Biology

Fourth Edition

Brad R. Batdorf

Elizabeth A. Lacy

bju press®

Greenville, South Carolina

BIOLOGY
Fourth Edition

Brad R. Batdorf, MAEd
Elizabeth A. Lacy, MEd

Consultants
Brenda Ball, EdD
Ginger Ericson, MAT
Amy Tuck, PhD

Contributing Writer
Lynne Woodhull

Bible Integration
Bryan Smith, PhD
Benjamin Lewis

Project Managers
Ted Williams
Franklin S. Hall

Project Editor
Adelé Hensley

Cover Design
Elly Kalagayan

Concept Design
Andrew Fields

Composition
Black Dot Group

Technical Consultant
Patricia Tirado, DZign Associates

Text and Photo Permissions
Sylvia Gass
Brenda Hansen
Joyce Landis
Rita Mitchell
Susan Perry

Illustration
Julie Arsenault
Matt Bjerk
John Cunningham
Aaron Dickey
Justin Gerard, Portland Studios
Cory Godbey
Courtney Godbey
Preston Gravely
Jim Hargis
Brian D. Johnson
Joyce Landis
Dave Schuppert
Del Thompson

Thomas E. Porch served as co-author of the third edition.

The front cover and spine photos are of the red-eyed tree frog, *Agalychnis callidryas*. The back cover features the Atlas moth, *Attacus atlas*.

© 2011 BJU Press
Greenville, South Carolina 29609
First edition (student text) © 1980 BJU Press
First edition (lab manual) © 1981 BJU Press
Second edition (both) © 1991 BJU Press
Third edition (both) © 2005 BJU Press

ISBN 978-1-60682-017-9

15 14 13 12 11 10 9

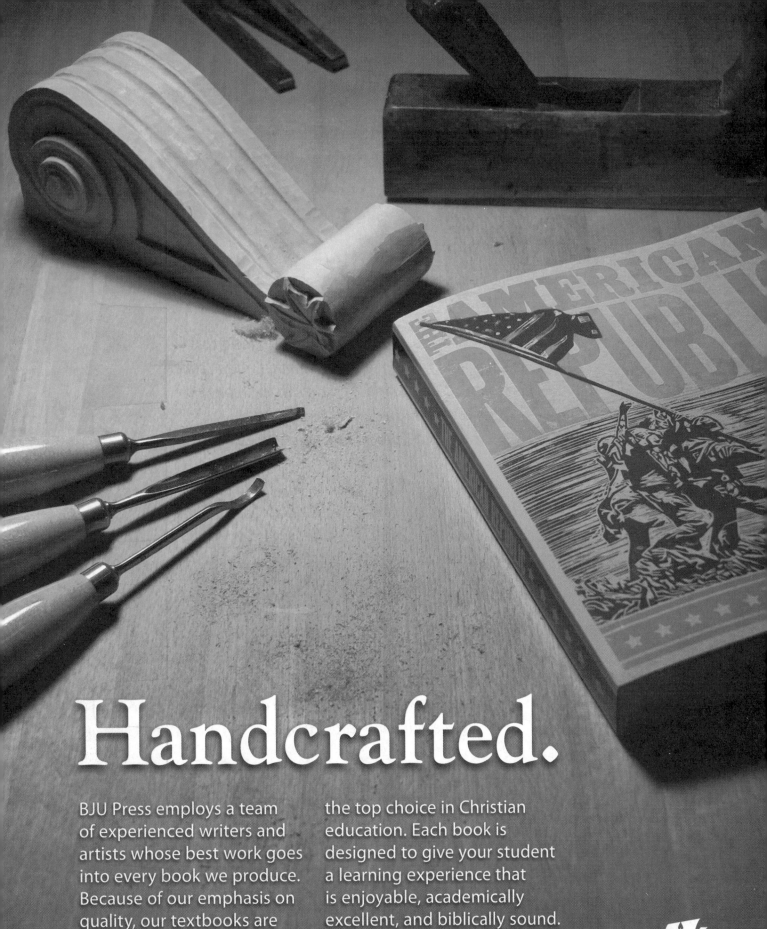

Handcrafted.

BJU Press employs a team of experienced writers and artists whose best work goes into every book we produce. Because of our emphasis on quality, our textbooks are the top choice in Christian education. Each book is designed to give your student a learning experience that is enjoyable, academically excellent, and biblically sound.

This is
My Father's World

This is my Father's world, and to my listening ears
All nature sings, and round me rings the music of the spheres.
This is my Father's world: I rest me in the thought
Of rocks and trees, of skies and seas;
His hand the wonders wrought.

This is my Father's world, the birds their carols raise,
The morning light, the lily white, declare their Maker's praise.
This is my Father's world: He shines in all that's fair;
In the rustling grass I hear Him pass;
He speaks to me everywhere.

This is my Father's world. O let me ne'er forget
That though the wrong seems oft so strong, God is the ruler yet.
This is my Father's world: the battle is not done:
Jesus Who died shall be satisfied,
And earth and Heav'n be one.

Maltbie D. Babcock

Contents

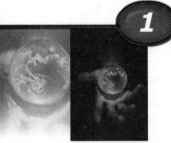

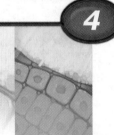

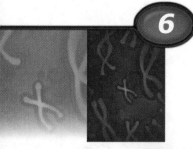

Introduction for the Student

You are about to embark on a journey that you may find fun, fascinating, or, in a few cases, even frustrating or frightening. Biology is the study of life, and with all the forms of life on this planet, each with its own unique characteristics, it is a subject that seems limitless. We live in a world of remarkable diversity, but all living things, from algae to zebras, share certain key characteristics. Behind this unified yet diverse creation is the God every Christian knows and worships. From the complex inner working of the cell to the beauty and balance of nature, the believer sees the hand of a Creator Who is not only mighty and powerful but also wise and good.

Format of the Text

Units and Chapters

This book is divided into three units. *The Science of Life* deals with topics basic to all biological studies. Much of this material concerns philosophy and theory, and it covers the areas in which many of the biological discoveries that most affect our lives are made. This material is foundational to the two units that follow. *The Science of Organisms* is a survey of the major groups of living things on our planet. This unit takes the traditional classificatory approach to studying organisms. Its final chapter shows how these living things relate to each other and to the nonliving aspects of the environment. *The Study of Human Life* includes sections on human anatomy and physiology and on Christian philosophy related to our physical, mental, and spiritual selves.

Each of these units is divided into chapters. Each chapter is divided into sections, which are lettered (for example, 1A, 1B, 1C, 2A, 2B). The chapter sections are further divided into subsections (for example, 1.1, 1.2). Finally, the subsections are divided into parts with several levels of headings, just as this introduction is divided. These headings should give you insight into what you are going to read. Do not skip over them; use them to help organize your thoughts and construct a framework for understanding.

Objectives

Each chapter subsection begins with one or more **objectives**. These tell you what you should know or be able to do after reading and studying the subsection. These objectives are also a valuable review tool at test time. Preview them before you read a subsection, and then review them after reading to see if you have grasped the most significant concepts.

Key Terms

An important part of any study is the vocabulary of the subject. Several vocabulary learning aids are incorporated into *BIOLOGY*. If you use these aids, you will not only make better grades, but you will also glean material useful in other areas of life. Usually the terms considered most important are printed in dark type, called **boldface**. These terms are also listed beneath the objectives at the start of each subsection. Once a term has appeared in boldface, it will sometimes appear in *italic type* if it is stressed again in that chapter or in following chapters. Many biological terms of secondary importance appear in italic type. Italic type is also used for emphasis and for scientific names.

Every subsection begins with a list of the boldface terms that will appear in that subsection. This list of **key terms** is arranged in the order that the terms appear in the text. After you have read a subsection, you should be able to define the terms in this list. For concise definitions of many terms used in this

Objectives

- Become acquainted with the organization of the text
- Recognize the significance of text formatting
- Read about some of the helpful features of this text

Key Terms

objective	review question
boldface	thought
key term	question
glossary	appendix
etymology	figure
Facets of	table
Biology	illustration
box	

Pronunciation Key

The pronunciations given in this text are designed to be self-evident and should give the average reader an acceptable pronunciation of the word. For precise pronunciations, consult a dictionary. This sample pronunciation key may help those who have difficulty with the symbols used.

Stressed syllables appear in large capital letters. Syllables with secondary stress and one-syllable words appear in small capital letters. Unstressed syllables are in lowercase letters. Most consonants and combinations of consonants (*ng, sh*) make the sounds normally associated with them.

Examples:

a	cat = KAT, laugh = LAF	**o**	potion = PO shun
a_e	cape = KAPE, reign = RANE	**o_e**	groan = GRONE
ah	father = FAH thur	**oh**	own = OHN
ar	car = KAR	**oo**	tune = TOON
aw	all = AWL, caught = KAWT	**oo**	foot = FOOT
ay	neigh = NAY, paint = PAYNT	**ow**	loud = LOWD
e	jet = JET	**oy**	toil = TOYL
ee	fiend = FEEND	**th**	thin = THIN
eh	rebel = REH bul, care = KEHR	**th**	then = THEN
eye	ivory = EYE vuh ree	**u,uh**	above = uh BUV
i	women = WIM un	**ur**	person = PUR suhn
i_e	might = MITE	**wh**	where = WHEHR
ih	pity = PIH tee	**y**	mighty = MY tee

book, consult the **glossary**, which begins on page 640. If the term has multiple meanings, make sure that you choose the definition that applies. Unfamiliar terms used in this text often have a pronunciation immediately following their appearance in the text. The pronunciation key on this page should help you read these pronunciations properly.

Etymologies

Etymology (ET uh MAHL uh jee) is the study of the history of words. It is useful because after you have learned a number of the common root words, you should be able to figure out the meaning of some words you may not be familiar with. This text gives etymologies for certain words, and in each case the word will appear as blue text. The etymology is found either in the margin or in the box or facet that contains the term. The first time a root appears in a term, the combining form and its origin language are given along with the definition. (If the language is the same for multiple parts, it is not repeated.) After that, only the definition is given. Appendix D is a glossary of the most commonly used word roots. Most roots that appear more than once can be found there.

Root Languages

The following abbreviations of languages are used in the etymologies:

E.	English
Fr.	French
Ger.	German
Gk.	Greek
L.	Latin
ME	Middle English
OE	Old English

Facets of Biology and Boxes

Additional information is presented in the **Facets of Biology**, which you will find scattered through the book. Facets contain material that is related to the content of the chapter. Some of them contain interesting examples, and some deal with practical or biblical applications of scientific material. Some facets contain terms that are listed in the key terms of the subsection that the facet is in.

Material may also be presented in **boxes**. Boxes are found throughout the text and sometimes inside facets. Generally, boxes contain additional information about or examples of the scientific material being discussed in the text. The "Closer Look" boxes contain more in-depth information related to

etymology: etymo- (Gk. *etumos*—true) + -logy (*logia*—word)

material in the regular text. Your teacher will inform you of what facets and boxes you are expected to read and which ones contain terms or other material you may be tested over.

Other Study Helps

At the end of each subsection, you will find a list of **review questions**. After carefully reading the preceding material, you should be able to answer the review questions. Only when you can answer these questions have you really comprehended the significant material. Some students find it profitable to read the review questions before they read the material. This alerts them to the most important points. When preparing for tests, you will find the review questions and the list of key terms valuable in directing study.

Some of the questions at the end of subsections may be marked with a red bullet (⊙) or a blue bullet (⊙). A blue bullet indicates that the question is drawn from a facet within the chapter. Those with a red bullet are **thought questions**. These may require you to have more knowledge than the text presents. Often you must draw from material covered in other chapters. You may need to use the index at the end of the book. Other times these questions will require the use of other sources such as encyclopedias, other textbooks, and, frequently, the Bible. The thought questions may also ask you to state your opinion and then defend it.

In the back of the book are several **appendixes** with which you should become familiar. These provide an overview of the classification of organisms, describe the world's major biomes, list common units of measurements and abbreviations, and furnish an index of many of the word roots used in the etymologies. Although you may not need their help every time you read this text, using these materials will often make your study of biology easier.

The **figures**, **tables**, and **illustrations** in this book have been designed to help you. Consider these carefully as you read the text. Some of the material in the tables and figures is reference material. You will not, of course, be expected to memorize this type of material. Your teacher will point out which tables and figures contain material you must remember and which are reference material. Many of these summarize related structures, processes, or concepts and are valuable study aids.

Use of Numbers

In this book, numbers longer than four digits are expressed in a way that may be unfamiliar to you. In the place where you would normally expect to find a comma, there is a small space. For instance, 2,345,600 would be written 2 345 600. This system is also used for numbers with more than four digits after the decimal. While it may at first seem awkward, this method is more scientifically acceptable. In many other countries and in scientific applications, spaces like this are used in place of commas. This new way of expressing long numbers is less confusing for those in other countries who are using this text.

Review Questions

1. List the topics of the three major units of this text.
2. In this text, what is the significance of (a) boldface type and (b) italic type?
3. Why is the study of etymologies significant?
4. List the three different types of questions that appear in each chapter.
⊙5. Write a single paragraph explaining why you are studying biology.

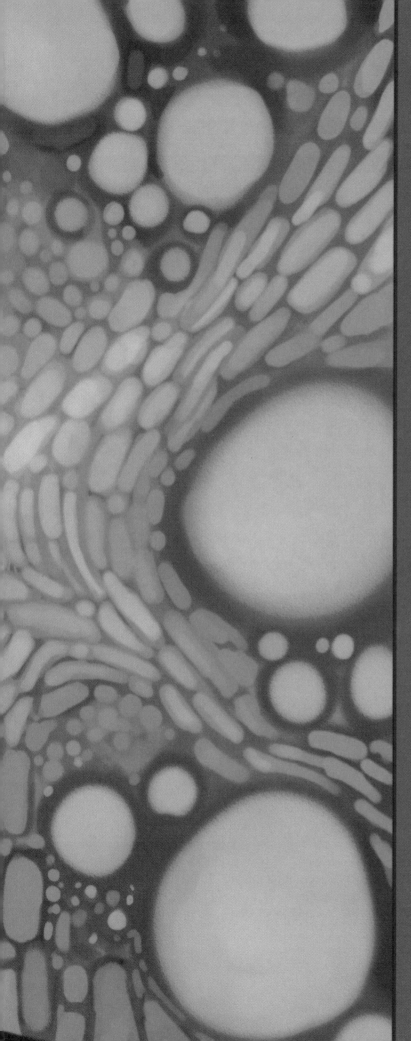

Unit 1
The Science of Life

1

The Science of Life and the God of Life

Facets

Few fields of study have the potential to affect your life the way that biology does. Do you ever wonder how breeders develop new types of cattle or what causes heartburn? How do leaves at the top of a tree make food and feed roots deep underground? How can doctors detect potential diseases before a baby ever leaves the womb? All of these questions are within the scope of biology.

Today, we know more than ever about how life functions, from the breeding behavior of penguins to how the cells in our own bodies communicate with each other. Advances in technology allow us to view things that only a few years ago were shadowy mysteries. These breakthroughs bring great potential for solving problems that have plagued mankind for centuries. Diseases, environmental concerns, and a growing population all call for solutions that are reasonable and practical.

Christians share the added responsibility of using this knowledge in a way that glorifies God and shows concern for humanity. To properly apply the tools of science, we must first submit ourselves to God. Through this submission, we can learn to see all of life—biology included—through the corrective lens of the Word of God.

1A—God and Science

What does God have to do with science? The Bible doesn't have a book titled "Ecology" or "Invertebrate Biology" or "Human Reproduction." True. But this does not mean that God and science have nothing to do with each other. In fact, they are bound together at key points—so bound that they cannot be separated. If people try to separate them, great damage will be done—damage that will distort our view of science and God.

The most obvious intersection between God and your study of biology comes in the opening chapters of God's Word (Gen. 1–2). Here, we have an authoritative account of how all life was *created* in six literal days by a direct act of God. For His own glory, He designed all living things to reproduce after their kind and created humans to enjoy and care for this wonderful new world. Adam and Eve were charged with exercising wise dominion over all of creation, a mission that is often called the **Creation Mandate** (Gen. 1:28).

Sadly, this perfect creation that started off with such promise soon was altered in detrimental ways because of Adam's sin. This sin and the curse that it led to are often referred to as the *Fall*. Through the Bible, we learn that suffering and death are natural evidences of God's judgment on sin. Diseases, genetic problems, and imbalances in the environment are all constant reminders that we live in a physical world that is broken. In addition, even man's mind is damaged by the Fall. Although our fallen minds make it difficult for us to understand the world around us, God, in His grace, does allow us to uncover some scientific knowledge that can improve our lives. Even so, man's fallen mind often misuses this information. Some, through the acquisition of knowledge, begin to doubt God and His Word, relying instead on their own abilities.

Although the Fall and its results paint a bleak picture of the world, the story does not end there. Scripture also reveals that God has a plan to *redeem*, or restore, this world, including the physical realm, to its original glory and goodness (Rom. 8:20–23). This will be accomplished through the work of Christ, but Christians also play a role in this mission. By properly applying biology, we may develop life-saving medicines or life-enhancing procedures. Life can also be improved through better agricultural practices and improved wildlife management. By properly applying the tools of biology, we become instruments of God's redemption. Just as Jesus worked to bring people to Himself by physical healing and proclamation of the gospel, so we can live redemptive lives by proclaiming the gospel in the context of serving people with the science of biology.

Biology is a powerful tool with great potential for good. To use it properly, we must know what it can and cannot do. The basis for that is grounded in understanding the nature and role of science.

1.1 What Is Truth?

Science is one of those words that we use regularly but have great difficulty defining. There are many possible definitions (and we will offer our own in the next subsection). On the popular level, people tend to think of science as somehow related to *truth*. If someone wants to assert that a certain idea should be disregarded, he may say, "That's not what science tells us." Or if people

1-1

Adam and Eve were charged with exercising wise dominion, or stewardship, over God's creation.

1-2

Smallpox and other more common diseases are constant reminders of the effects of the Fall.

1.1

▶ Objectives

- Define *Creation Mandate* and give examples of its application
- Define *truth*
- Recognize the errors that led to the acceptance of the Doctrine of Humors, the Doctrine of Signatures, and spontaneous generation
- Explain the difference between inductive and deductive logic

▶ Key Terms

Creation Mandate
Doctrine of Humors
Doctrine of Signatures
spontaneous generation
infusion
microbe
logical reasoning
inductive reasoning
deductive reasoning
Truth

want to cast doubt on a particular practice, they may say, "That's not very scientific." But is science really about truth? To answer that question, we'll have to think about what truth is—and isn't.

Truth: What Everybody Believes?

Some facts are established according to what everybody does. The rules of grammar and spelling and the meanings of words are based on how educated people commonly write and speak. But rules of language are man-made and man-governed. Is truth simply what everybody believes? An examination of an idea that everybody believed to be true for hundreds of years will help answer that question.

Hippocrates (hih PAHK ruh TEES), a Greek physician who lived about 350 BC, promoted the **Doctrine of Humors**, which states that living things are composed of four fluids, or "humors." These humors supposedly have certain properties and are produced by specific organs. According to this doctrine, a person is healthy and happy when these four humors are correctly proportioned and well mixed in the body. However, if a person has too much or too little of one or more of these humors, his temperament (humor) is affected. For example, a person with too much blood would be warm, ruddy, friendly, and happy. A person with too much black bile would be cold, sad, and melancholy.

It was thought that each individual tends toward one of these four humors. Hippocrates believed that the fluids were affected by the season, winds, temperature, sunshine, food, manner of living, age, and many other factors. In winter, phlegm (FLEM) would dominate. If a person tended to have too much phlegm, he would be sluggish and indifferent. Severe cases of too much phlegm were thought to produce disorders such as pneumonia, dropsy, diarrhea, or dysentery. Hippocrates, who is still called the Father of Modern Medicine, believed that counteracting the predominant fluids would cure a sick person. If moist phlegm entered the blood, resulting in chills, the person was to be kept warm and dry.

The Doctrine of Humors was the accepted medical practice for centuries. Today the Doctrine of Humors seems quaint or silly, yet it described human health satisfactorily for the Greeks. Until the mid-1800s, the practice of medicine was based on this doctrine—almost every educated person believed it. But is it true?

Truth: A Hunch That Works?

In a mathematics class, using the right method to solve a problem is sometimes considered more important than merely finding the right answer. If a student performs a series of mathematical steps with given numbers and produces the correct answer, he may assume that he followed the proper procedure. He may, in fact, have made errors that were compensated for by other errors. He arrived at the correct answer but did so by mistake. The incorrect steps actually worked. Does this mean that the incorrect steps are a valid method of discerning truth?

The ancient Babylonians used the **Doctrine of Signatures** to prescribe remedies for various ailments. In the fourteenth

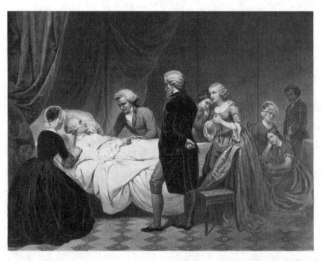

1-3
Hippocrates (left) is credited with devising the Doctrine of Humors (right). Shown are the four humors and the characteristics they were believed to cause. Hippocrates influenced medical practice for thousands of years.

Hippocrates: Medicine Becomes a Science by Robert Thom, UMHS.7

BRAIN — *which makes phlegm*
LIVER
WET COLD
HOT DRY
which makes yellow bile
SPLEEN — *which makes black bile*
HEART — *which makes blood*

1-4
In 1799, George Washington's physicians believed that too much blood caused his fever. They prescribed bleeding several pints of blood. Because they believed in the Doctrine of Humors, the first president of the United States was bled to death.

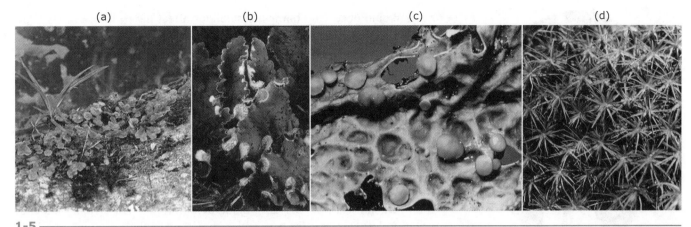

(a)   (b)   (c)   (d)

1-5

All of these organisms were used to treat human disorders according to the Doctrine of Signatures: (a) liverwort, (b) dogtooth lichen, (c) lungwort, and (d) hairy cap moss. Can you guess what each was supposed to be good for?

and fifteenth centuries, Europeans promoted this doctrine to its height. They believed that when God cursed man with disease, He mercifully left in the physical world signs (signatures) of cures for these diseases. Scriptural basis for this belief does not exist.

Medieval doctors prescribed a yellow papery lichen (a small plantlike growth found on rocks) as a cure for a yellow condition of the skin called jaundice. The similar yellow appearance of the lichen was believed to be God's sign that it was the remedy. A small plant that looks like the lobe of a liver was used to treat liver infections. This plant still has the name liverwort ("liver plant"). In 1540, Dorstenius gave directions for preparation of lungwort, a lichen that looks like a lung, as a cure for diseases of the respiratory system.

The Doctrine of Signatures sounds absurd to the modern mind, but the medical books of the Middle Ages indicate that these remedies had cured patients (usually in "other countries" or "olden times"). It is understandable that a person who was desperately ill would try some remedy if there were reports that it had cured people in the past. Besides, it was the only thing he could do, and to him it was probably better than doing nothing.

Scientists point out that maybe the people who were reportedly cured were never really sick or had a less serious disease that they would have recovered from without the treatment.

In the 1940s, however, it was found that some lichens do manufacture antibiotics that are effective in stopping the growth of some tuberculosis bacteria. Lichen antibiotics have been marketed commercially and have been found effective in treating wounds and burns. Is it possible that some people in the Middle Ages were actually aided by the antibiotics recently discovered in some lichens? If so, this aid could have strengthened a mistaken belief in the Doctrine of Signatures. The fact that something works, or appears to work, does not make it true.

Truth: Repeated Observations?

Some people think that ideas are established as true by repeated observations. If a phenomenon happens over and over, they feel it is safe to assume that it will happen again.

If a pencil is picked up and released a few inches above a table, it will fall. This falling, which happens every time, is the effect of gravity. If a person were to watch a magician seemingly levitate an individual and repeat this trick several times, would this force him to alter his belief about the truth of gravity? Would observing blimps or satellites affect the truth of gravity? Are repeated observations a sure guide to truth?

The **spontaneous generation** of life from nonliving materials was believed, on the basis of observations, as early as 800 years before Christ. The ancients were of course familiar with birds hatching and the birth of many animals, but some of the less familiar animals were believed to generate spontaneously whenever conditions were acceptable. In the Middle Ages many thought that frogs and fish were formed during storms and then "rained" to earth. Insects supposedly came from the soil, while maggots and worms developed from dead and decaying flesh. A fallen leaf of certain plants was believed to develop into a fish or snake, depending on where it landed.

If an **infusion** (in FYOO zhun), made by boiling animal or plant material in water, is left standing for a period of time and is then examined under a microscope, the examination will reveal quantities of small organisms called **microbes**. According to most scientists from the 1700s until the mid-1800s, the microbes found in infusions spontaneously generated as the mixture began to ferment (spoil).

Some scientists held the theory that microbes came from the air; therefore, they believed if the air were cut off, the microbes would not develop. In 1749, John Needham, an English priest and scientist, performed a set of experiments to prove the spontaneous generation of microbes. He filled a number of glass vials with an infusion of mutton, heated them "violently" in hot ashes, and then stoppered them tightly. He reasoned that the stoppers would prevent airborne microbes from entering the vials and that the heat would destroy any microbes already in the meat infusion. If there were any organisms found in his infusions, they would have had to generate spontaneously. After a few days, each of his vials was teeming with microbes.

| Infusion of mutton placed in stoppered jars | Infusion heated "violently" | Microbes found in the infusion a few days later |

1-6

John Needham (top) experimented with infusions to explore the concept of spontaneous generation.

Over and over scientists were able to demonstrate (they thought) that organisms, especially smaller ones like microbes, would just spontaneously appear if conditions were right. A pond may contain fish without anyone ever observing eggs being laid. The presence of fish without evidence of their origin does not support spontaneous generation. Nor did Needham's "scientific" experiments, however often he repeated them, prove spontaneous generation. He had made an error in reasoning that was not obvious to him. (Needham's errors are discussed along with other aspects of spontaneous generation later in this chapter.)

Is something true simply because it has been repeatedly observed? Senses are often fooled, and observations may be based on faulty experiments.

Truth: That Which Is Logical?

Logical reasoning constitutes truth to some people. If it makes sense, or if it can be figured out, they think it is true. Since mathematics is logical, some would say it is true. For example, it is said that $2 + 2 = 4$. If 2, $+$, $=$, and 4 can be defined, the truth of this mathematical statement is demonstrated. But numbers can be made to give inaccurate descriptions of what exists. By "cooking the books," a company can look profitable when it is nearly bankrupt. What may appear to be logical mathematically may not be true.

Logical reasoning is usually classified as either inductive or deductive. **Inductive reasoning** begins with a number of observed facts and uses them to derive a general conclusion. The principle of gravity can be induced by observing that most objects will fall to the earth when released. Inductive reasoning often leads to the statement of a scientific principle that is true in all or most cases. After seeing many feathered animals flap their wings and fly, you might induce that all feathered creatures can fly. However, repeated observations of natural processes or even controlled experiments may be faulty and lead to a false conclusion. For instance, both penguins and ostriches are covered with feathers, but neither of these types of birds can fly. Something can be reasoned inductively and still not be true.

Deductive reasoning begins with premises assumed to be true and draws conclusions about particulars. One of the first recorded illustrations of deduction is attributed to Aristotle, describing his teacher:

Major Premise: All men are mortal.
Minor Premise: Socrates is a man.
Conclusion: Socrates is mortal.

If, however, the statements in logical reasoning are not exclusive enough, the conclusion may exhibit faulty logic and be false. Which premise below leads to a faulty conclusion?

Major Premise: All feathered creatures can fly.
Minor Premise: Ostriches have feathers.
Conclusion: Ostriches can fly.

Obviously, the major premise above is not true. There are many types of birds that, although they are feathered, cannot fly. Even a single exception would falsify the premise and thus could lead to a false conclusion.

In the early 1600s, Jan Baptista van Helmont, a Belgian chemist and physician, experimented to prove that soil changed into the material that makes up plants. A large pot was filled with exactly 200 lb of dry soil. He planted a 5 lb willow tree in the pot and covered the pot with a shield that admitted water but excluded dust and new soil.

1-7

Ostriches can't fly, even though faulty reasoning may lead to the conclusion that they can fly.

At the end of five years, the tree was carefully removed. Its final weight was 169 lb 3 oz, a gain of 164 lb 3 oz. Next, van Helmont dried and weighed the soil and found that only 2 oz was missing. Obviously the entire 164 lb 3 oz gained by the plant had not come from the 2 oz of soil. He concluded that the additional 164 lb 1 oz that the tree had gained must have come from the only other substance the plant could get—water. Van Helmont's logic, although somewhat faulty, proceeded along these lines:

Major Premise: Plants, when growing in only soil and water, gain weight.
Minor Premise: Plants gain much more weight than is taken from the soil.
Conclusion: The weight gain of the plants comes almost entirely from water.

Van Helmont's premises are true, but his conclusion is false. His logical, well-designed, and carefully conducted experiment resulted in a faulty conclusion, even though the numerical information obtained was correct. Van Helmont did not know that plants use carbon dioxide and water to make sugar, the basic material from which plants make other substances. The process by which plants convert water and carbon dioxide into sugar was not discovered until long after van Helmont's death. Something that is logical and that seems to make sense is not necessarily true.

He observed that air is a mixture of gases and isolated CO_2 but failed to recognize it in this experiment.

Start
Tree: 5 lb
Soil: 200 lb

End
Tree: 169 lb 3 oz
Soil: 199 lb 14 oz

1-8

Jan Baptista van Helmont's conclusions were inaccurate in his tree experiment, yet he still contributed much to science.

Truth: That Which Is Accepted by Faith?

What a person believes can be called his *faith*. It has been said that whatever a person believes to be true will be truth for him. A person acts because of his

1-9

In the early nineteenth century, many scientists measured their subjects' skulls with devices like this. It was thought that the shape of a person's head determined his intelligence and personality. Did their firm belief in this theory make it true?

faith—what he believes. When a person crosses a bridge, he must have faith that the bridge will hold him. If he doubts the bridge's strength, he lacks faith in it and probably will not cross it. However, his faith in a bridge does not mean that it is strong enough to hold him. Men may have faith in something that is not true and may act wrongly in light of their belief. If they do, they must suffer the consequences of believing something that is false.

A person going to a medical doctor has faith in that doctor. But if the doctor practices the Doctrine of Signatures, the patient may die of the prescribed cure. The patient has faith, but he believes in something that is wrong. A person may be affected by his faith or lack of faith, but does faith or lack of faith affect truth? If a person does not believe there is a hell, does hell no longer exist because of his disbelief?

Truth: The Word of God

Our God is the God of truth. Christ said, "I am . . . the truth" (John 14:6) and later called the Holy Spirit the "Spirit of truth" (John 14:17, 15:26, 16:13). What the God of truth says must be true. Thus Scripture, the Word of God, is Truth (John 17:17). Logic, observations, common beliefs, or personal faith cannot disprove these claims of the Bible.

It is true that a person must accept the truth of the Bible by faith, but that faith is not blind. The physical world around us testifies to the power of God, Who created and sustains it. Thus repeated, accurate observations of the world help support the truth of the Bible (Ps. 19:1). The historical accuracy and fulfilled prophecy of Scripture also testify to its truth.

Our God is a God of logic. Although His ways are often beyond human understanding, that does not mean that they are without purpose or logic. "God is not the author of confusion" (1 Cor. 14:33).

To some people these testimonies to the truth of God's Word are not conclusive enough to justify their faith. Such people often spend time picking apart what the Bible says, looking for errors. But to those who are willing to have faith in God and His Word, these testimonies help support their belief that God's Word is **Truth**.

➥ *Review Questions 1.1*

1. List and give applications of the three key concepts of Scripture that relate to a modern study of science.

2. According to the Doctrine of Humors, what conditions were necessary for a person to be happy and healthy? What could happen to a person to make him ill? What could be done to cure him?

3. On what assumption is the Doctrine of Signatures based? According to the Doctrine of Signatures, what could be done to cure an illness?

4. What evidence did those who believed in spontaneous generation offer to support their belief?

5. Give (a) an example of inductive reasoning and (b) an example of deductive reasoning.

6. Describe van Helmont's tree experiment and tell why his conclusions were false.

⊙7. "The Bible does not deal with modern science; therefore, science is of the Devil." Give evidence to either contradict or support this statement.

1.2 What Is Science?

If the Bible is the source of settled truth, where does that leave science? Is science unimportant and unnecessary for those who believe the Bible? As God's Word, the Bible gives us God's view of the world. It gives us the orienting perspective we need for understanding every aspect of our lives. It tells us that the physical world is made by God and is something worth studying.

It also tells us that we glorify God by managing that world with wisdom and skill. But there is much that the Bible does not tell us. It does not tell us how trees produce sugar from water and carbon dioxide or how a broken bone heals itself. If we are to manage God's world wisely—as His Word commands—we will have to make use of science. But to use science properly, we must understand what science is—and isn't.

Modeling

Science is a very powerful tool for studying and managing God's world. But scientific study is not able to tell us the truth about God's world. Truth is unchanging; it is fixed and needs no revision or adjustment. Science, however, is always changing. One reason for this is that science is concerned with *modeling* the world around us.

A **model** is a simplified depiction of a far more complex object or concept. A model may be a drawing or representation of how it is thought something is constructed (e.g., the cell membrane), or it could be a step-by-step explanation of how a process occurs (e.g., blood clotting). Models such as these help scientists study the world because they simplify structures and processes that are too complicated for the human mind to understand—or even imagine. But these models are, by their very nature, limited. They may be valuable in explaining certain aspects of physical phenomena, but inadequate in others. As a result, models are subject to change.

Scientists devise models to explain the data they have obtained. A series of different experiments can often reveal much about a process. Putting together all the experimental information so that it "works" logically can produce a model. Further experimentation and testing of the model often reveal more information. If the results of these additional experiments agree with the model, they give greater validity to the explanation. If the results do not support the model, then either those data are erroneous, or the model must be greatly changed and retested or even discarded.

Despite the fact that models are often changed, they are the only method man has to describe how something functions. It is often crucial for man to know how a process operates in order to properly manage it. As the various models in this book are presented, however, it is important to recall that any explanation is merely man's guess at how God ordained something to operate.

Defining Science

As you read in Subsection 1.1, a Christian's understanding of truth is different from that of an unbeliever. However, there are some similarities. Both would recognize the organizing value of theories but would also realize they are subject to change as new information challenges them. They would also agree that theories may be regarded as fallacies if scientific evidence contradicts them. Finally, certain concepts must be true by definition for science to proceed and for its applications to lead to improvements. For instance, the mathematical relationships between numbers are accepted as true by definition, and numbers are the language of science.

Just as a Christian would have a different but related understanding of truth, he should also have a modified view of science. How then would a Christian define *science*, keeping in mind man's responsibilities to God and to his fellow man? Consider the following:

> **Science** is the collection of observations, inferences, and models produced through a systematic study of nature that enables humans to exercise proper dominion over God's world. The word *science* can

⇒ Objectives

- Explain the relationship between God and science
- Describe the necessity of modeling and its relationship to the nature of science
- Define *science*

⇒ Key Terms

model
science

1-10

God left it to man to learn practical uses of fire.

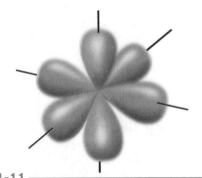

1-11

Physicists have offered this model for the location of electrons in the *p* sublevels of atoms.

also refer to the systematic methods that produce the observations, inferences, and models.

Several words in this definition may be unfamiliar to you. These will be clarified in the following sections of this chapter. Notice that this definition presents science as both a collection of knowledge and a process used to acquire that knowledge. As you read this textbook, you will find *science* used both ways.

Review Questions 1.2

1. Why is it important for a Christian to systematically study the physical world?
2. Why are models important in science?
3. Why can no model be accepted as a completely valid explanation of a phenomenon?
4. Defend the following statement: "Science is not able to tell us the truth about God's world."
5. Explain the meaning of this statement: "Science is a process as well as a product."

1B—The Scientific Method

For some people the thought of a scientist using the scientific method produces visions of a man in a white coat, clipboard in hand, surrounded by a complex system of tubes and flasks. He is looking at dials and flashing lights on a large console. Most people who have these mental pictures have great respect, even awe, for scientists who can use the scientific method.

Although scientists may sometimes wear white coats, use elaborate equipment, and perform experiments that are difficult to explain, none of these make a scientist any different from other skilled persons. The scientific method is not a "magical," miracle-working formula that will solve all man's problems. It is a method of reasoning that anyone can use daily.

The **scientific method** is simply a logical procedure for choosing an answer to a question. For example, after examining hundreds of ripe oranges, a person could use inductive reasoning to say, "Ripe oranges are orange in color and nearly round." Such reasoning is characteristic of what scientists often do and of what anyone does to employ the scientific method.

1.3 Using the Scientific Method

Although there is no set order of activities, a person will usually do all the following preliminary steps at one point or another when using the scientific method of collecting information.

✦ *Define the problem.* Before the actual experiment begins, he must state the specific question or problem to be solved. It must be limited sufficiently so that it can be dealt with conveniently. The question "What makes plants grow?" is too broad and indefinite, whereas "Do geraniums grow better in dim or bright light?" is limited enough to be answered by experimentation.

✦ *Do preliminary research.* Once the problem has been stated, he should **research** the problem to become familiar with the relevant area of science. Perhaps someone else has already solved the problem; if so, experimentation may be unnecessary. In any case, research, whether conducted from library sources or from conversations with knowledgeable people, will help provide an understanding of the problem and prevent the repetition of mistakes.

✦ *Form the hypothesis.* Once the question has been stated and researched completely, he formulates a **hypothesis**—a simple, testable statement that predicts the results expected from a research study. Next, he devises activities

1.3

Objectives

- List and describe the significant steps of the scientific method
- Explain the difference between a control group and an experimental group
- Explain the difference between a dependent variable and an independent variable

Key Terms

scientific method	independent variable
research	control group
hypothesis	experimental group
experiment	
survey	dependent variable
data	
controlled experiment	biogenesis

to see whether the hypothesis is correct. A good hypothesis will help define and direct the research.

Steps of the Scientific Method

The activities used to test a hypothesis can be an **experiment** or a **survey**. If an experiment is to be used, it must be tailored to answer the problem precisely. Should the problem ask what exists in a particular area or what is common practice, a survey is necessary. For example, determining what kind of tree is most common in a certain region or which pain remedy doctors recommend would be accomplished by surveys, not experiments.

Once the experiment or survey has been constructed, the person will go through the following steps:

✦ *Observe* the experiment or survey carefully.

✦ *Collect* the information from the experiment or survey and accurately record it. The recorded information is the **data** that will be used to solve the problem.

✦ *Classify* the data into a logical order or into logical groups.

✦ *Analyze* the data to determine what they reveal about the problem.

✦ *Choose* the solution that best answers the question. In some cases the data may point to only one solution; in others, several possible solutions may be implied. If the data seem to imply a conclusion that is different from the hypothesis, then the hypothesis may need to be changed or even discarded.

✦ *Verify* the chosen answer by repeating the experiment. The more often a well-designed experiment is repeated and produces similar results, the more valid and reliable the answer is.

✦ *Predict* what will happen in similar situations. The goal of using the scientific method is to be able to draw conclusions that can be applied to similar cases.

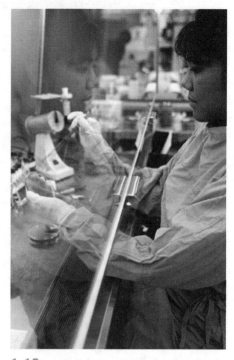

1-12

The scientific method can be used in the laboratory or at home.

Controlled Experiments

An ideal scientific experiment is sometimes called a **controlled experiment**. In a controlled experiment there are two groups—the control group and the experimental group. Both of these groups are identical except for a single factor. This difference between the two groups is called the experimental or **independent variable**.

The group not exposed to the independent variable is the **control group**, and the group exposed to the independent variable is the **experimental group**. During an experiment, the researcher measures a factor in both the control and the experimental groups—the **dependent variable**. It is called the dependent variable because it results from, or is dependent on, the independent variable. The experiment is designed to determine whether there will be a measurable difference in the dependent variable between the control and the experimental groups.

For example, if a company wants to evaluate a new hair-growth drug, it would need to design an experiment to test only the effects

The independent variable is the hair-growth drug that is dissolved in the experimental group's water.

Hair growth is the dependent variable and has been positively affected by the new drug. The dependent variable is unchanged in the control group.

1-13

Hair-growth drug experiment

Spontaneous Generation and the Scientific Method

Until the mid-1800s, many people considered spontaneous generation—the concept that organisms come to life from nonliving substances—an acceptable scientific theory. Two reasons why this belief held on for so long are examined here.

In Genesis 1 God spontaneously created living organisms out of the earth. Many people assumed that God continued to do so. Those who did not believe in spontaneous generation were accused of doubting God's Word, bordering on heresy. But Christians must conform their beliefs to the Bible, not conform the Bible to their beliefs. The supporters of spontaneous generation ignored the Scripture passages that say that God finished creation (Gen. 2:1–3) and that organisms reproduce after their own kind (Gen. 1:24–25).

Other people continued to believe in the spontaneous generation of some organisms because of their faulty use of the scientific method or because of their incorrect conclusions. Evaluating some examples of the use and misuse of the scientific method will promote a better understanding of the method and its proper application. The following examples describe some of the more interesting experiments dealing with spontaneous generation as well as experiments that support **biogenesis**, the concept that only living things can generate living things.

Jan Baptista van Helmont

Jan Baptista van Helmont (ca. 1600) proposed that mice could be spontaneously generated in at least twenty-one days by putting a sweaty shirt and grains of wheat in a dusty box. The sweat supposedly supplied the active principle that would cause the wheat grains and dust in the box to become mice. Every time van Helmont conducted the experiment, within twenty-one days he found mice that appeared to be gnawing out of the box.

The design of the experiment, however, was faulty because he did not perform a controlled experiment. By not excluding mice from the experiment, van Helmont failed to take into account that the mice might be gnawing into rather than out of the box. He was attempting to support his widely accepted belief. Since his results supported his belief, he saw no need to restructure his experiment.

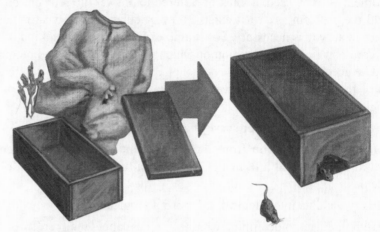

Van Helmont's demonstration of the spontaneous generation of mice

What could van Helmont have done to limit the number of variables?

Francesco Redi

Francesco Redi (ca. 1600), a poet and physician, also considered the concept of spontaneous generation. Aristotle's statements that maggots and flies spontaneously arose from rotting flesh and slime were contrary to what Redi had observed in nature. Redi believed that the decaying matter and slime served only as the material in which the organisms were growing and that the animals thought to be spontaneously generated were actually the results of biogenesis.

To support his belief, Redi devised a set of experiments. Redi placed dead fish, veal, dead eels, and pieces of snake in eight jars. He then placed paper covers securely over one set of jars and left the other set open. The two sets of jars were alike except for a single variable: the covers that prohibited the entrance of flies. In a few days, flies and maggots were found in the open jars, but no flies or maggots were found in the sealed jars. Redi's experiment supported his hypothesis.

Had Redi been satisfied and stopped his experiment at this point, he would have been criticized. Supporters of spontaneous generation would have said that since sealing the jars had cut off air (they believed air was necessary for spontaneous generation), Redi had prevented spontaneous generation. Redi's experiment had a double variable: the presence of flies and the presence of air. To eliminate one of the variables, Redi devised and conducted other experi-

ments. He set up the jars as before, but this time, rather than sealing the experimental jars, he covered them with "fine Venetian net," which would permit air but not flies into the jars. In a few days, the open jars had flies and maggots in them. Although flies were attracted to the closed jars and had even laid eggs on top of the net, no maggots or flies were inside the jars.

Redi repeated his experiment in different locations, using different substances in the jars, but he always obtained the same results. Evidence against the spontaneous generation of flies was shown by these simple, well-defined experiments. Nothing was conclusively proved; no absolute truth was gained; but the experiment produced valid support for biogenesis.

First experiment:
Redi used a set of open and a set of closed jars.

Second experiment:
Redi used a set of jars covered with nets.

Redi's experiments to disprove the spontaneous generation of flies

John Needham and Lazzaro Spallanzani

John Needham's experiments, which supported his ideas about spontaneous generation, are described earlier in this chapter. The growth of microbes in infusions that Needham assumed to be sterile (without living material present) was held to be proof of the spontaneous generation of these tiny organisms. After all, he reasoned, where could the microbes have come from? They were not like Redi's flies; they could not have traveled into the infusion.

About twenty-five years after Needham's experiments, the Italian scientist Lazzaro Spallanzani (ca. 1700) pointed out their weaknesses. Needham had not boiled the broth long enough to destroy all the microbes that were already in his flasks, and the stoppers were not tight enough to prevent airborne microbes from entering. Spallanzani was convinced that these two oversights were enough to account for the microbe growth seen in Needham's infusions.

To support his opinion, Spallanzani conducted a series of experiments. He placed various infusions in glass flasks and sealed the flasks by heating and then fusing the tops. He then placed the sealed flasks in boiling water for one hour to sterilize them. After several days, he opened each flask, and none of the infusions contained living organisms.

Spallanzani's experiment, however, was challenged. Since he had boiled the infusions for a long time, some who still believed in spontaneous generation said that he had destroyed the "active principle" in the infusions. Spallanzani then devised another set of experiments. He made several infusions and divided them into four identical groups. One group he boiled for half an hour; another group, for one hour; the third group, for an hour and a half; and the last group, for two hours. All of them he left open. If boiling destroyed the active principle, he should find the fewest organisms in the infusions that were boiled the longest.

After eight days he examined the infusions. Microbes were present in all of them, but the numbers and types of organisms varied. In all but one of them, he found that the infusions that had been boiled the longest had the most organisms. What had happened? Rather than destroying the active principle, longer boiling had dissolved more of the solid material in the infusion, making it better for the growth of microbes.

For the next century, those who believed in spontaneous generation of microbes ridiculed those who believed that microbes in an infusion came from microbes in the air or from their "seeds," as Spallanzani had called them. Since microbes in the air are invisible, most people assumed that there were none there.

Most of the respected scientists of the 1600s and 1700s believed in spontaneous generation, but few, if any, devised experiments to support it. They were content to believe in spontaneous generation and quoted one another and the ancients to support their beliefs.

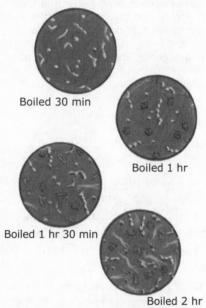

Boiled 30 min

Boiled 1 hr

Boiled 1 hr 30 min

Boiled 2 hr

Spallanzani's experiment to demonstrate that boiling did not hamper the growth of microbes involved boiling infusions for different lengths of time.

Louis Pasteur

Louis Pasteur (ca. 1800), a French chemistry professor, became interested in the spontaneous generation controversy. Pasteur believed that spontaneous generation was a myth, so he devised a series of experiments that to this day are of unquestioned validity.

One series of experiments was designed to support the idea that microbes are carried in the air. Pasteur prepared several sets of infusions and sealed them in flasks. He then sterilized the infusions by boiling. He opened one set of infusions along a dusty road, another set in a forest, and other sets in different places. Later, Pasteur examined the infusions and found that

Pasteur: The Chemist Who Transformed Medicine by Robert Thom, UMHS.32 (detail of original)

Louis Pasteur working with a swan-necked flask

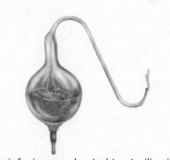

The infusion was heated to sterilize it.

Air entered but microbes were trapped in the neck. The infusions remained microbe free for years.

When the flask was tilted, microbes in the neck entered the infusion.

The infusion illustrated biogenesis.

The Science of Life and the God of Life

those opened in dusty places contained abundant and varied microbes. Those that were exposed to "cleaner air," like the set opened on a mountain, had fewer and different microbes. These results supported his belief that microbes are carried in the air.

Pasteur's famous swan-necked flask experiment dealt the final blow to the belief in the spontaneous generation of microbes. Pasteur prepared a sugar-and-yeast infusion, put it in a flask, and then bent the neck of the flask so that it formed a long S-shaped curve. He then heated the infusion to sterilize it. The boiling forced air and water vapor out of the flask; however, once the heat was removed and the flask cooled, air was able to return freely through the open neck. Pasteur reasoned that the dust and microbes that entered with

the air would be trapped in the lower bend of the long neck.

His reasoning was correct, for he found the infusion prepared in this way remained sterile (no growth of microbes) for over a year. However, once the flasks were tipped so that the dust particles trapped in the neck could enter the infusion, microbes developed quickly. Although the infusion was still capable of supporting life, no life spontaneously generated in the infusion.

Spontaneous Generation Today

After Pasteur's swan-necked flask experiment and thousands of other experiments supporting biogenesis, do people today still believe in spontaneous generation? Yes. Anyone who believes in evolution believes that

spontaneous generation has occurred. Evolutionists state that nonliving chemical compounds were randomly joined, resulting in the formation of a primitive life form that is the precursor to all of the various organisms seen today.

Many scientists today are trying to obtain data to support the spontaneous generation of life—the very thing that has been repeatedly disproved. They would be quick to agree that spontaneous generation is not taking place today, but they are working hard to try to make it happen in a laboratory. If they can create life, they think they can support their belief in life's beginning without God. But spontaneous generation has never been successfully demonstrated, not even in a laboratory.

of the drug. To exclude all other influences, the experimenters could start with two sets of mice grown in identical cages in the same room with the same diets. If one group has the special hair-growth drug dissolved in its water, it would be the experimental group, and the drug would be the independent (experimental) variable. The other group of mice would be the control group. The factor to be measured (the amount of hair growth) is the dependent variable.

Normally, multiple variables in a controlled experiment are not desirable. Why? If an experiment has more than one independent variable, it is difficult to know which variable was responsible for the observed results. If the experimental group was given larger cages than the control group, the added space (or both the added space and the drug) might have produced the results rather than just the drug.

For this reason, experiments with fewer variables tend to give more *valid* (accurate and reliable) results. That is, if the experiment is repeated, the results will be the same (or very similar). If the results are otherwise, they are not considered valid. To ensure valid results, scientists often repeat experiments or have scientists in other laboratories repeat their experiment.

At the conclusion of the mouse experiment, it is observed that 95% of the one hundred mice in the experimental group grew longer fur than the mice in the control group, while 5% of the mice in the control group grew longer fur. In experiments with living things, results that are 100% in support of or against a hypothesis do not often happen.

Since some of the data do not support the hypothesis, is the experiment useless? Scientists might consider the unexpected outcome to be the result of some uncontrolled factor such as mouse health problems or genetic differences. Such factors would introduce multiple variables in the affected mice that might account for data that do not appear to support the hypothesis. To remove the significance of such differences and to verify the results, scientists repeat the experiment. If after many repetitions the amount of data against the hypothesis remains low, scientists would accept the results as *workable*, even if not completely valid.

1. A scientific idea that has not been tested is a
   A. variable.   C. control.
   B. hypothesis.   D. bias.
2. List and describe the activities involved in the scientific method.
3. What types of cases are best investigated with a survey rather than an experiment?
4. Why must an experiment or a survey be controlled? What is the difference between the control group and the experimental group in a controlled experiment?
5. Why is it important to have a single variable between a control group and an experimental group? What happens if there is more than one variable?

1.4 The Limits of Science

The very nature of the scientific method greatly limits science. Although some people attribute godlike capabilities to it, science is actually little more than what man can sense around him. No matter how often an experiment or survey is repeated, no matter how carefully observed, no matter how accurately recorded, no matter how well it works or how valid it appears, a scientific "fact" can be, and often has been, wrong. To realize why, the limitations of science must be understood.

Limitations Inherent in the Scientific Method

Scientific investigations must deal with physical phenomena (sing., phenomenon) because experiments or surveys must have observable, measurable data to support a conclusion. A phenomenon is any observable object, property of an object, or process. A problem investigated by the scientific method is generally stated so that it can be answered with a yes, a no, or a number (such as a percentage or ratio). Questions asking how or why are not measurable and are therefore beyond the scope of science. The scientific method cannot explain a phenomenon.

Measurable, observable results require the use of most if not all of man's senses. The phenomenon must be seen, felt, heard, tasted, or smelled. Man's senses, of course, are often not accurate and can be fooled. If the phenomenon cannot be observed, a device capable of measuring the phenomenon must be used. For example, man cannot see, feel, touch, or taste x-rays, but photographic plates and other devices can measure their presence. However, using more devices to make observations introduces more variables and increases the possibility of mechanical errors.

The beginning of life, the future, and spiritual concepts such as heaven, angels, man's soul, and hell cannot be observed or measured; they are beyond the domain of science. All that can be known about these things is revealed in Scripture. These things are a matter of faith.

Another problem scientists have is biased thinking. A **bias** is what someone wants to believe. A scientist employed by a tobacco company to research the effects of smoking might tend to emphasize results that would please his employer and overlook results that would show tobacco to be harmful. Since a scientist must choose an answer, his answer will likely reflect his bias.

Limitations of the Results of the Scientific Method

Since it is nearly impossible to completely limit the variables in any experiment or survey, the exact answer (absolute answer) is not obtainable. Even if a person did obtain the exact answer, it would be so hidden in a group of other answers that it would be unrecognizable. Scientists then must choose the best answer to a problem. A valid choice would be an answer that is usable—one that is supported by repetitions of the experiment and that can be used to predict answers for similar situations in the future. But just because an answer is valid does not mean it is correct.

1.4

⇒ *Objectives*
- Recognize the limitations of science
- Describe how bias may influence research
- Explain the difference between valid and workable results

⇒ *Key Terms*
phenomenon
bias
workability

phenomena: (Gk. *phainesthai*—to appear)

1-14

Do you think that the water consumption of Chicago (left) residents is the same as that of San Antonio (right) residents?

The Limitations of Science

- Science must deal with observable, measurable phenomena.
- Science can only describe, not explain.
- No experiment can be completely controlled.
- Observations may be faulty.
- Man's beliefs affect his judgment.
- Science must deal with repeatable results.
- Science cannot deal with values or morals.
- Science cannot prove a universal statement.
- Science cannot establish truth.

What scientific statements could be made about these fossils?

Information gained by the scientific method is workable if it can be used in other circumstances. If the information cannot be used to predict outcomes in similar situations, it lacks **workability**. For example, suppose that controlled experiments performed in a laboratory support the hypothesis that a certain type of plant needs "chemical X" to produce fruit. However, if these plants produce fruit equally well in the countryside with or without "chemical X," the results of the experiment are not workable. The experimental results may be valid for those plants grown in the laboratory, but in the field some other variable in the soil, water, air, or some other unknown caused the plant to produce fruit without "chemical X." In this case the scientist has a piece of information that may be valid but cannot be used to predict the plant's production of fruit in other situations: the information lacks workability.

Suppose it is necessary to determine the amount of water people in the United States drink each day. This information will be used to help project the country's water needs for the next twenty years. A survey is taken of one hundred people living in the Chicago area. If the survey is taken several times in that area with similar results, the results can be considered valid. But would the results from the survey answer the question—would they be workable? No one can be sure that water consumption in Chicago is typical for the whole country without performing many similar surveys in other areas. Therefore, highly valid results of a survey or experiment may be unworkable if the survey or experiment is not typical or is too narrow in scope or if it included conditions that are not measured.

Limitations of the Use of the Scientific Method

People sometimes expect to be able to use science to make their decisions. Scientific research has developed compounds that, if put into a city's water supply, would kill everyone who drank the water. However, science cannot decide to poison the water supply. Men, not science, must make value judgments and moral decisions.

Scientific investigations cannot prove a universal statement. A universal statement includes terms such as *all*, *always*, *no*, or *never*. The use of these words makes the statement impossible to prove through experimentation. To proclaim that all woolly mammoths are extinct cannot be proved. Every place in the world would have to be checked simultaneously since the woolly mammoths could change locations. A more limited statement such as "There were no living woolly mammoths in the park at noon yesterday" may be provable, depending on the size of the park and how many people were observing.

Science cannot be used to establish truth. The fact that man can repeatedly observe something and measure it does not prove his hypothesis. The results of the scientific method may only support or not support a hypothesis. The more support a hypothesis has, the more valid it is assumed to be. But no amount of scientific experimentation, or any other type of work, can prove any statement beyond all doubt.

➡ *Review Questions 1.4*

1. Why is workability an important criterion for scientific knowledge?
2. Why can science not prove a universal statement?
3. What is the difference between the term *valid* and the term *proved*? (Hint: Scientists are seeking predictable and workable answers to problems.)
4. List and describe the nine limitations of science.

5. List two areas of human concern in which the scientific method is useless and explain why.
6. Why is a large problem more difficult to deal with than a small one when using the scientific method?
7. Science can only describe, not explain. Why is this statement true?

1.5 Science and the Christian

In Subsection 1.2, we offered a Christian definition of science. Throughout this chapter, we have expanded that definition through examples, clarifications, and even warnings about the limits of science. The important thing is not to have a series of words memorized to recite when asked what science is but to have a working understanding of what science is, what it can do, and what it cannot do. It is important for the Christian to realize what science is and to recognize the limitations of science so that he can see the proper relationship between science, God, and His Word.

Scripture's Scientific Challenge

No matter how much a Christian looks forward to heaven, he must accept the fact that God has placed him on the earth to faithfully do His will while the Lord Jesus prepares his eternal home. Part of doing His will is to "preach the gospel to every creature" (Mark 16:15). Other tasks involve worshiping and praising God, praying, and believing in His name. But the first recorded duty of man was one that required at least some scientific knowledge and skill. In Genesis 1:28 God told man to subdue the earth and to have dominion over it.

There are two basic methods of subduing and having dominion over anything. If a ruler wants to subdue and dominate a city, he can send his army to destroy the entire city. The city will be subdued, and he will dominate it. But what would he rule? A mass of rubble has no great value. It is better to subdue and dominate without destroying. The better method may take more time, but it can be just as effective and may produce better returns.

Second to the Bible, science is man's principal source for the knowledge needed to subdue and have dominion over the earth without destroying it. Mankind will never completely dominate

➡ **Objectives**
- Describe a Christian philosophy of science
- Explain the difference between applied and pure science
- Give examples of the research and technical methods in practice

➡ **Key Terms**
pure science
applied science
research method
technical method
worship

1-15

Block cutting is a technique used to harvest lumber without destroying the entire forest.

the earth because science has many limitations and man is a sinful creature unable to perfectly rule the earth. Not until Christ returns for His thousand-year reign will the earth be completely subdued and dominated. This does not mean that mankind should not try to properly subdue the earth; man should do the best that he can while realizing his limitations.

For instance, timber companies could get more lumber in less time if they leveled whole forests and moved on to continue the process. However, this type of domination is not wise from an ecological or even an economical standpoint. By judiciously harvesting selected trees or tracts of land and replanting to replace those trees that are taken, the forester helps preserve the forests, the forestry industry, and even his own job. When, where, and how to plant which trees for quickest production of lumber is not information given in Scripture. Yet men, using God-given intelligence, can determine how best to grow trees.

As the world population continues to grow, however, mankind could run out of wood despite reforestation. Other sources of energy, such as natural gas, oil, solar power, and electricity made from flowing water or nuclear power, have been used to heat homes in many countries. Less than a hundred years ago, these heating methods were only ideas. Today, through scientific endeavors, they are commonplace.

God Expects Man to Use Science

The physical world is made of substances that operate under God-ordained laws. Scripture teaches that God created the world and sustains it and that man is to subdue and have dominion over it. If he uses his God-given intelligence, he can subdue and have dominion over the world without destroying it. If he ignores what science can teach, he will have wasted two God-given gifts: the earth and his intelligence.

God has revealed in the Bible all the spiritual knowledge man needs. But God did not reveal all the scientific knowledge mankind would need in order to exist on the earth until His return. God, according to His timing, permits man to discover scientific knowledge. For example, beasts of burden, such as horses and donkeys, cannot do the things trains can do. Moreover, trains by themselves are not able to do what trucks and cars do today. God has permitted man, through science, to discover the knowledge he needs to build things he can use.

Some people feel that science is basically bad and attempt to return to what they consider more "natural" methods of doing things. They may even feel that returning to old ways is somehow godlier. But the old way of doing things is not necessarily the best way. For example, the fact that previous generations used candles and torches for light does not make those light sources better than electric lights. Before electricity was widely available and affordable, candles and torches were the best methods that man's knowledge had devised, but scientific investigation has since supplied additional knowledge about illumination. There is nothing wrong with using that information. Running from scientific ideas and returning to old methods of doing things is not turning

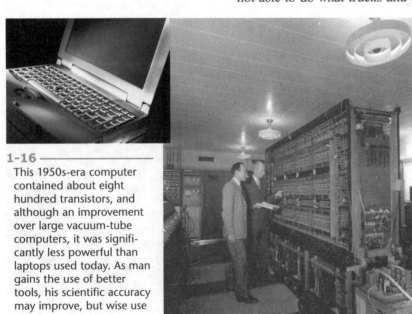

1-16

This 1950s-era computer contained about eight hundred transistors, and although an improvement over large vacuum-tube computers, it was significantly less powerful than laptops used today. As man gains the use of better tools, his scientific accuracy may improve, but wise use of information happens only when he bases decisions on God's Word.

Science: Pure or Applied

Scientific activities can be classified as either pure or applied science. This is not a value judgment regarding rightness but a judgment of how workable that science has become in our society.

Pure science is knowledge that scientific activities have produced. A scientist may, through experimentation, discover that a particular plant requires a certain nutritive element in order to produce fruit in maximum number and size. That knowledge, as well as the work behind its discovery, is pure science.

Applied science is using knowledge gained through scientific activities to solve problems. Once a person knows a plant's nutritive requirements, he can apply the proper amount of the nutrient to a field to obtain a more abundant harvest. The knowledge gained through pure science has now become applied science.

Two different types of scientific activities were involved in the above example. The first scientist used the **research method**. He recorded, classified, and analyzed data; chose an answer and verified it; and then made predictions about the nutritive element and the plant. He is a research scientist and has used the scientific method of thinking to discover knowledge.

The person who follows the instructions of the research scientist in order to apply the correct amount of nutrient to a field of plants is a technician. A technician, in this case, uses the techniques prescribed by the research scientist to predict the nutrient needs of a particular field. The use of prescribed techniques to obtain information about a particular example is the second type of scientific activity, called the **technical method**.

A research scientist may devise a test to determine the presence of a certain chemical in human urine. The person in the laboratory who follows the instructions of the research scientist to test human urine samples is a laboratory technician. Both research scientists and technicians are necessary. Without research, scientific knowledge is not gained. Without technicians, scientific knowledge is not used.

Study these pictures. Can you tell which of these people are using the research method and which are using the technical method? Can you tell which illustrates pure science and which illustrates applied science? The answers are on page 22.

back to God. Instead, it is the burying of scientific talents, much like the burying of spiritual talents spoken of in Matthew 25:14–30.

Some Improper Attitudes of Christians Toward Science

One wrong attitude toward science is to believe that science is anti-God. Since science is the discovering of workable information about God's creation, science is not inherently bad. Scientific information can be used for good or bad purposes, but science does not decide how information is used—men do. Science is not evil because men have abused the use of scientific knowledge.

It is interesting to note that when cars were first becoming widely used, some Christians condemned them, saying that these newfangled scientific inventions were going to cause man to sin. A person can sin by using a car, but simply *having* the car is not sin; it is man's *sinful use* of the car that is wrong. A car, rightly used, can be a tool of blessing. The same is true of many modern conveniences, like Internet technology, cell phones, and laptops.

It is also wrong for a Christian to think that scientific achievements can replace faith in God. Science is often portrayed as having "all the answers." Although science cannot supply all answers, and the answers it currently provides may need to be reevaluated as more information is obtained, science can and does supply workable answers. This belief may cause some people to place their faith in science rather than in God. They look to science for all the answers, but it is impossible for science to supply them. Science supplies information to solve physical problems. Man's real problem cannot be answered by science: man is guilty because of sin and is in need of salvation from an eternity in hell. Only faith in the Lord Jesus Christ and the salvation that He offers can solve man's sin problem. A person who believes that man's wisdom can solve this problem has put his trust in something Scripture says is foolishness (1 Cor. 3:19).

A Proper Christian Attitude Toward Science

Since science is a tool used by man to discover useful knowledge about God's creation, science itself is not good or bad. Men use or abuse knowledge gained by scientific investigations; therefore, it is wrong for a Christian to fear science, to condemn it as being anti-God, or to believe science can destroy man's faith in God. It is also wrong for a Christian to ignore science. Man's intelligent use of God's creation is part of subduing and having dominion over the earth.

The psalmist states in Psalm 19:1, "The heavens declare the glory of God; and the firmament sheweth his handywork." God's power and majesty are revealed in what He has created. The person who looks carefully at creation can know more about the God Who created it, and knowing God leads to worshiping Him.

True **worship** can be defined as man's recognizing his insignificance and turning his thoughts in love, reverence, adoration, and obedience to the almighty God. Worship does not require a church building, a preacher, or soft organ music. All too often, worship does not take place even with these helps. Worship takes place when a person submits every area of his life (including thoughts, words, and deeds) to the authority of God and seeks to know and obey God. Worship can take place when men look at creation and see the God Who designed, created, and sustains it. True scientific knowledge can enhance worship by giving more insight into the wonders of God's creation.

1-17
Even some of God's smallest and most fragile creations can lead us to worship Him.

Review Questions 1.5

1. Which of the following is *not* considered a limitation of modern science?
   A. Man cannot totally depend on his observations.
   B. God's Word does not cover many scientific points.
   C. Man is prone to bias in his interpretation of facts.
   D. No scientific experiment can be completely controlled.

2. What is the relationship between pure science and applied science? Give an example of each.

3. What is the relationship between the research method and the technical method of science?

⊙4. Some people look to science as a god. Since scientists have supplied the knowledge to heal many diseases and to produce many conveniences, many people look to science to solve all their problems. Why is this view of science unacceptable to a Christian?

⊙5. Some Christians turn away from anything that appears scientific. Since they may not understand science and cannot find most scientific facts in the Bible, they believe that science is anti-God. Give five biblically based reasons that these Christians should reconsider their opinions.

1C—Biology and the Study of Life

Biology is often defined by its Greek word origin: *bios*, which means "life" or "living," and *logos*, which literally means "word" but has come to mean "study" or "science of." The previous sections of this chapter discuss the term *science,* the *logos* of biology. Now the discussion turns to the other portion of the definition of *biology*—living. What does it mean to be alive?

1.6 The Attributes of Life

Rather than presenting definitions, this study will look first at a list of the attributes of life. This list does not include all the attributes, but it does present the major characteristics of living things. Since these attributes describe a very complex phenomenon—the **living condition**—they necessarily contain overlapping ideas.

Not everything that possesses a few of these attributes is alive. Living things must have all these attributes. Nonliving things, such as machines, computers, light bulbs, crystals, and chemicals, may have one or several of these attributes, but only those things that have all of them are alive.

✦ *Exhibits movement.* Normally most people think of movement as motion from one place to another. Actually, moving from place to place is *locomotion.* Plants, most fungi, many bacteria, and even some animals cannot carry on locomotion; nonetheless, they are alive and do exhibit movement.

Movement may be **internal movement** as well as locomotion. Even when you are perfectly still, blood is moving through the blood vessels. From birth to death, small structures are constantly moving inside the cells of the body, even when it is at rest. Plants move water internally, dissolve minerals, and process foods. Even a stationary single-celled organism carries on internal movement.

✦ *Achieves growth.* Though stalagmites or buildings may be said to grow, they are nonliving things and grow only in the sense that more material is added to them. But to increase the size of a brain, you cannot similarly add more brain substance directly into the skull. Living things do not grow that way. Organisms grow by **assimilation** (uh SIM uh LAY shun)—the assembling of the component parts that make up their living material. Growth is achieved by an organism, not done to an organism.

Growth does not always affect the size of an organism. Once an organism reaches maturity, growth is often the replacement of worn-out cells. An adult, as a rule, does not continue to grow in size but does grow new cells and parts of cells to replace older cells until he dies.

✦ *Reproduces.* At one time or another during its life span, every normal living thing is capable of reproduction. **Reproduction** is the making of another organism that has characteristics and limitations similar to the original. Splitting a rock in two is not reproduction since the two halves cannot become like the original. Some small organisms do reproduce by splitting in two. But unlike the rock, each new organism has the capability to become like its parents, just as a puppy becomes like its parents. Puppies are not "half dogs." They are small dogs that have inherited a dog's characteristics and limitations.

Although a puppy has been nourished by its mother's body, very little of the puppy (only the original egg and sperm cells) was actually part of the mother and father. After growing, the puppy will be similar to its parents.

Reproduction of living things is not making duplicates from a mold. A mother dog is not a mold for her puppies. She and the father dog have given the puppies the genetic information necessary for them to grow into dogs.

✦ *Comes from similar preexisting life.* As a result of careful observations of controlled experiments and natural conditions, biogenesis is now generally accepted.

1.6

✦ **Objectives**
• List and describe the attributes of life
• Differentiate between food and energy

✦ **Key Terms**

biology	variation
living condition	organic
internal	inorganic
movement	irritability
assimilation	food
reproduction	life

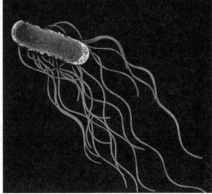

1-18

Although these stalactites and stalagmites (top) "grow," they are not alive. This bacterium (bottom), however, exhibits all the attributes of life.

Oak trees, like all living things, reproduce after their kind.

The idea that life comes from similar life is important. God created humans and all of the other kinds of organisms with the ability to reproduce after their own kind (Gen. 1:12, 21–28); therefore, humans reproduce humans, oak trees reproduce oak trees, and cats reproduce cats. The idea that all life forms descended from a common ancestor cell that originated from nonliving chemicals is contrary to the teaching of Scripture and is therefore contrary to a truly Christian worldview. It is true that no kitten will be exactly like its parents. Some kittens may be slightly larger or smaller or have different colors, depending on the genetic information given to them by their parents. These differences, some minor and unobservable and some very noticeable, are termed **variations**. But variations are limited. The eventual adult size of a kitten, for example, is limited to a certain range. The genetic information inherited by a kitten (along with its diet, exercise, and physical surroundings) will determine what size within the genetic range it will attain.

✦ *Has similar chemical makeup.* All living things are made of the same basic chemical elements in similar compounds. Since all organisms move, grow, reproduce, and come from preexisting life, the idea that all things have a similar chemical makeup is not surprising.

The substances that living things produce contain varying amounts of carbon and are called **organic** chemicals. Much of the composition of an organism is living material that is constantly changing. Many organisms, however, produce substances that are nonliving and often remain long after the living material has deteriorated. These nonliving substances are organic but are not alive. The shell of a snail, the hair of your head, and the outer bark of a tree are examples of dead organic materials. Of course, the body of an organism that has just died can also be considered dead organic material before it decomposes. Those things that are not alive and have never been alive, such as rocks, water, steel, and glass, are considered **inorganic**.

✦ *Is composed of cells.* Every living thing is made of either a single cell or many cells. Cells, (discussed in Chapters 3 and 4) the units of life, are limited by the membranes they manufacture.

✦ *Exhibits irritability.* All living things respond somehow to various forces in their environment. **Irritability** is the capacity of an organism to respond to stimuli. For example, human beings are highly sensitive to light and respond not only to its intensity, direction, and color but also to its source. An earthworm, on the other hand, does not have eyes but merely light-sensitive areas. Though human beings need to respond to what they see, an earthworm merely needs to know whether it is in the light or not; in its burrow seeing is not important. Sensing light, however, the earthworm is aware that it is not in its protective hole and can do something about it. God designed each organism to be sensitive to those conditions in the environment that affect that organism.

✦ *Requires energy.* All living things require a constant supply of energy. Humans obtain energy from food. Plants and some other organisms obtain their energy directly from the sun and store it for future use. **Food** can be defined as an organic, energy-containing substance. The energy that plants store is in the form of a food. Since many organisms, like plants, do not eat food, it is not accurate to list "eating" or "requires food" as an attribute of life. Organisms require energy.

✦ *Maintains a high level of organization.* Being alive requires that a collection of molecules be highly organized. This organized complex of molecules forms the structures of living things. Energy is required to keep these molecules in that organized condition that permits life. As soon as energy is no longer expended to maintain the organization, the molecules fall into disorder, and the organism begins to die. Think of it this way: in part, a person is a self-assembled mass of what he has eaten; he is being held together in complex order by the energy he has obtained from his food.

✦ *Faces death.* All things that are alive will eventually die. This fact is true for everything from petunias to people. We may be able to avoid death for several decades, but we are not able to avoid it entirely. The saddest truth we learn from biology is that nothing stays alive forever. All that lives will die. We have gone astray from our God, and He has caused our world to go astray from us.

Physical life is not some special element or compound that some objects have and others do not. Life is a condition, a state in which something exists. **Life** can be defined as a highly organized cellular condition that is derived from preexisting life; that requires energy to carry on processes such as growth, movement, reproduction, and responses; and that faces death. It is more important, however, to understand what life is and to recognize it than to memorize a definition.

1-20

This gerenuk, an African antelope, gets its energy from the plants it eats. Where do the plants get energy?

⇒ *Review Questions 1.6*

1. List and briefly describe the attributes of life.

2. Why is life best described as a condition?

1.7 The Study of Life

Studying life has always been difficult. Analyzing the chemicals that make up a living body does not reveal what life is. Diagrams of a watch can tell an engineer how the watch works; but careful descriptions of the size, shape, color, and even chemical makeup of the various structures of living things tell little about life processes. A complete set of diagrams of the human body is simply inadequate to describe how it works.

The next section discusses some tools and techniques scientists use to study living things and some of the difficulties they experience as they study.

The Microscope: A Tool for Biological Study

Although a description of an organism's structures is not adequate by itself to explain life, knowledge of the structures is essential. Unassisted human vision, however, cannot see millions of microbes or the parts of cells that make up an organism. **Microscopes** aid in studying these tiny structures.

The first microscopes were probably made in the first century AD by filling glass balls with water. Glass lenses that were used to magnify structures were introduced in the Middle Ages. A single lens constitutes what is called a simple microscope. Around 1590, a Dutch eyeglass maker, Zacharias Janssen, mounted two lenses in a set of adjustable tubes. The lens closer to the observer's eye magnified the already enlarged image of the lens closer to the object being observed. Such a microscope, consisting of two lenses or two sets of lenses, is called a compound microscope. Because of the distance necessary between the lenses, early compound microscopes were often over 0.6 m (2 ft) long.

1.7

⇒ **Objectives**

- List and compare the units used to measure microscopic objects
- Describe the difference between light and electron microscopes
- Identify the parts of a light microscope
- List and describe some of the difficulties when studying living organisms

⇒ **Key Terms**

microscope	body tube
compound light microscope	electron microscope
objective	reflection
ocular	refraction
stage	resolution

microscope: micro- (Gk. *mikros*—small) + -scope or -scopy (Gk. *skopein*—to view)

1-21
Zacharias Janssen made one of the first compound microscopes.

The best known microscopist is probably Anton van Leeuwenhoek (LAY wun HOOK), a Dutch merchant who made a hobby of grinding lenses in the early 1600s. Leeuwenhoek used simple microscopes rather than the cumbersome compound microscopes of his day. Some of Leeuwenhoek's lenses were smaller than a pinhead, and he had to push the lenses almost into his eye in order to see anything.

Leeuwenhoek made over two hundred different microscopes, which had various mountings to view many types of specimens. Some of his discoveries include circulation of fluids in the tiny vessels of a living fish's tail, thousands of microbes, blood cells, sperm, and even various human tissue structures. For most of his long life, Leeuwenhoek sent his observations of "very many wretched little beasties all cavorting about very nimbly" to the Royal Society of London. Leeuwenhoek, who is called the Father of Microscopy, never saw an object magnified more than 200 times. Today, even the microscopes commonly used in high schools can magnify 400 times ("400 times" can also be written "400×," in which the "×" means "times larger").

Leeuwenhoek and the "Little Animals" by Robert Thom, UMHS.15

1-22
Anton van Leeuwenhoek and one of his single-lens microscopes. The specimen was placed on the tip of a pointed screw and observed from the lens to the right.

The Compound Light Microscope

The **compound light microscope** is commonly used in classrooms and laboratories today. It consists of one set of lenses to magnify an object and another set to serve as a telescope to further enlarge the image and permit the observer to look at it from a convenient distance. In a light microscope the two sets of lenses are located in the **objective** and the **ocular** (AHK yuh lur), or eyepiece. Many microscopes have a revolving *nosepiece* that contains several objectives with different magnifications.

Many of the specimens that are observed using a microscope are permanently mounted on a glass slide. If a person examines the slide carefully, he may be able to tell that the specimen is colored. Special stains that react with specific chemicals in the structures of the specimen are used. Different colors react with different chemicals, so some specimens may have various colors. Thus, various structures are easily identified in a stained specimen.

When a person looks through the microscope, he sees light that has passed through the specimen on the **stage**, through the objective lens, through the **body tube**, and through the eyepiece lens. Actually, he is seeing the "shadow" of the specimen since the observed light has passed through the specimen rather than being reflected from its surface. Moving the objective lens closer to or farther from the specimen brings the image into focus.

Many modern microscopes have two eyepieces. Some of these microscopes (especially higher-powered ones) give the same image to both eyes. Others, however, are stereoscopic microscopes that provide two slightly different views of the same object, similar to the images you get when looking at an object with both eyes. These slightly different views result in depth perception. A number of these stereoscopic microscopes are low-powered and use light reflected from the specimen, which is placed relatively far from the objective lens. Microscopes of this nature are frequently used for dissecting small specimens.

1-23
Early compound microscopes

ocular: (L. *oculus*—eye)

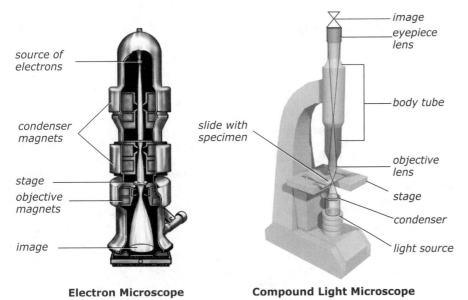

source of electrons

condenser magnets

stage

objective magnets

image

Electron Microscope

image

eyepiece lens

body tube

slide with specimen

objective lens

stage

condenser

light source

Compound Light Microscope

1-24

A comparison of an electron microscope and a light microscope

The Electron Microscope

One of the most useful tools for biological investigation developed in the twentieth century is the **electron microscope**. The electron microscope replaces light rays with electrons that are exceptionally small and that travel in a straight line. Because more electrons than light waves can pass through a given area at a given time, electron microscopes can magnify many times greater than light microscopes. Since electrons are smaller than light waves, electron microscopes have a much greater capacity for resolution.

There are two basic types of electron microscopes—transmission and scanning. The transmission electron microscope (TEM) produces an image similar to a light transmission microscope. Instead of light passing through the specimen, the TEM uses electrons. However, electrons do not pass through

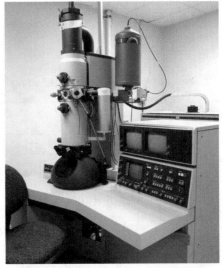

1-25

An electron microscope

Principles of Light Magnification

Normally the **reflection** of light from the surface of an object makes the object visible to the eye. If, however, a lens is placed between the object and the eye, the light waves will be bent. The wavy patterns seen through some window glass or the apparent bend in a drinking straw placed in water are the results of light waves being bent as they pass through the transparent substances. This bending of light as it passes through a substance is called **refraction**. The refraction of light as it passes through a curved lens either enlarges or shrinks an image.

There are limits to magnification. Since only a certain amount of light reflects from or passes through a specific point at one time, no image is visible if these waves are spread too far apart. The ability to distinguish fine detail is called **resolution**. The unaided human eye can distinguish two points 0.1 mm apart, but points closer than that appear as a single point. If two points 0.1 mm apart are magnified 100×, however, they will appear 10 mm apart. When an image is magnified, the resolution increases.

There is, however, a limit as to how close two points can be and still appear as two individual points when mag-

Refraction causes the pencil to appear enlarged and bent.

nified with a light microscope. In part, the ability of a lens to clearly separate two points depends on the quality of the lens. Due to the size of light waves, points too close together become distorted as they are magnified. Therefore, even the highest quality lens has resolution limits.

Biological Measurements

Most light microscopes magnify things 100× to 1000× larger than they really are. Some electron microscopes magnify objects 1 000 000×, but about 250 000× is the highest practical magnification. One can quickly get into hundredths, thousandths, millionths, and billionths of an inch. Although the English measurement system was once used all around the world, the United States is the only remaining country that primarily uses the English units of measure, which include inches, feet, miles, and yards. In order to communicate more effectively, scientists use a single standard measurement system—the Système International d'Unités (International System of Measurements), or SI. The SI has seven fundamental base units of measurement; however, the units most used in this biology course are those measuring distance, size, and mass. The table in Appendix C shows many of the SI-to-English conversions.

Why does it make any difference, since a person can always convert from one system to the other? An example best answers that question. In 1999, the Mars Polar Lander was set to explore a portion of Mars that had never been explored. Excitement abounded in the control room until all communication with the Polar Lander was lost. After several intense days of attempting to contact the Polar Lander, the mission was scrubbed—everything was lost. What had happened? One part of the program for the Lander had been done in English units while the remainder was in SI units; the oversight caused the mission to fail.

Since this text uses the International System of Measurements, a close examination of the SI is needed. The meter (m) is the basic unit of distance in the SI (1 m = approximately 39 inches). All of the other SI measures of distance are based on equal divisions of the meter. If a meter is divided into 100 equal parts, each part is called a centimeter (cm). Likewise, if a centimeter is divided into 10 equal parts, each part is a millimeter (MIL uh MEE tur) (mm). A millimeter scale is often shown on rulers, and a single millimeter unit can easily be seen with the unaided eye. Some large microbes can be measured in millimeters.

Metric Conversions

	Meter (m)	Centimeter (cm)	Millimeter (mm)	Micrometer (μm)	Nanometer (nm)	Angstrom (Å)
Meter (m)	1	0.01	0.001	0.000 001	0.000 000 001	0.000 000 0001
Centimeter (cm)	100	1	0.1	0.0001	0.000 0001	0.000 000 01
Millimeter (mm)	1000	10	1	0.001	0.000 001	0.000 0001
Micrometer (μm)	1 000 000	10 000	1000	1	0.001	0.0001
Nanometer (nm)	1 000 000 000	10 000 000	1 000 000	1000	1	0.1
Angstrom (Å)	10 000 000 000	100 000 000	10 000 000	10 000	10	1

To appreciate the smallness of these measurements, consider the following: If 1 meter were enlarged to equal the distance between Spokane, Washington, and Washington DC, then
- a millimeter would be the distance around the Indianapolis Motor Speedway,
- a nanometer would be the thickness of two nickels, and
- an angstrom would be the thickness of a paper clip.

meter: (Gk. *metron*—measure)

centimeter: centi- (L. *centum*—hundred) + -meter (measure)

millimeter: milli- (L. *mille*—thousand) + -meter (measure)

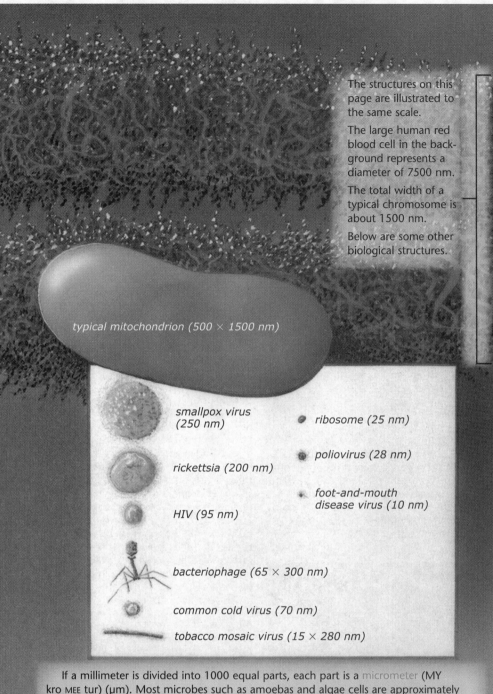

The structures on this page are illustrated to the same scale.

The large human red blood cell in the background represents a diameter of 7500 nm.

The total width of a typical chromosome is about 1500 nm.

Below are some other biological structures.

typical mitochondrion (500 × 1500 nm)

smallpox virus (250 nm)

ribosome (25 nm)

rickettsia (200 nm)

poliovirus (28 nm)

HIV (95 nm)

foot-and-mouth disease virus (10 nm)

bacteriophage (65 × 300 nm)

common cold virus (70 nm)

tobacco mosaic virus (15 × 280 nm)

substances easily. Specimens viewed on an electron microscope must be sliced extremely thin and treated with relatively harsh chemicals to allow the electrons to pass through. The specimen in an electron microscope must also be placed in a vacuum because air molecules will affect the path of the electrons. A typical TEM can clearly resolve two dots just 0.21 nm apart.

Electrons cannot pass through glass lenses; therefore, electromagnets of various charges and strengths are used to pull or push on the electrons, either condensing or separating them. The electrons are focused onto photographic plates, where they are detected by electron-sensitive devices, magnified, and transferred to a screen similar to a television for viewing. Electron micrographs, the pictures produced by an electron microscope, have no color but only light and dark areas where electrons passed or did not pass through the stained specimen. Computer technicians often add color to enhance the image and make it easier to visualize the structures.

The scanning electron microscope (SEM) is used to obtain surface images of structures or organisms rather than internal structures. After the specimen is coated with an electron-dense substance, the beam of electrons moves back and forth across the surface of the specimen. The coating on the specimen emits its own electrons that are picked up by a special detector. The detector transfers the energy from the electrons to a screen, where an image of the specimen's surface is displayed. The SEM can produce images up to 20 000×.

The electron microscope can magnify from 1500× to 250 000×. In other words, it starts at the highest magnification possible on a light microscope and can magnify that image about 160×. Throughout the text are micrographs produced by light microscopes, TEMs, and SEMs.

If a millimeter is divided into 1000 equal parts, each part is a micrometer (MY kro MEE tur) (µm). Most microbes such as amoebas and algae cells are approximately 100 µm long. To see things of this size, one must magnify the image at least 100× on a light microscope. Many human body cells are about 20 µm, and human red blood cells average only 7.5 µm in diameter. To make these structures clearly visible, 400× is necessary. Most medium-sized bacteria are about 1 µm. These require 1000× on the light microscope to be seen.

If a micrometer is divided into 1000 equal parts, each part is a nanometer (NAN uh MEE tur) (nm). Some small bacteria are about 100 nm, and the mumps virus is approximately 115 nm. This virus appears as a mere dot using the highest power of a standard light microscope but is easily seen with the electron microscope. The poliovirus—one of the smallest viruses—is only about 28 nm and cannot be seen using a light microscope.

Although not an SI unit, the angstrom (Å) is often used in scientific literature. Many structures inside the cell are measured in angstroms. If a meter is divided into 10 billion equal parts, each part is an angstrom, and 10 Å equal 1 nm. The membrane surrounding a cell is usually 75–100 Å thick. Structures larger than 5 Å are theoretically visible with an electron microscope if they can be properly treated. Hydrogen, the smallest atom, is about 0.5 Å in diameter.

micrometer: micro- (small) + -meter (measure)

nanometer: nano- (L. nanos—dwarf) + -meter (measure)

micrograph: micro- (small) + -graph or -graphy (Gk. graphein—to write)

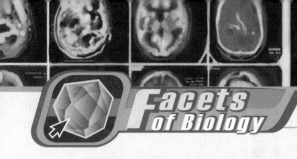

Today's scientists have many sophisticated tools that allow them to observe many objects that would have been impossible to see in the past. Some of the techniques described below are used in the medical profession to aid in diagnosing diseases. Others are used primarily in researching the anatomy and function of a particular organ or structure.

Radiology

Radiology (x-ray technology) was the first development that enabled scientists and physicians to see inside the body and was first used extensively during the First World War in treating combat injuries. X-rays pass easily through some materials but are either absorbed or reflected by denser materials. When a structure is exposed to x-rays, those rays that pass through the structure will expose photographic film to produce an image. Thus, scientists can learn about internal structures without cutting into the person or organism. Since many structures in living things are similar in density, scientists often must use substances that make the structures denser. To view the structures of the digestive system, people are often asked to drink an x-ray dense substance (a substance which blocks x-rays). The x-ray pictures then show the digestive system's structures.

Computerized Axial Tomography Scan (CT Scan)

Since the 1970s, CT scanning has become the most significant medical imaging technique used to screen for and diagnose diseases. In tomography, x-rays are passed through a single area from many different angles and are picked up by a detector. The information is fed into a computer, compared, and made into a picture of the anatomical structure. A CT scan can be

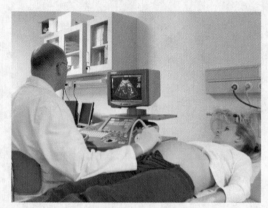

Ultrasound is a common and safe method used to visualize an unborn child.

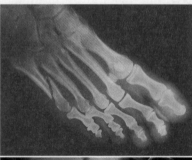

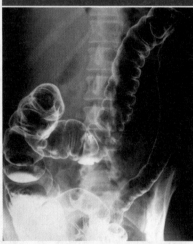

X-rays can be used to evaluate hard structures such as bone (top—note broken bone) and some soft structures such as the intestine (bottom), using special liquids that are swallowed to enhance the image.

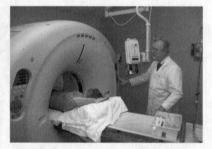

Patient receiving a CT scan

used in many different ways to study various internal body parts. A CT scan machine is about one hundred times more sensitive than a conventional x-ray machine and does not use x-ray dense material to identify structures. The information in the computer can also be used to create a three-dimensional image of the structure.

Ultrasonography

Ultrasonography (UHL truh suh NAWG ruf fee) is much like the sonar used by ships. A device releases high-frequency sound waves and then records the echoes of the sound waves as they bounce off internal structures. It is pressed directly on and then moved slowly over the surface of the body, where the anatomical structures being examined are located. The resulting data is fed into a computer that produces an image based on the speed and quality of the echo. The pictures are not as accurate or precise as CT scans or x-rays but are quite useful in observing unborn children, who can be harmed by x-rays. Ultrasound pictures are also useful in evaluating the heart, which is often difficult to observe with x-rays because heart muscle is soft tissue.

Magnetic Resonance Imaging (MRI)

An MRI produces incredibly clear, detailed pictures of the anatomy of organisms from any plane. Rather than using x-rays or sound waves to create an image, MRIs create a magnetic field into which the patient is placed. By altering the magnetic field with a specific radio frequency, the magnetic field in the tissue is altered. (The protons of the body's water molecules align with the direction of the magnetic field.) The MRI machine can detect the changes in the magnetic field, and computers translate the signals into images. Patients have to be closely screened to ensure that they do not have metal objects in their body that the MRI machine might dislodge or that might distort the image. For example, patients with pacemakers or other metallic implants, such as orthopedic devices, may not be able to be "scanned" safely since a patient may be harmed by the movement of the metallic devices by the powerful magnetic forces of the MRI machine.

Positron Emission Tomography (PET)

The PET scan produces three-dimensional images or pictures of physiological activities in the body that are more detailed than CTs, ultrasound, or MRIs. A radioactive substance that emits positrons (a tracer) is injected into the organism or person. As the positrons are emitted, they collide with electrons, and gamma rays are formed. Special screens detect the gamma rays, and computers generate an image. PET scans are used to detect the extent of cancer and heart disease and to aid in diagnosing Alzheimer disease. Researchers are continuing to find diagnostic uses for this technology.

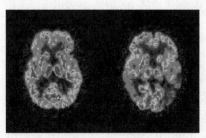

The PET scan of a human brain on the left illustrates a normal brain. The brain scan on the right reveals cocaine use.

Cellular Fractionation

Cellular fractionation (FRAK shuh NAY shun) is used to remove structures from inside a cell in sufficient quantity to be analyzed chemically. After a cell has been physically broken up and placed in a solution in a tube, it can be put in a centrifuge (SEHN truh FEWJ). The centrifuge spins the material at extremely high speeds. This causes the pieces to settle in layers. As the denser materials settle to the bottom of the tube, the lighter materials move toward the top. Between these extremes are many layers of cellular fragments of various densities. By taking a layer of material from the centrifuged tube and repeating the process several times, an almost pure collection of structures of a certain density can be obtained and then chemically analyzed. This process permits scientists to determine what chemicals compose cellular structures.

Endoscopy

Endoscopy (en DAH skuh pee) is the term for a procedure that uses a hollow tube or a fiber-optic cable to look inside a structure. Often the tube will have other devices in it that permit the scientist to supply light to the dark area for greater visibility, remove specimens for biopsy or analysis, help clarify diagnoses, perform surgery, or pump air into the structure so that it

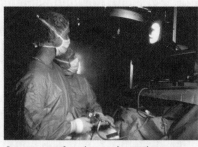

Surgeon performing endoscopic surgery on the esophagus. The tiny camera scope is inserted through the nose and directed into the esophagus.

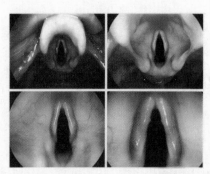

This is an endoscopic view of a healthy, rinsed esophagus.

can be seen more clearly. Different endoscopes are designed for looking at various structures. Some endoscopes are inserted through natural body openings such as the mouth or nose; others enter through incisions made by a physician or scientist.

Benefits of Advanced Research Techniques

Many scientists over the last two hundred years have contributed to the advancement of research techniques, equipment, and diagnostic medicine. These innovative technologies help scientists and physicians visualize and study the anatomy and physiology of humans and other creatures with less cutting and loss of blood and life. These men and women, using their God-given abilities, have benefited mankind by giving earlier and more accurate diagnoses of diseases and by developing more effective and less invasive surgical procedures and treatments, increasing the rate of patient survival and quality of life.

Variables in Biological Studies

Because God designed life to function in a certain range of environmental conditions, it is difficult to experiment with living things. When a person wants to measure the amounts of oxygen and hydrogen needed to produce a certain amount of water, the experiment can be repeated over and over with the same results. However, when performing experiments with living organisms, variables are harder to identify and difficult to control.

Identifying Biological Variables

An experiment to determine how much weight a horse can pull would be affected by many variables, including the type of horse, its age, health, and size. Using the same horse for all experiments could control some of these factors, but then fatigue and boredom would have to be considered. The results of such an experiment could not be applied to all horses, only to the horse that was tested. Thus, the results would not be workable.

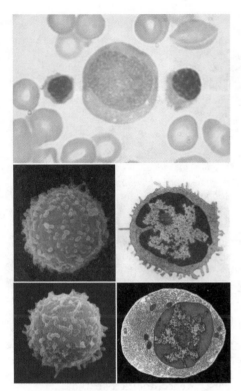

1-26

Photomicrographs of a white blood cell. Light microscope (480×) (top). SEM and TEM (15 500×) before (middle) and after (bottom) color enhancement

Suppose a student conducting an experiment for a science fair project put hamsters into two groups of five hamsters each. His hypothesis was that hamsters on a diet of seeds and carrots would gain weight and be healthy, while hamsters being fed cake and cookies would lose weight and become sickly. After several months, all the seed-and-carrot hamsters remained the same weight and appeared healthy. Although most of the cake-and-cookie hamsters also stayed the same weight and appeared healthy, two of them gained a considerable amount of weight.

Because the science fair was close at hand, the student began to draw conclusions about cake, cookies, and hamsters. The day of the fair, though, he found two litters of baby hamsters in the cake-and-cookie group along with two considerably slimmer hamsters. His conclusions had been wrong because there was a factor he had failed to take into consideration.

This story is amusing; unfortunately, eminent, qualified scientists have made the same type of mistake over and over again. All the variables involved in a living condition are almost impossible to control; therefore, reliable experimentation on living things is difficult.

The Validity of Biological Studies

From this discussion about some of the tools and techniques biologists use to study life, one thing should be evident: studying life itself is not easy. There are far too many variables to control.

- Errors can be made by the biologist in his observations.
- Errors can result if the specimen is atypical.
- Errors can be made when the techniques used are crude.

To overcome these problems, a scientist testing a living organism must usually repeat his work many times with a high percentage of similar results before considering a conclusion valid.

The conclusion must also agree with other scientific information about the topic before it can be considered valid. A scientist performing electron microscope observations may conclude that a certain molecule is box shaped. Another scientist performing experiments on the chemical makeup of the same molecule may conclude that it is ball shaped. There are several possible explanations for these contrasting results.

- The molecule may change shape as it is prepared for electron microscope viewing or as it is analyzed chemically.
- One or both scientists may be wrong in their observations or conclusions.
- The molecule may be both box and ball shaped.

In any case, further study is necessary.

Review Questions 1.7

1. Why would it be appropriate to call Anton van Leeuwenhoek the Father of Microscopy?
2. Describe how a lens magnifies an image.
3. Why must you use a thin specimen when viewing with a light microscope?
4. What structures of an electron microscope are comparable to the following requirements for the light microscope: (a) light waves, (b) lenses, (c) viewer's eye, (d) mirror or light source?
5. Why is a description of the structures found in a living organism not adequate to describe life?
6. Why is studying living organisms difficult?
7. List several problems that can affect the validity of experiments with living things. How do scientists overcome these problems?
⊙8. Based on an analysis of the attributes of living things, can a robot with a highly sophisticated computer attached to it be considered alive? Why or why not?

2

The Chemistry of Life

Facets

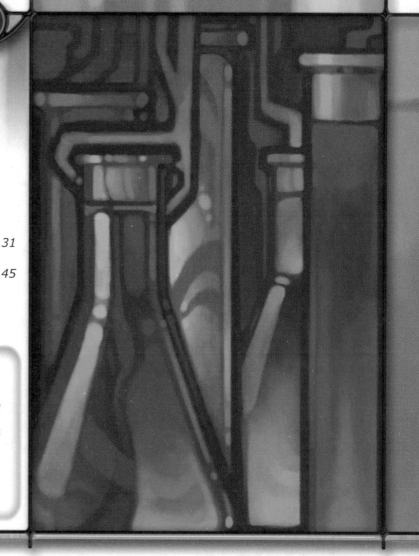

2.1

➤ Objectives

- Define *matter* and *energy*
- Describe the structure of an atom
- Distinguish between atom, molecule, compound, and mixture
- Describe covalent and ionic bonding
- Differentiate between chemical and physical changes

➤ Key Terms

matter	ionic bond
element	covalent bond
atom	hydrogen bond
neutron	molecule
proton	compound
electron	chemical change
isotope	physical change
ion	mixture

Biology was once almost entirely natural history, which is learning to recognize organisms and discovering where they live. Then came descriptive biology. Scientists, armed with hand lenses and microscopes, described in minute detail the structures of everything that grew, crept, or flew, as well as a few things that were not even alive.

With the use of new technologies, biologists started describing smaller and smaller parts of life forms. Now they are describing the basic substances of life. A scientific revolution happened because biology needed the models of another branch of science to better describe life. For this reason you must first understand *chemistry* (the study of substances and the rules that govern their behavior) in order to understand biology. Whether in a test tube or in a living cell, all substances behave according to God-ordained rules.

2A—Basic Chemistry

2.1 Matter

It is not always easy to distinguish the two basic components of the physical universe—matter and energy—because they exist and work together. For the sake of simplicity, this study of basic chemistry will consider each one separately.

The simple definition of matter (although scientists are modifying it as they learn more about the physical universe) suffices here: **matter** is anything that occupies space and has mass. Normally there are three *states of matter* associated with life: solid, liquid, and gas. These states of matter are the result of varying amounts of energy possessed by that matter. If thermal energy is added to solid water (ice), the water will enter the liquid state; if more thermal energy is added to the water, it becomes a vapor, the gaseous state.

Elements and Atoms

Matter is composed of about 90 naturally occurring elements. Occasionally, scientists synthesize other elements in the laboratory, but usually these elements are highly unstable. **Elements** are pure substances that cannot be further broken down into simpler substances by ordinary chemical reactions.

According to the atomic theory, if a piece of an element such as pure gold is cut in half and then in half again and again and again, in time there would be a piece that could not be cut again and still be gold. That piece would be an **atom**—the smallest unit of an element. Although nuclear reactions can break atoms apart, the products of these reactions are no longer the original element. A pure sample of an element consists entirely of the atoms of that element.

The *periodic table of the elements* is an arrangement of the elements according to the structure of their atoms. This table displays the *element symbols* (usually the first letter or letters of the chemical names in Latin or English). The common elements found in living organisms are listed along with their symbols in Table 2-1. The first five elements listed are essential in large

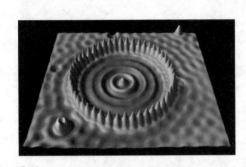

2-1

Researchers used a scanning tunneling microscope to move forty-eight atoms on the surface of this sample to create an "arena."

Periodic Table of the Elements

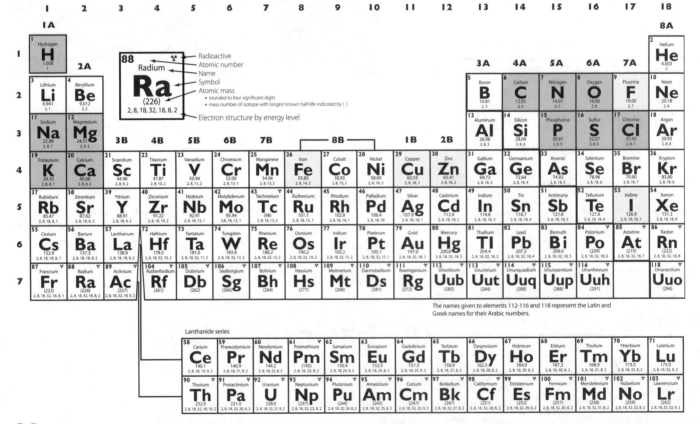

2-2

The shaded elements are found in living things.

amounts. The next six elements are essential in smaller amounts. The rest of the elements in the table are *trace elements*, which are essential to some organisms only in tiny amounts.

An atom is not solid like a marble. One model pictures an atom somewhat like the solar system: a central part and outer moving parts with much empty space in between. In this model, the center of the atom is the *nucleus*, composed of **neutrons**, which have no charge, and of positively charged **protons**. **Electrons**—particles with a negative charge—constantly circle the nucleus within orbits or shells at varying distances from the nucleus. Although there are other kinds of subatomic particles, they are not discussed here.

The electron shells are relatively distant from the nucleus. If a single oxygen atom were the size of a football field, the nucleus would be the size of a tennis ball in the center, and the electrons would be the size of pinheads circling the field at the distance of the goal posts. Most of matter, then, is empty space between the nucleus and the electrons of atoms. The matter that makes this book is quite solid, even though it is made of atoms that are mostly space. Strong forces hold the atoms of this book together so that they form a solid substance.

What forces hold electrons and protons together as atoms, and what forces bind atoms together? Scientists believe that the opposing charges of the electrons (–) and the protons (+) cause them to stay together as an atom. To understand what binds one atom to another, it is necessary to know more about the atom's composition.

Atoms of different elements vary in their number of electrons, protons, and neutrons. In a single atom, the number of electrons is always the same as the number of protons, and the positive and negative charges cancel each other, making the net charge of the atom zero—a neutral atom. If there is one proton in the nucleus and one electron in its shell, the atom is hydrogen. An atom with eight protons (with the proper number of neutrons) in the nucleus and eight electrons is oxygen. All atoms of a particular element have the same number of protons. The element's *atomic number* indicates the number of protons in one atom.

If the number of protons in an atom decreases or increases, the atom becomes a different element. Normally, the number of protons does not change except in nuclear reactions, as in atomic bomb explosions, nuclear power reactors, or stars.

Many elements consist of atoms that, although they must have the same number of protons, can vary in their number of neutrons. Because neutrons have no charge, the atoms would have the same charge but different masses. These atoms, which have the same number of protons but different numbers of neutrons, are called **isotopes** (EYE suh TOPES). Isotopes of an element have the same chemical properties because they have the same number of electrons. Carbon, one of the most common elements found in living things, has three different isotopes. Each isotope is identified by its mass number, which is the sum of the protons and neutrons in the nucleus. The three isotopes of carbon are carbon-12, carbon-13, and carbon-14.

Chemical Bonding: Ionic

Most atoms are unstable and will normally bond with other atoms to become more stable. An atom is most stable when its outermost electron shell is full. The innermost electron shell, which is the first to be filled, can contain only two electrons. The second shell can hold eight electrons. The next shell has

2-1	Elements Common to Life	
Symbol	**Element**	**Abundance in human body (% weight)**
O	Oxygen	65
C	Carbon	18
H	Hydrogen	10 98%
N	Nitrogen	3
Ca	Calcium	2
P	Phosphorus	0.5–1.0 each
K	Potassium	
S	Sulfur	0.1–0.5 each
Cl	Chlorine	
Na	Sodium	
Mg	Magnesium	
Cu	Copper	Less than 0.1 each
F	Fluorine	
Fe	Iron	
I	Iodine	
Zn	Zinc	

☐ Essential   ☐ Trace

hydrogen (H)

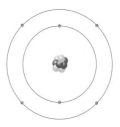

carbon (C)

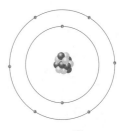

oxygen (O)

2-3 ——————

These three atoms contain different numbers of protons, neutrons, and electrons.

isotope: iso- (Gk. *isos*—equal) + -tope (Gk. *topos*—place); thus, atoms having the same place (on the periodic table)

the potential to hold eighteen electrons but acts as if it were full when it has eight. The shells farther out hold larger numbers, but all behave as if they are full when they have eight.

Atoms with naturally full shells are quite stable, such as the noble gases—helium, neon, argon, and so forth—located in the last column on the right side of the periodic table. Other atoms, especially if their outermost shells lack one or two electrons or if they have only one or two electrons, will give, take, or share electrons to achieve a full outermost shell. Atoms that give, take, or share electrons to become more stable are *chemically active*.

For example, an atom of chlorine, a chemically active nonmetal, has seventeen electrons: two in the first shell, eight in the second shell, and seven in the third shell. If it gets one more electron, it will have a full outermost shell. Sodium, a chemically active metal, is just the opposite. It has eleven electrons: two in the first shell, eight in the second, and one in the third. If it gives up its outermost electron, it will have a full second shell. When these two elements meet, sodium gives its outermost electron to chlorine, which takes it readily to fill its outermost shell. When this occurs, the two elements combine, forming a new substance, sodium chloride—common table salt. Sodium chloride is more chemically stable than the two elements that form it.

Since one atom gives and the other takes an electron, the atoms of sodium and chlorine in salt each have unequal numbers of electrons and protons. An atom with a charge (having an unequal number of protons and electrons) is called an **ion**. Since the sodium atom in salt is missing an electron, it has more protons than electrons, making it a positively charged ion. The chlorine atom becomes a negatively charged ion because it has more electrons than protons. Since positive and negative charges attract each other, the sodium ion and the chloride ion are attracted to each other. This attraction between the oppositely charged ions is called an **ionic bond**. Since all cells contain and are surrounded by ions, the study of ions and ionic bonding is significant to biological studies.

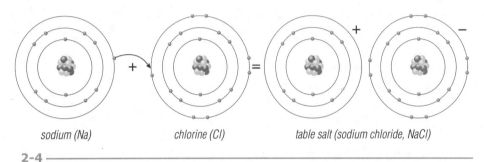

sodium (Na)   chlorine (Cl)   table salt (sodium chloride, NaCl)

2-4

Ionic bonding combines sodium and chlorine to form sodium chloride, or table salt.

Chemical Bonding: Covalent

Another type of chemical bonding involves the *sharing* of electrons. The formation of a water molecule is a good example. Hydrogen has only one electron and requires another to fill its first shell. Oxygen has eight electrons: two in its first shell and six in its second shell. Two more electrons would complete the second shell of the oxygen atom. One oxygen atom will therefore combine with two hydrogen atoms. The oxygen shares one electron from its second shell with each of the two hydrogen atoms so that each hydrogen atom has a filled shell. The two hydrogen atoms share their electrons with the oxygen so that the oxygen has a filled outer shell.

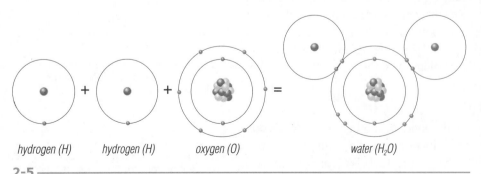

hydrogen (H)   hydrogen (H)   oxygen (O)   water (H₂O)

2-5

Covalent bonding combines hydrogen and oxygen to form water.

Two hydrogen atoms bonded to an oxygen atom form a water molecule. These atoms share the two pairs of electrons involved in this bond. This sharing of electrons is called a **covalent** (koh VAY lunt) **bond**. Covalent bonds are relatively strong and do not separate easily.

Some ionic bonds are weaker than others, and likewise, not all covalent bonds are equally strong. Some chemical bonds are not purely ionic or covalent. They are somewhere between the two and often form substances that behave like both *ionic compounds* and *covalent compounds*.

Molecules and Compounds

In the examples given previously—water and salt—atoms of different elements combined to form molecules. A **molecule** is a distinct group of atoms bonded together. Most molecules, such as those of water, carbon dioxide, sugar, and fat, are composed of atoms from different elements. Some molecules, however, contain atoms of only a single element. The oxygen in the air is not in the form of single oxygen atoms, but rather two oxygen atoms bonded together covalently. *Molecular formulas* express the number and type of atoms in a molecule. H_2O indicates that there are two atoms of hydrogen and one of oxygen. $C_6H_{12}O_6$ is the molecular formula for the simple sugar glucose. There are twenty-four atoms represented in this formula: six carbon atoms, twelve hydrogen atoms, and six oxygen atoms. A **compound** is a pure substance that is made of two or more elements that are chemically combined. A molecule is the smallest unit of a compound that is still that compound in a natural state.

A structural formula is often more important in the study of a compound and its reactions with other compounds than the molecular formula. In a *structural formula* scientists see where the atoms are actually located and the arrangement of the chemical bonds. Figure 2-6 shows that there are two different structures that a molecule of glucose can take. Galactose is another simple sugar that has the same molecular formula as glucose because it has the same number and kinds of atoms. However, in galactose, those atoms are bonded together in a different configuration. Its structural formula, therefore, is different. Because of its different structure, galactose has properties slightly different from the properties of glucose.

Chemical and Physical Changes

A **chemical change** takes place when the atoms of a substance bond (i.e., undergo a chemical reaction) with different atoms or compounds. In the examples of water and salt, the new substances have completely different characteristics, or properties, than the individual elements have. Pure sodium is a highly reactive metal that reacts violently with water, producing flammable hydrogen. Chlorine is a very poisonous gas which, when dissolved in water, is used to kill bacteria in swimming pools. Separately, sodium and chlorine are harmful to living things. But bonded together, they form a crystal that is essential to most living organisms. Hydrogen is a highly combustible gas, and oxygen is a gas necessary for combustion. However, as a compound, they form water, which is used for putting out fires!

The following are characteristics of all chemical changes:

- Chemical changes take place in definite proportions. Ten hydrogen atoms can make only five water molecules, no matter how many oxygen molecules are present.
- New compounds are formed, and/or there is a release of elements.
- Energy is involved.

ring form

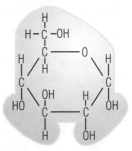

linear form

2-6 —————————
Two structural formulas of glucose

2-2	Compounds Common to Life
Name	**Molecular formula**
carbon dioxide	CO_2
carbon monoxide	CO
glucose	$C_6H_{12}O_6$
hydrochloric acid	HCl
nitrogen	N_2
oxygen	O_2
sodium chloride	$NaCl$
water	H_2O

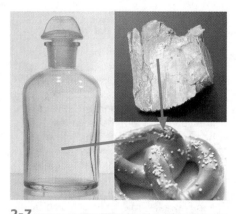

Chlorine gas and sodium metal combine to form sodium chloride—table salt.

Not all changes in matter are chemical changes. Adding thermal energy to ice merely causes it to change its state. Such a change is a **physical change**—the process of altering the state of something, its appearance, or its combination with substances, without involving change in electron sharing or giving.

A **mixture** is formed when two substances are combined without chemical bonding. Forming a mixture involves only a physical change. Since they are not chemically bonded, mixtures can be separated by heating or by mechanical methods. Water and sand do not have definite proportions when mixed together as do chemically bonded compounds. There can be a lot or just a little sand in that mixture. Mixtures maintain the individual properties of the components of which they are made.

Review Questions 2.1

1. Why is an understanding of chemistry important for an understanding of biology?
2. Define the two components of the physical world.
3. What elements are essential to life? What are their chemical symbols?
4. Describe an atom. What is the significance of its electron configuration?
5. Name and describe the two primary types of chemical bonding.
6. Define and compare the following: an atom, an ion, an isotope, a molecule, a compound, a mixture.
7. What things can a scientist determine (a) by studying a molecular formula of a compound and (b) by studying a structural formula of a compound?
8. List the characteristics of a chemical change and a physical change.

2.2

Objectives

- Differentiate between kinetic and potential energy
- Explain how kinetic energy affects the physical states of matter
- Describe how catalysts work
- Explain how enzymes work

Key Terms

energy
kinetic energy
potential
  energy
entropy
reactant
product
activation
  energy
catalyst
enzyme
active site
substrate

kinetic: (Gk. *kinein*—to move)
thermodynamics: thermo- or therm- (Gk. *therme*—heat) + -dynamics (*dunamis*—power)

2.2 Energy

One of the attributes of life discussed in Chapter 1 is that all living things require energy. But what is energy? **Energy** can be defined as the ability to do work or to cause a change. For the purposes of this study, there are two types of energy: kinetic (kih NET ik) and potential. **Kinetic energy** is the energy of motion, such as falling, heat, light, and electricity. **Potential energy** is stored energy, like the energy found in a rock sitting at the top of a cliff, in a log waiting to be burned, or in a battery ready to be connected to a light bulb.

In each of these examples, potential energy can be converted into kinetic energy. Pushing the rock, igniting the log, or connecting the battery will start the change from potential to kinetic energy.

Kinetic energy can also be converted into potential energy. One of the most important conversions to take place on the earth is the one in which green plants absorb light energy from the sun and convert it into the potential chemical energy that is stored as sugar. Later, the plant, humans, or other organisms will convert the stored chemical energy in sugar into the other forms of energy needed to carry on life.

Thermodynamics

The laws of thermodynamics (THUR moh dye NAM icks) are man's statements of how energy changes occur. The *first law of thermodynamics*, sometimes called the *law of conservation*, states that in any process, energy is neither created nor destroyed. Energy can change from one form to the other, but there is always as much at the end as there was at the beginning of the process. The *second law of thermodynamics* states that whenever energy is used (changed from one form to the other), some of it is rendered unusable (though not destroyed). That is, not all energy put into something is still there at the end of

Potential energy

Kinetic energy

2-8

The potential energy of the unlit logs is converted to kinetic energy as the logs burn.

the transfer. Some of the energy radiates out of the object, usually as heat or light. Eventually this unused energy goes into space. But because of the vastness of space, all the energy that has gone into it has not yet diminished its darkness or even raised its temperature to near the freezing point of water.

A natural corollary to the second law of thermodynamics is the *law of degeneration*: in all natural processes there is a net increase in disorder and a net loss of usable energy. For reactions such as burning a piece of wood, it is easy to see that the end products (carbon dioxide, water vapor, and ashes) contain much less energy and that the molecules are much more random than before the burning took place. This increase in randomness and loss of usable energy is termed **entropy** (EN truh pee).

Some reactions, however, appear to contradict the law of degeneration. Green plants use light energy to make sugar from water and carbon dioxide. This conversion of kinetic energy to potential energy and of simple substances to a larger, more orderly substance requires the plant to use *more* energy than is contained in the new molecule and is a result of the law of conservation. This process permits the growth and reproduction of plants.

At first glance, the first and second laws of thermodynamics appear to be in contradiction. Actually, however, they work in God-ordained harmony. For example, living things build complex substances according to the law of conservation. In time, the living organism will either store the energy in molecules or release the energy in the form of heat or motion. When the organism dies, energy is released as its molecules decompose. The use or release of the energy by the organism is in keeping with the law of degeneration. This balance is discussed in more detail throughout the text.

Kinetic Molecular Energy

It may be surprising to some to learn that each of the molecules that make up this book is wiggling. All of the molecules that make up every substance have a certain amount of kinetic energy. Even solid substances are composed of moving molecules.

The amount of thermal energy in a substance determines how fast its molecules move and subsequently its physical state. In ice, for

2-9

The energy stored in this ear of corn is less than the energy required to produce it.

2-10

Increasing the thermal energy changes water from a solid to a gas. Is this a chemical or a physical change?

example, molecular motion is slight; the individual water molecules barely move. However, in liquid water the movement is greater because it contains more thermal energy. The increase in thermal energy causes some of the bonds between the water molecules in ice to be broken. As the water molecules are heated to 100 °C (212 °F)—the boiling point of water—they become so active that many leave the surface of the fluid as water vapor, a gas.

Potential Energy of a Molecule

Living things are made of many types of molecules. The potential energy that fuels the processes of life is stored in the bonds between these molecules' atoms. Energy changes occur during the reactions that form these molecules as well as those that break them down.

Chemical reactions fall into two basic groups: those that require or absorb heat are called *endothermic* and those that give off heat are called *exothermic*. For example, heat is required to make cake batter solidify into the cake that is eaten; this is an endothermic reaction. A piece of wood gives off thermal and light energy as it burns; this reaction is exothermic.

Let the letters A and B represent two elements or compounds before they chemically react with one another—the **reactants**—and let C and D be the compounds that are the results, or **products**, of the chemical reaction. The reaction would be written like this:

$$A + B \longrightarrow C + D$$

This chemical equation is read "A plus B yields the products C plus D." An example of a chemical reaction is the formation of sugar by photosynthesis. Its simplified chemical equation is written this way:

$$6H_2O + 6CO_2 \longrightarrow C_6H_{12}O_6 + 6O_2$$

The water (H_2O) and the carbon dioxide (CO_2) are the reactants, and the sugar ($C_6H_{12}O_6$) and the oxygen (O_2) are the products of this reaction.

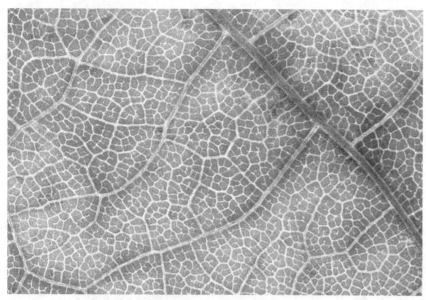

2-11

Can you determine where this leaf gets the two reactants it needs for photosynthesis?

endothermic: endo- (Gk. *endon*—within) + -thermic (heat)

exothermic: ex- or exo- (Gk. *exo*—outside of) + -thermic (heat)

The energy given off or absorbed in a reaction is often represented in the chemical equation. If the reaction between A and B is exothermic—giving off thermal energy—the reaction is written like this:

$$A + B \longrightarrow C + D + energy$$

Assume that the reaction between W and X requires thermal energy to become Y and Z. This endothermic reaction is written this way:

$$W + X + energy \longrightarrow Y + Z$$

But where does the energy required for baking a cake (or for any other endothermic reaction) go? In the endothermic reaction above, energy does not appear on the product side of the arrow. Does all the energy therefore escape into the atmosphere? Not at all. The products contain some of the energy. Where does the energy given off by a burning piece of wood (or from any other exothermic reaction) come from? In an exothermic reaction the energy released is contained in the substances that enter the reaction.

Whenever atoms take, give, or share electrons, energy is involved. Combining smaller molecules to form larger molecules requires energy. Generally, in living things, when the molecules combine by sharing electrons, energy is supplied to some of those electrons. With the exception of the energy that escapes, the energy supplied is still there in the form of chemical bond energy.

In the previous chapter, the need for a constant supply of energy is listed as one of the attributes of life. One of the reasons for this requirement is that living organisms must constantly make large, complex molecules. These molecules are the very substance of the organism itself. The food you eat supplies you not only with the energy you use to move but also with the energy necessary to build the complex molecules that form cells.

Catalysts and Enzymes

Many reactions are spontaneous; that is, whenever the proper substances come into contact with one another, the reaction immediately takes place. For instance, the combining of sodium and chlorine is spontaneous. Other reactions, however, require energy to start them. Paper does not burn when only oxygen is present. The molecules in a piece of paper must be heated to a certain point before the reaction will start. Once started, this reaction is exothermic and supplies the thermal energy necessary for other molecules to carry on the reaction as well. The energy necessary to start a reaction is called the **activation energy**. Activation energy is often in the form of thermal energy, but for other reactions it may be electrical, light, nuclear, or some other form of energy.

Some reactions, such as burning wood, happen very rapidly, even explosively. Others, such as the combining of iron and oxygen to form rust, happen slowly. To control reactions, scientists may use catalysts (KAT ul ists)—substances that affect the rate of a reaction but are not themselves changed in the reaction. Catalysts reduce the amount of activation energy required for the reaction.

Specific chemicals serve as catalysts for specific reactions, so there is not one catalyst for speeding up all reactions and another for slowing down all reactions. Many chemical reactions have no known catalyst. In a chemical equation, the catalyst is indicated over the arrow. A chemical reaction with its catalyst is written this way:

$$A + B \xrightarrow{\text{catalyst}} C + D + \text{energy}$$

All living things must carry out thousands of chemical reactions, each one either requiring or releasing energy, and must repeat these reactions many times. Many of these reactions occur too slowly or have activation energies that are too high to be practical for the organism. **Enzymes** are the compounds

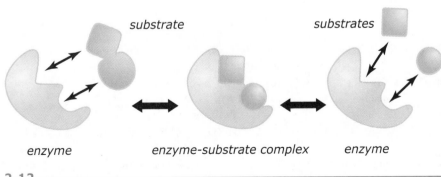

substrate   *substrates*

enzyme   *enzyme-substrate complex*   *enzyme*

2-12
Enzyme action

catalyst: cata- (Gk. *kata*—down) + -lyst (*luein*—to loosen)

that serve as *organic catalysts* that speed up chemical reactions, sometimes over a million times faster. How does this happen? Enzymes lower the activation energy of the reaction.

Figure 2-12 shows how an enzyme operates. From left to right, the drawing represents the kind of enzyme that breaks a molecule apart. One part of the surface of the enzyme is called the **active site**. The active site fits with the shape of the **substrate**, which is the molecule that the enzyme will bind with. The shape of the active site of the enzyme is specific for the substrate—it will fit no other. The combination of the enzyme and substrate is called the *enzyme-substrate complex*. While they are together, the shape of the enzyme is slightly altered and weakens some of the chemical bonds in the substrate, which is how scientists believe that the activation energy is lowered. The changed substrate is called the product. Notice that the enzyme is unchanged at the end—it has merely helped the reaction to take place.

Characteristics of Enzymes

- *Enzymes are proteins.* Although there are exceptions, most enzymes are proteins.

- *Enzymes are highly specific.* The enzyme that catalyzes the reactions between two sugar molecules will work only with those two sugar molecules or with molecules that are almost identical to them.

- *Many enzymes require energy to perform their functions.* Without the proper supply of energy, enzymatic reactions cease; therefore, life ceases.

- *Many enzymes release energy as they perform their functions.* Normally, the amount of energy released in a single enzymatic reaction is small. This energy can therefore be trapped and temporarily stored for use in reactions that require energy. Because the amount of energy released during enzymatic reactions is controlled, cells are not destroyed by too much heat energy being released at one time.

- *Enzymes often require coenzymes.* All coenzymes are substances that fit into or affect the active site so that it accepts the substrate properly. Without an adequate supply of coenzymes, enzymes do not function. Some important coenzymes are vitamins.

- *Enzyme action is affected by heat, radiation, pH, and chemicals.* Any of these can cause the enzyme to function improperly or not at all, thereby causing illness or death.

- *Enzymes most often work in series.* A molecule of sugar, for example, is broken down by dozens of different enzymes in succession. Each one may either break off, add an atom or molecule, or change the shape of the substrate.

➡ Review Questions 2.2

1. Name, describe, and give examples of two basic forms of energy.
2. What are the first and second laws of thermodynamics?
3. Give an example of the law of conservation and the law of degeneration working against each other.
4. Does the death of an organism result in increased or decreased entropy? Explain.
5. Give two reasons living things need energy and describe how they illustrate the law of conservation.
6. Describe how heat affects the molecules of a substance.
7. According to the heat energy involved, what are the two basic types of chemical reactions?
8. List two attributes of a catalyst.
9. Differentiate between an active site and a substrate in an enzymatic reaction.
10. Why are enzymes essential to life?

2.3 Solutions

Living things are mostly water—the average human body contains approximately 45.4 L (12 gal) of water. The other substances of life are in some way dissolved or suspended in water. The characteristics of the watery combinations that make up living things are vastly different, depending on what substances are in the water.

A **solution** is a homogeneous (HOH muh JEE nee us) mixture of one or more substances within another substance. A homogeneous mixture is one that is the same throughout. The kinetic energy of the molecules in the solution keeps the solution uniformly mixed. When sugar is dissolved in a glass of water, it forms a solution—the molecules of sugar are equally distributed throughout the water, and as long as the sugar does not settle out, the sugar water is a solution. The substance that is dissolved (sugar) is called the **solute**, and the substance in which it is dissolved (water) is called the **solvent**. Solutes are generally made of ions or small molecules, and often several different solutes will be in a single solution. An example is blood plasma—the noncellular portion of blood. It contains water (the solvent) and many dissolved solutes such as gases, molecules, and ions, as well as many other substances.

The composition of solutions is not fixed; the proportions of the solutes in the solvent can vary. The **concentration** of a solution is the ratio of the solute in the solvent. Concentration percentages are often found on bottle labels. For example, a 10% glucose ($C_6H_{12}O_6$) solution contains 10 g of glucose dissolved in enough water to make 100 g of solution. The more solute in the solution, the greater the concentration.

Since living things are mostly water, most of the biologically important chemical reactions occur in an *aqueous solution*—a solution for which the solvent is water. The solutions around and within cells are aqueous solutions, and the chemical reactions necessary for life occur in these watery solutions.

Diffusion

Diffusion is one of the most important physical processes that affects organisms. The proper concentration of solutes inside and outside the cell is highly critical to its continued existence. **Diffusion** is the net movement of molecules from an area of higher concentration of a substance to an area of lower concentration of that substance. *Net movement*, in this definition, means the number of molecules that moved from the higher concentration area minus those that moved back into the higher concentration area. When there are no longer places of higher and lower concentrations, a state of **equilibrium** (EE kwuh LIB ree um) has been reached, and diffusion stops.

2.3

▶ Objectives
- Define *solution*, *solvent*, and *solute*
- List the important molecular characteristics of water
- Differentiate between diffusion and osmosis
- Explain the difference between acids and bases

▶ Key Terms

solution	concentration
solute	gradient
solvent	semipermeable
concentration	membrane
diffusion	acid
equilibrium	base
polar molecule	pH
cohesion	buffer
adhesion	osmosis

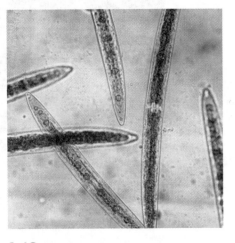

2-13
These aquatic organisms as well as all other living things are dependent on solutions.

Suspensions and Colloids

One mixture often seen in biological studies is a suspension. *Suspensions* are mixtures in which the kinetic energy of the molecules is not sufficient to maintain the large solute particles evenly dispersed within the solvent. They may or may not be homogeneous. If left undisturbed, the particles of a suspension will eventually settle out due to gravity. Blood is an example of a suspension. A portion of blood, blood plasma, is a solution of water with dissolved ions, but larger objects such as blood cells are suspended and will settle out if the blood ceases to circulate.

Colloids are similar to suspensions in that the dispersed particles are too large to completely dissolve. However, colloidal particles (usually defined as 1–1000 nm in size) are smaller than those in a suspension and do not settle out quickly.

Colloids may exist in two states. In the *sol phase*, the colloid is fluid, and there is little attraction between the colloidal particles. In the semisolid *gel phase*, weak bonds form between the dispersed colloidal particles. The material within a cell is a complex type of colloid, alternating between the sol and gel phases.

---

homogeneous: homo- or homeo- (Gk. *homoios*—same) + -geneous (*genos*—stock, race, or kind)

solute: (L. *solutus*—loose)

diffusion: dif- or di- (L. *dis*—apart) + -fusion (L. *fundere*—to pour)

equilibrium: equi- (L. *aequus*—equal) + -librium (*libra*—balance)

Water—Life's Essential Molecule

Water is so ordinary and plain that a person rarely thinks about it unless he is hot and thirsty or looking for a pool on a hot summer day. It is colorless, tasteless, and odorless, yet God uniquely designed water to sustain all of His creation. The human body is about 65% water; without water, a person would die within just a few days. The chemical reactions that maintain life take place in the watery environment of cells.

Without water in its liquid form, life as we know it could not exist.

Polarity
Water is composed of two hydrogen atoms covalently bonded to one oxygen atom. Although these atoms are sharing the electrons, the sharing is unequal—the oxygen atom pulls the shared electrons a little closer to its nucleus—making the area around the oxygen atom more negatively charged than the area around the hydrogen atoms. This uneven charge around the water molecule gives it a distinctive positive area, or pole, and a negative pole. A molecule with charged poles is said to be a **polar molecule**.

Hydrogen Bonding
The polar characteristic of the water molecule attracts not only other molecules and ions but also other molecules of water. The positive pole of one water molecule attracts the negative pole of another water molecule. This *hydrogen bonding* occurs as the hydrogen of one molecule is attracted to the negative region of another water molecule.

Hydrogen bonds give water some other biologically important characteristics. Water is often referred to as the *universal solvent* because it will dissolve so many different compounds, especially other polar compounds. When sodium chloride dissolves in water, the positive sodium ions are attracted to the negative portions of the water molecules, and the negative chloride ions are attracted to the positive poles of the water molecules. Substances that are not polar, such as vegetable oil, will not be dissolved by water.

Hydrogen bonding is also responsible for two other important characteristics of water. **Cohesion**, the attraction between molecules of the same substance, causes water droplets to bead up and also allows some insects and spiders to skate on the surface of water. The attraction between the molecules of different substances, known as **adhesion**, is one of the forces that helps water move through the vessels of stems and leaves. As long as the other substance is polar, water molecules will be attracted, or adhere.

Water and Temperature Change
Everyone knows that solid water (ice) floats. But what happens when another liquid, alcohol, is frozen? Unlike solid water, solid alcohol sinks. In fact, most liquids contract and sink when they reach a solid state. God designed water with its unique molecular structure and hydrogen bonding so that as it solidifies, it expands. This expansion makes ice less dense, allowing it to float; therefore, water freezes from the top down. The ice that forms on the surface of a lake acts as a barrier and shields the relatively warmer water below, an important fact for the survival of fish and other aquatic organisms.

Why does it seem to take forever for a pan of water to boil? A property called *specific heat* requires a large amount of thermal energy to bring about a temperature change in water. The hydrogen bonds are responsible for this property. When water is initially heated, most of the thermal energy is used to break the hydrogen bonds between the molecules. Only after these bonds are broken does the kinetic energy increase enough to cause an increase in temperature. This property is important biologically. Large bodies of water can hold enough thermal energy to affect the temperature of the environment around them. Warm ocean currents can moderate the temperature of coastal communities when more inland areas are much colder. At the cellular level, the ability of water to help moderate temperature changes aids the cell in maintaining temperature equilibrium with its surroundings.

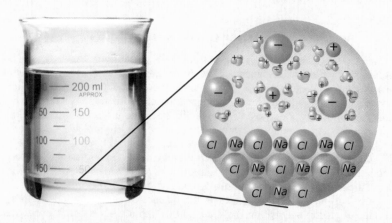

The polarity of the water molecules "pulls" the sodium and chloride ions out of a sodium chloride crystal.

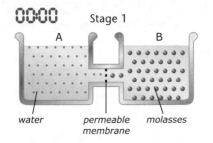
Stage 1

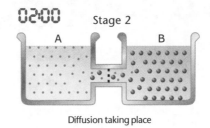
Stage 2

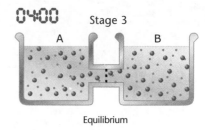
Stage 3

water   permeable   molasses
         membrane

Diffusion taking place

Equilibrium

2-14

Diffusion progresses through a permeable membrane to reach equilibrium.

In Figure 2-14 a tube connects two containers (A and B), and a cloth barrier has been placed in the tube to separate the two containers. Next, water is placed in container A and molasses in container B. Because the water side has no molasses molecules and the molasses side has relatively few water molecules, a concentration gradient (GRAY dee unt) is established. A **concentration gradient** is the difference between the numbers of one type of molecule in two adjacent areas. In this example, the concentration gradient is high for water on side A and high for molasses on side B. A high concentration gradient indicates that the difference is very great; a low concentration gradient indicates that there is little difference. The two areas of high concentration will seek to establish equilibrium, and the molecules will tend to "flow down" the concentration gradient from an area of high concentration to one of lower concentration.

The *diffusion pressure*, which is based on the relative concentrations of the adjacent areas (in this case the water and molasses molecules), is very high in Stage 1 of Figure 2-14 because the molasses and water are each strongly concentrated and no mixing has occurred. Diffusion occurs rapidly when the diffusion pressure is high. Since the molecules of both water and molasses can easily pass through the cloth barrier, diffusion happens in both directions. A barrier that allows passage of all molecules is said to be *permeable* (PUR mee uh bul).

After a period of time, the concentration gradient decreases as the number of water molecules and the number of molasses molecules on each side even out, and diffusion slows. The individual molecules are not slowing down, though; they continue to move because of the kinetic energy they possess. As long as the temperature remains the same, the molecules will continue to move at the same rate.

As diffusion continues, the number of molecules moving out of the areas of high concentration approaches the number of molecules moving *into* the areas of high concentration. That is, about the same number of molasses molecules are entering and leaving side B in Stage 3. The same is true for the water molecules. Diffusion, it should be remembered, is the net movement. Although these molecules continue to move, the gain in number on the lower concentration side is not as great as it was. As the concentrations of water and molasses get closer, diffusion pressure decreases, and diffusion slows down. In time, equilibrium will be reached.

Osmosis

If instead of a cloth, a membrane having very small pores is placed between the molasses and water, the results are different. If the membrane's pores are so small that they will permit the water molecules to pass through but will not permit the larger molasses molecules to pass through, it is a **semipermeable** (SEM ee PUR mee uh bul) **membrane** (or selectively permeable membrane). It is permeable to some things but not to others. Living organisms have many selectively permeable membranes.

permeable: per- (L. *per*—through) + -meable (*meare*—to pass)

semipermeable: semi- (L. *semis*—half) + -permeable (to pass through)

Acids, Bases, and Buffers

Most substances, when dissolved in water, are either acids or bases. Most chemical reactions, especially those that take place in living things, are affected by whether they are taking place in an acid or a base and by how strong the acid or base is.

Acids and bases are really opposites which, when they get together, do not fight; but they do destroy each other. An **acid** is a compound that donates hydrogen ions (H^+) when dissolved in water. A **base** is a compound that releases hydroxide ions (OH^-) when it is dissolved in water.

When an acid and a base are put together, they *neutralize* each other, forming a salt and water. For example, sodium hydroxide and hydrochloric acid will combine to form sodium chloride (table salt) and water.

$$NaOH + HCl \longrightarrow NaCl + H_2O$$

The *acidity* (amount of hydrogen ions) or *alkalinity* (amount of hydroxide ions) of a solution is usually expressed in terms of a value called **pH**. For biologists, the useful pH scale goes from 0 (very acidic) to 14 (very alkaline). A pH reading of 7 is neutral, meaning the solution has no excess H^+ or OH^- ions.

Living Things and pH

The pH of the environment surrounding living things is crucial. Some fish that normally live in water with a pH of 6.5 will begin to show signs of distress when placed in water with a pH of 6.0, and if placed in water with a pH of 5, they die.

The internal pH of living things is also important. Human stomach enzymes work best if the pH of the stomach juices is between 1.6 and 2.4. Sometimes a person's stomach may secrete too much acid for too long, causing a "sour stomach." The acid attacks the lining of the digestive system, causing a person to have a burning sensation (often called "heartburn" even though it has nothing to do with the heart). A common over-the-counter treatment for this acidic heartburn is the use of antacids. These neutralize the excess stomach acids, increasing the pH. Stronger drugs, available by prescription, actually limit the amount of hydrochloric acid the stomach cells can secrete.

In many living systems there are buffers that keep the pH within a tolerable range. A **buffer** is a substance that will combine with either H^+ or OH^- ions—whichever is in excess. Human blood, for example, maintains a pH between 7.35 and 7.45 using buffers. If an acidic substance enters the blood, the buffers will quickly pick up the excess H^+ ions. The reverse would happen if an alkaline substance gets into the blood.

There are, of course, limits. If someone gets too much of an acid in his blood, the buffers may not be able to handle all the ions. His blood pH would go down, and he could end up in a coma; if it drops too far, he would die. The brain cells can tolerate only small pH fluctuations beyond the normal range.

Common Acids and Bases

Acids:
- hydrochloric acid—HCl
- sulfuric acid—H_2SO_4
- carbonic acid—H_2CO_3

Bases:
- sodium hydroxide—NaOH
- calcium hydroxide—$Ca(OH)_2$

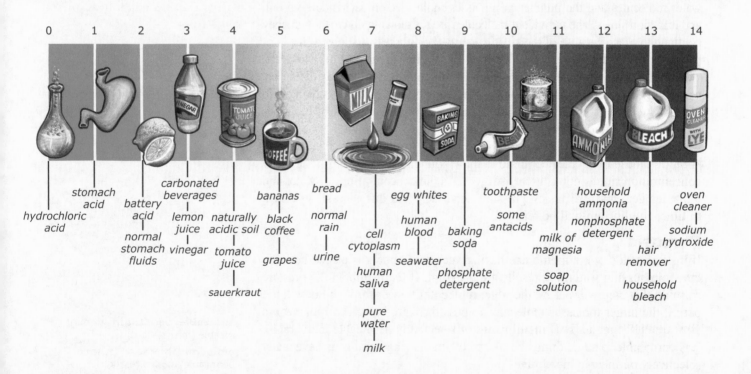

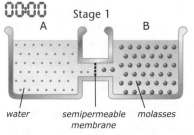 Stage 1

water   semipermeable   molasses
         membrane

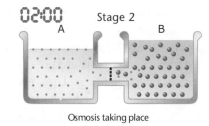 Stage 2

Osmosis taking place

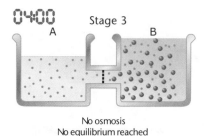 Stage 3

No osmosis
No equilibrium reached

2-15

Osmosis occurs through a semipermeable membrane.

If a semipermeable membrane separates water and molasses having high concentration gradients, the membrane will allow only water molecules to pass through and move into the molasses. Diffusion of water molecules through a semipermeable membrane is **osmosis** (ahz MOH sis). The large molasses molecules, however, are not able to move into the water because of the small size of the pores in the semipermeable membrane. As the water moves across the membrane to the molasses side, the water level falls on side A, forcing the level higher on side B.

Will this unequal movement of water molecules into the molasses ever produce an equal concentration of molasses and water molecules on each side? No, that result is impossible in this case. Nor can the water molecules continue going through the membrane until all the water has moved to the molasses side, even though a continuous concentration gradient exists, because the molasses cannot enter the water. Eventually, the net movement of water molecules into the molasses will decrease. At that point osmosis—the net movement of water molecules through a semipermeable membrane—will cease, even though the concentrations are not equal. Molecules of water will continue to move back and forth across the membrane, but there will be no net gain or loss of water.

osmosis: osmos- or osmot- (Gk. *osmos*—thrust)

⇒ Review Questions 2.3

1. What is the difference between a solute and a solvent?
2. Describe diffusion and give two examples of how it occurs in the home.
3. List two main differences between diffusion and osmosis.
4. Where does the energy needed for the movement of molecules in diffusion and osmosis come from?
5. List and compare the characteristics of a solution, a suspension, and a colloid.
6. List several examples of osmosis.

7. When a substance has diffused and is at the state of equilibrium, is it displaying an increase in entropy? Explain.
⊙8. How does the polarity of a water molecule relate to hydrogen bonding and the properties of adhesion and cohesion?
⊙9. Contrast acids and bases, and then explain what happens when the two are mixed together.
⊙10. What scale describes the degree of acidity or alkalinity of a substance? What range of values indicates an acidic solution? An alkaline solution? A neutral solution?

2B—Organic Chemistry

Chemistry can be divided into two very broad categories—*organic* and *inorganic chemistry*. These terms are actually holdovers from earlier times when it was thought that organic substances came from only animals and plants, while inorganic substances came from mineral sources, a belief which is not totally accurate.

Organic compounds are those that contain covalently bonded carbon atoms, with a few exceptions. Their unique properties are based on the carbon atom and its unusual bonding characteristics. Every organic compound has carbon in it, usually in conjunction with large quantities of hydrogen and frequently with oxygen.

Living organisms make organic compounds that are required for life. Many of these compounds are very complex and large molecules. Though some of the simpler ones have been synthesized in laboratories, under normal circumstances most organic compounds are the result of biosynthesis, the putting together of substances by living things. Biosynthesis is the process by which assimilation is accomplished and is essential to life.

Occasionally, headlines say that life has been created in a laboratory. These misleading headlines are written to get attention. A close reading of such an article (if it is written accurately) reveals that the new development falls far short of creating life in a test tube. Usually scientists have simply manufactured another biochemical compound by use of an elaborate apparatus. To accomplish even this, they must use large amounts of energy, special procedures, and elaborate equipment. Each time scientists employ sophisticated techniques to synthesize these compounds, they actually indicate how unlikely it is that life could spontaneously generate in a test tube, much less in a prehistoric pond. To "create" life, not only must one organic compound be made, but literally thousands of different ones must be made in large quantities, all at the same time (for many of them are very unstable on their own), and all in the same place.

Next, these materials have to be given life. Will life just happen if all the chemicals are right? Even those scientists who attempt to create life in a test tube admit that their chemicals need some special "starting" force. Some evolutionists have suggested that a bolt of lightning in a prehistoric pond started life. Regardless of thunderbolts or special jump-starts, life is more than the proper alignment of the correct chemicals. Even unbelieving scientists must admit that there is a great contrast between a living organism and that same organism after its life has ended. With our fallen minds, we cannot fully comprehend the secret to life, but a Christian worldview requires us to acknowledge that the Creator plays a key role in both the original creation as well as the continuance of life. Whether God chooses to use only physical processes to accomplish this or instead uses supernatural forces beyond our comprehension, it is still in His hands.

Is it possible for a living thing to be assembled from nonliving substances in a laboratory? At the time of this writing, it has not been done. Life, the Scriptures tell us, is the creation of God (Isaiah 42:5). But the Bible does not tell us that humans will not be able to put together a living thing. If, by using elaborate equipment and special man-made chemicals, scientists do form a structure that can be considered alive, have they supported spontaneous generation and evolution? No. The use of the equipment, the man-made chemicals, and human intelligence to form the living structure actually support the concept of a special creation of life by God, not a happenstance origin of life. Also, any life that humans could make in a lab would be made differently than the way God made life. God made life out of nothing, while humans have to use materials already created by God.

2.4 Organic Compounds

A carbon atom has four electrons in its outer shell that will readily form covalent bonds. Carbon is unusual because it will readily bond with other carbon atoms, forming long straight chains, branched chains, or even rings of carbon atoms. The chain or ring upon which the remainder of the molecule is built is called the carbon backbone. As Figure 2-17 shows, a great variety of organic molecules can be formed.

Carbon will also share more than one electron with other atoms, either another carbon or a different atom. If a carbon shares two electrons, it is called a

2-16

Some evolutionary biologists propose that the first life arose in a habitat similar to the one around this deep-sea hydrothermal vent. Could this be possible?

biosynthesis: bio- (Gk. *bios*—life) + -syn- (Gk. *sun*—together) + -thesis (*tithenai*—to put)

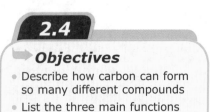

2.4

→ **Objectives**

• Describe how carbon can form so many different compounds
• List the three main functions of organic compounds in living organisms

→ **Key Term**

organic compound

double bond. A triple bond occurs when three electrons are shared with another atom. These different types of bonds will affect the shape of the organic molecule, and shape is very important in organic chemistry. Glucose and galactose, introduced in Subsection 2.1, have identical molecular formulas, but it is their structural formulas—their shapes—that determine their unique characteristics.

Organic compounds perform various functions for living organisms and can be categorized as follows:

- *Structural*—compounds that are used as building blocks of a cellular or extracellular structure
- *Enzymatic*—compounds that are enzymes or that help enzymes in their functions
- *Storage*—compounds that store energy, other substances, or information for future use

Many of the organic compounds perform combinations of these functions. The large number and great variety of organic compounds make grouping them difficult. They are often classified into four groups: carbohydrates, lipids, proteins, and nucleic acids.

C — C — C — C — C — C
straight

C — C — C — C — C — C
         |
         C
branched

ringed

C — C   single bond

C == C   double bond

C ≡ C   triple bond

2-17
The lines between the carbons represent bonds.

Review Questions 2.4

⊙ 1. If a scientist were to assemble a living thing from nonliving parts, would this disprove the Bible? Why or why not?

2. List three characteristics of the carbon atom that are important in forming organic compounds.

3. List the three primary functions of organic compounds.

4. What chemical element is found in all organic compounds?

2.5 Carbohydrates and Lipids

No matter where they are found—in plants, pigs, or people—carbohydrates and lipids are the same, although their amounts may vary. Both are organic compounds, both store energy, and both are important molecules in many structural components of organisms. Molecules of carbohydrates and lipids can be either quite small or very large, consisting of thousands of atoms.

Yet, they are also different. Many of the less complex carbohydrates dissolve in water; most lipids do not. Gram for gram, lipids store more energy than carbohydrates.

Carbohydrates

Carbohydrates (KAR boh HYE drates) are organic compounds that contain carbon, hydrogen, and oxygen. Generally, the carbohydrates found in living things have twice as many hydrogen atoms as they do oxygen atoms.

Monosaccharides

The basic units of carbohydrates are monosaccharides (MAHN nuh SAK uh rides), or simple sugars. A monosaccharide may contain as few as three carbon atoms. The monosaccharides that contain five or six carbon atoms are important building blocks of other organic compounds. A common 5-carbon sugar is *ribose*, a component of nucleic acids.

Glucose is a 6-carbon sugar manufactured by plants. It is also the sugar that the blood transports to cells in humans and in many animals. Two other simple 6-carbon monosaccharides are important in biology: *galactose* and *fructose*.

Disaccharides

Monosaccharides can be joined together in chemical reactions to form a disaccharide (dye SAK uh RIDE). In Figure 2-18 a glucose and a fructose

2.5

Objectives

- Describe the basic structures of carbohydrates and lipids
- Discuss the various functions carbohydrates and lipids have in living organisms
- Differentiate between saturated and unsaturated fats
- Compare the properties hydrophilic and hydrophobic
- Explain the relationship between monomers and a polymer

Key Terms

carbohydrate	fatty acid
monosaccharide	hydrophilic
glucose	hydrophobic
disaccharide	triglyceride
polysaccharide	saturated
starch	unsaturated
glycogen	steroid
cellulose	monomer
lipid	polymer

monosaccharide: mono- (Gk. *monos*—one) + -saccharide (L. *saccharum*—sugar)

disaccharide: di- (Gk. *di*—two) + -saccharide (sugar)

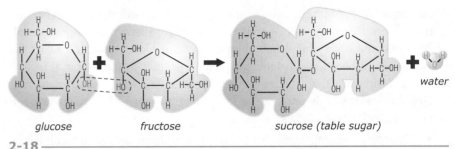

glucose   fructose   sucrose (table sugar)   water

2-18

Formation of a disaccharide

molecule combine to form the disaccharide *sucrose*, a common sugar found in many foods and often called table sugar. Because a water molecule is released in the synthesis reaction that forms a disaccharide, it is sometimes called a *dehydration synthesis* reaction.

When a cell needs the monosaccharides from which a disaccharide is made, enzymes will combine the disaccharide and a water molecule to separate the disaccharide into monosaccharides. A reaction such as this in which water is added to split a molecule is called *hydrolysis*.

Polysaccharides

A **polysaccharide** (PAHL ee SAK uh ride) is a large molecule of monosaccharide units. Using specific enzymes, living things can build up or break down polysaccharides. Three polysaccharides are particularly significant.

Starch is one of the primary substances that plants store as food. Starch from corn, potatoes, wheat, rice, and other grains constitute the major energy source for humans.

Glycogen is often called animal starch. The starches a person eats are broken down to simple sugars and then taken, along with the other sugars in his diet, to the liver where they are made into glycogen for temporary storage. Glycogen is a branching chain of glucose molecules.

Cellulose (SEL yuh LOHS) molecules are much larger than starch molecules. They are made of long chains of glucose molecules bonded in an alternating arrangement. The long cellulose molecules account for the strength of plant cell walls. Plant fibers that form cotton, wood, paper, and many other products contain large quantities of cellulose.

Lipids

Lipids are a group of organic substances that are only slightly soluble in water but are very soluble in organic solvents such as alcohol, ether, acetone, and chloroform. Lipids are most often structural, but they also store energy. Because lipids occupy less space than starches, humans and animals store their excess energy as lipids. Plants, however, add to their size by storing starches.

2-19

Sugars and starches (such as this pasta) are primary sources of energy for living things.

Cellulose and Digestion

Cellulose is indigestible to humans and most animals because they do not have the proper enzymes for breaking apart the bonds between its glucose molecules. Therefore the energy contained in the glucose molecules of cellulose is not directly available. Cellulose is a primary part of the indigestible portion of the diet called bulk or roughage. Although indigestible, bulk is needed for the digestive system to work properly. Certain bacteria, fungi, and protozoans, however, do produce enzymes for digesting cellulose.

sucrose: sucr- (Fr. *sucre*—sugar) + -ose (sugar)

hydrolysis: hydro- or hydra- (Gk. *hudro*—water) + -lysis (Gk. *lusis*—a loosening)

polysaccharide: poly- (Gk. *polus*—many) + -saccharide (sugar)

lipid: (Gk. *lipos*—fat)

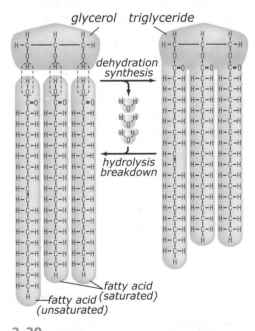

glycerol   triglyceride

dehydration synthesis

hydrolysis breakdown

fatty acid (saturated)

fatty acid (unsaturated)

2-20

The formation of a triglyceride (fat) from fatty acids and glycerol

The most abundant lipids are **fatty acids**, which are unbranched chains of fourteen to twenty-eight carbon atoms that have a carboxyl group (–COOH) added to one end. The carboxyl end of the molecule is polar and is attracted to water: it is **hydrophilic** (HYE druh FIL ik). The other end is repelled by water and is termed **hydrophobic** (HYE druh FOH bik). This unusual characteristic permits fatty acids to be used in specific ways in cells. The importance of these characteristics will become clearer as this study progresses.

Fatty acid molecules can be broken down by enzymes into many 2-carbon molecule pieces. Each of these molecules may eventually release usable cellular energy, making fatty acids a good energy source.

The body's most abundant type of lipid is *triglycerides*, or fats. Triglycerides are formed by combining three fatty acid molecules to a molecule of glycerol (glycerin), a 3-carbon alcohol.

If each of the carbon atoms in a fatty acid molecule (except the end ones) has two hydrogen atoms attached to it, the fatty acid is said to be **saturated**. However, should one or more of the carbon atoms be double bonded, there are fewer hydrogen atoms attached to the carbon backbone. This type of fatty acid molecular structure is called **unsaturated**.

The straight chains of a saturated fatty acid allow them to stack tightly together; therefore, at room temperature a saturated fatty acid will be a solid. Examples include butter and lard. The double bonds in the unsaturated fats cause bends in the chain and prevent a tightly stacked arrangement. These molecules will exist as liquids at room temperature. Lipids that are liquid at room temperature are called oils. Corn oil, olive oil, peanut oil, and coconut oil are examples of unsaturated oils as are most oils found in fruits, vegetables, and fish.

Phospholipids are composed of two fatty acid molecules attached to a glycerol molecule. The third carbon atom of the glycerol molecule has a phosphate-containing group attached to it. Phospholipids have a hydrophilic end that orients itself toward water and a hydrophobic end that avoids water. This characteristic is essential in the formation and structure of the cell membrane, which you will study in Chapter 3.

Steroids have a carbon backbone of four different carbon rings plus a side chain of carbons. They are classified as lipids because they are hydrophobic, but they differ from other lipids in their function and structure. Many steroids circulate through the body and function as chemical messengers. The sex hormones, estrogen and testosterone, are two such steroids. *Cholesterol*, a common but important steroid, is found in the membranes of human and animal cells.

2-21 —————
The molecular structure of triglycerides determines many of their physical characteristics. Which is the saturated triglyceride?

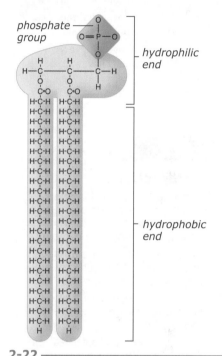

2-22 —————
A phospholipid

hydrophilic: hydro- (water) + -philic or -phile (Gk. *philos*—loving)

hydrophobic: hydro- (water) + -phobic (Gk. *phobos*—fear)

Waxes

Waxes result when several fatty acids join to alcohols made of long chains of carbon. Honeybees produce the waxes that make up a honeycomb. A thin layer of wax commonly covers plant leaves and stems to help deter pests and prevent water loss.

2-23 —————
A cholesterol molecule showing the carbon backbone in red

Macromolecules

Many of the molecules found in the cells of organisms are very large and are called *macromolecules*. Macromolecules are constructed from hundreds or even thousands of smaller molecules. The smaller units that make up a macromolecule are called **monomers**. When the monomers are assembled into the much larger molecule, it is a **polymer**. In some, but not all, polymers, the monomers are identical.

➡ Review Questions 2.5

1. What substances make up carbohydrates? What characteristics do carbohydrates have?

2. What are the primary functions of carbohydrates?

3. List several monosaccharides, a disaccharide, and several polysaccharides.

4. What substances make up a lipid? What characteristics do lipids have?

5. Describe the difference between a hydrophobic and a hydrophilic molecule.

6. Differentiate between saturated and unsaturated fatty acids. How are the physical characteristics affected?

7. List several lipids.

8. A large polysaccharide is made of many monosaccharide units linked together. What other terms can be used to represent *monosaccharide* and *polysaccharide*?

2.6

➡ Objectives

- Describe the formation and structure of proteins using the terms *amino acids*, *peptide bonds*, and *polypeptide chains*

- Recognize DNA as the primary nucleic acid and hereditary molecule in most organisms

- List the three parts of a nucleotide and describe their arrangement

- Compare and contrast DNA replication and RNA transcription

➡ Key Terms

protein
amino acid
polypeptide
nucleic acid
DNA (deoxy-
ribonucleic
acid)

double helix
nucleotide
replication
RNA (ribonu-
cleic acid)
transcription
RNA polymerase

2.6 Proteins and Nucleic Acids

Whereas carbohydrates and lipids are the same in different organisms, proteins, although sometimes similar in structure and function, are different. For example, human insulin, a protein that helps in food metabolism, is virtually identical to the insulin found in animals. Human hair protein, on the other hand, is different from the protein found in animal hair. Even among people, although one person's hair proteins are similar to another's, they are not identical. Proteins, it may be said, are what make a person unique.

The uniqueness of an organism's proteins is the result of nucleic acids and enzymes. Nucleic acids contain the code to manufacture the proteins; enzymes, which are themselves proteins, help the nucleic acids function properly. Thus, proteins and nucleic acids are codependent.

Proteins

Proteins consist primarily of carbon, hydrogen, oxygen, and nitrogen; but they sometimes include phosphorus, sulfur, and a few other elements. They come in a wide variety and are used either as enzymes or as structural building blocks. Some are very small, but some protein molecules are large enough to be seen under high-powered microscopes.

The basic building blocks of a protein are **amino** (uh MEE noh) **acids**. There are about twenty different amino acids. An amino acid contains an *amino group*

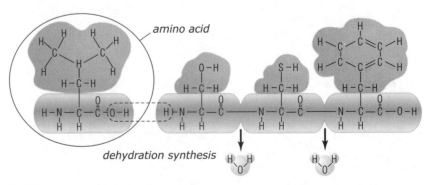

2-24
Formation of a polypeptide chain from amino acids

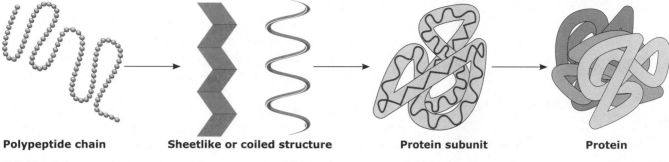

| Polypeptide chain | Sheetlike or coiled structure | Protein subunit | Protein |

2-25

Four different levels of organization are apparent in most proteins.

(–NH₂) and a *carboxyl group* (–COOH), both attached to a carbon backbone. Different side groups (some simple, others complex) attach to the carbon of the amino acid. Some proteins are composed of only a hundred connected amino acids, but several thousand can line up to make some of the larger proteins.

When two amino acids line up and the proper enzymes are present, a dehydration synthesis reaction occurs, removing a water molecule and forming a *peptide* (PEP tide) *bond* between the amino acids. A peptide bond links the carboxyl group of one amino acid to an amino group of another amino acid. A protein, then, begins as a **polypeptide** (PAHL lee PEP tide) chain of amino acids.

When the proper sequence of amino acids is arranged, the chain twists and then folds back upon itself. Certain amino acids bond together in a specific pattern that is determined by the sequence of the acids. The polypeptide chain of amino acids now has a three-dimensional arrangement. It is this shape that forms the active sites of enzymatic proteins.

As many as four different levels of organization can be seen in the structure of proteins.

1. The lowest level is the basic order of amino acids in the polypeptide chain.
2. The polypeptide chain may then be coiled or even folded in a sheetlike arrangement.
3. The coiled and/or folded structure itself is folded and held in a precise shape.
4. Several of these reconfigured polypeptide chains (called *subunits*) may lock together to form a large protein.

If some of the amino acids are out of sequence, the shape of the protein molecule may be altered. Should this change of shape affect the active site of an enzyme, the enzyme may not react properly with its substrate. Heat, radiation, ion concentrations, and other environmental factors can also affect the bonding of amino acids to other parts of the amino acid chain. When the bonds break between the amino acids that give the protein its three-dimensional shape, the shape of the protein molecule may change. If this change affects the active site of an enzyme, it may no longer function as an enzyme. These changes in shape may also affect the ability of a cell to use the protein as a building block for other structures.

Proteins often work together with other proteins or other substances. Enzymatic proteins may require coenzymes in order to function. Some proteins combine with lipids (lipoproteins), carbohydrates (glycoproteins), or even other proteins to form a structural component. Many hormones are lipids joined with proteins.

peptide: (Gk. *peptein*—to digest)
polypeptide: poly- (many) + -peptide (to digest)

Watson and Crick in the 1950s

Rosalind Franklin's Clue

Rosalind Franklin (1920–1958), a young molecular biologist working in a London laboratory, had obtained some significant x-ray diffraction images of the DNA molecule. While the pictures did not show exactly what DNA looked like, they provided the essential clues for Watson and Crick to develop their model. Franklin thus played a significant pioneering role in this breakthrough. Sadly, she died of ovarian cancer in 1958, four years before Watson and Crick and another biologist, Maurice Wilkins, were awarded the Nobel Prize for their work.

Nucleic Acids

Although scientists have long known that proteins are different in various organisms, they did not discover until just before World War I that **nucleic** (noo KLEE ik) **acids**, not proteins, are the hereditary material. In other words, in order to transmit its characteristics to another cell, a cell must pass nucleic acids to that cell.

Nucleic acids, in the form of DNA and RNA found in the cell nucleus, direct the activities of a cell by guiding the formation of both structural and enzymatic proteins. Because they contain the information necessary for the manufacture of the organism's proteins, nucleic acids determine its characteristics. The way the nucleic acids direct protein manufacture and the exact method of transmitting this information to the next generation of cells are important concepts discussed in later chapters. This section examines what nucleic acids are.

The Structure of DNA

DNA, **deoxyribonucleic** (dee AHK see RYE boh noo KLEE ik) **acid**, is the primary nucleic acid in most organisms. It contains the information that is needed for almost all of a cell's activities. By the time of World War II, scientists had analyzed the components of DNA but were confused about their exact arrangement. To understand how DNA functions, it was necessary to form a working model of its components. In 1953, James D. Watson, an American biologist, and F. H. C. Crick, a British biophysicist, assembled information about DNA's components. Using information obtained from x-rays of DNA, they created a DNA model. Watson and Crick received a Nobel Prize in 1962 for their accomplishment. Although it has been slightly modified since then, the basic model is still consistent with all the information known about DNA.

Watson and Crick's model is a **double helix**, or spiral, consisting of two strands attached at regular intervals. Visualizing a flexible ladder that is twisted on its long axis gives a good idea of what a DNA molecule looks like.

An understanding of the DNA structure must begin by mentally untwisting the DNA molecule and examining its component parts. DNA is a double-chain polymer of **nucleotides**. A nucleotide (NOO klee uh TIDE) has three components:

- *Sugar.* The sugar in DNA is the 5-carbon sugar deoxyribose.
- *Phosphate.* The phosphate in a nucleotide is a small molecule made of phosphorus and oxygen atoms.
- *Base.* Each nucleotide will have one of these four bases: adenine, thymine, guanine, or cytosine.

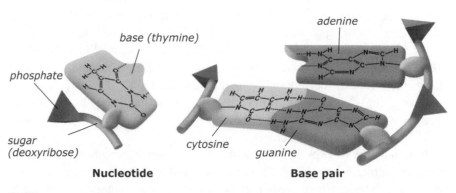

Nucleotide

Base pair

A nucleotide and a base pair, cytosine and guanine. The dotted lines in the base pair represent hydrogen bonds.

nucleus: nucle- or nucleo- (L. *nux*—central part)

2-28
As the bases pair, the DNA molecule forms a double helix.

Just as in the formation of polysaccharides and polypeptide chains, the nucleotides are put together by a dehydration synthesis reaction. The nucleotides are arranged so that the sugars and phosphates form one long sidepiece of the ladder with the bases pointing off to one side, forming one-half of each rung (step) of the ladder. The other half of the ladder is likewise formed by nucleotides.

Figure 2-27 illustrates DNA base pairing. Look carefully at the bases. Each adenine is bonded with a thymine by two hydrogen bonds, and each cytosine is bonded with a guanine by three hydrogen bonds. When there is a thymine on one side, there is an adenine on the other; and when there is a guanine on one side, there is a cytosine on the other.

DNA Replication

The sequence of the bases in the DNA molecule can be compared to the sequence of letters in this sentence. In one order the letters make words and have meaning. In another order they may make different words and mean something else. In a different arrangement the letters may make words in another language, or they may be completely senseless. The sequence of bases in DNA forms a specific code that directs the sequence of amino acids in the protein molecules that the cell produces.

The message encoded within the sequence of bases in DNA must be copied so that it can be passed to the next generation of cells. This copy must be an exact duplicate of itself, even down to the sequence of bases. If the base sequence is not correct, an improper protein may be manufactured—or possibly no protein at all.

DNA uses itself as a direct blueprint in duplicating itself. Visualize this process as the opening of a zipper. An enzyme moves down the DNA molecule, breaking the hydrogen bonds between the bases that hold the two sides of the DNA molecule together. The two sides do not close like a zipper, however. Instead, a different enzyme attaches free nucleotides to rebuild the missing half of the molecule. The final result is two identical molecules of DNA. Once complete, the sequence of bases is exactly the same in both strands of DNA. Producing two new DNA molecules from an old one is called **replication**. The enzyme systems necessary for replication are vast and intricate.

RNA—Another Nucleic Acid

RNA, **ribonucleic** (RYE boh noo KLEE ik) **acid**, is the basic nucleic acid for a few viruses; however, it is important for all living things. There are three important structural differences between DNA and RNA. RNA is a single strand

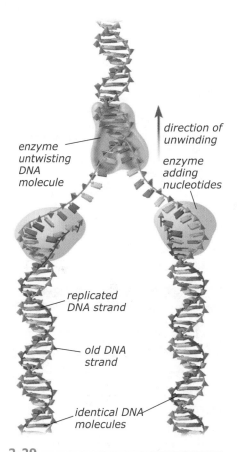

enzyme untwisting DNA molecule

direction of unwinding

enzyme adding nucleotides

replicated DNA strand

old DNA strand

identical DNA molecules

2-29
The replication process

replication: re- (L. *re*—back) + -plication (*plicare*—to fold)

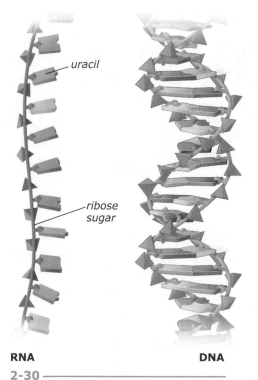

uracil

ribose sugar

RNA　　　　　　　　　　**DNA**

2-30

The RNA molecule is single stranded and contains the sugar ribose and the base uracil.

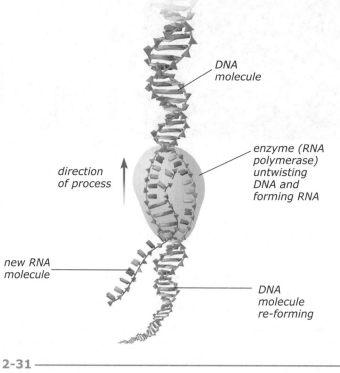

DNA molecule

enzyme (RNA polymerase) untwisting DNA and forming RNA

direction of process

new RNA molecule

DNA molecule re-forming

2-31

RNA transcription

of nucleotides rather than a double strand as in DNA. In RNA the sugar is ribose, which has one more oxygen atom than the deoxyribose in DNA. Also, in RNA the base uracil replaces the thymine found in DNA.

RNA is formed by DNA in a process called **transcription**. An enzyme called **RNA polymerase** (puh LIM uh RACE) starts at a specific section of the DNA molecule called the *promoter site*. From this point, the RNA polymerase moves down the DNA molecule, causing a section of it to separate. Other enzymes pair the proper RNA nucleotides to the corresponding DNA, forming an RNA chain. As the enzyme continues to move down the DNA molecule, the RNA chain separates from the DNA and the separated DNA strands re-form. This process is discussed in more detail in Chapter 4.

2-3 Functions of Major Organic Compounds					
Organic compounds	**Functions**				
	Structural	**Enzymatic**	**Storage**		
			Energy	**Material**	**Information**
carbohydrates	√		√	√	
lipids	√		√	√	
proteins	√	√		√	
nucleic acids					√

transcription: trans- (L. *trans*—across) + -scription (*scribere*—to write)

➡ **Review Questions 2.6**

1. Describe the relationship (a) between proteins and nucleic acids and (b) between amino acids and proteins.
2. How does the order of the amino acids contribute to a protein molecule's three-dimensional shape?
3. What are the two primary nucleic acids?
4. Describe the structure of a DNA molecule.
5. What is unique about the replication of a DNA molecule?
6. What are the structural differences between DNA and RNA?
⊙7. Carbohydrates are carbohydrates, and lipids are lipids. But nucleic acids and the proteins they manufacture truly constitute the organism. Support or disprove these statements.

Introduction to Cells

Cytology Part I

Facet

salm 19 encourages us to examine God's handiwork in order to learn more about Him. While the vastness and majesty of God can be appreciated by gazing on expansive mountain vistas and distant galaxies, His power and attention to detail are even more clearly seen in a careful examination of the minute details He planned and executed as He created life.

The cell has turned out to be far more complicated than anyone could have imagined just a century ago. What was once regarded as a box of organic compounds is now better described as a self-replicating, high-energy, extremely efficient factory. No one can delve into cellular biology without sensing that there is a God and that He has infinite power (Rom. 1:20). Early evolutionary theories on how the first life arose have been revised and replaced repeatedly as scientists have had to contend with the marvelous complexity of even the simplest cell. The minute details, so often overlooked, are where God's power as Creator and Sustainer are most clearly seen.

It is easy to become bogged down as you study cellular structures, functions, and reproduction (the topics of this chapter and the next few chapters). Christians, however, should rejoice in the fact that such a study reveals more of God. If He is so powerful and careful in His creation of the minutiae of cells, how much more is He concerned about eternal souls (Matt. 6:25–34). Truly, God is in the details.

3A—The Structure of Cells

Chemicals are the basic components of all matter. Atoms may combine to form molecules, and molecules can combine to form various cellular structures. When first proposed, the idea that living things were made up of cells seemed radical. As microscope technology advanced, it became evident that the cell was the basic unit of living things and that cellular components, although often vastly different, were based on similar patterns in most cells. Just as atoms combine to form molecules with different properties, so the organic molecules of living things are organized in different ways to form different cellular structures. This chapter on cellular structure shows how the chemistry learned in Chapter 2 can be applied to both the structure and function of the many entities that compose a cell.

3.1 Cell Theory

Cytology (sye TAHL uh jee), the study of cells, can be traced back over three hundred years to the English scientist Robert Hooke. In 1665, he published *Micrographia*, a report of his observations using a simple compound microscope that could provide a magnification power of approximately thirty times (30×). By contrast, a typical modern high-school microscope may magnify the image of the specimen as much as one thousand times. In Observation XVIII of *Micrographia*, Hooke described his observation of a thin piece of cork.

> I could exceedingly plainly perceive it to be all perforated and porous, much like a Honey-comb. . . . [T]hese pores, or cells, . . . were indeed the first microscopical pores I ever saw, and perhaps, that were ever seen, for I had not met with any Writer or Person, that had made any mention of them before this.

Because the neat rows of little boxes reminded him of the rows of rooms in a monastery, he named the structures **cells**. Later he studied sections of carrots and ferns and described similar findings.

In the specimens that Hooke studied, he saw only the walls of rather uniformly shaped "boxes." Because all plant cells have sturdy cell walls, Hooke was looking at the walls surrounding dead cells. Even though scientists now know that the "boxes" are found in vastly different shapes, sizes, and arrangements, they are still called cells. The first person to describe living organisms under a microscope was Anton van Leeuwenhoek in 1674—an amateur lens maker who had been intrigued after reading Hooke's *Micrographia*.

However, for the next two hundred years, microscopes were considered to be exotic

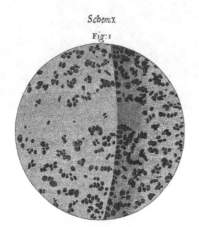

3-1

Sketches of cork "cells" from Hooke's *Micrographia*

cytology: cyto-, -cyte, or -cytic (Gk. *kutos*—hollow or cell) + -logy or -logo (*logos*—word; hence, the study of)

cell: (L. *cellae*—chambers)

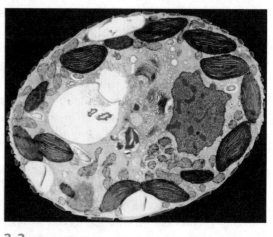

3-2

Compare the complexity of this cell as seen through an electron microscope with the empty cells Hooke observed and drew.

instruments. It was not until the nineteenth century that the use of microscopes became more commonplace. Gradually, as microscopes improved, scientists began to realize that a cell is not empty but is filled with minute structures. Although it had been recognized earlier, Robert Brown described the nucleus in 1833 as a special structure found in all cells. His and other studies attached significance to the role of the nucleus during cellular division and to the nuclei of egg and sperm cells.

The year 1838 brought a startling and daring statement from a German botanist, Matthias Schleiden (SHLY den). After studying much botanical material under a microscope, he stated that all plants are composed of cells. The next year Theodor Schwann (SHVAHN), a German physiologist, made a similar statement about animals. In 1855, Rudolf Virchow proposed that cells arise only from preexisting cells. The observations of these scientists have been combined to form what is called the **cell theory**. Continued research over the years by many scientists using more and more technically advanced instruments confirmed the validity of this theory, which is composed of three basic principles:

- The cell is the basic unit of all living things.
- Cells perform all the functions of living things.
- New cells come from the reproduction of existing cells.

The Cell as the Basic Unit of Life

The size of most organisms does not indicate the size of its cells but rather the number of its cells. For example, although a mouse and an elephant have cells of about the same size, their quantity of cells is quite different.

Not all of an organism is made of cells. Those parts of an organism that are not composed of cells are made of materials that cells manufacture. An insect's body covering is a secretion of some of the insect's cells. Many of the materials composing bones are cellular secretions.

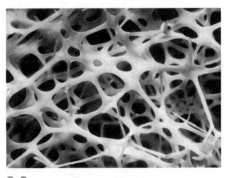

3-3

Much of the material in our skeletons is not made of cells but of cellular secretions.

Cell Size

How large can a single cell grow? The maximum size a cell can attain is limited by several factors. First, the message encoded in the DNA in the nucleus governs the size of the cell. Another factor is the ratio between the surface area of the cell and its volume. The nutrients a cell requires must pass from the outside of the cell, through the membrane, and to the inside. As the cell grows, its volume increases faster than does the surface area. At a certain size, the cell will need more nutrients than can pass through the membrane. Compare these three hypothetical cube-shaped cells. In the smallest cell, the surface area (6 mm^2) is six times as large as its volume (1 mm^3). The largest cell has a surface area (54 mm^2) that is only twice its volume (27 mm^3). In other words, the bigger a cell gets, the less adequate the surface area of the cell is to provide nutrients to the interior portions of the cell.

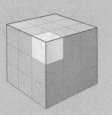

Nutrition	absorption	transport of dissolved substances into cells to serve as building blocks or energy sources
	digestion	enzymatic breakdown of substances to obtain building materials or energy
Internal functions	synthesis	assembly of organic compounds from smaller units obtained from digestion, absorption, or some other synthesis reaction in the cell; results in a cell's growth, secretion, or replacing worn out cellular parts
	respiration	breakdown of food (usually glucose) with the release of energy
	movement	movement of the cell itself (locomotion) or movement of substances and structures inside the cell (internal movement)
	irritability	ability to respond to external factors that affect the operations of the cell; in other words, the response or reaction of the cell to its environment
Release of materials	excretion	removal of soluble waste from the cell
	egestion	elimination of nonsoluble waste from the cell
	secretion	synthesis and release of substances from the cell
Continuation of existence	homeostasis	ability to maintain a steady state in the cell
	reproduction	formation of new cells

The Cell as the Functional Unit of Life

Cells are responsible for all the functions of any living thing. Because the functions are so various and overlapping, it is difficult to describe them. The cellular functions given in Table 3-1 are more a list of the processes of life than a list of what cells do. Not all cells perform all these functions, but in order for an organism to be alive, all of these functions must be carried out. In complex organisms like human beings, some cells perform a few of these functions almost to the exclusion of other functions.

The Cell as the Reproductive Unit

For most cells the biosynthesis of materials results in maintenance and growth of the cell. The science fiction cell "that grew so large it swallowed New York," however, is not possible. When most cells reach a particular size, they either slow down their synthesis of materials, secrete certain synthesized materials, or divide. The built-in control of a cell (the DNA) can control only so much; therefore, cell division after a limited amount of growth is essential for most cells.

Cells reproduce by dividing. Cellular division results in a new organism (if the organism is unicellular), more cells in the same organism, or a sex cell, like an egg or sperm. Processes of cellular reproduction are discussed in Chapter 5.

digestion: di- (apart) + -gestion (L. *gerere*—to carry)

excretion: e- or ex- (L. *e-*—out) + -cretion (L. *cernere*—to separate)

egestion: e- or ex- (out) + -gestion (to carry)

⇒ Review Questions 3.1

1. List the concepts of the cell theory.
2. Describe how the surface area of a cell can influence the maximum size a cell might attain.
3. Trace the development of the modern cell theory, listing the scientists and what they contributed to the theory.

3.2 Levels of Cellular Organization

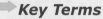

Whether they are one-celled organisms floating in a pond or gigantic sequoia trees, living things are composed of cells. Those organisms that are composed of just one cell are called **unicellular organisms**. Bacteria, as well as many protozoans, some algae, and some fungi, are unicellular.

Multicellular organisms are made of many cells. Some algae and fungi are multicellular, as are all animals, humans, and plants. Some multicellular organisms are colonial. A **colonial organism** is a collection of similar cells living together. Virtually any of a colonial organism's cells can carry on all the life processes of the organism. Thus, if they are separated from the colony, they can grow, divide, and become a new organism. Colonial organisms can vary in complexity from what appears to be a clump of cells to a highly organized mushroom.

Most multicellular organisms are made of cells that are organized into tissues. A **tissue** is a group of similar cells that work together to carry out a specific function. The muscle cells of the body compose tissues whose primary function is movement. The layer of cells that line your digestive tract is another example of a tissue.

Many multicellular organisms have **organs**, which are composed of several types of tissues working together to perform a specific function. The stomach is an example of an organ. It is made of muscle tissue that churns the food, a special tissue that lines the inside of the stomach to protect it from acid, nerve tissue that directs the movement of the stomach, connective tissue that holds the stomach together, and cells embedded in the stomach lining that produce acid and other enzymes. All these tissues work together to help digest food. An **organ system** is made up of a group of organs that work together to accomplish life functions. The digestive system includes the mouth, stomach, intestines, and other organs that work together to digest food and eliminate waste so that the body can function properly.

Objectives

- Differentiate between unicellular and multicellular organisms
- Distinguish between colonial organisms and organisms containing organ systems

Key Terms

unicellular organism
multicellular organism
colonial organism
tissue
organ
organ system

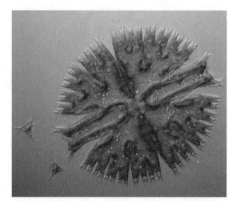

3-4 —————
This unicellular organism is a type of alga called a desmid.

Evolutionary Bias

Sometimes unicellular or colonial organisms are referred to as "first" or "early" organisms. Use of these terms should alert the reader of evolutionary bias, implying that the later, more complex organisms evolved from the "early" organisms. However, these terms are not correct since all life was created during the week of Creation.

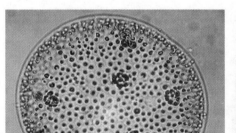 

3-5 —————
Volvox, an alga, and mushrooms, which are fungi, are both colonial organisms.

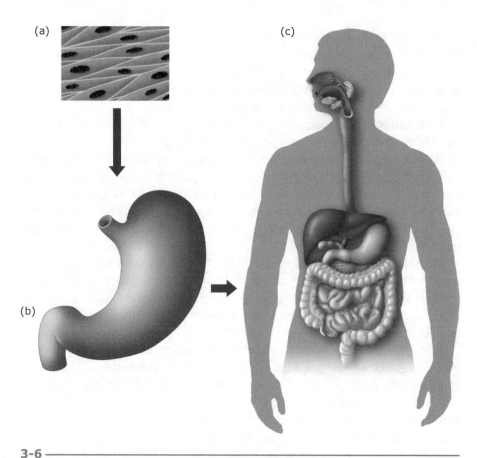

3-6

Smooth muscle (a), along with other tissues, forms the stomach (b). The stomach is one organ in a multiorgan system—the digestive system (c).

However, none of the specialized systems of tissues can survive alone. They depend on other tissues and organ systems to supply their needs. Muscles provide movement, but they depend on the digestive system to provide the energy needed for movement, the circulatory system to bring them food and oxygen, and the nervous system to tell them when to move.

Review Questions 3.2

1. Why is it accurate to describe a unicellular organism as both simple and complex?

2. Explain the difference between a colony and a tissue.

3. Give three examples of (a) organs in the human body and (b) organ systems in the human body.

⊙4. Is it acceptable to call unicellular organisms "early organisms"? Why or why not?

3.3 Cellular Anatomy

Cells are so diverse in their functions and structures that finding or describing a "typical" cell is impossible. While some structures are common to all cells, their form or function may vary. Other structures may be found only in a particular type of cell. As cytological techniques have improved, theories of cellular structure and function have changed, and they may change again as more information is discovered. Currently cell anatomy can be divided into three broad categories:

- A boundary that encloses the cell
- The enclosed area, containing various structures and molecules
- The nucleus that contains DNA and other materials

Two Types of Cells

Cells can be easily divided into two major classifications based on the structure of the nucleus and the organelles it has. An organelle (OR guh NELL) is a structure inside the cell that performs special functions in the cell. Organelles can be compared to the various organs in the body—each one performs special tasks essential for the survival of the cell. Organelles with membranes around them are called *membrane-bound organelles*; those lacking membranes are termed *non-membrane-bound organelles*.

One of the two kinds of cells is eukaryotic (yoo KEHR ee AHT ik). Eukaryotic cells have a membrane-bound nucleus and both membrane- and non-membrane-bound organelles. Humans, plants, animals, and several other organisms are composed of eukaryotic cells.

Bacteria, blue-green algae, and some similar organisms are termed prokaryotic (proh KEHR ee AHT ik) cells. These cells lack a membrane around their nuclear area and contain only non-membrane-bound organelles. Differences between these two types of organisms are so significant that eukaryotes and prokaryotes are placed into separate kingdoms. Prokaryotic organisms and their cellular structures are discussed in greater detail in Chapter 10, and the remainder of this chapter is devoted primarily to eukaryotic cells.

Cell Boundaries

Technically the outer boundary of the cell, which is still part of the cell, is the cell membrane. However, many cells have structures outside their cell membrane that are not actually part of the cell. These structures—cell walls and capsules—are also discussed as part of cell boundaries.

Objectives

- Explain the difference between eukaryotic and prokaryotic cells
- Describe the structure and function of the cell membrane
- Describe the difference between a cell membrane and a cell wall
- List the major organelles in a cell and describe their function
- Describe the nucleus and its structures

Key Terms

organelle	centrosome
eukaryotic	microfilament
prokaryotic	flagellum
cell membrane	cilium
lipid bilayer	plastid
cell wall	chloroplast
cytoplasm	vacuole
mitochondrion	vesicle
ribosome	turgor pressure
endoplasmic	nucleus
reticulum	nuclear
Golgi apparatus	envelope
lysosome	chromatin
cytoskeleton	material
microtubule	nucleolus

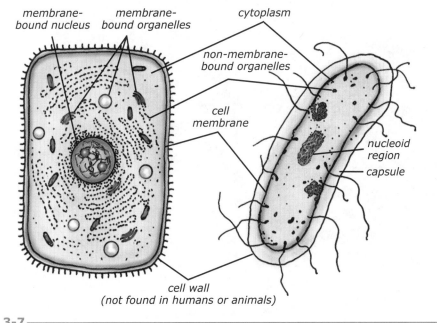

membrane-bound nucleus
membrane-bound organelles
cytoplasm
non-membrane-bound organelles
cell membrane
nucleoid region
capsule
cell wall
(not found in humans or animals)

3-7 A comparison of a eukaryotic cell and a prokaryotic cell

Evolutionary Bias Again

The term *prokaryotic* was chosen by evolutionary scientists because it basically means "before the nucleus." In evolutionary theory, prokaryotic cells are primitive cells that evolved to become eukaryotic cells, which have better developed nuclei. Even though these terms are evolutionary, we still use them because we cannot communicate scientifically with others without them.

organelle: (L. *organella*—little organ)

eukaryotic: eu- (Gk. *eu-*—good or true) + -karyotic (*karuotis*—with a central part)

prokaryotic: pro- (Gk. *pro*—before) + -karyotic (with a central part)

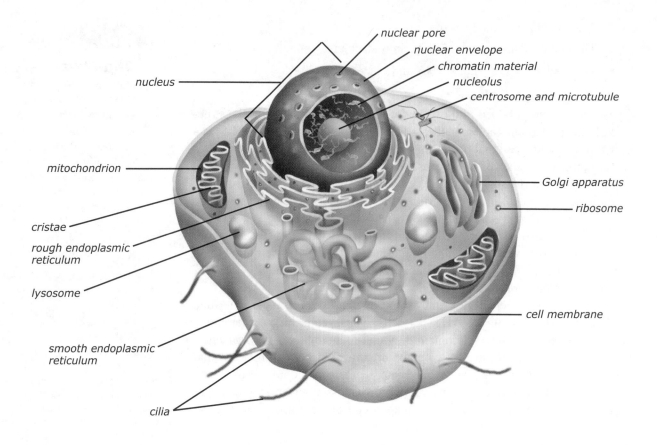

nuclear pore

nuclear envelope

chromatin material

nucleus

nucleolus

centrosome and microtubule

mitochondrion

Golgi apparatus

ribosome

cristae

rough endoplasmic reticulum

lysosome

cell membrane

smooth endoplasmic reticulum

cilia

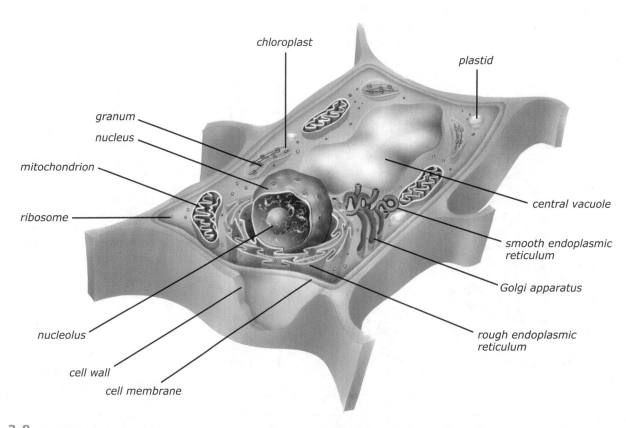

chloroplast

plastid

granum

nucleus

mitochondrion

central vacuole

ribosome

smooth endoplasmic reticulum

Golgi apparatus

nucleolus

rough endoplasmic reticulum

cell wall

cell membrane

3-8

Structures in a typical animal cell (top) and typical plant cell (bottom)

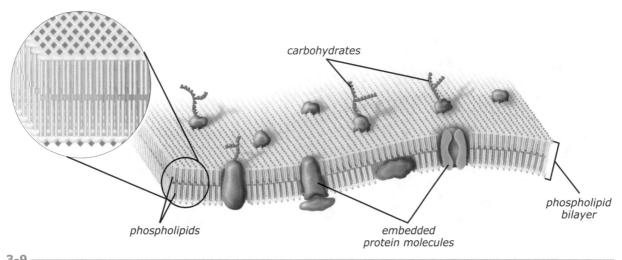

carbohydrates

phospholipids

embedded
protein molecules

phospholipid
bilayer

3-9

Cell membrane section

Cell Membranes

Every cell, whether prokaryotic or eukaryotic, has a **cell membrane** (sometimes called the plasma membrane), which serves as the outermost boundary of the cell itself. The cell membrane is the structure that separates the cell from its environment. At the same time that the cell membrane is protecting the cell from the environment, it must also allow the cell to have access to the environment—nutrients must be absorbed, wastes removed, and substances produced by the cell must be secreted.

One thing that membranes have in common is that they contain phospholipids. The fact that phospholipids have hydrophobic and hydrophilic ends and do not dissolve in water gives cell membranes their basic structure. Because

3-2 Cellular Components and Their Functions	
cell membrane	controls movement of substances into and out of the cell
mitochondrion	transforms energy stored in sugars to usable cellular energy
ribosome	synthesizes proteins
endoplasmic reticulum	synthesizes proteins and sterols and transports materials within the cell
Golgi apparatus	prepares substances to be secreted by the cell
lysosome	breaks down ingested substances, old organelles, and cytoplasmic molecules
cytoskeleton	provides structure for the cell; necessary for movement and reproduction
flagella and cilia	provide locomotion or move substances over the outer surface of the cell
vacuole	provides storage
vesicle	provides storage; may contain enzymes for specific reactions
contractile vacuole	collects and pumps water out of the cell
nucleus	stores DNA; produces RNA
cell wall*	protects and supports cell
plastid*	houses pigments and stores starches (Special plastids, chloroplasts, are the site of photosynthesis.)
central vacuole*	provides storage; regulates turgidity of cell

*These structures are found in plant cells and some other eukaryotes but not in human or animal cells.

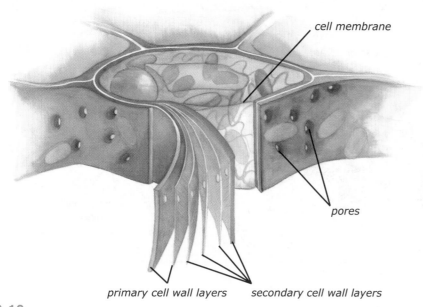

cell membrane

pores

primary cell wall layers   secondary cell wall layers

3-10
Plant cell wall

water is found inside and outside the cell, the phospholipids align themselves into a **lipid bilayer**. This is a double layer of phospholipids with the hydrophilic ends adjacent to the water environment and the hydrophobic ends toward each other, away from the water.

While phospholipids are found in all cell membranes, the exact composition of the cell membrane differs between cells. One difference is the combination of other lipids, such as cholesterol, and proteins that are embedded in the membrane. These molecules impart unique properties to the membrane.

Cell Walls

In addition to the cell membrane, plant cells and some similar kinds of cells are surrounded by a cell wall. The **cell wall** is a rigid or nearly rigid structure that is located on the outside of the cell membrane.

Plant cell walls are made primarily of cellulose and other carbohydrates. They are multilayered, providing strength and rigidity. Such strong walls would seem to make the cell an isolated island; however, this is not the case. Cell walls have pores through which the cell can communicate with other cells and take in and export substances.

Capsules

Many unicellular or colonial organisms have a *capsule* in addition to (or occasionally in place of) a cell wall. Sometimes this covering is called a *slime coat*. Capsules are made of cellular secretions, such as polysaccharides and lipids, of varying thickness. Unlike cell membranes and cell walls, capsules have no structural organization. Capsules give bacterial and algal colonies their shiny appearance and slimy feel. These structures protect the organism because the material must first either dissolve the capsule or pass through it before entering the organism.

Cytoplasm

The term **cytoplasm** (SYE tuh PLAZ uhm) refers to all of the structures and materials inside the cell membrane, excluding the nucleus. The actual fluid portion of the cytoplasm (minus the organelles) is called the *cytosol*. Some molecules in the cytoplasm are used to rapidly build structures to move organelles around the inside of the cell in a process called cytoplasmic streaming. In some plant cells, *cytoplasmic streaming* moves the organelles that carry on photosynthesis throughout the cell so that they can be positioned to take best advantage of the available light.

Mitochondria

Nicknamed the "powerhouse" of the cell, **mitochondria** (MYE tuh KAHN dree uh) (sing., mitochondrion) are typically bean-shaped organelles that are responsible for the reactions that transform the chemical energy stored in sugars into usable energy for the cell. The number of mitochondria contained within a cell will vary, depending on the type of cell and its energy needs.

Cell Walls and Plant Growth

The *primary* cell wall is formed when a plant cell is young and still growing. Containing cellulose fibers, a jellylike substance called pectin, and water, it is flexible enough to allow continued growth.

The *secondary* cell wall forms when the cell has reached full size. This layer between the cell membrane and the primary cell wall contains more cellulose and is arranged in alternating layers like plywood. Once the secondary wall forms, cell growth ends.

---

cytoplasm: cyto- (cell) + -plasm (Gk. *plassein*—to mold)

Two membranes surround the mitochondria—a smooth outer membrane and an inner membrane that has many folds. These inner folds, or *cristae*, not only greatly increase the surface area inside the mitochondria, but they also contain enzymes and other proteins embedded in the membrane. Many of the chemical processes used during cellular respiration occur at these sites.

The *mitochondrial matrix*—the fluid inside the mitochondria—also contains many enzymes needed for respiration. Depending on the usual energy demands of the cell, mitochondria may have many cristae, as in muscle cells, or they may have only a few cristae.

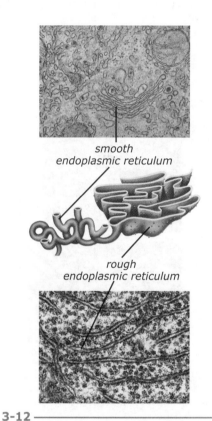

3-11
A TEM of a mitochondrion (a) and an illustration showing the outer membrane and the cristae of the inner membrane (b)

Ribosomes

A **ribosome** is a non-membrane-bound organelle found in both prokaryotic and eukaryotic cells. These small organelles are composed of two basic units—proteins and multiple strands of RNA. The primary function of ribosomes is to line up the amino acids during the production of proteins. Ribosomes can be either free-floating in the cytoplasm or attached to the endoplasmic reticulum.

Endoplasmic Reticulum

The **endoplasmic** (en doh PLAZ mik) **reticulum** (rih TIK yuh lum)—often abbreviated ER—is a system of interconnected folded membranes inside the cell. In some cells the ER is continuous with the membranes around the nucleus. There are two types of ER contained in most cells. *Smooth ER* is so named because it lacks any attached ribosomes. Smooth ER is found in cells that secrete sterols and in liver cells where toxic substances are broken down. Some ER is dotted with ribosomes and is called *rough ER*. Rough ER is especially common in cells that produce proteins to be secreted by the cell. The ribosomes send some of the newly formed proteins into the ER, where the protein may be further modified before it is secreted. The ER also provides a transportation network for compounds inside the cell and helps maintain the cell's shape.

Golgi Apparatus

In 1898, Camillo Golgi (GOLE jee) first described a cellular structure that has since been found in one form or another in most eukaryotic cells. The **Golgi apparatus** appears as flattened, curved, membrane-covered sacs. It is important in the final processing and packaging of many complex substances

3-12
The smooth ER lacks ribosomes, whereas the rough ER is dotted with ribosomes.

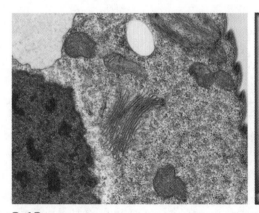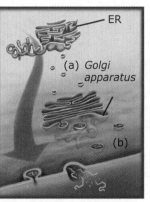

3-13
A TEM of a Golgi apparatus (center). The ER sends sacs of molecules to the Golgi apparatus (a), where they are prepared for release by the cell or used inside the cell (b).

endoplasmic: endo- (in) + -plasmic (to mold)
reticulum: (L. *rete*—net)

3-14
The cytoskeleton fills the cytoplasm and gives structure to the cell.

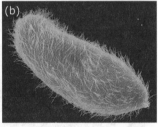

Dr. Richard Kessel & Dr. Gene Shih
Visuals Unlimited, Inc.

(a)

(b)

3-15
SEM of a euglena with a single flagellum (a). Cilia completely cover this parame-cium (b).

lysosome: lyso- (Gk. *lusis*—a loosening) + -some (Gk. *soma*—body)

microtubule: micro- (small) + -tubule (L. *tubule*—tube)

centrosome: centro- (L. *centrum*—center) + -some (body)

microfilament: micro- (small) + -fila-ment (L. *filium*—thread)

flagellum: (L. *flagellum*—little whip)

cilium: (L. *cilium*—eyelid)

leucoplast: leuco- (Gk. *leukos*—clear, white) + -plast (to mold)

chromoplast: chromo- (Gk. *khroma*—color) + -plast (to mold)

produced by the cell. Once the Golgi apparatus receives these substances, they are sealed in small sacs whose membranes appear to be made by pinching off the membrane of the Golgi apparatus as it bulges outward. The Golgi apparatus works in close association with the ER.

Lysosomes

Lysosomes (LYE so SOHMS) are small, irregularly shaped membrane-bound organelles filled with digestive enzymes. They are very common in the cells of protozoans, fungi, humans, and animals. The enzymes contained in the lyso-somes can digest invading bacteria and viruses or food substances. Lysosomes are also important in breaking down old or nonfunctional cellular structures. This process permits the recycling of materials for the formation of new or-ganelles and cell growth.

Cytoskeleton

The cytoplasm is not just an unorganized region with organelles float-ing about in the cytosol. Like the body, eukaryotic cells have a structural component that maintains their shape. Under an electron microscope, tiny fibers can be seen crisscrossing the inside of the cell. This microscopic system of fibers is called the **cytoskeleton**.

Two major types of fibers compose the cytoskeleton. The **microtubules** are straight, hollow tubes of proteins. They help maintain the cell's shape and pro-vide tracks for the movement of organelles and substances. The **centrosome** (SEN tro sohm) is a region located near the nucleus that is important in the production of microtubules. The **microfilaments** are thinner, solid strands of protein. They are especially important for cells such as muscle cells that con-tract and for cells that exhibit crawling movements.

Microfilaments and microtubules, along with special proteins attached to them, form a system of scaffolding, ropes, pulleys, and engines that control the cell's shape, give it strength, and move structures around inside the cell. They also play a role in cellular locomotion.

Flagella and Cilia

A **flagellum** (fluh JEL um) (pl., flagella) is a long tubular extension of the cell membrane surrounding a special arrangement of microtubules. Flagella are often longer than the cell and usually exist singly; however, they may occur in pairs or groups of three, four, or five. **Cilia** (sing., cilium) have an internal organization similar to that of flagella, although they are much shorter and frequently cover the entire cell or an entire section of a cell.

Cilia and flagella in eukaryotic cells beat in specific patterns. Although both of these structures are used in cellular locomotion, they can also function in a different way. Instead of moving small organisms through watery environ-ments, they can move the environment past a cell whose position is fixed. For example, the surface of the cells that form the lining of the respiratory tract is covered with thousands of small cilia and a thin layer of mucus. As the cilia beat, mucus is carried up and away from the lungs along with the dust and bacteria that the mucus has trapped.

Plastids

Plastids are found in cells of plants, algae, and a few other organisms, but not in humans or animals. They are classified as either *leucoplasts*—colorless structures used as storehouses—or *chromoplasts*—structures that contain pig-ments and usually function in synthesis processes. Leucoplasts are found in the fleshy storage areas of plants. Potatoes, root crops, and many fruits have

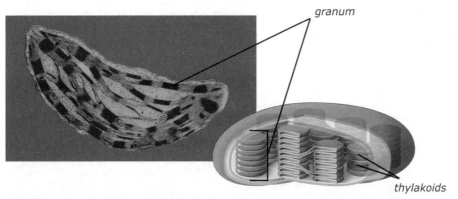

granum

thylakoids

3-16
Chloroplast TEM (a) and cutaway view (b)

leucoplasts that contain starches and occasionally oils. Chromoplasts contain pigments that give the plant its coloration.

The **chloroplast**, one of the most familiar chromoplasts, is a green organelle in which light energy is converted into organic compounds. Some chloroplasts are cup shaped and nearly as large as the cells that hold them. Others are flat or even spiral shaped. The most common chloroplasts are round or football shaped. Internally, the chloroplast contains a complex system of flattened sacs called *thylakoids*. The thylakoids are arranged into stacks called *grana* (sing., granum). The membranes of the grana contain the green pigment chlorophyll, which catches sunlight for the manufacture of sugar in the process of photosynthesis.

Vacuoles and Vesicles

Both vacuoles and vesicles are multipurpose membrane-bound organelles, and the difference between them is often hard to discern. Although vacuoles are found in all types of cells, they are more common in plant cells.

Vacuoles (VAK yoo ohlz) can be compared to various household containers—different boxes, jars, and bags—that are all used to hold various substances. Vacuoles are membrane-bound organelles that contain food, water, wastes, or other materials. They tend to be stationary in the cell or to move very slowly.

Vesicles (VES ih kulz) are also membrane-bound organelles. Although there are no distinctly defined size differences between them, vesicles are usually smaller and more mobile than vacuoles. The best way to learn about the two structures is to examine some of the more common and important types.

✦ *Food vacuole.* Food vacuoles are formed when the cell takes in food, a process called *ingestion*. Lysosomes fuse with the vacuolar membrane, releasing their digestive enzymes into the vacuole. The enzymes break down the food into substances that will pass through the membrane into the cytoplasm, where they will be used by the cell.

✦ *Waste vacuole.* After the lysosomal enzymes have completed their task, the indigestible materials remain in the vacuole, which is now termed a waste vacuole. Cytoplasmic streaming moves the waste vacuoles to the inner surface of the cell membrane, where the vacuole fuses with it, pushing the wastes outside the cell. This method of waste release is called *egestion*.

✦ *Central vacuole.* Plant cells often have a central vacuole. The central vacuole can be quite large—often 90% of the total volume of the cell—pushing the cytoplasm and other organelles out to the cell membrane.

Because of the high concentration of solutes in the fluid of the vacuole, water from outside the cell moves into the vacuole by osmosis. As the vacuole expands,

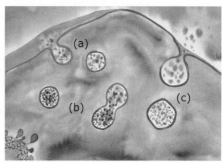

3-17
Cellular digestion: A food vacuole is formed (a). A lysosome from a Golgi apparatus fuses with the food vacuole (b). The waste vacuole fuses with the cell membrane, where the waste products are expelled (c).

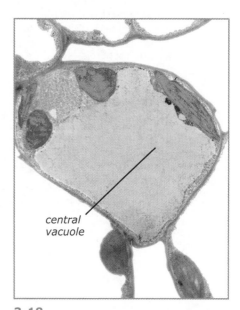

central vacuole

3-18
TEM of a plant cell showing the large central vacuole

chloroplast: chloro- (Gk. *khloros*—green) + -plast (to mold)

thylakoid: (Gk. *thulakos*—a sack)

granum: (L. *granum*—seed)

vacuole: (L. *vacuus*—empty)

vesicle: (L. *vesicula*—little bladder)

ingestion: in- (L. *in*—in) + -gestion (to carry); hence, to carry in

Cytoskeleton—Cellular Support and Movement

The cytoskeleton is an elaborate network of protein fibers inside the cell that extends from the nucleus to the cell membrane. As the *-skeleton* portion of the name implies, the cytoskeleton gives support to the cell much like the bony skeleton supports the body. Unlike a bony skeleton, however, the cytoskeleton is constantly being reorganized as the cell changes its shape, moves, and rearranges its cytoplasmic contents. Some components of the cytoskeleton provide for the structural integrity of the cell, others form tracks on which organelles are moved around, and still others move the cell itself.

The cytoskeleton is composed of two primary structures—microtubules and microfilaments—each composed of different protein subunits. Thousands of protein subunits are needed to make up a single microtubule or microfilament.

Microtubules are the thicker of the cytoskeleton components and are very important in the organization of the cell, in cell division, and in cellular movement. They are long, somewhat stiff tubular structures that can be rapidly assembled, disassembled, and reassembled as needed by the cell. Not only do these microtubules give shape

and support to the cell, but they also can provide tracks to transport organelles and other cellular substances inside the cell. Special "motor" proteins bond the microtubule to specific cytoplasmic components so they can travel up and down the microtubule.

In most animal and human cells, microtubules originate from the centrosomes and extend out to the cell membrane. The centrosome is important in cellular division and is discussed in Chapter 5. Inside the centrosomes of human cells and some animal cells are two structures called *centrioles* (SEN tree ohlz); however, scientists are unsure of the centrioles' function in the centrosome.

Microtubules are also an important part of cilia and flagella. Both flagella and cilia are projections of the cell membrane that have a core of microtubules. The internal structures of cilia and flagella are the same: an extension of the cell membrane covers a bundle of nine microtubule doublets (pairs) arranged in a ring, and two unpaired microtubules are in the middle of the ring.

Although cilia and flagella are structurally very similar, the movements of cilia and flagella are different. Cilia have coordinated movements similar to the rowing of oarsmen. Flagella make wavelike movements along their length, just as if a piece of rope were being shaken rapidly up and down.

Microfilaments are the thinner of the cytoskeleton elements. Each microfilament is composed of two molecules of the protein *actin* twisted around each other. Like microtubules, they can be assembled and disassembled as needed by the cell, and many different kinds of proteins can bind to the microfilament, depending on the function to be performed. Some of the microfilaments are located just under the cell membrane and, like other cytoskeleton components, provide a latticelike network that gives the cell mechanical strength and imparts shape. During cell division (discussed in Chapter 5), microfilaments help form the contractile rings that pinch the cell into two separate cells.

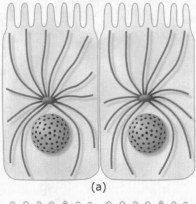

(a)

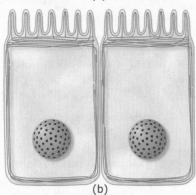

(b)

The two basic elements of the cytoskeleton: microtubules (a) and microfilaments (b)

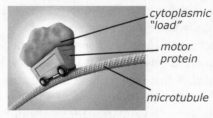

cytoplasmic "load"

motor protein

microtubule

Movement of cytoplasmic contents is coordinated by special motor proteins and microtubules.

(a)

(b)

The 9 × 2 microtubule arrangement in flagella: TEM (a) and illustration (b)

Flagella beat with a whiplike motion.

Cilia have a repetitive cycle that consists of a power stroke and a recovery stroke—much like someone rowing a boat.

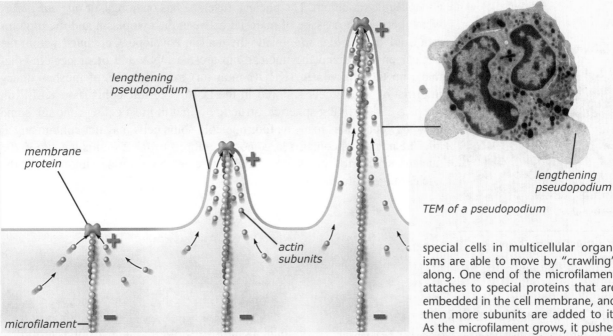

lengthening
pseudopodium

membrane
protein

actin
subunits

microfilament—

Microfilament forming a membrane protrusion, or pseudopodium

*lengthening
pseudopodium*

TEM of a pseudopodium

Muscle cells contain large numbers of microfilaments that are linked to filaments of another protein called *myosin*. When these two different fibers are activated, they slide along each other, causing the cell to contract. The coordinated movement of millions of fibers sliding along each other causes muscular movements.

Microfilaments are also important in another type of cellular movement. Some unicellular organisms and special cells in multicellular organisms are able to move by "crawling" along. One end of the microfilament attaches to special proteins that are embedded in the cell membrane, and then more subunits are added to it. As the microfilament grows, it pushes out the leading edge of the cell membrane, forming extensions. As these lengthen, the opposite side of the cell is pulled in by contraction of the microfilaments. Using this kind of movement, some white blood cells are able to slip through blood vessel walls and crawl through body tissue to hunt down disease-producing organisms.

pressure is exerted on the vacuole membrane, which pushes the cytoplasm and its contents against the rigid cell wall. This cellular fullness caused by the water pressure in the central vacuole is called **turgor** (TUR gur) **pressure**. The difference between crisp and wilted lettuce is the result of the plant cells' being turgid (crisp) or not being turgid (wilted).

✦ *Contractile vacuole.* Many unicellular organisms that live in water environments do not have cell walls. These cells, it seems, should burst, since water molecules quickly come through the cell membrane to equalize the concentration difference between the water environment and the cytoplasm inside the cell (osmosis). Under normal circumstances the cell collects these water molecules in the contractile vacuoles. After reaching a certain size, the contractile vacuole fuses with the cell membrane and ejects the water.

✦ *Secretion vesicle.* These vesicles are usually formed by the endoplasmic reticulum or Golgi apparatus. Vesicles transport the materials made by these structures to the cell membrane. When the vesicle membrane and the cell membrane unite, the materials are secreted.

The Nucleus

The **nucleus** is sometimes called the control center of the cell and is usually the most prominent structure seen when observing a stained cell with a microscope. The nucleus is the site where DNA replication and RNA transcription take place. The **nuclear envelope** is a double membrane that completely surrounds the nucleus. In some cells, the nuclear envelope is continuous with the

turgor: (L. *turgere*—to be swollen)

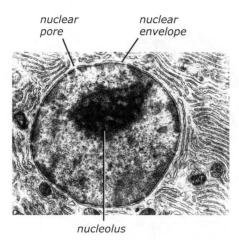

nuclear pore   nuclear envelope

nucleolus

3-19
TEM showing structures of the nucleus

chromatin: chroma- (color)

endoplasmic reticulum. The nuclear envelope has openings, or *nuclear pores*, which permit the passage of material between the cytoplasm and the protein-rich fluid, the *nuclear sap*, inside the nuclear envelope. Contained within the nuclear envelope and the nuclear sap are the DNA and other proteins that make up the **chromatin** (KROH muh tin) **material**. All of the hereditary information of the cell is stored in the DNA. The **nucleolus** (noo KLEE uh lus) is a dark-staining spherical structure found in most cells. Although some cells may contain as many as four nucleoli, some cells may not contain any at all. The nucleolus contains large concentrations of RNA and is the site in the nucleus where ribosomes are partially assembled before passing through the nuclear pores into the cytoplasm.

▶ Review Questions 3.3

1. Name the three categories of eukaryotic cell anatomy.
2. What is an organelle?
3. How does a prokaryotic cell differ from a eukaryotic cell?
4. Describe the structure and functions of the cell membrane.
5. List the cellular organelles and describe their functions.
6. Describe the two functions of cilia.
7. Why do you think that the nucleus is often called the "control center" of the cell?

8. Describe the structure of the plant cell wall. Is it living? Why or why not?
9. What are the two main components of the cytoskeleton? Discuss the structure and function of each.
10. What is the difference between a leucoplast and a chromoplast?
⊙11. In what ways are cell membranes, cell walls, and capsules only the outer limits of a cell and not merely the cell's boundaries?

3B—Cells and Their Environment

Rather than being passively isolated, a cell interacts with its environment. This section discusses how the environment can affect a cell and how the cell can respond to changes in its environment.

3.4 Homeostasis

Being alive and being surrounded by living things may cause you to lose an appreciation for the narrow range of conditions necessary to maintain the homeostasis (HOH mee oh STAY sis) of life. **Homeostasis** means "steady state," but in reality, it is a *dynamic equilibrium*—the organism is constantly interacting with the surrounding environment to maintain the delicate balance of conditions for life. When the surrounding conditions are altered, the organism must adjust to maintain the homeostasis of life.

Optimal Point and Range of Tolerance

Scientists call the precise conditions at which something functions best its *optimal point* for each condition. The optimal point for temperature in human muscle cells is 37 °C (98.6 °F).

Most organisms have an *optimal range*. Within the limitations of the optimal range, an organism's or cell's performance is stable. Most human muscle cells

3.4

▶ Objectives

- Relate homeostasis to the living condition
- Explain optimal range, optimal point, and range of tolerance
- Explain the dynamic nature of homeostasis
- Describe the effects of hypotonic, hypertonic, and isotonic solutions on a cell

▶ Key Terms

homeostasis
isotonic
hypotonic solution
hypertonic solution

homeostasis: homeo- (same) + -stasis (Gk. *stasis*—a standstill)

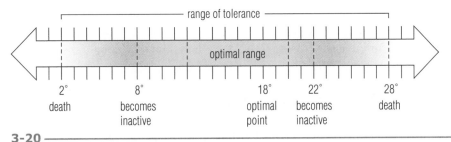

range of tolerance

optimal range

2°	8°	18°	22°	28°
death	becomes inactive	optimal point	becomes inactive	death

3-20

Temperature (in °C) for a soil microbe

will function equally well in temperatures within a degree or two of 37 °C. The optimal range for a muscle cell is quite narrow when compared to the range of temperatures that are optimal for the body. A person can function well and comfortably in rooms with temperatures of 16–27 °C (60–80 °F), depending on the humidity, air currents, clothing, and activities being performed. Throughout this range of temperatures, the body works hard to maintain a constant temperature (homeostasis) for its muscle cells.

If the body's temperature increases above the optimal range, a person begins to sweat and his body sends more blood to the arms, legs, and skin. The evaporation of the sweat works to decrease the temperature of the skin and helps cool the blood. If his body's temperature continues to rise, the cells in his body begin to malfunction and he becomes ill. On the other hand, if he becomes chilled, he may begin to shiver, since shivering is the body's attempt to generate heat. To conserve heat, the body will decrease the amount of blood sent to the arms, legs, and skin. If the person is not warmed to the proper temperature, his body begins to stop functioning. The change in temperature forces the body to alter its normal operations to restore homeostasis.

The same principle is true for individual cells—the environment affects the functioning of the cell. In the case of temperature, enzymes may be destroyed or rendered nonfunctional and cellular processes may occur too fast, too slow, or not at all. If the temperature gets too high or continues for too long, the cell will die.

When a cell or organism gets warmer or cooler than the limits of its optimal range, it enters the *range of tolerance* for temperature. The range of tolerance is that range of temperature in which a cell or organism will remain alive but will not function properly. The further it is outside the optimal range, the more poorly it functions. When it reaches its *limit of tolerance*, it dies.

The range of tolerance for heat and cold for a muscle cell is only a few degrees, quite narrow when compared to the range of tolerance for a skin cell. God designed skin to deal with the more extreme temperature fluctuations in the environment. Nerve cells, on the other hand, have a very narrow range of tolerance; it is almost the same as their optimal range. Brain cells are extremely specialized nerve cells and must be kept within their optimal range to maintain proper functioning.

All living things were designed with ranges of tolerance that enable them to survive in their normal environment. This is why banana trees grow well in rainy tropical areas but grow quite poorly or not at all in deserts or on chilly mountaintops. Designers of botanical gardens spend hundreds of thousands of dollars to provide environments within the optimal range for the many different species of plants they display. However, some organisms (and cells) live in environments that change drastically in the course of a year. God designed some organisms whose environment normally fluctuates beyond its

3-21

Man-made rainforest and desert environments must maintain the proper temperatures, humidity, and light levels.

optimal range to enter a period of *dormancy* in order to survive unfavorable conditions. The adjustments made by some trees to live through the winter are examples of dormancy. Nonetheless, if the conditions become too harsh and exceed the limits of their tolerance, they will die.

The Solutions Around Cells

Whether a cell is a unicellular organism or is part of a multicellular organism, it is in a solution. One of the most critical influences upon the existence of a cell is the concentration of the substances in the solution around it. The cell membrane is the structure that stands between the surrounding environment and the cell, working to maintain homeostasis. To illustrate this, consider red blood cells and the solution around them.

When the concentration of solutes outside the cell is the same as the concentration inside the cell, the solution is said to be **isotonic** (EYE suh TAHN ik). The red blood cells in the bloodstream are a good example of cells in an isotonic solution—the cytoplasm of the red blood cells and the solution that surrounds the cells (plasma) have identical concentrations. Although the solutes in the red cells may be different from those in the plasma, the ratio of water to solutes is the same. Since there is no concentration gradient, there is still movement of water molecules across the semipermeable membrane but no *net* osmosis.

If red blood cells are placed into pure water, a concentration gradient is established. When the solution outside the cell has a higher concentration of water molecules and a lower concentration of solutes than the solution inside, the cells are said to be in a **hypotonic** (HYE poh TAHN ik) **solution**. The solutes inside the cell are unable to move through the membrane into the water around the cell; however, the water molecules can move through the membrane into the cell. This sets up a one-way flow of water molecules into the cell, causing the cell to swell and eventually burst. The bursting of a cell from internal water pressure is called *cytolysis* (sye TAHL ih sis).

If red blood cells are placed in a fluid with a concentration of solutes higher than that inside the cell, the cells are in a **hypertonic** (HYE pur TAHN ik) **solution**. In hypertonic solutions, water molecules will diffuse out of the cell. As the cell loses water, it shrinks.

3-3 Direction of Osmosis		
External conditions	**Net movement of water**	
hypotonic	into the cell	
hypertonic	out of the cell	
isotonic	no net movement	

Cells in Hypotonic Solutions

Since fresh water has fewer solutes than the cytoplasm of a cell, a unicellular organism that lives in fresh water exists in a hypotonic solution. The unicellular and colonial organisms found in such aquatic environments must either deal with the concentration gradient that is around them or die.

Many cells that live in aquatic environments have rigid cell walls that prevent cytolysis. As water diffuses into the cell, it fills the central vacuole. As the central vacuole expands, it presses the cytoplasm against the inside of the cell wall, increasing the pressure inside the cell. The turgor pressure inside the cell effectively counters the pressure of the water trying to diffuse into the cell, thus preventing cytolysis.

Other organisms that live in hypotonic solutions must have a flexible membrane. Because a rigid cell wall would restrict movement and food-gathering capabilities of protozoans, God designed other mechanisms that enable them to survive a hypotonic environment. Many such organisms have contractile vacuoles. As water molecules diffuse into the cell, they are collected by the

dormancy: (L. *dormire*—to sleep)

isotonic: iso- (equal) + -tonic (*tonos*—tone or tension)

hypotonic: hypo- (Gk. *hupo*—beneath) + -tonic (tension)

cytolysis: cyto- (cell) + -lysis (to loosen)

hypertonic: hyper- (Gk. *huper*—over or beyond) + -tonic (tension)

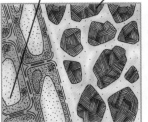
central vacuole   soil

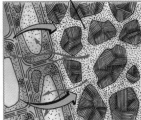
fertilizer

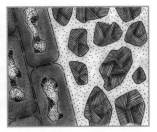

3-22

Fertilizer burn: Normal conditions (left). The addition of fertilizer increases the solute concentration outside the cell, resulting in the movement of solutes into the cell [red arrows] and water out of the cell [blue arrows] (middle). The movement of solute and water can result in plasmolysis and possibly death of the plant (right).

contractile vacuole and then released to the environment. If the contractile vacuole cannot establish and maintain equilibrium with the incoming water, cytolysis will eventually occur.

Cells in Hypertonic Solutions

Most cells are not designed to deal with hypertonic solutions as well as they deal with hypotonic solutions. With the exception of cells that live in seawater, most cells do not naturally live in hypertonic environments. Humans, however, occasionally put cells in hypertonic solutions and then wonder why they do not thrive.

For example, some gardeners think that "if a little is good, then a lot must be better," and they put more fertilizer on the soil around their plants than the directions recommend. As the fertilizer dissolves, it soaks into the soil. If too much has been applied, the root cells may be surrounded by a hypertonic solution. Ions from the fertilizer enter the cells, and water leaves the cells in an attempt to equalize the concentration gradient. As the water continues to leave the cell, it first loses its turgidity, and then the cell membrane pulls away from the cell wall. This process is called *plasmolysis* (plaz MAHL uh sis), and if it continues, the cell will die. If enough root cells are involved, the entire plant may die.

Plant cells are sometimes temporarily short of water. Hot days may cause water to evaporate from leaves more rapidly than the roots can replace it. The results are the same as if a hypertonic solution surrounded the cells—there is a net loss of water from the cells. As the water molecules exit the leaves and enter the atmosphere, the cells lose their turgidity and the plant wilts. Under normal conditions a cooler evening and night will permit the plant to restore turgidity by obtaining water from the soil faster than it is lost by evaporation. If the wilting continues because of a lack of water or damage to the plant, the cells will experience plasmolysis and die.

plasmolysis: plasmo- (to mold) + -lysis (to loosen)

▶ Review Questions 3.4

1. Why is the homeostasis demonstrated by osmosis in a cell called an equilibrium? Why is this homeostasis described as dynamic?

2. Compare and contrast optimal point, optimal range, range of tolerance, and limit of tolerance.

3. Explain the difference between plasmolysis and cytolysis.

4. Compare and contrast isotonic, hypotonic, and hypertonic solutions.

5. List two methods God designed for some cells to withstand hypotonic environments, discussing important structures and how they function.

- Compare and contrast passive transport and active transport
- Describe the differences between carrier proteins and channel proteins
- Distinguish between facilitated diffusion and active transport
- Compare and contrast endocytosis and exocytosis

Key Terms

passive transport
facilitated diffusion
transport protein

active transport
endocytosis
phagocytosis
pinocytosis
exocytosis

3.5 Transportation Across the Membrane

Most of the previous discussion about substances entering or exiting cells can be explained as simple diffusion through the semipermeable membrane based on concentration differences inside and outside the cell. Substances such as water, carbon dioxide, oxygen, and many fat-soluble molecules cross the membrane in this manner. However, many substances will not diffuse directly across the membrane. In order to maintain homeostasis, cells must often move substances against the concentration gradient.

Passive Transport

Many substances cross the cell membrane with no energy expenditure by the cell. **Passive transport** is the movement of molecules across a membrane with the concentration gradient without the expenditure of chemical energy. The kinetic energy of the molecules, not the cell, supplies the force that drives passive transport. Both diffusion and osmosis are examples of passive transport; not all molecules, however, are able to diffuse freely across a membrane. The rate of passive transport is influenced both by solute factors and by the particular composition of the membrane.

There are several factors besides the concentration of a substance that affect the rate of passive transport across the membrane. One factor is the size of the solute molecule. Some very small molecules, such as water and gases (oxygen, carbon dioxide, and nitrogen), seem to pass easily through almost any cellular membrane. Some other very small molecules and ions pass quite slowly, so there must be other factors besides size. The shape of the molecule will also affect its ability to cross the membrane. Some long, thin molecules pass slowly through a membrane, while round molecules of approximately the same size and mass may not pass at all. The polarity, or electrical charges, of a molecule also seem to be a factor. Fat-soluble molecules appear to dissolve the phospholipids of a membrane and penetrate more quickly than molecules that are not fat-soluble. In addition to the properties of the solute molecules, the membrane's composition is an important factor in determining the ability and the rate at which a substance will pass through a membrane.

Small molecules such as monosaccharides, amino acids, fatty acids, and glycerols penetrate slowly through cell membranes. Larger molecules such as disaccharides and proteins penetrate even more slowly, while lipids barely penetrate membranes at all.

Facilitated Diffusion

For years scientists were baffled that some cells permit the passage of certain large molecules (usually specific carbohydrates or proteins) but exclude other molecules of similar size and shape even when a favorable concentration gradient exists across the membrane. After much research, scientists determined that a factor contained within the membrane helped, or *facilitated*, the passage of certain molecules. Further study revealed that the factors that facilitated the transportation across the membrane were actually proteins that are embedded within the membrane structure. Since this type of diffusion requires a facilitator or helper, it is termed **facilitated diffusion**, and the embedded proteins are called **transport proteins**.

Scientists believe that there are two types of transport proteins involved in facilitated diffusion. Both are highly specific, allowing only certain substances to cross the membrane. The first is

3-4	Factors Affecting Passage of Molecules Through a Membrane
Concentration of the molecule (diffusion pressure)	
Size and weight of the molecule	
Shape of the molecule	
Charge of the molecule	
Fat-solubility of the molecule	
Composition of the membrane	

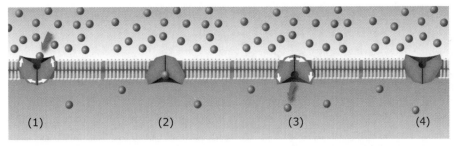

3-23

In facilitated diffusion, molecules are moved across the membrane with the concentration gradient. (1) Molecule is bound to a transport (carrier) protein. (2) Carrier protein changes shape as the molecule is transported across the membrane. (3) Molecule is released on the other side of the membrane. (4) Carrier protein returns to its original shape.

called a *channel protein* and forms tiny pores (or holes) in the membrane through which specific substances pass. The other transport protein is termed a *carrier protein*. The carrier protein has a specific receptor site that binds to the molecule it is designed to carry. As it binds to the substance, the carrier protein changes shape, and the molecule passes through the membrane. Once the molecule has passed through the membrane, the carrier protein reverts to its original shape.

Active Transport

Many times the movement of molecules or solutes is against the concentration gradient, from an area of lower concentration to an area of higher concentration. Just as in facilitated diffusion, carrier proteins are utilized to help the molecule cross the membrane. However, the difference is that energy is required. When energy is required and the substance is moved across the membrane against the concentration gradient, the process is called **active transport**. Some types of active transport systems are termed *pumps*, because they use cellular energy to pump substances against their concentration gradient.

In active transport, the substance that is to be transported binds to a carrier protein. Once that occurs, the carrier protein uses cellular energy to change its shape, and the molecule is transported across the membrane. Once the molecule is across the membrane, the carrier protein returns to its original shape. The processes by which cells store and use energy are discussed in the next chapter.

Bulk Transport: Endocytosis and Exocytosis

Large substances like starch and protein molecules are too big for any of the transport systems studied thus far. Some cells are able to eat large molecules and other cells whole. There are other mechanisms that cells use to transport substances in bulk across their membranes—endocytosis and exocytosis—both of which require energy and so are types of active transport.

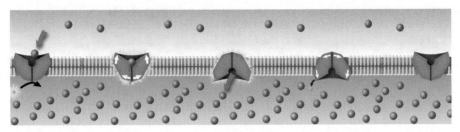

3-24

Active transport requires cellular energy and moves molecules against the concentration gradient.

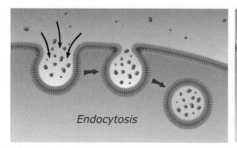
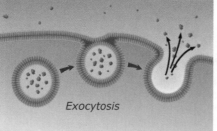

Endocytosis

Exocytosis

3-25

Endocytosis and exocytosis are opposite processes.

Endocytosis (EN doh sye TOH sis) is the process a cell uses to transport substances in bulk across the membrane and into the cell. When a cell contacts the substance to be transported, the cell membrane folds inward until it completely surrounds the substance. Once the substance is completely surrounded, the membrane is pinched off into the inside of the cell. There are two types of endocytosis, depending on the kind of substance that is transported.

The term **phagocytosis** (FAG uh sye TOH sis), referring to the process of a cell ingesting another cell, applies to the movement of any bulk solid material across the membrane. **Pinocytosis** (PIN oh sye TOH sis) is the movement of bulk fluids or solutes across the membrane. Sometimes phagocytosis is called "cellular eating," and pinocytosis is called "cellular drinking."

The reverse of endocytosis is exocytosis. **Exocytosis** (EK soh sye TOH sis) is the process whereby vesicles or vacuoles in the cytoplasm fuse with the cell membrane and release their contents into the solution outside the cell. Cells use exocytosis to secrete materials produced by the cell, such as hormones or proteins, or to release nonsoluble, indigestible wastes from food or phagocytic vacuoles.

endocytosis: endo- (in) + -cyto- (cell)
phagocytosis: phago- (Gk. *phagein*—to eat) + -cyto- (cell)
pinocytosis: pino- (Gk. *pinein*—to drink) + -cyto- (cell)
exocytosis: exo- (outside) + -cyto- (cell)

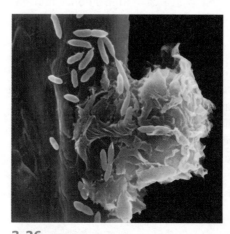

3-26

SEM of a white blood cell (yellow) phagocytizing bacteria (small red structures)

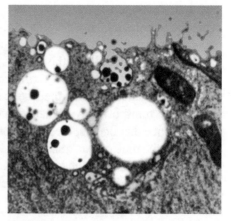

3-27

TEM of exocytosis

➡ Review Questions 3.5

1. Explain the difference between simple diffusion and facilitated diffusion.
2. Explain the similarities and differences between facilitated diffusion and active transport.
3. Explain the differences between the two types of transport proteins.
4. Describe the process of endocytosis.

4

Cellular Processes

Cytology Part II

Facets

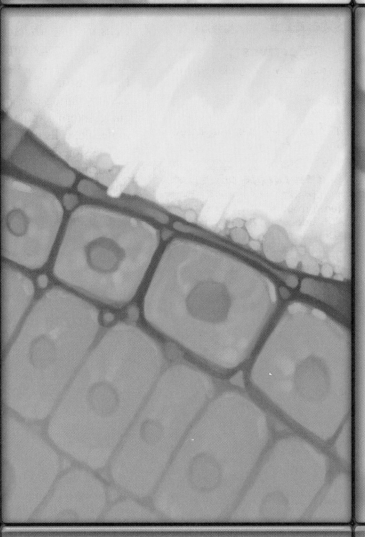

I magine a factory that produces not one, but thousands of different products. It operates hundreds of assembly lines, 24/7, at blazing speeds with never a breakdown. Some of its raw materials are shipped in from outside sources, but the factory also recycles some of its old products into new ones. Now imagine that this highly organized production facility is so tiny that a hundred of them could easily fit on the period at the end of this sentence. You have just imagined a typical cell—the unit that God designed to be the building block as well as the manufacturer of all living things.

4A—Cellular Energy

Although cells depend upon many conditions in the environment, a constant supply of energy is essential for them to perform all of their cellular processes. Even in the most favorable environments, cells need energy to maintain their homeostasis. They use energy to manufacture needed substances and to tear down others. In fact, when a cell stops using energy, it is dead. Expenditure of energy is a primary attribute of life.

Objectives

- Explain why energy is required for life
- Define *autotroph* and *heterotroph*
- Describe how energy is stored in ATP molecules

Key Terms

autotroph
heterotroph
ATP (adenosine triphosphate)
ADP (adenosine diphosphate)

4-1

Heterotrophs depend on autotrophs for their survival.

autotroph: auto- (Gk. *autos*—self) + -troph (*trophe*—nourishment)

heterotroph: hetero- (Gk. *heteros*—other) + -troph (nourishment)

4.1 Energy Relationships

Cells use amino acids and other substances again and again in the process of making many large and different molecules. Energy, however, is a one-time commodity; every time it is used, some escapes and becomes unusable. Cells must use energy to make energy-storing molecules. Since some energy escapes during every energy transaction, more energy is needed to build an energy-storing molecule than is stored in the molecule. To make any large molecule, cells need large quantities of energy.

The stored energy in large molecules is of value to the cell primarily when it is being used. Every time a large molecule is broken apart by the cell to get this stored energy, some of the energy escapes. As a result, the amount of usable energy in an energy storage molecule is considerably less than the total amount of energy in the molecule.

Cells constantly battle to maintain a supply of energy. Energy—crucial for all life—is the one commodity that is stored (but not without considerable loss), is used (but never reused), and escapes constantly. Yet all living cells need energy at all times.

Autotrophs and Heterotrophs

Organisms can be classified into two groups according to how they obtain their energy. Organisms that make their own food are called **autotrophs**. Plants, algae, and some other organisms are able to capture light energy and produce their own food. They are the primary autotrophs. Autotrophs are also called *producers*. Organisms that depend on other organisms for their energy are called **heterotrophs**, or *consumers*. These include humans, animals, fungi, and most bacteria.

Photosynthesis, the process used by most autotrophs in storing energy (making their food), is discussed in the next subsection. Cellular respiration, the process all organisms (including autotrophs) use to release energy from storage, is discussed in Subsection 4.3.

ATP: The Energy Currency in Cells

When a person wants something from a vending machine but finds only a $20 bill in his pocket, he cannot get what he wants. He has enough money for the item that he wants, but he cannot get it because he does not have the amount in a usable form.

Similarly, a cell may have many units of stored energy in starch or lipid molecules, but there is too much energy in these molecules to be used all at once. If all the energy in a molecule of starch or lipid were released in a cell at one time, that area of the cell would be destroyed. The energy from these molecules must be converted into smaller, usable units. In all known living organisms, from bacteria to humans, that smaller, usable unit is the energy stored in a molecule of **ATP—adenosine triphosphate** (uh DEN uh SEEN • try FAHS FAYT). It could be said that ATP is the "coin" of cellular energy currency.

Examining its structure will explain how energy is made available in ATP. The discussion of nucleic acid

high-energy bonds

adenosine

phosphate   phosphate   phosphate

4-2

ATP molecule

structure in Chapter 2 introduces all three of the ATP molecule components. The backbone of the ATP molecule is adenosine, which is made of a simple sugar and adenine, one of the bases found in DNA. The other component of ATP is a chain of *three phosphate groups* attached to the adenosine backbone.

The key to how ATP stores energy is found in the bonds between the three phosphate groups; they are bound to each other by unstable, high-energy covalent bonds. When the bonds are broken, a useful amount of energy is released (an exothermic reaction) and is immediately available for use in any cellular function that requires energy (an endothermic reaction). Occasionally two or more ATP molecules supply the energy for a single reaction. Active transport, biosynthesis of molecules, cellular movement, and locomotion are all processes powered by energy released from ATP molecules.

In most reactions, it is the phosphate group on the end of the chain that is broken off. This reaction produces a molecule of **ADP—adenosine diphosphate** (ATP minus one phosphate group), a free phosphate group, and energy. The reaction can be summarized as follows:

$$\text{ATP} \longrightarrow \text{ADP} + \text{P} + \text{energy}$$

An ATP molecule that has given off its energy can be reused. With the proper enzymes and an adequate supply of energy, the ADP and a phosphate can recombine to form ATP. The reaction can be summarized as follows:

$$\text{ADP} + \text{P} + \text{energy} \longrightarrow \text{ATP}$$

Muscles use energy from ATP in order to contract; however, ATP cannot be brought to muscles through the blood. Why? Since the ATP molecule is very unstable, the flow of blood would break ATP down to ADP long before it reached the muscles. Thus, ATP is a temporary, unstable energy-storage molecule and must be constantly manufactured by every cell as long as it is alive.

A normal adult male uses about 2800 kilocalories (or 2800 Cal) of energy each day, requiring about 150 kg (328 lb) of ATP each day. However, there is only about 0.005 kg (0.175 oz) of ATP contained in a person's body at any given

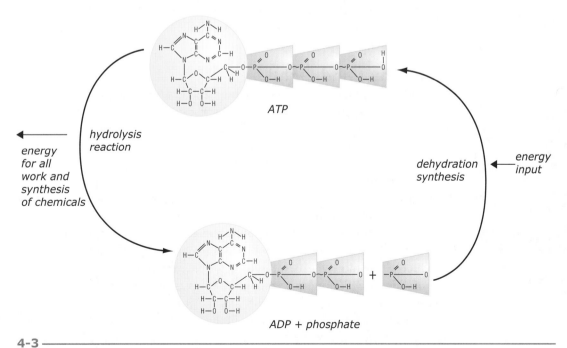

4-3

ATP is a renewable resource.

time. If a person is to stay alive, his body must constantly make ATP from ADP and energy. God has graciously designed cells to be astonishingly efficient at recycling ADP back to ATP. In a working muscle cell, 10 million ATP molecules are used and regenerated every second! In 60 seconds the same cell will have regenerated all of its ATP. This rapid recycling ensures that the cell has all of the energy it needs to function at optimal levels. We truly are fearfully and wonderfully made (Ps. 139:14)—down to the molecular level.

Review Questions 4.1

1. What is the primary need of all cells?
2. What are the two major groups of organisms based on how they obtain their energy?
3. List the two components of an ATP molecule. What is ATP's function?
4. Describe how ATP stores energy.
5. What is the difference between ADP and ATP? Which holds more energy?
⊙6. Why must cells have a continuous supply of energy?

4.2 Photosynthesis

On the earth, the sun is ultimately the source of energy for virtually all living things. One of the things the sun does is heat the earth. But warmth, though essential, is not the energy needed to keep living things alive. Chemical energy is required for the reactions that maintain life.

Since the sun's energy does not come as chemical energy, it must be changed into a usable form. There are certain organisms that are capable of absorbing light energy and converting it into stored chemical energy in a process called **photosynthesis**. Green plants and algae—the primary photosynthetic organisms—perform this essential energy transformation not only for themselves but also in large enough quantities to supply stored chemical energy for almost every living thing. Because it is the essential step between solar energy and life, photosynthesis is one of the most important biological processes. Photosynthesis is important for another reason—it produces oxygen. Human beings as well as animals, plants, and most other organisms require oxygen in order to live. The following study explains how oxygen and stored chemical energy are produced by photosynthesis.

Chlorophyll and Light

Chlorophyll (KLOHR uh fil), a green pigment, is the primary catalyst of photosynthesis. **Pigments** are special light-absorbing molecules. The chlorophyll molecule has a ring-shaped head portion containing a single magnesium atom and a tail containing a carbon chain. Although the chlorophyll molecule does not contain iron, it is necessary for chlorophyll's formation. In fact, chlorophyll is made only when the cell has a supply of iron and is exposed to light. Plants grown in reduced light or in iron-poor soil lack chlorophyll and appear pale yellowish or whitish.

The two primary pigments of photosynthesis, chlorophyll *a* and chlorophyll *b*, are found in the grana of the chloroplasts in photosynthetic eukaryotic cells. The chlorophyll molecule is located in the membranes of the grana. The *a* form of chlorophyll is more abundant and serves as the primary catalyst for photosynthesis.

The two forms of chlorophyll differ in the wavelengths of light that activate them. White light has wavelengths of all the colors of light. When something appears red, it is actually reflecting red light waves and absorbing all the

4-4

Control plant (left) and plant grown in reduced light and iron-deficient soil (right)

photosynthesis: photo- (Gk. *phos*—light) + -synthesis (to put together)

chlorophyll: chloro- (green) + -phyll (Gk. *phullon*—leaf)

others. This fact explains which wavelengths, or colors, of light are necessary for photosynthesis. The green color of most photosynthetic organs of plants indicates that wavelengths other than in the green range are being absorbed.

The Process of Photosynthesis

Photosynthesis takes place in the grana of the thylakoids of the chloroplast. Chlorophyll and other pigment molecules are located in clusters of one hundred to two hundred in the thylakoid membrane. Experiments have revealed that photosynthesis actually occurs in two separate, yet closely related, phases—the light-dependent phase and the Calvin cycle.

The Light-Dependent Phase

The first phase is called the **light-dependent phase**, or *light reactions*. As the name implies, light energy is required. Light energy is absorbed by various pigments and is used to energize electrons in a chlorophyll *a* molecule. The energized electrons leave the chlorophyll *a* molecule and pass through a series of protein molecules called the **electron transport chain**, which is embedded in the thylakoid membrane. As the excited electrons pass through the electron transport chain, some of their energy is used to form ATP.

In order to replace the energized electrons that left the chlorophyll *a* molecule, photolysis occurs. *Photolysis* is the splitting of a water molecule into hydrogen ions, electrons, and oxygen. The electrons replace those lost when chlorophyll became energized, the hydrogen ions remain in the thylakoid, and the oxygen is eventually released into the atmosphere or used by the cell.

After an electron has passed through the electron transport chain, it is still highly energized. To capture the remaining energy, the electrons bind in pairs to a special electron carrier molecule, *NADP+ (nicotinamide adenine dinucleotide phosphate)*. When NADP+ picks up these two electrons and a hydrogen ion (H^+), it becomes NADPH. This molecule stores the energy for later use.

In short, the light-dependent phase uses light energy to form NADPH and ATP. It also produces oxygen gas.

The Calvin Cycle

The **Calvin cycle**, or *light-independent phase*, is the second phase and takes place in the **stroma** of the chloroplast, where the necessary enzymes and other reactants are located. Although light is not required, the Calvin cycle depends upon the products of the light-dependent phase (ATP and NADPH) and carbon dioxide (CO_2) from the atmosphere. The Calvin cycle is sometimes called the carbon fixation cycle because during this phase carbon is fixed, or assimilated, in an organic compound.

To start this phase, a molecule of carbon dioxide diffuses from the cytoplasm of the cell into the stroma of the chloroplast, where it binds to a 5-carbon sugar called *ribulose biphosphate (RuBP)*. Through a stepwise series of reactions, the energy stored in the ATP and NADPH that was produced during the light-dependent phase is used to produce a 3-carbon sugar (*phosphoglyceraldehyde*, or *PGAL*). Some of the PGAL molecules are used to make glucose and some to form more RuBP so that the Calvin cycle can continue to function.

To summarize, the Calvin cycle produces sugar using the atoms from CO_2, energy from ATP, and the electrons and hydrogen ions provided by NADPH.

4-5

Chlorophyll *a* molecule. Note the magnesium (Mg) and nitrogen (N). The circled side group is replaced to make the different kinds of chlorophyll.

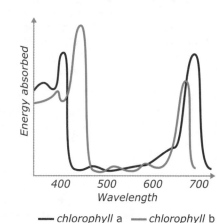

400 500 600 700
Wavelength

—— *chlorophyll* a —— *chlorophyll* b

4-6

Wavelengths of light absorbed by chlorophyll *a* and *b*

photolysis: photo- (light) + -lysis (to loosen)

The **stroma** is the dense fluid within the chloroplast, surrounding the thylakoid membrane.

The Steps of Photosynthesis

The clusters of chlorophyll molecules are called photosystems. There are two types of photosystems—photosystem I (PI) and photosystem II (PII)—based on the geometric arrangement of the chlorophyll molecules. As with many biological processes, photosynthesis can most easily be studied in a stepwise fashion.

The Light-Dependent Phase

Step 1: Light energy is collected by the various pigments and transferred to a chlorophyll *a* molecule, where electrons are excited in photosystem II. At the same time, an enzyme splits a water molecule into electrons, oxygen, and hydrogen ions in a process called *photolysis*. The electrons from the water molecule replace those lost by photosystem II, the hydrogen ions stay inside the thylakoid, and the plant eventually releases the oxygen to the atmosphere.

Step 2: An excited electron leaves the chlorophyll *a* molecule and is picked up by an electron carrier molecule, which is the beginning of a series of molecules called the electron transport chain.

Step 3: As the electron passes through the transport chain, some of the energy is used to pump hydrogen ions across the thylakoid membrane.

Step 4: After the electron has passed through the electron transport chain, it is accepted by a molecule of chlorophyll in photosystem I. Simultaneously, light energy excites the newly arrived electron.

Step 5: The excited electron in photosystem I is passed to an electron carrier in a different electron transport chain.

Step 6: This electron transport chain carries the electron to the stroma side of the thylakoid membrane, where it combines with a hydrogen ion and NADP+, an electron carrier molecule, to form NADPH.

Step 7: The hydrogen ions produced by photolysis and those pumped into the thylakoid during Step 3 begin to form a large concentration gradient between the inside of the thylakoid and the stroma of the chloroplast.

Step 8: As the hydrogen ion concentration builds up inside the thylakoid, the ions pass from a higher concentration to a lower concentration through an enzyme—**ATP synthase**—very similar to the process of osmosis. This movement of hydrogen ions caused by the concentration gradient is called *chemiosmosis*. As the hydrogen ions pass through the ATP synthase molecule, it converts the potential energy of the H+ concentration gradient into chemical energy as it combines ADP with a phosphate group to produce ATP.

The chemical reaction that summarizes the light-dependent reactions is

$$12H_2O + 12NADP^+ + 18ADP + 18P + \text{light energy} \xrightarrow{\text{chlorophyll}} 6O_2 + 12NADPH + 18ATP$$

It should be remembered that the overall reaction for photosynthesis is

$$6H_2O + 6CO_2 + \text{light energy} \xrightarrow{\text{chlorophyll}} C_6H_{12}O_6 + 6O_2$$

A comparison between the products of the light reactions and the products in the overall chemical equation for photosynthesis shows similarities. The oxygen produced in the light-dependent reactions is the same oxygen that is seen in the overall photosynthesis reaction. The ATP and NADPH produced by the light-dependent reactions are used in the Calvin cycle.

The Calvin Cycle

Step 1: A carbon dioxide molecule diffuses from the cell's cytoplasm into the stroma of the chloroplast and is fixed, or attached, to RuBP (ribulose biphosphate). RuBP is a 5-carbon sugar with 2 phosphates bonded to it. The resulting 6-carbon sugar is very unstable and breaks apart to form 2 molecules of PGA (phosphoglyceric acid). PGA is a 3-carbon sugar with a phosphate attached to it.

Step 2: PGA is converted into another 3-carbon molecule called PGAL (phosphoglyceraldehyde). PGA is energized by the energy from a phosphate bond in an ATP molecule and then receives hydrogen and electrons from the carrier molecule

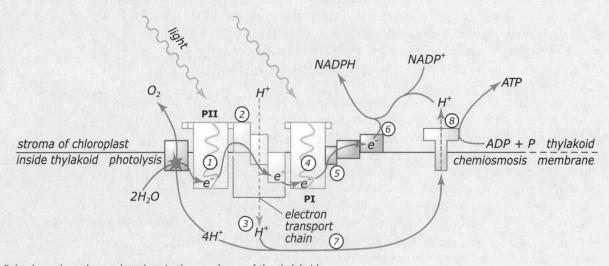

The light-dependent phase takes place in the membrane of the thylakoid.

NADPH. The NADP⁺ and ADP released during the formation of PGAL may be used again in the light-dependent reactions to synthesize additional ATP and NADPH.

Step 3: Some of the molecules of PGAL never make it to Step 3—they are used in the synthesis of other organic molecules such as glucose. However, in Step 3 many of the PGAL molecules are converted into RuBP, utilizing the energy from a molecule of ATP to "recharge" the cycle. The RuBP returns to the start of the Calvin cycle.

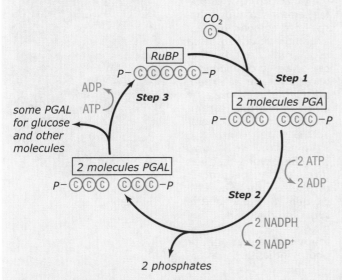

The Calvin cycle takes place inside the thylakoid.

Photosynthesis Recap

Photosynthesis is divided into two major phases—the light-dependent phase and the Calvin cycle—and each of these proceeds in a stepwise fashion. The PGAL produced at the end of the Calvin cycle is used as the building block of organic compounds such as glucose. How much energy is needed to produce 1 molecule of PGAL? Every spin of the Calvin cycle

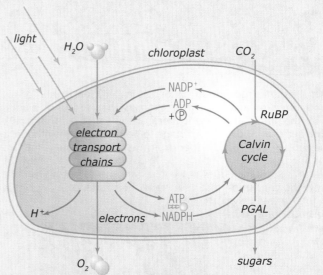

Overview of photosynthesis. Light energy, water, and carbon dioxide enter the chloroplast, where the light-dependent phase and Calvin cycle work together, using ATP, ADP, and the electron carrier molecules. Oxygen and PGAL, which is used to make sugars, are the final products.

fixes 1 CO_2 molecule, and since PGAL has 3 carbon atoms, it takes 3 spins to produce 1 molecule of PGAL. Two molecules of ATP and 2 of NADPH are used in Step 2, and 1 ATP molecule is used in Step 3, using a total of 3 ATP and 2 NADPH molecules per spin. Therefore, the total energy requirement for producing 1 PGAL molecule is 9 molecules of ATP and 6 of NADPH.

The overall chemical equation for photosynthesis is this:

$$6CO_2 + 6H_2O + \text{light energy} \xrightarrow{\text{chlorophyll}} C_6H_{12}O_6 + 6O_2$$

Glucose is not produced in the actual photosynthesis reactions; it is the molecules of PGAL that are used to make glucose and other organic molecules.

Conditions for Photosynthesis

Adequate quantities of the right wavelengths and intensity of light are necessary for photosynthesis to occur. If they do not absorb enough energy, chlorophyll *a* molecules cannot become sufficiently energized. On the other hand, the rate of photosynthesis will reach a plateau and will go no faster if all of the chlorophyll *a* molecules are functioning at maximum capacity.

The cell must also be able to absorb sufficient carbon dioxide. This is seldom a problem since carbon dioxide normally accounts

Chemosynthesis: Other Autotrophs

There are a few bacteria capable of obtaining energy from inorganic chemicals that they break apart in a process called **chemosynthesis** (KEE moh SIN thih sis). These bacteria can use that energy to synthesize sugar. Since chemosynthetic bacteria do not depend on other organisms for their food, they are autotrophs.

Chemosynthetic bacteria often grow in dark areas with little organic matter in their environment. Some of these bacteria change ammonia to nitrites and nitrites to nitrates; others use iron compounds. Probably the best-known chemosynthetic bacteria use sulfur compounds and produce sulfur dioxide. A walk through a swamp may disturb some collections of this gas that are trapped underwater. The rotten-egg odor of sulfur dioxide that bubbles to the surface may have been produced by chemosynthetic bacteria.

The deep-sea hydrothermal vent shown in Figure 2-16 (p. 46) also harbors chemosynthetic bacteria. These tiny organisms are the basis of a whole food chain that exists in total darkness.

Sulfur-producing bacteria are chemosynthetic.

chemosynthesis: chemo- (chemical) + -synthesis (to put together)

A Brief History of Man's Knowledge of Photosynthesis

Man has long recognized his dependence upon plants for his food. Even meat comes from animals that consume plants. From the time of the ancient Greeks, people believed that plants obtained their substance from the soil. Simply put, they thought the roots sucked up the food from the soil. Jan van Helmont's famous tree experiment (see p. 7) demonstrated that most of a plant's substance does not come from the soil. Van Helmont drew the conclusion that virtually all the substance of a plant is water. He was only partly right.

Joseph Priestley

Portrait of Joseph Priestley, Attributed to Ozias Humphrey (1742–1810), Oil on canvas, Gift of Chemists' Club Collection, Chemical Heritage Foundation Collections, Photo by Will Brown

About one hundred years later, in 1772, Joseph Priestley conducted experiments with air. He discovered that if he burned a candle in a closed jar, the resulting "impure air" would kill a mouse that had also been put into the jar. Priestley also noted that if a living sprig of mint was placed in a jar containing "impure air," the air was "restored" by the mint. How did he know? A mouse that was placed in the jar along with the mint did not die. Another piece of the puzzle had been discovered—living plants add something to the air.

Several years later, Dutch physician Jan Ingenhousz studied and reproduced Priestley's results. Based on his findings, Ingenhousz began to manipulate the variables in the experimental groups and determined that the air was restored only when the green leaves of a plant were exposed to sunlight. Using the results of his experiments, he proposed a new model: green leaves use sunlight in some process to break apart carbon dioxide (CO_2) into carbon and oxygen. He further proposed that the carbon combined with water to produce carbohydrates. The basic information was clear: green plants use water, affect the air, require light, and make organic matter.

carbon dioxide + water + light energy   carbohydrate + oxygen

Over the next several decades, scientists used refined techniques to uncover new facts about the process of photosynthesis. Initially it was believed that the process of photosynthesis happened all at once as the equation indicates.

$$CO_2 + H_2O + \text{light energy} \longrightarrow C_6H_{12}O_6 + O_2$$

or

carbon dioxide + water + light energy $\longrightarrow$ glucose + oxygen

Work by Dr. Melvin Calvin in the 1950s, however, indicated that photosynthesis is actually a series of steps. Calvin exposed unicellular algae to radioactive isotopes of carbon in the form of carbon dioxide. He exposed the algae to varying periods of light and then killed the algae during differing stages of growth.

Using the radioactive carbon as a tracer, his research group was able to discover the exact path that carbon takes during photosynthesis, from entering the leaf as carbon dioxide to being used in the plant as other carbon-based compounds.

In another set of experiments, Calvin exposed the algae to normal carbon dioxide but used radioactive water containing oxygen-18. He found that the oxygen that was given off by the process of photosynthesis was all oxygen-18. None of the sugar produced had any oxygen-18 in it. Note the following equation for photosynthesis:

Jan Ingenhousz

$$12H_2O^* + 6CO_2 + \text{light energy} \xrightarrow{\text{chlorophyll}} C_6H_{12}O_6 + 6H_2O + 6O_2^*$$

An asterisk (*) indicates the oxygen-18 that Calvin used. The 12 molecules of water on the left supply the 12 atoms of oxygen (6 oxygen molecules, O_2) given off. The water on the right side of the equation is newly formed water. The formula cannot be reduced by canceling 6 water molecules from each side since the water molecules on one side are not the same water molecules on the other side. Later studies demonstrated that the oxygen in the glucose molecule came from the carbon dioxide. Thus water is actually a byproduct of photosynthesis.

What has happened to the light energy put in on the left side? No energy is expressed on the right side. Some energy, of course, has been lost. Energy was used to make the new water and to put together the glucose molecule. In other words, the energy that did not escape is now contained in the molecules made by photosynthesis. The energy in the products of water and oxygen is of little use to the organism; however, the energy stored in the glucose molecule will be used to carry out cellular processes.

Much of our present knowledge of the process of photosynthesis is the result of Calvin's work, for which he was awarded a Nobel Prize in 1961.

for about 0.03% of the atmosphere by volume. Just as with light intensity, the rate of photosynthesis will reach a plateau when all of the chlorophyll *a* molecules are at maximum capacity.

Proper temperatures for photosynthesis vary from plant to plant. For most plants, room temperature (21 °C, or 70 °F) is about right. However, above 32 °C (90 °F), many plant enzymes do not function properly for photosynthesis. At midday during the hot summer months, many plants do not carry on photosynthesis. The lower temperature limit for photosynthesis is near freezing for some plants, but others require more moderate temperatures. In most plants the functioning of chlorophyll is temperature dependent. A greenhouse with improper ventilation or with drastic temperature changes can affect the supply of carbon dioxide, and photosynthesis may cease.

Occasionally a lack of water will cause photosynthesis to cease. On hot, dry summer days the roots of a plant may not be able to absorb sufficient water and therefore cannot transport enough water up the stem to maintain photosynthesis in the leaves.

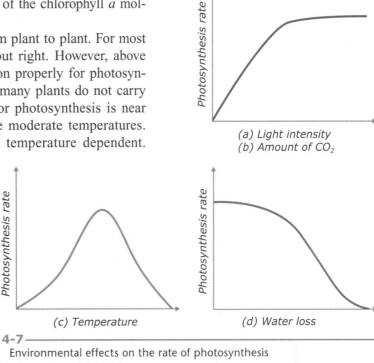

(a) Light intensity
(b) Amount of CO_2

(c) Temperature

(d) Water loss

4-7 Environmental effects on the rate of photosynthesis

Review Questions 4.2

1. What is the primary purpose of photosynthesis?
2. List (a) the primary materials necessary for and (b) the materials produced by photosynthesis.
3. Describe chlorophyll *a* and tell its function.
4. Write a chemical equation for photosynthesis.
5. What inputs are needed for and what products are made by the light-dependent phase of photosynthesis?
6. What inputs are needed for and what products are made by the Calvin cycle of photosynthesis?
7. Tell the function in relation to photosynthesis of the following substances: chlorophyll, hydrogen and electron carrier molecules, PGAL, and RuBP.
8. Tell how the light-dependent phase and the Calvin cycle of photosynthesis are dependent upon each other.
9. Describe how changes in light, temperature, water, and carbon dioxide affect the rate of photosynthesis.
10. What are the two types of autotrophs? List an example of each.

4.3 Cellular Respiration

Respiration is commonly thought of as breathing—air being taken into and forced out of the lungs. Breathing is only the first step in the body's respiration. Oxygen obtained by breathing is transported from the lungs to the cells where cellular respiration takes place. **Cellular respiration** is the breaking down of a food substance into usable cellular energy in the form of ATP. In most cells glucose is the primary food source of cellular energy; however, lipids, other monosaccharides, and even proteins can be used.

Cellular respiration may be **aerobic** (requiring oxygen) or **anaerobic** (not requiring oxygen). Most cells carry on aerobic respiration. Many cells that normally perform aerobic respiration can, if necessary, operate anaerobically. There are some bacteria and fungi that carry on only anaerobic respiration. The primary processes of aerobic cellular respiration are discussed first, and then two important anaerobic respiration processes are presented.

4.3

Objectives

- Differentiate between aerobic and anaerobic respiration
- Summarize the major events in glycolysis, the Krebs cycle, and the electron transport chain
- Compare and contrast alcoholic and lactic acid fermentation

Key Terms

cellular respiration	glycolysis
	Krebs cycle
aerobic	acetyl CoA
anaerobic	fermentation

aerobic: aero- (Gk. *aer*—air) + -bic (life)
anaerobic: an- (Gk. *an*—without) + -aero- (air) + -bic (life)

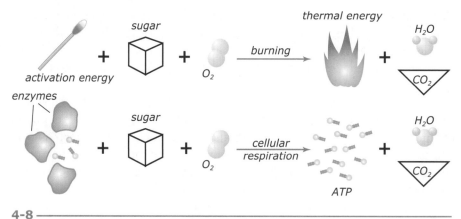

4-8

A comparison of burning and cellular respiration

Comparing Aerobic Respiration and Photosynthesis

Aerobic cellular respiration can be compared to burning. If a lighted match is held to a sugar cube, the match will supply the activation energy necessary to set the sugar on fire. The burning sugar releases large amounts of thermal and light energy. Carbon dioxide and water vapor are also released, although they cannot be seen. If the flame were put in an airtight chamber, the fire would soon be extinguished because of a lack of oxygen.

In aerobic cellular respiration, the sugar glucose requires a small amount of activation energy to begin a series of enzyme-controlled chemical reactions that release the energy contained in ATP molecules. Just as the burning sugar cube did, cellular respiration releases carbon dioxide and water molecules. If cells do not receive oxygen, they cannot carry on aerobic respiration, and in time many will die.

Aerobic cellular respiration is the opposite of photosynthesis. Photosynthesis combines water, carbon dioxide, and light energy to form glucose. Aerobic cellular respiration breaks down glucose to form water, carbon dioxide, and energy. The reactants that are needed for the one are the products of the other.

Photosynthesis can be generalized thus:

$$6H_2O + 6CO_2 + \text{light energy} \xrightarrow{\text{chlorophyll}} C_6H_{12}O_6 + 6O_2$$

Cellular respiration can be generalized thus:

$$C_6H_{12}O_6 + 6O_2 \longrightarrow 6H_2O + 6CO_2 + ATP \text{ (energy)}$$

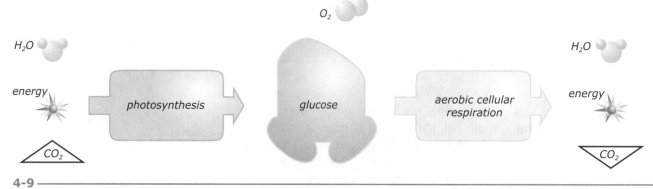

4-9

A comparison of photosynthesis and aerobic cellular respiration

These processes, however, are not the exact reverse of each other. For example, chlorophyll is not used in cellular respiration. Although some chemicals function in both processes, they proceed down different enzyme paths. These differences are necessary because the purposes of the two processes are different.

Photosynthesis captures light energy and converts it to stored chemical energy. Respiration takes stored chemical energy and converts it to a ready-to-use chemical energy carrier (ATP).

The light energy harnessed by a chlorophyll molecule during photosynthesis is made useful to all cellular processes by cellular respiration. The atoms of oxygen, carbon, and hydrogen are essentially the same after these processes, but the energy involved has been changed several times, and much of it has been lost.

The Process of Cellular Respiration

Cellular respiration may follow several different pathways, depending on the organism and the environment. Each process is discussed in steps to show the logical progression of the reactions. All types of cellular respiration begin with a common process—glycolysis.

Glycolysis

Glycolysis (glye KAHL uh sis) involves the breakdown of glucose into pyruvic (pie ROO vik) acid, hydrogen ions, and electrons. It takes place in the cytoplasm, which contains the enzymes necessary for this series of reactions. Pyruvic acid can then be used in the different forms of cellular respiration. Because glycolysis does not require oxygen, it is considered anaerobic. The products of glycolysis are shuttled to the mitochondria to continue cellular respiration.

Step 1: Two molecules of ATP supply the activation energy necessary to start glycolysis. Using the energy and 2 phosphates from the molecules of ATP, glucose is broken down into 2 molecules of PGAL. Notice in Figure 4-10 that each molecule of PGAL contains 3 carbons and 1 phosphate. (This is the same compound seen in photosynthesis.)

Step 2: Each molecule of PGAL releases electrons and hydrogen ions to an electron carrier molecule (NAD^+). Each PGAL molecule also has a phosphate attached to it, forming 3-carbon compounds, each now with 2 phosphates attached.

Step 3: Each phosphate group added in Steps 1 and 2 is removed and added to a molecule of ADP, forming a total of 4 ATP molecules. The reaction in this step produces 2 molecules of pyruvic acid. This pyruvic acid holds most of the energy that was found in the original glucose molecule.

The total energy production of glycolysis is 4 molecules of ATP; however, 2 ATP molecules were needed to initiate the process. The net energy production is 2 molecules of ATP. In addition, 2 NADH molecules have been produced. What happens next depends on the type of cell and whether oxygen is present.

Aerobic Respiration

Aerobic respiration can be divided into two phases—the Krebs cycle, or citric acid cycle, and the electron transport chain. In eukaryotic cells both of these phases take place in the mitochondria; in prokaryotes they occur in the cell's cytoplasm.

✦ *Krebs cycle.* The pyruvic acid produced by glycolysis diffuses across the mitochondrial membrane into the matrix of the mitochondria. In the **Krebs cycle** the pyruvic acid reacts with an enzyme that removes a carbon from the pyruvic acid to produce acetyl (uh SEET uhl) coenzyme A, or **acetyl CoA**, CO_2, hydrogen ions, ATP, and electrons.

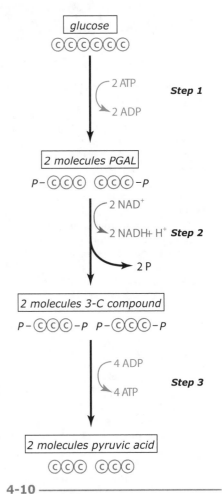

4-10
Glycolysis has three main steps.

glycolysis: glyco- (Gk. *glukus*—sweet) + -lysis (to loosen)

The Steps of the Krebs Cycle

The Krebs cycle can be broken down into five steps. Each step is illustrated at the right.

Step 1: After the pyruvic acid from glycolysis has been converted to acetyl CoA, The 2-carbon acetyl CoA combines with a 4-carbon compound, *oxaloacetic* (AHKS uh low uh SEET ik) *acid*, to form citric acid, a 6-carbon compound. This reaction also regenerates coenzyme A.

Step 2: Citric acid loses a carbon to form CO_2. It also loses a hydrogen atom and electron, which are picked up by NAD^+ (nicotinamide adenine dinucleotide) to produce a molecule of NADH. These electron carrier molecules will become important later in the electron transport chain.

Step 3: The 5-carbon compound from Step 2 releases another hydrogen atom and electron, forming another molecule of NADH. ATP is also produced during this reaction. A carbon atom is lost in forming another CO_2 molecule.

Step 4: The 4-carbon compound releases another hydrogen and electron, this time to a different carrier molecule, FAD (flavin adenine dinucleotide), forming $FADH_2$. The electron acceptor molecule may be different, but the purpose is the same. These electron carrier molecules will also become important later in the electron transport chain.

Step 5: During this reaction, the 4-carbon compound from Step 4 releases another hydrogen and electron to form another NADH and oxaloacetic acid. Just as the Calvin cycle had to continuously produce RuBP, the Krebs cycle has to continuously produce oxaloacetic acid in order to keep running.

What has the Krebs cycle produced? One glucose molecule will cause two turns of the cycle to produce the following: 6 NADH, 2 $FADH_2$, 2 ATP molecules, and 4 CO_2 molecules. The

ATP can be used to provide energy for other cellular processes, the CO_2 is a waste product that will be eliminated by the cell, and the NADH and $FADH_2$ will be used in the next phase of aerobic cellular respiration. These two carrier molecules contain the high-energy electrons that drive the electron transport chain.

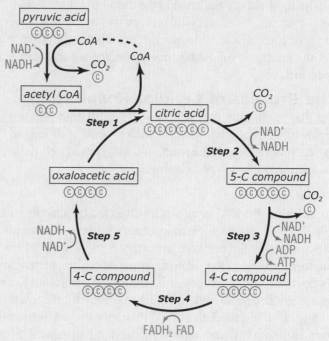

The Krebs cycle begins with acetyl CoA and proceeds in a stepwise fashion.

The CO_2 diffuses out of the mitochondria and will eventually leave the cell. The hydrogen ions and electrons will combine with electron carrier molecules to be used in the second phase of aerobic respiration. These hydrogen and electron carrier molecules are carrying high-energy electrons that are essential to the next phase of aerobic respiration.

✦ *Electron transport chain.* The majority of the ATP energy is produced in the electron transport chain. Chapter 3 explains that the mitochondria have infoldings of the inner membrane called cristae. The molecules that form the electron transport chain are located in this membrane. The production of energy by the electron transport chain in mitochondria is similar to the workings of the electron transport chain in the chloroplast.

During this very productive phase of aerobic respiration, the high-energy electrons from the electron carrier molecules (produced by the Krebs cycle) are used to generate huge amounts of ATP, the energy coinage of respiration. In a cascade of reactions that shows remarkable design, each pair of electrons moves through the steps of the electron transport chain and, on average, gives off enough energy to convert 3 molecules of ADP to ATP. In total, 1 molecule of glucose yields 32 molecules of ATP just from the electron transport chain.

✦ *Aerobic respiration energy output.* Looking at the entire process from glycolysis to the electron transport chain, what is the final energy production in terms of ATP molecules from 1 molecule of glucose? Glycolysis had a net production of 2 molecules of ATP, and the Krebs cycle also produced 2 ATP

The Electron Transport Chain

Steps 1 and 2: The electron and hydrogen carrier molecules—FADH₂ and NADH—release their high-energy electrons to an electron carrier molecule, which passes the electron through the electron transport chain. Some of the energy that is released is used to pump hydrogen ions to the outside of the membrane.

Step 3: At the end of the chain, the electrons are released into the mitochondrial matrix, where they combine with oxygen and hydrogen ions to form water.

Step 4: As the hydrogen ions (H^+) from the carrier molecules are being pumped to the outside of the cristae membrane, the concentration gradient of H^+ between the inside and the outside of the cristae membrane builds up. By the process of *chemiosmosis*, the hydrogen ions pass through a molecule of ATP synthase, releasing energy that powers the production of ATP from ADP and phosphate.

Step 5: The electrons from Step 3 and the incoming hydrogen ions from Step 4 combine with oxygen to form water. The

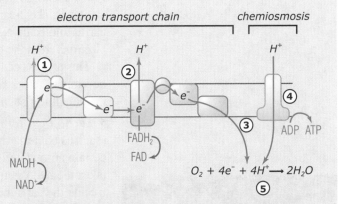

The electron transport chain

oxygen in this step is the *rate-limiting* factor. In other words, if there is no oxygen, the hydrogen and the electrons will not combine, and the entire process will stop functioning.

molecules. The electron transport chain is the big producer of ATP. It used the electron carrier molecules to produce 32 molecules of ATP. Table 4-1 shows the maximum results from 1 glucose molecule—36 molecules of ATP.

The aerobic cellular respiration of glucose converts approximately 40% of the energy contained in a glucose molecule into the form of ATP molecules. Although a 60% loss of energy may seem wasteful, this breakdown of sugar is one of the most efficient energy processes known. In most man-made machines (cars, for example), only a small amount, about 20%, of the energy available in the fuel is used. Most of the rest radiates as heat and is unusable.

Anaerobic Respiration

In the cellular respiration just described, oxygen is a key reactant. However, some cells exist in environments where oxygen is not available. Many bacteria in the lower layers of swamps, lakes, and oceans do not have oxygen available in their environment, and some organisms may actually be destroyed by exposure to free oxygen. Other cells that normally operate best with a supply of oxygen can occasionally function without it. For example, when a person engages in strenuous physical activity, he may begin to breathe heavily but still be unable to supply enough oxygen for his muscle cells to function properly. Without enough oxygen, his muscles cannot carry on sufficient aerobic cellular respiration to supply the needed ATP. If cells in these circumstances are to supply themselves with usable cellular energy, they must rely on anaerobic cellular respiration.

Fermentation is the breakdown of food (usually glucose) without oxygen. There are two main pathways of fermentation, depending on how the cell modifies pyruvic acid. Why pyruvic acid and not glucose? The answer is that glycolysis is the starting point for both aerobic *and* anaerobic respiration—one glucose molecule is broken down into two pyruvic acid molecules. Depending upon the organism and the enzymes available, the pyruvic acid is usually converted to either an alcohol or lactic acid.

✦ *Alcoholic fermentation.* In *alcoholic fermentation*, the pyruvic acid molecule is changed to a molecule of ethyl alcohol in a two-step process.

4-11
Alcoholic fermentation by yeast causes this bread dough to rise.

Step 1: The pyruvic acid gives off a carbon dioxide molecule, converting it into a 2-carbon compound.

Step 2: The 2-carbon compound takes the hydrogen ions and electrons from the carrier molecule NADH to make ethyl alcohol. The use of the hydrogen and the electrons frees the carrier molecule so that it can be used again in glycolysis.

This process occurs in certain bacteria and yeasts and is used to make alcoholic beverages. Yeasts are also used in breads. The carbon dioxide production causes the dough to rise, and baking evaporates the alcohol from the bread dough.

✦ *Lactic acid fermentation.* This process is used by some microorganisms (such as bacteria that form yogurt and cottage cheese) and, when necessary, by certain animal cells. *Lactic acid fermentation* has only one step—pyruvic acid takes the hydrogen and electrons from the NADH produced during glycolysis, freeing the carrier molecule to be used again in glycolysis to produce ATP. The lactic acid gives yogurt and cheeses their tart taste.

Muscle cells are one type of cell that can perform either aerobic or anaerobic respiration. During normal activity, muscle cells obtain energy through the aerobic pathway. However, during prolonged physical exertion or very strenuous activity, such as running at top speed, aerobic respiration cannot supply energy rapidly enough for muscles. During that period of intense need, the muscles switch over to lactic acid fermentation. The lactic acid will accumulate in the muscle cells and will eventually be removed by the bloodstream. If the lactic acid continues to accumulate in the muscle cells, it can cause pain and cramping of the muscles.

✦ *Fermentation energy output.* Fermentation supplies no ATP energy beyond that obtained from glycolysis. Cellular fermentation, therefore, produces a net gain of only 2 ATP molecules per glucose molecule.

4-12

Lactic acid fermentation during cheese production imparts the tart taste.

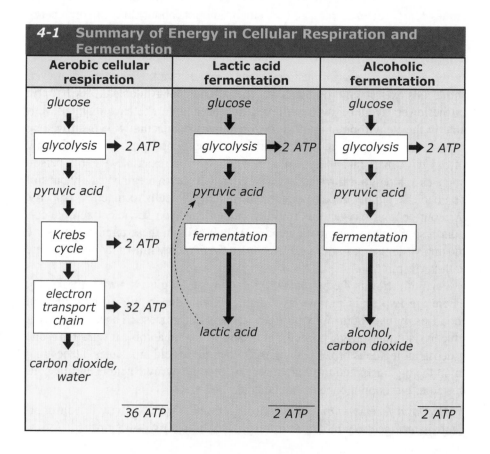

4-1	Summary of Energy in Cellular Respiration and Fermentation	
Aerobic cellular respiration	**Lactic acid fermentation**	**Alcoholic fermentation**
glucose ↓	*glucose* ↓	*glucose* ↓
glycolysis ➡ 2 ATP ↓	*glycolysis* ➡ 2 ATP ↓	*glycolysis* ➡ 2 ATP ↓
pyruvic acid ↓	*pyruvic acid* ↓	*pyruvic acid* ↓
Krebs cycle ➡ 2 ATP ↓	*fermentation* ↓	*fermentation* ↓
electron transport chain ➡ 32 ATP ↓	*lactic acid*	*alcohol, carbon dioxide*
carbon dioxide, water		
36 ATP	2 ATP	2 ATP

Review Questions 4.3

1. What is the difference between cellular respiration and breathing?
2. Name the two types of cellular respiration and list organisms that carry on each type.
3. Give a chemical equation for aerobic cellular respiration.
4. Where does each of the following occur: glycolysis, Krebs cycle, electron transport chain?
5. What is required to start glycolysis and what is produced?
6. What is required to start the Krebs cycle and what is produced?
7. What is required to start the electron transport chain and what is produced?
8. List two forms of fermentation. Give the beginning substances and the end products of each.
⊙9. Compare and contrast aerobic cellular respiration with the following: (a) photosynthesis, (b) alcoholic fermentation, and (c) lactic acid fermentation.

4B—Cellular Metabolism and Protein Synthesis

The **metabolism** (muh TAB uh LIZ um) of an organism is the sum of all its life processes: photosynthesis, movement, respiration, growth, and everything else the organism does. Different organisms, of course, carry on different kinds and different quantities of the various metabolic pathways. For example, a human cannot carry on photosynthesis. The sugars the human body needs to build carbohydrates are supplied by another metabolic pathway: digestion. Plants, on the other hand, do not carry on digestion to obtain sugars.

For years scientists have carefully studied cellular metabolic pathways. By understanding the chemical reactions, scientists hope to be able to correct metabolic problems that either cause disease or do environmental harm. For example, if scientists determine that a disease is caused by certain cells making too much of a chemical, they have a better chance of developing a treatment if they know how the cell makes that chemical. Suppose high temperatures and abundant nitrogen in a pond cause algae to produce a chemical that kills fish. Scientists have a better idea of how to control the chemical being produced if they understand the metabolic processes cells use to make it.

The understanding of cellular metabolism has permitted scientists to encourage certain processes that are profitable to us and to discourage those that are harmful. At one time scientists had hoped to greatly alter a cell's metabolic pathways and even introduce different metabolisms to various cells. The cell's metabolism can be slightly altered by this tinkering process, but the complexities and the interrelations of metabolic pathways in the organism and its cells have prohibited major changes.

Before a discussion of how some of the metabolic pathways relate to each other must come a discussion of what some scientists consider the most important metabolic pathway, the one that controls and permits all the others: protein synthesis.

4-13

The various metabolic processes that allow this athlete to compete are dependent on proteins.

metabolism: meta- (Gk. *meta*—change) + -bolism (*ballein*—to throw)

Objectives
- Describe the structure and function of the three kinds of RNA
- Differentiate between introns and exons
- List and describe the steps in translation

Key Terms

metabolism	anticodon
codon	ribosomal RNA
messenger RNA	(rRNA)
(mRNA)	translation
transfer RNA	exon
(tRNA)	intron

4.4 Protein Synthesis

The discovery of how proteins are made was a major breakthrough in the understanding of life. An organism's design to manufacture proteins enables it to carry on its life processes. Since what a cell can and cannot do depends upon its enzymes, cells are controlled by enzymes, which are proteins. In addition, almost every cellular and extracellular substance made by a cell is or contains a *structural protein*.

The mechanism for protein manufacture needs to be exceptionally flexible, able to produce the several thousand different proteins found in every cell. But the process must also be stable. When the amino acids of a particular protein are lined up, they must be in the same sequence every time. If the sequence alters as much as one amino acid, the protein will often not function properly. Thus the process of protein synthesis must manufacture a vast number of different proteins, but it must also manufacture thousands of exact copies of these proteins.

The Code of Life

The sequence of the four bases found in cellular DNA determines the sequence of the approximately twenty different amino acids found in that particular cell's proteins. It is important to understand the code before studying the mechanism responsible for the lineup of amino acids.

Anyone who speaks, speaks in a code. Each word in his language is a symbol for something. The twenty-six letters of the alphabet make hundreds of thousands of English words. These words express even more ideas when arranged into sentences.

The code of life uses the same basic format. The four bases—*adenine* (AD uh NEEN), *thymine* (THY MEEN), *guanine* (GWAH NEEN), and *cytosine* (SY toh SEEN)—can be called the alphabet of life. These "letters" are arranged in groups as "words" that may be interpreted as amino acids. All the "words" in the DNA code are three letters long. Three bases then become the code for a particular amino acid. The code of life, then, is formed from triplets of bases called **codons**. When many of these codons are in a particular sequence, they are read to form a chain of amino acids.

A series of words in a sentence has more meaning than just the words alone. The polypeptide chain of amino acids also becomes more than just a string of amino acids. It twists and bonds and combines with other amino acid chains or other substances to become functional proteins.

A Common Language

Scientists have discovered that nearly all living things, from bacteria to people, share this amino acid codon system. Someone with an evolutionary worldview might claim this is solid evidence for the descent of all living things from a common ancestor who passed on these codons. How would you respond to this?

First letter	Second letter				Third letter
	U	**C**	**A**	**G**	
U	phenylalanine	serine	tyrosine	cysteine	**U**
	phenylalanine	serine	tyrosine	cysteine	**C**
	leucine	serine	stop	stop	**A**
	leucine	serine	stop	tryptophan	**G**
C	leucine	proline	histidine	arginine	**U**
	leucine	proline	histidine	arginine	**C**
	leucine	proline	glutamine	arginine	**A**
	leucine	proline	glutamine	arginine	**G**
A	isoleucine	threonine	asparagine	serine	**U**
	isoleucine	threonine	asparagine	serine	**C**
	isoleucine	threonine	lysine	arginine	**A**
	(start) methionine	threonine	lysine	arginine	**G**
G	valine	alanine	aspartate	glycine	**U**
	valine	alanine	aspartate	glycine	**C**
	valine	alanine	glutamate	glycine	**A**
	valine	alanine	glutamate	glycine	**G**

4-2 Messenger RNA Codons for Amino Acids

The first base of the messenger RNA codon is found on the left; the second base, in the middle; the third, on the right. Thus, UGG codes for tryptophan, the only codon for that amino acid. Many codons code for the same amino acid. For example, GGU, GGC, GGA, and GGG all code for glycine. AUG, methionine, is the beginning amino acid; UAA, UAG, or UGA is the last codon, indicating a stop of the polypeptide chain.

In DNA each of the three characters that make a word (codon) can be any one of the four letters representing the four bases. Therefore, the total number of possible combinations of four letters making a three-letter codon is sixty-four. Because in most living things there are only twenty amino acids, some amino acids are coded with more than one codon. There are some special codons that signal the start and stop of the process of lining up a sequence of amino acids. They may be considered punctuation marks in the DNA code.

Types of RNA

Three different types of RNA are necessary to translate the sequence of bases in a DNA molecule into the sequence of amino acids in a polypeptide chain. DNA is capable of manufacturing RNA by *transcription*. Transcription places the DNA code into the sequence of bases in an RNA molecule.

One of the RNA molecules that is produced by the transcription of the DNA molecule is called **messenger RNA (mRNA)**. The mRNA contains the code for a polypeptide chain of amino acids and carries the code from the DNA in the nucleus to the ribosomes in the cytoplasm, where the code is "read." The mRNA contains the proper sequence of codons that will be interpreted as a chain of amino acids.

The amino acids are located in the cytoplasm of the cell, but they are not assembled into proteins directly by the mRNA. Instead, the amino acids are attached to **transfer RNA (tRNA)** molecules, the second type of RNA. The tRNA molecules are about eighty nucleotides long and are formed in the nucleus. The chain of nucleotides forms several loops, taking a cloverleaf shape. At one loop of this configuration are three unattached bases called the **anticodon**. The anticodon has complementary base pairs for specific codons in the mRNA. If mRNA has the codon GAC, the tRNA that lines up on that codon will have the anticodon CUG. Anticodons, then, consist of bases that are the complements of the bases in the codon with which they align.

The tRNA end opposite of the anticodon usually has the three nucleotide bases CCA. Using a high-energy bond, an amino acid attaches to the final base (adenine) of this sequence. The bonding of an amino acid to tRNA requires the energy of one ATP molecule.

Ribosomal RNA (rRNA), the third type, is also manufactured by the DNA of the nucleus. The rRNA combines with various proteins in the cytoplasm to form a *ribosome*. Some of the proteins located around rRNA form three binding sites that are essential for assembling amino acids. One site binds to the mRNA so that its codons can be read. The other two sites hold two tRNA molecules in place as their anticodons pair with the mRNA codons. Amino acids are lined up to manufacture a polypeptide chain only at the point where mRNA is in contact with the ribosome.

Translation—the Manufacture of an Amino Acid Chain

So far, this study has discussed the tools needed to assemble a polypeptide. The actual process of assembling the polypeptide from information coded in mRNA is called **translation**.

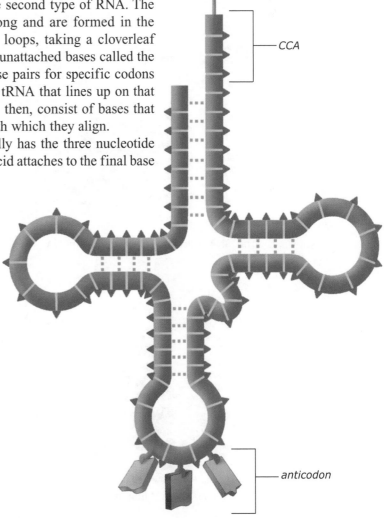

CCA

anticodon

4-14
Transfer RNA molecule

The Manufacture of Messenger RNA

Messenger RNA was once believed to be a simple copy of one side of a DNA molecule. Today scientists believe that in eukaryotic cells the transcription process is only the first step. The initial mRNA manufactured by transcription, called *pre-mRNA*, is actually longer than the number of nucleotides needed to code for the polypeptide chain of amino acids it produces. The transcribed RNA contains exons and introns. The **exons** are the sections of mRNA that are actually read as codons when making the protein. The **introns** are sections of transcribed RNA that do not code for amino acids and

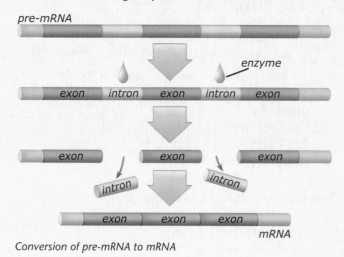

Conversion of pre-mRNA to mRNA

therefore cannot be translated into proteins. Since they are not read when making the protein, it appears that introns are "snipped out" and the exons are rejoined as the final mRNA molecule is formed. This process occurs before the mRNA leaves the nucleus.

These introns, as well as repetitive sections of DNA, have been regarded by many scientists as "junk" DNA. In keeping with an evolutionary worldview, secular scientists believe them to be useless leftovers from a long evolutionary history. While scientists still cannot find purposes for *all* of the non-coding sections of DNA, recent research shows that many of the sequences play a role in how genetic traits are expressed. Their function seems to be turning on and off certain genes. In time, we will probably learn how important this "junk" really is!

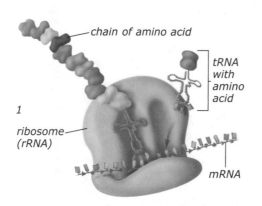

1 The tRNA anticodon aligns with the mRNA codon.

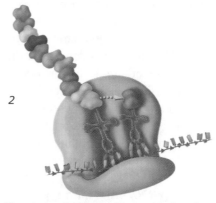

2 The chain of amino acids is added to the amino acid brought in by the tRNA.

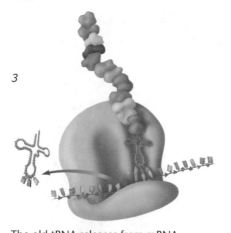

3 The old tRNA releases from mRNA.

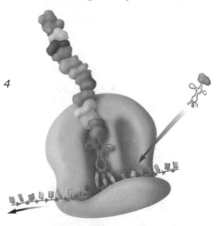

4 Another tRNA brings in a new amino acid as the mRNA shifts.

4-15

Translation: the process of manufacturing a specific polypeptide chain

Translation begins as mRNA leaves the nucleus with DNA's triplet code by passing through a pore in the nuclear envelope. A ribosome lines up with the codon AUG on the proper area of mRNA. The AUG codon, which codes for methionine, is called a *start codon*. The ribosome uses one of its binding sites to hold an mRNA codon at the proper place so that a tRNA anticodon can line up with it. Since the mRNA codon is AUG, a tRNA with the anticodon UAC will line up at that position and will have the amino acid methionine attached to its opposite end.

Once the codon and the anticodon have lined up, the ribosome moves down the mRNA, and the next codon is held in place to be read. If the next codon is UUU, the tRNA with anticodon AAA will join it and be held in place by one of the binding sites on the rRNA. The amino acid phenylalanine will now be right next to methionine. As they are held in place, they will bond together. Methionine and phenylalanine are now in the proper sequence for this polypeptide chain. Their sequence was dictated by the sequence of bases in the mRNA.

Ribosomal enzymes cause the amino acids lined up on the mRNA to form a peptide bond. The energy to form the bond holding the amino acids together was in the bond between the first amino acid, methionine, and the tRNA that brought it to the ribosome.

After the bonding of the amino acids (methionine to phenylalanine in the example), the first tRNA is no longer bonded to an amino acid. That tRNA molecule now leaves the ribosome and enters the cytoplasm as the ribosome slides down the mRNA to reach the next codon. (Soon another amino acid bonds with the tRNA, and it can be used again.) The process of reading new codons, bonding amino acids, and releasing empty tRNAs continues forming a polypeptide chain of amino acids.

Certain codons do not code for any amino acids. The codons UAA, UAG, and UGA are termed *stop codons* since they do not correspond to any amino acids. When the ribosome reaches one of these codons, the translation process ends. The mRNA is released from the binding site on the ribosome and the polypeptide chain is completed.

The new polypeptide chain is twisted and looped to form a protein. Several of the amino acids in this chain may be bonded to other specific amino acids within the chain. Thus, proteins have a specific shape. Not all proteins are a single chain of amino acids. Some proteins are composed of several different amino acid chains bound together. Lipids, carbohydrates, or inorganic materials combine with some amino acid chains to form functional proteins.

Mitochondria—a Different Genetic Code

Not all of the genetic information found in a cell is located in the nucleus—mitochondria have their own DNA, as do the chloroplasts in plant cells.

Mitochondrial DNA (mtDNA) is a loop of DNA that manufactures RNA, including tRNA and rRNA, and appears to be responsible for the manufacture of some of the proteins used in mitochondria. The code, however, is slightly different. UGA is usually a stop codon, but in the mtDNA of many organisms it is translated as the amino acid tryptophan. In plants, however, UGA is a stop codon in mtDNA.

New mitochondria are made in the cytoplasm by division of old mitochondria. In order to properly function, each new mitochondrion must get a copy of the mtDNA from an old mitochondrion, which contains five to ten copies of mtDNA. Mitochondrial DNA does not have the code for all the proteins needed for a mitochondrion to grow and reproduce and function; therefore, it must also have proteins made by the DNA of the nucleus.

In humans and animals, 99.99% of the mtDNA is inherited from the mother. This is true because nearly all mtDNA arises from the original mitochondria that are found in the ovum. There are very few mitochondria in a sperm cell, and these rarely penetrate the egg at fertilization. This fact has led researchers to attempt to use this "pure" line of DNA as a genetic marker for particular races. An assumption that mutations in this mtDNA occur at a known, constant rate has also led many scientists to suggest how long ago various human migrations occurred and even how long man has been on the earth. Their estimates are based on so many assumptions that there is no strong consensus. In fact, many scientists believe that ages established by mtDNA dating are so fraught with errors that they are worthless.

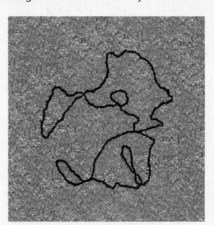
SEM of mitochondrial DNA

Some evolutionists have suggested that mitochondria were once prokaryotic cells (like bacteria) that were "taken in" by eukaryotic cells, and over millions of years the two somehow became one organism and evolved into the various forms of life that contain mitochondria. They base this on the similarity of the prokaryotic DNA loops to the form of mtDNA. To the evolutionist, this may explain the existence of mtDNA. However, it does not explain many details about mitochondria: why are they so different in different cells—even different cells of the same organism? How did mitochondria get into all eukaryotic cells? Also, mitochondria cannot live outside the cell membrane because they lack many of the features that prokaryotic cells have to live out in the open.

The evolutionist's explanation of mtDNA is both unnecessary and unacceptable to one who believes in the biblical account of Creation. Those who believe the evolutionists' models have an evolutionary bias in their interpretation of the physical phenomenon of mtDNA. What is actually known about mtDNA so far? Scientists know that it definitely exists, and they know what it is like and much of how it functions. But that is not enough to justify any explanation of mtDNA's origin or its value as a molecular clock.

 Facets *of Biology*

Interrelated Metabolisms—Putting It All Together

From this chapter's discussion of photosynthesis, cellular respiration, and protein synthesis, it should be apparent that these very different processes have some molecules in common. For example, both photosynthesis and respiration deal with glucose; PGAL is an intermediate product in glycolysis and is also produced during the Calvin cycle of photosynthesis. ATP is used or produced in most of these reactions. Do these processes fit together? What about those that you don't find in other processes—like amino acids? Where do they come from?

Most cells have the ability to transform substances by moving them from one process to another process. By the interrelationship of cellular processes, cells can change whatever organic substance they have in abundance into almost any other organic substance that they normally

Plants produce carbohydrates by photosynthesis and then use them to manufacture the lipids and proteins they need.

produce. This is essential for cells that live in an environment that cannot provide all the substances they need.

Plant cells are a good example. They produce abundant carbohydrates but obtain few or no lipids or proteins from their environment.

Therefore they must be able to convert the sugars they produce into the lipids and proteins necessary to form their cellular structures. Another example: humans do not have to eat fats to get fat (store excess lipids). A person can eat too many carbohydrates and end up storing the energy as lipids around the waist. (It is easier, however, to gain excess "energy" by eating lipids because fats have twice as much energy per unit of weight as carbohydrates.) The next paragraphs identify the compounds that can be used to go from one process to another, and their pathways are shown in the chart.

Carbohydrate and Lipid Metabolisms

One of the most important anabolic processes is photosynthesis, which combines water, carbon dioxide, and energy to form PGAL. With the proper enzymes, PGAL can be used to produce glucose, which can be easily changed into different monosaccharides. Other enzymes can convert monosaccharides into disaccharides, starches, cellulose, or any other natural carbohydrate. Some cells do not have the enzymes to make every one of these carbohydrate conversions because the cells do not normally require every type of carbohydrate. Cells usually have the enzymes necessary to convert the carbohydrates they have to glucose and the glucose back to carbohydrates. One notable exception, however, is cellulose in plant cell walls. Once cellulose is made, most plant cells cannot break it down.

PGAL, the first stable product of photosynthesis, is involved in glycolysis. PGAL can also be converted to glycerol, one of the components of most lipids. Through a series of steps, simple sugars can be converted into fatty acids. Most cells, therefore, can make carbohydrates into fats and fats into carbohydrates. If a human diet does not include sufficient fats, the liver can synthesize fats from carbohydrates. Fats, however, are not as easily converted to sugars. Generally this conversion occurs only when a person is dieting or starving.

Cellular respiration changes glucose and oxygen to water, carbon dioxide, and ATP. However, that is not all that cellular respiration can do. Fatty acids can be synthesized from acetyl CoA, and the reverse is also true: most

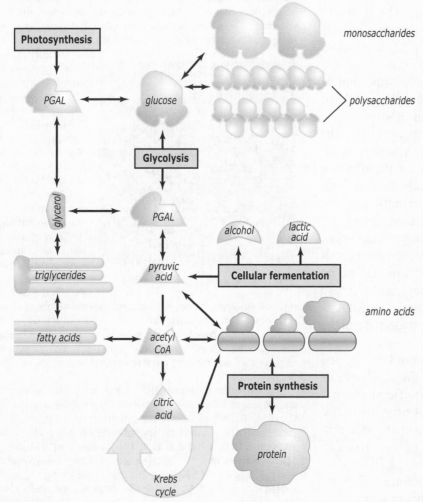
A comparison of major metabolic pathways

96 *Chapter 4*

cells can break long fatty acid molecules into 2-carbon molecules. These 2-carbon molecules can be used to produce an acetyl CoA that can be used in the Krebs cycle.

For each acetyl CoA entering the Krebs cycle, there is a net gain of 16 ATP molecules. An 18-carbon fatty acid can produce 128 ATP molecules by way of the Krebs cycle and the electron transport chain. Using the same processes, 18 carbons in the form of sugar (3 glucose molecules) can produce only 108 ATP molecules.

Protein Metabolism

A cell may obtain the amino acids necessary for protein synthesis by digestion and absorption, or it may manufacture them. Certain amino acids can be made from substances at certain steps in the Krebs cycle, from pyruvic acid, or from acetyl CoA; however, some organisms cannot manufacture all the amino acids they require. Such amino acids are called *essential amino acids* because organisms must obtain them. Not all organisms require the same essential amino acids—those essential for one organism may be easily synthesized by another.

Amino acids can also enter the Krebs cycle and be broken down to carbon dioxide, water, and ATP energy. Some amino acids can be converted to acetyl CoA or to pyruvic acid and then be broken down as energy sources. Long before a person starves to death, his body will begin to convert proteins, including muscle proteins, into usable ATP energy.

Nucleic Acid Metabolism

Nucleic acids are formed from nucleotides. Usually a nucleotide can be used repeatedly. As mRNA falls apart with use, its component nucleotides can be resynthesized into more RNA or DNA. If a cell is to grow and expand its protein-manufacturing abilities or if the cell is to replicate its DNA and divide to form two cells, it will need additional nucleotides. A nucleotide is composed of a sugar, a phosphate group, and a nitrogen-containing base. The sugar can be supplied by carbohydrate metabolism. The phosphate is an inorganic substance that most cells easily absorb. The nitrogen-containing base is similar to those found in amino acids. A cell with the proper enzymes can synthesize nucleotides.

⇒Review Questions 4.4

1. Compare the structures that make up the code of language (letters, words, sentences, punctuation) to corresponding parts of the code of protein synthesis.

2. List the three types of RNA. Describe each and give its function.

3. What are exons and introns?

4. Describe the process of making an mRNA molecule.

5. The following processes take place either in the nucleus, in the cytoplasm, or on the ribosome. Indicate where each process takes place.
   (a) transcription of a DNA molecule
   (b) reading of an mRNA molecule
   (c) lining up of a codon and an anticodon
   (d) replication of a DNA molecule
   (e) attachment of two amino acids to each other
   (f) attachment of an mRNA molecule to a ribosome
   (g) attachment of an amino acid to a tRNA molecule

6. What are start codons and stop codons?

7. Determine the codons and anticodons for the sequence of bases on the mRNA molecule given below. Then, using Table 4-2, determine the sequence of amino acids in the section of a protein for which this mRNA would code:

   AUGUUCGUUAACGACCAAAUUUAA

8. Other than the nucleus, where is DNA located in eukaryotic cells?

9. What is unusual about mitochondrial DNA?

⊙10. Considering the steps of protein synthesis and the results, list several attributes of this process that are critical for its success. Explain why each attribute is necessary for the continuation of life.

⊙11. Since proteins make the differences between various organisms and individuals, discuss why DNA is considered the genetic material instead of proteins.

⊙12. Describe how carbohydrate and lipid metabolisms relate to each other.

⊙13. How does protein metabolism relate to carbohydrate and lipid metabolisms?

⊙14. Tell where the various components of a DNA nucleotide come from.

4.5 Metabolism and Homeostasis

An organism does not constantly face exactly the same environmental situation, so the metabolic activities of cells must change to maintain the organism's homeostasis. For example, cells require ATP energy constantly. If the environment is favorable, the cell may grow rapidly and therefore use large quantities of energy to carry on the necessary processes to supply substances for growth. The same cell in an unfavorable environment will not grow but may still use the same amount of ATP to maintain its life. The metabolic *rate* under these two conditions will be the same; the metabolic *processes* will be different. In each situation, homeostasis—that is, the steady state of life—will be maintained.

4.5

⇒**Objectives**
- Differentiate between anabolism and catabolism
- Discuss the three types of cellular digestion

⇒**Key Terms**
intracellular digestion
extracellular digestion

Metabolic Rate

Some organisms have a higher metabolic rate than others do. A rapidly growing young organism, be it a child or a bacterial cell, usually builds large quantities of substances and therefore has a high metabolic rate. At the point of maturity, the metabolic rate will slow down. Plant cells in sunlight carry on photosynthesis and respiration, but in darkness they carry on only respiration. During the day, then, the rate of metabolism is higher.

Members of athletic teams force their bodies to engage in higher metabolic activity than a person who spends most of his out-of-school time reclining in front of a television or playing video games. Rates of metabolism vary even during sleep. Illness causes metabolism to be higher as the body reacts to the disease. However, resting metabolisms are not the same even among healthy people.

For some organisms the rate of metabolism depends on the environment. The amount of available light or water may determine the rate of photosynthesis in plants. The presence or absence of certain vitamins (to serve as coenzymes) may affect a person's health. Biological reactions seem to happen twice as fast for each increase of 10 °C. Most enzymes of living systems, however, have optimal ranges. At higher or lower temperatures, they begin to function improperly rather than just faster or slower.

Increasing a human's internal temperature 10 °C is usually fatal. The desert pupfish can exist in temperatures from 10 to 40 °C but does not function well at either extreme. Increasing a fish's body temperature will increase its metabolism, but there is a limit. The Antarctic fish, for example, thrives in water from −2 to +2 °C. As the temperature reaches 0 °C, the fish is active. As the temperature rises, its metabolism slows down. Around +2 °C it

4-16

In which photo is the tree experiencing a higher rate of metabolism?

4-17

A desert pupfish (left) and an Antarctic fish (right)—neither could survive in the other's environment.

becomes immobile. If temperatures continue to increase, the fish will die. God enables this fish to survive the constant near-freezing water temperatures of the Antarctic; its metabolism is specialized. At the other extreme are bacteria that thrive around geothermal vents on the floor of the Pacific Ocean. They grow readily at 121 °C, and some can even survive a few hours at 130 °C.

Anabolism and Catabolism

Metabolism can be divided into two types:

- Anabolism (uh NAB uh LIZ um)—processes that build molecules and store energy
- Catabolism (kuh TAB uh LIZ um)—processes that break down molecules and release energy

Many cellular processes involve both anabolism and catabolism, but there is usually a net gain or loss. In glycolysis, 2 molecules of ATP are

anabolism: ana- (Gk. *ana*—up) + -bolism (to throw)

catabolism: cata- (*cata*—down) + -bolism (to throw)

Cellular Digestion

One of the important methods by which cells obtain substances is digestion. Many cells obtain materials for digestion by *phagocytosis* and *pinocytosis*. Those materials are then broken down by enzymes in **intracellular digestion**. The enzymes, which are produced by the ribosomes and secreted by the Golgi apparatus as membrane-bound organelles, such as lysosomes, break down the materials in food vacuoles into soluble carbohydrates, fatty acids, amino acids, salts, and other substances. These products of digestion diffuse through the vacuolar membrane into the cytoplasm and are available for other cellular processes.

Not all cells carry on digestion in food vacuoles; sometimes the food substance remains outside the cell. The enzyme-containing vesicles of this type of cell fuse with the plasma membrane, and the enzymes are released (secreted) outside the cell. The enzymes then digest the food substance. This process is called **extracellular digestion**. The soluble products of extracellular digestion are then absorbed through the membrane by the cell. Most cells in multicellular organisms do not carry on digestion but rely on a supply of predigested food in their environment. The cells of the human digestive system carry on extracellular digestion and supply the nutrients in a soluble form to all the body's cells.

Some cells carry on a "remodeling" process called *autophagy* (aw TAHF uh jee). The cell forms a membrane around a damaged, worn out, or no longer needed cellular structure, forming a vacuole. Lysosome membranes join with the autophagic vacuole's membrane and empty their enzymes into the vacuole. The cellular substances within the autophagic vacuole are then broken down to soluble materials that are absorbed into the cytoplasm. Thus, some substances from nonfunctional cellular structures can be recycled.

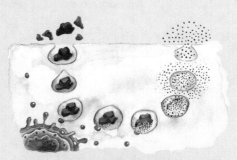

intracellular digestion

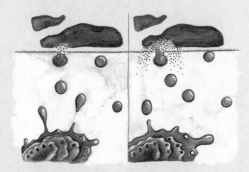

extracellular digestion

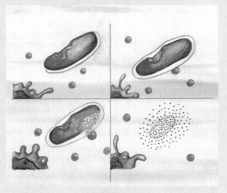

autophagy

intracellular: intra- (L. *intra*—within) + -cellular (pertaining to a cell)

extracellular: extra- (L. *extra*—outside) + -cellular (pertaining to a cell)

autophagy: auto- (self) + -phagy (to eat)

consumed in the production of 2 molecules of PGAL, and a total of 4 ATP molecules are formed when pyruvic acid is formed from PGAL—a net gain of 2 ATP molecules. Since molecules are broken down and energy is released, the process of glycolysis is considered a catabolic process. On a larger scale, a growing cell must break some substances down in order to build others up. Since there is a net gain of molecules, growth is usually referred to as anabolism.

The cell is an awesome chemical factory. Not only can it take raw materials and put together the substances it needs, but it can also take its own substances and re-form them when necessary. It can, within limits, change the course of its operations by keeping a constant supply of needed materials and by maintaining the internal homeostasis that permits its life. It is truly amazing what God has wrought in even the tiniest cell!

Review Questions 4.5

1. Relate the terms *homeostasis* and *metabolism*.
2. List several factors that may affect metabolic rates.
3. What are the two basic types of metabolism? Briefly describe each.
4. What are the two basic ways that cells can digest a substance? Describe them.
5. For maturing into a frog, a tadpole relies heavily on the process of autophagy. Explain how this process is useful in the change of the frog's physical structures.
6. The interrelationship of the life processes (briefly outlined in this chapter) is carefully balanced while the organism is alive. Explain how this balance illustrates the first two laws of thermodynamics and the laws of conservation and degeneration.

Genetics

The Continuity of Life
Part I

od showed His glory mightily when He created each living thing with a massively complex code of DNA to direct its design. Then He gave humans dominion over creation, including genetics information. Chapter 7 will deal more with humanity's dominion of genetics. Unfortunately, humanity's fall into sin affected genetics. Now there are genetic mutations and mistakes that imperil health and life. Sin has also warped humanity's dominion of genetic material. Humans now use genetics for sinful things such as human embryonic stem cell research. Christians can, however, hope for Resurrection day when Christ will come and give us new bodies that will not be affected by genetic mutation, errors, or sin ever again (1 Cor. 15:19–54).

5A—Genes, Chromosomes, and Cell Division

Humans resemble one another. They usually have two eyes, two ears, a nose, and a mouth. Their hands are basically the same. Internal human features such as stomachs, livers, kidneys, and bones are about the same size, shape, and location in different humans. Even human cells have basic chemical and structural similarities. For example, most human cells have forty-six chromosomes. Characteristics that every member of a species possesses are called *species characteristics*.

But humans also have some distinguishing features. Characteristics such as eye, hair, and skin color, body build, intelligence, and other features that make a person unique are called *individual characteristics*. Who a person is depends in part on the species characteristics and in part on the individual characteristics that he inherits.

Does a person inherit all his characteristics from his parents? Do the twenty to thirty thousand genes he inherits so control him that he just mechanically lives out what he inherited? Both questions must be answered yes and no. Genes do determine species characteristics and individual characteristics. But the extent to which these characteristics will be expressed depends in many ways upon the *environment*, the second major physical factor that determines who a person is. For example, if a person inherited genes for being over 6 feet tall but had a poor diet, insufficient rest, and little exercise as he matured, he would likely not reach his genetic potential. This often happens in countries where malnutrition is rampant. Intelligence potential, most geneticists believe, is also inherited. However, if an individual's environment gives little stimulation, his intellect will not become what it could be.

Today many people have wrongly argued that their genetics and their environment have caused them to sin. They blame their homosexuality, their addictions, and their violent actions on their genes and their environment. But a human is more than just the sum of inherited genetic potential and environment. Humans are not enslaved to either their genetics or their environment but to sin. Jesus by His death on the cross has the power to free people from their sins and to free them to pursue His will. As a person yields his will to God, God directs his environment and uses his genetic potential to develop in him the fruit of the Spirit to glorify Himself (Rom. 8:28; Phil. 4:13; Heb. 13:21).

5.1 The Mechanism of Heredity

Genetics, the study of heredity, is both one of the oldest and one of the newest of the biological sciences. The ancient Babylonians and Assyrians used selective breeding of plants during the time of King Hammurabi—approximately 1750 BC. Although the ancients were unaware of the mechanism of inheritance, they were able to improve their plants and animals. Still common today, this type of selective breeding is actually *applied* genetics.

Men have always been interested in how the characteristics of one generation are passed on to the next. The ancient Greeks believed that each offspring was a mixture of elements from the parents' bodies. Aristotle proposed the *particulate theory of reproduction*: particles of the parents' blood mix and then join to form the offspring. The terms "pure blood" and "blood relative" came from this belief of circulating particles in the blood.

During the seventeenth century, a group of people called *preformationists* believed that there are tiny, completely formed organisms in the sperm that, when planted in an egg, merely grow up. Most preformationists also believed that inside the tiny person in that sperm are other sperm that contain even tinier people, which contain sperm with even smaller people *ad infinitum*. In other words, inside Adam was all of mankind, just waiting to grow up. The impossibility of such decreasingly tiny people was part of the "miracle" of reproduction. As the lenses in microscopes were improved, this theory was disproved.

In the middle of the 1800s, Gregor Mendel (1822–1884), who is discussed later, performed a series of experiments using garden peas. Mendel advanced

► Objectives

- Describe the impact of both genetics and environment on an individual
- Define *genetics*
- Differentiate between a gene and a chromosome
- Describe the structure of a chromosome
- Explain the difference between a haploid and a diploid chromosome number

► Key Terms

genetics	karyotype
gene	homologous
chromosome	chromosomes
histone	homologue
chromatid	diploid
centromere	haploid

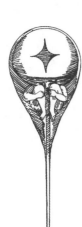

5-1

A drawing of a human sperm by a preformationist. Preformationists cited Scripture out of context to support their theory.

the understanding of genetics by proposing that there are pairs of *factors* in organisms and that each parent gives a single set of these factors to its offspring. It was later discovered that these factors are located on DNA. In 1953, Watson and Crick described the structure of the DNA molecule, opening the way to modern genetics.

Today genetics is studied at the molecular level and has given rise to knowledge and procedures that were once considered science fiction. All of this may seem very far removed from high-school biology class. However, in order to understand and participate in discussions about information in the headlines, on the Internet, or on television, you need to understand the basics of genetics. The study of genetics begins with learning about genes, chromosomes, and cell division.

Genes

The common term *gene* is hard to define. Mendel called these particles, which are involved in passing on inherited characteristics, *factors*; they are now called genes. But what exactly is a gene?

Scientists now define a **gene** as a section of DNA that produces a particular polypeptide chain of amino acids (a protein or a section of protein) that causes a particular trait. Not all traits are as visible and discernible as blue eyes or brown hair. Many genes produce structural or enzymatic proteins that perform functions that are never seen. For example, the production of the enzymes used in the Krebs cycle is the result of traits encoded in the DNA of genes. If a person lacked these genes, he would not be alive. It is important to note that this definition applies only to *protein-coding* genes. Many other sections of DNA do not actually produce a protein. Some of these sections function to control the activation and deactivation of protein-coding genes.

It is hard to believe that all the information needed to produce a person's physical traits is carried in the genes on his DNA. DNA is tightly compressed; the DNA from all a person's cells would fit into a 1-inch cube. However, if the DNA in just one human cell were uncoiled and stretched out, it would form an extremely thin string about 2 yards long. The information found in the DNA in just one human cell equals the information contained in one thousand books, each having six hundred pages—that is the amount of information in this textbook multiplied by one thousand! The nucleus of one human cell is a huge library of chemical information.

Characteristics of a Gene

✦ Genes are fundamentally identical in both type and amount in the cells of an organism. For example, virtually all a person's cells have the same number and types of genes.

✦ Genes are fundamentally identical in both type and amount in the cells of each organism of a species. For example, a normal human has the same number and types of genes as other normal humans.

✦ Genes are chemicals that can function as individual units—they are different chemicals, each of which can do a different thing.

✦ Genes are able to carry information for the formation of organic chemicals—they are responsible for biosynthesis.

✦ Genes are able to reproduce themselves—they are able to make copies of the information they have.

✦ Genes can be passed on to the next generation.

✦ Genes are found in the cell's nuclear material.

Chromosomes

Each DNA strand in a nucleus actually consists of many genes. These long strands of DNA are called **chromosomes**. In an active nondividing cell, most of the chromosomes appear in the nucleus as a fuzzy, tangled mass called the *chromatin material*. Some of the genes are active, producing RNA. Others are not being used and are often tightly coiled.

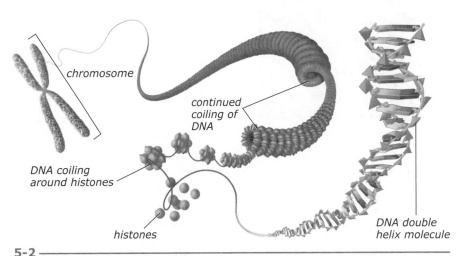

chromosome

continued coiling of DNA

DNA coiling around histones

histones

DNA double helix molecule

5-2

A chromosome consists of tightly coiled DNA and proteins called histones.

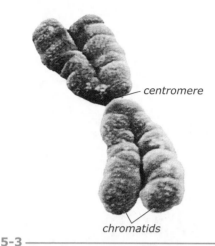

— centromere

chromatids

5-3

A duplicated chromosome ready for cell division consists of two chromatids joined at the centromere.

5-1	**Diploid Chromosome Numbers in Some Organisms**		
Plants	**Number (2n)**	**Animals**	**Number (2n)**
pea	14	fruit fly	8
red clover	14	house fly	12
onion	16	honey bee	32
cabbage	18	cat	38
corn	20	mouse	40
watermelon	22	rat	42
lily	24	rhesus monkey	42
tomato	24	rabbit	44
white pine	24	cow	60
cotton	26	donkey	62
African violet	28	horse	64
adder's tongue fern	1400+	crayfish	200

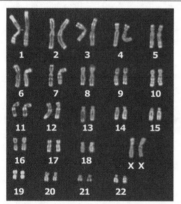

5-4

This karyotype from a human cell shows that humans have 46 chromosomes arranged in 23 pairs.

homologous: homo- (Gk. *homos*—same) + -logous (*logos*—word)

diploid: diplo- (Gk. *diploos*—double) + -oid (Gk. *eidos*—shape or form)

haploid: haplo- (Gk. *haploos*—single) + -oid (shape or form)

Proteins called **histones** help support and protect the long, thin strands of DNA. The histones also help maintain the shape of the chromosome and play a role in tightly coiling and packing the DNA.

The shape of a chromosome that is most familiar is one that is already duplicated and ready for cell division. A chromosome in this stage has two identical halves. Each half is called a **chromatid**, and the chromatids are joined at a constricted area called the **centromere**. The centromere holds the two chromatids together until they are ready to separate.

The number of chromosomes in a cell is a species characteristic. Humans have forty-six chromosomes in the nuclei of most of their cells. Fruit flies have eight; certain species of goldfish have one hundred; crayfish have two hundred. However, it should not be supposed that the number of chromosomes determines the species—lilies, tomatoes, and white pines all have twenty-four. It is the genes on the chromosomes that make the organism. But every normal cell in a particular organism, as well as every normal organism of that type, has the same number of chromosomes.

If evolution were true, it would be logical to assume that the smaller species with the least complex structures would have smaller and fewer chromosomes. But this is not the case, as Table 5-1 shows. Evolutionists have come up with several hypotheses, but they still have no way to explain chromosome numbers.

Figure 5-4 shows a complete set of chromosomes from a human cell. They were obtained from a cell that was stained to reveal chromosomes and then treated chemically to separate the individual chromosomes. They were then photographed, cut out, and arranged into pairs according to their length and centromere location. The result is called a **karyotype** (KEHR ee uh TYPE).

The karyotype shows that chromosomes occur in pairs. There are two number one chromosomes, two number two chromosomes, and so forth. Humans actually have twenty-three pairs of chromosomes. These pairs are called homologous (hoh MAHL uh gus) **chromosomes**. Each member of a homologous pair of chromosomes is called a **homologue** (HAHM uh LAWG). In a karyotype, homologues are arranged in homologous pairs.

When a cell has homologous pairs of chromosomes, it is said to be diploid (DIP LOYD) and is abbreviated 2n. Most common organisms are diploid. The diploid number of an organism is its number of chromosomes. The human diploid number is forty-six (2n = 46).

Both homologues in a pair have genes for the same characteristics. In other words, humans have two of almost every gene, one on each homologue of the pair. Assume, for instance, that humans have a gene for tongue length on chromosome number one. Since humans have a homologous pair of chromosome number one, everyone has two genes for tongue length. A later discussion shows the significance of having pairs of genes.

Some cells have only one set of chromosomes—they are not paired. These cells are termed haploid (HAP LOYD) cells and have half the number of chromosomes as diploid cells. The haploid number is abbreviated n.

1. What are the three factors that contribute to the makeup of a person?
2. List (a) several species characteristics and (b) several individual characteristics of dogs.
3. List seven characteristics a gene must possess.
4. Describe a chromosome. Distinguish between a chromosome and a chromatid.
5. What problem does the chromosome number of various organisms present to evolutionists? Give an example to support your answer.
6. What is the difference between a diploid cell and a haploid cell?

5.2 Cell Division

When a eukaryotic cell divides, it is not essential that an equal amount of cytoplasm and an equal number of organelles be in each of the new cells. If the new cell has a complete set of genes, it can manufacture what is missing. Many students think that cell division and mitosis are the same thing; however, each process is a distinct phase in the life cycle of a cell.

The Cell Cycle

The repeating cycle of events in the life of a cell is called the **cell cycle**, which is divided into three major phases—interphase, mitosis, and cytokinesis. Interphase is a time of growth for the cell. Mitosis is the process whereby the copies of the genetic material are separated into two sets. The final phase—cytokinesis—is the actual division of the cell, and it occurs differently in plants than in humans or animals. Although we discuss these phases as separate steps, they can overlap somewhat. For example, cytokinesis may begin before mitosis has completely finished.

Interphase

As Figure 5-5 shows, the cell spends most of its time in **interphase**, the major growth phase of the cell cycle. Interphase is further divided into three stages: G_1, S, and G_2. The *G_1 phase* (the *G* stands for *gap*—the time between cell division and DNA replication) is the first stage of interphase. Here the cell undergoes a major growth spurt, not only increasing in size but also making new organelles and synthesizing proteins and other needed molecules. This stage ends when the cell has reached maturity.

During the *S phase* (*S* stands for *synthesis*), the DNA is replicated. At the beginning of the S phase, each chromosome has a single copy of its DNA. By the end of this phase, all of the DNA molecules have been replicated and the chromosomes consist of two duplicate copies of the DNA. Although chromosomes have been duplicated, they are still in the form of chromatin material—not yet the tightly coiled chromosomes as seen in the karyotype in the previous section. Although these DNA duplicates are attached to one another by the centromere, the fuzzy appearance of the chromatin material hides these structures during interphase. Just outside the nucleus in the cytoplasm, a small region called the *centrosome* divides, and the *centrioles* become visible within the centrosome. The centrioles produce fibers that aid in separating the duplicated chromosomes during mitosis. It is not known exactly what other roles they play in cell division since not all animal or human cells contain centrioles; plant cells do not have centrioles.

5.2

⇒ **Objectives**

- List the phases of the cell cycle
- Describe the G_1, S, and G_2 phases of interphase
- List the stages of mitosis and the major changes of each
- Describe how cytokinesis in plant cells differs from that in human and animal cells

⇒ **Key Terms**

cell cycle   anaphase
interphase   telophase
mitosis   cytokinesis
prophase   cell plate
mitotic spindle   asexual
metaphase   reproduction

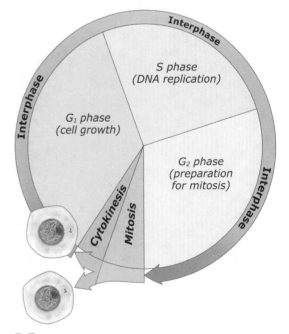

5-5

The cell cycle consists of three phases—interphase, mitosis, and cytokinesis.

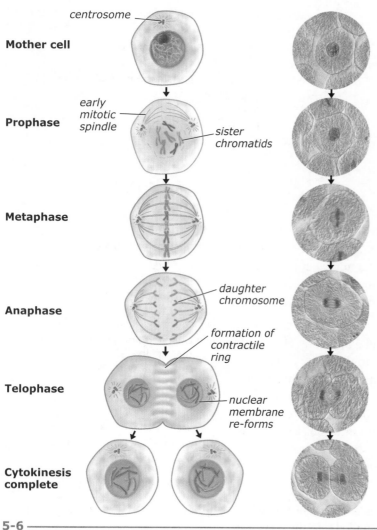

Mother cell — centrosome

Prophase — early mitotic spindle, sister chromatids

Metaphase

Anaphase — daughter chromosome

Telophase — formation of contractile ring, nuclear membrane re-forms

Cytokinesis complete

5-6

Mitosis begins after interphase. Follow the diagram while reading the description in the text. Mitosis in whitefish embryo cells is shown to the right of the illustrations.

The final phase of interphase is called G_2. It lasts from the end of the S phase to the start of mitosis. Just as in G_1, the G_2 phase is one of intense cellular activity in which the cell is producing the proteins and other molecules needed for mitosis. A cell that is ready to begin mitosis is called a *mother cell*.

Mitosis

The division of the nuclear material so that each new nucleus has a complete identical copy of the genetic information from the mother cell is called **mitosis** (mye TOH sis). Scientists have divided the process of mitosis into four phases:

✦ *Prophase.* The first phase of mitosis is **prophase**, during which there are changes in both the nucleus and the cytoplasm. In the cytoplasm, the centrosomes migrate to opposite poles of the nucleus. This migration is not haphazard; they are carried to their new positions by motor proteins on microtubules in the cytoplasm. Once in place, they begin to organize their own microtubules that form the **mitotic spindle**—special microtubules that will "direct" the movements of the chromosomes during mitosis.

Inside the nucleus, the nuclear chromosomes get short and thick as they coil up, becoming visible as individual chromosomes composed of two sister chromatids that are pinched together at the site of the centromere. The nuclear membrane disintegrates, and the nucleolus disappears.

✦ *Metaphase.* By the end of prophase, the chromosomes are positioned at the center of the cell. This signals the start of metaphase. **Metaphase** is the stage when the centromeres are aligned on the *equatorial plane*—an imaginary line bisecting the spindle. During metaphase the sister chromatids appear to repel each other, forming the familiar X shape of a chromosome.

✦ *Anaphase.* At the onset of **anaphase**, enzymes break down the proteins in the centromeres, allowing the two chromatids to separate. Once the chromatids separate, they are considered individual *daughter chromosomes*. As anaphase continues, the spindle fibers pull the chromosomes toward opposite poles of the cell. This separates, or segregates, the new chromosomes.

✦ *Telophase.* When the new chromosomes reach the ends of the spindle, **telophase** begins. A new nucleus begins to re-form around the chromosomes, forming two daughter nuclei. Inside the nuclei, the chromosomes begin to uncoil and the nucleoli reappear. Outside in the cytoplasm, the mitotic spindle disappears, leaving the centrosomes just outside the nuclear membranes. Once the nuclear membrane has re-formed around the nucleus, mitosis is complete.

Cytokinesis

Once the duplicated chromosomes have been separated into two sets, the cytoplasmic contents—organelles, proteins, membranes, and cytoskeleton—must be divided. This final step of the cell cycle is called cytokinesis (SYE toh kih NEE sis). Cytokinesis proceeds differently in human and animal cells than in plant cells. In plant cells, small membrane-bound vesicles formed by the Golgi apparatus begin to align and fuse, forming a membrane-bound **cell plate**. Later, cellulose and the other components of the cell wall are produced and transported to the cell plate to form a mature cell wall separating the two cells.

In human and animal cells, a *contractile ring* begins to divide the cell. As the contractile ring tightens, it causes the plasma membrane to pinch in, or *invaginate* (in VAJ uh NATE). As the ring gets smaller and smaller, the opposing membranes touch and fuse to form two *daughter cells*. Each of the daughter cells has received a complete copy of the mother cell's DNA as well as some cytoplasmic contents from the mother cell. The daughter cells now begin interphase, and the cell cycle starts again.

Variations in Mitosis

Some cells—for example, certain unicellular organisms—will go through mitosis but not cytokinesis. This action will result in a single cell with multiple nuclei.

The length of time necessary for mitosis differs for various types of cells. Some prokaryotic cells are able to complete a cell division in about ten minutes and are ready to divide again after an additional ten minutes of growth. Most plant cells require about thirty minutes for mitosis. Some animal cells require about three hours to pass from interphase to daughter cells.

Interphase may be extremely short for some cells. Embryos, for example, must form many cells rapidly. Almost as soon as the nuclei in the daughter cells have formed, the two nuclei may begin another prophase, even before cytokinesis is complete. Later these cells will grow and specialize.

Most cells in the body will divide to replace damaged or dead cells. Skin cells, for example, divide constantly to replace the skin rubbed and washed off. Bone cells can divide, although much more slowly than skin cells, and grow to repair broken bones. Some cells, once they have reached maturity, will never carry on mitosis again. Scientists believe that most nerve cells in the human brain, for example, may grow in size and replace worn-out or damaged cell parts, but they will not divide. When brain cells die, they usually are not replaced.

Uses of Mitosis

Mitosis with cytokinesis results in two cells that have the same DNA. In other words, they both contain identical genetic information. The most obvious uses of mitosis are growth, repair, and replacement of cells in multicellular organisms. For example, a human's 10 trillion cells are the result of repeated mitosis of the original one-celled embryo. Also, injured or worn-out parts of the body are repaired by the growth of new cells to replace the damaged ones. Often cells are needed to replace those that wear out naturally.

Many organisms are able to carry out reproduction by undergoing mitosis. Any form of reproduction that involves only mitotic cell divisions is called **asexual reproduction**. For example, if a unicellular organism goes through mitosis, the organism has reproduced asexually. Asexual reproduction produces offspring that are genetically identical to the parents; there is no genetic variety in the offspring.

Many colonial organisms are able to reproduce asexually by *fragmentation*. Breaking a simple colony in two and then permitting additional mitosis to

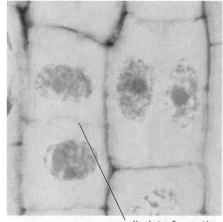

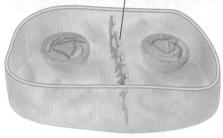
cell plate formation

5-7

In plant cell cytokinesis, a cell plate forms that divides the cell into two daughter cells.

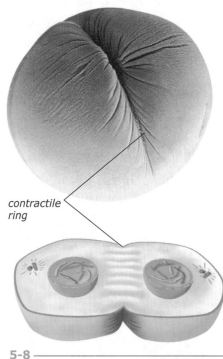
contractile ring

5-8

In animal and human cells, a contractile ring forms that eventually divides the cell.

cytokinesis: cyto- (cell) + -kinesis (to move)

5-9

Hydras (discussed in Chapter 15) can reproduce asexually by budding. The newly formed offspring will separate from the parent and live independently.

5-10

The new strawberry plant is genetically identical to its parent.

replace missing cells will result in asexual reproduction. Many algae, fungi, and some bacterial colonies reproduce asexually by fragmentation.

Some multicellular organisms can, by producing a large number of cells in a certain area, produce a new small organism on the side of the parent. This is called *budding*. Some plants, like the strawberry, carry out a similar process and produce small plants on the ends of special stems.

Many multicellular organisms are able to reproduce asexually by forming spores. A spore is a cell (sometimes cells) with a hard protective covering. Often a species is kept alive because the organism forms spores that live in an inactive state through a dry, cold, or otherwise unfavorable period. Spores of some molds can remain alive for over fifty years. Many organisms such as mosses and fungi reproduce and are spread predominantly by spores.

spore: (L. *spora*—seed)

Review Questions 5.2

1. List the phases of the cell cycle and describe each.
2. Describe the major changes that occur in each of the three phases of interphase.
3. In proper sequence, list the phases of mitosis and describe each.
4. What is the difference between mitosis and cytokinesis?
5. Compare and contrast the purpose of mitosis in unicellular and multicellular organisms.
6. List several types of asexual reproduction and describe each.

5.3

Objectives

- Describe the function of meiosis
- List and describe the phases of meiosis
- Summarize the processes of spermatogenesis and oogenesis

Key Terms

zygote
meiosis
tetrad
gamete
fertilization
sperm
ovum
spermatogenesis
oogenesis
sexual reproduction

5.3 Meiosis

Belgian scientist Edouard van Beneden discovered the chromosomal process of meiosis (mye OH sis) in 1883 while studying the reproduction of a particular type of worm. He noticed that reproductive cells (sperm and ova) of the worm had only half the number of chromosomes as were contained in cells of mature worms. Based on these findings, he proposed that some sort of process occurred that reduced the number of chromosomes in the sperm (the male reproductive cells) and eggs (the female reproductive cells). Once an egg and sperm were united at fertilization, the proper number of chromosomes was restored.

In a discussion of meiosis in human beings, simple mathematics indicates that if a father's sperm contains twenty-three pairs of chromosomes (46) and a mother's ovum contains twenty-three pairs of chromosomes (46),

5-11

Edouard van Beneden discovered the process of meiosis.

the number of chromosomes in their baby's cells will be forty-six pairs (92). Normal human cells, however, have only twenty-three pairs of chromosomes. As the parents' sperm and ovum were forming, the chromosome number of each was reduced to the haploid number. (Recall that a haploid cell is one that has only one chromosome from each pair.) In humans the haploid number is twenty-three ($n = 23$). When the two haploid cells (n) from the parents unite, a diploid cell called a zygote ($n + n = 2n$) is formed. After countless mitotic divisions, their baby enters the world.

The cellular division necessary to form a haploid cell is not a random dividing of chromosomes. A haploid cell must have one of every homologous chromosome pair. Under normal circumstances a gamete does not have two of one pair of chromosomes and none of another pair. If it did, the zygote it formed might not be diploid even though it would have the right number of chromosomes. It might have none, one, three, or four of a homologue rather than a pair. If a zygote does not have all its chromosomes in pairs, it may lack genes that are on the chromosomes for which it has only one homologue. Or it may have too many of the genes that are on the chromosomes for which it has three or four homologues. Either case can be harmful or even fatal.

To form a haploid cell, meiosis must occur. **Meiosis** is the reduction of a cell's chromosome number from diploid to haploid by two consecutive cell divisions. To keep the divisions straight, it is helpful to call the first division *meiosis I* and the second division *meiosis II*. Within each division, the cells still pass through prophase, metaphase, anaphase, and telophase. These phases will also be designated as I or II, depending on whether they occur in meiosis I or meiosis II.

Cells that are preparing to begin meiosis still must undergo the G_1, S, and G_2 phases of interphase—the cells grow and the DNA is duplicated. As the steps of meiosis are discussed, it is helpful to follow the progress of the duplicated chromosomes that are present in the initial cell.

Meiosis I: The First Division

Meiosis I proceeds similarly to mitosis, with just a few changes that are pointed out along the way.

✦ *Prophase I.* Just as in mitosis, the chromosomes tightly coil and the mitotic spindles begin to form. Next, the duplicated homologous chromosomes pair up, forming a structure called a **tetrad**. This is a major distinction from mitosis. In mitosis, only the sister chromatids form pairs joined at the centromere. However, in meiosis, not only do the sister chromatids pair up, but the homologous chromosomes pair up as well, forming the tetrad. During this process, the chromatids are very close together, and sometimes genetic information can be transferred from one chromatid to another. This process allows the exchange of genetic material between maternal and paternal chromosomes.

✦ *Metaphase I.* During metaphase I, the tetrads line up on the equatorial plane.

✦ *Anaphase I.* Rather than the sister chromatids separating as in mitosis, the homologous pairs separate, and the sister chromatids of each pair remain together. The homologues then travel to the ends of the spindle.

✦ *Telophase I and cytokinesis.* The telophase of the first division of meiosis is the same as in mitosis—the chromosomes arrive at the poles of the cell; however, the chromosomes usually do not uncoil. Then the cell undergoes cytokinesis, and the two new cells enter directly into the second division of meiosis.

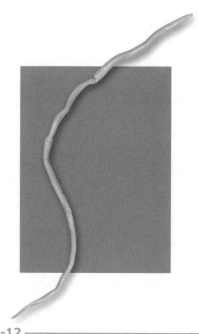

5-12

Ascaris, the roundworm studied by van Beneden

zygote: (Gk. *zugon*—yoke)

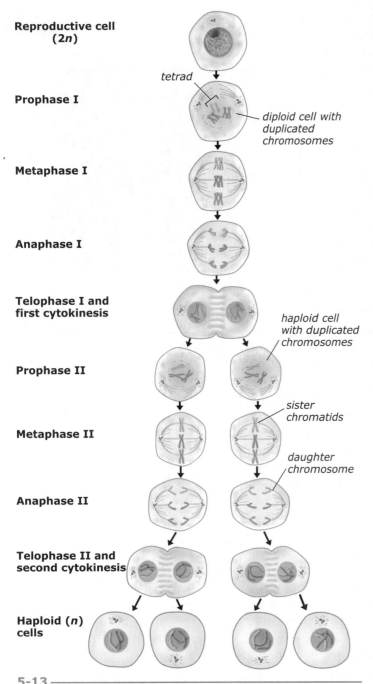

Reproductive cell
(2n)

Prophase I

tetrad

diploid cell with
duplicated
chromosomes

Metaphase I

Anaphase I

Telophase I and
first cytokinesis

haploid cell
with duplicated
chromosomes

Prophase II

sister
chromatids

Metaphase II

daughter
chromosome

Anaphase II

Telophase II and
second cytokinesis

Haploid (n)
cells

5-13

The final products of meiosis are haploid gametes.

Meiosis II:
The Second Division

At the start of meiosis II, each daughter cell has a haploid (*n*) chromosome number. It should be remembered, though, that the chromosomes are still duplicated. Meiosis II will produce four daughter cells, each with one unduplicated chromosome of each pair.

✦ *Prophase II.* During prophase II, a mitotic spindle re-forms and the spindle fibers begin to move the chromosomes toward the equatorial plane of the cell.

✦ *Metaphase II.* The chromosomes are aligned on the equatorial plane. Compare this stage to metaphase I. Note that now there is only one member of the homologous chromosome pair on the equatorial plane.

✦ *Anaphase II.* The sister chromatids separate, and the resulting daughter chromosomes move toward opposite poles.

✦ *Telophase II and cytokinesis.* The nuclei are re-formed and each of the four new cells has a haploid chromosome number.

Comparison of Mitosis and Meiosis

Mitotic cell division produces additional cells for growth, repair, and replacement of cells and tissues in multicellular organisms. The two cells that are produced have the same number of chromosomes as the parent cell. Meiosis, on the other hand, produces four haploid cells that have only one member of each homologous pair of chromosomes.

Whether cells undergo mitosis or meiosis, they all go through interphase. Interphase is the period when the chromosomes are replicated; therefore, in both mitosis and meiosis, the chromosomes are duplicated only once. The events that are unique to meiosis all occur during meiosis I. In prophase I, chromosomes pair to form tetrads. The tetrads are lined up on the equatorial plane in metaphase I, as opposed to the individual chromosomes in mitosis. During anaphase I and telophase I, the tetrads separate, and homologous pairs (two homologous pairs form a tetrad) of chromosomes move to the opposite poles of the cell; in mitosis the sister chromatids are pulled to the opposite poles.

The steps in meiosis II are almost the same as in mitosis. The homologous pair of chromosomes moves to the equatorial plane, where they separate, and then they are pulled to opposite poles of the cell. The resulting four daughter cells have the haploid number of chromosomes.

Gamete Formation

Gametes are those haploid cells that, when they unite, form a diploid cell called a zygote, the first cell of a new individual. The process of forming a zygote—the union of gametes—is called **fertilization**.

gamete: (Gk. *gamos*—marriage)

Some organisms produce gametes that are all alike, called isogametes (EYE so GAM eets). Usually isogametes move by cilia or flagella. Fertilization occurs when two isogametes unite. Many algae and fungi produce isogametes.

Humans, all animals, many plants, and a few other organisms produce heterogametes (HET uh ro GAM eets). A heterogamete is usually either a **sperm** formed by a male or an **ovum** (pl., ova) formed by a female. Sperm are usually smaller than ova and are able to move on their own. The ovum, if it moves, must be moved by structures around it. (An ovum is sometimes called an egg. However, the term *egg* properly applies to an ovum with accessory structures—like a shell—such as a bird egg.)

The forming of sperm is called **spermatogenesis** (spur MAT uh JEN ih sis). Spermatogenesis produces four functional gametes. Following the chromosomes in Figure 5-14 reveals that there are actually two pairs of identical sperm. In human sperm, as well as in sperm of other species, the cells formed by meiosis lose most of their cytoplasm and form a flagellum. The tightly coiled DNA in the nucleus of the sperm is then moved toward the ovum by the flagellum.

The forming of an ovum is called **oogenesis** (OH uh JEN ih sis). During the cytokinesis of meiosis I, one of the two cells receives most of the cytoplasm. The cell that receives the smaller amount of cytoplasm is called the *first polar body*. When the larger cell divides at the end of meiosis II, again only one of the two cells gets the majority of the cytoplasm and forms the ovum. The cell that gets very little cytoplasm is the *second polar body*. The first polar body often goes through the second meiotic division, forming two additional polar bodies. Oogenesis therefore results in one ovum and three polar bodies.

In humans and animals, the polar bodies soon disintegrate. The large quantity of cytoplasm in the ovum is necessary for the development of the zygote. The stored food and the cellular structures of the ovum's cytoplasm will be used during the period of rapid cell division at the beginning of the embryo's life.

Sexual Reproduction

The union of haploid gametes, resulting in a diploid zygote, is **sexual reproduction**. It results in offspring that are not genetically identical to either parent. In sexual reproduction one member of every homologous pair of chromosomes comes from one parent and the other homologue comes from the other parent. The offspring, therefore, have one chromosome of every pair from each parent.

Inevitably one parent does not have exactly the same genes for a particular characteristic as the other parent has. Since each organism receives one gene of each set from each parent, the offspring will not have exactly the same genetic makeup as either parent.

5-14

To form male gametes, the original cell undergoes spermatogenesis, producing four haploid sperm cells.

5-15

Oogenesis produces only one functional haploid ovum.

isogamete: iso- (same) + -gamete (marriage)

heterogamete: hetero- (other) + -gamete (marriage)

sperm: sperm-, sperma-, or spermato- (Gk. *sperma*—seed)

ovum: (L. *ovum*—egg)

spermatogenesis: spermato- (seed) + -genesis (Gk. *genesis*—beginning or birth)

oogenesis: oo- (Gk. *oion*—egg) + -genesis (beginning or birth)

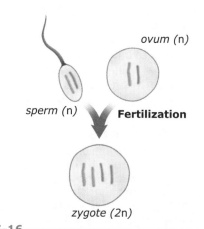

5-16

In sexual reproduction, two haploid cells unite to form a diploid zygote.

1. What happens to the chromosomes during (a) the first division of meiosis and (b) the second division of meiosis?

2. Describe the differences between mitosis and meiosis.

3. Name the two types of meiosis that are carried out by humans, and describe the difference between them.

4. Explain the difference between heterogametes and isogametes.

5. What type of gamete is used during sexual reproduction (haploid or diploid)? How do those gametes affect the genetic makeup of the offspring when compared to the parents?

5B—Basic Genetics

Gregor Mendel was born in 1822 on a farm in Heinzendorf, Austria. At the age of twenty-one, he entered the Augustinian order of the Roman Catholic Church. As a monk he studied science at the University of Vienna and became an excellent mathematician. Later, as a schoolteacher, he engaged in many scientific activities. He recorded sunspots, read Darwin (with whom he disagreed), and maintained fifty beehives in which he tried to mate various European, American, and Egyptian types of bees. In 1857, Mendel began a program of selective breeding of peas in a small plot in the vegetable garden of the St. Thomas Monastery near Brünn, Moravia.

Mendel's experience in breeding and raising plants and animals on his father's farm, along with his mathematical and experimental science background from the university, equipped him well to investigate heredity. After nine years of raising and classifying thousands of pea plants and recording many pages of notes, Mendel wrote a paper that presented a set of conclusions now called *Mendelian genetics*. In 1868, Mendel became the abbot of the monastery, and political problems forced him to give up most of his scientific work. In 1884, he died of a kidney disorder.

Gregor Mendel's paper on heredity in peas was published in 1865 but lay unnoticed in libraries for about thirty-five years. In 1900, after scientists had learned much about the cell, they rediscovered the paper and recognized its worth. Mendel became known as the Father of Genetics. What Mendel discovered about heredity is important to the understanding of modern genetic theories.

5.4 Mendelian Genetics

Mendel ordered thirty-four varieties of pea seeds, planted them, and observed their characteristics. From those varieties he chose seven sets of opposing characteristics. For example, he noted that pea plants are either about 6 ft or about 2 ft tall. The tall and short characteristics compose a set of opposing characteristics. Pod color is either green or yellow, another set of opposing characteristics. Peas are also either round or wrinkled.

The flower of the pea plant made it ideal for genetic experimentation. The petals are arranged so that the pollen (which contains the male gamete) naturally fertilizes the pistil (which contains the female gamete) of the same flower. This is called **self-pollination**. If Mendel wanted to **cross-pollinate** the plant with another, he had to open the petals and remove the pollen sacs before they matured. When the pistil was to be fertilized, he could supply pollen from another pea flower of his choice.

5-17 ⎯⎯⎯⎯⎯⎯⎯⎯⎯⎯⎯⎯
Gregor Mendel and the St. Thomas Monastery where he developed his theories of heredity

5.4

⬆ **Objectives**

- Describe three of Mendel's concepts
- Explain the difference between recessive and dominant traits
- Perform monohybrid crosses using Punnett squares
- Explain the use of a test cross

⬆ **Key Terms**

self-pollination
cross-pollinate
dominant trait
recessive trait
phenotype
genotype
locus
allele

homozygous
heterozygous
monohybrid cross
Punnett square
test cross
pedigree

	tall plants	axial flowers	green pods	inflated pods	yellow peas	round peas	colored seed coat
Dominant trait X	X	X	X	X	X	X	X
Recessive trait	short plants	terminal flowers	yellow pods	constricted pods	green peas	wrinkled peas	white seed coat
F₁ generation	all tall plants	all axial flowers	all green pods	all inflated pods	all yellow peas	all round peas	all colored seed coats
F₂ generation	787 tall: 277 short (2.84:1)	651 axial: 207 terminal (3.14:1)	428 green: 152 yellow (2.82:1)	882 inflated: 299 constricted (2.95:1)	6022 yellow: 2001 green (3.01:1)	5474 round: 1850 wrinkled (2.96:1)	705 colored: 224 white (3.15:1)

Mendel began his experiments with peas that had been self-pollinating and breeding true: in other words, the tall plants always produced tall plants, the short plants always produced short plants, and so forth. He called these the parent plants and used the symbol P_1 to represent them. One of Mendel's experiments involved cross-pollinating a tall pea plant with a short pea plant. He called the offspring of this cross the *first filial generation* (F_1). All the F_1 plants were tall. He next allowed the F_1 plants to self-pollinate and produce the *second filial generation* (F_2). Of the 1064 plants in Mendel's F_2 generation, 787 of them were tall and 277 of them were short. So here was Mendel's problem. True-breeding tall plants crossed with true-breeding short plants yielded all tall offspring. Yet, when these tall offspring reproduced, both tall and short offspring were produced with predictable and repeatable results. What mechanism could account for these results? (Keep in mind that chromosomes and meiosis were unknown to Mendel.)

Mendel's Concepts

To explain the outcome of this experiment, as well as the similar results he obtained when crossing peas with other sets of opposing characteristics, Mendel proposed several concepts. These concepts have been validated as scientists have observed similar results in other organisms and the cellular structures responsible for heredity. The following concepts are illustrated in the crosses described in Figure 5-19.

The Concept of Unit Characteristics

Mendel stated that an organism's characteristics are caused by units that he called *factors* (now called genes) that occur in pairs. In pea plants the tall parent has two genes for being tall, which are represented by *TT*, and the short parent has two genes for being short, *tt*. It should be remembered that (1) genes (composed of DNA) are responsible for inherited characteristics, (2) genes are located on chromosomes, and (3) most organisms have homologous pairs of chromosomes. Thus most organisms have pairs of genes in their cells.

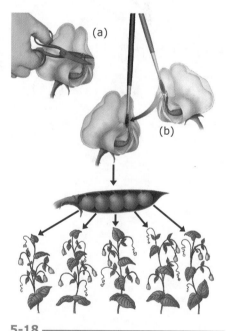

5-18 —
Mendel controlled the pea plant breeding by (a) removing the pollen-producing structures from one plant and then (b) transferring pollen from a plant with different characteristics to the original plant.

filial: (L. *filius*—son)

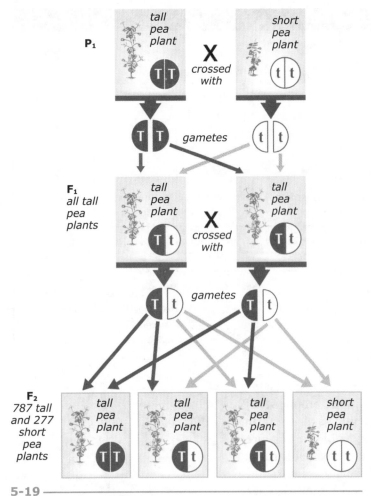

P₁

tall pea plant **TT** X crossed with short pea plant **tt**

T T gametes **t t**

F₁ all tall pea plants

tall pea plant **Tt** X crossed with tall pea plant **Tt**

T t gametes **T t**

F₂ 787 tall and 277 short pea plants

tall pea plant **TT**

tall pea plant **Tt**

tall pea plant **Tt**

short pea plant **tt**

5-19

A summary of Mendel's crosses of tall and short pea plants

The letter for the genotype is taken from the dominant trait. For example, *T* is used for the tall/short characteristic because *tall* is the dominant trait.

phenotype: pheno- (Gk. *phaino*—showing) + -type (*tupos*—impression)

genotype: geno- (beginning) + -type (impression)

The Concept of Dominant and Recessive

Since the short plant of the P_1 could give only a short gene (*t*), and the tall plant only a tall gene (*T*), the F_1 generation was *Tt*. But rather than being medium-sized as Mendel expected, all the F_1 generation were tall. However, the short plants reappeared in the F_2 generation in a ratio of three tall to one short plant. Something caused one factor to hide the other factor in the first generation, but somehow it reappeared in the second generation. Mendel called a trait that is expressed when factors for two opposing traits are present the **dominant trait** (caused by a *dominant gene*). The trait that is masked (hidden) when two genes for opposing traits are present is the **recessive trait** (caused by a *recessive gene*).

The Concept of Segregation

Mendel reasoned that when a cell forms gametes, the genes segregate (separate) so that there is only one gene for each characteristic in each gamete. What is known of the behavior of chromosomes in meiosis confirms that Mendel's description of gamete formation was accurate.

These concepts are also illustrated by the self-pollination that produced Mendel's F_2 generation of tall and short plants. Since all the F_1 generation were *Tt*, half of their pollen should contain a *T* gene and the other half a *t* gene. Also, half of the ova they produce should contain a *T* gene, and the other half a *t* gene. This illustrates the concept of segregation.

The recessive gene (*t*) was not destroyed or altered when the dominant gene masked it in the F_1 generation; therefore, in the F_2 generation there are some short pea plants. Whenever both parent plants gave the recessive gene, the offspring expressed the recessive trait. This also illustrates the concept of unit characteristics.

It is important to notice the possible unions of the various gametes of the F_1 as illustrated in Figure 5-19. Three-fourths of the possible gamete combinations in the F_2 have at least one dominant gene; only one-fourth of them have two recessive genes. This, combined with the fact that Mendel's F_2 generation results were 787 tall to 277 short (which is a ratio of about 3:1), lends support to the concept of unit characteristics.

Genetic Terminology

An understanding of the following terms is necessary for a discussion of genetic principles.

✦ **Phenotype** (FEE nuh TYPE)—the physical expression of an organism's genes—what an organism is like (tall, green, constricted). However, not all genes result in a visible trait. For example, every person inherits genes for digestive enzymes that are not seen but must work for him to be healthy. Expressing these genes is part of his phenotype.

✦ **Genotype** (JEN uh TYPE)—the specific genes that an organism contains—its genetic makeup. The genotype is often expressed by letters such as *TT*, *Tt*, *tt*, and so forth.

* **Locus**—the specific site on a chromosome where a particular gene is located.

* **Allele** (uh LEEL)—an alternate form of a gene that occupies the same locus on homologous chromosomes. They are often expressed by letters—for example, *T* or *t*.

* **Homozygous** (HOH moh ZYE gus)—the condition where both alleles in an organism are the same—for example, *TT* or *tt*.

* **Heterozygous** (HET ur oh ZYE gus)—the condition in which both alleles in an organism are not the same—for example, *Tt*.

* **Monohybrid** (MAHN oh HYE brid) **cross**—a cross between individuals that deals with only one set of alleles, that is, with one set of opposing characteristics. The cross between short and tall pea plants is a monohybrid cross.

In humans, free, or unattached, earlobes are dominant over attached earlobes. Suppose a man with the dominant phenotype for earlobes (free earlobes) knows that he is homozygous for the trait. His genotype is *FF*. His wife has the recessive phenotype. What kind of earlobes does she have? The only genotype a person with the recessive phenotype can have is *ff*. (If she had a dominant allele, her phenotype would be different.) All the man's sperm contain the dominant allele (*F*), and all the woman's ova, the recessive allele (*f*). All their children, therefore, would have the heterozygous genotype (*Ff*). What would be the phenotype of these children?

Punnett Squares
Sometimes the arrows in figures like Figure 5-19 can be very confusing, especially for a problem with more than one trait. To make things easier, geneticists and high-school students often use a diagram called a **Punnett square** to depict genetic crosses and to determine the probability of the offspring's particular genotype or phenotype. The possible female gametes are listed across the top of the Punnett square; the possible male gametes are placed down the left side. The gametes are then combined in each of the boxes within the square to give the possible gamete combinations (the possible genotypes) of the offspring. The resulting Punnett square shows the *probability* of an offspring having each of the possible genotypes; it does not necessarily guarantee that an offspring will have a certain genotype.

Figure 5-21a depicts the Punnett square for the man and woman with the free and attached earlobes described previously. This would be a homozygous dominant × homozygous recessive cross (*FF* × *ff*). All their offspring, no matter how many they have, will have the father's phenotype, even though none of them have his genotype.

Now assume that one of the sons from this family married a woman with attached earlobes. The man's genotype must be *Ff*, and the only genotype the woman could have is *ff*. This would be an example of a heterozygous × homozygous recessive cross. Figure 5-21b shows the Punnett square for this couple.

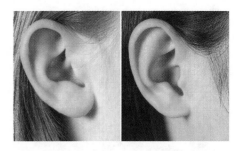

5-20
Free (left) and attached (right) earlobes

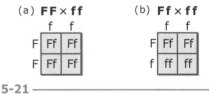

(a) **FF × ff**

	f	f
F	Ff	Ff
F	Ff	Ff

(b) **Ff × ff**

	f	f
F	Ff	Ff
f	ff	ff

5-21
Punnett squares showing (a) a homozygous × homozygous cross and (b) a heterozygous × homozygous cross

locus: (L. *locus*—place)
allele: (Gk. *allos*—other)
homozygous: homo- (same) + -zygous (yoke)
heterozygous: hetero- (other) + -zygous (yoke)
monohybrid: mono- (single) + -hybrid (L. *hybrida*—mongrel; hence, offspring of different parents)

5-22
The arrows show how to properly place the gametes in a Punnett square.

f   f

F Ff   Ff

F Ff   Ff

5-23

Punnett squares are named after Reginald Punnett, one of the pioneer genetic researchers in the early 1900s.

(a) **P P × p p**

	P	P
p	Pp	Pp
p	Pp	Pp

(b) **P p × P p**

	P	p
P	PP	Pp
p	Pp	pp

Genotypic ratio: 0:4:0 Genotypic ratio: 1:2:1
Phenotypic ratio: 4:0 Phenotypic ratio: 3:1

5-24

Punnett squares for corn kernel color

5-25

This corn cob has the phenotypic ratio of 3 purple: 1 yellow.

Test Cross #1:
P P × p p

	p	p
P	Pp	Pp
P	Pp	Pp

If you received these F_1 results, you would know that your unknown genotype was *PP*.

Test Cross #2:
P p × p p

	p	p
P	Pp	Pp
p	pp	pp

Two purple and two yellow phenotypes in the F_1 would indicate a *Pp* genotype for your unknown.

5-26

Test crosses for the unknown genotype—is it *PP* or *Pp*?

Now suppose that two individuals who are heterozygous for earlobe attachment (heterozygous × heterozygous) marry and have a child. What would be the possible genotypes and phenotypes of their child? Make a Punnett square for this cross. (Check your square with the answer on page 126.)

Genetic Ratios in Monohybrid Crosses

The likelihood of offspring having or not having a particular allele or trait is usually expressed as a ratio. Corn kernel color clearly illustrates genetic ratios since on each ear of corn there are enough individual kernels to present an accurate ratio. In corn the allele for purple kernels (*P*) is dominant over its allele for yellow kernels (*p*). If pollen from a homozygous yellow plant is applied to the developing ear of a homozygous purple plant, the ear of corn will develop all purple kernels. The Punnett square in Figure 5-24a indicates four boxes with the heterozygous genotype (*Pp*) and none with any other genotype. The genotypic ratio for the F_1 generation is 0:4:0, or zero homozygous dominant (*PP*) to four heterozygous (*Pp*) to zero homozygous recessive (*pp*). In this case the phenotypic ratio is 4:0, or four purple to zero yellow. (Often such ratios are reduced to 1:0, but both are accurate.) In phenotypic ratios the number of dominant individuals is listed first.

The F_2 generation of this cross, however, will produce different ratios. When two heterozygous corn plants are cross-pollinated, the F_2 phenotypic ratio will be 3:1; that is, three purple kernels will occur for every one yellow kernel. The genotypic ratio, however, will be 1:2:1—that is, one homozygous dominant, two heterozygous, and one homozygous recessive. See Figure 5-24b. (Always write genotypic ratios in the following order: the number of homozygous dominant, followed by heterozygous, and then homozygous recessive.)

Try to determine the phenotypic and genotypic ratios when corn heterozygous for purple kernels is mated with one homozygous for yellow kernels. The results can be seen in the second test cross in Figure 5-26.

Test Crosses and Pedigrees

How can it be determined whether the genotype of a purple kernel of corn is *PP* or *Pp*? For many organisms, one of the best methods is to run a test cross to determine the genotype. A **test cross** involves mating an organism that has the dominant phenotype but an unknown genotype with another plant that has the homozygous recessive phenotype.

The genotype of a purple kernel of corn can be either *PP* or *Pp*. If a test cross is performed, the unknown genotype can be determined by examining the ratios of the phenotypes in the F_1 generation. Following the test crosses in the Punnett squares in Figure 5-26 will be helpful in understanding the following paragraph.

If the kernel was a homozygous purple kernel plant (*PP*) and was crossed with a yellow kernel plant (*pp*, homozygous recessive), what would be the color of the offspring? All would be purple. However, if the kernel was a heterozygous purple kernel plant (*Pp*) mated with a yellow kernel plant, how would the phenotypic ratio of the offspring change? Half would be purple and half would be yellow. Thus, the genotype of the dominant parent can be determined by looking at the offspring of a test cross (if there are enough offspring).

A **pedigree** is a chart that geneticists use to trace the presence or absence of a trait in a number of generations. Pedigrees often use symbols to indicate sex, marriage (or matings of animals and plants), offspring, and other related

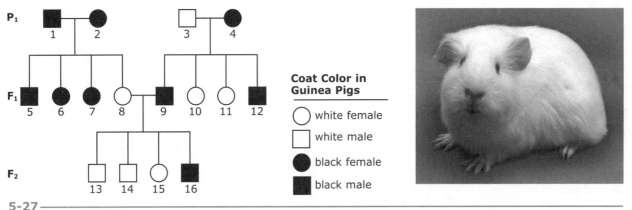

5-27

This pedigree illustrates the inheritance of coat color through three generations of guinea pigs. What is the genotype of each individual?

Coat Color in Guinea Pigs

○ white female
□ white male
● black female
■ black male

factors. Symbols and the use of colors or shading indicate different phenotypes. For example, in the guinea pig pedigree, the males are designated by a square and the females by a circle. Geneticists create pedigrees to predict the possibilities of certain traits in offspring.

> **Review Questions 5.4**

1. Who is known as the Father of Genetics? What did he do to earn this title?
2. Describe (a) the concept of unit characteristics, (b) the concept of dominance, and (c) the concept of segregation.
3. Give the differences between the genotype and phenotype of an organism.
4. The gene for yellow peas is dominant over the gene for green peas. What is the difference between homozygous yellow peas and heterozygous yellow peas?
5. What is the difference between a gene and an allele?

6. Two organisms heterozygous for a single trait are crossed. What is the expected genotypic ratio of the offspring? Explain using a Punnett square.
7. What is the phenotypic ratio of the cross described in Question 6? Explain.
8. What can a biologist learn by looking at a Punnett square for a cross?
9. Describe a test cross. What would a test cross be used to determine?
10. What information can a geneticist learn by looking at a pedigree for a cross?

5.5 Variations of Mendelian Genetics

If the Mendelian concepts studied thus far were the only factors involved in heredity, the prediction of probable phenotypes would be relatively easy. Although genetics is based largely on the concepts illustrated in a monohybrid cross with simple dominant and recessive characteristics, there are many variations. These variations, although simple, make predictions of an organism's phenotype quite difficult. Many human traits, including sex, eye color, height, skin color, hair color, blood type, and intelligence, as well as a number of disorders, are determined by these genetic mechanisms.

Incomplete Dominance

Not all genetic traits are exhibited as purely dominant or recessive. Many alleles express incomplete dominance. **Incomplete dominance** occurs when two alleles are both expressed, resulting in a phenotype that is intermediate, or a blending, of the two traits. Flower color in snapdragons and other common garden flowers demonstrates this condition. When homozygous red and homozygous white snapdragons are crossed, all of the heterozygous offspring are pink.

5.5

> **Objectives**

• Explain the effects of incomplete dominance and codominance on offspring
• Use a Punnett square to predict dihybrid cross results
• Describe the difference between multiple alleles and polygenic inheritance

> **Key Terms**

incomplete dominance
codominance
multiple alleles
dihybrid cross
independent assortment
polygenic inheritance

5-28

Pink snapdragons result because neither the red nor the white color allele is completely dominant.

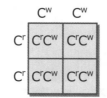

The F_1 generation

5-29

Crossing of red (C^rC^r) and white (C^wC^w) snapdragons produces all pink snapdragons.

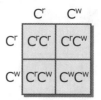

The F_2 generation

5-30

Crossing pink snapdragons produces a ratio of 1 red to 2 pink to 1 white.

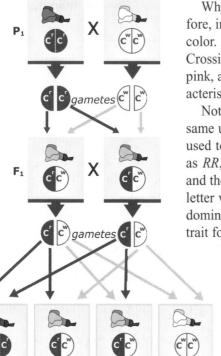

5-31

Incomplete dominance in snapdragon color

Why? In snapdragons neither red nor white is completely dominant; therefore, in a heterozygous flower both alleles are expressed, resulting in the pink color. Although the F_1 offspring are pink, the alleles have not been altered. Crossing two pink F_1 snapdragons yields an F_2 generation of 25% white, 50% pink, and 25% red snapdragons—the alleles maintained their individual characteristics.

Note that in the Punnett square showing incompletely dominant alleles, the same uppercase letter C is used for both alleles and a superscript r and w are used to denote the different colors. Using upper- and lowercase letters (such as RR, rr, and Rr) might make it appear that one of the alleles is dominant and the other recessive. Therefore, it has become conventional to use a single letter with different superscripts to demonstrate the crossing of incompletely dominant alleles. In the example of the snapdragons, C^rC^r will represent the trait for red, C^wC^w for white, and C^rC^w for pink.

Many human traits appear to be the result of incomplete dominance. People who suffer from brachydactyly (BRAK ih DAK tuh lee) lack a bone in each finger and toe and therefore have abnormally short fingers and toes. A person with brachydactyly is heterozygous (that is, he has one abnormal allele and one normal allele); people with normal fingers and toes are homozygous for the normal allele. Individuals who are homozygous for the brachydactyly gene are severely crippled, with a complete lack of fingers and toes.

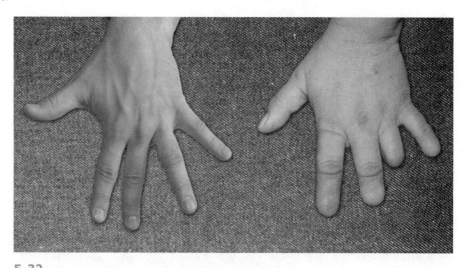

5-32

Normal hand and fingers (left) and hand and fingers of a person with brachydactyly (right)

The Key to Genetics—Fruit Flies

In the early twentieth century, the tiny fruit fly captured the attention of the scientific world—but not because of major crops it destroyed or diseases it carried. What caused some of the most brilliant scientists at major universities to begin spending countless hours and millions of dollars studying the fruit fly was the ease of studying its genetics.

For several reasons scientists found *Drosophila melanogaster* (droh SAHF uh luh • MEL uh noh GAS tur), the tiny fly often found around ripe fruit, ideal for genetic investigations.

✦ *Abundant offspring.* Just one mating will produce hundreds of offspring, enough to show reliable genetic ratios.

✦ *Short life span.* Fruit flies reach maturity in about sixteen days. In many organisms a scientist must wait years for a characteristic to be expressed.

✦ *Ease of keeping and handling.* Their size and temperament permit hundreds of fruit flies to be kept in a small bottle. A small amount of mashed fruit will feed them for their entire lives.

✦ *Noticeable differences.* Unlike the sexes of many small organisms, fruit fly sexes have different appearances. They also have many contrasting traits that can be observed easily.

✦ *Four pairs of chromosomes.* A small number of chromosomes makes the manufacture of karyotypes and the study of chromosomes and their behavior relatively easy.

Normal fruit fly

Codominance

Codominance occurs when two alleles for a gene are both expressed in a heterozygous offspring. This may sound the same as incomplete dominance, but there is a distinct difference. In incomplete dominance, there is a *blending* of the characteristics in the heterozygous offspring—red + white = pink. In codominance, both alleles are expressed with no blending. For example, hair color in many mammals is a codominant characteristic. If a horse that is homozygous for red hair is crossed with one homozygous for white, a color pattern termed *roan*—white hairs intermingled with red hairs—is produced.

Multiple Alleles

The traits discussed thus far have only two contrasting alleles. In peas there is the allele for the tall trait and the allele for the short trait. In snapdragon color there is the allele for red color and the allele for

5-33

Another example of codominance is the blue Andalusian chicken, which is a cross between a homozygous white-feathered chicken and a homozygous black-feathered chicken. Although at a distance the heterozygous offspring appears blue, a close examination reveals that both black and white feathers are clearly visible.

5-34

A roan horse exhibits codominance.

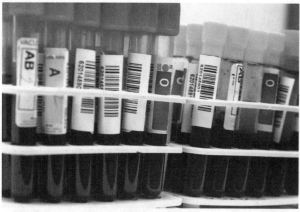

5-35

After blood is donated, it is tested to determine its type using the ABO system.

	I^A	i
I^B	I^AI^B	I^Bi
i	I^Ai	ii

This cross results in a genotypic and phenotypic ratio of 1:1:1:1.

What are the blood types produced by this cross?

5-36

Cross between heterozygous type A (I^Ai) and heterozygous type B (I^Bi)

dihybrid: di- (two) + -hybrid (offspring of different parents)

white color. It should be remembered that the place on a chromosome where a particular gene is located is the gene's *locus* (pl., loci). The alleles for the traits of a monohybrid cross appear at the same loci on the homologous chromosomes. Since chromosomes in peas and humans and all diploid organisms occur in pairs, there are two of most of their alleles. One of these two will be at each locus.

Sometimes, however, there may be **multiple alleles**—that is, one of several alleles—at a given locus. A diploid organism, of course, can have only two of the possible alleles—one on each of the homologues.

A simple multiple allele condition in humans is the ABO blood types. There is one allele for the factor causing blood type A and another allele for the factor causing blood type B, both of which are dominant. Additionally, there is a recessive allele for no factor. In diagramming this multiple allele cross, it is traditional to use an uppercase letter I to represent the chromosome and a superscript to represent the dominant alleles, I^A and I^B. The recessive allele, which is neither A nor B but recessive to both, is represented by the lowercase letter i.

A person with blood type A could be homozygous (I^AI^A) or heterozygous (I^Ai); someone with type B could be I^BI^B or I^Bi. A person heterozygous for both dominant alleles (I^AI^B) has blood type AB, and an individual homozygous for the recessive trait (ii) has type O.

The heterozygous A and B cross makes it apparent that there are two inheritance patterns occurring in this cross. First, there is a pure dominant/recessive inheritance pattern when either the dominant I^A or I^B allele pairs with the recessive allele, i. Second, in the genotype I^AI^B, both dominant alleles are being expressed—codominance.

Dihybrid Crosses

A **dihybrid** cross deals with two pairs of contrasting traits at the same time. Figure 5-37 shows a cross between a homozygous green inflated-pod pea plant (*GGII*) and a homozygous yellow constricted-pod pea plant (*ggii*).

Since there are two characteristics to deal with, there are four alleles to consider, two for each characteristic. When gametes are formed, however, the alleles segregate (as stated in Mendel's concept of segregation), and only one allele for each characteristic will be in each gamete. Since chromosome pairs separate,

F₂

Phenotypic ratio:
9 green inflated (*G_I_*):
3 green constricted (*G_ii*):
3 yellow inflated (*ggI_*):
1 yellow constricted (*ggii*)

5-37

A dihybrid cross of yellow and green inflated and constricted pea pods

there will normally not be a gamete containing two alleles of one gene and none of the other. The forming of the F_1 by the uniting of the gametes results in individuals that have green inflated pods. They are, however, heterozygous for both traits.

The formation of gametes in the F_1 results in four different possibilities: *GI*, *Gi*, *gI*, and *gi*. The procedure for finding the possible gamete combinations for the two traits is much the same as for finding the possible combinations in a monohybrid cross: list the possible gametes from the male on the side of a Punnett square and the possible gametes from the female on the top. Since there are four possible gametes from each parent, the Punnett square for this cross will have sixteen blocks. The phenotypic ratio of individuals within the Punnett square is nine green inflated, three green constricted, three yellow inflated, and one yellow constricted.

Assume that a plant breeder is interested in obtaining a homozygous strain of green constricted peas. In the F_2 Punnett square there are three boxes that contain individuals of this phenotype; however, there are two different genotypes—*GGii* and *Ggii* (one homozygous and one heterozygous for green). Can the plant breeder tell the difference between them just by their physical appearance? No, the two genotypes produce peas with identical phenotypes. How could you help the plant breeder with his problem? One way he can find out is to test cross the peas in question. Draw Punnett squares for the two possible test crosses to see what you would expect from each of the two genotypes. The answer is on page 127.

Mendel's Concept of Independent Assortment

Mendel himself used dihybrid crosses, and the results of those crosses led him to formulate his concept of **independent assortment**: the segregation (separation) of one set of alleles during gamete formation is not affected by the presence or segregation of other sets of alleles. When Mendel was observing two sets of characteristics in the same cross, not only did he note that the predicted ratios occurred, but he also saw that the various characteristics were randomly mixed with other characteristics. For example, the segregating of the green-yellow alleles did not affect the segregating of the inflated-constricted alleles. Green color will not always accompany just constricted pods or just inflated pods. The traits mix and match freely.

Polygenic Inheritance

A single gene does not always cause a single characteristic. Sometimes two or more genes working together result in a single trait, a process called **polygenic** inheritance. One simple example of this is the variation in shape of the combs of some types of chickens. The comb is the fleshy crest found on the heads of some birds, especially poultry.

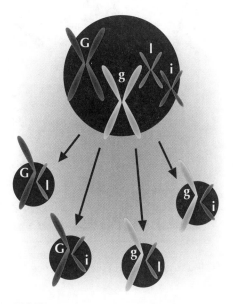

5-38

The *G/g* alleles are on one chromosome and the *I/i* alleles are on another; thus, they segregate independently of each other.

polygenic: poly- (many) + -genic (beginning)

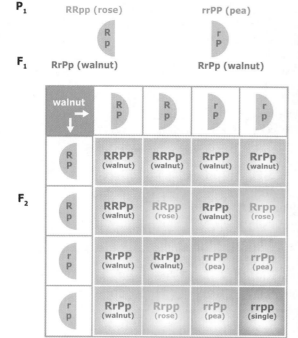

P₁   RRpp (rose)   rrPP (pea)

F₁   RrPp (walnut)   RrPp (walnut)

F₂

walnut →	RP	Rp	rP	rp
RP	RRPP (walnut)	RRPp (walnut)	RrPP (walnut)	RrPp (walnut)
Rp	RRPp (walnut)	RRpp (rose)	RrPp (walnut)	Rrpp (rose)
rP	RrPP (walnut)	RrPp (walnut)	rrPP (pea)	rrPp (pea)
rp	RrPp (walnut)	Rrpp (rose)	rrPp (pea)	rrpp (single)

5-39

The combs of chickens exhibit polygenic inheritance.

rose comb
(*RRpp, Rrpp*)

pea comb
(*rrPP, rrPp*)

walnut comb
(*RRPP, RRPp,*
RrPP, RrPp)

single comb
(*rrpp*)

In this case, two genes interact to form a different type of comb. One gene, *R*, produces a rose comb; its recessive form, *r*, produces a single comb. A third type of comb—pea—is formed by a different gene, *P*. The recessive form of this gene, *p*, produces a single comb. As the two genes are crossed, they interact with each other to produce different comb types.

Having a dominant gene at one locus and two recessive genes at the other locus (*RRpp* or *Rrpp*) causes a rose comb. A pea comb is caused by exactly the opposite condition (*rrPP* or *rrPp*). A walnut comb—this is a new type caused by the interaction of the genes—results from having at least one dominant allele in each pair (*RRPP, RRPp, RrPP,* or *RrPp*), and a single comb is the result of having homozygous recessive alleles at both loci (*rrpp*).

Most human traits have some degree of polygenic inheritance. For example, five or six gene pairs at different loci determine human skin color. The Punnett square for such a large number of gene possibilities requires over one hundred different blocks and has dozens of different genotypes and phenotypes. Some of the gene pairs seem to be incompletely dominant, and some work as *inhibitors* (genes that prevent the expression of other genes). It is also possible that two or more of these genes are on the same chromosome and thus further increase the complexity of this type of cross.

Humans produce very few offspring, and relatively poor records have been kept of the exact shade of skin color for more than a few generations. Knowing a person's genotype and predicting his child's phenotype for skin color are almost impossible.

Hair color and eye color are similar but probably involve fewer gene pairs. Polygenic inheritance determines, in part, a person's height, body build, intelligence, and many other human characteristics.

5-40

The wide variation in human skin color is a result of polygenic inheritance.

Review Questions 5.5

1. Compare simple dominance with incomplete dominance. List several examples of incomplete dominance.
2. Compare incomplete dominance with codominance, using examples as support.
3. Describe the condition of multiple alleles. What human characteristic is controlled by multiple alleles?
4. Describe the concept of independent assortment.
5. Describe polygenic inheritance and distinguish between it and multiple alleles.

⊙ 6. Distinguish between a dihybrid cross and a cross involving multiple alleles.

⊙ 7. In what way is incomplete dominance an exception to the concept of dominance? Could multiple alleles be an exception to the concept of dominance? Why or why not?

⊙ 8. Explain how polygenic inheritance may have confused early researchers who expected traits to be inherited by simple dominant and recessive monohybrid mechanisms.

5.6 Sex Determination and Sex-Linked Traits

5.6

King Henry VIII of England is remembered for his disappointment at having only one male heir. When a wife did not bear him a son, he disposed of her and married again. But Henry VIII should probably have been more upset with himself; it is the man's genes that determine the child's sex.

Sex Chromosomes and Autosomes

Karyotypes reveal that normal people have twenty-two pairs of autosomes and one pair of sex chromosomes. The **sex chromosomes** are those chromosomes that determine the sex of an individual—in humans they are designated X and Y. The **autosomes** are the other non-sex-determining chromosomes and are traditionally numbered. In humans the Y chromosome is considerably smaller than the X chromosome. A person with one X and one Y chromosome is male; a person with two X chromosomes is female.

Crosses of X and Y chromosomes can be determined on a Punnett square. Note that only the father can give a Y chromosome to his offspring. Since it takes a Y chromosome to form a male offspring, the father thus determines the sex of the child.

5-41

Henry VIII of England and his only male heir, Edward VI

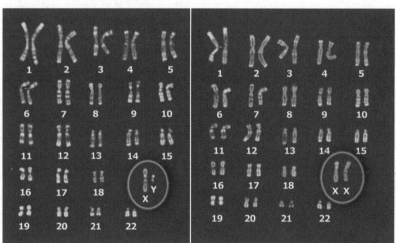

X X × X Y

female

male	X	X
X	XX	XX
Y	XY	XY

5-42

5-43

Male karyotype (left) and female karyotype (right). The circled pairs of chromosomes are in the 23rd position.

Sex-Linked Traits

The alleles found on autosomes are in pairs. Since autosomes are in pairs, the allele found at a particular locus on one will have a companion allele at the same locus on the homologue. At that locus there may be the exact same gene (homozygous condition) or a variation of that gene (heterozygous condition). But on sex chromosomes conditions may vary. In humans, for example, the X chromosome is much larger than the Y chromosome and thus can contain many more genes. While the X and Y pair up during meiosis, these chromosomes are not homologous.

Traits that have their genes on the X or Y chromosome are called **sex-linked traits** because they are linked to the sex of the individual. The Y chromosome, found only in males, contains genes for some male characteristics. The X chromosome, however, is found in both males and females and contains genes for other traits. These are called *X-linked traits*. *X*-linked traits can be studied by examining the trait of red-green colorblindness.

About 5% of the white males of northern European ancestry have a reduced amount of a chemical in the retina of their eyes that causes them to be red-green colorblind. To them, most greens appear tan, olive greens appear brown,

Objectives

- Differentiate between autosomes and sex chromosomes
- Explain how sex-linked traits affect inheritance of traits
- Use a Punnett square to demonstrate sex-linked crosses

Key Terms

sex chromosome
autosome
sex-linked trait
hemophilia
carrier

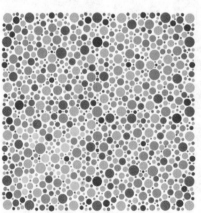

5-44

Physicians often test for colorblindness during a physical examination by having a patient look at several colored images and asking him to identify what is hidden within the image. You should see a pink X and pink triangle in a field of gray dots.

Used with permission of Richmond Products. This image is a reproduction and is not accurate as an eye test.

Hemophilia (HEE muh FIL ee uh) is sometimes called the "bleeder's disease." People who have this genetic disorder lack an enzyme in the blood that is essential for normal blood clotting. Even a small cut can result in severe bleeding or death. Bumping into something, which for most people results in simple bruises, can cause severe internal bleeding, swelling, and possibly death for hemophiliacs. In times past, hemophiliacs usually died very young. Today, medication can supply the missing blood chemical.

Hemophilia has been found to be an *X*-linked recessive trait. Although there are very old records of people who are believed to have been hemophiliacs, the best known historical record of the disease involves the royal house of Great Britain. Hemophilia has thus been called "the royal disease." Queen Victoria (reign 1837–1901) was a carrier of this sex-linked trait.

The pedigree shows only some of Queen Victoria's offspring. Her son Leopold was a hemophiliac. At a time when most hemophiliacs died as infants or young children, he was exceptionally protected and lived to maturity. Although his brothers joined and commanded British military regiments, Leopold was not even permitted to wear a uniform. A record of Leopold's life is one of "falling ill," as his mother put it, and spending long days in bed recovering.

After thwarting most of his attempts to be married, the Queen relented and arranged for Leopold's marriage to the German princess Helena of Waldeck. After a year Alice was born. A Punnett square for Leopold and Helena shows that their daughters could not be hemophiliacs but would be carriers. This fact is also demonstrated by Alice's offspring.

The second child of Leopold and Helena was a normal son, Charles Edward. Since Leopold must give the *Y* chromosome to any son, he could not give the gene for hemophilia to any of them. Leopold, however, did not know this and also did not know whether his second child was a hemo-

$$X^hY \times X^HX^H$$
(Leopold)(Helena)

	X^H	X^H
X^h	X^HX^h	X^HX^h
Y	X^HY	X^HY

philiac or not. At age thirty-one, before his son was born, Leopold injured his knee and died of internal bleeding within a few days.

Queen Victoria was quoted as saying that hemophilia "was not in our family," and she was correct: no one on her side of the family expressed the recessive characteristic. She supposed that her husband, Albert, introduced the disease into their lineage, but she was wrong. It is believed that Queen Victoria, or possibly her parents, actually introduced the gene by mutation.

In the pedigree, Leopold is the only hemophiliac to have children. The others died before they married. Today, however, hemophiliacs can receive treatment and live nearly normal lives. Also in the pedigree there are no hemophiliac females. Female hemophiliacs would only be possible if a hemophiliac male married a woman who had the hemophiliac gene (a carrier). Today there are hemophiliac females, but in times past, because the hemophiliac men died young, there were none.

Hemophilia is not really a royal disease because it does exist in other families. Also, the present British royal family is free of the gene for hemophilia because they are the descendents of Edward VII, a healthy male. Queen Victoria's carrier daughters and hemophiliac son, however, exported the gene to various royal houses of Europe.

The British royal family (about 1860) with carriers indicated. Left to right: Alfred, Prince Albert, Helena, Arthur (in kilt), Alice, Beatrice (infant), Queen Victoria (seated), Princess Victoria, Louise, Leopold (facing camera), and Edward.

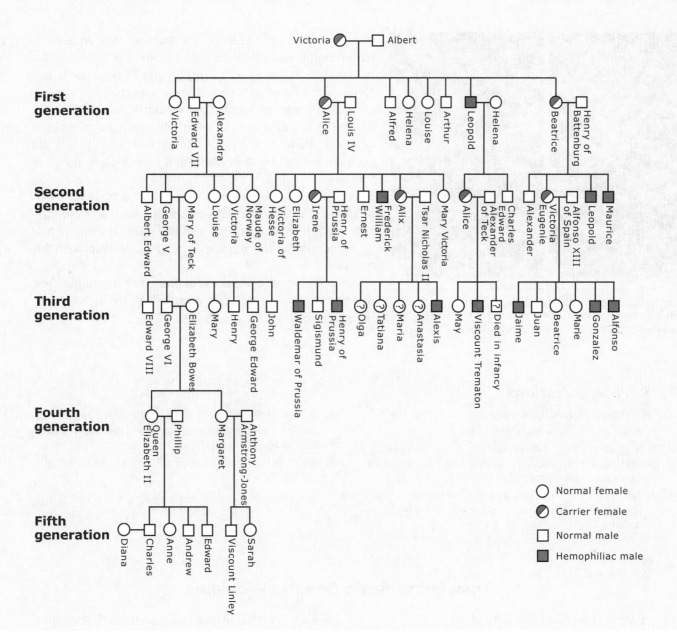

First generation
Second generation
Third generation
Fourth generation
Fifth generation

Victoria — Albert

Victoria, Edward VII, Alexandra, Alice, Louis IV, Alfred, Helena, Louise, Arthur, Leopold, Helena, Beatrice, Henry of Battenburg

Albert Edward, George V, Mary of Teck, Louise, Victoria, Maude of Norway, Victoria of Hesse, Elizabeth, Irene, Henry of Prussia, Ernest, Frederick William, Alix, Tsar Nicholas II, Mary Victoria, Alice, Charles Edward Alexander of Teck, Alexander, Victoria Eugenie, Alfonso XIII of Spain, Leopold, Maurice

Edward VIII, George VI, Elizabeth Bowes, Mary, Henry, George Edward, John, Waldemar of Prussia, Sigismund, Henry of Prussia, Olga, Tatiana, Maria, Anastasia, Alexis, May, Viscount Trematon, Died in infancy, Jaime, Juan, Beatrice, Marie, Gonzalez, Alfonso

Queen Elizabeth II, Phillip, Margaret, Anthony Armstrong-Jones

Diana, Charles, Anne, Andrew, Edward, Viscount Linley, Sarah

○ Normal female
◐ Carrier female
□ Normal male
■ Hemophiliac male

and most reds appear reddish brown. The condition is a recessive characteristic that is found on the X chromosome; therefore, it does not have a companion allele on the Y chromosome.

In order to diagram these crosses and keep track of the sex of the individuals, the X and Y will be used to indicate the chromosome, and superscripts will indicate the genes. Thus X^G is for a female sex chromosome that has the gene for normal vision; X^g is for a female sex chromosome that has the gene for red-green colorblindness. The Y is for the male chromosome, and since it does not have a gene for this trait, no G or g is added; however, the X chromosome in the male genotype will have a G or g.

The Punnett square in Figure 5-45 shows the cross of a female who lacks the recessive trait with a male who is red-green colorblind. Note the sex and genotypes of the offspring.

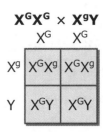

$$X^G X^G \times X^g Y$$

	X^G	X^G
X^g	$X^G X^g$	$X^G X^g$
Y	$X^G Y$	$X^G Y$

5-45

A normal female crossed with a colorblind male

5-46

The left picture represents what someone with normal vision would see. The right picture has been altered to simulate what a red-green colorblind person would see.

None of the sons have the trait because they get their X chromosome from their mother and their Y chromosome (which does not carry genes for this trait) from their father. None of the daughters are red-green colorblind either, because they are heterozygous for a recessive trait ($X^G X^g$). The daughters, however, are carriers of the red-green colorblindness trait. A **carrier** for an X-linked trait is an individual that does not exhibit the characteristic but does carry the gene for the trait.

Is it possible for a man to pass X-linked traits on to his son? No. A man will always give to his son the Y chromosome, which lacks genes for X-linked traits. If a boy is red-green colorblind, he inherited the characteristic from his mother.

Is it possible for a woman to be red-green colorblind? Yes. She must, however, be $X^g X^g$. What will be the genotypes of her parents? The answer is on page 127.

Review Questions 5.6

1. Distinguish between autosomes and sex chromosomes.
2. Explain how sex is determined in humans.
3. Why are X-linked traits passed from father to daughter but not father to son? Can a mother give an X-linked trait to her son and to her daughter?
⊙4. Why is the garden pea a good organism for studying genetics? Why is the human a poor organism for studying genetics?
⊙5. What are the characteristics of hemophilia?
⊙6. What is an X-linked characteristic, and how can you be sure hemophilia is one?
⊙7. Why were no hemophiliac females born until relatively recently?

Answers to Basic Genetic Problems

Free or Attached Earlobe (p. 116)

Ff × Ff

	F	f
F	FF	Ff
f	Ff	ff

Whose Baby Is Whose? (p. 121)

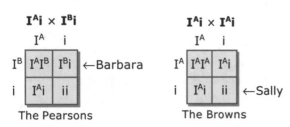

$I^A i \times I^B i$

	I^A	i
I^B	$I^A I^B$	$I^B i$ ←Barbara
i	$I^A i$	ii

The Pearsons

$I^A i \times I^A i$

	I^A	i
I^A	$I^A I^A$	$I^A i$
i	$I^A i$	ii ←Sally

The Browns

Since Sally is blood type O, she could be the Pearsons' child. If both Pearson parents were heterozygous (Mr. Pearson $I^B i$ and Mrs. Pearson $I^A i$), then it would be quite possible for them to have a homozygous recessive (*ii*) child who would have blood type O. (See Pearsons' Punnett square.)

Barbara (type B), however, could not be the Browns' child. Neither Mr. Brown nor Mrs. Brown has the *B* allele; therefore, they could not give it to their daughter. The Browns' children must be either blood type A or blood type O. (See Browns' Punnett square.) Since the Pearsons could have a child with blood type B, Barbara must be their daughter, and Sally the Browns' daughter.

How can we be sure that the parents' genotypes are the ones used above? Their phenotypes and their daughters' phenotypes can be used to show that there are no other possible genotypes for any of the parents, except Mr. Pearson. He could be heterozygous B ($I^B i$) or homozygous B ($I^B I^B$) and still be Barbara's father.

Dihybrid Cross of Peas (p. 121)

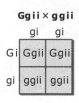

A test cross mates an unknown genotype with a known genotype (the homozygous recessive), which in this case is *ggii*. The test cross of a pea that is homozygous for green constricted pods (*GGii × ggii*) will produce all green constricted individuals in the F_1. If this happens, the farmer will know that his parent pea is homozygous for those traits. If a heterozygous green pea is test crossed (*Ggii × ggii*), half the offspring will be yellow constricted. If this happens, the farmer knows that his parent pea was heterozygous for pod color.

Red-Green Colorblind Female (p. 126)

A red-green colorblind girl ($X^g X^g$) can be born to a red-green colorblind man ($X^g Y$) and a carrier woman ($X^G X^g$). The father must be red-green colorblind, because if he has normal vision he can give only a normal gene for this trait to his daughter. If she gets a normal gene, she will not be red-green colorblind. The mother must be a carrier ($X^G X^g$) or be colorblind herself ($X^g X^g$) in order to pass the recessive trait to her daughter.

Advanced Genetics

The Continuity of Life
Part II

6A—Chromosomal and Genetic Changes

The inheritance mechanisms discussed thus far could be considered normal. In most cellular divisions, gamete formations, and gamete combinations (fertilizations), the principles of Mendelian genetics hold true. There are, however, many irregularities that can occur. Some of these are merely a reshuffling of genes that happens quite often in most organisms without any noticeable effects. Because of the curse for man's sin, however, other irregularities are lethal, killing the organism, or cause abnormal (weak or deformed) organisms.

There are two basic types of genetic changes:

- **Chromosomal changes**, which involve either the number of chromosomes or the arrangement of genes on a chromosome, and

mutation: (L. *mutare*—to change)
- **Gene mutations**, which change the sequence of bases in a gene.

6.1 Changes Affecting the Numbers of Chromosomes

6.1

Objectives
- Explain the difference between euploidy and aneuploidy
- Discuss the advantage of polyploid organisms
- Describe how nondisjunction affects the chromosome number

Key Terms

chromosomal change
gene mutation
euploidy
polyploid
aneuploid

nondisjunction
Down syndrome (trisomy 21)
trisomy
monosomy

Most organisms people are familiar with are diploid in their adult stage. Diploid organisms must carry on meiosis in order to form haploid gametes. At fertilization, the haploid gametes combine to produce a diploid zygote.

Some organisms are haploid even as adults; they form gametes without going through meiosis. Mosses, algae, and many fungi, for example, are haploid. Their diploid zygote, however, usually undergoes meiotic divisions before the organism begins to grow.

In the insect world there are other naturally occurring haploid organisms. The queen and worker bees are diploid, but the drones (males) are haploid, developing from unfertilized eggs. Scientists call the development of an unfertilized egg *parthenogenesis* (PAR thuh noh JEN ih sis).

Euploidy

Euploidy (YOO PLOY dee) is the condition of having a chromosome number that is an exact multiple of the haploid number for that organism. Euploid organisms have one or more complete sets of chromosomes. The bee drones mentioned in the previous paragraph would be considered euploid—they have one complete set of chromosomes.

Any cell or organism that has three or more complete sets of chromosomes is a **polyploid**. A polyploid cell would also be considered euploid because its chromosome number is now $3n$, $4n$, $5n$, and so on. There are two basic types of polyploids: those that have multiples of the same chromosome set ($4n$) and those that have multiples of different sets of chromosomes ($2n + 2n$).

Triploids

If a diploid gamete is fertilized by a haploid gamete, the resulting zygote is *triploid* ($3n$), having three complete sets of chromosomes. During mitosis, all the chromosomes replicate and divide normally. When a triploid organism enters meiosis, however, and the chromosomes line up, the third homologue prevents successful replication, and unfertile (sterile) gametes are produced.

Living triploids do not occur in humans or in the animal kingdom; however, they occasionally appear in other kingdoms. Triploid plants are usually taller and stronger, have more and larger leaves, and produce larger fruits than their diploid counterparts. Some triploids are therefore cultivated as ornamental plants. Triploid grapes, oranges, and other fruits are common on today's market shelves. Because they lack seeds, triploids must be reproduced asexually.

Tetraploids

Tetraploid (TET ruh PLOYD) organisms have four complete sets of chromosomes and are common in plants but rare in humans and animals. While no humans are completely tetraploid, some cells of the human liver are tetraploid. For some reason, and by an unknown mechanism, some human

parthenogenesis: partheno- (Gk. *parthenos*—virgin) + -genesis (birth)

6-1
Queen bee ($2n$) surrounded by drones (n)

6-2
Triploid grapes are larger than their diploid counterparts.

Euploidy (complete chromosome sets)	haploid (*n*)		gametes, mosses, algae, fungi, bee drones
	diploid (2*n*)		man, animals, most plants
	triploid (3*n*)		seedless plants (watermelons, bananas)
	tetraploid (4*n*)		Irish potato, alfalfa, some rare plants
	tetraploid (2*n* + 2*n*)		corn, wheat
Aneuploidy (missing or extra chromosomes)	monosomy (2*n* – 1)		various types of plants, Turner syndrome in humans
	trisomy (2*n* + 1)		various types of plants, Down and Klinefelter syndromes in humans

The Mule

The mule is a well-known exception to many of the genetic "rules." A mule is a cross between a female horse and a male donkey. The horse has sixty-four chromosomes per body cell nucleus, and the donkey has sixty-two. The union of a horse gamete (*n* = 32) and a donkey gamete (*n* = 31) results in a mule with sixty-three chromosomes (*n* + *n* = 63). Because the horse chromosomes and the donkey chromosomes are different, they fail to pair properly during meiosis. At the same time, the mule's uneven chromosome number forces one chromosome to be without a pairing partner. This presumably causes sterility.

Someone has rightly said that a mule is an animal with an ignoble past and a hopeless future; however, one in ten thousand female mules can conceive and bear young. It seems that for some reason the gametes she forms have either all the horse or all the donkey chromosomes. Depending upon the kind of mule being bred, the offspring of a female mule are usually horses or donkeys, rarely mules. No fertile male mules have been known to exist.

The mule, with its unusual characteristics, was the ideal animal to meet many of man's needs before the invention of cars, trucks, and certain types of machinery. Today man breeds very few mules.

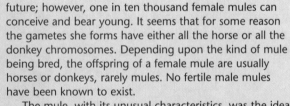

The mule (right) is a cross between a donkey and a horse.

Man's Use of Polyploids

Man has often looked at the plants and animals he grows for food and wished that one organism could have the desirable traits of two. The Russian plant geneticist Karpechenko is remembered for his desire to have a plant that produced the large edible leaves of a cabbage and the large edible root of a radish. In 1927, he crossed a radish (2*n* = 18, genus *Raphanus*) with a cabbage (2*n* = 18, genus *Brassica*) and eagerly awaited the results. He even invented the word *Raphanobrassica* as the scientific name for this new plant.

Some seeds were formed, but the plants that grew were sterile. Since the chromosomes were from two different sets of chromosomes, the nine radish chromosomes did not pair with the nine cabbage chromosomes during meiosis. This plant had the diploid chromosome number (2*n* = 18), but its genetic makeup was really two different haploid sets of chromosomes (*n* + *n* = 18).

In a few plants, however, the chromosomes did double, forming a tetraploid (2*n* + 2*n* = 36). These plants were fertile, each of the chromosomes having a partner during meiosis and producing seeds that grew new plants. *Raphanobrassica* was hailed as a new man-made species. Although it cannot grow in the wild, it can be cultivated. Today people can buy *Raphanobrassica* seeds, but the plant is not grown for food. To Karpechenko's dismay, it has radish leaves and cabbage roots.

Crossing a grapefruit and a tangerine (above) forms a tangelo (inset).

Karpechenko's cabbage-and-radish cross (bottom) was not the successful food crop he had envisioned.

Most of man's attempts to cross organisms from different genetic backgrounds have met with a similar lack of success. Occasionally some crosses between similar species do produce valuable offspring. A cross between a grapefruit and a tangerine produces a tangelo. Once a plant with desirable traits is obtained, asexual reproduction is used to keep the organism's line alive.

Many valuable crops are polyploids: corn, wheat, cotton, grapes, alfalfa, bananas, potatoes. Humans had nothing to do with developing these polyploids—they are believed to have occurred naturally. However, these plants would also have *ended* naturally were it not for their cultivation by man. Many polyploids, such as grapes and bananas, either are sterile or produce seeds that carry inferior characteristics due to problems in meiosis. Although wild counterparts of these plants do exist in nature, the polyploid strains require man's cultivation. Even many fertile polyploids, like corn, wheat, and cotton, would become extinct within a few years if man did not cultivate them.

Man has been greatly involved with the genetics of these cultivatable plant polyploids since they were formed. This is an excellent example of humanity carrying out the Creation Mandate of using the earth's resources to better human life (Gen. 1:29). These plants have been crossed and recrossed, forming many different strains. It appears that their extra chromosomes have permitted a greater flexibility in the expression of their genes. Consequently, many polyploid crops can grow in different areas and produce variations of their products. Consider all the different kinds of corn, wheat, and potatoes that are available. Also polyploid plants tend to be bigger than the wild variety. Humans have used these characteristics to feed larger and larger populations. Without this use of dominion over polyploid plants many people would have starved.

Some Christians have speculated that God permits these genetic conditions for the benefit of man but does not permit similar genetic variations on a widespread basis. Although various polyploids have apparently formed naturally, polyploidy does not happen often. Man has not actually observed the forming of polyploids except when drugs (such as colchicine) or other experimental techniques are used. Do you think Christians should favor using polyploids in this way?

6-3

The Irish potato: $4n = 48$

liver cells completely duplicate their chromosomes without dividing. These cells, then, have four sets of chromosomes ($4n$). Since liver cells do not form gametes, this unusual chromosome number does not affect the offspring.

The white or Irish potato is a tetraploid. Its haploid chromosome number is twelve ($n = 12$), and the diploid state has twenty-four chromosomes ($2n = 24$). The cells of an Irish potato, however, have forty-eight chromosomes ($4n = 48$). Just as in triploids, when the four homologues line up to form the tetrad during meiosis, the chromosomes do not segregate properly, and few fertile gametes are formed. Instead, farmers use the eyes of the potato to reproduce this crop asexually.

Tetraploids can also be formed when a diploid gamete ($2n$) from one organism is crossed with a diploid gamete ($2n$) of a different organism ($2n + 2n = 4n$). Although this may seem unusual, many food crops, including corn and wheat, have been developed by this type of breeding.

Aneuploidy

Aneuploids (AN yoo PLOYDZ) either lack chromosomes or have extra ones. Most aneuploids are believed to be the result of an error during meiosis. For some unknown reason, a chromosome pair will occasionally fail to separate during meiosis. This is called **nondisjunction**. Two of the resulting gametes therefore have an extra chromosome ($n + 1$), and two lack a chromosome ($n - 1$).

Aneuploidy in Humans

The genetic makeup of humans (as well as animals) appears to be quite fixed when compared to plants—many of the chromosomal changes that are common in plants do not occur in humans or animals. However, in humans there are several well-known aneuploid conditions.

Down syndrome (trisomy 21) is caused by a trisomy of the twenty-first chromosome. Individuals with Down syndrome may be male or female and are characterized by low mentality, short stature, stubby hands and feet, an extra fold of skin on the eyelids, as well as other defects. Down syndrome occurs in one out of every seven hundred people born in the United States.

Occasionally someone is born with a trisomic condition of one of the other smaller human chromosomes. In these cases, however, the individual is often severely deformed, and the life expectancy is only a few weeks.

Various aneuploids of the sex chromosomes also occur. Possible conditions are these:

✦ *Turner syndrome (XO).* Females lack a second sex chromosome (denoted by the *O* in the genotype). The result is underdevelopment of the female sex glands, short stature, and other deformities.

Some people with Down syndrome are more intellectually impaired than others. In the past they were all assumed to have a very low intelligence. Proper education helps them develop to their full potential.

✦ *Klinefelter syndrome (XXY).* This male has an extra *X* chromosome. He often appears normal but is typically infertile.

✦ *Trisomy X (XXX).* Various symptoms may develop in these individuals. Many appear normal, but some resemble women with Turner syndrome, and some are so-called "super females," having a tendency toward male characteristics. Many of these individuals are sterile.

Aneuploids of other human chromosomes have been found in naturally aborted fetuses. Aneuploid conditions of most human chromosomes are therefore believed to be lethal before birth. Only the sex chromosomes (which are believed to carry relatively few genes) and a couple of the autosomes can be trisomic or monosomic and not kill the individual. Almost all human aneuploids that survive birth are sterile, and many die before they reach maturity.

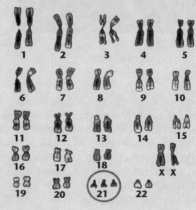

Down syndrome karyotype showing the extra twenty-first chromosome

---

If the gamete that has an extra chromosome unites with a normal gamete, the result is a **trisomy** (try SO mee) of one of the chromosomes ($2n + 1$). All the chromosomes are in the diploid state except one, which has three chromosomes instead of two. If the gamete that lacks a chromosome unites with a normal gamete, the result is **monosomy** ($2n - 1$). All the chromosomes are in the diploid state except for one, which is single. Other numbers of missing or additional chromosomes do exist but are rare.

Aneuploids are common in the plant kingdom, especially in cultivated crops. Some of the various types of wheat, for example, are aneuploids of a complex polyploid. Aneuploids in the animal kingdom and among humans are known to exist, but the trisomy or monosomy condition can happen only with a few of the chromosomes. It seems that having too few or too many chromosomes is usually lethal in animals and humans.

Review Questions 6.1

1. Name the two major types of genetic changes and briefly explain the difference between them.
2. Explain why most tetraploid organisms are fertile but triploid organisms are sterile.
3. In what ways are polyploid plants advantageous to humans?
4. It has been said that although many polyploid plants appear strong, as a species they are weak. What basis could be used to support this statement?
5. Distinguish between euploid and aneuploid organisms.
6. Describe how nondisjunction can cause aneuploid organisms.
7. Describe the genetic condition that results in Down syndrome.

6.2 Mutations

In the broadest sense of the word, a **mutation** is any change in the DNA of an organism. More specifically, it may involve a rearrangement of a chromosome or a change in a single nucleotide in the DNA sequence—either way the DNA has been altered. The effects of a mutation can vary. The mutation may affect only a few cells within the organism, or it may affect the entire organism. The mutation may or may not be passed to the offspring of the mutated organism. The effects of the mutation may not be noticeable at all, or they may be lethal. A *lethal mutation* is one that causes the death of the organism, either before or after birth. This section discusses the various ways mutations can occur and then examines the biological effects of mutations. Just as in Subsection 6.1, understanding the information previously covered on meiosis is extremely important.

Changes Within the Chromosome

Subsection 6.1 discusses changes in the numbers of chromosomes within an organism. There can also be changes within the chromosomes themselves. Chromosomes are actually long chains of many genes and can be visualized as a string of pearls or beads. Just as it is possible for a string of pearls to break and be repaired, chromosomes can also break. Such breakage actually happens quite often. Most of the time the chromosome is repaired without any problems; however, there can be some unexpected results if the "gene-pearls" are strung together in a different sequence or end up on a different strand altogether.

Translocation

Occasionally, two *nonhomologous* chromosomes exchange genetic information during meiosis. This type of chromosomal change is called a translocation and can occur by one of two methods. In the first method, broken pieces of one chromosome attach to a chromosome from a different

6.2

Objectives

- Define *mutation*
- Compare and contrast the changes within a chromosome and changes within a gene
- Discuss the differences between somatic and germ cell mutations

Key Terms

mutation	frame shift
translocation	mutagen
deletion	somatic
inversion	mutation
point mutation	germ mutation

translocation: trans- (across) + -location (L. *locare*—to place)

Crossing-Over and Genetic Recombination

Although it is not a mutation, crossing-over is similar to translocation. Crossing-over occurs when *homologous* chromosomes exchange information during meiosis. Suppose an organism has two chromosomes—one with the genes *ABCDE* and one with the genes *abcde*. Since genes *ABCDE* are all on one chromosome and genes *abcde* are on the other chromosome, the gametes would contain either the *ABCDE* genes or the *abcde* genes. When crossing-over occurs, there is a change in the expected genes in the gametes. Two of the gametes have chromosomes with the expected genes *ABCDE* or *abcde*; however, the other two gametes have different genes on their chromosomes—*ABCDe* and *abcdE*. This new mixture of genetic material is called *genetic recombination*.

Crossing-over sometimes occurs at two or more places within a single chromosome pair. If the chromosomes join together so that there is a complete set of genes in each chromosome, there are no bad effects. This crossing-over merely shuffles the genes that are on the chromosome and thus permits greater genetic variation.

It is important to understand that if genetic recombination occurs throughout many generations, there may be vast differences between the first parents and the latest offspring. Even though there may be many changes, this in no way offers any support for evolution. Why? There has been no *new* genetic information produced, only various recombinations and shuffling of the original genetic material.

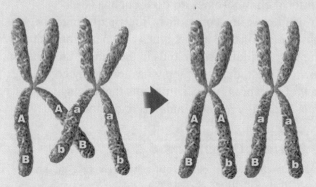

Crossing-over occurs when chromosomes that form a tetrad exchange genetic information.

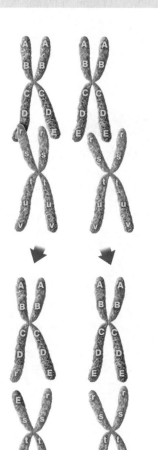

6-4

In translocation the chromosome fragment reattaches to a nonhomologous chromosome.

homologous pair. In the second type of translocation, a segment of chromosome attaches to the end of a nonhomologous chromosome that has not broken.

Some gametes made from cells that underwent translocation will be normal; however, some will have extra genes, and some will lack certain genes.

About 4% of the people suffering from Down syndrome have a translocation of genes from the twenty-first chromosome onto one of their other chromosomes. This translocation causes the zygote to have three sets of the genes on the translocated segment of its twenty-first chromosome. A person with this condition will have Down syndrome even though he has only one pair of chromosome twenty-one. The three sets of genes cause the syndrome.

Deletion

Occasionally a segment of the chromosome is left out after it breaks. This complete loss of a segment of the chromosome is called **deletion**. The chromosome piece usually winds up outside the nucleus, where it disintegrates.

If this happens, the offspring will lack certain genes. You might think that this would not be a problem for a diploid organism because the genes on the homologous chromosome could supply the missing information. That is a logical conclusion; however, if the homologous chromosome is carrying harmful recessive traits, they will be expressed. Why? The deleted segment that carried the dominant, "good" traits is no longer present to override the harmful, recessive genes.

6-5

In deletion the chromosome fragment fails to reattach.

A person lacking genes may suffer very harmful effects. An example of a harmful deletion is seen in an individual suffering from *cri du chat* (cry of the cat), which results from the deletion of part of chromosome number five. When a baby has this deformity, he may have mental retardation, and his cry sounds like a cat's screeching because of a malformation of the larynx (voice box).

Inversion

In **inversion**, a segment of the chromosome breaks off and reattaches at the same position. However, the genes on that segment are in the reverse order, or inverted, from their original sequence. Inversions are less likely to cause serious conditions because the genes are still present; they are just located in a different order on the chromosome.

Gene Mutations

The term *mutation* can often refer to a *gene mutation*—the alteration of an individual gene. A gene is a section of the DNA molecule that contains the genetic code, the specific sequence of nucleotides that directs the manufacture of an amino acid chain. If the nucleotide sequence is changed by even one nucleotide, the codons of RNA can be affected. When only one or a few nucleotides are changed, it is called a **point mutation**; and even though changing just one nucleotide seems to be a very insignificant change, the effect on the organism can be profound. Point mutations can occur in one of three ways—substitution, addition, or deletion of nucleotides.

✦ *Substitution* occurs when a nucleotide in the DNA sequence is removed and replaced with a different nucleotide or when two nucleotides are inverted. If the substitution makes an RNA codon that codes for the same amino acid (for example, GCA and GCC both code for alanine), the mutation will not be noticeable. In the substitution illustrated, however, the new codon calls for a different amino acid, but all of the other amino acids in the polypeptide chain of this protein will be the same. The disease sickle cell anemia is caused by substituting adenine for thymine in the DNA molecule, resulting in a defective form of hemoglobin.

✦ In *nucleotide addition*, an extra nucleotide is placed in the DNA sequence. Since the codons contain three nucleotides, adding a nucleotide to the DNA shifts all the nucleotides down by one. This is called a **frame shift** because the nucleotides in the codons (the frame) have been shifted. The chance is quite small that the new codons will code for a polypeptide chain of amino acids that is at all similar to the original.

✦ *Nucleotide deletion* is just the opposite of an addition—a nucleotide is removed from the DNA. A deletion has the same results as an addition. When the nucleotide sequence changes, a frame shift mutation occurs.

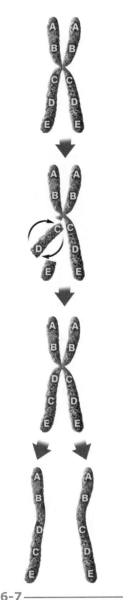

6-7
During inversion the chromosome fragment reattaches in reverse order.

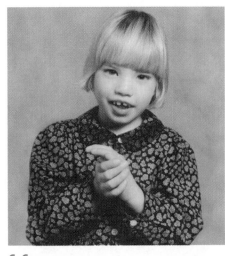

6-6
Child exhibiting *cri du chat*—a result of a deletion

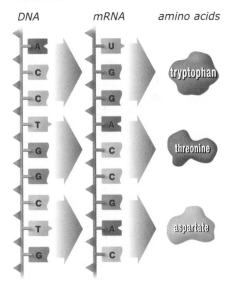

6-8
Normal transcription and translation process

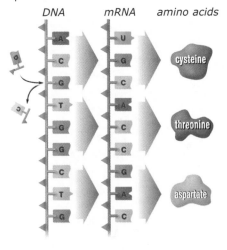

6-9
Substitution of a different nucleotide in the DNA results in a different amino acid.

Advanced Genetics   135

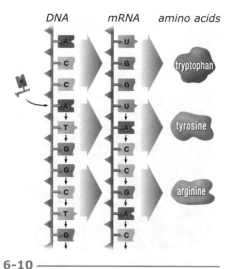

DNA | mRNA | amino acids

tryptophan

tyrosine

arginine

6-10

A frame shift mutation. The addition of a nucleotide causes the codons to be re-arranged.

These alterations of genes can have the following effects:

- *Major effects.* They produce no protein or vastly different proteins that can be lethal or cause severe genetic diseases.
- *Minor effects.* They produce a protein that is only slightly different or is not important to the functioning of the organism.
- *No effect.* They do not really change the codon, or the affected codon codes for the same amino acid as the nonmutated codon.

Generally, only substitution mutations result in minor effects, while addition and deletion mutations cause more severe disorders. However, if the addition or deletion mutations occur near the terminal end of the gene, they may cause only minor changes to the polypeptide chain. Some scientists believe these minor mutations may cause some of the different alleles that have been observed.

Biological Effects of Mutations

Mutations that happen naturally are called *spontaneous mutations.* Mutations can also be induced by mutagens (MYOO tuh juhnz). A **mutagen** is anything that causes a mutation to occur, such as various chemicals, viruses, or radiation. According to some estimates, every human has thousands of mutations in his body. Most of these mutations, however, will not affect the person or his offspring.

A **somatic** (soh MAT ik) **mutation** is a mutation that occurs in the nongamete, or body, cells of the organism. A mutation in a somatic cell will typically do one of three things:

✦ *It may produce an odd protein.* Some odd proteins either decompose or are given off as waste products. Since most human cells are diploid, most mutations of this sort do not drastically affect the cell—the other allele will still function and will usually counteract some of the effects of the mutation. This type of mutation may account for certain blemishes or deformities such as moles and some tumors.

✦ *It may have no effect.* The mutation of a gene that was not operating in a particular cell has no effect until that gene is turned on. In some human cells certain genes are never turned on. Also, many substitution gene mutations have no effect because they do not significantly change the codon.

If a single blood-forming cell in an adult mutated to the sickle cell gene, would that person suffer from sickle cell anemia? No. Even though this one cell may produce the odd protein, many other cells are producing the normal protein. Most mutations have no effect because they are somatic and in multicellular organisms.

✦ *It may kill the cell.* The accumulation of an odd protein that is lethal to the cell or the lack of a needed protein results in the death of the cell. Since there are thousands of every type of cell in the human body, the loss of one cell is not noticed. This would be a lethal mutation, however, if it occurred in a single-celled organism.

A **germ** **mutation** is a mutation in a gamete or a cell that forms gametes. Since a germ mutation produces gametes with some genetic change, it will affect every cell in the new individual. Some somatic mutations, which cause either no effect or simply the production of an odd protein, would be lethal as germ mutations. A mutation that kills the cell will, of course, be lethal as a germ mutation.

somatic: (Gk. *soma*—body)
germ: (L. *germen*—seed)

1. Describe a human condition that results from a change within a chromosome.
2. Compare and contrast the three types of mutations caused by changes within the chromosome.
3. Contrast crossing-over and translocation.
4. List several causes (mutagens) for gene mutations.
5. Gene mutations can affect the code for an amino acid in what three ways?
6. Which type of change would be more likely to have severe consequences—substitution or deletion? Why?
7. What are three possible results that could be expected from a gene mutation in a somatic cell chromosome?
8. Why is a germ mutation more significant than a somatic mutation?

6.3 Gene Expression

Genes have specific jobs to be performed at specified times; therefore, not all the genes on a chromosome are turned on, or active, all the time. For example, some genes are turned on only during the initial stages of fetal development. Once their particular task is completed, they are turned off and may not be used again until that gene has been passed to the developing fetus of the offspring. **Gene expression** is the activation of a gene that results in its transcription and the production of a specific protein.

Gene expression begins at the very moment that the gametes unite to form a zygote and then directs the production of various proteins. These proteins cause cells to undergo specialization to form all the different cell types that the organism needs to grow and function. This specialization of cells is called **cellular**

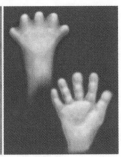

6-11

A human zygote (left). By 6 weeks, tiny hand buds have formed. Within another week, tiny fingers are clearly visible. These changes are the result of gene expression and cellular differentiation.

differentiation. As growth continues, these specialized cells form various tissues and organs. The one-celled zygote grows and begins to take on the form of the organism. As a human embryo grows, the body visibly develops—little hands, feet, eyes, and ears appear where before they were unrecognizable. The change in form that an organism undergoes is called *morphogenesis*; it is controlled by the expression of the genes.

Master control genes called **homeotic genes** regulate cellular differentiation and morphogenesis. These genes determine when and where the cells will change and what the cells will change into. For example, scientists have discovered a homeotic gene in fruit flies that controls eye development. If a mutation causes the expression of this gene in the wrong cells, eyes can develop on the flies' legs, antennae, and even wings.

Scientists are working to discover the mechanism for cell specialization in embryos as well as in the cells of the adult organism. In their attempt to understand gene expression, scientists have gone back to the DNA molecule itself to search for the mechanisms that control the genes. Although scientists have learned a great wealth of information, the exact mechanism of embryonic development is still one of the mysteries of the body, for it is "fearfully and

6.3

⮕ **Objectives**

- Define *gene expression* and *cellular differentiation*
- Compare and contrast molecular and environmental controls on gene expression
- Explain the relationship between cancer and the functions of proto-oncogenes and tumor suppressor genes

⮕ **Key Terms**

gene expression
cellular differentiation
homeotic gene
tumor
benign
malignant
carcinogen
cancer

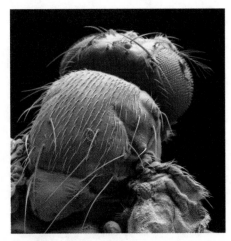

6-12

A mutation in the fruit fly's homeotic gene that controls eye development can lead to the presence of eyes in some strange places.

morphogenesis: morpho- (Gk. *morphe*—shape) + -genesis (birth)

wonderfully made" (Ps. 139:14). As humans search out more of the beautiful complexity of the development of the embryo, we can see more clearly the beauty and wisdom of the embryo's Creator.

Some Controls of Gene Expression

Scientists have studied the DNA of some bacteria and have developed models of gene expression; however, most bacteria have only one chromosome. The control of gene expression in eukaryotic cells is a much more complex process, and no one knows all the answers. What is known is that there are certain external environmental factors, such as temperature, that cause genes to be expressed and that there are internal environmental factors, such as hormones or certain nutrients, that do the same.

Molecular Factors

Gene expression actually starts with the transcription of the DNA molecule. The DNA in the nucleus can be either unwound in very thin threads or tightly coiled; only the areas that are uncoiled can be transcribed. Each gene has its own set of regulatory proteins that bind to specific control sequences on the DNA and initiate the transcription process that forms mRNA.

The proteins produced by translation of the mRNA may be used within the cell's cytoplasm or organelles, transported to other cells, or returned to the nucleus. If a cell has an abundant supply of a particular protein, the DNA stops manufacturing the mRNA responsible for that protein until the excess is depleted.

Sex-limited characteristics (characteristics that occur only in males or only in females) are the results of gene expression. The presence or absence of sex hormones turns on or off the genes that code for sex-limited characteristics. The feathers of some male birds, brightly colored when compared to those of the female of the same species, are a sex-limited characteristic.

Sex-limited characteristics in humans include growth of body and facial hair, beard growth, breast development, and milk production. Both male and female humans have genes for all these characteristics. The male reproductive hormones stimulate the growth of body hair and beards in young men, while the female hormones stimulate breast development in young women.

Environmental Factors

An example of an external environmental control is the Himalayan rabbit. During the winter the Himalayan rabbit is normally white with black ears,

6-13

The female duck (right) has the genes for the brightly colored feathers of the male (left), but the expression of these genes is limited by the sex hormones.

nose, tail, and feet. If, however, the hair on his ears and tail is removed and these body parts are kept in heated mufflers while his winter coat regrows, the fur will regrow white. If an area of the rabbit's back is kept cold while his hair is growing, the hair on his back will be black. In this example, temperature turns on or off the gene for coat color.

6-14

An experiment with the Himalayan rabbit demonstrates the effect of temperature on certain genes.

Gene Expression and Cancer

An abnormal mass of cells produced by abnormal cell division is called a **tumor**. Tumors can be divided into two broad categories: benign and malignant. **Benign** (bih NINE) tumors contain cells that stay within the body of the tumor and do not spread to other parts of the body. A mole is a good example of a benign tumor. Malignant (muh LIG nunt) tumors are cancerous; they contain cells that may spread, or *metastasize*, beyond the boundaries of the original tumor to other parts of the body. Cancer can spread by local infiltration of adjacent tissue or by entering the bloodstream or lymphatic system, where cells of the tumor break off and are transported to other tissues.

A carcinogen (kar SIN uh jun) is any substance that increases the risk of cancer. Carcinogens can be chemicals (such as the tars in cigarette smoke), certain viruses, or ionizing radiation (such as x-rays and ultraviolet radiation from the sun or from the lights in tanning booths). Most carcinogens are also mutagens, causing mutations to occur in the genetic code. Whether someone develops cancer depends on many factors. Researchers have also determined that the type of carcinogen, the amount, and the duration of exposure play important roles in the potential for development of cancer. For example, someone who has never smoked is much less likely to develop lung cancer than someone who does smoke. Even exposure to the sun can increase the possibility of skin cancer. One bad sunburn is not likely to cause cancer, but multiple sunburns (especially before the age of 20) or exposure to the sun over the course of many years certainly increases the possibility of skin cancer.

But what causes cancer cells to grow so wildly, and why? One good definition of cancer that also helps explain the *why* is that **cancer** is unrestrained cell growth that has escaped the normal controls of the cell cycle. Yet, as good as this answer is, it raises still another question: what causes a cell to disregard the cell cycle controls?

Now the discussion can move to the molecular level. Researchers have identified two types of genes that regulate the cell cycle—the proto-oncogenes and the tumor suppressor genes. The *proto-oncogenes* (PRO toh ON kuh jeenz) code for proteins that stimulate cell division or that affect the

6-15

Moles are examples of benign tumors.

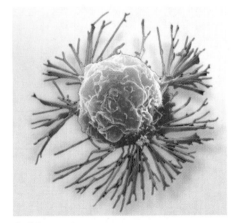

6-16

A cancer cell

malignant: (L. *malignus*—harmful)

carcinogen: carcino- (Gk. *karkino*—crab, cancer) + -gen (born)

proto-oncogene: proto- (Gk. *protos*—first) + -onco- (*onkos*—tumor) + -gene (*genea*—family)

6-17

Most people know that the carcinogens in cigarette smoke can cause cancer, but many do not realize that long-term sun exposure can increase the risk of skin cancer.

cancerous lung

synthesis of growth factors. *Tumor suppressor genes* produce proteins that prevent uncontrolled cell growth. Working together, the proto-oncogenes and the tumor suppressor genes properly control the cell cycle. Any mutation that suppresses or inactivates the function of the tumor suppressor genes or that overactivates the proto-oncogenes can increase the likelihood of cancer formation. If a proto-oncogene mutates, an oncogene is formed. An *oncogene* causes cells to begin to divide uncontrollably, and if the cells overwhelm the tumor suppressor genes, a tumor starts to grow. With this knowledge of oncogenes scientists are now working on cancer treatments that target and reverse the effects of the oncogene. It is hoped that one day these treatments will replace tiring and dangerous treatments like radiation and chemotherapy.

Cancer development is certainly more complicated than this presentation, but from what researchers have learned, the initial definition of cancer can be modified as follows: cancer is a genetic disorder in which damaged genes are unable to control cell growth. The devastating effect of cancer on the human race is staggering. Sadly, many people have used cancer (and other similar maladies) to argue against the existence of the Christian God. They say that if a good God exists, then He would not have made a world

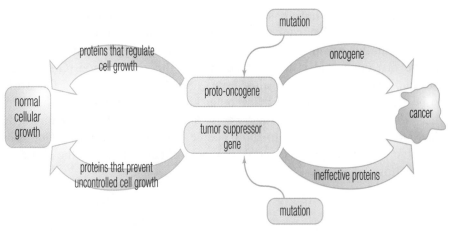

6-18

The relationship between proto-oncogenes, tumor suppressor genes, and mutations

Human Genetic Disorders—Treatments and Cures

A *genetic disorder (inherited disorder)* is any malformation or malfunction of an individual's body that is caused by a gene or group of genes. Some genetic disorders can be quite severe (such as Down syndrome), while others are barely noticeable. Even within the disorder itself, there are degrees of severity. For example, the severity of Down syndrome can range from mild to profound mental and physical retardation. Some disorders are so mild that they go undetected until they are discovered at a routine physical examination. Other genetic disorders are even accepted as variations of normal—some types of birthmarks are considered to be a lesser type of genetic disorder.

For most genetic disorders there are no cures. When an individual has a genetic disorder, every cell in his body has the genetic information for that disorder. In order to cure the disorder, the responsible gene or genes would need to be corrected or replaced in all the cells of the body in which that gene is expressed. An alternative would be to replace all the cells that are using the defective gene. Gene therapy, which is discussed in the next chapter, is an attempt to cure genetic disease. Many genetic disorders, however, can be treated. Treatments of genetic disorders relieve only the symptoms; they do not cure the disease.

Many genetic disorders are apparent at birth. People with pure white hair, chalky complexions, and sometimes even pink eyes have *albinism*—that is, they lack a dark-colored pigment called melanin, which is necessary to make most of

A birthmark is considered a minor genetic disorder.

the pigments that color various parts of the body. Their eyes and skin can appear pink because the red of their blood shows through the skin. An albino (al BYE noh) has two recessive, mutant alleles for the albino trait. This is an example of one gene controlling several characteristics.

Some genetic disorders are not apparent at birth. A person with Huntington chorea, for example, usually will not express this dominant gene until he is over thirty years old. A man named Huntington first described the symptoms of this disorder: the slow loss of mental ability, the loss of voluntary control of muscular activity, and occasional spasmodic

A person with albinism. Most people with albinism live normal lives, but they must avoid long exposure to sunlight. Lacking melanin, their cells can be harmed by strong light. Some albino persons seek to improve their appearance by coloring their hair, wearing glasses or contact lenses, and using makeup.

movements. People with this genetic disorder know they will die after an extended, humiliating period of uncontrolled muscular activity and loss of mind. At present there is little that can be done for people with Huntington chorea.

Apparently, there are genetic disorders (such as certain types of diabetes) caused by the turning off of a particular gene. Scientists are searching to find ways to cure genetic disease at the molecular level, either to prevent the gene from turning off or to turn it back on. Even then, such a technique is a treatment, not actually a cure. Because the defective gene still exists in a person, he may pass it on to his offspring.

albino: (L. *albus*—white)

where humans—including small children—suffer from such an awful disease. The fact that cancer exists, they say, helps prove that humans are alone in this world. The problem with this thinking is that it fails to recognize that our world is fallen. God created this world good (Gen. 1:31). But humans have rebelled against Him. Therefore, God allowed the world to be filled with cancer and many other kinds of suffering and pain (Gen. 3:17–19). Cancer reminds us that we are sinners, not that God is absent. In fact, God often uses cancer to show people His mercy and grace. God uses this kind of suffering to lead people to repent of sin and believe the gospel of Jesus Christ.

Review Questions 6.3

1. Compare gene expression, cellular differentiation, and morphogenesis.
2. Since every cell in your body has the same genes, why is every cell in your body not like every other cell?
3. Under normal circumstances the beard genes are not activated in women and the milk production genes are not activated in men. What are these gene expressions called, and how are these genes activated?
4. Give an example of how the environment can influence a genetic trait.
5. Describe how a mutation of either the proto-oncogenes or the tumor suppressor genes could result in cancer.
6. Is the existence of cancer and other devastating diseases evidence that God does not exist? Explain.
7. Why is it difficult to cure a human genetic disorder?
8. List several human genetic disorders and describe the treatments used to overcome the symptoms of the disorders.
9. List several human genetic disorders for which there are no effective treatments at this time.

6B—Population Genetics

The first part of this chapter discusses how changes can occur in both the chromosomes and the genes as well as some of the effects of genetic changes on individuals. In this section the text examines the effects of genetic changes in large groups of individuals. While understanding how genes work and are passed from one generation to the next is essential, the role of genetics in producing the variability and changes seen in groups of organisms is equally important. While there are those who teach that these genetic changes are the driving force behind evolution, these evolutionary beliefs are examined and refuted in Chapter 8.

6.4 The Gene Pool

A group of individuals of the same species that live in the same area is called a **population**. Since all members of a population can interbreed, they share a common group of alleles. Geneticists often refer to the sum of all the alleles that all members of a species of an organism can conceivably possess as the **gene pool** for that organism. Since most species are distributed over a wide area comprising many different populations, an individual would not have access to the whole gene pool for its species. Every normal individual has a complete set of genes from the gene pool; however, that complete set does not include every possible allele of every gene. For example, all the possible alleles that Mendel's peas could have had form the gene pool for peas, but every pea plant does not have all the possible alleles of every gene—a short pea plant (homozygous recessive) does not have the allele for being tall.

6.4

Objectives

- Explain the difference between *population* and *gene pool*
- Discuss how mass selection, hybridization, and inbreeding affect the gene pool
- Describe desirable as well as undesirable effects of hybridization

Key Terms

population
gene pool
variation
mass selection
Hardy-Weinberg principle
genetic equilibrium
genetic drift
hybridization
inbreeding

Genetic Variation

The genotypic differences between individuals from the same gene pool are called **variations**. For example, the human gene pool contains alleles for hemophilia, but most people do not have this allele. If a person's parents do not have the allele for hemophilia in their genotype, there is no chance of his having it (unless a mutation occurs). An individual has access to the gene pool only through his parents, who are two separate accesses to the gene pool. In other words, one parent brings to his children one set of genes, and the other parent brings another set. Some genes in a cross are homozygous, each parent giving the same allele as the other. For other genes, the parents may give different alleles, thereby producing offspring that are not exactly like themselves. The alleles that the parents give to their offspring are determined by the segregation of the chromosomes during meiosis as the gametes are formed.

Natural variations in the genotype can occur in three basic ways: (1) by the random fusion of two gametes from all of the possible gamete combinations, (2) by the segregation and possible exchange of information between homologous and nonhomologous chromosomes that occur during meiosis, or (3) by altered copies of genes that result from mutations.

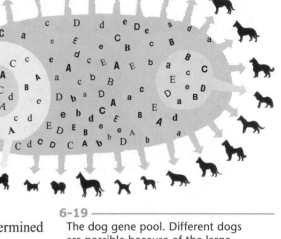

6-19

The dog gene pool. Different dogs are possible because of the large number of alleles in this gene pool. Smaller gene pools exist as "puddles" of this larger pool. Occasionally, "puddles" homozygous for certain genes exist (usually the result of selective breeding), resulting in pure strains of dogs (e.g., collie, dachshund). These purebreds can mate with other members of the gene pool, resulting in hybrids.

Mass Selection

The gene pools of organisms contain both desirable and undesirable traits. For thousands of years the breeding of animals and plants for desirable traits has been a profitable activity for man. The objective of such breeding is to produce offspring that are superior to the parents. Superior characteristics might include greater weight gain, greater milk production, greater resistance to disease, better coat or fur quality, or the ability to survive in an area of low rainfall, late spring frost, or early autumn freezes. Actually, the breeder may want any number of desirable characteristics. He can use any of several techniques to select from the gene pool those genes that will give good traits to the offspring.

Ancient peoples selected the most desirable animals and plants from the herd or field as breeding stock. This method of selecting breeding stock, called **mass selection**, continues to be used today. The loblolly pine, for example, is a valuable lumber tree that grows throughout the southern United States. Unfortunately, southern pine beetles often attack forests of the loblolly pine and in a few weeks can kill thousands of trees. Often, only a few trees remain alive after an epidemic. The remaining trees apparently possess something that repels the beetles. Seed that is harvested from those trees is being used in a breeding program to develop a strain of loblolly pine that is resistant to the southern pine beetle.

It is important to notice that mass selection attempts to cultivate an already existing trait, not to develop a new trait. Breeders seek to produce a pure line that emphasizes a characteristic already in the gene pool, not to expand the gene pool. Creationists and evolutionists agree that mutation is the only way to introduce new genes into the gene pool. Changes caused by mutations, however, do not prove evolution (see pp. 177–78, 185–87).

6-20

Pine trees (top) destroyed as a result of pine beetle (bottom) infestation

Studying Populations

With all of the various alleles in a gene pool and the possible changes in chromosomes and genes, you might wonder what the long-term effect on a given population of individuals might be. Do they stay the same? Do they change, and if so, how much? Consider the illustration of the dog gene pool and look at all the different kinds of dogs—most Creationists believe the dogs on the ark had a sufficiently large gene pool to account for all the kinds of dogs seen today. There has obviously been a great amount of change since Noah's time! However, man has caused much of the change through the breeding of domesticated dogs to select certain "desirable" alleles.

At the start of the 1900s, when the study of genetics was just beginning and Mendel's work had been rediscovered, scientists began to wonder how gene pools and populations might be affected over time. One of the questions scientists wanted to investigate was, "Will the dominant forms of an allele eventually replace the recessive forms?" Two researchers working independently of each other—German physician Wilhelm Weinberg and English mathematician Godfrey Hardy—were able to demonstrate that dominant alleles will not automatically replace the recessive alleles. Their findings are known as the **Hardy-Weinberg principle**: the frequency of alleles and the ratio of heterozygous to homozygous individuals in a given population will remain constant unless the population is affected by factors outside the original population. In other words, the Hardy-Weinberg principle is true only for a population that is in genetic equilibrium.

But what exactly is a genetic equilibrium? A **genetic equilibrium** exists when the allele frequencies within a population do not change. Allele frequencies are the number of times an allele occurs in a given gene pool compared with the number of times other alleles are expressed. There are five factors that must be met for a genetic equilibrium to exist and for the Hardy-Weinberg principle to remain true.

- There is random mating.
- The population is very large.
- There is no movement of individuals into or out of the population.
- Mutations do not alter the gene pool.
- All individuals in the population must have an equal chance to survive and reproduce.

It is not very likely that a genetic equilibrium would exist naturally unless a population lived in isolation, say at the top of a mountain or in a controlled laboratory. Studying a population in genetic equilibrium helps researchers understand how specific changes may affect the population.

Using genetically pure populations of lab rats, researchers can study the effects of specific changes.

Changes in Genetic Equilibrium

Any violations of the factors required for a genetic equilibrium will cause a change in the gene pool and result in changes in the population. Some changes that will affect the gene pool include the following:

✦ *Nonrandom mating.* Most species do not mate randomly. There may be geographic barriers, or they may select mates that have similar phenotypes. Mating with relatives—inbreeding—as is sometimes performed when breeding animals for a particular trait, will reduce the number of heterozygotes in the population, although the gene frequency will remain unchanged.

✦ *Small population.* In a small population any change will produce amplified results. Such genetic change due to chance is called **genetic drift**. For example, consider a population of 100 000 deer. This population has a set allele frequency; if by random chance four deer fail to reproduce (0.004% of the population), there would be minimal if any change in the allele frequency of the gene pool. But if the population is only ten deer and by chance four fail to reproduce (40% of the population), there will be a profound change in the allele frequency. As the population recovers, the new allele frequency will be based only on the six deer that were able to reproduce, and there will be a decreased amount of variability between individuals in the population.

✦ *Migration.* Most populations experience migration of individuals either into (immigration) or out of (emigration) the population. This constant incoming or outgoing of genetic information is called *gene flow*.

✦ *Mutation.* Mutations are the only mechanism by which new alleles are made. In the example of sickle cell anemia, the substitution of a single base results in an altered allele.

✦ *Selection.* In nature not all members of a population survive to reproduce. They may be sick, they may have some characteristic that makes them less likely to survive, or predators may kill them. The loss of individuals will alter the alleles in the gene pool.

Isolated populations of cheetahs and elephant seals have resulted in less genetic variability, making them more susceptible to harmful mutations. Researchers have cross-bred individuals to increase their genetic variability.

Many factors can cause a population to change over time. Some people would take this fact and use it as evidence to support the idea of evolution. However, they should think again about these population changes. These genetic changes involve either the rearrangement of already-existing alleles or the loss of genetic information; they do not create the new information necessary for molecules-to-man evolution. Genetic change does not equal evolution! In the deer example the deer are still deer; in the photographs the elephant seals and cheetahs are still seals and cheetahs. Creationists and evolutionists would agree that mutations are the only mechanism that creates new alleles; however, most mutations are neutral or harmful, and none have created a new type of organism. Population genetic studies are useful in studying how populations change, but they offer no support for evolution.

Hybridization

Hybridization (HYE brid uh ZAY shun) is the crossbreeding of two genetically dissimilar individuals. The offspring of such a cross is called a hybrid. Hybridization often involves two varieties of the same species. Crossing two strains of corn, crossing two different kinds of apples, and crossing two different varieties of chickens are examples of hybridization of two members of the same species.

The hybrid offspring of animals or plants often have characteristics superior to either parent. This superiority is called *heterosis* (HET uh ROH sis), or hybrid vigor, and may involve such things as greater resistance to disease, larger bodies, more milk production, or more crop yield per acre.

Plant breeders have used heterosis to improve crop yield. Corn, a wind-pollinated crop, is usually heterozygous for many traits. Since each corn kernel is fertilized by a different pollen grain, a single ear may have kernels with many different genotypes. Before careful breeding, 120 bushels per acre (bpa) was an average corn yield. **Inbreeding**—the mating of an organism with itself or with close relatives—can, in time, produce pure strains. *Pure strains* are organisms that are homozygous for various traits. For example, consider four different pure strains of corn, labeled *A*, *B*, *C*, and *D*. These pure strains are of low quality. Some of them produce only one small ear per plant, and the ears have very small kernels. The average yield is only about 60 bpa.

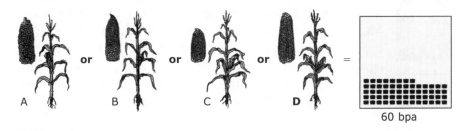

60 bpa

6-21 —————————————————————————————————
Corn strains *A*, *B*, *C*, and *D* yield 60 bpa.

When breeders crossed strain *A* with strain *B*, heterosis produced a crop yield of about 140 bpa. The same was true of the cross of the *C* and *D* strains.

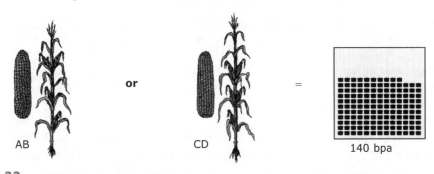

140 bpa

6-22 —————————————————————————————————
Corn strains *AB* and *CD* yield 140 bpa.

hybrid: (mongrel; hence, the offspring of different parents)

But mating the offspring of the *A* + *B* cross with the offspring of the *C* + *D* cross produced a superior corn with a yield of about 200 bpa. This type of hybrid corn is the type most often grown in the United States. Using controlled breeding techniques, seed companies supply hybrid corn seed to farmers. Farmers who try to grow their own corn seed without careful breeding techniques get poor quality corn, much like the crops before controlled breeding.

 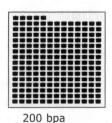

ABCD = 200 bpa

6-23

Corn strain *ABCD* yields 200 bpa.

6-24

Clydesdale horses measure 5.5–6 ft at the shoulder, while some miniature breeds stand only 18–24 in. tall.

In animals, close inbreeding (mating of parents to their own offspring) often results in inferior offspring. Although organisms that are homozygous for good traits are desirable, close inbreeding may also result in organisms that are homozygous for poor traits, including physical deformities. Careful inbreeding for generations has produced relatively pure strains of certain dogs, cattle, chickens, horses, and other animals. These domesticated animals possess characteristics man considers desirable. For instance, the longhorn cattle that ranged across the Southwest and fueled the great cattle drives have been nearly replaced by short-horned breeds that are much more practical in crowded trucks, train cars, and feedlots. Man's preference for animals with certain traits and his indifference about other traits have determined the lineage of animals such as cows, sheep, and chickens. These animals are now in some respects inferior to the wild stock from which they came in that many of them are no longer able to live without man's care.

➡ Review Questions 6.4

1. Explain why a single organism cannot have one of every allele from its gene pool.
2. Can a breed be improved after many generations of mass selection? Why or why not?
3. Why does inbreeding often produce inferior organisms?
⊙4. A purebred strain of dogs often has weaknesses (poor traits) in some of the offspring, while crossbreeds produce more and healthier pups. Give an explanation for this.
⊙5. List five factors that must be met for a genetic equilibrium to exist.
⊙6. Explain why it is so unlikely for a natural genetic equilibrium to exist.
⊙7. Is the fact that populations change over time a valid support for evolution? Support your answer.

6.5 Eugenics

The term **eugenics** (yoo JEN iks), meaning "good origins" or "well born," is traditionally applied to efforts to improve the human gene pool. This term was coined by an Englishman, Sir Francis Galton (1822–1911), who was a cousin of Charles Darwin. After reading *The Origin of Species* he became a believer

eugenics: eu- (good) + -genics (birth)

Objectives
- Define *eugenics*
- Describe current methods of genetic screening
- Describe two methods of artificial reproduction

Key Terms
eugenics
genetic screening

in evolution. Soon he began to apply the rules of selective animal breeding to human reproduction in an attempt to improve the evolution of the human race. He wanted people of "superior" stock to breed more and people of "inferior" stock not to breed. He did not propose any specific means for running this human breeding program; that job was left to the next generation. In America during the late 1800s and early 1900s, eugenists got the eugenic program rolling. Eugenics became part of the social gospel movement and became a matter of religious fervor. States passed laws that allowed them to forcibly sterilize people who may have passed along mental and physical problems in their genes. Sadly, over 60 000 people in America were forcibly sterilized.

Probably one of the most widespread and horrific eugenic endeavors was conducted in the 1930s by Adolf Hitler. The German dictator's systematic killing of millions of Jews, gypsies, and other selected peoples resulted from his desire to eliminate what he believed were "inferior races." At the same time in special camps, scientists of the Third Reich were inbreeding humans of "superior" German stock to form a "super race."

Consider this opinion:

> I wish very much that the wrong people could be prevented entirely from breeding; and when the evil nature of these people is sufficiently flagrant, this should be done. Criminals should be sterilized and feebleminded persons forbidden to leave offspring behind them. . . . The emphasis should be laid on getting desirable people to breed.†

A Nazi did not make this statement in the 1930s; the American patriot Theodore Roosevelt made it in 1913—seventeen years before Adolf Hitler. Although Theodore Roosevelt was a great leader in other ways, he, with most of the American upper class of the early 1900s, thought that the many immigrants from Europe and the Orient were a threat to American society.

However, additional genetic experiments with animals revealed that most of the practices of eugenic leaders of the early 1900s were based on faulty ideas and would not eliminate the "bad blood" (actually bad genes) and encourage "good blood." They also failed to recognize that man's environment and spiritual nature are major influences on human behavior and well-being.

The Second World War, with Hitler's atrocities, put a stop (at least temporarily) to most major eugenic activities in the United States. But today, advanced techniques are making eugenics not just a possibility but in fact a subtle reality in present American life. The remainder of this chapter examines some of the techniques that scientists and researchers are currently using to alter the human gene pool. The next chapter discusses further advances in biotechnology that are both promising and frightening.

Genetic Screening

Man-made **genetic screening** is used to determine an individual's genetic makeup. In America today, there are hundreds of genetic counseling centers that use genetic screening to supply individuals with information about themselves, their relatives, or their future children. Even using the most modern techniques, however, people learn about only a few of their genes. At present there are three basic genetic screening categories that are widely used.

---

† Theodore Roosevelt, *The Works of Theodore Roosevelt* (New York: Charles Scribner's Sons, 1926), XII, p. 201.

6-25

Sound waves and computer analysis are used to produce a three-dimensional sonogram of the unborn child. Sonograms can be used to evaluate whether the child is developing properly.

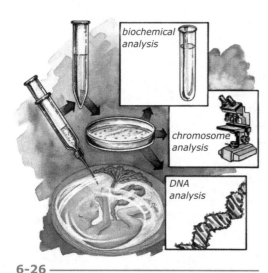

6-26

The fluid removed by amniocentesis contains cells and chemicals produced by the child. The sex of the child, abnormal chromosomal conditions, and some metabolic disorders can be diagnosed before the child is born. Sometimes treatment can begin by changing the mother's diet or even by transfusing the unborn child's blood. Amniocentesis is not without risk; it should be performed only if the physician feels it is necessary.

Pedigree Analysis

By obtaining information about the individuals and their families, genetic counselors construct pedigrees. Sometimes people can be tested to determine their genetic makeup regarding a particular condition. If the inheritance pattern of the trait is known, the counselors can then make accurate predictions regarding the probability of individuals having the trait. Many genetic disorders, for example, are simple recessive traits. By constructing pedigrees, a genetic counselor can tell prospective parents the percentage of their offspring that should have the trait, carry the trait, or be free of the trait. Prospective parents can use this information as they plan their families.

Analysis of the Unborn

Today there are several methods of learning about a child before he is born. Ultrasonic scanning, or *ultrasonography*, and *fetoscopes*, fiber-optic devices inserted into the womb, are used to obtain images of the unborn child. The images can often reveal genetically related deformities.

Amniocentesis involves removing some of the fluid that surrounds the unborn child. The child's cells in this fluid can be used to produce karyotypes and to look for chromosomal defects such as Down syndrome. The fluid can be chemically analyzed to determine other genetic conditions.

A fourth testing method is called *chorionic villus sampling*, or CVS. The chorionic villi are tiny projections that are part of the placenta, and they have the same genetic makeup as the fetus. The small sample of cells obtained by the physician can be examined for any chromosomal abnormalities.

Surgery on unborn babies has even been used to correct some defects that were discovered using ultrasound and fetoscopy.

6-27

Fetal surgery—operating on an unborn child—may be possible to correct some abnormalities discovered by amniocentesis, CVS, or ultrasonography.

By knowing that a child has a genetic disorder, parents and physicians can begin proper treatment at birth, if not before. Tragically, some parents use genetic screening as a deciding factor to abort a child with a genetic abnormality. All children, even those with incurable genetic disorders, are precious in God's sight because all are made in His image. God's image bearers should not be destroyed just because their life would be hard. To make such a decision is to violate one of the most basic commands to the human race: "Thou shalt not kill" (Exod. 20:13).

Analysis of the Newborn

A baby undergoes a number of tests soon after he is born. Some tests are conducted visually; other tests require blood or urine specimens. The results of these tests give the physician information needed to treat those infants who have certain genetic disorders.

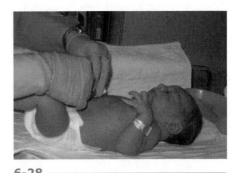

6-28

Technological advances make it possible to screen newborns for over 50 disorders. Some states test for only about 10 disorders, while others test for 50 or more.

Artificial Reproduction

Artificial insemination (in SEM uh NAY shun) is the mechanical injection of sperm into a female's body. It has long been practiced on animals in laboratories and has been used in livestock breeding for many years. Human artificial insemination has been routinely practiced since the 1950s. Each year thousands of women are artificially inseminated with sperm from their husbands, most because of medical reasons.

Women can also be inseminated with sperm from men who are not their husbands. Human sperm can be frozen and stored. Women have had children whose *biological fathers* (the men who supplied the sperm) have been dead for years. Some would contend that it is immoral for a woman to be artificially inseminated with the sperm of a donor who is not her husband. Those who accept this form of artificial insemination argue that true purity is not violated by this clinical technique.

Ova have also been removed from a human female's body, fertilized in a laboratory by human sperm, and then implanted back into the woman's body. This process is called *in vitro fertilization*, or "test-tube" fertilization, and has been used to make childbearing possible for women who lack certain reproductive structures. However, in vitro fertilization has a major ethical problem. The doctor normally fertilizes several ova and then observes them to ensure that they are growing. He may perform genetic screening tests to determine whether the embryos are carrying any genetic disorders. Then several of the tiny embryos are implanted into the female in hopes that at least one will survive. The ethical question is this: what is done with the embryos that are not implanted into the female? If they are destroyed, this action is no different from killing embryos during an abortion.

In some instances of in vitro fertilization, the *biological mother* supplies the ovum but the zygote is implanted in another woman who serves as the *surrogate mother*. Some Christians object to biological mothers serving as ovum donors. The arguments regarding ovum donors are about the same as those for sperm donors. Today some women serve as surrogate mothers for other women who cannot or do not wish to become pregnant. Many Christians object to surrogate mothers because the biblical principles of the family are ignored. Many feel that the one-flesh union spoken of in Genesis 2:24 and compared with Christ's relationship to the church in Ephesians 5:31–32 is violated by the introduction of outside parties into the relationship.

Eugenics: Dream or Nightmare?

A world populated with people who are strong, healthy, and intelligent does sound like a dream—no more sickness or disease, and the finest minds solving the remainder of mankind's problems. Social planners and the medical profession have been working toward this goal for hundreds of years. Some people see genetic screening as a beneficial form of eugenics. If those individuals likely to pass major genetic deformities on to their children do not reproduce, the percentage of genes causing the deformities in the gene pool would decrease. Such knowledge, however, has many potential problems.

Christians must be aware of the technology that God has allowed man to develop. They should also be aware of biblical principles that apply to man's use of the technology and should prayerfully consider what their attitudes and responsibilities in these areas ought to be.

Man's power over animals, plants, and even his physical self, however, cannot make him truly good. Only Christ can save him and give him a new heart. No matter how friendly, strong, intelligent, and disease- and disorder-free man is, his righteousness is nothing but filthy rags unless he has a regenerated, cleansed soul. "For all have sinned, and come short of the glory of God" (Rom. 3:23).

The Problems of Eugenics

Eugenics may be unable to produce for mankind the improvements it promises. Remember that artificial domestic breeds develop weaknesses that make them dependent on man. Likewise, scientists may someday successfully breed man for intelligence, strength, or other traits, but what traits will be sacrificed? If eugenics is practiced widely, the gene pool will shrink as "bad" traits are eliminated. In that shrinking, man's resistance to disease and his adaptability to different environments may weaken. Man might become like his domesticated animals—dependent on a controlled environment.

Since eugenics—like all sciences—cannot make decisions, man will have to make the difficult decisions about which traits to eliminate. Should sickle cell anemia be eliminated? The trait is disadvantageous in the United States but is of some value where malaria abounds. Strength and intelligence are desirable, but should everyone be a muscular genius? Diversity has encouraged human achievement. Many people with undesirable traits, in trying to compensate for them, have excelled in their fields. Paul's "thorn in the flesh" (2 Cor. 12:7) was perhaps one such undesirable trait, but it promoted Paul's spiritual progress.

Christian Consideration of Eugenics

Some Christians condemn modern genetic technologies because they consider these techniques "unnatural." But like x-rays, surgery, and all of modern medicine, these genetic techniques are tools that humanity can use to show love to one's neighbor by helping to prevent and cure sicknesses (Mark 12:31). But, just like any other tool, these genetic practices can be misused.

Can Christians use any eugenic principles? Only in a limited sense. A person can use genetic screening to see if he is likely to pass on a defective gene. However, it is that person's personal responsibility to take action, and every case is different.

Some Christians may be tempted to use genetic screening and selective breeding to keep up with others who use these technologies. Other people may use abortion and selectively develop "superior" embryos while destroying the "inferior" embryos to have "genetically advanced" children. Christians, however, should not be willing to abandon biblical morality in order to advance genetically. If that means having children "genetically inferior" to the genetically advanced designer babies of others, then so be it. In the end, we are all accountable to God, not to people with "superior" offspring.

In addition to the dangers discussed above, the Christian should temper his approval of eugenics with one other qualification: disagreement with secular scientists who believe that eugenics can solve all of man's problems. Secular scientists think that all of man's problems have physical causes and thus must have physical solutions. The Bible teaches otherwise: man's suffering is the result of man's sin (Gen. 3:17–19). Eugenics will not establish a perfect race, for God's curse on the earth will prevail until Christ returns (Rom. 8:19–21).

As genetic research continues to expand (see Ch. 7), Christians will have to make difficult judgments about its use. In doing so, Christians must be sure that their position is always in accordance with biblical principles and is not merely a personal opinion or prejudice. Although the Bible does not deal with specific medical or scientific techniques, the Bible does deal with such principles as the value of human life (see Ch. 25, Sec. B). Christians can never violate biblical principles and still consider themselves obedient to God. A conclusion based on feeling and not on the Bible is merely an opinion.

⮞ Review Questions 6.5

1. List three categories of genetic screening and give examples of each.

2. List and describe four methods currently used to learn about the unborn.

3. What is meant by *test-tube baby*? In what way is this term misleading?

4. Differentiate between a biological mother and a surrogate mother.

⊙ 5. Discuss several factors that a Christian must consider when making decisions regarding eugenics.

⊙ 6. Tell whether you agree or disagree with the following statement: artificial reproduction should more accurately be called man-assisted reproduction. Support your opinion.

Biotechnology

Modern technology has allowed many advances in biology. We are now able to describe the building blocks of life, reaching down to the molecular level. This advance in biology has opened the door to a new kind of dominion. Humans can now use their knowledge to produce technologies that manipulate many aspects of life. As with all products of human knowledge, biotechnology holds great potential for both good and evil. People can use it for God's glory in such uses as finding new cures for diseases and growing better plants. But people can also use biotechnology for ungodly purposes, such as embryonic stem cell research or abortion. As you read about the techniques presented in this chapter, ask yourself these questions: Can a Christian support this technique? How could this technique be used for evil? How could the technique be used for God's glory?

7A—Biotechnology

Biotechnology is the use of technology to enhance living organisms and processes. This chapter examines three broad areas of biotechnology—cloning, genetic engineering, and stem cell technology—and includes applications of each. Two important questions to keep in mind are these: How does the

Creation Mandate apply to these technologies? How should Christians respond to the research and application of these technologies?

7.1 Clones and Cloning

When the subject of clones comes up, many people automatically think of Dolly the sheep, CC the cat, or even news stories stating that successful human clones are being developed. One would think that cloning is a newly developed technology; however, many clones are common in nature. Many man-made clones, as well as naturally occurring human clones, have existed since the time of Adam.

How can that be possible? A **clone** is merely an exact genetic duplicate of a cell or organism. If an organism reproduces only by mitosis, the result is a clone. For example, if a unicellular protozoan or alga goes through mitosis and forms two cells, both of those cells are identical genetically and are clones of the parent organism. Therefore, clones have identical genomes. The **genome** of an organism is the hereditary information encoded in its DNA.

Natural Clones

In addition to reproduction by mitosis, any asexual reproduction produces natural clones—clones that occur without man's intervention. Multicellular organisms that reproduce by fragmentation, budding, or asexually produced spores form natural clones. Some organisms in the animal kingdom, such as jellyfish, some worms, and some insects, can reproduce asexually and thus produce natural clones. Most of the more familiar animals, like fish, birds, and mammals, normally reproduce sexually and thus do not form natural clones of themselves.

In the plant kingdom natural clones are more common and are even commercially beneficial. If a strawberry grower finds a strawberry plant that grows well in his climate and has the qualities of taste, texture, and color he wants, he can clone the plant. Strawberries naturally reproduce asexually by stolons (thin stems that grow tiny plantlets on their ends). In time the farmer would be able to fill his field with clones of the desirable strawberry.

Artificial Clones

Many economically important plants do not naturally reproduce asexually but can be either helped or forced to do so. For example, the "eyes" of desirable varieties of the Irish (or white) potato are used to produce clones of that kind of potato. Eventually there can be fields full of genetic duplicates of this potato. Since man cloned this organism and helps the clone to continue, it is said to be an *artificial clone*.

Seedless banana plants are propagated by planting the cut-up pieces of their large underground stems. The genetic change that resulted in a seedless banana has been kept alive by cloning. Today most people have never seen a seed-filled banana.

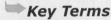
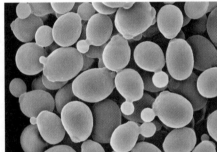

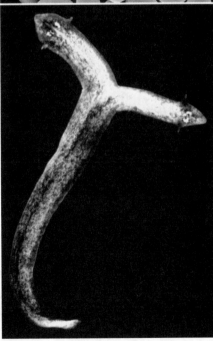

7-1

Budding in yeast (top) and regeneration in planarians (bottom) produce natural clones.

7-2

The wild variety of banana is filled with seeds and would not be much fun to eat.

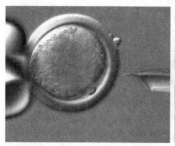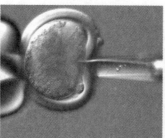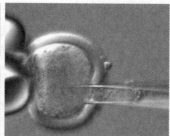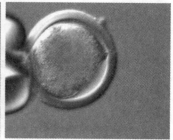

7-3

Needle being used to remove nuclear contents from a cell

Frogs were used in some of the original successful experiments with cloning from the zygote stage. When the zygote became two cells, scientists separated the cells. The cells then developed into two genetically identical frogs.

Because frogs produce large ova, they were also used as experimental animals in another form of artificial animal cloning. Scientists are able to use ultrafine needles to remove the nucleus from a frog's ovum without destroying the cytoplasm's structure. They can then take a nucleus from a differentiated cell of another frog and implant it into the empty, or *enucleated*, ovum. A *differentiated* cell is one that has become specialized to perform a particular function, such as muscle or nerve cells. The artificial zygote can then develop into a frog. The process of transplanting the genetic material from one cell into an enucleated ovum is called *somatic cell nuclear transfer*.

In 1997, Scottish scientist Ian Wilmut was the first to successfully develop a clone of a mammal—Dolly the sheep—using a fully differentiated adult cell. Prior to Dolly, most scientists believed that cloning of mammals was possible using only undifferentiated cells from either embryos or fetuses. Wilmut removed the nucleus from a sheep ovum and fused the enucleated ovum with a fully differentiated cell from a different sheep's udder. After the cell began to divide, it was placed into the uterus of a third sheep where it continued to develop. Since the birth of Dolly, several types of mammals have been successfully cloned, including pigs, rabbits, cats, mice, and cows.

Some scientists are now pursuing a new method of cloning called *direct reprogramming*, or de-differentiation, which does not require the use of enucleated egg cells. In this method, scientists attempt to stimulate differentiated adult cells to develop like embryonic cells.

As the techniques for cloning improve, many more types of animals will be cloned. Why is cloning important? Those who perform clone research might answer that producing cloned animals will aid in medical research. If the research animals are genetically identical, the differences between animals resulting from genetic variables are eliminated. Other scientists have suggested that cloning may be an acceptable method to save endangered species of animals. In agriculture, once the desirable characteristics have been developed in crops and livestock, scientists could produce clones that would always produce those characteristics. Some secular scientists believe that cloning is a promising method of producing cells or substances for the treatment of disease.

There are also drawbacks to using cloned organisms. For example, although endangered species might be saved, cloning would decrease their genetic variability. This could cause them to be more susceptible to diseases or to pass on genetic disease. If entire populations of organisms—plants or animals—are genetically identical, infection by a single viral or bacterial strain could destroy them all. Cloned laboratory organisms will decrease experimental

How Old Was Dolly?

How old is a clone? There could be two answers—one based on the organism's birth date and the other based on the age of the chromosomes. In February 2003, Dolly the sheep was "put to sleep." Although just six years old, she had begun to suffer from arthritis and a type of lung disease usually seen only in sheep twice Dolly's age.

But was she really six years old? Since Dolly was cloned from a six-year-old sheep, some researchers consider Dolly to be twelve years old genetically. No one can really be sure because of the scarcity of cloned animals. Some researchers think that aging is revealed in the telomere section of the chromosome. The telomere is located at the end of a chromosome and is made of thousands of repeating bases. At each cell division, the telomere is shortened; the older the chromosome, the shorter the telomere. Not all researchers agree with using the telomere as an indicator of age.

The aging of cloned animals is an area of intense study. If a forty-year-old man donated genetic information for a clone, would the child really be almost forty-one years old at birth? This is a question to which no one knows the answer.

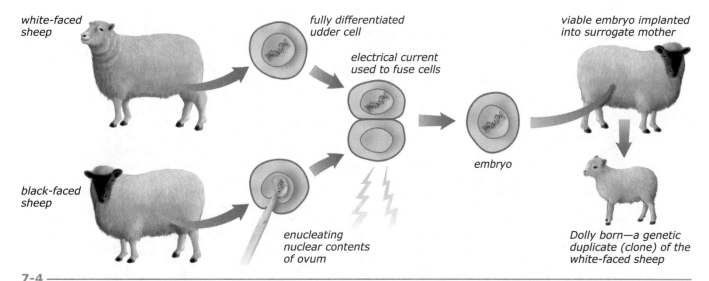

white-faced sheep

fully differentiated udder cell

electrical current used to fuse cells

viable embryo implanted into surrogate mother

black-faced sheep

enucleating nuclear contents of ovum

embryo

Dolly born—a genetic duplicate (clone) of the white-faced sheep

7-4

Diagram of somatic cell nuclear transfer process used to produce Dolly the sheep

variables, but any conclusions the researcher makes to answer his hypothesis will be true only for that particular genetic population. He is less able to draw conclusions regarding a more diverse population.

Researchers still have much to learn about producing and using clones. Many people question the morality of cloning animals for use in research. Others, however, believe that such cloning would not be outside man's biblical call to manage God's creation (Gen. 1:28; James 3:7) if the cloning is performed within scriptural guidelines for the benefit of humanity (Prov. 12:10).

Human Cloning

Human clones already exist; identical twins are naturally occurring clones. Their hereditary material is nearly identical because they are the result of one fertilized egg dividing into two cells that separated and developed into two individual embryos.

The successful artificial cloning of humans is the next major cloning event in the future. The question Christians need to consider is this: *Should* humans be cloned? Although some look favorably on the possibility of human cloning, the reasons for doing so are questionable, and the presently available techniques make human cloning objectionable to a Bible-believing Christian. Hundreds of animal embryos died before any clones were successfully produced. It took over two hundred attempts for Ian Wilmut to successfully produce Dolly. That God has allowed man to develop cloning techniques does not mean that he is free to perform experiments using human life. But does this mean that human cloning is an acceptable form of research? To answer this question, one must consider both the technology and the scriptural principles involved.

Clones and Souls

Would an artificially cloned human have a soul? Although some Christians may disagree, many Christian scholars believe that artificially cloned humans would have souls. Identical twins are natural clones, and they do not share one soul, nor is one of the twins soulless. Although they are genetically identical, they sin independently and must be saved individually. Using this logic, it appears that if man develops the technology to produce a human clone, each would still have an individual human soul in need of salvation just like everyone else.

Why Clone Humans? Therapeutic versus Reproductive Cloning

Scientists have divided human cloning research into two basic categories: reproductive cloning and therapeutic cloning. One goal of reproductive cloning is to provide a baby to certain couples, for example, those who have no children or those who have lost a child. Using the somatic cell nuclear transfer method, an embryo would be produced, implanted into a woman, and carried to birth. Proponents of this type of cloning say that, morally, it is no different from in vitro fertilization (IVF) procedures. The difference is that the child is a genetic clone of one selected individual: he would not possess DNA from both the mother and father.

Therapeutic cloning also uses somatic cell nuclear transfer to create a clone of an adult. However, the goal of therapeutic cloning is to produce cells for treatment of disease or research. In other words, it develops embryos to harvest cells for the production of "replacement parts." Since the harvested cells are an exact genetic match to the cloned adult, the cells could be transplanted without the need for a tissue match. Some scientists say that many diseases that are now untreatable could potentially be cured through cloning techniques.

Would anyone ever consider creating a life to be sacrificed to harvest its cells? Consider this example. An individual named Betty has a fatal disease that can be cured only by transplanting cells that match her tissue type. Since no relatives can be found with an acceptable match, Betty's parents allow scientists to produce a number of healthy embryos by cloning. The embryos are then destroyed to harvest their stem cells. Scientists implant the stem cells into Betty's body, and her health is much improved.

Was that harmful? The stem cells gathered were able to help Betty live. However, consider the fate of all the embryos—they were destroyed. In this case, humans attempted to save the life of one human created in the image of God by wrongfully taking the life of many other image bearers of God (cf. Gen. 9:6).

Mass-producing human embryos to be killed during research is unacceptable from a Christian point of view; it is murder. God has called man to exercise wise and good dominion over the earth and its creatures (Gen. 1:26–28). Killing one human in order to save or improve the life of another human cannot be considered wise or good (Gen. 9:6). It is the modern-day equivalent to human sacrifice and cannibalism. Christians need to stand against any individual, institution, or government policy that allows the pursuit of scientific knowledge using unbiblical methods regardless of how "wonderful" or "humanitarian" the results might be. In the future, cloning techniques may change and become much safer. Depending on the techniques, Christians may support cloning then. At the present, though, cloning techniques would cause much human death and therefore should not be pursued now.

Review Questions 7.1

1. Describe the process of somatic cell nuclear transfer.
2. List some advantages and disadvantages of using cloned organisms in research.
3. In light of the Creation Mandate, is the cloning of humans a valid field of research at this time? Explain.

7.2 Genetic Engineering

7.2

Objectives

- Define *genetic engineering*
- Describe the steps to produce recombinant DNA
- Describe how recombinant DNA is used in bacteria to produce human insulin

Key Terms

genetic engineering
recombinant DNA
restriction enzyme
plasmid
transgenic

Genetic engineering is the manipulation of genes by methods other than normal reproduction. There are several methods of genetic engineering that are currently being used and others that are being developed. One of the most common genetic engineering techniques joins specific sections of DNA from two different sources. The modified DNA is called **recombinant DNA**.

Scientists have found certain enzymes that can snip, or cut, DNA strands. These **restriction enzymes** snip DNA at specific base sequences, leaving several nucleotides unpaired. These unpaired bases form what scientists call a *sticky end*. Another DNA molecule that has been snipped with the same restriction enzyme will have a complementary sticky end. The two complementary sticky ends can combine, forming a recombinant DNA molecule.

Using restriction enzymes, scientists can snip genes out of chromosomes and insert them into other chromosomes or even into chromosomes of other organisms by using a vector. A *vector* is an agent that is used to carry the new gene from one organism to another. Vectors are often, but not always, plasmids. A **plasmid** is a small, circular piece of bacterial DNA, similar to a chromosome.

An example of the use of recombinant DNA is the manufacture of the protein insulin, a chemical essential for the proper use of sugars. Insulin is normally produced in the pancreas. People suffering from diabetes mellitus either do not produce insulin or cannot make adequate amounts of insulin and thus require insulin injections.

Prior to recombinant DNA techniques, most insulin used to treat those who suffered from diabetes mellitus was obtained from pig or cow pancreases. Animal insulin works well for most diabetic patients; however, some people are allergic to it.

To overcome these problems, scientists used restriction enzymes to snip out the insulin gene from human cells. The human insulin gene was then inserted into a bacterial plasmid. The plasmid served as the vector carrying the recombinant DNA back into the bacteria.

Bacteria grow and divide rapidly. As the bacterial cells divide, the plasmids also replicate and divide. Therefore, all of the bacteria contain the human

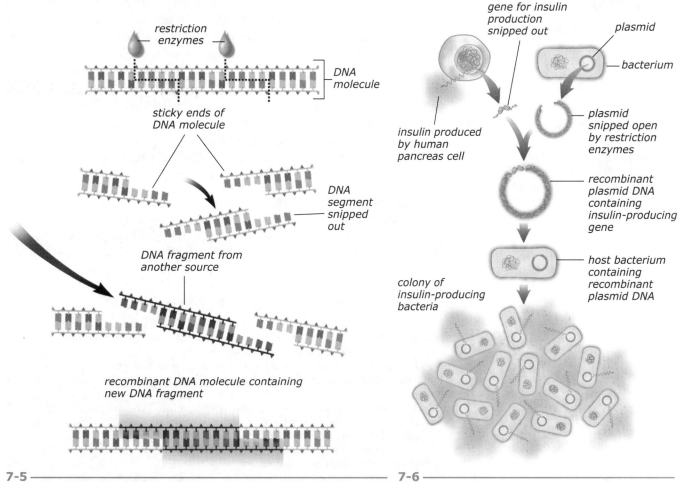

7-5

Restriction enzymes are used to open the DNA molecule so that a gene can be removed or added.

7-6

Bacteria can be altered to produce human insulin by inserting the human insulin gene into the bacterium's genome.

insulin–producing gene and can produce human insulin. When the bacteria are killed, the insulin is collected and purified. Although bacterial cells are used to produce the insulin, the protein produced is human insulin since a human gene directed its production.

Organisms such as the human insulin–producing bacteria in the example above are called **transgenic** organisms because they contain genes from more than one species. Transgenic microorganisms, animals, and plants have all been produced. Scientists are researching ways in which transgenic organisms may be used to benefit man; areas of interest include agriculture and treatment of disease. For example, transgenic mice have been produced that have enough human DNA to enable them to contract HIV. These mice are very useful for testing medicines that could help people with AIDS.

transgenic: trans- (L. *trans*—across) + -genic (relating to genes)

⮕ Review Questions 7.2

1. What is genetic engineering?

2. Explain the role of restriction enzymes, vectors, and plasmids in genetic engineering.

7.3 Stem Cell Technology

7.3

⮕ Objectives

• Define *stem cell*

• Explain the difference between somatic and embryonic stem cells

⮕ Key Terms

stem cell
embryonic stem cell
somatic stem cell
sequencing

Stem cell technology holds much promise in the treatment of disease and the repair of injuries. But what exactly are stem cells, and where do they come from? **Stem cells** are cells that have the ability to divide almost indefinitely and to give rise to, or differentiate into, specialized cells. For example, a stem cell in the tip of a plant root can become any of the types of cells found in plants. Due to environmental and hormonal cues, the cell develops into a particular type of cell; it becomes differentiated or specialized. Similarly, animal embryos contain stem cells that differentiate to become the various tissues and organs of the animal. The same is true for human embryos. Since much of the controversy regarding stem cells relates to using human stem cells, that context is used in this section.

Types of Stem Cells

When a sperm cell fertilizes an ovum, a zygote is formed. That one cell has the potential to give rise to all of the cell types in the human body; it is *totipotent*. When the zygote divides into two cells, each of these cells can also become

How Do Stem Cells Differentiate?

The totipotent cell of the zygote has all of the instructions needed to develop into a complete adult organism—in this discussion, a human. As the embryo grows, portions of the genetic information are either switched on or off at very specific times and in the proper order. Much of the switching instructions are coded in the genes; however, some are located outside the genes. These instructions outside the genes are called *epigenetic* controls.

Many researchers consider epigenetic controls to be responsible for differences such as hair color and patterns, eye color, blood chemical concentrations, and other variations between clones and their parent organism. Epigenetic controls, as well as environmental and nutritional effects, make it impossible to produce a clone that is identical to the parent organism in every respect.

The elaborate instructions, with their specific sequencing and on/off switching, allow only one reasonable conclusion: an all-powerful and all-knowing God is the Creator.

any of the cell types in the body. If these two cells should separate and continue to grow separately, they would form identical twins.

As the zygote continues to grow by cell division, the cells begin to differentiate into the parts of the early embryo. They can give rise to many, but not all, cell types; they are *pluripotent*. In other words, these cells have lost some of their potential to specialize. As these pluripotent cells continue to divide, they form stem cells that will give rise to particular tissues such as blood and skin. Now they are committed to a specific type of tissue but still are not fully specialized. For example, bone marrow stem cells can still give rise to all of the various types of human blood cells.

Sources of Stem Cells

The larger controversy over stem cells is not whether scientists should be performing stem cell research but rather concerns the source of the stem cells used in research. There are two basic sources for stem cells—embryonic and adult.

As the name states, human **embryonic stem cells** (ESCs) come from human embryos. In order to obtain embryonic stem cells, the embryo is destroyed. Since life begins at conception (fertilization of the ovum by a sperm), human lives are killed in order to harvest embryonic stem cells for research (Ps. 51:5; 139:13–16).

A more descriptive name for adult stem cells is **somatic stem cells** since they are obtained from the differentiated somatic, or body, tissue of adults or children. Bone marrow stem cells have been successfully used for more than thirty years in the treatment of leukemia. More recently, scientists have isolated somatic stem cells in a variety of tissues, such as nerves, skin, and blood vessels. Although they are more specialized than embryonic stem cells, researchers are finding that somatic stem cells can be made to differentiate into many types of tissue cells.

Uses of Stem Cells

Scientists are looking to stem cells for new treatments and cures for many diseases and disorders. For example, in insulin-dependent diabetes, researchers theorize that if stem cells can be harvested, differentiated into the proper kind of pancreatic cells, and then transplanted into the diabetic's pancreas, the newly differentiated cells will continue to grow and produce insulin. If successful, the person would no longer need to inject insulin to treat his diabetes. Many other conditions are being studied to determine whether they can be successfully treated with stem cells.

At this time, only somatic stem cells have been used to successfully treat human diseases. In fact, somatic stem cells are already being used to treat over 70 different diseases, including diabetes, Parkinson disease, and several forms of cancer. Recent studies continue to find that somatic stem cells can be differentiated into many cell types and have some advantages over embryonic stem cells. Somatic stem cells are easier to grow and to differentiate into specific cell types.

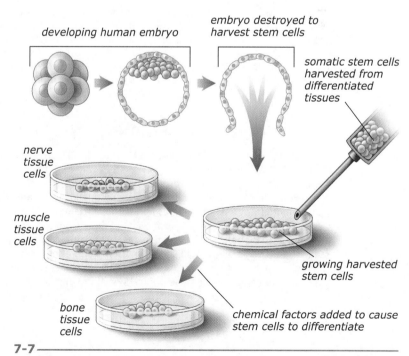

developing human embryo

embryo destroyed to harvest stem cells

somatic stem cells harvested from differentiated tissues

nerve tissue cells

muscle tissue cells

bone tissue cells

growing harvested stem cells

chemical factors added to cause stem cells to differentiate

7-7

Embryos are killed to harvest embryonic stem cells; no one is killed to harvest somatic stem cells.

Advances in biotechnology have allowed scientists to decode the entire genome of certain organisms. Determining the order of nucleotides in an organism's DNA is called **sequencing**. Scientists began with relatively small genomes, such as that of the bacterium *Escherichia coli*. The effort to decode the human genome began in 1990 as the *Human Genome Project*, or HGP, headed by Dr. Francis Collins. The project involved scientists from sixteen laboratories in six countries, and the projected date of completion was 2005. The announcement that the first draft had been completed (actually only 90% complete) was made in June of 2000. Finally, it was announced in April 2003 that the remainder of the genome had been decoded. However, a few areas of the genome remain unsequenced because of the difficulty of decoding them. The finished sequence is 99% complete and 99.99% accurate. The HGP provided the DNA base sequence on each of the 24 human chromosomes (22 autosomes, X and Y sex chromosomes). It did not reveal what traits each sequence codes for.

Researchers were surprised by some of the knowledge gained from the project. Previously it had been estimated that there were around 100 000 human genes; however, now it is estimated that the human genome contains only 20–25 000 genes—about the same number as an ear of corn.

However, a man is much more complex than an ear of corn. Scientists are now rethinking old "truths." Prior to this discovery, it was thought that each gene coded for a specific protein. If this were true, the 20–25 000 genes could not account for the hundreds of thousands of parts of the human body. Scientists now believe that it is how the DNA code is interpreted in combination with other genes, along with the influence of many epigenetic factors, which determines the product. They concede that having a list of the sequence of nucleotides is far from being able to explain how that list works.

A second surprise: only about 1.5% of the genome codes for proteins; the remaining parts are noncoding. In fact, researchers originally called the noncoding portion of the genome *junk DNA* because it was considered useless. They have since found that this area of the DNA contains important nucleotide sequences that affect genetic expression, turning particular genes on or off as needed. No one knows what additional knowledge will be learned from this portion of the genome.

The data from the HGP have been made public (via the Internet) and is freely available for other scientists to use in their research. One new scientific endeavor that is rapidly developing since the completion of the HGP is the field of proteomics—the study of

Dr. Francis Collins announces that the decoding of the human genome is complete.

proteins. Scientists are attempting to determine the structure and function of the thousands of proteins found in various organisms. One major application of the field of proteomics is the treatment of disease. The hope is that the knowledge gained through these studies will allow doctors to focus treatments on the underlying causes of the disease, rather than merely treating the symptoms.

How should Christians view the information being learned from the Human Genome Project? Throughout history, scientists have gained knowledge about the functioning of the human body that has provided new methods to counteract the effects of the Curse. The completion of the HGP is no exception. The Creation Mandate has not been revoked, nor are advances in knowledge (or the pursuit of knowledge) evil; man chooses whether to use knowledge for good or evil purposes. Christians need to be rooted in the Bible, for without God's moral absolutes, it is easy to be swayed by the opinions of others. As researchers interpret the human genome, Christians need to evaluate the findings by using a biblical framework, and they need to point out any inconsistencies that are identified.

As new information is discovered, it becomes more and more obvious that the message contained in the human genome points to an omniscient, omnipotent Author—the mighty God of the Bible, the Creator of the entire universe.

Junk DNA

The presence of so-called junk DNA has been used as an argument against Creation and for evolution. Some evolutionists argue that these regions are "leftovers" from earlier organisms in our evolutionary past.

However, the fact that we do not know the function of a particular DNA sequence does not mean that it has no function. Scientists are with further research discovering the functions of many of these DNA sequences.

Some of these noncoding DNA sequences may also introduce genetic variation that could allow rapid speciation (formation of new species). This would fit well into a Creationist model that calls for changes within kinds after the Flood, resulting in the diverse populations of organisms we see today.

Using somatic stem cells also eliminates the possibility of rejection of the new cells by the immune system. If researchers plan to implant embryonic stem cells from one individual into a second, the tissues of the two individuals must be matched just as in organ transplants. Tissue typing is not a problem with somatic stem cells because these stem cells are harvested, differentiated into the desired tissue type, and reimplanted into the same individual.

The most important advantage of somatic stem cells is that no embryo is killed to harvest the cells. If researchers should ever determine that embryonic stem cells are superior to somatic stem cells, would that justify their use? Those who support embryonic stem cell research argue that it is acceptable to sacrifice human embryos to advance scientific and medical knowledge for the benefit of all humanity. They refuse to accept that human embryos are human beings from the moment of conception, created in the image of God.

The use of the knowledge that God has allowed man to discover must always be guided by scriptural principles. The Bible states that killing an innocent person is never justified (Gen. 9:6). Doing evil so that some good might result is also condemned (Rom. 3:8). Therefore, even if all of humanity's suffering could be relieved, Christians must stand against all research and treatments that are in opposition to scriptural principles.

⮕ Review Questions 7.3

1. Describe the differences between embryonic stem cells and somatic stem cells.
2. Give biblical references that prohibit the harvesting of stem cells from a human embryo.
⊙ 3. List several surprising discoveries revealed by the Human Genome Project.

7B—Practical Uses of Biotechnology

The production of insulin by inserting the human gene into a bacterial genome was just the beginning of DNA technology. Researchers continue to develop new techniques and uses in many areas, including medicine, pharmaceuticals, agriculture, and forensics.

7.4 Gene Therapy

Chapter 5 defines a gene as a section of DNA that produces a particular polypeptide chain of amino acids (a protein or a section of protein) that causes a particular trait. **Gene therapy** is a technique to correct a defective gene that has been identified as the cause of a specific disease. Once the defective gene has been corrected, researchers hope that it will reverse the course of the disease.

The most common approach to gene therapy uses the techniques developed for recombinant DNA (see Subsection 7.2). The abnormal gene is cut out using restriction enzymes, and a corrected gene is inserted. The corrected gene can be delivered to the recipient in one of two ways. First, cells with the corrected gene could be grown in the lab and then injected into the patient's body. In the second method, a vector, often a virus that has been altered to carry human DNA, is used to deliver the new genetic information directly

7.4

⮕ **Objectives**
- Define *gene therapy*
- Discuss how gene therapy could be used in humans

⮕ **Key Term**

gene therapy

Problems with Gene Therapy

✦ *Short duration.* In many of the gene therapy experiments, the benefit of the corrected DNA is lost after several cell divisions. Patients must undergo many treatments of gene therapy to maintain the desired result.

✦ *Viral vector problems.* Viruses are the most common vectors for gene therapy; however, they present potential problems. Although researchers try to remove the disease-causing portion of the virus, it is possible for the virus to revert to the disease-causing form and harm the host. Also, once the virus is introduced into the body, there is no certainty that it will infect only the targeted cells.

✦ *Immunity issues.* When a foreign substance enters the body, the immune system fights it. The first treatment may not cause a problem because the immune system has not been sensitized to the virus vector. Subsequent treatments can be ineffective because the body recognizes the foreign substance and successfully destroys it.

✦ *Multiple gene interactions.* Gene therapy is most effective against diseases caused by one defective gene. Since most genetic disorders are caused by changes and interactions among several different genes, gene therapy is not effective in many instances.

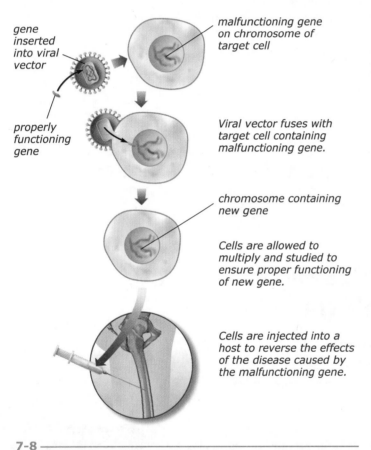

gene inserted into viral vector

properly functioning gene

malfunctioning gene on chromosome of target cell

Viral vector fuses with target cell containing malfunctioning gene.

chromosome containing new gene

Cells are allowed to multiply and studied to ensure proper functioning of new gene.

Cells are injected into a host to reverse the effects of the disease caused by the malfunctioning gene.

7-8
One goal of gene therapy is to replace a malfunctioning gene by inserting a properly functioning gene. This diagram illustrates the first method of delivery described in the text.

to target cells in the body. When the virus infects the target cells, the corrected DNA is incorporated into the DNA of the target cell to produce the proper protein and hopefully reverse the disease.

Although this sounds quite simple, there are many potential problems associated with gene therapy. These include short duration of the benefits from treatment, problems with the viral vectors, immunity issues in the recipient, and difficulty in treating diseases caused by multiple gene interactions.

In order to be successful, gene therapy must overcome these obstacles and treat the disease. Researchers are continuing to evaluate gene therapy for use against many different diseases. Gene therapy experiments have been successful in treating some diseases, including certain types of inherited blindness and cancers. However, gene therapy is still experimental; at the time of this printing, the FDA has not approved any human gene therapies. Currently there has been only limited success; however, most researchers are confident that gene therapy will eventually be successful.

➡ *Review Questions 7.4*

1. Describe the potential advantages and disadvantages of gene therapy.

⊙2. A researcher has successfully developed a long-lasting gene therapy. How might the results be different if the therapy were applied to germ cells rather than somatic cells? (Hint: The difference between somatic and germ cells [gametes] is discussed in Chapter 6.)

7.5 DNA Fingerprints

Fingerprints have been used as a method of identification for over a hundred years. So far, no two individuals have been found to have identical fingerprints—even identical twins have different fingerprints. Just as everyone has unique fingerprints, everyone has a DNA "fingerprint" that is exclusive to that person. A **DNA fingerprint** is a distinctive pattern of bands composed of fragments of an individual's DNA.

There are more than three billion nucleotides in the DNA sequence of a human. Researchers have found that about 99.9% of the sequence is the same among individuals, leaving only a 0.1% difference. Although 0.1% seems like a very small amount of variation, it represents about three million nucleotides that are unique to that person.

Using DNA Fingerprints

Since 1987, hundreds of court cases have been decided with the help of DNA fingerprinting, and some people have been released from prison because DNA fingerprinting proved them innocent. Pursuing justice is certainly one important and effective use of DNA fingerprint technology. Another important use is the identification of lost family members or of the remains of people killed in disasters such as airplane crashes or earthquakes.

DNA fingerprinting can also be used to diagnose inherited disorders in newborns. A DNA fingerprint is made of the individual and compared to a DNA fingerprint of a known disease. If the two match, it is likely that the baby is susceptible to the suspected disease. Adults can also be tested to see whether they have or carry a particular gene. Some of the diseases that can be tested for include cystic fibrosis, hemophilia, sickle cell disease, Huntington disease, and some types of Alzheimer disease.

As with all technology, DNA fingerprinting must be used wisely. In the case of disease detection, how much of an individual's genetic information should be public knowledge? An insurance company that knows a person may develop a genetic disease might elect to deny health insurance coverage. Employers might deny jobs to those who, based upon DNA testing, might be out sick more than other employees.

DNA Fingerprinting—Who Is It?

Suppose a group of hikers stumble onto a plane crash high in the mountains. As they peer inside, they see the decayed remains of the pilot, totally unrecognizable and without identification documents. When the crash investigators arrive, they carefully take samples that will be processed in the lab to produce a DNA fingerprint. Other investigators get information from the plane and search through missing persons records. Unfortunately, they find that the pilot did not file a flight plan, and there is some confusion at the local airport about who actually rented the plane. Finally the investigators narrow the identity of the pilot to two possible individuals. Fortunately, family members of the two potential victims are able to supply the lab with usable hair samples.

DNA fingerprints are produced using the hair samples from the crash site and the samples provided by the families of the two missing individuals. The three DNA fingerprints are compared, and a match is found, confirming that the crash victim is one of the missing individuals and providing positive identification.

Objectives
- Define *DNA fingerprint*
- Describe some uses of DNA fingerprint technology

Key Terms

DNA fingerprint
polymerase chain reaction (PCR)
gel electrophoresis

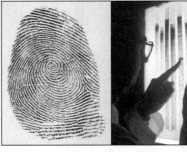

7-9

Fingerprints have long been used as a means of identification. DNA fingerprints can provide identification from a drop of blood or from some strands of hair.

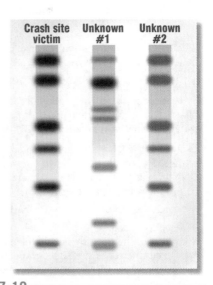

7-10

Comparison of DNA fingerprints of unknowns to crash victim. Which unknown was the crash victim? The answer is listed below the review questions.

Step 1: isolation of sample

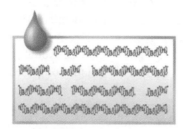

Step 2: use of restriction enzymes

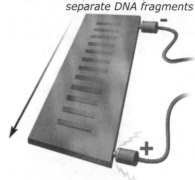

Step 3: electrophoresis to separate DNA fragments

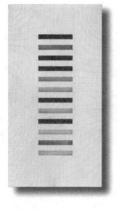

Step 4: transfer of fragments from gel to nylon sheet

Step 5: addition of radioactive probes

Step 6: DNA fingerprint

7-11

Production of a DNA fingerprint

Making DNA Fingerprints

Producing a DNA fingerprint involves isolating DNA from a sample, cutting it into pieces, and sorting it by length. There are six basic steps in the process.

✦ *Step 1*: The DNA is isolated from a sample containing cells, such as blood or skin cells.

Production of a DNA fingerprint requires the amount of DNA found in a drop of blood (approximately 1 mg), in a nickel-sized blood stain, or in several hairs. Unfortunately, the amount found is often much less. Scientists sometimes use a technique called **polymerase chain reaction (PCR)** to produce millions of copies of DNA in just a few hours.

In PCR, the DNA is heated, separating the two strands. Next, primers—short pieces of DNA that are complementary to the ends of the DNA strands being replicated—attach to the ends of each single-stranded DNA. These primers provide a starting point for DNA replication. The resulting copies of the DNA strand serve as templates for more copies, allowing millions of copies of DNA to be produced in a relatively short time.

✦ *Step 2*: Restriction enzymes are used to cut the DNA at specific sites. (See Subsection 7.2 for a review of restriction enzymes.)

✦ *Step 3*: The segments of DNA are sorted by a process called **gel electrophoresis**. This process separates the DNA fragments according to their size by sieving them through a gel that contains millions of tiny pores. The sample is placed into wells in the gel, and an electric current is passed through the gel. Since DNA is negatively charged, the fragments will move toward the positive end of the gel. As the DNA migrates, the fragments are sorted by size—the smaller the DNA fragment, the faster it will migrate through the gel.

✦ *Step 4*: The DNA fragments are transferred to a sheet of nylon (some processes use filter paper) by laying the sheet on the gel surface and allowing it to soak overnight.

✦ *Step 5*: The DNA fragments are marked by radioactive probes. These probes attach to the segments of DNA.

✦ *Step 6*: Finally, the DNA fragments are exposed to photographic film. The radioactive probes form lines on the film. The DNA fingerprint can then be analyzed by technicians and computers. The final DNA fingerprint looks something like the barcode seen on store items.

Review Questions 7.5

1. List the steps in producing a DNA fingerprint.
2. What is the purpose of polymerase chain reaction (PCR)?

⊙3. What are some potential uses and abuses of DNA fingerprint technology?

The answer to the DNA Fingerprinting—Who Is It? question is Unknown #2.

7.6 Genetically Modified Plants

DNA technology is being applied to plants as well as humans. The same basic research techniques of genetic engineering studied in the previous sections can be used in plant research. Some of the characteristics that researchers are working to achieve by genetic engineering include increased flavor and nutrient content, herbicide and pest resistance, and easier transportation. Plants that have undergone genetic engineering are called **genetically modified (GM) plants**.

Pest-Fighting Crops

Farmers are constantly fighting pests that can destroy their crops. Often this fight requires multiple applications of poisons that may also kill helpful insects and could contaminate nearby water supplies. Researchers found that the spores from a soil bacterium named *Bacillus thuringiensis* contain a crystalline protein with an interesting property. Once ingested by the insect, the protein breaks down and releases a toxin that causes perforations in the gut wall, killing the insect. The researchers were able to identify the location of the bacterial gene, snip it out, and insert it into the genome of corn. The Bt corn, as it is called, produces the protein, and when an insect begins to eat the corn plant, the insect is killed. There are also Bt varieties of cotton and soybeans.

In addition to insect pests, farmers also fight plant pests, or weeds. They must select herbicides that will spare the crop while killing the noncrop plants. Using recombinant DNA technology, researchers have developed soybeans and corn that are herbicide-resistant. Farmers can plant these strains of corn or soybeans that will survive the herbicides used to control weeds.

GM Plants as Food

Are GM plants being used to produce GM food for humans? Yes, primarily from two crops: corn and soybeans. Ingredients from these crops are widespread in processed food. It is estimated that over 60% of the food on the grocer's shelves has some GM food component. There are currently fewer than 20 GM plants that are approved for commercial production in the United

7.6

Objectives
- Explain how genetic engineering is used to modify plants
- Describe some potential adverse environmental impacts of genetically modified plants

Key Term
genetically modified (GM) plant

7-12
Genetically engineered Bt cotton (left) effectively prevents damage from many pests, increasing the crop yield per acre.

Alternative Fuel Sources

Research is being done to develop genetically modified plants that will produce alternative sources of fuel. Some plants, including corn and wheat, are already being used to produce ethanol. This alternative fuel is available at some gas stations today.

Scientists have also produced genetically engineered bacteria that are capable of converting plant cellulose into ethanol, which could then be used as a fuel source. Further research is being done in this area, and scientists hope to make more of these alternative fuels commercially available in the near future.

GM Animals

Several genetically modified plants are already being sold in grocery stores. What about GM animals? At the time of this printing, there are no genetically modified animals approved for human consumption. However, research is being done on several types of GM animals. One example is GM salmon (back), which have been reported to be faster-growing than non-GM salmon (front).

States. Since GM and non-GM crops are often not separated during processing, it is possible that food containing ingredients from any approved crops might contain some GM components. Table 7-1 shows where GM products might be found in food.

The Future of GM Plants and Foods

Since the production and use of GM plants is relatively new, there are no long-term studies that document their safety and their impact on the environment. In the United States, several government agencies are responsible for ensuring the safety of GM plants. They include the Environmental Protection Agency (EPA), the Department of Agriculture (USDA), and the Food and Drug Administration (FDA).

Much of the concern about these products is their potential impact on human health and the environment. For example, some people fear that the new proteins produced in GM plants might be toxic to humans or that the traits that produced herbicide-resistant plants could be transferred to wild varieties, producing "super weeds." The bacteria that are used as vectors are often resistant to certain antibiotics. If these bacteria remain viable, some fear they will lead to a more rapid emergence of antibiotic-resistant bacteria. For these reasons, some countries have banned the importation of any GM plants or foods.

Another concern is the ownership of GM plant technology. Companies that spend millions of dollars on the development of new GM crops will certainly want financial success. If there is no demand for the product, there will be little incentive to plant the crop, even though it requires less of the chemical pesticides and herbicides or even though it has potential health benefits. On the other hand, a demand for the product might price the seeds beyond what farmers could afford to pay.

7-1 GM Versions of Food Crops		
Food type	**Likely to be found in food products**	**Comments**
Soybeans	Yes	These are mostly herbicide-resistant. Some estimate that over 50% of the soybean crop worldwide is of a GM variety (85% in the U.S.), typically found as soybean oil, flour, or protein extracts in processed foods.
Corn	Yes	Over 30% of field corn (corn used for animal feed) is GM. Growers do not separate GM and non-GM corn; therefore, if a product includes corn, it is likely to contain some GM corn. Sweet corn (that which is sold as fresh ears) is much less likely to be GM—probably around 3%–5%.
Canola	Yes	Around 60% is GM. Canola oil is used in vegetable cooking oils, processed cheese, many nondairy products, cosmetics, soap, chocolate, and salad dressings.
Cotton	Yes	Over 70% of crops are GM varieties. Cottonseed oil is used in many products, such as peanut butter, pastries, chips, crackers, and cookies.
Tomato	No	Although the Flavr Savr tomato was on the market in the mid-1990s, it was pulled due to poor performance. Other approved varieties have not made it to consumers.
Potato	Unlikely	These comprise a very small portion of crops. Poor sales caused the developer to stop production.
Squash, sugar beets, rice	Unlikely	GM varieties have been developed, but either farmers are reluctant to plant them due to consumer concerns (European markets are closed to GM foods), or they are awaiting approval by the appropriate government agencies.

Considering the potential blessings and dangers in genetically modified plants may seem unimportant compared to the other aspects of biotechnology discussed in this chapter. However, these issues are no less a part of the Creation Mandate than stem cell research or gene therapy. Investigating and weighing the effects of GM plants on the environment and on other humans are a part of fulfilling man's duty to exercise wise and good dominion over the earth under God. In fact it is an important part of dominion because GM plants have increased the food supply to feed an ever-growing human population. All of God's creation is precious to Him, its Maker. Likewise, it should be precious to us, its stewards.

Review Questions 7.6

1. Describe two potential environmental and health problems that might result from using GM plants.

2. How is the use of GM crops a valid application of the Creation Mandate?

The History of Life

Imagine that you were hiking high in the Rocky Mountains and discovered a slab of rock containing shell fossils along a peak. This might lead you to speculate on when and how they were deposited there. In a meadow at the foot of the same mountain you watch, in amazement, a beautiful swallowtail butterfly uncoil its long tongue to drink nectar from deep within a blue wild-flower. The design and behavior of the butterfly, as well as its relationship to the flower, might cause you to wonder how they came to be. Ultimately, an exploration of life naturally leads to the question "How did it all start?"

Is life the gift of an all-powerful, righteous, loving God? Or is life the result of a random natural process? Are humans merely highly evolved primates whose characteristics and behaviors are related to those of animals, or did God make man in His image? The answers you give to those questions profoundly affect how you will deal with the many moral issues of biology.

8A—Worldviews and the History of Life

Biologists face a great challenge when they attempt to understand the history of life. Normal methods of biological research do not apply—not in the same way at least. Electron microscopes have not found words written on DNA explaining how life came into being. We cannot travel back in time and observe where life came from. And it would be impractical to study the evolution of a species at the slow rate that it is proposed to occur.

So how do biologists study life's history? They do this, in part, by using their powers of observation along with careful scientific analysis. In addition to these, however, something else plays a role in all biological study and dominates the study of the history of life. That something is the biologist's *worldview*.

8.1 What Is a Worldview?

A **worldview** is the overall perspective enabling a person to see and interpret the world; it is composed of basic beliefs about life and the universe. To understand this important word, we need to consider each part of this definition.

Understanding the Term

A worldview, first of all, is about life and the universe. It is, in other words, something comprehensive in scope. There is no part of a person's life that is not somehow affected by his worldview.

It is also important to note that a worldview is composed of beliefs. By *beliefs* we do not mean opinions or feelings but rather commitments or convictions. A worldview is made of ideas that people are committed to, and these commitments are accepted by faith. They are important ideas that science cannot prove but that form the foundation for science.

These ideas are often called *presuppositions*, statements or ideas that people reason from, not ideas they reason to. For example, most scientists assume (or presuppose) that the world we can see is real and that it will continue to respond in predictable ways into the future. If they did not believe this, they would be unable to say that the results of an experiment are repeatable. No scientific experiment can prove this claim, but unless we assume this claim is true, no scientific experiment is worth doing. Obviously, both believing and nonbelieving scientists share some core presuppositions, but they differ on others. Can you think of some Christian presuppositions about the nature of man?

A worldview is, at its core, a perspective. It is a way of looking at things, and it helps us orient ourselves in this world. For this reason people often describe a worldview as a lens through which ideas are seen, a blueprint for understanding how our lives should be lived, or a roadmap showing us where our ideas will lead us.

We should also note that this perspective usually takes the form of a grand, overarching story or narrative. Our worldview does not orient us through a detailed outline. (If that were the case, few of us would have a worldview because we would not be able to remember the outline!) It guides us through what we believe is the unifying story of the universe. This plot-line tells where our world came from, why it exists, what part we play in it, and where this world is headed.

Finally, a worldview enables a person to see and interpret the world. Because a person has beliefs about life that

8.1

Objectives
- Describe the nature of a worldview
- Defend the position that all humans have a worldview

Key Term
worldview

8-1

Each of these objects represents the role of a worldview in our lives.

come from his sense of the world's history, he interprets things in a certain way. For example, if a scientist believes that all organisms have come from one single-celled organism (over millions of years), then he will interpret similarities between a chimpanzee and a human as proof that they are related. But if a scientist believes that God made humans and chimpanzees on the same day, then these similarities are seen as proof that both were made by the same God.

Does Everyone Have a Worldview?

When you think about the term *worldview*, it's natural to wonder how common worldviews are. Understanding this concept is difficult. Are worldviews just for very intelligent people? It's true that if you asked your friends to describe their worldview, you would probably get blank stares. But their beliefs about life and the universe would quickly tumble out if you asked them what they thought of capital punishment, divorce and remarriage, or homosexual rights. If you were then to ask them to defend their beliefs, they would probably try to show that these beliefs come from their understanding of where the world came from, why humans exist in this world, and where the world is headed. Having a worldview is part of being human—we cannot live without an overall perspective that enables us to make sense of the world we live in.

Review Questions 8.1

1. What are presuppositions and why are they important?
2. A worldview is an overall perspective that is usually expressed in what form?
3. What does a worldview enable a person to do?
4. Does everyone have a worldview? How do you know?

---

redeem: (L. *redimere*—to buy back) *Redeem* means that God has purchased the salvation of humankind through the death and resurrection of His Son and will reclaim both humankind and creation for Himself.

8.2 Which Worldview Should We Use?

There are many different worldviews. In this chapter we will focus on just two—the two that matter most to the Christian who is trying to understand how biology deals with the history of life.

A Christian Worldview

The Bible claims to be the Word of God (2 Tim. 3:16), and the Christian accepts this claim. Christians derive their worldview from the Bible because its grand, overarching story is written by the Person Who created and sustains the universe. Human reason and scientific research are highly valued by Christians. But they recognize that human reason and science are not sufficient to understand the world and the history of life. God's Word is the foundation for knowing these things. Christians are happy to build on that foundation using their powers of reason and study. As they study the world in this way, they repeatedly discover that the data of the physical universe fits the worldview of the Bible.

As we discussed in Chapter 1, the basic components of a **Christian worldview** are as follows:

1. God made the world and placed humans at the center of that world for the purpose of glorifying Himself;
2. the world has fallen into a broken condition because of human sin;
3. God is working to redeem this world to Himself.

Christians seek to orient themselves in this world by fitting their experiences—including their study of the history of life—into the events of *Creation*, the *Fall*, and *Redemption*.

The Bible's teachings about Creation tell us where life has come from and why humans exist. Genesis 1–11 reveals that God called this world into existence just a few thousand years ago. Scripture indicates that God created the world out of nothing, or *ex nihilo*. The world He made was also "very good" (Gen. 1:31). This description suggests that the universe, immediately after it was created, was beautiful and fully functioning. In these accounts we also learn that humans are God's great masterpiece. He made humans in His image (Gen. 1:26–27). Because we are in His image, we are like Him in many significant ways. Therefore, human life is precious—far more precious than the lives of the other living things God made. Also, because we are created in God's image, we have the ability to study and understand the world around us. The Bible also reveals that God has called us to exercise dominion over the things He has made (Gen. 1:28). Because of this command, called the *Creation Mandate*, we have a calling from God to figure out ways to maximize the usefulness of the many resources He has placed in this world.

The Bible's teaching regarding the Fall tells us how things have gone wrong in God's world. We have rebelled against the rule of God, and He has caused our world to rebel against our work of dominion (Gen. 3:17–19). Suffering, injustice, and death are not normal events in a Christian worldview; they are proof that God has cursed our world because of our sin. But God has also cursed our entire being. We do not love as we should, and we do not think as we should (Gen. 3:12; 1 Cor. 2:14; Eph. 4:17–18). This inner brokenness is the reason that humans cannot understand reality—including the question of life's history—unless the light of God's Word shines on them (Ps. 119:130; Prov. 1:7). Also, because of this brokenness, humans tend to see this world as normal, not fallen. They also tend to reject one of the most important events in the history of life—the great Flood of Noah's day (Gen. 6–8; 2 Pet. 3:3–6).

But these tragic truths are not the end of God's grand, overarching story. God has promised to redeem this fallen world. God's Son, Jesus Christ, came to earth to die for the sins of the human race and to rise from the dead (Gen. 3:15; Matt. 1:21; Luke 24:46–47). All who trust in Him are forgiven of their sins. God also does a work in their hearts so they can understand His Word and His works in the world. And in the age to come, these believers will live in a redeemed earth, a world where the effects of the Fall will be entirely removed (Rom. 8:18–23; Isa. 11:1–9; Rev. 21–22).

We should also note that Christ revealed He would use believers in this mission. He has commanded all who believe in Him to "go . . . and teach all nations" (Matt. 28:19). This teaching involves proclaiming the gospel, but it also involves acts of mercy—acts that require some knowledge of biology. When Christ sent out His disciples, He did not send them to preach only; He also sent them to help people (Luke 10:9). Christ indicated that it is not just the words of the Christian that are to testify to God's redemption but also the Christian's deeds: "Let your light so shine before men, that they may see your good works, and glorify your Father which is in heaven" (Matt. 5:16). Thus, according to a Christian worldview, biology is an important study. It is a noble pursuit that helps a person meet the needs of God's fallen, but precious, image bearers. Biology is important to us as humans because God created us in His image and has given us the responsibility to care for His creation.

Results of Believing Evolutionary Theory

Consider the following implications of a secular worldview in relationship to science. Can you think of examples from current events that demonstrate these statements?

✦ Man is not responsible to God because He does not exist.

✦ Man does not need a savior because there are no moral absolutes and thus no sin.

✦ Man is a highly evolved animal, not a special creation of God.

✦ Man's religion should be scientism.

ex nihilo: ex (out of, from) + nihilo (L. nihil—nothing)

8-2

The origin of the universe is explored in the study of cosmology.

cosmological: cosmo- (Gk. *cosmos*—universe) + -logical or -logy (the study of)

A Secular Worldview

Most biologists believe that a Christian worldview has no place in scientific study. They believe science must be kept **secular**; that is, it must be free from religious influence. The key presupposition of a secular worldview is faith in **scientism**—the belief that the only things we can know with confidence are the things we learn through scientific study. If a question cannot be answered through scientific study, then it cannot be answered and is not even worth asking. This assumption has produced a worldview that takes a very high view of science but a very low view of humans. For example, many in society readily accept the "scientific" claims about endangered species or embryonic stem cells and are determined to take actions in spite of the impacts they would have on human life.

When it comes to biology and the history of life, most secularists construct their grand, overarching story using the **theory of evolution**. This is a very broad term and includes many variations, but there are three major components of evolutionary theory that evolutionists have in common:

✦ *Theory of cosmological beginnings.* Man's attempts to explain how the universe, the earth, and even matter and energy came into being are theories of cosmological beginnings. Since those who believe in the theory of evolution reject the truth of God's Word, they often propose that the big bang began the process of evolution or even that the universe has always existed.

What Kind of Change?

Creationists do not dispute the fact that the kinds of genetic changes described in Chapter 6 can cause the traits of a species to change over time. In fact, these changes might even be significant enough that scientists declare that a new species has developed. However, these changes do not result in a new kind of organism. It is simply a reshuffling of the genes that were already present and fits in a biblical worldview of variation within a God-created kind (Gen. 1:24–25).

Throughout this chapter, we are not trying to discredit these minor changes that obviously occur, sometimes even very rapidly. Instead, we are refuting the belief that these random changes can, over time, produce new kinds of organisms, and that a branching lineage of these kinds of living things has given rise to all forms of life today. That is, if you could trace it back far enough, there would be a common ancestor from which all life arose. This molecules-to-man view of evolution is clearly a contradiction of Scripture.

The differences between these two woodpecker species can be explained by slight genetic variation. Thus, a Creationist might view them as members of the same created woodpecker kind.

Downy woodpecker

Hairy woodpecker

◆ *Theory of biological evolution.* Biological evolution proposes that over time less complex organisms give rise to more complex organisms, which in turn produce even more complex offspring. Humans are not special beings made in God's image. They are simply more highly evolved animals. The things that humans value most—things like love, joy, peace—are not gifts of the eternal God. They are the results of complex chemical reactions in the brain. And since the evolutionist does not believe in the fall of man into sin, suffering and death are not evils. They are the regular rhythms of an evolving world. Pain and even criminal activity are unpleasant, but the evolutionist cannot call them evil. In the end, there is no difference between the person who is kind and the person who murders. Science can distinguish pain from pleasure; it cannot, however, distinguish good from evil.

◆ *Philosophy of evolution.* If there is any redemption in this worldview, it is found in what may be called the philosophy of evolution. This philosophy suggests that all things are progressing toward a future perfection and that things are currently improving. This basic assumption helps the evolutionist justify his mission in science. When asked for examples of this progress, an evolutionist may point to the "improvements" in man's knowledge and in human society. It is at this point that this worldview clearly conflicts with itself. For science to make significant strides in the struggle against disease and death, great sacrifices of time and energy will be required. But why should these sacrifices be made for humans who are, in their estimation, merely animals? Why not let the forces that have produced man have their way? Christians believe they are called to properly use science to help humans, the image bearers of God. The progress comes from God's hand, however, not blind processes of evolution.

When the term **evolutionist** is used in this book, it refers to a person who believes these three ideas. The term **Creationist** refers to those who believe the basics of a Christian worldview: the Bible is the inspired Word of God, and the Genesis account is a literal description of God's Creation of all things.

In the sections that follow, we will examine how a secular, evolutionary worldview has dealt with the question of the history of life. Then we will examine how a Christian worldview has dealt with the same issue. Throughout these sections it will be clear—as it has been already—that the secularist lives by faith just as much as the Christian. The difference between the secularist and the Christian is where they put their faith. Secularists put their faith ultimately in human reason and scientific study. Christians put their faith ultimately in God and His Word. It will also be clear that whereas one worldview fits as an explanation of the world we live in, the other worldview produces a picture of the world that is meaningless and without hope.

A Difference in Worldviews

Looking at a reconstructed skeleton of a large reptile in a museum, a student who believes in evolution may say to his friend, "These dinosaurs were huge. They lived during the Mesozoic period, between 70 million and 200 million years ago. Dinosaurs became extinct about 70 million years ago, probably as a result of a change in the earth's environment that was likely caused by one or more huge meteors striking the earth."

Looking at the same skeleton, a Christian student might reply, "You're right; they were huge! But the Bible says that God created them, along with the rest of creation, in six days. They lived prior to Noah's Flood, which many Bible scholars agree occurred about 4000 years ago. Some scholars believe that Noah may have taken some dinosaurs on the ark and that most dinosaurs became extinct some time after the Flood because of environmental changes. The Flood and the environmental changes that probably resulted from it can easily account for the formation of these fossilized bones."

Such a response can easily lead to a discussion of the three components of a Christian worldview—Creation by God, the fall of man into sin, and the redemption of man by Jesus Christ.

The existence of dinosaur skeletons cannot be denied, but their interpretation is based on a person's worldview.

1. Why do Christians derive their worldview from the Bible?

2. What role does human reason and scientific study play in a Christian's approach to studying the history of life?

3. What are the basic components of a Christian worldview?

⊙4. What is the nature of the emotion *love* in a secular, evolutionary worldview?

5. In what way does the evolutionary worldview clearly conflict with itself?

8B—Biological Evolution

Before the 1700s most Europeans believed that the earth was created by God about 6000 years ago. However, during the period of history referred to as the Enlightenment, many began to doubt what the Bible said. They started to formulate ideas about nature without any reference to the statements of Scripture. Looking at God's world through man's wisdom rather than God's Word, they saw a very different world than they once had seen.

8.3 The Beginnings of Evolutionary Theory

One of the few direct evidences of life in the past is the fossil record. A **fossil** is any direct or indirect evidence of a once-living organism that is embedded or preserved in the earth's crust. A dead body or a footprint can therefore be considered a fossil. A dead body usually decomposes quickly; a footprint, unless made in cement or some similar material, is also temporary. However, the term *fossil* normally refers to remains of long-dead organisms that have been naturally preserved.

As early scientists began to examine the fossil record, they noticed that some of the fossil organisms were similar to—yet different from—organisms that were then alive in the same regions. Scientists began to wonder why this was. They theorized that the organisms had changed, or evolved, over time. These observations formed the basis for the early theories of evolution. However, for the story of evolution to be possible, either evolution would need to have occurred rapidly or the earth would have to be far older than people had once assumed.

The Age of the Earth

Until the 18th century, most people looked to their Bibles to find the age of the earth and agreed that it was around 6000 years old. But there were some scientists who were willing to ignore the Bible as they attempted to determine the earth's age. Most of their models produced ages of the earth thousands of years older than the Bible. This was not by accident. Most of these geologists were Deists (people who believe a god created the earth but then left it alone), and they wanted science to back up their unbelief in the God of the Bible.

Two geologists were especially influential during this period: James Hutton

---

8.3

⮞ **Objectives**

- Understand the developments of ideas about an old earth
- Relate changing views of geology to the emergence of evolutionary theories
- List Lamarck's theories

⮞ **Key Terms**

fossil
apparent age

Creation with Apparent Age

People often fail to consider that the earth as well as animals, plants, and humans must have been created with **apparent age**, or *fully mature*. Adam was not an infant on the seventh day of Creation. Even though he was just a day old, he had the body of at least a young adult who was able to care for himself. The same was true of the animals. The Garden of Eden had fruit-bearing plants, not just seeds and seedlings.

The earth itself must also have had some apparent age. Although streams, soil, and many other features of the earth take time to form today, they are described as being fully formed in God's Creation. If a person had measured the soil depth in Eden the day after Creation week and had calculated the age of the garden based on the length of time required for soil production, he undoubtedly would have come up with an age for Eden considerably older than it actually was.

This creation with apparent age throws off some of the buildup and decay clocks that evolutionists use to date the earth. For instance, scientists who measure how rapidly certain chemicals are building up in an ocean and then compute backward to a point where none of these chemicals were yet in the ocean obtain an apparent age but not an actual age. Why? First, they assume that the original seas did not contain any of those chemicals, and second, they assume that the rate of deposition of the chemicals was constant. It is quite possible that the newly formed seas already contained some of these chemicals. If a scientist knew exactly the concentration of certain chemicals actually in the sea at Creation and knew all the factors (including the Flood) that could affect their further buildup, he could more accurately determine the age of the earth. But this information is not available.

(1726–1797) and Charles Lyell (1797–1875). Hutton theorized that continents had formed out of the sea and fallen back into the sea repeatedly over millions of years. Charles Lyell developed the doctrine of *uniformitarianism*. He said that theories of the earth's age should be governed by the observation of the rates of gradual geologic processes of the earth today. According to this view, catastrophes such as the Flood were out of the question when looking at geologic features.

Although some natural processes do proceed at fairly predictable rates, there are many exceptions. For example, stalactites and stalagmites can form rapidly under certain conditions, and large meteorites can cause sudden, widespread changes. Uniformitarians also rely on these slow, steady rates to explain how the earth came into existence. If they do believe in a god, he is one that is bound by these natural laws, and supernatural events are impossible. The theories promoted by Hutton and Lyell provided the foundation for doubting the Bible and accepting unbiblical views on the age of the earth and, ultimately, origins.

8-3

The concentrations of various minerals and salts in the ocean are an unreliable source of data for determining the age of the earth.

Lamarck's Theory of Evolution

In an attempt to explain how organisms change over time, Jean Baptiste Lamarck, a French biologist, introduced his theory of biological evolution in 1801. At the core of his theory were three concepts.

✦ *Theory of need.* For an organism to evolve a structure, it must need the structure.

✦ *Theory of use and disuse.* If an organism continues to use a particular structure, the structure will continue to evolve. If an organism stops using a structure, the structure will degenerate and disappear.

✦ *Theory of inheritance of acquired characteristics.* If an organism acquires a characteristic, it can pass this characteristic on to its offspring.

The most common example of Lamarck's theory is the proposed evolution of the giraffe. He asserted that the giraffe's ancestors were deerlike animals on the African plain. A long drought forced these early giraffes to stretch their necks to reach higher tree leaves (theory of need). Their necks became longer with stretching (theory of use and disuse). Adults who gained slightly longer necks produced offspring with slightly longer necks (theory of inheritance of acquired characteristics). After many generations, the giraffe's long neck developed through a series of slight changes.

8-4

Jean Baptiste Lamarck and his proposed evolution of the giraffe

Evaluation of Lamarck's Theory

Although Lamarck was very influential, his theory was eventually discredited. Scientists discovered that most giraffes do not need to stretch their necks for food. In fact, their long necks are actually an obstacle for drinking. According to Lamarck's theory, giraffes' necks should now be getting shorter.

Most evolutionists today agree that the theory of need is not an actual factor in the evolutionary process. All organisms need structures (most humans could use radar, better hearing, third hands, or eyes in the back of their heads), but the DNA, which determines what structures an organism has, is not affected by need.

Modern scientists have also disproved the inheritance of acquired characteristics. Children inherit genes from their parents, who inherit their genes from their parents, and so on. Although a parent may have developed a certain characteristic in himself, his child will not inherit it. If a man has his arm amputated, will any of his children be born without an arm? No, the DNA of the father contains the information for two arms.

A child may inherit the genetic potential to develop a characteristic that is well developed in one or both parents, but the actual characteristic is not inherited. For example, if a boy's father lifts weights for years and develops a strong, muscular body, the boy does not inherit his father's build. The son may inherit the genes to permit him to develop such muscles, but he will have to exercise to build his body. The same is true of musical and artistic abilities and similar characteristics.

The theory of use and disuse is unacceptable based on modern genetics. Genes can be turned off and on, but use or disuse does not alter or destroy them for the next generation.

8-5
If this tree were to produce seeds, the plants that grew from them would be of normal shape and size, not in the form of this manipulated bonsai tree.

Review Questions 8.3

1. List reasons why dating the earth by decay or buildup methods can be unreliable.
2. How does belief in the Flood of Genesis contradict a belief in uniformitarianism?
3. List the three basic tenets of Lamarck's theory of evolution and describe each.
4. List several objections to Lamarck's theory of evolution.

8.4 Darwin's Theory of Evolution

Charles Darwin was born February 12, 1809, into a religious household. Even as a child, he was interested in natural phenomena. He was sent to medical school, but he soon lost interest in his studies there. Then he went to Cambridge to study to be a pastor, but he spent most of his time there hunting for and studying beetles. Also, during his college years he studied the Bible's account of creation and arguments for design in nature.

Voyage to the Galápagos Islands

In 1831, Darwin signed up to be a naturalist on the HMS *Beagle*. The trip, which lasted five years, took Darwin to South America and then to the Galápagos (guh LAH puh gus) Islands off the west coast of Ecuador. The Galápagos Islands are the result of volcanic activity and have a variety of habitats. Some are very desertlike and even have cactus "forests," while other areas have humid forests.

A group of birds Darwin observed reminded him of the finches he had seen in England and other parts of the world. Normally finches are seed-eaters. Although some of the finches on the Galápagos Islands did eat seeds, others

8.4

Objectives

- Explain the two main aspects of Darwin's theory of evolution
- Evaluate Darwin's theories in light of the Bible's teachings on origins and nature

Key Terms

theory of descent with modification
common ancestor
artificial selection
theory of natural selection
survival of the fittest
fitness
extinct
environmental determinism

fed on the fleshy parts of cacti. One group even used the spine of a cactus held in their beaks to obtain insects from under the surface of the cacti. Darwin was confused by the birds he saw. In fact, he did not even recognize that they were all finches but felt they were unique species from several major bird groups. As a naturalist, he collected specimens for later study.

Publication of *The Origin of Species*

After returning to England, Darwin diligently studied his observations and began to develop his theory of evolution. During this period, his beloved ten-year-old daughter died. Darwin had already been troubled by the suffering he had seen in the animal world. Now with the death of his own daughter, he questioned whether God was involved in the natural order at all. How could God be good and yet design and govern a world filled with pain—for animals as

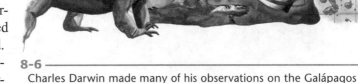

8-6

Charles Darwin made many of his observations on the Galápagos Islands.

well as for humans? Perhaps, Darwin reasoned, all living things experience suffering and sorrow because God has little or nothing to do with this world. In 1859 Darwin published his most important work, *On the Origin of Species by Means of Natural Selection.*

The first element of Darwin's theory discussed in *The Origin of Species* is the **theory of descent with modification**. This states that newer forms of organisms are actually the modified descendants of much older, similar organisms. A colleague of Darwin examined the bird specimens he had collected and declared that they represented thirteen different kinds of finches. Darwin reasoned that all the finches he studied on the Galápagos Islands had descended from one type of finch that had originally populated the islands. Using that thought as a model, he theorized that all living things descended from a single life form, or **common ancestor**. Darwin used this theory to account for similarities between organisms found in the fossil record and living organisms. For example, he found fossils of the giant ground sloth along the coast of Argentina. The extinct sloth was similar to the surviving three-toed sloth he saw hanging in the trees, but much larger. Some were the size of elephants!

Long before Darwin began his observations, scientists had recognized variations in organisms. Plant and animal breeders, for example, had practiced controlled breeding to obtain desired strains of existing variations. Darwin referred to this as **artificial selection**. Dozens of varieties of pigeons, for example, have been bred from the wild rock pigeon. Artificial breeding is man's choosing individual organisms to breed in order to obtain offspring with certain characteristics. Darwin proposed that the finches he observed, as well as other groups of organisms, had experienced the same type of breeding without man's help. Darwin called this *natural selection.* Darwin's **theory of natural selection** states that the environment may cause certain characteristics to become dominant in a population by allowing organisms with the most suitable traits to survive long enough to reproduce. This idea has become a major tenet of evolutionary theory.

Almost every population of organisms has natural variations. According to Darwin, organisms struggle with one another to find food, shelter, and other necessities of life because of overpopulation. The individuals having the best characteristics for success in the struggle survive, reproduce, and pass those characteristics on to the next generation. This principle

8-7

The great variety seen in the different dog breeds is the result of many generations of artificial selection.

8-8

Variations in shell shape among different Galápagos tortoise populations influenced Darwin's thinking on the origin of species.

became known as the **survival of the fittest**, and it is a part of the theory of natural selection.

The populations of finches that Darwin observed during his visit to the Galápagos Islands were similar to each other but showed variations in coloring, size, and beak length. It seemed that these slight changes made each population more suitable for its particular island or habitat. In a similar way, he noticed that the large Galápagos tortoises had shells that varied in shape from one island to another. When Darwin reflected on these differences after his trip, he theorized that there must have been one original type of finch and tortoise, but over time differences developed, causing different races of animals to form.

The concept of fitness is at play here. An organism's **fitness** is measured by the production and survival of its offspring. Those individuals that make a more significant input on the next generation by leaving a greater number of offspring are considered to have greater fitness than those that leave fewer. Some populations have such a low level of fitness that they eventually all die out. They become **extinct** because their traits and abilities are not a good match for the environmental conditions.

Publication of *The Descent of Man*

In 1871, Darwin wrote *The Descent of Man*. In this book he applied his theories of descent with modification and natural selection to human beings. He proposed that man—like all other living things in his theory—originated (descended or evolved) from lower forms of life. It was at this point that Darwin made it clear he did not believe God made humans in His own image through a direct act of Creation. To Darwin, man was simply a highly evolved animal.

Evaluation of Darwin's Theory

Although many Christians are rightly critical of Charles Darwin because of his views, it is important to recognize that some of his conclusions are useful and valid. The Bible recognizes that the world is a place that is filled with violence as organisms compete for limited resources. Death and suffering are an inevitable result of the Fall. This reality fits well with Darwin's description of the struggle for survival. In addition, nothing in Scripture contradicts Darwin's observations of variations within a group of organisms.

Effects of Darwin's Theories

Darwin's theory of improvement through struggle (survival of the fittest) was just what many of the people of his day wanted. A contemporary of Darwin, Karl Marx, was in favor of evolutionary concepts because they supported his political ideology—communism—which teaches that through collective effort man can improve his existence. More importantly, Darwin's theories appeared to give scientific evidence against the reality of God. Communistic philosophy denies the existence of God. In the 1930s, Adolf Hitler adopted the ideas of survival of the fittest and superior races in his political ideology of Nazism.

Darwin's theories have been used to support **environmental determinism**, the concept that the environment determines what an individual is. This concept blames the world around a person for the sin he commits. According to environmental determinism, a drunkard is not a sinner but rather a victim of his environment; his environment, this belief says, drove him to drink.

The Christian view of man—that man was created perfect but is now a wicked sinner who is responsible for his sin—is by comparison unappealing. The Bible places the blame for sin on the individual. It says that man cannot save himself but is in need of the Savior.

Darwinian evolution promises great things in the future if man works at it. Darwinism is, therefore, contrary to God and His Word because it blames natural causes for man's present state and hopes in natural causes for man's future. In contrast, the Bible holds man accountable for his condition and presents God as man's only hope.

Adolf Hitler's Nazism was an outgrowth of Darwin's principle of the survival of the fittest.

On the other hand, there are *many* points where a Christian should disagree with the teachings of Darwin. His theory clearly violates the Bible's account of the origin and history of life. God created distinct kinds of living things and gave them the ability and a command to reproduce after their kind. Nothing in the Bible speaks of a single common ancestor or the development of new kinds of living things. To Darwin, man is just another product of evolution as are mosses, millipedes, and monkeys. But the Bible teaches that God created humans in His own image by a special act of Creation.

Darwin's theory robs God of His glory in Creation. Darwin attempted to explain the origin and development of life on the earth excluding God. But the Bible presents God's works of Creation and His governing of all living things as something that declares His own glory.

A Christian worldview sees everything through the themes of Creation, the Fall, and Redemption. God made this world a good place, without pain and suffering. Because of human sin, however, God has cursed this world. The pain that troubled Darwin does not call into question the integrity of God. It calls into question our own integrity; it is God's way of reminding us that we are sinners. Darwin's teaching denies the biblical theme of Creation and the Fall and, in so doing, denies the hope of Redemption. This world is a tragic place, but it will not always be so. God will one day restore all things to their former beauty and holiness. In that day all of His creatures will glorify Him for creating, sustaining, and redeeming this world. But in Darwin's theory, the glories of redemption are lost.

Darwin's theory has also damaged humans' relations with each other. Many people, including politicians, psychologists, and ministers, have applied Darwin's ideas to humans and societies. These dangerous extensions (highlighted in the margin box, "Effects of Darwin's Theories") illustrate the impact of ungodly thinking.

Scientifically, the theories of Darwin rest on an unstable foundation. Darwin never offered scientific proof for his most significant claim—that all organisms descended from a common ancestor by means of natural selection. Although his observations and conclusions support the fact that small changes may occur in populations over time, he was unable to connect the dots and show that one kind of organism ever developed into another kind. In fact, no clear evidence has ever been found in the many years since Darwin proposed it.

The molecules-to-man type of evolution Darwin tried to defend is no more illustrated by the changes in finches and their separate populations than it is by the different types of cattle, sheep, chickens, or corn man has been able to develop. The generalized finch gene pool has always contained the genes to produce the variations Darwin noted on the islands. However, the change of frequency of a characteristic in a population is not biological evolution. Nothing new was made; a more complex organism did not develop. Natural selection merely sorted through the characteristics that were available. In fact, more recent research on Darwin's finches has revealed that in a relatively short period of time, some of the finches are reverting to a generalized form with more average beaks. Biological evolution requires new characteristics, not just a slight change in or regrouping of existing characteristics.

The natural selection of the fittest organisms may appear to be quite logical. It has drawbacks, however. First of all, overpopulation is not always a valid concern. Some organisms, like predatory birds, have behavior patterns in their life cycles that prevent overpopulation. When food is scarce, these parents produce fewer young so that there will not be overpopulation.

Another drawback to Darwin's survival of the fittest is that in nature, it is not always the fittest that successfully reproduce. For example, an animal that can run fast and thereby escape predators may be considered the fittest. If, however, the genes that cause its running ability also cut down on its reproductive ability, the "fittest" may survive, but it may not reproduce. Consider what would happen if speed were genetically linked to lower intelligence, poor digestion, susceptibility to disease, or other inheritable disorders. Although the organism has particular genes enabling it to survive certain environmental conditions, it may have other undesirable genes. In other words, the best all-around organisms are not always the ones to survive, nor are they always the ones to reproduce.

Evolutionists, of course, define *fittest* as those organisms that survive regardless of their qualifications and would not accept this argument. However, by using the concept of "fitness," evolutionists are actually damaging their own argument and saying nothing more than "the organisms that survive are the fittest; therefore, the fittest are the organisms that survive."

Survival of the fittest as evolutionists explain it is not a method of developing new characteristics. It works by simply selecting for or against characteristics that already exist. A method of increasing genetic information, which is the means to develop new characteristics, is essential to a theory of biological evolution.

Another part of Darwin's theory attempts to explain how changes occur. Darwin believed that each organ of the body produces *pangenes*, which travel through the blood to the reproductive organs to be given to the offspring. Darwin thought that pangenes are affected by the organ where they originate and therefore result in the inheritance of acquired characteristics. Darwin borrowed this concept from Lamarck but claimed it as his own.

Today scientists know that the inheritance of acquired characteristics by pangenes is false. When introduced, however, Darwin's pangenes sounded interesting and scientific. The abundant examples of natural selection he gave in his works were accepted as adequate proof of pangenes. Examples of natural selection, however, do not support the inheritance of acquired characteristics through pangenes.

Review Questions 8.4

1. List the two elements of Darwin's theory of evolution. What are some objections to these concepts?

2. What mechanism did Darwin propose to supply the changes needed for evolution? What are some objections to this mechanism?

3. Why is Darwinian evolution so compatible with environmental determinism?

8.5 Later Developments in Evolutionary Theory

After Darwin's theory was published, scientists started to add ideas to support the theory. They theorized about a history of life on the earth, from tiny single-celled organisms to the current biosphere. Also, as biological knowledge grew, scientists added genetics to the idea of natural selection.

Origin of Life

Before life can evolve, there must be something alive to start with. The origin of the first living cell is one of the biggest obstacles and oft-debated topics in the story of evolution. The origin of life from nonliving substances,

abiogenesis, has been pondered for over a century but is still a mystery to evolutionists. They have proposed many theories—from crystal formation to deep sea vents to space aliens seeding life—but none has been demonstrated scientifically.

New scenarios are proposed every few years for how life originated. One of the most popular theories is that life emerged from the primordial sea. According to this evolutionary theory, when life first started the early earth was no Garden of Eden; it was more like hell. Volcanoes covered the face of the earth, meteors struck the earth constantly, and the atmosphere was mostly methane and ammonia. Somehow the chemicals for life were mixed together in the right way in the early ocean, and then with energy from a bolt of lightning, life began. The Miller-Urey experiment was once used to support this theory.

In 1952 Stanley Miller and Harold Urey made a mixture of gases simulating what they believed to be the early earth atmosphere. Then they sparked the artificial atmosphere with electricity, and a few amino acids, the building blocks of proteins, were formed. This was widely acclaimed as evidence for how the first life may have originated. Many secular scientists now dispute the significance of this experiment because they have new explanations on what the "early" atmosphere might have been like. Based on studies of rocks and other evidence, they believe that the "early" atmosphere contained much more oxygen than was thought before. This is a major problem for the evolutionary origin of life because oxygen destroys the chemical compounds that are needed to form life. Their experiment also required manufactured equipment and the careful control and manipulation of conditions. All of these depend on the intelligence and expertise of man. Today, evolutionists are as much in the dark about the origin of life as they were in Darwin's day.

Evolutionary History of Life

Evolutionists developed their theory of the history of life based on what they thought the fossil evidence showed. The deepest fossil layers seem to start out with "simple" marine life, and then the layers above them are filled with increasingly complex organisms. Many scientists interpret this as evidence that simple life forms existed first and eventually gave rise to larger and more complex organisms. Elaborate diagrams and terms are used to describe this geologic evidence. We will look at a Creationist response to this interpretation of the fossil layers in Subsection 8.7.

The Geologic Column

After Hutton and Lyell began to challenge the biblical view of the age of the earth, many geologists and paleontologists felt the freedom to propose increasingly older ages for the earth. Based on the rates at which sediments were collecting at that time, they surmised that it must have taken millions of years to accumulate the thicknesses of sedimentary layers of rock they were finding. By using common fossils and other clues, they eventually developed a **geologic column**—a sequence of rock layers that supposedly provided a permanent record of the history of the earth and the sequence and rate at which life evolved through its various forms. By 1854, they published this scale with names for eras, periods, and smaller subdivisions of these. Over the years, the time scale has been expanded to include older dates, but little change has been made to the names. This geologic sequence with its corresponding time scale appears in most books that discuss evolution and fossils. Most secular geologists believe the earth is approximately 4.6 billion years old.

Objectives

- Describe the development and use of the geologic column
- Explain the differences between Darwinism and Neo-Darwinism
- Give arguments against mutations as the means of increasing genetic variation
- Explain why homologous structures do not support evolution
- Contrast punctuated equilibrium with earlier evolutionary theories

Key Terms

abiogenesis
geologic column
radiometric dating method
Neo-Darwinism (mutation-selection theory)
genetic load
punctuated equilibrium
homologous structure
convergent evolution
vestigial structure

Dating Fossils

A person who counts the number of layers above a fossil and then attributes a certain age to each layer is assuming that fossils were formed progressively. But as discussed in this chapter, fossils are probably the record of one major catastrophe, Noah's Flood. Even if they are a record of several major catastrophes, it would be difficult, if not impossible, to determine which layers belong to which catastrophe and what period of time to assign to each layer.

Evolutionists often use another method to determine the age of a fossil: they consult an evolutionary timetable of when organisms supposedly developed. When consulting such a chart, the evolutionist is assuming not only that biological evolution occurred but also that the fossil record is progressive.

Evolutionists often use index fossils to tell the age of other fossils in the same area. An *index fossil* is the remains of an organism (usually an extinct animal) that supposedly lived during only a certain time in evolutionary history. When an index fossil is found in a layer of rock, evolutionists believe they can consult evolutionary timetables and thereby accurately date that layer as well as the fossil layers above and below it.

By analyzing the use of index fossils, you can see that dating the fossils by the fossils themselves is primarily guesswork. It often winds up with one scientist quoting another scientist as the authority for his dates. If scientists have merely guessed at the date based on assumptions, it is impossible to scientifically determine the age of a fossil. Consequently, many scientists look to other dating methods to determine the age of fossils.

Dating by Decay or Buildup

If a person examined a palm tree, measured its breadth and height carefully, and then made the statement that the tree is 10.5 years old, it would be appropriate to ask him how he knows that is true. He might say that in the environment in which he found it, a tree of that species would need to grow exactly 10.5 years to reach its current size.

All his statements may be accurate and even testable. However, several assumptions make his statement pure guesswork rather than thorough science. He is assuming that the tree sprouted from a seed in the place where it now grows. In fact, it might have started growing elsewhere under different conditions and then been transplanted to its current location. He must also assume that the tree has grown at a known, constant rate. One or all of these assumptions may not be true. An error in any of them could affect the accuracy of the estimation, perhaps by as much as several years.

The same principle applies to using the decay or buildup of substances to determine the age of the earth; underlying these methods are many speculative assumptions.

Although highly favored by evolutionists because they often give very old dates, radiometric (RAY dee oh MET rik) dating methods are losing scientific credibility because of faulty assumptions and measurement problems. A **radiometric dating method** measures the decay of a radioactive substance into a nonradioactive substance. Probably the best known of the radiometric dating methods is the *uranium-lead method*. Other radiometric dating methods examine the ratios between potassium and argon and between rubidium and strontium.

At present there is some question about whether decay rates changed significantly in the past.

A scientist can measure (with a degree of accuracy) the amount of uranium and the amount of lead in a rock sample. If, however, the scientist then makes a statement regarding the age of that rock based on the ratio of lead to uranium in it, he is assuming several things.

Various species of trilobites, extinct arthropods, have been used as index fossils.

- When the rock was formed, it contained all uranium and no lead.

- The rate of decay has been constant from the time the rock was formed.

- None of the uranium, its subproducts, or lead escaped from or was added to the rock since it was formed.

In essence, the scientist is assuming the same things as the person who guessed how old the tree was. Good evidence indicates, however, that these assumptions are not valid.

Because many of the products and subproducts of radioactive decay easily escape from rocks by exposure to water or even air, the accuracy of dates obtained is even more questionable. To use this method, a scientist must know all the conditions under which the substance has been kept since it was formed.

To account for these variables, scientists try to set the radiometric clock by dates from other sources. For example, when they test a rock found near an index fossil, and the rock yields a certain ratio of the radioactive element to its end product, they often

radiometric: radio- (L. *radiare*—to emit beams) + -metric (measure) assign the date of the index fossil to that ratio.

Since other rocks in the area have been exposed to similar conditions, scientists assume that they can date the other rocks by obtaining their ratios and then comparing them to the ratio of the rocks near the index fossil. In other words, they are setting the radiometric clock by the assumed date of the index fossil. The date of the index fossil, as discussed earlier, is merely an evolutionary guess and contradicts scientific knowledge and the Bible. Setting the radiometric clock by index fossils therefore yields erroneous dates.

The Radiocarbon Dating Method

The *radiocarbon dating method*, or carbon-14 dating method, is a radiometric procedure used to date substances that were once alive. Most of the carbon on the earth and in the atmosphere is the nonradioactive isotope carbon-12. About 6 miles up in the atmosphere, carbon-14 is formed by cosmic radiation bombarding nitrogen. As soon as carbon-14 is formed, it begins to degenerate into nitrogen. The half-life of carbon-14 is about 5730 years (figures vary as much as 200 years).

Carbon-14 in the atmosphere combines with oxygen, forms carbon dioxide (CO_2), and spreads throughout the air. Only a small amount of atmospheric carbon dioxide is made of carbon-14. Carbon dioxide (both carbon-12 and -14) enters the photosynthetic process of plants and in time is passed to all other living things.

When an organism dies, it stops taking in carbon-14. Scientists can measure the carbon-14 and carbon-12 in a specimen and obtain a ratio. By comparing this ratio to the ratio of carbon-14 and carbon-12 in currently living things, scientists date the specimen, finding the time that its organic elements were last in a living thing.

The radiocarbon dating method can give reliable dates within the framework of several thousand years. Since the ratio of carbon-14 to carbon-12 on the earth is very small, the tiny ratios obtained when measuring the carbon-14 of a supposedly very old organic substance are doubtful.

Scientists realize that if the amount of radiation coming to the earth from outer space was at any time significantly different from the present amount, the dates obtained by radiocarbon dating would be unreliable. Evidence shows that the rate of radiation from the heavens has not always been the same.

Before accepting carbon-14 dating, a person must consider its several assumptions:

- The method assumes that the amount of carbon-14 forming in the atmosphere is now, and always has been, constant. Current measurements indicate that the carbon-14 to carbon-12 ratio is increasing. More is being formed than is degenerating. Differences in cosmic radiation (of which there is evidence) could be the reason.
- The method assumes that carbon-14 has reached

equilibrium: the amount of carbon-14 forming equals the amount decomposing into nitrogen. Assuming present conditions, this equilibrium would take about 30 000 years to reach. Present data shows that the carbon-14 ratio is increasing, indicating that the ratio has not yet reached equilibrium (as those who believe in a young earth are not surprised to learn).

- The method assumes that the carbon-14 decay rate is constant. It can be demonstrated, however, that the decay rate can be inconstant.
- The method assumes that the amount of carbon-12 available to organisms is constant. The abundance of carbon fossils (coal, oil) indicates that the amount of carbon-12 available has probably not always been the same.

The radiocarbon dating method is probably the most reliable method of dating organic materials of unknown age, but it obtains reliable dates only within certain limits. For relatively short periods of time (up to a few thousand years), the method can accurately tell the age of some specimens. Dates that are very old or are based on specimens that may have been kept in unusual conditions should not be trusted. Radiocarbon dating is not infallible. There are a number of specimens for which the method has given wrong dates. Shells taken from living clams have been dated thousands of years old.

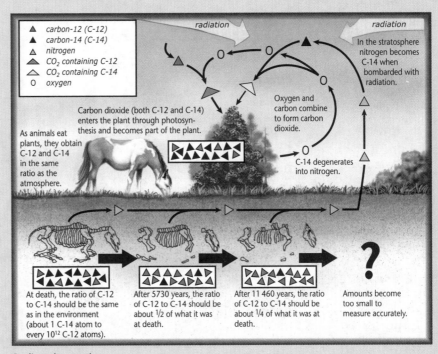

Legend:
▲ carbon-12 (C-12)
▲ carbon-14 (C-14)
△ nitrogen
⬟ CO₂ containing C-12
△ CO₂ containing C-14
O oxygen

In the stratosphere nitrogen becomes C-14 when bombarded with radiation.

Carbon dioxide (both C-12 and C-14) enters the plant through photosynthesis and becomes part of the plant.

As animals eat plants, they obtain C-12 and C-14 in the same ratio as the atmosphere.

Oxygen and carbon combine to form carbon dioxide.

C-14 degenerates into nitrogen.

At death, the ratio of C-12 to C-14 should be the same as in the environment (about 1 C-14 atom to every 10^{12} C-12 atoms).

After 5730 years, the ratio of C-12 to C-14 should be about 1/2 of what it was at death.

After 11 460 years, the ratio of C-12 to C-14 should be about 1/4 of what it was at death.

Amounts become too small to measure accurately.

Radiocarbon cycle

How should a Creationist view this scale and its suggested ages for the earth? The account of the Flood (to be discussed more fully in Subsection 8.7) certainly provides the type of cataclysmic changes that could deposit vast layers of sediment in a short period of time. In fact, the general sequence of fossils can be explained just as easily by the chronology of the Flood as it can by a gradual sequence of evolutionary development. You must also realize that there is no place on earth where the complete column is found in sequence. The ideal order of layers upon which the geologic column was developed is actually a combination of layers from many different sites. For reference purposes, below is a table summarizing the geologic time scale with a diagram showing when certain groups supposedly originated.

Neo-Darwinism

In the early 1900s, a Dutch botanist, Hugo De Vries, made some observations in his garden. He noticed various sudden changes, which he called *mutations*, in the offspring of an evening primrose, an American plant. In his book *Species and Varieties, Their Origin by Mutation*, De Vries suggested that these sudden inheritable characteristics were the means of obtaining variations that could result in evolution. The characteristics which De Vries observed, it was later learned, were nothing more than the sorting of genetic information already existing in evening primroses.

The concept that mutation supplies the variations and that natural selection determines which variations will survive and breed is called **Neo-Darwinism**, or the **mutation-selection theory**, and is accepted by many evolutionists today.

8-9

Both the rock layers and the carving of the Grand Canyon are a testament to the power of water, during and just after the Flood.

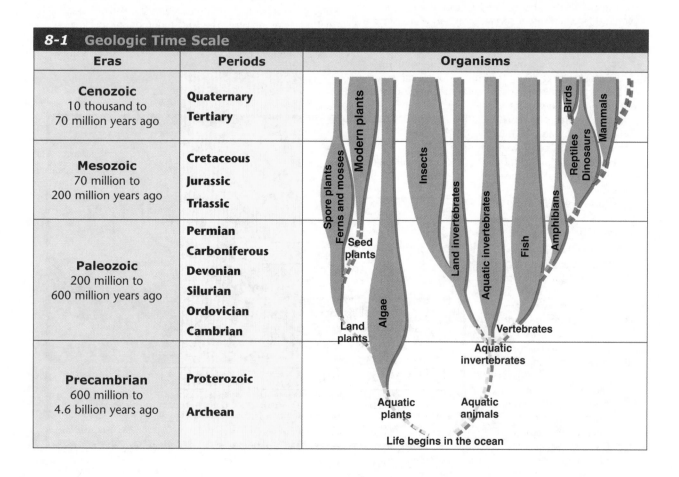

Eras	Periods	Organisms
Cenozoic 10 thousand to 70 million years ago	**Quaternary** **Tertiary**	
Mesozoic 70 million to 200 million years ago	**Cretaceous** **Jurassic** **Triassic**	
Paleozoic 200 million to 600 million years ago	**Permian** **Carboniferous** **Devonian** **Silurian** **Ordovician** **Cambrian**	
Precambrian 600 million to 4.6 billion years ago	**Proterozoic** **Archean**	

8-1 Geologic Time Scale

According to Neo-Darwinism, giraffes did not evolve because of a need to eat higher tree leaves. Some early short-necked giraffes just happened to produce offspring with mutations that caused them to have longer necks. At the time the long-neck mutations were happening, a drought began. The scarcity of food favored those giraffes with the longer necks, since they could reach leaves on higher branches. The long-necked giraffes survived to breed and thereby passed on their long-neck genes; the short-necked giraffes were unfit and died without breeding.

This theory might seem possible until the adolescent giraffe is considered. It was too old for its mother's milk but not yet tall enough to reach the treetops and take advantage of its inherited long-neck genes. Unable to nurse and not tall enough to eat leaves, the natural selection caused by the drought could easily wipe out these giraffes, eliminating them before they could evolve into a new species of taller giraffes.

Mutations and Evolution

For evolution to occur, new genes must form and enter the gene pool. A gene mutation is the only known method that could form new genes in nature. Gene mutations, however, are random and have never been shown to increase the amount of genetic information. Although scientists can increase the rate of mutations, they cannot direct which genes will mutate.

Some scientists estimate that genes mutate once in 10 million cell divisions. In a large population (say a billion individuals) the possibility of having a number of mutant genes is rather high. The necessity that these mutations be in the reproductive cells (germ mutations) and that they produce fertile offspring, however, limits the possibility of many mutations affecting the gene pool. The problem with Neo-Darwinism is not that mutations cannot occur. The problem is that there would need to be an accumulation of millions of related "good mutations" in one line of organisms in order to develop a workable new gene—if such a thing is even possible.

The odds of flipping two coins and having both of them land heads up is the product of multiplying the odds of getting heads up with one coin by the odds of getting heads up with the other coin. Assume that, when flipped, a coin lands heads up half of the time. If two coins are flipped at the same time, the odds of getting heads up on both coins is ½ × ½, or ¼. In fact, the more times the two coins are flipped, the more likely it is that both coins will show heads one-fourth of the time.

If the odds of a mutation happening are 1 in 10 million cell divisions, the odds of two mutations happening to one gene are 1 in 100 trillion cell divisions. The question, then, is this: How many mutations are necessary for evolution? To take an existing gene and make it into another gene may require hundreds of gene mutations. Surely the number of mutations necessary to go from one type of organism to another must be in the hundreds of thousands, millions, or billions.

Evolutionists, believing that the earth is billions of years old, have been able to say, "In that length of time anything could have happened." Yet statistical evidence has shown that even with billions of years, biological evolution could not occur.

The Harmful Nature of Mutations

A mutation must alter an existing gene. Since gene mutations are random and can occur in any gene, the vast majority of noticeable mutations are harmful. They often result in deformity because they change the gene for a needed characteristic. A random change in a highly organized system (and life's

8-10

Hugo De Vries and an evening primrose, the flower in which he first observed his supposed evolution by mutation

Observed Mutations—Expected Mutations

If evolution were to take place today, it would most likely be in an organism that has had an increased mutation rate and therefore opportunity to evolve by mutations. The fruit fly (see Ch. 6) has been exposed to mutation-inducing chemicals and radiation in experiments for many years. Thousands of different mutant fruit flies have been produced. Few of these mutations, if any, could possibly be considered helpful to the fruit fly. Most nonlethal mutations induced in fruit flies produce characteristics like curly wings, deformed eyes, or missing parts. In the wild, flies with these mutations would quickly be eliminated by natural selection. A fruit fly can carry only a certain number of mutations. If evolution were true, organisms would have to be able to carry thousands of nonbeneficial mutations.

The number of mutations an organism (or sometimes a gene pool) has is often referred to as its **genetic load**. Too much of a genetic load kills the organism. Since mutations are usually recessive to the allele from which they come, a light genetic load is counterbalanced in heterozygous individuals by normal alleles. If, however, the genetic load is too heavy, some homozygous individuals for the mutation develop. For this reason close inbreeding often produces an inferior stock of organisms. Such organisms produce few offspring, many of which are deformed.

You might think that since evolutionists believe that mutations are the means whereby evolution takes place, they would encourage mutations to occur. The exact opposite is true. If a substance is shown to increase the mutation rate, scientists quickly speak against it. For example, many scientists believe that decreasing the ozone layer in the atmosphere would permit more of the sun's mutation-inducing radiation to come to earth. But scientists predict that diminishing the ozone layer would produce not the evolution of man into a super race, but rather thousands of new cases of cancer and other dire consequences.

Against All Odds?

The possibility of evolution has been compared to the possibility of a printed, bound novel resulting from an explosion in a print shop. Even if the calamity happened every minute for 4.6 billion years, it is statistically impossible for a book to have formed under such circumstances. The same is true of evolution by mutations.

genetic code is probably the single most highly organized system known) will most likely be a change toward disorder.

A mutation could be compared to randomly selecting one letter in one word on this page and changing it to a letter or punctuation mark that was also randomly selected. The chances of introducing a spelling or grammatical error are extremely high. Such an error would be considered harmful to the meaning of this page. If more random letters were changed, the probability of errors would rise. Soon the page would contain many meaningless letter groups, and the meaning of this page would be lost.

Making even one random letter change per second, it would take trillions of years before anyone could expect to have a page with no spelling errors on it. An even greater amount of time would need to pass before a page would be error-free and sensible.

Those who believe in evolution by mutation are, in essence, contending that books (DNA) can be rewritten to form other books (different DNA) by random letter replacements. When writing a new book, authors carefully select letters to make words that have meaning when put together; they do not write by randomly substituting letters in other books. The genetic material found in living things was designed by God, not by random substitution of nucleotides in the DNA of other living things. The order and functionality of creation is a testimony to the greatness of God.

Punctuated Equilibrium

For years nonevolutionists have pointed out that there is no observable example of evolution happening today. Since Darwin's time, evolutionists have been able to answer that evolution is a long, slow process, and man has just not observed closely or long enough to see evolution. In the hundred years since Darwin's death, scientists have been looking very carefully for an example of the origin of new kinds of organisms. But there is no concrete, observable evidence of this kind of evolution happening today.

The fossil record also raises questions about slow, gradual evolution. In many cases, complex structures and body forms appear abruptly in the geologic column with none of the early developmental stages scientists might expect to uncover.

As evolutionists have recognized these problems, some have added a new twist to evolutionary theory: **punctuated equilibrium**. This is the belief that there were periods of time when evolution happened rapidly, followed by periods of time when almost no evolution took place.

In other words, they say that there were times when, for unknown reasons, mutations and natural selection happened rapidly and organisms evolved. Then, for other unknown reasons, mutation rates and natural selection slowed, and evolution virtually ceased. According to this theory, most organisms are currently in a nonevolutionary phase, which explains why, according to them, there are no present examples of evolution. Some evolutionists try to explain missing links by saying that they appeared during a period of rapid evolution and quickly died off. Since they were not around long enough to have made fossils, they are truly missing, and no one should be concerned about their absence.

Because punctuated equilibrium seems to explain some of the arguments against evolution, some evolutionists have adopted it. This new twist, however, does not change the basic scientific arguments against evolution. Even when scientists have tried to make evolution happen in a laboratory by increasing the mutation rate (as they assume must have happened during the evolutionary periods), there has been no observable evolution.

Even with punctuated equilibrium, the statistical improbability of evolution is still great. Although mutations may occur rapidly, the possibility of "good mutations" and their coming together are not increased by punctuated equilibrium. Periodically increasing the rate at which evolution is supposed to have happened does not make evolutionary theory any more scientifically acceptable.

In recent years, it has become increasingly difficult to describe the beliefs of evolutionists as a group. In the face of ever-increasing scientific data, many evolutionists are finding their beliefs challenged and are at times turning to unorthodox evolutionary beliefs. Today there are almost as many different beliefs about evolution as there are evolutionists.

Other Arguments Used to Support Evolutionary Theory

In times past some evolutionists have used certain "scientific observations" to "prove" evolution. For a long time these arguments held significant sway, especially among nonscientists who were deceived by the words and scientific sound of the arguments. Additional observations and less biased conclusions deflate these arguments.

Homologous Structures

In science classes students can dissect certain animals to learn about human anatomy. This is possible because some animals have many structures in common with humans. Many organs that are similar in structure and function between two different organisms are called **homologous structures**.

Evolutionists assume that if there are many homologous structures found in two organisms, they must be related by a common ancestor. They say that since a bird's wing and a whale's flipper have a similar number of bones and muscles that are used in movement, somewhere in their evolutionary past the bird and whale must be related. The observation of bones and muscles may be accurate, but the conclusion is based on bias.

Creation shows design. The whale's flipper works as well at moving the whale through water as the wing does at moving the bird through air. There may be similarity between these two vastly different organisms, but this similarity points less to a common ancestor and far more to a common Creator.

Molecular Similarity

Just as evolutionists claim that homologous structures support common ancestry, they also believe that similarity in organic molecules such as proteins and DNA are evidence for common ancestry. The enzymes needed for cellular

The Answer to Everything?!

Not every similarity between two organisms is attributed to homology. Because very similar body plans often appear in two species that evolutionists believe came from entirely separate evolutionary lines, they must offer a different explanation. They describe these shared traits as examples of **convergent evolution**. Similar environmental pressures, they assert, caused similar or identical traits to arise in two unrelated organisms. The great anteater of South America and the smaller spiny anteater of Oceania are commonly used as examples of convergent evolution. They share strong digging claws, a long, sticky tongue, few teeth, and a digestive tract ideal for digesting ants and termites, yet they are said to have developed independently. The spiny anteater's unusual egg-laying method of reproduction puts it in a completely different lineage from the great anteater that gives birth in the more common mammal fashion.

Thus, evolution is used to explain why "related" as well as "unrelated" animals share common traits. It really is the answer to everything!

Human Structures Once Considered Vestigial

✦ *Tonsils.* The tonsils are a part of the body's defense against diseases. Occasionally the tonsils can become overly infected (just as any part of the body can) and need to be removed. However, properly functioning tonsils are advantageous.

✦ *Coccyx (tailbone).* The coccyx, a set of small fused segments of backbone at the base of the spine, was once thought to be a vestigial tail. This tiny bone serves as a place of attachment for leg and lower back muscles.

✦ *Pituitary gland.* This gland was once thought to be either a degenerating or developing third eye because it is located under the brain, between where the two optic nerves leave the brain and go to the eyes. The pituitary gland, however, produces many hormones that carefully regulate the body's metabolism. Without these hormones, a person would die.

✦ *Appendix.* This organ can sometimes become infected and must be removed. A person can function normally without it. Scientists now believe that the human appendix does function in digestion and in the body's defense against infection.

✦ *Thymus gland.* This mass of tissue in a child's chest atrophies (goes away) as he matures. The thymus gland is now known to produce antibodies and various cells involved with immunity to infectious diseases. Without it, a child needs special medical attention to grow to adulthood.

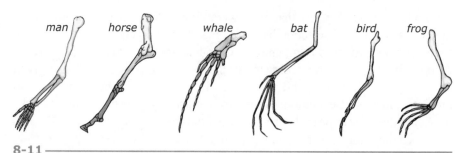

man   horse   whale   bat   bird   frog

8-11
The homologous bones in the appendages of man and some animals tell little about how these organisms came into existence.

respiration may be similar, or even the same, among many types of organisms, and the DNA needed to code for the enzymes may have a similar nucleotide sequence. However, that does not mean that aerobic bacteria have a common ancestor with any other organism that also performs aerobic respiration. Rather than being evidence for a common ancestor, this is evidence of the God-directed design in His creation.

Vestigial Structures

If evolution by mutation were true, then scientists would expect to find **vestigial** (veh STIJ ee uhl) **structures** in evolving organisms. A vestigial structure (vestigial organ) is one which no longer has a function. These are organs that supposedly were useful to ancestors but have now become nonfunctional through mutations. They are viewed as homologous structures that have outlived their usefulness and no longer serve a purpose.

At one time the list of human vestigial structures had over 180 entries. Gradually scientists discovered that these structures do have functions. Indeed man cannot live without some of these structures once thought vestigial. Man can survive without others but functions best if he has them.

At present there are no structures in the human body that are believed to be vestigial. However, the functions of many structures in plants and animals are not known. It is safe to assume, though, that with further study scientists will discover the functions for these supposedly vestigial structures. Because of the intricate design of creation, Christians can recognize that God is not a wasteful Creator.

▶ *Review Questions 8.5*

1. Why was the Miller-Urey experiment not a strong support for abiogenesis?

2. How would you answer a person who says that the geologic column gives support for an earth that is at least 4.6 billion years old?

3. Describe the basic difference between Darwinism and Neo-Darwinism.

4. List several reasons why gene mutations could not be the means whereby evolution occurs.

5. In what way does the concept of a genetic load support the Creationist's views?

6. Describe punctuated equilibrium. Explain why many evolutionists have adopted the theory, and give a Creationist's response to it.

7. Briefly describe the evolutionary argument based on homologous structures. Explain why this argument is not valid.

8. Name several human structures that were once considered vestigial. Why are they no longer considered vestigial?

⊙9. Discuss the statement "Evolution by mutations and natural selection is possible but improbable."

⊙10. Describe how scientists use fossils to date fossils. Why is this an unacceptable method?

⊙11. Describe the radiometric dating methods.

⊙12. What assumptions must be made when using radiometric dating methods?

⊙13. What technique is used to compensate for variables in radiometric dating methods?

⊙14. What assumptions must be made when using the radiocarbon dating method?

8C—Biblical Creationism

Jesus claimed, "The scripture cannot be broken" (John 10:35). This important statement applies just as much to Genesis as it does to Psalms or Romans. If a Christian is serious about growing in all aspects of Christianity, he will study the history of life using Scripture as his corrective lens. In the previous section, we examined how the history of life has been viewed by those who ignore or reject the Bible. In this section we will examine how to view the history of life from the perspective of a Christian worldview.

8.6 Interpretation of Genesis

After the emergence of the theory of evolution, many Christians' faith in the Genesis narrative was shaken. Some responded with disbelief in Genesis as actual history. Others tried to interpret Genesis in a way that allowed for billions of years. All of these nonliteral interpretations strayed from the straightforward reading of Genesis that had been the accepted interpretation for ages. Although these views are under separate headings, they are not all mutually exclusive. In other words, some of these views share common ideas, and a person's interpretation of Scripture might include elements of several of these.

Nonliteral Interpretations

Theistic Evolution

Theistic evolution is a mixture of the Bible and evolution. A theistic evolutionist tries to interpret biblical statements to support evolution as God's method of Creation. However, the Bible teaches that God created by direct act and that man did not arise from any preexisting life form (Gen. 2:7). To deny this is to deny the plain truth of God's Word. A person who believes that God directed evolutionary processes must also believe in millions of years as well as the existence of suffering and death before Adam's sin and the curse, which resulted in death.

Day-Age Theory

Some Christians interpret the seven days in Genesis 1 as seven ages. This interpretation is sometimes called the **day-age theory** (long day theory). Second Peter 3:8 ("One day is with the Lord as a thousand years, and a thousand years as one day") is often cited as support for the long day theory. This passage, however, simply teaches that God is eternal and timeless.

Many people who hold to the day-age theory are attempting to put enough time into the Genesis account of Creation for evolution to take place. In so doing, they attempt to prove that God either directed or permitted the evolution of the universe rather than actually creating it as He did.

Progressive Creationism

Progressive creationism tries to harmonize the Creator with an old universe. Proponents of this theory believe that the earth is billions of years old and that at particular moments in time, God created new creatures. Although this is not an evolutionary model (God is the creative force), it attempts to fit the evolutionary time scale of an old earth with Scripture. Progressive creationists also claim that the Flood was a local, not a worldwide, catastrophe.

Framework Hypothesis

The *framework hypothesis* tries to interpret Genesis in a literary light. This hypothesis claims that chronology is not as important to the passage as the literary framework in which the passage is set. According to the framework hypothesis, in the first three days God created the realm where the things

8.6

Objectives
- List and summarize the most common nonliteral interpretations of Genesis
- Give biblical support for a literal interpretation of Creation
- Explain why a literal interpretation must apply to all portions of the Bible

Key Terms

theistic evolution
day-age theory
progressive creationism
intelligent design
irreducible complexity
specified complexity

created in days four through six would be. The stars are placed in the heavens, the fish in the sea and the flying things in the sky, and man and land animals on the land. The framework hypothesis says nothing about the timing of creation; it could have happened in one second or 13.7 billion years.

Gap Theory

According to the *gap theory*, there was a long period of time between Genesis 1:1 and 1:2. (Some people place the "gap" even before verse 1.) Gap theorists believe there was a "first creation" that was most likely destroyed by the fall of Satan described in Isaiah 14:12–17. This period ("first creation" and "gap") may have been thousands, millions, or even billions of years long. The gap theory attempts to harmonize the evolutionists' claim of an "old earth" that was millions of years old with the Creation account in Genesis.

There are two major problems with the gap theory. First, it assumes a great deal more about the wording of Genesis 1 than the context allows. According to this view, one of the most important details regarding the origin of the universe has been omitted from and must be inferred between the two opening verses of the Bible. Second, it contradicts the time of the first sin and death. According to Romans 5:12, "by one man sin entered into the world, and death by sin." If the gap theory were correct, an entire civilization would have had to die out before Adam's creation, much less his sin. It is impossible for both the gap theory and Romans 5:12 to be correct.

A Literal Interpretation

The account of Creation given in Genesis 1 and 2 states that God created the universe in six days. (See Figure 8-12.) The Genesis account gives no indication that the days described are long ages of millions of years or that the

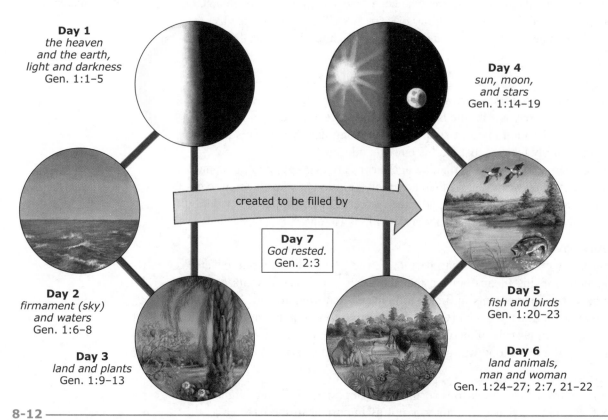

Day 1
the heaven and the earth, light and darkness
Gen. 1:1–5

Day 4
sun, moon, and stars
Gen. 1:14–19

created to be filled by

Day 7
God rested.
Gen. 2:3

Day 2
firmament (sky) and waters
Gen. 1:6–8

Day 3
land and plants
Gen. 1:9–13

Day 5
fish and birds
Gen. 1:20–23

Day 6
land animals, man and woman
Gen. 1:24–27; 2:7, 21–22

8-12

The days of Creation can be illustrated by this diagram. The materials created on days 1–3 correspond to the materials created on days 4–6.

days are twenty-four hours long, separated by long ages of millions of years. The Bible gives a straightforward timetable of the Creation events—six literal days, each one approximately twenty-four hours in length. The word *literal* means conforming to or limited to the most obvious meaning of a word or phrase; therefore, the belief that God created the universe in six twenty-four-hour days is based upon a literal interpretation of Scripture. Those who hold to this view are often called *young-earth Creationists*.

Creation cannot be proved scientifically since no one was there to observe and record it. Knowing why or how God created things the way He did is not only beyond the scope of science (for it cannot be observed or tested) but also beyond the comprehension of the human mind (Job 38:1–40:5).

Scriptural Support for a Literal Interpretation

The word for *day* in the Hebrew language is *yom*; and as many scholars have pointed out, *yom* can mean a twenty-four-hour day or an unspecified period of time. However, when *yom* is modified by certain words, the length of time can be determined. For example, when evening and morning are used to modify *yom* in Scripture, it always describes a literal day. Also, when *yom* is used in combination with a number or ordinal, such as the "fifth" day, a twenty-four-hour day is meant. If Moses, writing under inspiration, wanted to convey a long period of time, there are several other Hebrew words that he could have chosen. One such word is *olam*, which like *yom* can have several shades of meaning. However, *olam* is most often used to mean a very long period of time.

The Creation account in Genesis can also be interpreted using other Scripture. Moses states, "For in six days the Lord made heaven and earth, the sea, and all that in them is, and rested the seventh day" (Exod. 20:11). In this example, the word *six* modifies *yom*, making the interpretation six twenty-four-hour days. A similar example can be found in Exodus 31:17.

The Importance of a Literal Interpretation

Psalm 119:160 says, "Thy word is true from the beginning." The Bible is the cornerstone of the Christian faith. If the Bible is fallible, if it must be amended to conform to scientific theory or man's worldview, then Christians are without hope.

People who do not believe that the Bible is the inerrant Word of God choose what parts they want to believe and ignore those passages they do not want to believe. If it is permissible to disbelieve the biblical account of Creation, then a person might disbelieve other parts as well. He might dismiss God's condemnation of sin and the need of a blood atonement simply because these doctrines do not appeal to him. In fact, both of these doctrines of salvation are introduced in the third chapter of Genesis. A person who rejects any portion of the Bible has placed himself above God's Word.

Men begin to have trouble when they impose their ideas onto Scripture. There are certainly portions of the Bible that people disagree about. Even Peter states that some parts of Paul's letters are hard to understand (2 Pet. 3:15–16). There are other places, though, such as the account of Creation, where the words are clear. They become confusing only when someone tries to twist their meanings to make the Bible fit nonliteral and antibiblical theories.

Christians who accept other interpretations when the Bible clearly teaches a literal Creation are saying that a section of the Bible is untrue. The question of whether the Bible or human speculation is true then becomes a matter of choice, open for debate. Those who choose human speculation that contradicts the Bible are actually questioning the inerrancy and authority of God's Word.

8-13

The Bible is the foundation of the Christian faith. If it is undermined, the entirety of the faith and worldview begins to fall.

Intelligent Design

Appearing in the 1980s, the **intelligent design** (ID) movement offers a relatively new theory on origins. This movement was largely initiated by Philip Johnson with the publication of his book *Darwin on Trial*. The leaders in the ID movement are interested in proving how life and the universe show evidence of design that cannot be totally explained by the random processes of Darwinian evolution. Although the ID movement uses some of the same arguments and evidence that Creationists have used for years, most of its proponents come at the issue from a different perspective. They are not interested in using the Bible in arguing with evolutionists. They prefer to meet the evolutionists on the "common ground" of the evidences in nature. The two key ID arguments concerning biology are irreducible complexity and specified complexity.

Irreducible Complexity

The biochemist Michael Behe set off a firestorm in Darwinism with his book *Darwin's Black Box*. In it, he argues that Darwinism will be undone by the "black box" of the cell. The black box metaphor represents a device whose actions can be observed but whose inner details and workings are not visible and can only be inferred. The cell in Darwin's time was thought to be little more than a sac of organic compounds, but today science has revealed that the cell is a complex chemical machine, baffling evolutionary explanations. Behe focuses his argument for design on **irreducible complexity** in cellular systems.

Behe presents irreducible complexity with the illustration of a mousetrap. A mousetrap is made up of specific parts acting in a coordinated manner to perform a specific purpose—to catch mice. If one part of the mousetrap is taken away, the mousetrap will not work.

Many biological systems, like mousetraps, have so many interdependent parts and steps in their processes that they could not operate unless all the components were present simultaneously in their finished state. They could not have developed in gradual steps by evolutionary processes. Some of the best examples include the blood clotting system, the immune system, and some cellular transportation systems.

The bacterial flagellum is an oft-cited example of irreducible complexity. The bacterial flagellum is the biological equivalent of a power boat. It is made up of a motor, which powers the flagellum, a hook, which holds the flagellum in place, and a propeller, which is the flagellum itself. Each of the systems that make up the flagellum is very complex. In total the flagellum is made up of more than 40 different proteins. If any of the flagellum's parts are removed, it will not work, and the bacteria cannot move.

Evolutionists have tried to answer irreducible complexity by returning to the example of the mousetrap. They argue that a mousetrap which is missing a part can be used for other purposes, such as a doorstop. Against the example of the bacterial flagellum, they point out that another bacteria uses its flagellum as a needle to inject poison into its prey. These objections miss the point of irreducible complexity. The slow processes of Neo-Darwinism cannot explain how life developed these complex systems which are needed for life to survive in the first place. How did an animal get by before an immune system developed? How did organisms not bleed to death before the advent of the blood clotting system? Evolutionists have yet to give clear answers for these problems.

Specified Complexity

William Dembski, a mathematician and philosopher, developed the argument of **specified complexity** to answer this question: how can we know something is designed? It is easy to see that a car or building is designed and made by man, and specified complexity claims that it is just as easy to see that life is designed by an intelligent agent.

Specified complexity is defined in two parts. A designed system must be specific, or in a particular pattern. A designed system must also be complex, or highly improbable. Specificity and complexity are both present in designed systems. Dembski uses the example of language to illustrate specified complexity. A single letter *a* is specified because it is a particular letter, but it is not complex because it is probable that it is just a random letter. The string of letters *ajkdfahdg-kajsdkgjkasdj* is complex because it is an unlikely combination of letters, but it is not specified because it does not follow a pattern. The quote "the quick brown fox jumps over the lazy dog" is both specific and complex because it follows a pattern and is highly unlikely that it is a random assortment of letters. This series of letters expresses meaning and shows evidence of design.

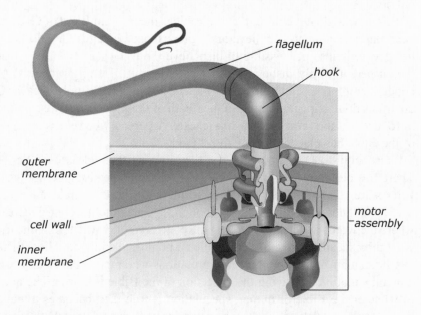

The bacterial flagellum functions as a well-designed machine, demonstrating irreducible complexity.

DNA is a good example of specified complexity. Many evolutionists believe that DNA formed by random chemical reactions in a primal pond 3.8 billion years ago, but because DNA shows such a high level of specified complexity, that cannot be the case. The chemical combinations that form DNA do not happen spontaneously in nature, so that rules out DNA forming according to natural laws. DNA is also improbably complex, consisting of thousands of base pairs even in "simple" organisms. Lastly, DNA is set in a certain meaningful pattern that carries the code for the manufacture of proteins. Therefore, DNA has all the elements of design.

ID and Biblical Creationism

Biblical Creationists can appreciate the work of the ID movement in arguing for design against evolution. Many of their arguments are useful to the Creationist cause. The research and scholarship of the ID camp can certainly provide critical ammunition for the fight against the threats of evolution. A few of their writers are not only sympathetic with but even promote a literal interpretation of Genesis. However, ID is not the same as biblical Creationism.

There are a number of cautions for Creationists who are attracted to ID. First, ID proponents do not look at the Bible as the primary authority. They believe that man's reason and science are the authority—not the Bible. Second, many of them believe

in an old earth. In light of the fact that many ID proponents feel that evolution still occurs to a certain extent and lack an adherence to a literal interpretation of Scripture, it is natural that they come to this conclusion. Third, they do not identify the Designer of life. The designer could be the God of the Bible—or it could be Gaia, Mother Nature, or some unspecified life force.

Although there are some Bible-believing Creationists within its fold, people do not have to believe in creation by God to identify with ID. The ID movement is so broad in its ranks that its supporters range from biblical Creationists to agnostics. Biblical Creationists can cheer ID on against evolution, but they should also be clear on the differences between ID and Creationism.

Since the God Who created everything that scientists study is the same God Who gave the Bible, primary allegiance in scientific study should be to the unaltering standard of God's Word. This is the only consistent Christian position. All scientific facts and the interpretation of those facts, therefore, must fit into the model prescribed by the Word of God. A scientific "fact" that does not fit into the worldview outlined in the Bible either is in error (and therefore not really a fact) or is being misinterpreted.

Review Questions 8.6

1. List what was created on each day of the Creation week.
2. Why is a literal interpretation of the Bible important?
3. Check five Bible commentaries or study Bibles regarding nonliteral Creation theories. Note carefully the reasons they give. See if you can find flaws in the logic used either to support or to discredit them.
4. List at least three ways in which the intelligent design movement differs from young-earth Creationism.

8.7 Noah's Flood and Fossilization

One of the most challenging problems for young-earth Creationists has been how to account for the fossil record. The existence of such a large number of fossilized organisms seems to suggest that the earth is very old. However, those who take a literal interpretation of Genesis have found that the record of Noah's Flood accounts for the fossil evidence quite well.

Before and After the Flood

The Bible tells very little of the antediluvian (AN tih duh LOO vee un) (before the Flood) earth. However, it was in some ways similar to the present-day earth. There were seas, rivers, and mountains (Gen. 1:10; 2:10; 7:20), as well as animal and plant life. Some Bible scholars believe that there was no rainfall before the Flood, plants being watered by a mist that rose from the earth (Gen. 2:5–6).

To many people the story of Noah with the animals on the ark is a delightful children's tale that clearly teaches God's punishment for sin and His watch-care over His children (and animals). The details and implications of the biblical account, however, teach even more about God and His creation.

Based on the Genesis account and assuming a 0.46 m (18 in.) cubit, the ark was over 138 m (450 ft) long, 23 m (75 ft) wide, and 14 m (45 ft) high. It contained three internal decks with a total deck area of over 9500 m^2 (101 000 ft^2)—larger than twenty basketball courts.

The ark was probably box shaped rather than streamlined like a modern ship. Its purpose was simply to float through a severe flood, not to travel anywhere. A completely enclosed structure strengthened by the three internal decks, it would have to be turned more than halfway over to capsize. Made completely of wood and covered with pitch (probably a tarlike substance), only major structural damage could have sunk it.

Although Noah built a good vessel, the reason for the ark's successful passage through the Flood was that Noah followed God's instructions and that "God remembered Noah, and every living thing . . . that was with him in the ark" (Gen. 8:1). Such a great engineering feat as this speaks of the greatness of God and His knowledge of the world.

Many people have said that the ark could never have contained all the necessary animals. Noah was told to take seven of every kind of clean and two of every kind of unclean animal. (There were probably far more unclean than clean.) Most likely only land animals were on the ark (aquatic animals could have survived in the water). Today there are approximately 25 000 species of vertebrate land animals, such as mammals, birds, reptiles, and amphibians. Noah's ark, therefore, needed to carry about 50 000 vertebrate land animals.

Researchers believe that a sheep is about the average size of the animals taken on the ark. A double-decker

A modern artist's rendition of the ark

railroad stock car can carry 240 sheep. Only 208 stock cars would be needed to carry 50 000 ark animals. The only major group of invertebrates to have land-dwelling members includes the insects and a few similar organisms. An additional stock car or two could contain these. The ark had a total volume of over 530 stock cars, even though only the volume of about 208 was needed to hold its inhabitants.

Noah did not need to catch the animals. The Bible states that God brought the animals to the ark (Gen. 6:20; 7:9, 15). Some people ask whether there were dinosaurs or other now-extinct organisms on the ark. The deluge fossil formation theory (discussed on p. 197) assumes dinosaurs were alive in Noah's day. God possibly brought dinosaurs (perhaps young ones) to the ark. There was easily enough room for them.

It is also possible that God, knowing that the postdiluvian world would not be suitable for large dinosaurs, did not bring them and other now-extinct animals to the ark. Some Christians object to this idea, saying that Noah was to take two "of every living thing of all flesh" (Gen. 6:19), and dinosaurs must have been included. But the next phrase of the verse states "two of every sort." It is possible that smaller dinosaur-like reptiles represented the "sort" that included dinosaurs.

Others argue that God would not completely destroy organisms He had created. However, other organisms, for which God's purpose must have been finished, have become extinct since the Flood. God could have destroyed creatures outside the ark in the Flood as easily as He could have destroyed them after the Flood.

Some people see Noah as a harried, understaffed animal keeper with only seven helpers, running from cage to cage trying to keep things clean and the animals' food and water bowls filled in a rocking, floating zoo. Others wonder how Noah could supply meat for the carnivores without killing some of the animals onboard and plants for the vegetarians without a sizable storehouse. Still other people think that neither humans nor animals consumed meat until sometime after the Flood.

One possible answer is that the carnivores survived on a diet of plant protein. This diet can support nearly all carnivores. Others suggest that many of the animals entered into a state of dormancy (hibernation or other quiet state) for the year on the ark. Some people object, saying that this would require a miracle from God. But the Flood, the ark, and the animals coming to and entering the ark were all miracles. God finishes the work that He begins. It is not likely that God would have permitted His creatures to harm one another or to be overly demanding on Noah and his family.

Many researchers believe that when the fountains of the deep were opened up (Gen. 7:11), there may have been an outflowing of both liquid water and superheated steam from thousands of volcanoes. The steam and debris thrown up into the atmosphere would have condensed and fallen back to the earth as rain. Theories also include the shifting of the tectonic plates and the separation of the landmasses. The shifting of the landmasses supports the fossil evidence of tropical plants and animals seen in places such as Antarctica.

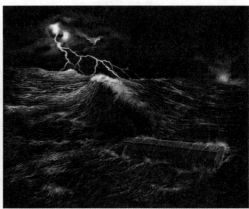

As the floodwaters receded, many scientists think that the mountains were raised up and ocean floors were lowered. Scientists theorize that the postdiluvian (after the Flood) mountains are higher and the oceans deeper than the antediluvian mountains and oceans. They also propose that the rapid movement of water running off the landmasses accounts for many of the geologic formations seen today that secular scientists say took millions of years to form. Some biblical Creation scientists also think that the earth and its environment continued to change after the floodwaters receded. This accounts for the evidence of large amounts of volcanic activity and an ice age, although the latter is thought to have been much shorter than secular scientists believe.

8-14

Creationists can only imagine what the antediluvian environment was like (top left) or the terror of God's judgment during the Flood (top right). The postdiluvian earth (bottom) is probably quite different from the antediluvian world.

The average human life span before the Flood (based on Genesis) was 912 years. After Noah, the life span quickly dropped to about 400 years and continued to decline. Some scientists believe this change in life span was due to postdiluvian changes in the atmosphere. Abraham, Isaac, and Jacob lived well over 100 years, but few men after their time have reached even 100 years. A cause for this decrease could be the loss of the filtering effect of the antediluvian atmosphere resulting in more direct exposure to solar radiation. Other Bible scholars believe this decrease in life span is a result of sin and its effect on the human gene pool.

The Bible also implies differences. Second Peter 3:6–7 says that "the world that then was, being overflowed with water, perished: But the heavens and the earth, which are now . . . are kept in store." Apparently, "the world that then was" (before the Flood) differed from "the heavens and the earth, which are now" (after the Flood).

A Worldwide Catastrophe

The Bible also states that the floodwaters covered the entire world; this fact can be established from the Genesis account and from the physical evidence. Evidence for a worldwide flood can be seen from the following examples.

♦ *God's revelation to Noah.* God states in Genesis 6:7 that He will destroy man "from the face of the earth." In Genesis 6:17, God tells Noah that the flood will "destroy all flesh, wherein is the breath of life, from under heaven; and every thing that is in the earth shall die." For everything on the earth to be destroyed, the flood had to be universal; otherwise, people and animals could have migrated beyond the area of a local flood.

♦ *Confirmation in the New Testament.* The testimony of Christ (Luke 17:27) and of Peter (2 Pet. 2:5; 3:6) is that the world was destroyed by the Flood.

Other Facts That Support a Universal Flood

1. The duration of the Flood was over a year (started Gen. 7:11; ended Gen. 8:14). Local floods do not last that long.

2. If the Flood were local,
   - the time spent building the ark (over 100 years) would have been wasted.
   - the raven and the dove would have been able to find places to rest.
   - there would have been no need to take most animals into the ark.
   - the prescribed size of the ark would have been unnecessary.

8-15

The rainbow is a frequent reminder of God's covenant with Noah.

8-16

Parts of organisms are usually preserved in minerals (top), but some small organisms are trapped in fossilized tree sap called amber (bottom).

✦ *The covenant of the rainbow.* God made a covenant with Noah that there would never again be such a flood. The rainbow symbolizes His promise (Gen. 8:21; 9:11–16). God has repeatedly broken the covenant if it referred to local flooding since there are both rainbows and local floods today.

✦ *The continental shelves and seamounts.* The shelves of land around the continents, as well as many mountains in the sea, appear to have been dry land at one time. They would have been above sea level before the Flood.

✦ *Universality of the fossil record.* The waters covered the highest mountains to a depth of over 20 ft (Gen. 7:19–20; 8:5). Since water seeks its own level, covering a mountain by this depth requires a universal flood. Fossils have been found around the world and on the tops of mountains; therefore, the mountaintops had to have been underwater at some point.

Fossil Formation

Fossilization (the forming of fossils) is not naturally happening to any appreciable extent today. When an organism dies now, it is only a matter of time before it decomposes, leaving no trace that it existed. For years scientists have puzzled over how existing fossils were formed. The following paragraphs discuss some types of fossils and some considerations of how these fossils were formed. As you read this section, consider how the conditions of the Flood might have promoted fossilization.

✦ *Preservation of parts.* The most abundant type of fossil is the preserved parts of animals (bones, teeth, shells) and plants (stems, leaves, seeds). These often lie in vast layers, sometimes hundreds of square miles long and many feet thick. In some areas even the soft parts have been so well preserved that scientists can determine types of chlorophyll, contents of stomachs, and shapes of muscles and even skin. Today when something dies, even its hard parts will usually deteriorate within a few months. For its hard parts to fossilize, an organism must have been quickly placed under pressure in rock-forming sediment. Many small creatures have been preserved in amber, fossilized tree sap. These types of fossils are not formed naturally today.

✦ *Preservation of carbon (coal).* Coal is believed to be a massive collection of mostly plant material that was placed under high temperatures and pressure. Experiments have shown that under proper conditions plant material can form coal in twenty minutes. It is estimated that to form a vein of coal 12 m (40 ft) thick, over 61 m (200 ft) of green plant material must be compressed. Although plant material does collect in the bottoms of swamps and bogs, Creation scientists do not believe that this slow accumulation of organic material is an acceptable model for coal formation. No coal is known to be naturally forming today.

✦ *Preservation of forms (casts or molds).* Casts or molds of organisms are believed to have formed when the organisms were placed in a substance that quickly solidified before the body decomposed. When the body of the organism dissolved, an empty or fluid-filled form, a mold, was left. If mineral-rich water filled this space, it might have filled the mold with rocklike sediments, creating a cast. This type of fossilization may occur today (though rarely) in tar pits. But no extensive casts or molds are now forming in sedimentary rock.

✦ *Preservation of tracks.* Scientists have found extensive areas of preserved animal tracks, usually in rock that was once clay or mud. Tracks of large reptiles and various other extinct animals are numerous. Even small animals such as crabs can leave tracks if the sediment is soft and fine enough. Apparently, the tracks were made in soft mud, and other sediments quickly filled the tracks and turned them to stone before the prints had a chance to be destroyed. Today footprints made in clay or mud are soon gone.

✦ *Petrifaction.* Usually only the hard parts of an organism are petrified. Petrifaction seems to happen when water containing dissolved minerals penetrates a structure. The water then leaves, and the minerals remain, collecting within the spaces of the structure and eventually turning them to stone. Very little petrifaction is happening today.

✦ *Frozen fossils.* Not all fossils are found in sedimentary rock. Some organisms, sometimes even humans, are preserved for thousands of years, locked in ice. Large numbers of animals have been found frozen in Siberia and Alaska. The freezing of so many large animals (many of which are not native to the arctic today) suggests some major catastrophe or change in the environment. Many of the animals appear to have been trapped in mudflows or ravines and then frozen when they could not escape.

Dinosaur tracks

Petrified tree stump

8-17
Occasionally, dinosaur tracks are discovered in rock strata (top). The petrifaction of trees (bottom) implies that the wood was submerged in water with minerals.

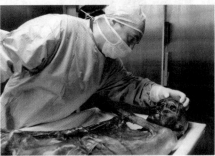

8-18
A frozen human in remarkable condition was found in the Alps in 1991. Because the body was found in the Ötzal area of the Austrian Alps, it was nicknamed Ötzi. Dating methods indicated he probably died around the time of the Flood. Here is an artist's rendition of how he may have appeared (top) and a photo of a scientist examining this rare find (bottom).

The Deluge and Fossilization

Many Christian scholars and scientists support the **deluge fossil formation theory**. Until about 150 years ago most written material—Christian and non-Christian—indicated that the Deluge (Noah's Flood) formed the fossils. Today, fossils are used widely by evolutionists to support their theories. Some Christians are uncomfortable discussing fossils, assuming that their presence somehow verifies a belief in an old earth and in evolution. However, research and experimentation by Creation scientists continues to produce support for the deluge fossil formation theory.

Water covering the entire earth, as in the Deluge, would contain many strong currents, possibly even stronger than the major currents found in oceans today. God's purpose for the Flood was to destroy the entire world, and so the Flood currents would have been strong enough to carry any material on the face of the earth that could be moved.

As the currents slowed down or changed course during the Deluge, they would have deposited the materials they carried. This **sedimentation** (settling out of materials) accounts for the layers of *sedimentary rock* that

deluge: de- (apart) + -luge
(L. *lavere*—to wash)

8-19

The erosion of the Badlands in South Dakota is presumed to have taken millions of years. However, the amount of water involved in the Flood could have formed these and similar structures quite rapidly.

contain most of the fossils and compose most of the **topography** (tuh PAHG ruh fee)—the shapes and contours of the earth's surface—that are seen today. Rock layers are visible where mountains have been cut away by rivers, as in the Grand Canyon, or by man, as in building roads. According to the deluge theory of fossil formation, Noah's Flood formed these rock layers. As the dry land emerged and ocean floors dropped, strong currents were again created, cutting channels and depositing sedimentation layers.

Several conclusions about fossilization can be made that support the deluge fossil formation theory and contradict the progressive fossil record that evolutionists propose.

✦ Abundant fossilization is not taking place today. Some time in the past, however, fossilization rates must have been vastly different to account for the fossils currently found.

✦ The abundance of fossils and the arrangement of the fossils indicate a major catastrophe.

✦ Fossilization of an organism's parts must be rapid; otherwise, organisms would decompose before fossilization could take place.

✦ Most types of fossils found today apparently required water movements for depositing them.

Some evolutionists, recognizing the problem of fossilization, suggest that there must have been a series of major catastrophes to account for the fossils. These catastrophes would have to have been separated by long ages, and each one could have formed only a few layers in order for the fossils to verify evolution. The fossils themselves present problems even for these evolutionists.

8-20

Fossils such as these fish (left) are frequently found in masses, an indication of a catastrophe. A *polystrate* fossil (right) is one that runs through several layers of rock, also indicating a rapid deposition of sediment.

Sometimes *human artifacts* (things man has made, such as gold chains or metal objects) or human fossils such as bones and footprints have been found in fossil layers that are thought to be millions of years old. This is a grave problem for evolutionists who believe that humans did not develop the ability to create those objects until millions of years later.

Review Questions 8.7

1. List seven biblically based reasons for believing the Flood was universal.

2. List several methods of fossilization.

3. Give several reasons why the deluge fossil formation theory is the most scientifically acceptable theory for fossilization. Is it also a scripturally acceptable theory?

⊙4. List several objections that people have to the scriptural account of the animals' being on Noah's ark and give answers to the objections.

⊙5. Compare the size of Noah's ark to the size of some familiar objects.

Unit 2
The Science of Organisms

The Classification of Organisms

One of the first tasks of dominion that God gave Adam was to name all the creatures that He brought to him (Gen. 2:19–20). Man is still naming organisms today, continuing his dominion. The process of naming organisms has changed since Adam's time. It has now become a scientific discipline using specialized language. Today man also has to give names to variations of creatures that did not exist at the time of Creation. These are populations that God has allowed to develop into separate species, but they are still within the limits of the original created kinds.

9A—The Necessity of Classifying

People group things together for convenience. All the spices in a kitchen are usually kept in the same cabinet. The pots are in one place, and the dishes are in another. Imagine the confusion that would be created if the spices and cooking utensils were put haphazardly into the nearest empty space when they were no longer needed.

Musical instruments, too, are not thought of only as individual instruments; they can be put into groups, such as woodwinds, strings, percussion, and brass.

It is easy to put the violin, viola, cello, and string bass together in the string group. They look alike and are played in essentially the same way. If someone mentions the viola da gamba and says that it is an antiquated instrument in the string group, it is not difficult to picture what type of an instrument it is.

Classifying some other musical instruments is not easy. The piano, harpsichord, and harp all have strings but are played very differently. Do they belong with the violin? The piano has strings that are struck by small hammers, not plucked or stroked by a bow. Because of this, some think the piano belongs with the drum, xylophone, triangle, and other percussion instruments. Others think that since the piano has a keyboard, it should be grouped with other keyboard instruments. Except for their keyboards, however, the piano and the organ have very little in common. Does the piano actually belong in all these groups? Maybe it should be classified by itself.

The same problems apply to **taxonomy** (tak SAHN uh mee) (or *systematics*), the science of classifying organisms into groups. The taxonomist (one who classifies organisms) has over 1.5 million different species of organisms to group, with 30 000 to 40 000 more being added each year.

Sometimes when a new organism is classified, so little is known about it that it is put into the group where it seems to "best fit." After more investigation, the organism may be placed into a different group. Occasionally, after learning more about a group of organisms, taxonomists may decide to divide the group. At other times they expand the definition of a group to include several groups. These acts are sometimes referred to as splitting and lumping, respectively.

Today's biological classification system suffers from many of the same problems as classifying the piano. But it does permit similar organisms to be put into groups so that the organisms and information about them can be dealt with conveniently. For instance, if a new bird species found in the rainforest was declared to be part of the macaw family, even without seeing it, you would have a general idea of its size, shape, beak type, and food habits, provided you knew something about macaws.

Were it not for a good classification system, most information about the organisms would be lost in a hodgepodge of facts. Grouping organisms by taxonomy helps organize this information. Attempts at classification also have the added benefit of clarifying communication among scientists. Taxonomy helps them "speak the same language."

9.1 The Classification Hierarchy

The Greek scientist and philosopher Aristotle attempted the first recorded classification of organisms. He classified living organisms into one of two major groupings—plants or animals. Each of these two groups was subdivided into three categories. Plants were subdivided based on the presence or absence of woody parts. *Herbs* had no woody parts, *shrubs* had several short woody stems, and *trees* had one large woody stem. Animals were grouped by where they lived—animals that could fly were *birds*, those that lived in water were *fish*, and those on land were *animals*.

9-1

Plants as classified by Aristotle: trees, shrubs, and herbs

taxonomy: taxo- (Gk. *taxis*—arrangement) + -nomy (*nomos*—a system of rules)

9.1

→ **Objectives**

- Explain the need for a classification system
- List the seven levels of taxonomic hierarchy
- List the major characteristics of the six kingdoms

→ **Key Terms**

taxonomy	genus
artificial classification system	species
	domain
kingdom	Eubacteria
phylum	Archaebacteria
class	Protista
order	Fungi
family	Plantae
	Animalia

According to Aristotle's system, there were three groups of animals: land animals (a), birds (b), and fish (c).

Because Aristotle's six groups were based on appearance, his system is called an **artificial classification system**—one that is based primarily on observable characteristics. In Aristotle's system the lobster and the tuna are both fish, even though the lobster has more physical characteristics in common with insects. Are butterflies land animals or birds? Surely they are more like insects than they are like robins or hawks.

Whenever physical characteristics are used for classification of organisms, there will be either many generalities—"most organisms that fly are birds, most that swim are fish"—or many exceptions—"birds include all flying organisms except butterflies, flies, and bats."

Although Aristotle's artificial classification system left much to be desired, it was used for almost two thousand years. During this time period world exploration was expanding, and many explorers were returning with new and never-before-seen plants and animals. Naturalists were having difficulty fitting these new plants and animals into Aristotle's system. In the mid-1700s, when Swedish naturalist Carolus Linnaeus (lih NEE us) set forth a new classification system in his two works *Species Plantarum* and *Systema Naturae*, scholars of the time readily adopted it.

The Linnaean system is also an artificial classification system. Yet because it has more flexibility than previous systems, the Linnaean categories and system of naming organisms are still used today, over two centuries after their development. As new species are found and as new research techniques are developed to study organisms, scientists recommend changes to the system. International scientific committees meet every few years to review any proposed changes and decide whether they should be adopted.

Today's System of Classification

Taxonomists today use a classification hierarchy, an arrangement of graded levels. This system has seven basic levels:

- **kingdom**
- **phylum** (pl., phyla; technically called *division* in the kingdom Plantae)
- **class**
- **order**
- **family**
- **genus** (pl., genera)
- **species** (pl., species)

A few of the *taxons* (taxonomic levels) in the animal kingdom and a few characteristics for each level are given in Figure 9-3. This figure can be used

KINGDOM ANIMALIA
1. heterotrophic
2. multicellular

9-3

In the 1700s, Carolus Linnaeus proposed the binomial classification system still used today. He believed that his species level was the same as the biblical kind (see p. 217). A few levels, characteristics, and examples of the kingdom Animalia are given.

Phylum Arthropoda
1. exoskeleton
2. jointed legs

Phylum Chordata
1. internal notochord for support
2. hollow dorsal nerve cord

Subphylum Cephalochordata
1. notochord in adults
2. gill slits for respiration

Subphylum Vertebrata
1. notochord in embryos replaced by vertebral column
2. usually two pairs of appendages

Class Chilopoda
1. head and many body segments
2. one pair of legs per segment

Class Arachnida
1. two body divisions
2. four pairs of legs

Class Insecta
1. three body divisions
2. three pairs of legs

Class Osteichthyes
1. fins and scales
2. two-chambered heart

Class Amphibia
1. smooth, moist skin
2. two-chambered heart that becomes three-chambered when adult

Class Aves
1. feathers
2. four-chambered heart

Class Mammalia
1. hair (fur)
2. four-chambered heart

Order Orthoptera
1. chewing mouthparts
2. heavy front wing, thin hind wing

Order Diptera
1. sucking mouthparts
2. one pair of thin wings

Order Lepidoptera
1. sucking mouthparts
2. scaled wings

Order Carnivora
1. claws
2. small incisor teeth, large canine teeth

Order Rodentia
1. two sets of incisor teeth
2. rotating elbow

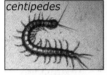

centipedes

spiders

ticks

scorpions

flies

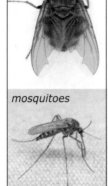

mosquitoes

bony fish

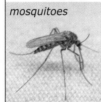

salamanders

frogs

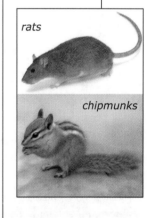

rats

chipmunks

wolves

cheetahs

crickets   grasshoppers

butterflies   moths

birds

to illustrate the following general rules for using the biological classification hierarchy:

♦ *Each group on one level of the hierarchy may be divided into several groups on the next lower level.* For example, the kingdom Animalia is divided into about thirty phyla (only two are shown in Figure 9-3). Each phylum is divided into several classes, the next lower level; each class is divided into several orders; and so on.

♦ *Each group in the hierarchy has various characteristics that all levels under the group possess.* For example, the phylum Arthropoda, according to Figure 9-3, contains organisms that have an exoskeleton and jointed legs. Occasionally, the effort to simplify or generalize a description results in the use of terms such as *usually* or *most*. The caption under subphylum Vertebrata reads, "usually two pairs of appendages." An obvious exception is a snake. Because this is an artificial classification system, such qualifying statements appear often.

♦ *Each level of the hierarchy can be divided into smaller units before reaching the next lower level.* For example, in the phylum Chordata there are subphyla that are divided into classes. Prefixes such as *sub-* (below), *infra-* (below), and *supra-* (above) are used to name these divisions.

A complete listing of all the taxonomic levels, their characteristics, and all their organisms would be as large as an encyclopedia. A listing of the levels studied in this book and a few others can be found in Appendix B.

The Significance of Classification

The classification of organisms into a hierarchy of groups is a necessary step in trying to organize our thoughts about God's world. In fact, it may even be an aspect of the image of God in man, enabling him to exercise wise dominion over the world. It is natural to use similarities in grouping organisms and to place those groups into larger groups based on more general characteristics. For example, all woodpeckers share certain characteristics that other birds lack. In a broader sense, woodpeckers and similar non-woodpecker songbirds have more in common with each other than they do with ducks and geese.

Man's inclination to group things makes it natural to organize living things into various levels of classification. Most science books present these in a tree structure, implying that one kind of organism gave rise to another. As you learned in Chapter 8, Charles Darwin believed that all life could be traced back to a single common ancestor, and that later common ancestors existed at key points when new kinds of organisms first developed and branches sprouted from the tree.

Obviously, this line of thinking contradicts the Bible's account of how the creation came into existence. Whereas an evolutionist points to many similarities as evidence for common descent, a Creationist can view the same characteristics in one of two ways. If two organisms are so similar that they probably belong to the same original *kind* of created organism (downy and hairy woodpeckers [p. 172], perhaps), then common descent is an acceptable explanation, within the divinely ordained genetic limits. Organisms that share a few features (downy woodpeckers and hummingbirds, for example) are better described as evidence of a common Designer that is God.

The Kingdoms

Aristotle's classification system placed his six groups of living things into two main divisions: plants and animals. Linnaeus also used two groups: kingdom

Domains: Higher Than a Kingdom

There is an addition to taxonomy that is gaining popularity in some areas. It is a category larger than a kingdom, called a **domain**. Currently, there are three domains named *Archaea* (includes only the kingdom Archaebacteria), *Bacteria* (includes the kingdom Eubacteria), and *Eukarya* (includes the four remaining kingdoms of Protista, Plantae, Fungi, and Animalia). This system is based on biochemical analysis and comparison of the RNA of organisms and attempts to group them according to those similarities. Although the thrust of the research behind this classification system is to find a common ancestor from which all other life forms arose, the addition of domains only demonstrates the frustrating challenge of finite man trying to organize the creation of an infinite God.

Plantae and kingdom Animalia. These two men, however, had no knowledge of the complexity and variety of microscopic organisms. Nor did they realize the vast difference in some of the groups such as the fungi and plants. They neatly tucked fungi into their systems and, to a great extent, ignored them. Because more types of living things are now known and more information about organisms is available, it has become necessary to recognize additional groups.

As scientists discovered new organisms and used more advanced technology to study the existing organisms, they all agreed that two kingdoms were not enough. In the 1860s, scientists proposed a third kingdom—Protista—that included bacteria, protists, and sponges. Taxonomic classification was again modified in the 1950s when the bacteria were removed from Protista and placed into the new kingdom Monera. In 1969, the number of kingdoms increased to five with the formation of a separate kingdom for fungi. Fungi had previously been included in the kingdom Plantae. The five-kingdom system has been accepted by biologists around the world.

However, science rarely stands still. New kinds of life forms have been found in places where no one ever expected to find life, and new techniques have allowed scientists to explore and maintain these life forms in the laboratory. In the late 1970s, the kingdom Monera was divided into two separate kingdoms—Archaebacteria and Eubacteria—making a total of six kingdoms. However, since the international committees responsible for classification have not agreed on the "official" number that should be established, other texts and reference materials may refer to either the five- or the six-kingdom taxonomic systems. We will use the six-kingdom organization in this text. Both systems are correct; the taxonomic system is merely a tool that man has devised to aid his understanding of a world that is remarkably diverse.

Kingdom Eubacteria

Members of the kingdom Eubacteria (YOO back TEER ee uh) are the most abundant organisms on the earth. This kingdom contains unicellular prokaryotic organisms—the kinds of bacteria with which most people are probably familiar. They are the cause of many diseases such as tuberculosis, food poisoning, and even dental decay. However, many are free-living in the soil and are essential for decomposition. These organisms contain peptidoglycan (PEP tuh duh GLY KAHN), large organic molecules containing protein and sugar, in their cell walls—one of the major characteristics that differentiates them from the other kingdom that contains prokaryotes.

Kingdom Archaebacteria

The kingdom Archaebacteria (AHR kee back TEER ee uh) also contains prokaryotic organisms, but these do not contain peptidoglycan in their cell walls. Many of these organisms are also called *extremophiles* because they live in environments that were once considered too harsh to support life. Archaebacteria have been found in thermal ponds and undersea volcanic vents at temperatures as high as 110 °C (230 °F) as well as in extremely acidic and salty environments. Others live in anaerobic environments, obtain energy by chemosynthesis, and release methane gas as a waste product. Still other archaebacteria live in habitats similar to those in Eubacteria, but they have chemicals in their makeup that are characteristic of Archaebacteria. Both of these bacteria kingdoms are discussed further in the next chapter.

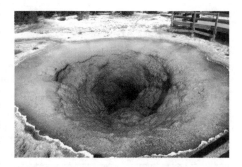

9-4

Some extremophiles live in thermal springs such as this one in Yellowstone National Park.

Eubacteria: Eu- (good) + -bacteria (Gk. *bakterion*—small rod)

Archaebacteria: Archae- (Gk. *arkhaios*—ancient) + -bacteria (small rod)

Kingdom Protista

The algae, protozoans, and some funguslike protists called slime molds and water molds compose the kingdom Protista (proh TIS tuh). Since protists are autotrophic and heterotrophic, mobile and stationary, unicellular and colonial (sometimes more than 30 m [100 ft] long), they are difficult to classify, and some scientists even place some protists in other kingdoms. Essentially, protists are eukaryotic organisms that are *not* animals, plants, or fungi. Examples of protists include seaweeds, amoebas, paramecia, and sporozoans of the genus *Plasmodium* that causes malaria.

Kingdom Fungi

The organisms in the kingdom Fungi (FUN jye) were originally in the kingdom Plantae. Fungi are all heterotrophic and feed on dead or decaying

Protista: Protista (Gk. *protistos*—the very first)

9-1 Classification and Characteristics of Living Organisms

Domain	Archaea	Bacteria	Eukarya			
Kingdom	Archaebacteria	Eubacteria	Protista	Fungi	Plantae	Animalia
Cell type	prokaryotic	prokaryotic	eukaryotic	eukaryotic	eukaryotic	eukaryotic
Cell structure	cell walls do not contain peptidoglycan; lack organized nucleus and membrane-bound organelles	most have cell walls; cell walls contain peptidoglycan; lack organized nucleus and membrane-bound organelles	organized nuclei, membrane-bound organelles; some have cell walls	organized nuclei, membrane-bound organelles; some have cell walls; cell walls contain chitin	organized nuclei, membrane-bound organelles, chloroplasts; cell walls contain cellulose	organized nuclei, membrane-bound organelles; no cell walls or plastids
Cellular organization	usually unicellular, may form colonies; no tissues or organs	usually unicellular, may form colonies; no tissues or organs	unicellular or colonial (some with specialized structures); no true tissues or organs	unicellular or colonial (some with very specialized structures); no true tissues or organs	all multicellular with true tissues and organs	all multicellular with true tissues and organs
Nutrition	autotrophic (usually chemosynthetic)	heterotrophic or autotrophic (photosynthetic or chemosynthetic); some perform fermentation	autotrophic (photosynthetic), heterotrophic, or both	heterotrophic; digest food externally and then absorb the nutrients	autotrophic (photosynthetic); some forms heterotrophic or both	all heterotrophic
Reproduction	asexual	asexual	asexual; some reproduce by conjugation	all reproduce asexually; some also reproduce sexually	all reproduce sexually; many also reproduce asexually	all reproduce sexually; many also reproduce asexually
Examples	thermophiles, acidophiles	*Streptococcus*, *E. coli*	*Amoeba*, kelp, slime molds	mushrooms, molds	trees, ferns, shrubs	worms, sponges, reptiles, birds mammals

organic matter. They may be unicellular like yeasts or colonial like mushrooms. Besides yeasts and mushrooms, the kingdom Fungi includes mildews and molds.

Kingdom Plantae

Most of the organisms recognized as plants belong in kingdom **Plantae**. They range in size from the giant sequoias to tiny mosses. Most plants are autotrophic and perform photosynthesis, but some, like mistletoe, have some heterotrophic characteristics. Plants are usually stationary when adults. They also contain cells surrounded by cell walls. The formation of true tissues is found only in the kingdoms Plantae and Animalia.

Kingdom Animalia

Kingdom **Animalia** contains heterotrophic, eukaryotic, multicellular organisms. Most of them have some means of locomotion during at least part of their life cycle. Insects, fish, worms, birds, and mammals are all part of the kingdom Animalia.

➥Review Questions 9.1

1. Give two reasons for having a biological classification system.
2. Why are Aristotle's and Linnaeus's groupings of living things considered artificial classification systems?
3. List from largest to smallest the seven basic levels in the modern classification hierarchy.
4. Describe three general rules for using the modern biological classification system.
5. List the six kingdoms of living things and give characteristics of each.
6. What is a domain?

9.2 Scientific Names

A snake common in the eastern United States is pictured in Figure 9-5. Frequently called the hognose snake because of its upturned, blunt snout, this nonpoisonous snake rarely bites. However, when provoked it will spread its neck, open its mouth, and hiss. This imposing bluff has given it a variety of other names, including "common spreading adder," "blowing viper," "spreading viper," and "hissing snake." Still other common names for this snake include "chunkhead," "spread moccasin," and more than sixty others. But it has only one scientific name: *Heterodon platyrhinos*.

Quite often a common name applies to several different organisms. The name "gopher" may refer to a salamander, a turtle, a frog, one of several snakes, or any of about fifty different types of rodents. Common names that apply to multiple organisms, as well as organisms with multiple common names, can easily lead to confusion.

The problem of which language to use to name an organism also arises. Is the bird in Figure 9-6 a redheaded woodpecker, *el pájaro carpintero de cabeza roja*, *der Rotkopfspecht*, or *le pic à tête rouge*? If one language is chosen to name an organism, those speaking other languages will have to learn a "foreign term." If all languages are acceptable, then a scientist will have to become a linguist just to be sure he is talking about the same organism his colleagues are discussing.

Binomial Nomenclature

To solve these problems, Carolus Linnaeus proposed and used a system of **binomial nomenclature** (bye NOH mee uhl • NOH muhn KLAY chur).

9-5

Although known by many common names, scientifically this snake is a *Heterodon platyrhinos*.

"What Is It?"—Using Dichotomous Keys to Identify Organisms

When a person sees an organism that he does not recognize at the zoo or in an aquarium, what does he do? He might look for the card or plaque that gives the organism's name. Or he might point his finger and ask the question biologists probably hear the most: "What is that?" To most people, being able to identify an organism by name is a worthy goal in itself. When scientists *identify* an organism, they determine to which taxonomic groups it belongs, including genus, species, and sometimes even variety.

But identification should not be confused with classification. To *classify* an organism, a highly trained scientist closely examines the organism and then assigns it to a particular group according to the identified characteristics. Usually the first person to find and describe the organism classifies and names it.

There are two common methods of identifying organisms. The first is to ask someone. If possible, the organism or a representative piece of it can be taken to a qualified person, who can perhaps give its name. Another method of identification is to use a *biological key*. Most of these are structured as *dichotomous keys*. A dichotomous key is a series of paired statements or characteristics about the specimen being identified. Only

An organism to identify

one statement from each pair can be true. This will either lead you to the identification of the specimen or refer you to another pair of statements to consider.

"A Simple Key to the Kingdoms" uses sets of descriptive statements, as do many dichotomous keys. By choosing the statement that describes the organism and then going to the set of statements indicated in the margin and repeating the process, you can identify the organism.

Follow along as "A Simple Key to the Kingdoms" is used to determine which kingdom the above specimen belongs to.

1. Begin with the first set of statements in No. 1. These ask about nuclear membranes and membrane-bound organelles. Although the picture does not reveal cellular structures, other characteristics in the

first set of statements indicate that large organisms with tissues fall under the second statement. Recognizing a tree as being comparatively large and probably having tissues, you decide that the statement "Nuclei of cells have membranes" best describes the organism. In the right-hand column this choice tells you to go to the statements in No. 3.

2. The statements in No. 3 again ask about tissues. The second statement fits best, and in the right-hand column you are instructed to go to the statements in No. 6.

3. The first statement in No. 6 best describes the organism. It also indicates in the right-hand column that the organism being identified is in the kingdom Plantae. There are no additional numbers; you can go no further using this key.

If you need to identify the phylum, the class, or the genus-species name of an organism, you must use additional keys. Some multivolume keys contain all the plants in a region of the United States. Other keys may deal only with pond algae, birds common in the eastern United States, or similarly limited groups.

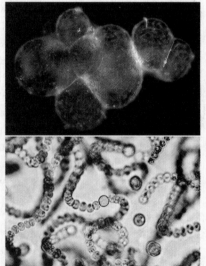

This ball-shaped colony of Nostoc *could be difficult to identify. When seen under a microscope (bottom) it becomes easier. In what kingdom does* Nostoc *belong? Use the index of this book to check your answer.*

dichotomous: dicho- (Gk. *dikho*—in two) + -tomous (*temnein*—to cut)

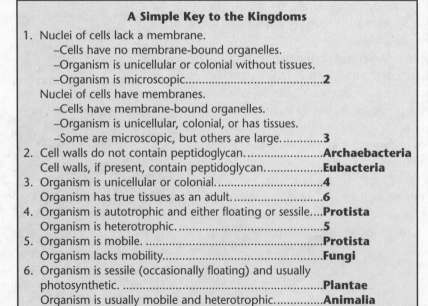

A Simple Key to the Kingdoms

1. Nuclei of cells lack a membrane.
    –Cells have no membrane-bound organelles.
    –Organism is unicellular or colonial without tissues.
    –Organism is microscopic...**2**
   Nuclei of cells have membranes.
    –Cells have membrane-bound organelles.
    –Organism is unicellular, colonial, or has tissues.
    –Some are microscopic, but others are large.............**3**
2. Cell walls do not contain peptidoglycan.......................**Archaebacteria**
   Cell walls, if present, contain peptidoglycan.................**Eubacteria**
3. Organism is unicellular or colonial.................................**4**
   Organism has true tissues as an adult.............................**6**
4. Organism is autotrophic and either floating or sessile....**Protista**
   Organism is heterotrophic..**5**
5. Organism is mobile...**Protista**
   Organism lacks mobility...**Fungi**
6. Organism is sessile (occasionally floating) and usually
    photosynthetic. ....**Plantae**
   Organism is usually mobile and heterotrophic...............**Animalia**

Keys are important to field biologists who must identify exactly what species, and even variety, they have found. Keys can also provide enjoyment on a hike or camping trip. Using one of the popular field guides (which are actually modified keys) to identify plants and animals makes the outdoors more interesting.

However, dichotomous keys are not foolproof. For example, a ball of *Nostoc* (NAHS ᴛᴀʜᴋ) scooped up from a pond would be almost impossible to identify by kingdom using "A Simple Key to the Kingdoms." Just by looking at it, a person could not tell if the first or second statement in No. 1 described it. A high-powered microscope and special stains would be necessary to note the nucleus or membrane-bound organelles. Even at this point, chemical tests would be required to note the presence or absence of pep-

tidoglycan for No. 2. Not being able to see a distinguishing characteristic makes using a key difficult. If a key for the plant kingdom uses only flower color or fruit size as characteristics on which to make one of the choices and the specimen being identified is not in bloom or has already dropped its fruit, the key is useless.

General keys (such as "A Simple Key to the Kingdoms") often add general statements to the primary statement to help the user if a problem arises. In "A Simple Key to the Kingdoms," the statements in No. 1 are of this type. Although easily misleading, they are often helpful. Some general keys make false generalizations. To avoid this, the authors of keys often find it necessary to use terms such as "usually," "often," "occasionally," and "some."

Another common problem with keys is not understanding what they

describe. If a person looks at a particular structure on a specimen and assumes that "this must be what the key is talking about" without really knowing, he stands a chance of reaching a wrong conclusion. His faulty assumption will make all his subsequent answers wrong.

Not only can the key or the user be inaccurate, but also the specimen can be atypical. If a flower normally having twelve petals is damaged so that it now has only ten, and if the key offers the choice of "twelve petals" or "fewer than twelve petals," then the damaged specimen is misleading. The key and the user of the key may be accurate, but the answer will still be wrong. The use of a key is only as accurate as the key, the user, and the specimen combined.

Binomial means "two-name," and *nomenclature* means "naming." A common system for naming people is binomial. The name "Jack Brown" identifies a person as belonging to the Brown family and that he is the particular Brown called Jack.

Linnaeus needed to choose a language for his system of naming organisms. He knew that languages that are being used by people as native tongues are constantly changing. New words are added as new things, ideas, or actions are devised. Words also change meaning. For example, the English word *charity* today means the giving of money or service. In the 1600s—when the King James Version of the Bible was translated—*charity* simply meant "love."

If Linnaeus had used his own language, Swedish, to name the organisms, his life's work would have become out-of-date as the meanings of the descriptive words he chose changed. Furthermore, Swedish is not a very familiar language to most scientists! Linnaeus therefore used Latin. No one spoke Latin as a native tongue; therefore, it was unchanging. In addition, most scholars knew Latin and would thus be able to understand the scientific name of the organism. Also, because Latin is a descriptive language, it suited Linnaeus's purpose quite well.

9-6

No matter what the language, this bird is *Melanerpes erythrocephalus*.

Genus-Species Names

The scientific name of an organism includes its genus and species names. In other words, the final two categories of the classification hierarchy for an organism are its binomial scientific name, or **genus-species name**. Since the genus-species name of an organism is a foreign term, it is printed in italics—if handwritten, genus-species names are underlined. Only the genus name is capitalized.

For example, the genus *Equus* is so called because *equus* means "horse" in Latin. *Equus caballus* is the common horse. This genus and species contains horses ranging in size from the Shetland pony to the Clydesdale. Other

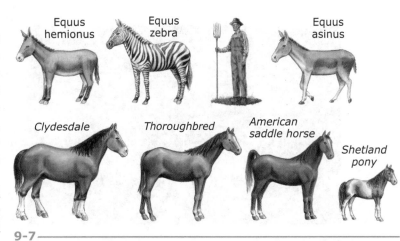

9-7

Various species within the genus *Equus* (top) and varieties within the genus-species *Equus caballus* (bottom). All are drawn to scale, including the farmer.

Varieties

A species may be a very large group of similar, but different, members. *Canis familiaris*, for example, is the genus-species name of the "familiar canine," the dog. In this particular species, however, there are many different *varieties*. Some of these varieties are purebreds, such as the collie, Irish setter, Chihuahua, Saint Bernard, and dachshund. Others may be crosses, like the peekapoo (a cross between a Pekingese and a poodle). And of course, in most neighborhoods there are many hybrids known as mutts.

Toy poodle   +   Pekingese   =   Peekapoo

members of the genus are *Equus asinus*, the donkey; *Equus zebra*, the zebra; and *Equus hemionus*, which includes the onager, a wild donkey-like animal native to central Asia.

A *genus* can be defined as a group of similar organisms. For some organisms the genus name and the common name are the same. For example, *Paramecium caudatum* is the genus-species name of an organism; however, there are a large number of related organisms that share the common name "paramecium." When "paramecium" refers to the genus name, the *p* should be capitalized and the word italicized or underlined.

The species name tells which specific organism of a particular genus is being described. Although the concept of a species is very useful, it is almost impossible to define, as the next section shows.

➡️ ## Review Questions 9.2

1. Give two reasons for having a system of scientific names.
2. What are two reasons for using Latin to name organisms?
3. What is the primary division of a species?
⊙4. What is the difference between classifying and identifying an organism?
⊙5. Describe a dichotomous key.
⊙6. List several problems that could make using a dichotomous key difficult or inaccurate.

9B—The Species and the Kind

Early in his career, Carolus Linnaeus thought that the number of species was established at Creation. While his publications were going through many editions to include the ever-growing number of organisms that were being found, Linnaeus wrote to a friend in 1749, "Am I to work myself to death, am I never to see or taste the world? What do I gain by it?" Later, he began to believe that those organisms present at Creation could change through the process of hybridization. By the end of his life, he and his students had described and named almost ten thousand plants and animals.

Today more than twice the number of species that Linnaeus classified in his entire lifetime is added *annually* to the lists of species. There are now over 1.5 million living species known to science, and taxonomists estimate there are another 5 to 10 million organisms yet to be discovered and classified. Why so many? Scientists are using increasingly sophisticated technology to seek out new organisms. Unmanned submersibles have gone to great ocean depths to explore, and researchers continue to invent new techniques to

9-8

This crane allows researchers to find new species in the forest canopy.

The discovery of new species is not limited to tiny insects or microscopic bacteria. Researchers often work with local residents to explore areas that have been previously off-limits or inaccessible—searching for animals that have been described but never officially recorded.

On the African continent, in remote mountains of Tanzania, a team from the California Academy of Sciences discovered and photographed the gray-faced sengi (*Rhynchocyon udzungwensis*) in 2005 and captured the first one in 2006. This animal is the largest known species of the giant elephant shrew group and eats invertebrates that it digs from the ground. It weighs about 700 g (1.5 lb).

The gray-faced sengi was not known to exist prior to 2005.

more closely examine previously studied environments. In addition, DNA analyses are causing some taxonomists to split existing species as molecular differences are detected.

9.3 The Species

"A group of similar organisms" is probably the only definition of *species* upon which all scientists would agree. This definition, however, is not adequate. All the trees in a forest are "similar" in some ways, but they are certainly not all one species. One problem with defining *species* is that the term has different meanings in different groups.

For example, a definition that applies to species of trees does not equally apply to species of fish, mammals, protists, or bacteria. Generally speaking, biologists agree that there are two major points of consideration for a definition of species.

1. Members of a species are structurally similar but do have a degree of variation.
2. Members of a species can interbreed and produce viable and fertile offspring under natural conditions.

Now the initial definition of *species* can be modified: a **biological species** is a group with members that resemble each other and that can generally interbreed to produce viable, fertile offspring. However, even this definition has problems, two of which are discussed below.

Problems with the Species Concept: Artificial Characteristics

Using physical characteristics to define the limits of a species often causes problems. For example, one variety of the tiger salamander, *Ambystoma tigrinum*, lives in some ponds in the western United States. In other ponds in the same region, researchers also found immature-appearing, salamander-like animals that looked similar to another type of amphibian called an axolotl (AK suh LAHT uhl) and gave it the name *Siredon mexicana*. Later it was found that if these axolotls were fed iodine-enriched material, they matured into tiger salamanders. The lack of iodine in their environment had caused them not to mature. Their appearance was so different from a mature tiger salamander's that they were initially classified as a separate species when really the difference was that the pond in which they grew was deficient in iodine. Scientists no longer use the name *Siredon mexicana*.

9.3

Objectives

- Describe the components that make a species
- Explain the problems with the species concept
- Respond to the assertion that "speciation causes new kinds of organisms to form"
- List and describe some of the factors that contribute to speciation

Key Terms

biological species
speciation
migration

geographic isolation
behavioral isolation
adaptation

9-9

Ambystoma tigrinum, previously known as *Siredon mexicana*

9-10

A coydog shows that the species concept is difficult to define.

Since an organism's environment can greatly affect its appearance, classification by physical characteristics is only an artificial method that may be prone to many errors.

Canis familiaris is the genus-species name of the common dog. It includes the toy poodle as well as the Great Dane. Along with their variety, all dogs have some similarities. They can interbreed and produce fertile offspring. All dogs have seventy-eight chromosomes. By the previously given definition, dogs are a species.

Canis latrans is the coyote. Although coyotes resemble dogs, they are placed in a different species within the same genus because they live in particular habitats and have certain characteristics—coat color, size, and others—that set them apart from most dogs. However, coyotes, like dogs, have seventy-eight chromosomes. Occasionally, in the wild or in captivity, a coyote mates with a dog, and a litter of *coydogs* is born. Coydogs are fertile and have characteristics of both parents.

Although the coyote has a certain set of dog characteristics and can interbreed with dogs, it is still classified as a separate species because it tends to live in areas where most dogs do not live and because it rarely mates with dogs. Why is the coyote not classified as a particular variety of the dog species, like a terrier or poodle? Maybe it should be. But it must be remembered that the classification of organisms was done by man for his convenience. So far, taxonomists consider it more convenient to ignore the fact that coyotes and dogs are so similar and leave them as two separate species.

Some evolutionists wrongly use the fact that there are varieties designed into creation. By seeing what *has been called* a species mate with another species and produce offspring with characteristics of both parents, some people conclude that evolution is occurring and that a new species is being formed. If two varieties of dogs were crossed, would the offspring be a new species? No. If a dog and a coyote produce offspring, is a new species developing? Some would say yes, because a coydog is a cross between two species and is "new." Again, it should be remembered that man classified dogs and coyotes in different species for convenience and that the characteristics of the coydog came from a dog and a coyote. The coydog is still just a kind of dog, drawing its characteristics from the gene pool supplied by the dog group.

Classification by physical appearance, habitat, or any other outward characteristic is artificial. In these cases similarities that cannot be seen but nonetheless exist are often ignored.

Problems with the Species Concept: Interbreeding

The sexual reproduction of a species with other members of the same species has, for quite a while, been considered a method of avoiding artificial classification of organisms on the basis of physical characteristics. If two organisms are capable of sexual reproduction and produce fertile offspring, then they should be classified together, no matter what they look like. This idea seems to make sense, but it should be examined closely.

First, there are thousands of organisms (bacteria and types of protozoans and fungi) that are not known to reproduce sexually at all. Obviously, some other criteria must be used to classify these organisms.

A second problem of using interbreeding as a criterion for classification is uncovered in another look at *Canis familiaris*. It is physically impossible for a male Saint Bernard to mate with a female Chihuahua. Even if the match were artificially made, at birth the pups would each be about the same size as their mother, a condition that would kill either the mother or the pups long before the pups were born.

Should this inability to interbreed put Chihuahuas and Saint Bernards in two separate species? If so, where should the species line be drawn? Chihuahuas can mate with dogs slightly larger than themselves, and these dogs with larger dogs, and finally, several generations and many sizes removed, a Chihuahua offspring may successfully mate with a Saint Bernard. Any line drawn has to be based on artificial distinctions.

God's primary purpose in creating different organisms was to have them perform specific functions in specific ways in specific places, not to make them easy to classify. Classifying organisms is not wrong just because it is not easy or does not always seem to work. Organisms are classified for convenience and as an aid in the study and understanding of creation.

9-11

Speciation

Where do new species come from? Linnaeus originally thought that all living species were populations that had been distinct since Creation. Today, we recognize that **speciation**, the formation of a new species, does happen. But speciation does not create new kinds of organisms. They are new species merely because taxonomists choose to define them as such. This type of speciation demonstrates that populations of organisms change from generation to generation. Eventually, these changes may reach the point where the now distinct populations no longer interbreed. Creationists have observed that populations change over time. They emphasize, however, that there is a limit to this change. Genetic information has merely been shuffled or lost. No new information is added. The molecules-to-man type of evolution requires the addition of information that is new, not just to the species, but to the whole creation.

It seems logical to assume that as the animals moved away from Noah's ark following the Flood (Gen. 8:15–19), they formed into loose groups, and in time they became distinct groups by natural selection. Various *isolating* mechanisms may have caused the distinctions that led to their being called separate species.

Evolutionists contend that migration, isolation, and natural selection are driving forces of evolution since, by their definition, a new species has developed. To the contrary, this rearrangement and selection of existing characteristics is not evolution, at least in the macro sense. A new, more complex organism has not evolved. Rather than new genes being formed, certain genes appear to be eliminated (not expressed) from a small population. Such a

Types of Isolation

✦ *Geographic isolation.* **Migration**, the moving of organisms from one area to another, can cause such a separation of traits that the groups of organisms could be considered two different species. Separation by water, mountain ranges, or even open space after a subgroup has migrated may lead to **geographic isolation**. At one time, all African elephants were considered the same species, *Loxodonta africana*. Today, the elephants of the savanna retain this classification, but the forest elephants of central and western Africa are placed into a different species, *Loxodonta cyclotis*. Taxonomists have agreed that differences in their genetics and physical characteristics demand this separation.

✦ *Behavioral isolation.* Two separated populations may become incapable of interbreeding due to conflicting reproductive behaviors or rituals, resulting in **behavioral isolation**. For instance, Eastern and Western meadowlarks may have been created as a single type of meadowlark, but they will no longer breed, even in areas where their ranges overlap. Their courtship (mating) songs are too different to attract each other.

Western meadowlarks (left) will not breed with Eastern meadowlarks (right).

Rapid Speciation

Evolution requires millions of years for enough mutations to occur to produce more genetic variability by adding new DNA. The new DNA, evolutionists believe, is passed on to new generations as the new organism interacts with the environment through the process of natural selection. This supposedly accounts for the thousands of types of organisms seen on the earth today. However, recent observations have shown many instances of rapid changes and the emergence of new species of organisms as a result of natural selection.

The discovery of new varieties and species should not be a stumbling block to Creationists. On the contrary, it is actually evidence for the many variations seen after the animals were released from Noah's ark. As the kinds of animals on Noah's ark were dispersed, the great variability designed into their gene pool allowed them to interact with the environment. The evidence shows that the emergence of new varieties and species can occur in some organisms in as little as five to ten generations—an observed fact that evolutionists cannot explain. Thus, the biblical model of earth history allows for rapid speciation.

change is not the biological evolution necessary to explain the origin of new types of life.

Adaptations are another force claimed by evolutionists to drive speciation and ultimately evolution. An **adaptation** is defined as any inheritable characteristic that gives a survival advantage to the organism. Adaptations could include structural changes such as a different color of fur, behavioral changes, or even changes to physiological processes. While these shifts in characteristics may result in changes significant enough to justify the naming of a new species, no new kinds of organisms have formed. A closer examination of the new species usually reveals a decrease in genetic potential. In other words, genetic information is lost as the organism becomes more specialized.

For instance, suppose that you have a population of bacteria living in a pond. The optimal pH range for these bacteria is 6.1–6.9. However, some bacteria possess a gene that allows them to survive down to a pH of 5.1. Assume that pollution results in the lowering of the pond's pH from 6.4 to 5.3. Can the bacteria still survive? Yes, the bacteria that have the gene for surviving in the lower pH can survive. After several generations, there will be a population of bacteria that express the gene, allowing them to survive in the lower pH. The bacteria have "adapted" to the change in their environment. Evolutionists believe that adaptation is the result of many beneficial genetic changes, resulting in new genetic information that improves an organism's chance for survival. However, in the bacteria example, no new genetic information was created. The population changed over time because some bacteria possessed characteristics that enabled them to survive better in their environment.

Changes in the behavior of organisms can also be viewed as adaptations. If some animals adapt to hunting an unusual food source when their preferred diet is no longer available, this behavioral adaptation would give them a survival advantage. The gathering of some animals into social groups may also be seen as an adaptation that increases the chances of survival. These types of adaptations merely show that the strongest, quickest, and most adaptable organisms survive. In many cases, these behaviors may revert to the original if environmental conditions return to the previous state. New or modified behaviors are not traits that can be transmitted through genes. Except for a few behaviors learned through modeling, offspring must learn new behaviors as their parents did. Also, you must recognize that changes in behavior do not result in different kinds of organisms developing. The fittest organisms may leave more offspring, but they are still essentially the same, even if they do behave differently.

➡ Review Questions 9.3

1. Give the two commonly accepted parts of the definition of a species.
2. Why are physical characteristics an inadequate basis for grouping members of a species?
3. List two reasons why interbreeding is a difficult basis to use for grouping organisms into species categories.
4. List and describe three factors that may lead to speciation.
5. Explain how migration, isolation, and adaptation might account for various groups of animals coming into existence after the Flood.
6. Respond to the statement, "Speciation is the force that drives evolution."

9.4 The Biblical Kind

Scripture does not include a detailed classification system, although it occasionally includes lists of organisms. What God created during the Creation week and which organisms were brought into Noah's ark have been suggested as classification systems. However, these are simply lists and are not meant to be classifications.

The Scriptures state that God created each organism to reproduce "after its kind" (Gen. 1:20–25). It appears that God established the **biblical kind** as the natural grouping of organisms and that the ability to reproduce is the criterion He established for classification into a biblical kind. Carolus Linnaeus, a Creationist, thought that he was distinguishing the kinds referred to in Genesis as he identified his species. He was the first to use the term *species*, which means "kind" in Latin, as a classification unit. Although Linnaeus thought that the species was the same as a biblical kind, he—like modern taxonomists—used various physical characteristics to determine his classifications and therefore had an artificial system.

The Dog Kind

Considering the species *Canis familiaris*, Linnaeus recognized the possibility of variation and did not put all the varieties of dogs into separate species. According to Linnaeus the coyote is a separate species since it does not naturally interbreed with dogs. Nonetheless, recent studies show that since dogs and coyotes are capable of interbreeding, the coyote is probably a member of the dog kind.

Are there other organisms that should also be placed in the dog kind? What about wolves? It is possible for dogs and wolves to produce fertile offspring. To determine all the organisms that should be placed into the dog kind, extensive breeding experiments would have to be conducted. The failing of a few matings should not significantly affect the conclusions.

Remember that Chihuahuas and Saint Bernards do not directly interbreed, but they can "indirectly." The expense, time, and know-how necessary to perform extensive breeding experiments make it probable that it will never be known how far the dog kind extends.

Linnaeus unknowingly hurt the cause of Creationism. By setting up his species and equating them with biblical kinds, he implied that his species could not interbreed and that new species would not develop. But species do interbreed (a fact that Linnaeus later recognized), and a new set of characteristics can occasionally occur in the offspring of the interbreeding of two species. The hybrid organism may be so different that, according to the currently accepted definition of species, a new species has developed. This type of development has often been used to demonstrate biological evolution. However, the expressing of old genes in new combinations within a biblical kind is *not* evolution.

Baraminology

Some Creationists, seeing problems with the current classification system, have developed **baraminology** (bah RAH min AH lo jee), the study of classification based on the idea of biblical kind. The word *baramin* is taken from the Hebrew of the Old Testament, and it means "created after its kind." Creationists have formulated several kinds under baraminology and use Scripture as well as science to organize organisms into groupings by their original created kind.

▶ **Objectives**
- Describe the criteria for a biblical kind
- Explain the basis for a natural system of classification

▶ **Key Terms**

biblical kind
baraminology
phylogenetic tree
clade

derived character
evolutionary classification

Noah's Flood and the Biblical Kind

Many animals and plants that were alive both before and after the Flood are extinct today. God apparently created a greater variety of organisms than what now remains. However, every extinct organism is similar to existing organisms. Dinosaurs were very similar to the reptiles that have survived. Trilobites (TRY luh BITES), a common fossil form, are easily classified because they are very similar to organisms alive today.

It is possible, but not probable, that although different groups within a biblical kind may have become extinct, either naturally or by man's efforts, there are still the same number of kinds existing on the earth.

In any case, the number of animals on the ark could have been far fewer than two of all the land-dwelling and seven of all the clean species. God brought kinds onto the ark. For example, since dogs are considered unclean based on Jewish ceremonial law, it can be assumed that there were only two of them on the ark. It is possible that several other species similar to the dog kind are the offspring of that one pair of dogs on the ark.

Trilobite fossil   *Horseshoe crab*

Darwin proposed one of the first evolutionary trees in his *Origin of Species* (1859). Since that time, the tree has had literally thousands of variations. Modern diagrams appear in every secular biology textbook and purport to show the relationships between modern organisms and the *common ancestors* that preceded them in evolution. These evolutionary diagrams are called **phylogenetic** (FYE loh juh NET ik) **trees**. *Phylogeny* is the study of the ancestry or evolutionary history of organisms. Those diagrams that attempt to depict the full path of evolution begin with a single common ancestor, and the trunk from this hypothetical species branches repeatedly to form a dense tree.

This phylogenetic tree is one evolutionary view on the origin of vertebrates.

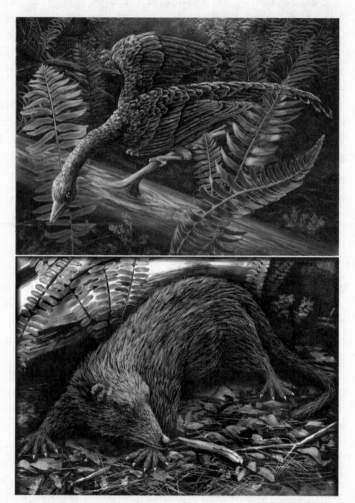

The archaeopteryx (top) was once thought to be a transitional link between dinosaurs and birds. Today it is recognized as being a bird. Its feathers, skeleton, and even brain case all support this conclusion. Many evolutionists propose that the genus Eozostrodon (bottom) contains the common ancestor for all mammals, including humans.

Evolutionists claim that organisms that are alike in some way, whether on the structural or the molecular level, may have had a common ancestor. This claim is based on their presupposition that all life arose from a common origin and over time gave rise to every living thing.

When a limb of a phylogenetic tree branches, a common ancestor is implied. A phylogenetic tree that contains many organisms and is relatively detailed usually has the known organisms (those alive today and those extinct organisms that are known only from fossils) on the ends of the branches. The common ancestors at the forks of the branches are nothing

more than guesses. None of them are living today. Various species known only from fossils have been proposed as candidates for some of these common ancestors, but there is no scientific basis for proving this.

In the early days, these phylogenetic trees were based on external artificial characteristics and assumptions related to the fossil record. By the early 1960s, with the recognition that all living things contain DNA, the development of new technologies provided a tool to decode this information in order to streamline phylogeny. Scientists began comparing the DNA and protein sequences of living organisms to

calculate how related they are to each other. They assumed that organisms that shared more traits on the molecular level must be more closely related and share a more recent common ancestor than those with fewer similarities.

Molecular analysis supports a phylogeny that views a group of related organisms as a clade. A **clade** represents a branch of a phylogenetic tree (for instance, the cat family) and may be part of a larger clade (e.g., the mammal class) if you move down to a more distant common ancestor (lower on the tree). The diagrams that result from these groups are called *cladograms*.

Scientists assign organisms to a particular clade on the basis of shared unique features called **derived characters**. The homologous structures discussed in Chapter 8 would be an example of derived characters. Increasingly, molecular markers are chosen as derived characters.

This modern phylogeny based on molecular biology, known as *cladistics*, did little to clarify the situation. In fact, it only complicated things more. Longstanding relationships between groups, based on artificial characteristics, were challenged and in many cases overthrown. For many years phylogenetic trees assumed that lizards, snakes, and crocodiles all branched from a common pre-reptile ancestor and thus were assigned to the same class, Reptilia. After all, they share many common features. Today, however, analysis of certain derived characters on the molecular level places the crocodiles in a common group with birds, rather than with the other reptiles.

With this ability to analyze molecules, you would assume that if all organisms were in fact truly related, all phylogenetic analyses would come to the same conclusions and produce similar trees. It turns out that early molecular phylogenetic trees based on ribosomal RNA (rRNA) are very different from later trees based on gene sequences. Other tests comparing particular proteins led to even more differences in the trees. The whole field of phylogeny is in such a state of crisis that some taxonomists have suggested it may be impossible to represent the history of life as a tree. Since God created distinct kinds of living things, that is exactly what would be expected.

Creationists have no difficulty with similarities between organisms since God designed each organism for a particular purpose. If four long legs are the right design for movement in one animal, it is likely that four long legs will be the best design for a similar movement in a similar animal. Neither do Creationists have difficulty with differences between organisms. God designed each of His creatures for a particular purpose, so differences are also due to His design. Molecular similarities and differences are no challenge either. It only makes sense to use similar blueprints and building blocks over and over.

Evolutionists, however, have great difficulty in accounting for differences between organisms since they assume a common ancestor. Similarities and differences come from God's design, not from a common ancestor.

Baraminology is useful for Creation science and to further our understanding of biblical kinds. It is also important to develop baraminology because evolutionists are trying to make a classification system to replace the current system. However, we will focus this study on the current system of classification because it so widely used and because we must be able to engage the culture in which we live.

An Evolutionary Classification System

Currently scientists are looking for a system that is not hindered by artificial boundaries such as the ones previously discussed. They have turned to genetic and biochemical research to develop **evolutionary** (or natural) **classification**—a system based on biochemical similarities, not on physical characteristics. Their hypothesis is this: the amount of similarity between one organism's biochemical makeup and another's is directly proportional to their degree of relatedness. This includes individual proteins as well as the nucleotide sequences in DNA and RNA.

For example, many organisms contain a hemoglobin-like macromolecule, and the amino acid sequences in the different hemoglobin molecules are similar. Human and frog hemoglobin molecules differ by sixty-seven amino acids; human and gorilla hemoglobin molecules by only one amino acid. Evolutionists who support a natural system of classification would say that humans are more closely related to gorillas than frogs and that frogs split off from the common ancestor much earlier than did gorillas.

How would a biblical Creationist counter that argument? God designed a means to carry oxygen in His creatures; thus, He created the macromolecule hemoglobin. The various creatures that were created have very different body forms as well as different oxygen requirements and different environments. God modified the basic hemoglobin molecule to fit the specific needs of His creatures. Would this argument convince an evolutionist? Probably not, but remember—the Bible is the source of all truth, and everything, not just science, must be evaluated based on Scripture. If a hypothesis or scientific model seems to make sense and all the evidence points to an answer that is contrary to the

Bible, then the evidence, not the Bible, must be reevaluated and the conclusions changed.

As researchers decode the genomes of various organisms (including the human genome), they are comparing the nucleotide sequences in the DNA and RNA. Each nucleotide in a gene is treated as a point of comparison, making possible millions of comparison points. Again, the closer the similarity of sequences between organisms, the closer they are thought to be related. According to evolutionists, genetic similarities, like other biochemical similarities, prove common ancestry.

Using these techniques, evolutionists are trying to classify organisms according to evolutionary relationships. The experimentation necessary to determine genetic similarities is advancing rapidly, and occasionally scientists discover a fact that they think helps in establishing the natural classification system. When this new system is all worked out, they expect to have an accurate evolutionary lineup of organisms (a *phylogenetic tree* [see p. 218]) that they claim will prove evolution.

At this point the discussion enters speculation, but speculation with scriptural backing: it would be interesting if scientists distinguished groups of organisms whose members are capable, although possibly indirectly, of interbreeding within the group but not with members of other groups. When this happens, science will not have arrived at what the evolutionist was looking for—the path evolution has taken—but science will be closer to realizing what God created—the biblical kind.

➡️ Review Questions 9.4

1. What is the primary characteristic that would place an organism into a biblical kind?

2. Compare an evolutionary and an artificial system of classification of living things.

⊙3. According to the cladogram on the right,

    a. which pair shares a more recent common ancestor, spiders and trilobites or spiders and centipedes?

    b. what is the number of the most recent ancestor common to both insects and millipedes?

    c. what group of organisms are in the clade arising from common ancestor #3?

⊙4. What is the basic concept used to justify the building of phylogenetic trees? What are some objections to building phylogenetic trees?

⊙5. Suppose a Christian objects to the modern system of classification, saying that since he does not see a basis for it in Scripture, it is wrong. Prepare an answer for this person.

⊙6. One of your friends has just read an article in a well-respected scientific journal that he says proves beyond a doubt that the close genetic similarities between the human genome and the chimpanzee genome prove descent from a common ancestor. What would be your response?

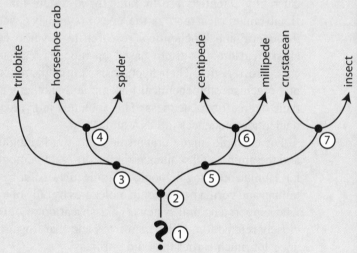

10

Bacteria and Viruses

Microbiology Part I

Although the particles discussed in this chapter are small, they play an important role in God's creation—for good and for ill. Bacteria help eliminate our planet's waste, enable us to digest our food, and make the air around us breathable. Viruses too have beneficial purposes, such as regulating bacteria populations and increasing diversity. These beneficial functions of bacteria and viruses remind us that God made this world good— even the smallest creatures help balance nature and contribute to human health and well-being. This statement is based on the assumption that bacteria and viruses were part of the original Creation and did not originate at the time of the Fall. While it is impossible to be dogmatic about exactly when God formed them, the benefits of many bacteria and viruses, as well as their interdependence with living things from peas to parrots, support the conclusion that they were present and functioning in harmony by the end of the sixth day.

Nevertheless, sometimes these organisms are very destructive, causing disease and death. This destructive function reminds us that our world is not just created but also fallen. Because we have rebelled against God, He has cursed our world, and that curse extends to the tiniest things on the earth.

If we are to please God with the dominion that He has called us to exercise, we will need to study the good and the bad aspects of these organisms. We need to study their destructive functions to control and counteract them, and we need to study their beneficial functions to make wise use of them.

10A—Bacteria

Bacteria—the organisms that are included in the kingdoms Archaebacteria and Eubacteria—outnumber all the members of the other four kingdoms combined. They include the tiniest known living organisms; many are visible only with the use of the higher powers of a light microscope and are probably best studied by using electron microscopy. Even so, it has been estimated that bacteria have a greater combined weight than all the other living things on the earth.

On average, bacteria are one thousand times smaller than most human cells. Although small, they are numerous—the bacteria that live on your skin and in your digestive tract outnumber the cells of your body.

Bacteria thrive in places where no other living organisms are known to exist. In the atmosphere 6.096 km (3.8 mi) high, they ride on dust particles. In hot springs at 98 °C (208 °F), in melted glacial water, and at hydrothermal vents deep in the ocean, bacteria multiply. A single gram of common soil contains from 1 to 100 million of them. They exist in every natural water supply, including the water that you drink. Every cubic meter of air contains from 100 to 20 000 of them. Despite the fact that bacteria are almost everywhere in astonishing numbers, scientists have only recently devised techniques for studying them and still have a great deal to learn about them.

10.1 Bacterial Classification and Morphology

As discussed in Chapter 9, bacteria were initially included as members of the plant kingdom. This placement seemed logical since some of these organisms can make their own food by photosynthesis, and many of their cellular characteristics do resemble plants. Later they were placed into the kingdom Monera; they are now split into two separate kingdoms—Archaebacteria and Eubacteria. Although members of both kingdoms are prokaryotic, newly discovered differences support two separate kingdoms. (See Appendix B for a summary of their characteristics.)

Today a *microbiologist* (one who studies microscopic organisms) uses various techniques to study and classify bacteria. In fact, microbiologists are discovering so much information so fast that there is no consistent, unified classification system for bacteria.

When most people think of **bacteria** (sing., bacterium), they think of germs that cause diseases. A **disease** is any change (except for those caused by injuries) that affects an organism's normal function. Although many bacteria cause diseases, the vast majority do not. Most bacteria are vitally important to living things, including humans, because of their effects on dead things.

Bacteria (along with fungi) are the primary **decomposer organisms** in soil and water. Decomposer organisms produce enzymes that break down proteins, starches, lipids, and almost every other known organic substance. Without this action by bacteria and fungi, in a few generations all the materials necessary for life would be contained in nondecomposing dead bodies. Think of what the world would be like if dead things did not decompose! It is easy to see that bacteria and fungi are essential to all life.

10-1

Without the action of bacteria and fungi, the earth would soon run out of raw materials for the formation of new organisms.

bacteria: (small rod)

Kingdom Archaebacteria

Taxonomists originally chose the name *archaebacteria* because, having an evolutionary worldview, they thought these organisms were probably the oldest life forms on the earth. *Archae-* comes from the Greek word *arkhaios*, meaning "ancient." That most of these organisms live in "extreme" environments also seemed to fit an evolutionary concept of the origin of the earth and life. But a study of these organisms will show that these evolutionary concepts are false. In fact, archaebacteria are more evidence of God's abundant creativity and design.

All bacteria are prokaryotic. So what makes archaebacteria sufficiently different to warrant a separate kingdom? One major difference is in the structure of the cell wall. The archaebacteria cell walls do not contain *peptidoglycan*. Another finding that sets them apart is that the structure of some of their genes resembles both eubacteria and eukaryotic cells. In fact, the resemblance to eukaryotic cells caused some evolutionary researchers to surmise that the archaebacteria were the precursor cells to eukaryotes; however, this did not "fit" their phylogenetic tree. Why? Because the genetic characteristics seen in the archaebacteria that were similar to eukaryotic cells would have to be present millions of years before the supposed evolution and appearance of eukaryotic cells. Now they are looking for other explanations.

Life in extreme environments was the early defining characteristic of these organisms. They were found in hot springs, salty lakes, and swamps: places where life was thought to be impossible. In fact, the current classification of archaebacteria is based on the environments where they live.

Thermoacidophiles (THUHR moh uh SID oh filez) live in highly acidic soils and hot springs where the temperature can reach 110 °C (230 °F). They also live around hydrothermal vents deep on the ocean floor. Hydrogen sulfide is often used by these organisms as a source of electrons for their metabolism.

The *methanogens* (meth AN uh jenz) live in anaerobic environments such as swamps, sewage, and the intestines of some animals. Most use hydrogen gas and carbon dioxide for anaerobic respiration and produce methane gas as a waste—hence their name, which means "methane-producer."

"Salt-lovers," or *halophiles* (HAL uh filez), live in areas with extremely high salt concentrations, such as the Great Salt Lake and the Dead Sea. These organisms metabolize the salt to produce ATP.

10-2 —
The methanogens in swamps release large amounts of methane gas.

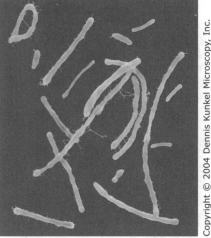

Copyright © 2004 Dennis Kunkel Microscopy, Inc.

10-3 —
Examples of archaebacteria: methane-producing *Methanosarcina mazei* (top) and a halophilic species (bottom)

thermoacidophile: thermo- (heat) + -acido- (L. *acidus*—sour) + -phile (loving)

methanogen: methano- (E. *methane gas*—CH₄) + -gen (Gk. *genes*—born)

halophile: halo- (Gk. *hals*—salt, sea) + -phile (loving)

10-1 Comparison Between Archaebacteria and Eubacteria		
	Archaebacteria	**Eubacteria**
Cell type	prokaryotic	prokaryotic
Cell membrane	various kinds of lipids	lipids different from archaebacteria
Cell wall	no peptidoglycan	contains peptidoglycan
Environment	extreme and nonextreme environs	nonextreme environs
Gene structure	contains introns and other similarities to eukaryotic cells; also similarities to eubacteria genes	no introns; distinctly different from eukaryotic cells

The Cyanobacteria (the Blue-Green Algae)

At one time the blue-green algae were considered plants. Today scientists classify them in the kingdom Eubacteria as a group within one of the bacterial phyla and call them the cyanobacteria. The major reason for this reclassification is that cyanobacteria are all prokaryotic.

The cellular structure of the cyanobacteria is essentially the same as that of bacteria. Blue-green algae lack flagella, but many of them float in the water, and some (such as *Oscillatoria*) even appear to move by a type of gliding. Although a few of the cyanobacteria are unicellular, most of them are colonial, forming long **filaments** (thin strands of similar cells) that are surrounded by a gelatinous sheath. In some these filaments are branched (*Nostoc*), while some form colonies shaped like discs or globs. Often these colonies share slime coats, and they sometimes form large gooey masses that are visible to the unaided eye.

Cyanobacteria are often attached by their sheaths to underwater surfaces or to places that are constantly wet. Although most live in bodies of water, some live in the soil. Cyanobacteria reproduce by simple cell division. Some species are known to form spores, but none are known to reproduce sexually.

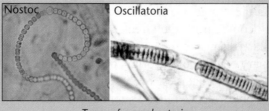

Types of cyanobacteria

Usually, within a colony all individual cells are the same. Some cyanobacteria, however, contain a *heterocyst*, generally a single cell which may have different functions in different species. Some heterocysts capture nitrogen gas from the environment and make it useful to the cyanobacteria. Occasionally a heterocyst is a dead cell that permits easy fragmentation of the filament. In some species the heterocyst is alive, can carry on photosynthesis, and can divide, forming a new filament.

All the cyanobacteria contain chlorophyll *a*, along with other accessory pigments. Thus they carry on photosynthesis similar to that of autotrophs in other kingdoms. Many cyanobacteria also contain a blue pigment (phycocyanin) and appear bluish green. About half of the cyanobacteria, however, are colorless, gray, green, yellow, orange, pink, purple, brown, violet, or red because of the presence or absence of other pigments.

Certain cyanobacteria, when in abundance, can affect the taste or odor of water. Other cyanobacteria produce poisons, making the water unfit to drink. If this happens to a farm pond or stream, animals that drink the water may die.

cyanobacteria: cyano- or -cyanin (Gk. *kuanos*—dark blue) + -bacteria (rod)

Kingdom Eubacteria

This kingdom is the larger of the two and contains all the remaining types of bacteria. The eubacteria are the organisms that most people think of when the word *bacteria* is mentioned (e.g., the bacteria that cause various diseases or that are used in food processing and industrial applications). Eubacteria have a wide range of morphology and live in many different kinds of environments. Historically, prokaryotes have been classified by their shape and by the results of a special process called Gram's staining. However, with the current emphasis on molecular taxonomy, many of the old phyla have been replaced or bacteria have been switched from one phylum to another. In fact, there is no consensus among scientists over exactly how many phyla this kingdom should contain or which bacteria should be in what phylum.

The remainder of this chapter examines the various characteristics of bacteria, and unless specifically stated, the information is applicable to the kingdom Eubacteria.

Bacterial Shapes, Sizes, and Colonies

Generally speaking, bacteria are one of three general shapes. Bacterial shape was one of the first means that scientists used to classify bacteria.

✦ *Coccus.* A spherical or oval bacterium, averaging about 1 μm in diameter, is called a **coccus** (KAHK us) (pl., cocci). The period at the end of this sentence could easily have five thousand cocci lined up across its diameter.

✦ *Bacillus.* A rod-shaped bacterium is called a **bacillus** (buh SILL us) and averages 1 μm in width and 2–10 μm in length.

✦ *Spirillum.* A **spirillum** (spy RILL um) is a spiral- or corkscrew-shaped bacterium that is slightly longer than the bacillus type.

Scientific names of bacteria often describe a characteristic of the cells. *Coccus, bacillus,* and *spirillum* are often found in genus names, such as

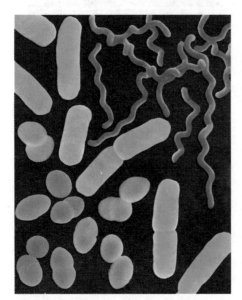

10-4

Bacterial morphologies (SEM): coccus (purple), bacillus (red), and spirillum (green)

coccus: (Gk. *kokkos*—berry or grain)

bacillus: (L. *bacillus*—small rod)

spirillum: (L. *spirillum*—a small coil or twist)

Streptococcus, *Lactobacillus*, or *Rhodospirillum*. Bacterial cells also form colonies, which are sometimes used in identifying and naming bacteria. Some bacteria form simple colonies, such as the diplococcus seen in Figure 10-4, while others may form long chains or grape-like clusters. When bacteria are grown in laboratory petri dishes, the colonies often exhibit specific growth patterns that are characteristic for a certain bacteria. Other types of bacteria actually form colonies that have stalks that produce spores. No wonder taxonomists continue to have difficulty in classifying bacteria!

Large bacteria can be seen when magnified about 400×. Most bacteria, however, are so tiny that they must be magnified 1000×.

Structure of a Bacterial Cell

Originally, prokaryotic cells were also thought to be very simple in structure, especially since they were believed to be a more ancient life form in the evolutionary timeline. However, as microscopic and biochemical techniques have advanced, so has appreciation of the complexity of the prokaryotic cell and God's intricate design.

Cell Walls and Capsules

Bacteria, like eukaryotic cells, have a bilayered phospholipid cell membrane whose major function is to control the flow of substances into and out of the bacterial cell. However, it does not contain cholesterol and many of the other molecules that are found in eukaryotic cell membranes.

Most bacteria have a cell wall immediately outside the cell membrane. Bacterial cell walls lack cellulose but do contain several chemicals found only in bacterial cell walls. One such chemical is a large molecule of carbohydrates and protein called peptidoglycan. This molecule is cross-linked with other molecules to give strength and shape to the cell wall. Another method used by taxonomists to classify bacteria is based on the content of peptidoglycan in the cell wall. A special staining technique called a *Gram's stain* will stain bacteria either purple (gram-positive) or pink (gram-negative) based on the amount of peptidoglycan found in their cell walls.

Outside the cell wall, many bacteria have a **capsule** made of gummy complex carbohydrates. Some capsules are very thin, while others may be several times the thickness of the cell. The capsule protects the cell from drying out during temporary dry periods and prevents certain substances from entering the cell. Any substance entering or leaving the cell must pass through the slimy capsule.

The thickness of a bacterium's capsule appears to affect an organism's ability to combat infection by that particular bacterium. For example, *Streptococcus pneumoniae* is a bacterium that comes in several strains, some with a thick capsule and some with almost no capsule. When the nonencapsulated strains are injected into laboratory mice, few, if any, symptoms develop. But when an encapsulated strain is injected, the mice develop pneumonia and die. The mouse's body is able to destroy the nonencapsulated variety but has little effect on the bacteria that have a thick capsule.

Cytoplasmic Structures

The DNA in a bacterium forms a single double-stranded circular chromosome. If this circular DNA were stretched out, it would be about one thousand

Uses of Bacteria

Because bacteria produce only certain chemicals as they break down their nutrients, many types of bacteria are useful in industry. Bacteria are used when making cheese, yogurt, silage, sauerkraut, and buttermilk. Natural products of bacterial fermentation, such as alcohol, vinegar (acetic acid), and lactic acid, are also useful since they are often the least expensive source for these substances. Other bacteria are useful sources of vitamins (B_{12}, C), amino acids, and most antibiotics.

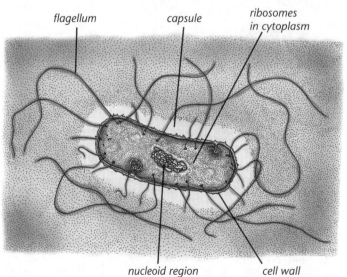

flagellum   *capsule*   *ribosomes in cytoplasm*

nucleoid region   *cell wall*

10-5 ———
Structure of a typical bacterium

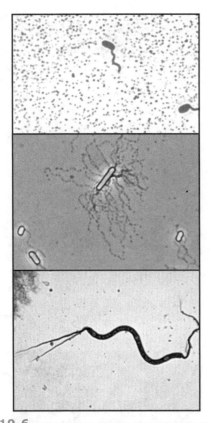

10-6

Bacteria have many different configurations of flagella.

times as long as the cell itself. Normally this chromosome appears in a non-membrane-bound nuclear area called the **nucleoid region**. Some bacteria also contain one or more smaller circular DNA chromosomes called *plasmids*, which you learned about in Chapter 7. Plasmids carry only a few genes and usually are not essential for the bacteria to survive. However, some plasmids may impart special characteristics such as resistance to antibiotics and the ability to withstand certain chemicals.

Bacteria do not have a cytoskeleton; their shape is maintained by the cell wall. Ribosomes, the non-membrane-bound organelles associated with protein formation, are abundant in bacterial cytoplasm. The cytoplasm also contains globules of fats, starches, ions, and proteins.

Locomotive Structures

Many spirilla, bacilli, and a few cocci have *flagella*. A bacterial flagellum is a long threadlike structure usually less than 0.05 μm in diameter. They may be either short or several times longer than the length of the cell. Bacterial flagella lack the 9 + 2 microtubule arrangement of the flagella in eukaryotic cells; instead, they are formed by a single strand of the protein *flagellin*. Spirilla with flagella usually have clusters of flagella on each end, while bacilli often have flagella scattered over the cell. Most bacteria with flagella are able to direct their movements: they usually proceed either toward or away from a particular stimulus. Photosynthetic bacteria, for example, swim toward the light source.

Bacterial cells without flagella appear to vibrate back and forth as water molecules bump into them. Even bacteria with no known form of movement seem to be capable of moving toward or away from certain stimuli. All bacteria are small enough to be easily carried along with even the slightest fluid current.

⇒ Review Questions 10.1

1. What is the primary function of bacteria in nature?
2. Compare and contrast the kingdoms Archaebacteria and Eubacteria.
3. What are the common cellular shapes and colonies of bacteria?
4. Describe the cellular structures of bacteria.
5. What is the primary reason the cyanobacteria were reclassified from algae to bacteria?
6. List several common characteristics that cyanobacteria share with other bacteria and several characteristics that separate them as a distinct group.
7. List several ways bacteria are used in industry.
8. What are two common methods of bacterial movement?

10.2

⇒ Objectives

- Describe bacterial reproduction
- Summarize the methods bacteria use to obtain energy
- List the conditions needed for bacterial growth
- Describe the methods bacteria use to obtain new genetic material

⇒ Key Terms

binary fission
conjugation
transformation
chemosynthetic
parasitic
saprophytic
endospore

10.2 Bacterial Reproduction and Growth

Under ideal conditions many bacteria are able to grow to full size and divide every thirty minutes. If these conditions persist, in about thirty generations (fifteen hours) a single bacterial cell would be multiplied to over one billion cells. An hour later the initial cell would have become about four billion cells. In less than twenty-four hours the mass of bacterial cells would weigh over 2000 tons. In another day there would be billions of tons of bacteria. Within a week the bacteria would weigh more than the earth.

With this phenomenal growth rate, it would seem that bacteria should have taken over the entire planet by now. However, even laboratory cultures of bacteria grown under optimal conditions cannot sustain such a growth rate for long. The bacteria on the outside of a colony may maintain a high metabolism, but those in the center do not receive enough of the materials necessary for growth. Soon the central cells are in a pool of wastes and dead cells—certainly not ideal growth conditions. In fact, the ideal range of conditions for high metabolism in bacteria is often quite narrow and difficult to maintain for a large quantity over an extended period of time.

Reproduction

Normally bacteria reproduce by simple **binary fission**. This type of asexual reproduction does not involve the normal steps of mitosis, but the result is the same. After the replication of the bacterial chromosome into two daughter chromosomes, an invagination of the cell membrane forms daughter cells. They contain identical DNA since there has been no exchange of genetic information.

Some bacteria can complete cell division in nine minutes. After about ten minutes of growth under ideal conditions, these cells will be full-sized and ready to divide again. Most bacteria grown in a laboratory can divide about every thirty minutes, but certain bacteria require three to four hours. Some disease-causing bacteria divide only once or twice a day.

Transfer of Genetic Material

The transfer of genetic material from one bacterial cell to another is believed to be a relatively rare occurrence naturally, although it is accomplished in the laboratory during genetic engineering procedures. The transferred genetic material may be a gene, a few genes, or a complete copy of the bacterial genome. Since there is no meiosis and no diploid cell or zygote formed, these genetic transfers are not considered true sexual reproduction. They do, however, result in genetic variation.

Conjugation in bacteria is a natural process of genetic transfer. Certain strains of bacteria are able to grow a hollow *conjugation tube*. Typically, it is plasmids that are transferred from one bacterium to the other through the conjugation tube.

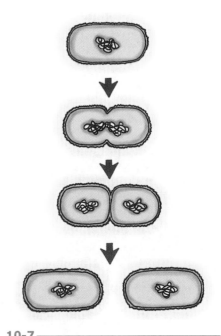

10-7
Asexual reproduction of bacteria

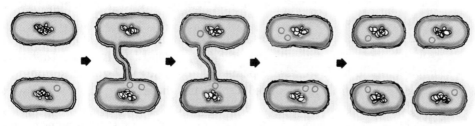
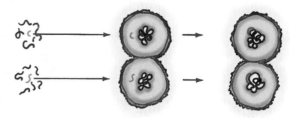

10-8
Conjugation: The plasmid (red) passes through the conjugation tube from one bacterium to another. Electron micrograph of bacteria undergoing conjugation (right).

In some other bacteria known to perform conjugation, the entire bacterial chromosome passes through the tube. The donor bacterium always gives its traits to the receiving bacterium. There is no mutual exchange of characteristics in bacterial conjugation.

In **transformation**, another form of bacterial genetic transfer, living bacterial cells take up other bacteria's DNA that is free in the environment. Often, this genetic material has been released from bacteria that have died.

A third but less common type of bacterial genetic exchange is transduction. *Transduction* is the transfer of genetic material from one bacterium to another by a virus called a *bacteriophage* that attaches to bacteria. The genetic material may come from another bacterium or even the virus itself.

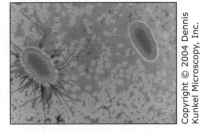

10-9
Transformation: The gene for capsule formation (red) is taken into living bacteria, causing them to change their genetic make-up and produce a capsule.

conjugation: con- (L. *com-*—together) + -jugation (*jugare*—yoke)

Nutrition

Of all known groups of living things, the bacteria kingdoms have the widest range of methods for obtaining usable energy and building organic substances. The most complex biosynthesis in bacteria takes place among the *autotrophs*, both the *photosynthetic* and **chemosynthetic** varieties. Some of these bacteria are almost completely independent, requiring only inorganic materials to live.

Some bacteria carry on photosynthesis that is considerably different from the photosynthesis carried on by other organisms. *Bacteriochlorophyll* is usually purple, red, or brown. These colors permit bacterial photosynthesis to be carried on in environments such as the depths of a pond, where the wavelengths of light necessary for green chlorophyll to function cannot penetrate. Recall that the photosynthesis carried on by green chlorophyll uses water as a hydrogen source; bacterial photosynthesis usually obtains hydrogen from hydrogen sulfide or other substances. Oxygen, therefore, is usually not a product of bacteriochlorophyll photosynthesis.

Although not a large group, the chemosynthetic bacteria perform some important conversions of inorganic materials. Chemosynthesis involves converting inorganic compounds that other living things cannot use into usable forms and capturing the escaping energy. Chemosynthetic bacteria use this energy to form materials (usually carbohydrates) for their own use.

The vast majority of known bacteria are *heterotrophic*—they obtain their energy by digesting organic substances. The bacteria that people usually think of are **parasitic** (feeding on a living host); however, most bacteria are **saprophytic** (feeding on dead organic matter). Saprophytic bacteria feed by secreting enzymes that digest external substances into soluble forms. These soluble materials then diffuse through the capsule and into the cell.

Some bacteria and similar organisms are *obligate parasites*; they require the presence of living tissue in order to grow. Parasitic bacteria lack many of the systems needed for the synthesis of organic substances that the saprophytic or autotrophic bacteria have.

Most evolutionists speculate that bacteria were the first living things. If evolution were true, the simplest organisms would have evolved first. The simplest would be the obligate parasites—the ones with the least complex biological systems. The autotrophic bacteria, those with the most complex biological systems, would have evolved from the simpler parasitic types. However, parasites cannot support themselves—they must have other organisms to survive. Therefore, evolution of another organism to serve as their host would be essential. But this necessity presents a problem: how can a parasite, which requires a living host to live, evolve before its host—a more complex organism?

Those with a biblical worldview do not have this problem of "Which came first?" God, as the omnipotent and omniscient Designer, fashioned all the organisms in His creation to work together. The fact that men, His creatures, do not understand the intricacies of creation only reveals their limited understanding.

Conditions for Optimal Growth

Several environmental factors must be within a bacterium's optimal range for it to grow and divide. These include the following:

♦ *Moisture.* Although many bacteria are protected by their capsules during temporary dry periods, all bacteria require moisture to grow. Some bacteria grow best submerged in fluids.

parasitic: para- (Gk. *para*—beside) + -sitic or -site (*sitos*—grain or food)

saprophytic: sapro- (Gk. *sapros*—rotten or putrid) + -phytic, -phyte, -phyta, or phyto- (*phuton*—plant)

✦ *Temperature.* Enzyme systems of different bacteria operate at different temperatures. As previously discussed, some bacteria grow in near freezing temperatures and some at high temperatures; nonetheless, most bacteria survive in temperatures from 27 °C to 38 °C (80 °F to 100 °F). Human bacteria obviously grow best at body temperature 37 °C (98.6 °F) and have a range of only a few degrees in which they can grow. Most bacteria have a range of about 30 °C in which they can grow.

✦ *pH.* Although many bacteria can grow in a nearly neutral condition, the pH of the environment is critical to their existence. Raising or lowering the pH can destroy one kind of bacteria while creating ideal conditions for another.

✦ *Nutrition.* The food source for heterotrophic bacteria not only must meet the energy needs but must also supply the materials necessary for bacterial biosynthesis. For some heterotrophs, especially the parasites, the proper food source is highly specific. Even though photosynthetic bacteria obtain their energy from light, they often require additional substances (inorganic and occasionally organic, such as vitamins) in order to grow. Chemosynthetic bacteria require specific chemicals as their energy source.

Although bacteria live in almost every known environment, what is advantageous for one species may kill another. Since each species of bacteria requires a specific set of environmental conditions to grow and reproduce, bacteria exhibit a high degree of *specificity*. In other words, they are highly specific regarding the environment in which they can grow and multiply.

During mildly unfavorable environmental conditions, many bacteria decrease their metabolic activities and wait for the return of favorable conditions. This reduced metabolism preserves certain soil bacteria during the higher temperatures of a summer afternoon and the cooler evening temperatures of spring or fall.

When some bacillus bacteria and a few other types are exposed to extended periods of unfavorable environmental conditions, they are capable of forming **endospores** in order to survive. Bacterial endospores differ from the spores of most other organisms in that the endospores form within the cell membrane. Several layers of materials harden around the nuclear area of the cell to form an endospore, and metabolism within the endospore is reduced to a very low level.

Some endospores can withstand twenty hours of boiling. Others can withstand freezing, dryness, extreme pressure, and even poisonous gases. When

Accelerating Mutation Rates

Saprophytic bacteria, when grown in the laboratory, require certain vitamins or other organic nutrients in their media. Experiments with mutation-inducing radiation or chemicals have produced bacteria whose offspring have even further reduced abilities to manufacture certain nutrients. The mutant offspring require these additional nutrient materials in their media that their "parents" were able to produce. If these mutant bacteria are subjected to the same type of radiation or chemicals, the next generation may have even more mutations and thus be capable of manufacturing even fewer nutrients. Therefore, each new generation needs to be supplied with even more organic nutrients.

If the evolution model were true, it would be expected that increasing the mutation rate would produce more complex organisms that are able to manufacture more and more substances. Theoretically, they should move from being saprophytic to being autotrophic—successive generations should require a smaller variety and amount of nutrient materials. They would become more complex; they would have evolved. However, in these experiments, increasing the mutation rate resulted in a loss of ability. It cannot be denied that the mutations did cause changes, but the mutations resulted in degeneration, not evolution.

Mutations in bacteria. The bacteria in the upper left can grow on media without any vitamins. When exposed to radiation, they change.

(c)

(a)

(b)

Unable to grow on media with no vitamins (a) or on media with only vitamin A (b), these bacteria can grow on media containing vitamins A and B (c). What conclusions can you make about the bacteria's reaction to radiation?

10-10
Endospore (top) within a bacterium

endospore: endo- (within) + -spore (seed)

Rickettsiae, Spirochetes, and Mycoplasmas

Kingdom Eubacteria contains several distinct groups of organisms. Three groups are significant—*rickettsiae*, *spirochetes*, and *mycoplasmas*—because of the pathogens they contain. With rare exceptions these are obligate parasites because they must have a living host.

The Rickettsiae

Dr. Howard T. Ricketts was the first person to describe a small bacterium in the blood of people suffering from Rocky Mountain spotted fever. In 1910, Dr. Ricketts saw a similar organism in the cells of people with typhus and in the lice that carry the disease to humans. Unfortunately while performing his studies, Dr. Ricketts contracted typhus and died. This group of organisms was named in his honor.

The rickettsiae (rih KET see EE) are *intracellular parasites*: they live inside cells. Rickettsiae can grow and divide rapidly, but they are highly specific, growing only in certain cells. They do not form spores and do not reproduce outside living cells. Most substances in the cytoplasm of their host cell easily penetrate their cell walls and membranes.

Rickettsial diseases usually cause fevers, rashes, and blotches under the skin. The blotches are the result of the rupturing of damaged small blood vessels. Once a person recovers from a rickettsial disease, he is usually immune to its harmful effects even though his cells may still contain the organism. Such a person may host the pathogen without being sick himself.

Typhus, caused by a rickettsial organism, is one of the diseases that has claimed the most human lives. Man is the natural host for this disease, and human body lice and head lice pass it from one person to the next. The typhus rash spreads over the body except for the soles of the feet and palms of the hands. Following a fever that makes the victim sluggish, death may occur within ten days. Drugs can control the disease, but control of the lice is the best preventive.

Rocky Mountain spotted fever was first recognized in the American West but is more common in the eastern states. Ticks carry it from small mammals (rodents, rabbits, dogs) to man. About two weeks following the tick bite, the person develops a fever, headache, chills, and rash; the disease is rarely fatal.

The Spirochetes

Most spirochetes (SPY ruh KEETS) are larger than the average bacterium, and many are corkscrew shaped. Some spirochetes are free living and normally inhabit the mouths, intestines, and reproductive organs of humans and animals. Some are believed to be nonpathogenic. Some spirochetes are found in sewage and in decomposing plant materials.

Spirochetes lack flagella but have flexible cell walls that permit them to move by contracting their coils. Many are highly specific as to their environment. None are known to form spores. Lyme disease, yaws, infectious jaundice, syphilis, and relapsing fever are a few of the diseases caused by spirochetes.

The Mycoplasmas

Mycoplasmas (MY koh PLAZ muhz) were discovered in the late 1890s as pathogens in the membranes around the lungs of cattle. Other mycoplasmas have since been found living harmlessly in the human mouth, nasal passages, and urinary tract. Several animal and plant diseases have been attributed to mycoplasmas. Atypical pneumonia, arthritis, and infections of the urinary tract in humans have been associated with mycoplasmas. Mycoplasmas lack a cell wall and thus can assume a variety of shapes. They are highly specific regarding the solute concentrations they can tolerate.

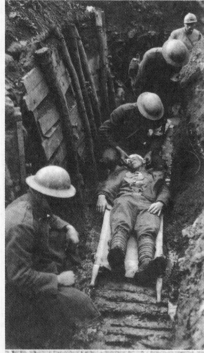

A rickettsial organism caused "trench fever" during World War I, so named because it was readily spread in the close quarters of the trenches (top) by body lice (bottom).

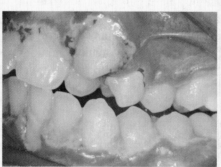

Spirochetes can be the cause of severe gum disease.

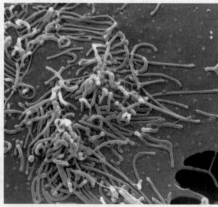

Mycoplasmas can cause many kinds of pneumonia.

favorable conditions return, the endospore covering (which may be in layers) weakens, and as bacterial growth resumes, the cell bursts out of the endospore case. Endospore formation is not a method of reproduction; it is a method of survival.

Bacteria and Oxygen

Louis Pasteur demonstrated that some bacteria can grow without (and are actually inhibited or destroyed by) free oxygen. Organisms that grow only in the absence of free oxygen are *obligate anaerobes*, and those that grow only in the presence of free oxygen are *obligate aerobes*. Many bacteria can grow with or without oxygen; they are *facultative anaerobes*.

Many bacteria in the kingdom Archaebacteria that grow in the depths of lakes and oceans or in the ooze of swamps are anaerobic bacteria, as are many of the eubacteria used in the production of cheese, yogurt, sauerkraut, buttermilk, and

Controlling Bacteria in Food

Although in some ways we rely heavily on bacterial metabolism and even encourage the growth of bacteria in some instances of food production, many of our dealings with bacteria are attempts to control them. Medical science and industry have invested much time and money in attempts to control food-borne disease-causing bacteria.

Since most bacteria decompose organic material, it should not be surprising that bacteria are among our primary competitors for food. Many foods would remain edible for months were it not for bacterial action fermenting carbohydrates, putrefying proteins, and turning fats rancid. There are two basic ways of controlling bacteria in food:

- Destroy the bacteria present and seal the food in a container to prevent the entrance of other bacteria.
- Place the foods in an environment that will not permit bacteria to grow or at least not grow rapidly. The bacteria may be present, possibly in a dormant form, but they either will not spoil the food or will spoil it less rapidly.

Some of the more common methods of food preservation are as follows:

✦ *Canning.* In canning, the food is heated enough to destroy all disease-causing bacteria and then sealed in a can or jar to prevent growth of the residual organisms and the entrance of new ones. Canned foods can keep for extended periods, but many foods lose desirable qualities in the process. Canning is primarily used for fruits, vegetables, meats, and prepared foods.

✦ *Preserves or jellies.* Foods that can be placed in a strong sugar concentration can be kept as preserves or jellies. The high concentration of sugar prevents most bacteria from growing.

✦ *Salt-curing.* Some meats can be preserved by placing large quantities of salt on them. The salt causes dehydration of active cells. Most salt-cured meats retain the heavily salted taste. Salt-cured meats were common before freezing and refrigeration were convenient.

✦ *Refrigeration.* Foods kept at low temperatures do not support rapid growth of bacteria or mold. Refrigeration, however, does not stop food spoilage.

✦ *Quick-freezing.* Those foods that can be frozen and still retain their desirable qualities after thawing can be stored by freezing. Freezing does not kill all the bacteria present, but it does greatly retard their growth.

✦ *Dehydration.* Removing the moisture from a food and then keeping it dry will prevent bacterial and fungal growth. Cereals and other naturally dry foods can be preserved by this method. Many foods can be dehydrated and kept for long periods (bananas into banana chips, grapes into raisins, meat into jerky).

✦ *Radiation.* Placing the food in an airtight plastic container and then exposing it to radiation will kill all living things in the container and prevent food spoilage. In this process the radiation passes through the food but does not remain in it. Sealed, radiation-preserved foods do not require refrigeration until they are opened because there is no living substance inside the container.

✦ *Pickling.* Acids are used to preserve some foods. Pickles are cucumbers placed in vinegar (2% acetic acid) with other flavorings (like sugar or dill). A few other foods (even meats) can also be pickled.

✦ *Chemical preservatives.* Added chemicals in many foods retard bacterial or fungal growth.

other food products. Some anaerobic bacteria cause diseases such as tetanus (lockjaw), gas gangrene, and food poisoning. Anaerobic bacteria carry on various types of fermentation and form different products, including lactic acid, alcohol, methane, carbon dioxide, hydrogen sulfide (a gas which smells like rotten eggs), acetic acid (vinegar), citric acid, and ammonia. Many bacterial names actually indicate their fermentation products (*Lactobacillus* makes lactic acid; *Acetobacter* makes acetic acid).

Review Questions 10.2

1. Describe asexual reproduction in bacteria.
2. Describe two methods that bacteria use to survive periods of unfavorable conditions.
3. Describe three methods of genetic transfer in bacteria.
4. List several differences between bacterial photosynthesis and photosynthesis in plants.
5. List the two types of autotrophic and the two types of heterotrophic bacteria.
6. Describe and compare obligate aerobes, obligate anaerobes, and facultative anaerobes.

7. When culturing a particular bacterium, what conditions must be considered to have it survive and grow?
8. What are the two basic methods of controlling food-spoiling bacteria? List several household or commercial practices and tell how they illustrate these methods.
9. Describe rickettsiae, spirochetes, and mycoplasmas. List several diseases caused by organisms in each of these groups.

10B—Viruses

Viruses have plagued mankind for a long time. Some of the earliest diseases and remedies to be documented were viral diseases and their prescribed treatments. After the germ theory of disease came into prominence during the late 1800s, various protozoans and bacteria were isolated and demonstrated to be the "germs" that caused various diseases. Many of the diseases that defied early attempts to isolate an organism are now known to be caused by viruses.

10.3 Virology

Both Edward Jenner, who dealt with smallpox, and Louis Pasteur, who studied rabies, successfully formulated vaccinations for these diseases. They thought they were working with bacteria or similar organisms that had not yet been isolated; however, that was not the case—smallpox and rabies are caused by viruses. Unknown to them, they were pioneers in the science of *virology*—the study of viruses.

The Discovery of Viruses

In 1892, a Russian biologist, Dmitri Iwanowski, was searching for the cause of tobacco mosaic, a disease that causes light green patches on tobacco leaves and stunts leaf growth. Iwanowski passed the juice from diseased leaves through an unglazed porcelain filter. Such a filter is so fine that it permits only dissolved substances to pass; not even the smallest bacteria can go through. When Iwanowski examined the filtered fluid under a microscope, he was unable to find any visible particles. Nonetheless, when this juice was placed on a tobacco leaf, it caused tobacco mosaic disease. Iwanowski assumed that the fluid contained a poison made by a bacterium. A few years later the term *virus*, the Latin word for poison, was assigned to the unknown agent that caused the disease.

Objectives

- Describe the structure of a virus
- Describe the five stages in the lytic cycle
- List several characteristics used to group viruses
- Compare and contrast the lytic and the lysogenic cycles
- Describe the difference between a persistent viral infection and a transforming virus
- Explain how viroids and prions differ from viruses

Key Terms

virus
capsid
envelope
bacteriophage
lytic cycle

retrovirus
viroid
prion
lysogenic cycle
vaccination

In the mid-1930s, Wendell Stanley isolated the actual virus. He reduced the juice from about 1 ton of infected tobacco mosaic leaves to about 1 spoonful of crystals. These crystals could be kept in a dry, airtight jar for extended periods of time. Since there was no evidence of metabolism, the crystals were thought to be an inert chemical. However, when placed in contact with a living tobacco leaf, they caused tobacco mosaic disease. Stanley was awarded the Nobel Prize in Chemistry in 1946 for isolating and purifying the *tobacco mosaic virus (TMV)*.

There is nothing unusual about a disease caused by a chemical. Various poisons and irritants can affect cellular metabolism, resulting in disease and death. A certain amount of cyanide (a chemical poison) given to a laboratory animal will kill it. Careful examination of the animal's body will reveal only the amount of cyanide given to the animal. There has been no increase in the amount of cyanide.

A virus, however, is very different. A tobacco leaf exposed to a tiny amount of TMV will develop the disease and die. Later, if the TMV is extracted from the leaf, much more will be found than was initially placed on the leaf. Somehow the TMV apparently grows and reproduces when placed on a tobacco leaf. But how? Is TMV alive, or is it a nonliving chemical? Does the leaf make more TMV? Does TMV alter the genes of the leaf cells? Does TMV supply genes to the cell to make more TMV? By performing controlled experiments and by making careful observations with an electron microscope, scientists have begun to understand the structures and life cycles of viruses and have found many of the answers to these questions.

10-11
Wendell Stanley isolated the TMV.

10-12
A tobacco leaf infected with TMV; TEM of TMV (inset)

The Structure of Viruses

A **virus** consists of two basic parts—a core of either DNA or RNA (never both) and a protein covering called a **capsid**. Some viruses have a membrane-like **envelope** surrounding the capsid that is made mostly of lipids. The lipids in the envelope are not formed by the virus but are taken from the host cell it infects. The surface of the envelope has several glycoprotein molecules projecting from its surface. These are used to identify and facilitate attachment of the virus to a host cell. Viruses are very selective in their targets. Plant viruses only attack plant cells. Some viruses will only attack cells of certain species or groups of related organisms.

Although viruses do have some elements of design and structure typical of living things, outside of a host they are lifeless. They have no cell membrane, cytoplasm, or organelles of their own. They cannot move on their own, and they can reproduce only by using the organelles and enzymes of a host organism.

Much of the early research on viruses was performed using a bacteriophage (back TEER ee uh fayje). A **bacteriophage** is a virus that infects only certain bacteria. One of the most commonly studied bacteriophages is the T4 bacteriophage that infects *Escherichia coli*, a common bacterium that lives in the human digestive tract.

The head portion of a bacteriophage is icosahedral in shape and contains about 200 000 base pairs in its DNA core. The tail structure, as well as the tail fibers, is important in the attachment and injection of the DNA core into a

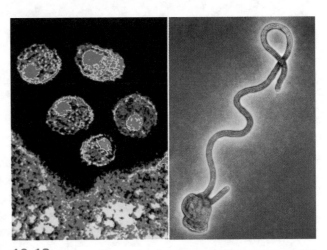

10-13
Viruses have varied shapes: HIV (green spheres) and Ebola virus (cylindrical).

bacteriophage: bacterio- (rod) + -phage (to eat)

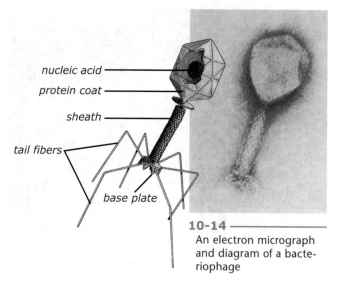

nucleic acid ——
protein coat ——
sheath ——
tail fibers ——
base plate ——

10-14 ——
An electron micrograph and diagram of a bacteriophage

host cell. The bacteriophage can produce dozens of different proteins, more than TMV, although the virus particles are about the same size. The more complex structure of the protein coat of the bacteriophage demands a wider variety of proteins than the simple recurring subunits of the TMV.

The Lytic Cycle

The ability of a virus to affect cells is called *virulence* (VIHR yuh lunce). If the virus does not affect a certain type of cell, the virus is nonvirulent for that cell type. TMV, for example, is nonvirulent to human cells. A person could eat TMV with little danger. Conversely, a virus highly virulent to humans can enter a cell, produce hundreds of viruses, and destroy the cell in less than an hour.

Scientists call the activity of a virulent virus the **lytic** (LIT ik) **cycle.** During the lytic cycle the virus invades the cell, uses the resources of the host cell to produce multiple copies of the viral nucleic acid, destroys the host cell, and releases new viruses into the environment. We will use a virulent T4 bacteriophage to illustrate the lytic cycle. Although slightly different for various kinds of viruses, the lytic cycle generally follows these steps.

Step 1: Attachment. The various capsid glycoproteins of some viruses aid in the attachment to the host cell. The bacteriophage will attach to a specific *receptor site* on the cell wall of the bacterium. If there is no receptor site, the bacteriophage will not attach. Receptor site specificity explains why viruses may attack only certain types of cells.

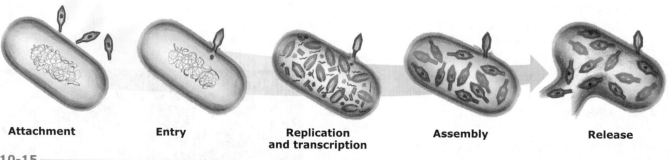

| **Attachment** | **Entry** | **Replication and transcription** | **Assembly** | **Release** |

10-15 ——
The lytic cycle

Step 2: Entry. Once the bacteriophage is attached to the cell wall, enzymes begin to break down the cell wall, and the core is injected into the bacterium. Often the empty protein capsid remains attached to the outside of the cell after the core has entered the cell. In some viruses the entire capsid may enter the cell and then release its nucleic acid.

Step 3: Replication and transcription. Using the cell's enzymes and organelles, the virus begins to produce multiple copies of the viral nucleic acid, mRNA, enzymes, and proteins.

Step 4: Assembly. The virus proteins form capsids around the replicated viral DNA (or RNA), forming new viruses.

Step 5: Release. The cellular organelles and enzymes used by the virus are not replaced, and eventually the host's cellular resources are depleted and the

lytic: (a loosening)

cell dies. Enzymes produced by the virus cause the cell to burst open and release the new viruses into the environment, where they seek out new cells to infect. The bursting, or disintegration, of the host cell is called *lysis*.

Certain bacteriophages, as well as the viruses that cause human polio and influenza, are relatively virulent, forcing the entire cellular metabolism to produce new viruses. When a virulent virus particle infects a cell, it can sometimes produce hundreds of viruses in as little as thirty minutes. Each of those new viruses can infect another cell and multiply into hundreds in each infected cell. Knowing this, it is easy to understand how the symptoms of certain viral diseases can appear very quickly.

The Classification of Viruses

Viruses are classified, or grouped, according to several characteristics. First, viruses can be classified in two large groups based on the type of nucleic acid they contain—RNA or DNA—and whether the DNA or RNA is single or double stranded. The type of viral nucleic acid also determines its mechanism of action. Once inside the host cell, a DNA-containing virus usually works in one of two ways. Either the viral DNA simply uses the enzymes of the host cell to produce RNA that then produces viral proteins, or the viral DNA is incorporated directly into the host DNA to make more viruses.

RNA-containing viruses work somewhat differently. Once the virus enters the host cell, the viral RNA is released into the cytoplasm. The viral RNA then takes over the host's ribosomes to produce viral proteins. A special type of RNA virus called a **retrovirus** contains the enzyme *reverse transcriptase*. Once inside the host, this enzyme uses the viral RNA as a template to produce viral DNA, which is incorporated into the host's DNA. The DNA that now contains the viral DNA is transcribed to produce new viral RNA that directs the host's ribosomes to produce viral proteins and new viruses. The enzyme is called *reverse* transcriptase because the usual order of transcription is reversed—RNA to DNA rather than the usual direction of DNA to RNA. The human immunodeficiency virus (HIV) that causes AIDS is a familiar retrovirus.

Viruses can also be grouped by their shape, the presence or absence of an envelope, or their method of infecting and replicating within the host cell.

The Lysogenic Cycle

Not all viruses begin to destroy the cell immediately after the entry phase as they do in the lytic cycle. Rather than immediately taking control and producing new viruses, during some viral infections, the virus that enters a cell may remain inactive (or *latent*) for

10-16
The human immunodeficiency virus (HIV)

reverse transcriptase enzyme
glycoproteins
envelope
RNA core
capsid

Retroviral Reproduction Review

1. Retrovirus enters host.
2. Reverse transcriptase uses viral RNA to produce viral DNA.
3. Viral DNA is incorporated into host's DNA.
4. Altered host DNA is transcribed to produce new viral RNA.
5. New viral DNA uses host's ribosomes to produce virus components.

Viroids

Even smaller than viruses are some other kinds of disease-causing agents—viroids. A **viroid** (VYE roid) is a short, single strand of circular RNA. It has no capsid or envelope, yet it is still able to replicate once it is inside a host. T. O. Diener first studied viroids in 1971 while investigating the cause of a disease that affects potatoes, causing them to grow into irregular shapes. It is thought that viroids disrupt ribosome assembly by interfering with ribosomal RNA. Although most viroids appear to infect plants, researchers now suspect that the causative agent of hepatitis D is a viroid.

T. O. Diener and a potato affected by potato spindle tuber disease

Prions

Prions (PREE ahnz) are even smaller than viroids and do not contain any nucleic acids. In fact, a prion is an abnormal form of a protein normally found in cells that is thought by some researchers to cause certain diseases. In 1982, Stanley Prusiner proposed the existence of prions and suggested them as the cause of certain infectious diseases that affect the brain. Although it is somewhat controversial, Prusiner proposed that prions interfere with the normal folding of proteins as they form their three-dimensional shape. The resulting malformed protein does not properly function and is thought to cause disease.

Prions were first studied in sheep infected with a disease called scrapie (SKRAY pee) that affects the brain. In the 1990s, thousands of cattle in England were destroyed in order to control an epidemic of "mad cow disease." Prions are also thought to cause a similar brain disease

Protective gear used to prevent the spread of mad cow disease (right); prion-infected brain tissue (left)

called Creutzfeldt-Jakob (KROITZ felt • YAK ohb) disease that affects humans.

prion: pr- (E. *proteinaceous*) + -i- (infectious) + -on (particle)

The "First Living Thing"?

Since a virus is a nonliving chemical structure that can utilize cellular parts to produce more of itself, some evolutionists have claimed that viruses are the step between living organisms and nonliving chemicals. Today, many research projects are underway that attempt to show that some RNA and peptides can self-replicate; in other words, they can duplicate themselves outside of a host. Most of these experiments have been performed in well-controlled and closely monitored environments that the researchers assume were present at the beginning of the earth.

There are many problems with these experiments. First and foremost is the frame of reference—they assume evolution is true—that biases their entire hypothesis. They also make certain assumptions about the environment that cannot be proved. Also, manipulating the chemical environment of their experiment indicates that some sort of intelligent designer (in this case the scientists) is needed for the success of the experiment.

long periods of time. Although this may appear to be a state of inactivity, during this time the viral genome is being incorporated into the genome of the host organism. Each time the bacterial genome is replicated during cell division, the viral genome also replicates and is also passed to the daughter cells. In time most of the bacterial population will contain the virus. This type of viral replication is called a **lysogenic cycle**. In animal and human cells it appears that many lysogenic viruses reproduce each time the cell divides so that all new cells have the virus in them.

When a certain stimulus (such as ultraviolet radiation, a temperature change, chemicals, or an unknown agent) is applied, the virus becomes virulent, enters the lytic cycle, and destroys the cells. Since the virus may be in a large number of cells before the stimulus is applied, a large area may be affected all at once. The viruses then enter other cells and, unless the stimulus is still present, become inactive again.

Lysogenic viruses are not limited to bacteriophages but can infect both animals and humans. The *herpes simplex* virus is such a virus. It remains latent

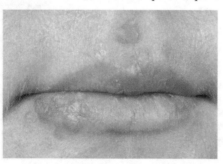

10-17
Herpes simplex is an example of a lysogenic virus life cycle.

inside nerve cells in the skin. Stress factors such as sunburn, windburn, illness (such as a cold), and even emotional stress can activate the virus. It produces small fluid-filled sacs, commonly called fever blisters or cold sores. These blisters, followed by scabs on the skin, usually form around the lips and always recur in the same area with each activation of the virus.

Smallpox: A Plague of the Past—or of the Future?

Smallpox plagues have affected man for centuries. But in the late 1700s in Europe, smallpox epidemics swept through cities and caused thousands of people to break out in red *pustules* (fluid-filled blisters). Most who contracted the disease died. Any who recovered had permanent scars but were *immune to* (would not again contract) smallpox.

Dr. Edward Jenner, an English physician, noted that individuals who lived in the country, and especially those who worked around dairy cattle, appeared to be immune to smallpox. He also observed that most farmers and dairy workers had previously experienced a mild disease called cowpox. This disease caused a small pustule, usually on the hands of those who milked the cows, which healed and left a small scar.

Edward Jenner inoculating James Phipps with matter from a cowpox pustule to develop immunity in the boy

In 1796 during a severe pox epidemic, Mrs. Phipps brought her son, James, to Jenner. She feared that James had contracted smallpox and had heard that Jenner was working on a cure. On May 14, 1796, Jenner took matter from a cowpox pustule on the hand of a dairymaid and inoculated James by applying the matter to two shallow cuts he made on the boy's arm. A pustule developed, formed a scab, and then disappeared, leaving only a scar.

Was the boy now immune to smallpox? In June of the same year, Jenner again inoculated his patient, this time using matter from a smallpox pustule. For two anxious weeks Jenner watched James for signs of the disease. None developed. Later in the year, he inoculated him again with material from a smallpox pustule, but James did not develop the disease.

Jenner wrote a paper describing what he called a "vaccination" against smallpox. Today it is known that a **vaccination** is not a cure for a disease but a method of developing an immunity by exposing a person to either a weakened form of the disease or a similar disease that prevents the first disease from occurring.

At the time of Jenner's work, most people did not understand the principle behind vaccinations. People organized antivaccination campaigns. Cartoons of the time showed people with cow heads, tails, legs, and hooves growing out of their bodies in the places where they had been vaccinated.

Jenner did not know what caused smallpox or cowpox. About 150 years after his vaccination of James Phipps, it was discovered that similar viruses cause both smallpox and cowpox. When a person develops immunity to one, he also becomes immune to the other.

About 200 years after Jenner's first vaccination, the World Health Organization (an agency of the United Nations) announced a plan to eradicate smallpox. Since there is an effective vaccine against smallpox and the only known place the virus can exist is in humans, it is theoretically possible to eliminate the disease. If everyone were vaccinated against the virus, there would be no place for the virus to exist, and that would be the end of smallpox.

The task of vaccinating over 4.5 billion people proved impossible for even the United Nations. Therefore they adopted the idea of vaccinating everyone near an outbreak of smallpox. This method of disease management is called the *ring method* or *encirclement method*. After a few years, there were no known major outbreaks of smallpox. In 1980, the World Health Organization certified that smallpox was the first major human disease to be completely eliminated. Routine vaccinations in the United States ended in 1972. Today a person

Fear of the unknown (vaccinations) often inspired cartoons showing cows emerging from vaccinated individuals.

does not need a smallpox vaccination to attend school or to travel to most foreign countries.

However, some scientists doubt that smallpox has been totally eradicated. To say that no one is suffering from smallpox is a universal statement—the kind of statement that science should not make and cannot support. Why? There may be some remote area where smallpox does exist, or it may exist in a nonvirulent form that could mutate into the virulent form. In addition to the possibility that smallpox exists somewhere in nature, there are also stores of smallpox virus in laboratories where it has been studied as a potential biochemical weapon. In the mid-1990s, officials in Russia admitted to having produced huge quantities of smallpox virus for potential use in biochemical weapons, and not all of it has been accounted for.

Many people fear that since most people have not been vaccinated, the release of the virus from a biochemical attack would have devastating results, especially in a major urban center. To counter such a threat, it has been suggested that a mass vaccination program be undertaken to prevent a major catastrophe. However, the cost of vaccinating so many people and the possible side effects of the vaccine have hindered such a program. The encirclement method works well in a somewhat isolated rural environment but would be almost useless in a large urban area. Only time will tell if smallpox will stay in the history books or once again become a major threat to human life.

vaccination: (L. *vaccinus*—of cows) ["of cows" because the first vaccine was prepared from the cowpox virus]

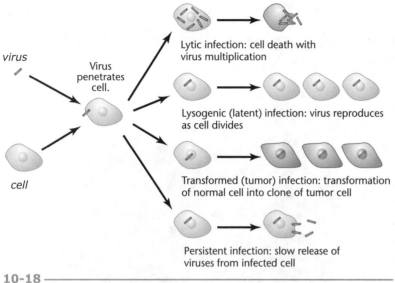

virus

Virus penetrates cell.

cell

Lytic infection: cell death with virus multiplication

Lysogenic (latent) infection: virus reproduces as cell divides

Transformed (tumor) infection: transformation of normal cell into clone of tumor cell

Persistent infection: slow release of viruses from infected cell

10-18
Ways different viruses affect cells

Other Kinds of Viruses

Viruses can also cause *persistent infections*. In a persistent viral infection the host cell does not go through lysis but slowly releases virus particles. This type of viral infection may not destroy the cell, but it does hamper the cell's metabolism. Even stress could harm the organism if enough cells contain the persistent-infection virus.

Some viruses are *transforming viruses*. These viruses transform their host cells by adding new genetic information. They significantly change the cell's metabolism but do not destroy the cell. The transformed cell, however, is not a productive cell in the organism's body.

An example of a transforming virus is one that causes certain warts in humans. The wart virus enters the skin cell and transforms it into a wart cell. As the transformed cell grows and divides, the wart grows. In time the wart reaches a certain size, and the cells stop growing and dividing.

▶ Review Questions 10.3

1. Give the contributions to virology made by (a) Dmitri Iwanowski and (b) Wendell Stanley.
2. What is one way a virus is different from a chemical poison?
3. Describe the structure of a virus particle.
4. Describe the lytic cycle.
5. What is the difference between a virulent virus and a lysogenic virus?
6. List and describe two effects, other than the lytic cycle, that some viruses may have on cells.

⊙ 7. Explain why an attack of a viral disease is often sudden and extensive.
8. Compare and contrast viroids and prions.
⊙ 9. What is the relationship between smallpox and cowpox?
⊙ 10. Who was Dr. Jenner and what is he credited with doing?
⊙ 11. Why do you think that the encirclement method of smallpox treatment would not be effective in a large urban setting?

10.4 Viral Diseases

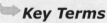

▶ Objectives

- Describe two methods of controlling viral diseases
- Compare and contrast attenuated and inactivated vaccines
- Describe three viral diseases that affect man

▶ Key Terms

interferon
emerging virus

Most viruses are highly specific regarding their host cells. It appears that a virus particle in contact with the wrong type of cell either lacks the mechanisms for entering the cell or, once inside the cell, lacks the mechanisms for affecting the cell's metabolism. This virus would most likely be destroyed by the cell or released as a waste product.

Most viruses are thus limited not only to one type of organism but also to one cell type in that organism. Smallpox, chicken pox, and measles usually affect skin cells. Rabies and polio attack cells in the nervous system. Influenza and the common cold are usually viral infections of the respiratory system. A few viral diseases, however, such as rabies and cowpox, can affect similar cells in different organisms.

The blood can carry various poisons released by virus-infected cells to other areas of the body. These transported poisons often cause the symptoms of the disease to be widespread, even though the actual infection may be in a limited area. Thus flu viruses in cells of the respiratory system can cause muscle soreness, headaches, and nausea, even though the cells of these areas are not infected with virus. The blood can also carry many types of viral particles from one area to another. A viral infection, therefore, is often found in areas of the same type of tissue throughout the body.

Control of Viral Diseases

Once a virus is inside a cell, it is virtually impossible to destroy the virus without harming the host cell. For this reason, developing medications to fight viral infections has been quite difficult. In addition, viruses are able to mutate quite rapidly, acquiring characteristics that may cause some medications to lose their effectiveness. The control of viral diseases has two major components—prevention by vaccination and antiviral drugs to halt the progression of the disease.

Vaccination Programs

The purpose of any vaccine is to activate the body's immune system so that it can readily recognize and rapidly respond to a virus, or other organism, and prevent disease. There are two basic types of vaccines against viruses—inactivated and attenuated.

To make an *inactivated vaccine*, researchers alter the virus so that it cannot replicate in a host cell; however, enough identifying characteristics of the virus remain so that the body can recognize and store the information in its immune system. Most inactivated vaccines do not require special handling, such as refrigeration, making them easier to use in rural or isolated areas and in mobile vaccination programs. Unfortunately, the immune response created by inactivated vaccines is relatively weak and may have to be repeated.

An *attenuated vaccine* is made from "live" viruses that can still replicate in a host cell. Unlike Jenner's vaccine that was made from unaltered cowpox virus, attenuated vaccines are made from viruses that are either grown under special conditions or genetically altered so that they are nonvirulent. One advantage of attenuated vaccines is a stronger immune response than with an inactivated vaccine. As with any vaccine or medication, there are possible side effects. Attenuated vaccines should not be given to those who have a compromised immune system, such as those taking immune-suppressing drugs, cancer patients, or people diagnosed with HIV. Also, there is a slight possibility that the virus could revert to a virulent form and actually cause the disease it was developed to prevent.

Antiviral Drugs

Once someone has a viral disease, the treatment must shift from prevention to control. Whereas antibiotics are quite effective in treating bacterial infections, they are ineffective in the treatment of viral disease. Researchers and physicians had to look elsewhere for effective medications.

In the 1950s, chemicals called **interferons** (IN tur FEHR ahnz) were discovered. An interferon is a protein that is produced by an infected host cell and released when the cell bursts. It then binds to receptors on other cells, causing them to produce enzymes that inhibit viral replication. Researchers thought that a cure for viral diseases was close at hand; however, interferons were difficult to produce in large quantities and did not live up to initial antiviral expectations. Although today interferons are produced by recombinant DNA technology and are used for some viral diseases such as hepatitis B, the effectiveness of interferons is still somewhat limited.

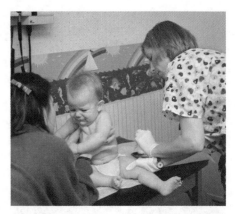

10-19
Vaccines have prevented many serious childhood diseases.

10-20
Some antiviral drugs are administered by nasal spray rather than injections or pills.

interferon: inter- (L. *inter*—between or among) + -feron (*ferire*—to strike)

10-2 Some Viral Diseases	
In humans	
• Ebola	• mumps
• hepatitis A, B, C, and D	• severe acute respiratory syndrome (SARS)
• herpes simplex (fever blisters, cold sores)	• warts
• mononucleosis, infectious	• yellow fever
In animals	
• cowpox	• hog cholera
• distemper (in dogs)	• monkeypox
• foot-and-mouth disease	• sacbrood (in bees)
• fowl leukemia	• sheep pox
In plants	
• "breaking" diseases (as in Rembrandt tulips)	• mosaic diseases (in cabbage, cucumbers, potatoes, sugar cane, and tobacco)

10-21

West Nile virus is an emerging virus transmitted by mosquitoes from birds to humans.

Emerging Viruses

Occasionally viruses can move from one species of host to an unrelated species, causing a disease in the new host. Although almost everyone has heard of the flu, few people may realize that the natural host of the influenza virus is birds. The influenza virus can spread directly from birds (often chickens) and cause disease in humans. Between 1918 and 1919, over 21 million people died worldwide from the flu.

Many researchers think that HIV originated in monkeys and was spread to humans. New viral diseases that arise this way are called **emerging viruses**. Many times these viruses are in geographically isolated areas, but with modern travel, diseases can potentially spread around the world in a matter of hours. One such emerging virus is the Ebola virus that was isolated in central Africa. There is no effective vaccine or treatment for Ebola, which has mortality rates as high as 90%. In the southwestern United States, hantavirus is considered an emerging virus. It occurs naturally in wild rodents, but it can be lethal in humans. West Nile virus and SARS—severe acute respiratory syndrome—are yet two more recent examples of emerging viruses. It is important to understand that emerging viruses are not examples of the evolution of new life forms. Because of genetic changes, these viruses can now enter more hosts, but they are still viruses.

Review Questions 10.4

⊙1. Rabies is a viral disease of both man and dogs. Man, however, is not susceptible to viral distemper that affects dogs. Why can dogs and man share certain viral diseases but not others?

⊙2. Immunity to cowpox also serves as an immunity to smallpox, but the two are separate diseases. What reasons can you give for one immunity protecting against two diseases?

3. Explain the differences between attenuated and inactivated vaccines.

4. Describe how interferons work.

5. List (a) several viral diseases of man and (b) several viral diseases that both man and certain animals can have.

6. What is an emerging virus?

10C—Diseases and Disorders

Anyone who has suffered or has seen someone suffer from a serious disease has wondered why there is such a thing and how it originated. Scripture teaches that there was originally no disease in God's creation. At the end of the Creation week, "God saw every thing that he had made, and, behold, it was very good" (Gen. 1:31). Today, however, "the whole creation groaneth and travaileth in pain together" (Rom. 8:22). What happened to destroy the peace and harmony originally inherent in God's creation? The answer is found in Genesis 3—man's sin, the Fall, and the Curse caused a host of degenerative changes in the biological realm, including disorders, diseases, aging, and death.

Thankfully that is not the end of the story. The accounts of New Testament miracles of healing indicate that Jesus was concerned with the relief of human suffering. Through these acts, Christ demonstrated that as God He had the power to one day restore creation to its original state, where sickness and death are no more (Luke 5:15; 1 John 3:8; Rev. 21:4). God still sometimes works miracles of healing, but most of the fight against sickness has come under man's responsibility in living out the Creation Mandate in loving and compassionate ways. God has blessed humanity with many ways to fight disease, and it is one of the missions of Christianity to help people who suffer (Luke 10:25–37). Sadly, there are some sicknesses that cannot be healed today, but it should be remembered that God can even use these dreadful diseases to work good in the life of the believer (Rom. 8:28). He can use diseases to turn His people back to dependence on Him (2 Cor. 12:7), and He can even use disease to turn unbelievers to Himself for salvation (2 Kings 5).

10.5 Infectious Disease

Many common ailments are called **infectious diseases** because they are caused by viruses or organisms. Such an agent that invades the body and causes a disease is called a **pathogen**. One example is a parasite that grows and reproduces within the body of its host, injuring it in the process. Bacteria and viruses are two important types of pathogens; viroids, protozoans, fungi, and some animals such as worms are also pathogens. Many scientists believe that the infectious proteins called prions should also be termed pathogens. While the previous section deals with viral diseases, this section deals primarily with bacterial pathogens. Diseases that are the result of other types of pathogens (protozoans, fungi, etc.) are discussed as these organisms are covered in later chapters.

Disease Detection

Before doctors can effectively treat and control a particular disease, they need to know what caused it: they need to know its *etiology* (EE tee OL uh jee). Men such as Edward Jenner were treating disease without knowing the causative agent or exactly how it was spread. Their work contributed to what is now known as the *germ theory of disease*. This is the theory stating that infectious diseases are caused by and transmitted from one individual to another by microorganisms.

It was not until the mid-1870s that a stepwise protocol was established to detect and confirm the cause of a particular disease. German physician Robert Koch was studying a cattle disease called anthrax that can spread to humans. In the cattle that had anthrax, Koch always isolated a particular type of bacteria. If he grew these in his laboratory and then injected them into a healthy cow, the cow developed anthrax. When Koch isolated bacteria from the second cow, it was the same type that he had found in the first cow with anthrax. From these results he concluded that this particular bacteria was the cause of anthrax.

Today, determining the etiology of a particular disease is based on the pioneering work of Robert Koch. His technique has four basic steps.

Step 1: The suspected pathogen should be found only in the body of a sick individual, not one that is healthy.

Step 2: The suspected pathogen should be grown in a laboratory culture.

Objectives
- Summarize Koch's Postulates
- Describe the methods by which diseases can be spread

Key Terms
infectious disease
pathogen
Koch's Postulates
toxin

10-22

Robert Koch: His postulates still form the basis for modern research on the causes of disease.

etiology: etio- (Gk. *aitia*—cause) + -logy (the study of)

Step 3: If a healthy individual is injected with the suspected pathogen from the laboratory, it should develop the same disease as the original individual.

Step 4: The pathogen from the second individual should be isolated and compared to the pathogen from the original. The two pathogens should be the same.

These steps are called **Koch's Postulates** and have been used by doctors and scientists around the world to identify the pathogens of thousands of diseases. Once the causative agent has been discovered, they can determine how the pathogen causes the disease, how it is spread, and how it can best be controlled and treated.

How Pathogens Cause Disease

How can a tiny microbe bring so much distress to a 150 lb human? During the *incubation period* of a disease, a few microbes multiply into millions of organisms. The incubation period is the time between contracting (becoming infected with) the disease and the appearance of the first symptoms. Once the pathogens have multiplied sufficiently, they can affect their host in two ways.

✦ *Tissue destruction.* Pathogens obtain their nutrition from the host's body, an action that often amounts to destroying host cells and ingesting their remains for food. For example, typhoid bacteria destroy portions of the intestinal wall, and tuberculosis bacteria destroy lung tissue. Viruses divert the cell's metabolic machinery from its own functions to producing the virus, and in time the cell is destroyed.

✦ *Toxin formation.* Many bacteria and other microorganisms have the ability to produce poisonous substances called **toxins**. Toxins can cause a malfunctioning of the cells, which in turn produces the symptoms of the disease. In many cases these toxins so disrupt the metabolism of the host's cells that large numbers of cells are destroyed.

Kinds of Toxins

There are two basic types of toxins:

✦ *Exotoxins.* Toxins that diffuse from the living pathogenic cell into the surrounding tissue are called exotoxins, or soluble toxins. Exotoxins are products secreted by the pathogen. The toxins formed by the tetanus, diphtheria, and rheumatic fever bacteria are examples of exotoxins. Exotoxins cause the symptoms of most infectious diseases.

✦ *Endotoxins.* Toxins that remain in the pathogen as part of its structure are endotoxins. Endotoxins become a problem to the host as the pathogen dies and disintegrates. Endotoxins can produce violent reactions in tissues around a dying pathogen if they are present in sufficient quantities. Some diseases caused by endotoxins are bacterial dysentery, bubonic plague, and typhoid fever.

exotoxin: exo- (out) + -toxin (L. *toxicum*—poison)

endotoxin: endo- (within) + -toxin (poison)

a Closer Look

Leprosy in Biblical Times and Today

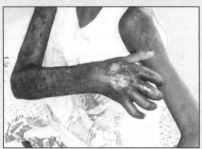

In ancient civilizations leprosy was a dreaded contagious disease. It started with minor white patches of skin and a numbness of the fingers and toes. It caused a progressive disfigurement and so weakened the body that other diseases would infect the person. In time the person died, not always of leprosy, but often of leprosy-related infections.

God told the Hebrews to have lepers live in quarantine outside the camp (Lev. 13:46), and if a leper came into a city, the Law required him to cry "unclean" so that others could avoid him (Lev. 13:45).

The disease known as leprosy today (caused by the bacterium *Mycobacterium leprae*) does not completely correspond to all the descriptions of leprosy given in Leviticus 13 and 14 and similar passages. The disease caused by *M. leprae* results in lumpy, discolored patches in the skin that become insensitive to cold, touch, and pain. In advanced cases the patient, if not treated, develops numbness of the hands, feet, and face. The muscles weaken, and the body becomes disfigured. Because of the numbness, the person can suffer cuts, burns, and even amputation without being aware of it.

In most industrialized countries leprosy is rare. Under normal conditions the leprosy bacterium is an obligate parasite of humans and is passed only through direct contact with an infectious leper. Once the disease is diagnosed, it can be treated and stopped before major damage occurs. The person is no longer infectious following treatment. Some lepers, however, continue drug treatments throughout their lives.

Today many countries do not require lepers to live in a leprosarium (special camp for lepers). Before modern drugs, however, leprosy was dreaded and quarantine was the only method of controlling its spread.

Most Bible scholars agree that when the Bible describes leprosy, it is speaking of the leprosy known today as well as several other skin diseases, some molds, and other fungi of cloth, animal hides, and masonry (walls). The spread of a good number of highly contagious diseases (between which people in Old Testament times had no way to distinguish) was prevented by the laws given in Leviticus.

10-23

Christ Healing the Blind Man (detail), Cornelis Cornelisz, van Haarlem, From the Bob Jones University Collection

Diseases and God's Providence

Occasionally God does use disease as a punishment for sin (1 Cor. 11:29–30); however, people must be careful not to jump to conclusions. God calls some of His choicest servants to endure serious illnesses for reasons that, for the present at least, only He understands. Job's affliction was not a result of sin (Job 1:8). On the contrary, it appears that he was selected for special testing because of his uprightness.

The account of a New Testament miracle of healing provides a general principle. In John 9:2 the disciples asked the Lord, "Who did sin, this man, or his parents, that he was born blind?" His answer was, "Neither hath this man sinned, nor his parents: but that the works of God should be made manifest in him." Diseases and disorders are clearly tools God uses to work providentially in the affairs of men.

In Bible times God used disease both to discipline (Num. 16:49; Josh. 22:17; 2 Sam. 24:15) and to avenge His people (Exod. 9:1–11; 1 Sam. 5:6–12). Occasionally, God used fatal diseases to remove evil rulers such as Jehoram (2 Chron. 21:18–19) and Herod (Acts 12:21–23) from office. He will send pestilences (some of which appear to be infectious diseases) as a sign of the end time (Matt. 24:7; Luke 21:11), and they will reach unprecedented proportions during the Tribulation (Rev. 16:2, 10–11).

Communicable Diseases

A disease that can spread from one person to another by either direct or indirect means is called a *communicable disease*. Most of the so-called childhood diseases such as chicken pox, measles, German measles, mumps, and whooping cough are communicable diseases that are highly *contagious*. A highly contagious disease is one that is easily spread to others.

The contagiousness of a disease often varies with the disease's stage of development in the patient. Although a person may be suffering the *symptoms* (effects) of a disease, he may be past the contagious stage and therefore incapable of infecting someone else. In many such cases, the pathogen is either no longer alive or no longer reproducing, but the effects of the damage the pathogen did are still evident in the host.

Some communicable diseases are so contagious that health authorities advise placing the patient in *isolation*. A person with meningitis, tuberculosis, scarlet fever, diphtheria, typhoid fever, and certain kinds of dysentery is often isolated from other people. At times people with some communicable diseases have been placed in *quarantine*, a strict isolation often enforced by law.

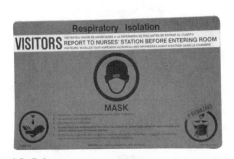

10-24

Isolation is one effective measure to control the spread of communicable diseases.

contagious: con- (together) + -tagious (L. *tangere*—to touch)

10-25
Many disease-carrying germs are made airborne by sneezing and coughing.

10-26
Hand washing is one of the most effective disease-prevention measures available.

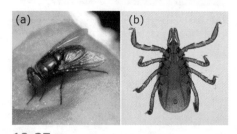

10-27
Common disease vectors include (a) the common housefly and (b) the deer tick, which spreads Lyme disease.

How Diseases Are Spread

One way of distinguishing between types of infections is by examining the various ways pathogens can enter the body. Knowing these ways can help a person observe certain simple precautions to safeguard good health.

✦ *Airborne infections.* Many diseases are transmitted by pathogens suspended in water droplets that remain aloft after humans cough or sneeze into the air. Airborne infections generally affect the respiratory tract, though some of them affect other parts of the body as well. Tuberculosis and diphtheria can be transmitted in this manner. Healthcare workers treating patients with these diseases and even family members of patients must wear masks to protect themselves from developing diseases spread in this manner.

✦ *Direct contact infections.* Some diseases are spread by direct contact with a sore or lesion on the skin or mucous membrane of an infected person. Bloodborne pathogens like hepatitis B and HIV fall into this category. Scarlet fever, colds, influenza, and measles can be transmitted by touch.

✦ *Indirect contact infections.* Touching a surface or object contaminated by a pathogen, such as eating utensils, can spread some diseases from one individual to another. Proper hand washing and dish washing, as well as refraining from eating or drinking after other people, can prevent this mode of disease transmission.

✦ *Contamination infections.* Pathogens that enter the body by way of contaminated food or water often afflict the digestive system. Cholera, typhoid fever, and many dysentery organisms invade and parasitize the intestines. The intestinal wastes of a person who has one of these diseases are highly infectious. Good sanitary procedures minimize spreading or getting these diseases. The disposal of body wastes prescribed in the Old Testament helped (at least in part) to maintain sanitary conditions (Deut. 23:12–13). Improper personal hygiene by food handlers—those who package, deliver, prepare, and serve—can also contaminate food and spread disease.

✦ *Wound infections.* Some pathogens enter the body through wounds. Even small cuts can be serious if they are not properly treated. *Staphylococcus* bacteria are among the most common pathogens of wound infections. *Streptococcus* infections are less frequent but more serious, for they are more likely to enter the bloodstream and spread to other parts of the body. Fortunately *Streptococcus* blood poisoning responds readily to antibiotics. Tetanus (lockjaw) and gas gangrene are serious wound infections.

✦ *Vector-carried infections.* Insects or other arthropods that carry pathogens from one host organism to another are *vectors*. Disease transmission can be accomplished in two ways: mechanically, as with food contamination by pathogens carried on the bodies of flies or roaches; or by injection, as with the bites of such organisms as mosquitoes, flies, and ticks, which inject the pathogen into the bloodstream of the host. In some cases, mammals may serve as temporary reservoirs for a pathogen that is then carried by a vector to another species. Vectors spread typhus, bubonic plague, and malaria.

✦ *Immune carriers.* People (and occasionally animals) sometimes spread pathogens to others but have no symptoms of the disease themselves. These healthy-appearing individuals are called *carriers.* The carrier can spread the disease by coughing or sneezing or by unclean hands. Often the carrier has had the disease previously and has developed immunity to it, yet he remains infectious. Diseases spread in this way include hepatitis, diphtheria, polio, scarlet fever, and typhoid fever.

Typhoid Mary

Perhaps the most famous immune carrier in medical history was a cook often called "Typhoid Mary." In the early 1900s, a number of cases of typhoid fever near Oyster Bay, New York, were found to involve people who had eaten at the place where Mary Mallon worked as a cook. Further study led the investigators to the carrier herself. She had never had symptoms of the disease and did not believe the New York State health officials that traced the typhoid outbreaks to her. At least fifty-one cases of typhoid (and three typhoid-related deaths) are attributed to her.

➡ *Review Questions 10.5*

1. List several possible reasons the Lord may have for permitting a person to suffer from a disease or disorder.
2. Describe the two ways pathogens can affect a host.
3. List and describe seven ways disease can be spread.
4. What causes the disease we call leprosy today? What are the symptoms of leprosy?

⊙ 5. In the Bible leprosy is often used to represent what spiritual concept? How does this idea parallel the disease?
6. Give examples from Scripture that show how God used diseases or disorders to accomplish His will.
⊙ 7. Give several reasons why man must continually seek new methods of conquering disease. Will man ever conquer all diseases? Why or why not?

10.6 Defense Against Infectious Disease

God has wonderfully equipped the human body to resist disease. The majority of us enjoy good health in spite of sharing an environment with pathogenic organisms. The body's system of defenses is a solid testimony against evolution. These defenses could not have developed over long periods; the entire population would have been killed off by disease at the beginning. The defenses had to work correctly the first time.

The Lord has also permitted man to develop drugs and medical techniques that can help overcome pathogens that invade the body. The first part of this section deals with the primary bodily defenses against infectious diseases. Then the text discusses antibiotics—what they are, how they work, and how resistance to antibiotics develops.

Structural Defenses

The *structural defenses*—the "first line of defense" against disease—prevent pathogens from entering the body. The skin forms an effective barrier against invading organisms. Moreover, the skin secretes certain fatty acids and salts that are believed to inhibit microorganisms.

The mucous membranes that line the respiratory, digestive, urinary, and reproductive tracts are composed of closely packed cells that also form a tight wall against invading organisms—sort of an "inside skin." The mucus secreted by these membranes traps microorganisms and other foreign materials such as dust and then disposes of them. For example, the mucous membranes of the nose, trachea, and lungs have cilia that move the mucus along with its trapped materials to the throat. The small amount of mucus continuously coming to the throat is normally swallowed. The digestive juices of the stomach are highly acidic and quickly kill most pathogens that are swallowed.

Certain nonpathogenic microorganisms that live in our intestines (the intestinal flora) are beneficial to us. Physicians do not fully understand how this defense works, but they do know that the intestinal flora occupy the areas

10.6

➡ Objectives

- Describe how bodily structures defend against disease
- Summarize the nonspecific defenses against disease
- Describe the differences between antibodies and antibiotics

➡ Key Terms

inflammation
fever
antibody
antibiotic

10-28
Human tears contain an enzyme that attacks bacteria, decreasing the risk of eye infections.

that pathogens would occupy if they could. Should the normal intestinal flora be destroyed by medications such as antibiotics, pathogenic bacteria could become established and cause disease.

Tear glands, in combination with blinking, not only keep the exposed surfaces of the eye moist and dust-free but also ward off certain pathogens. Tears contain *lysozyme*, a powerful enzyme that attacks the cell walls of bacteria.

Nonspecific Defenses

If microorganisms should get past the first line of defense, the second line of defense comes into action. This consists of the inflammatory response, phagocytic cells, the lymphatic system, and an elevated body temperature (fever). **Inflammation** is a condition characterized by an increased flow of blood that is caused by chemicals released from cells in response to the presence of a pathogen. One of those chemicals is called *histamine*; when it is released, it causes the blood vessels to increase in diameter (dilate) and to become more permeable. As a result more blood is delivered to the area, causing it to become swollen, tender to the touch, and warmer than the surrounding area. In the extremities or areas that are close to the surface, the overlying skin appears red.

Another result of increased blood flow to the area is that more white blood cells called *phagocytes* arrive at the site. The phagocytes are able to squeeze through the walls of the blood vessels to get to the pathogens in the tissue. How are the phagocytes able to distinguish pathogens from other body cells? You should recall from previous discussions of cell membranes, cell walls, and viral capsids that many kinds of molecules are embedded in these structures. When the phagocytes encounter these unknown substances, they treat them as invaders and begin to isolate them from the rest of the body by forming a barrier around the area. At the same time many white blood cells engulf and digest the pathogens with enzymes. During this process toxins kill many of the white blood cells. The dead bacteria and white blood cells, as well as the fluids that remain after the conflict, form *pus*.

phagocyte: phago- (to eat) + -cyte (cell)

The liquid part of the blood plays an important role in the defense system. Since the blood vessel walls are more permeable, the fluid leaks out into the spaces between the cells, washing many of the pathogens and other substances from the area. The fluid carrying the foreign substances enters the vessels and nodes of the lymphatic system. When in the lymphatic system, the fluid is called lymph. Once inside the lymph nodes, phagocytic cells help filter the lymph of pathogens and toxins. The lymph then reenters the bloodstream (see Ch. 22). Enzymes break down the materials that remain in the phagocytic cells of the lymph nodes into soluble, nontoxic substances that are eventually carried away by the blood as wastes.

Although most people think of a **fever** (a raised body temperature) as an adverse symptom of an

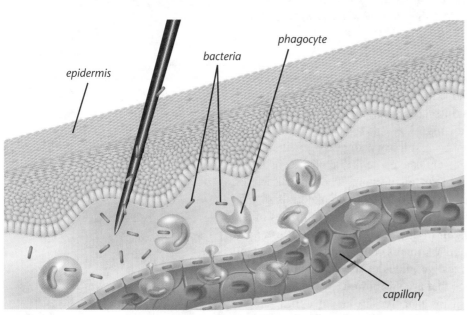

epidermis · bacteria · phagocyte · capillary

10-29
During inflammation, phagocytes leave the capillaries to attack and destroy bacteria.

infection, a fever is actually a defensive response. The higher temperature makes the environment less favorable for many invading organisms. Even a small increase in temperature can appreciably inhibit the growth of many pathogens. Also, many chemical reactions proceed more rapidly at higher temperatures, allowing the body not only to accelerate its defenses but also to use defenses that function only when the body temperature is elevated. However, if the body temperature goes too high or is raised for too long, damage to some body tissues (especially the brain) may result. In such cases a physician often prescribes measures to decrease the body temperature.

Specific Defenses

The third line of defense against disease is the immune system. The immune system has two components—chemicals called **antibodies** and several kinds of special cells, each with a particular duty. Antibodies are protein molecules made by specific cells in the blood. The antibodies circulate in the blood and are able to combat specific pathogens or their toxins. After an infection is over, other blood cells possess a "chemical memory" of that specific pathogen. If that pathogen ever infects the host again, these cells recognize it almost immediately and begin to produce antibodies and cells to fight that specific pathogen. The immune system is studied in detail in Chapter 22.

antibody: anti- (Gk. *anti*—opposite or against) + -body (O.E. *bodig*—body)

Some Human Infectious Diseases

Some people may not be familiar with all the diseases mentioned in this chapter or may have some incorrect information about them. The following gives some information or tells where in this book information about these diseases can be found. Other conditions that are discussed in this book can be located using the index. Viral infections are indicated by a *V*, bacterial infections by a *B*, and protozoan infections by a *P*.

✦ *African sleeping sickness.* (**P**) See page 262.

✦ *AIDS.* (**V**) See *HIV*.

✦ *Anthrax* (**B**) is caused by *Bacillus anthracis*, a gram-positive, spore-forming bacterium that usually affects hoofed animals. It can occasionally be transmitted to humans by contaminated food, through wounds, or by inhalation of spores. The inhalation type is the most dangerous; death can occur in just a few days after symptoms appear. It is also the type of anthrax infection that is used in biochemical weapons. There is a vaccine to prevent anthrax, but it is not recommended for the general public because of potential side effects; however, many military personnel have received the vaccine. Anthrax can also be treated with certain antibiotics.

✦ *Botulism* (**B**) is food poisoning caused by a bacterial exotoxin. It is one of the most powerful poisons known. The symptoms generally appear within twelve to thirty-six hours after ingestion and include fatigue, dizziness, and paralysis of the muscles of the eyes and pharynx. Paralysis spreads to the respiratory system. Botulism is often the result of improperly canned food. If the spores of the botulism bacteria (which can be found in the soil almost everywhere) enter the food and survive the canning process, they can grow anaerobically and produce the toxin. If an individual then consumes the contaminated food, he will contract the disease.

✦ *Chicken pox and shingles.* (**V**) The same virus causes chicken pox in children and shingles in adults who previously had chicken pox. One of the herpes DNA viruses, the chicken pox virus is very contagious and is probably spread by physical contact and airborne droplets from sneezing, coughing, or even shaking linens or clothes. It spreads in the bloodstream, causing a body rash and a mild fever. Adults may develop more serious symptoms from chicken pox. After the chicken pox stage, the virus can become latent in nerve cells. Years later the virus may move from the nerves to the skin and cause small, often painful sores called shingles. In 1995, a vaccine for chicken pox was approved for use.

✦ *Common cold.* (**V**) The common cold is a leading cause of illness. There are many different kinds of viruses that cause the common cold, and each one is different enough to prevent us from developing immunity or a vaccine. It appears that the common cold is not highly contagious; in one study only about 35% of those directly inoculated with the virus got a cold. Physical conditions (stress, amount of rest, other illnesses, etc.) seem to be significant factors determining whether a person gets a cold. Direct contact with the infected person or his secretions appears to be more important in spreading the disease than airborne droplets.

✦ *Cowpox.* (**V**) See page 237.

✦ *Dysentery* (**B, P**) is a condition characterized by abdominal cramps and bloody diarrhea. It may be caused by certain chemicals or by bacteria, protozoans, or worms. Bacterial dysentery is caused by several species. Some species cause mild symptoms; other species, however, cause high fever, chills, convulsions, and frequent bloody stools, resulting in dehydration. A spontaneous cure usually occurs within a few days. The bacteria invade the body through tainted food, water, or unsanitary practices (see also amoebic dysentery, p. 258).

✦ *Flu.* (**V**) See *influenza*.

✦ *Hepatitis A.* (**V**) This disease, also called infectious hepatitis, is spread by contaminated food or water. The virus can withstand mild heat, dryness, and chlorinated water. The early symptoms include fever and chills, loss of appetite, headache, muscle pains, jaundice (yellow color of the skin), and abdominal pain and swelling. Once the person recovers, he is immune. The disease causes the liver cells to lose their ability to function, but following recovery (which may take months), the liver may return to full function.

✦ *HIV (human immunodeficiency virus).* (**V**) This virus causes acquired immune deficiency syndrome, or AIDS, and was identified in 1983 after physicians began to diagnose some patients with very unusual types of infections that are normally prevented by the body's immune system. Once the virus enters the body, it attacks certain types of lymphocytes that are important in the immune system. As the lymphocytes are destroyed, the body is unable to fight off infections—even infections of organisms that normally present no problem to the host. Although researchers are working to inhibit HIV, the virus is difficult to control. Current drug therapy often consists of multiple drugs that must be taken in a specific order at specific times and that are quite expensive.

✦ *Influenza.* (**V**) In the United States, influenza, or the flu, is among the leading causes of illness. The flu is caused by a virus made of a protein coat covering up to eight pieces of RNA. It is transferred by airborne droplets. The infection begins in the upper respiratory tract; symptoms include soreness and redness of the nose and throat, a dry cough, a fever, muscle soreness, headaches, and nausea. Several kinds of flu exist. Each RNA fragment can mutate to form a new strain, and sometimes the RNA from different kinds appears to combine, forming new strains. These new strains may have different protein coats, the host cells they invade may be changed, or their virulence may be altered. When a person recovers from a particular flu, he develops a temporary immunity to that strain. But the body must develop immunity to each new strain. New influenza vaccines are developed for each season and are recommended especially for older adults, who are often at increased risk for flu complications.

✦ *Leprosy.* (**B**) See page 242.

✦ *Lyme disease* (**B**) is one of the most common vector-borne diseases in the United States. The bacteria that cause Lyme disease—*Borrelia burgdorferi*—live in the bloodstream of rodents and are transmitted to deer and humans by tick bites. Usually within seven to fourteen days after a tick bite, 80% of those bitten will have a red lesion that looks like the bull's eye on a target. The early symptoms include tiredness, fever, headache, and joint pain. If the disease is untreated, the heart and nervous system can become infected. In the early stages Lyme disease can be treated effectively with antibiotics, but if untreated there may be permanent damage.

✦ *Malaria.* (**P**) See pages 260–62.

✦ *Meningitis* (**V, B**) is any inflammation of the meninges (the membranes covering the brain and spinal cord). Some viruses and several species of bacteria can cause meningitis. The symptoms include headaches, spasms, stiff neck, and exaggerated reflexes. In severe cases, convulsions and death occur. Most forms of bacterial meningitis enter through the mouth and require close contact to be passed from one person to another. Vaccines are available for some forms.

✦ *Rabies.* (**V**) Caused by a single-stranded RNA virus, rabies attacks the nervous system of many different warm-blooded animals. Often associated with dogs, it is common in other carnivores as well, including skunks, raccoons, opossums, and coyotes. The virus is normally found in the saliva of an infected animal and enters the next victim as he is bitten. In the body the virus goes to the spinal cord, where it multiplies and destroys the nerve cells. Various symptoms develop as the virus spreads toward the brain. In dogs the virus is found in the saliva in about nine days, but serious symptoms begin to show only after twelve days and are not really noticeable for another day or so. The animal usually dies about sixteen days after becoming infected. If a human is exposed to rabies, the only known cure is to develop immunity using a vaccine.

✦ *Salmonellosis* (**B**) is an intestinal disorder caused by several members of the bacterial genus *Salmonella*. The organism lives in the intestine where it produces toxins that cause headache, chills, vomiting, diarrhea, and fever between eight and forty-eight hours after being consumed. The symptoms last a few days and are rarely fatal even without the use of antibiotics. Since some cases are mild, many people who contract this infection assume they just have a stomachache from something they ate.

✦ *Shingles.* (**V**) See *chicken pox and shingles*.

✦ *Smallpox.* (**V**) See page 237.

✦ *Syphilis.* (**P**) See page 622.

✦ *Tetanus* (**B**) is also called lockjaw. The spores of the anaerobic bacteria that cause tetanus are common in topsoil. If they enter a wound, they cause damage to the blood supply and then can grow in anaerobic conditions and produce toxins. These toxins affect the central nervous system by causing continual impulses to be sent to the muscles. The muscles contract and remain rigid. The muscles of the jaw are often involved. If not treated, the person often dies a painful death as more and more of his muscles are affected. Antibiotics can kill the bacteria, but the toxin must be treated by antitoxin. Temporary immunity can be achieved by injections of a toxoid, a form of the toxin that has its toxic properties removed but still stimulates an immune response for future protection.

✦ *Tuberculosis* (TB) (**B**) is still a dreaded respiratory infection. It may cause little damage in healthy people, but in others, for reasons not completely understood, the bacteria may cause fatigue, weight loss (hence it is also called consumption), and a persistent cough as it forms clumps of damaged lung tissue called tubercles. In advanced cases there is bleeding in the lungs as the bacteria destroy lung tissue, and the person coughs up blood. The bacteria are spread by airborne bacteria that can live in the air for hours. Treatment includes rest and antibiotics. It is difficult to wipe out the bacteria completely, and recurrent attacks are common. A species of tuberculosis bacteria was common in cattle and passed to humans in milk. Pasteurization of milk has virtually eradicated this type of tuberculosis in humans in the United States and Europe.

✦ *Whooping cough* (pertussis) (**B**) is caused by bacteria that a person inhales. After an incubation period of about a week, the initial symptoms of a sore throat and minor cough develop. A persistent cough develops after one or two weeks. The cough is accompanied by a long forced inhalation (or "whoop") and ends in the expulsion of colorless, sticky mucus. This stage may last four to six weeks and often leads to other complications.

Medical Control of Disease

If the body's defenses are unable to combat the pathogen, or if the physician feels it would be wise to help the body, there are several "chemical warfare" tools available. One of these is *chemotherapy*, the use of chemical agents to treat or prevent disease. Physicians select chemicals that will injure or kill specific pathogens without damaging the host's body. There may be side effects (symptoms caused by the chemotherapy). If side effects are minor, patients must tolerate them during the period of treatment.

Today the most used form of chemotherapy for infectious bacterial diseases involves **antibiotics**. These are chemicals which are either *bactericidal* (killing bacteria) or *bacteriostatic* (inhibiting growth of bacteria). The earliest antibiotics were produced from other living organisms, but many are now produced synthetically.

How Antibiotics Work

Bacteria produce toxins that can kill cells; antibiotics are toxins that man uses to kill certain kinds of bacteria. Since the metabolic pathways and enzymes of bacteria are different from those of a human, scientists can develop antibiotics that interfere with either bacterial metabolism or specific enzymes and that have little or no effect on the host. Almost all bacteria have a cell wall, and if the cell wall can be disrupted, the bacterium will die. Many antibiotics kill bacteria by blocking the assembly of peptidoglycan, an important component of the cell wall. Others interfere with protein synthesis—without the proper proteins, bacterial growth will be inhibited. One of the antibiotics used to treat tuberculosis inhibits RNA synthesis.

One drawback of antibiotics is that they are often not very specific in which bacteria they kill. For example, penicillin kills many different species of bacteria, not just one. Some beneficial bacteria that are critical in digestion may be eliminated by the overuse of antibiotics. Antibiotics that have little specificity are called broad-spectrum antibiotics. Some antibiotics are specific for either gram-positive or gram-negative bacteria, but few are specific for a single kind of bacterium.

Criteria for an Ideal Antibiotic

1. Selective toxicity: kills the pathogen and not the host
2. Narrow spectrum: kills only the pathogen and no beneficial organisms in the host
3. Minimal side effects for the host, such as allergic reactions, nausea, vomiting, diarrhea, sensitivity to light, or discoloration of developing teeth
4. Effective in small amounts
5. Low cost
6. Long shelf life with no special needs such as refrigeration
7. Simple dosing instructions

antibiotic: anti- (against) + -biotic (pertaining to life)

Sir Alexander Fleming and Penicillin

The properties of antibiotics were discovered by accident in 1929 by the British bacteriologist Sir Alexander Fleming. Fleming, returning to his laboratory after a short vacation, noticed a blue mold growing on a petri dish culture that had become full of bacteria. Around the mold there was a clear, circular area where the bacteria had been killed. The mold had produced a toxic substance that had diffused outward, killing the bacteria.

Fleming identified the mold as a species of *Penicillium* and named the substance it produced *penicillin*. Realizing the potential of this substance as a drug, he tested it, found that it was not toxic to most laboratory animals, and also determined that it was effective against many different kinds of bacteria. By 1942, scientists had produced pure penicillin—a yellow powder—that had a remarkable potency and was used successfully against many bacterial infections.

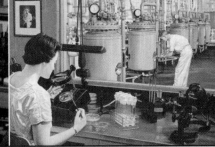

Although Fleming discovered penicillin in 1929, significant use of the drug did not occur until 1940 due to high production costs. Under the pressure of World War II, pharmaceutical companies mass-produced it, bringing the price to one-thousandth of its original cost.

The Era of Antibiotics by Robert Thom. Printed with permission of American Pharmacists Association Foundation. Copyright 2007 APhA Foundation.

Use of Antibiotics

To be useful, an antibiotic must have little effect on tissues of the host and at the same time be effective against the pathogen. When antibiotics were first available, they were truly miracle drugs—infections that had previously been fatal were easily cured. Indeed, antibiotics seemed to cure almost everything and were prescribed for almost anything. Over time, physicians began to see infections that were not cured by the usual course of antibiotics. Dosages were increased, but still the bacteria continued to multiply. Researchers began to look for different antibiotics to treat infectious diseases effectively.

Today the quest for new antibiotics continues. Some infectious diseases are not yet under control. Then, too, in any population of microorganisms, a few may be *naturally resistant* to an antibiotic. Although the antibiotic may kill off the nonresistant strains, the naturally resistant ones survive and multiply. As time goes on, the resistant organisms form a greater percentage of the total population of that pathogen. This makes it increasingly difficult to treat the disease with the same antibiotic. Since a pathogen resistant to one antibiotic can often be effectively treated using a different one, researchers and physicians are eager to increase the number of antibiotics available.

Antibiotic Resistance

Occasionally it is said that a pathogen has "developed" a resistance, implying that the pathogen is evolving. This "development" of resistant strains is usually an example of selection from a preexisting gene pool or the transfer of genes from one bacterium to another, not the development of new characteristics. Even if a pathogen did become resistant to an antibiotic because of mutation, this change is not evolution but merely a variation of an existing organism's genotype.

Review Questions 10.6

1. What are the body's three lines of defense against disease?
2. List and describe several of the body's structural defenses against disease.
3. Describe the working of the body's nonspecific defenses against disease.
4. What method can physicians use to help control a pathogen that is in the body?
5. What is the difference between bactericidal and bacteriostatic chemicals?
6. List four ways that antibiotics inhibit bacterial growth.

10.7 Disorders

For this study, a **disorder** is defined as any affliction not caused by a pathogen. Disorders can be grouped into three major categories:

✦ *Inherited disorders.* These disorders are either the direct result of an inherited gene (as in hemophilia, PKU, or sickle cell anemia) or an inherited tendency for a disorder (such as diabetes mellitus or heart disease).

✦ *Injuries.* Caused by physical damage to the body, injuries may be temporary (as a bruise or minor cut) or permanent (as the loss of a leg or an eye). Burns, broken bones, sprained joints, concussions (impaired activity of the brain caused by a severe jar or shock), and some hearing and vision defects are the results of injuries.

✦ *Organic disorders.* These conditions are not inherited or caused by injury. A *deficiency disease* results from improper nourishment such as the lack of a vitamin or mineral. *Chemical poisoning* and *radiation sickness* result from exposure to environmental factors. Strokes, ulcers, blood clots in the vessels (thrombosis), types of hardening of the arteries, kidney stones, gallstones, and many nervous disorders are caused by unknown or only partially understood factors.

10.7

Objectives

- Describe the three major groups of disorders
- Distinguish between benign and malignant tumors
- Summarize the three major methods of treating cancer

Key Terms

disorder
gerontology
clinically dead

The rest of this chapter includes a discussion of two organic disorders: benign tumors and cancer.

Benign Tumors

Occasionally a group of cells stops functioning normally and grows a structure different from the tissue of which it is a part. This abnormal growth of cells is called a tumor. If the growth is slow and localized, it is called a *benign tumor*. The body often walls off areas of benign tumors, preventing their spread.

Some examples of benign tumors are common (brown) moles and certain birthmarks. The tumor cells often closely resemble those of the tissue from which they originated. Some benign tumors grow to a certain size and stop, while others expand slowly and can exert pressure on the surrounding tissues. In some cases the pressure from a benign tumor can seriously impair the functions of an organ. For example, it may obstruct a secretion, cut off the blood supply to a region, or, as in the case of a brain tumor, cause serious disability or death. Benign tumors of the hormone glands can cause a gland to secrete either too much or too little of its hormones, a condition that will affect different areas or functions of the body.

For benign tumors that are causing such problems, surgical removal is usually a successful and lasting treatment in the vast majority of cases.

Cancer

If the growth of a tumor is rapid and chaotic, it is called a *malignant tumor*. A malignant tumor is often called a *cancer*. The nuclei in cancer cells are larger and often contain more DNA than normal tissue; cell growth and reproduction is also more rapid. The cells in a malignant tumor often develop a bizarre appearance and may even have an abnormal number of chromosomes. Cancer cells may *metastasize*, or separate from the parent tumor and travel to other parts of the body, starting new tumors.

Cancer cells are cells of the body that have changed their normal genetic expression. Cancer researchers point to a two-step process in the development of cancer: *initiation* and then *promotion*. Initiation is generally accomplished in one of three ways.

✦ *Carcinogenic chemicals.* Cancer-causing chemicals are called *carcinogens*. The best-known carcinogens are in the chemical residue from smoking or use of other tobacco products. A definite link between smoking and lung cancer was established many years ago. Other carcinogenic chemicals include formaldehyde, asbestos, and benzene.

✦ *Radiation.* Skin cancer can be caused by excessive exposure to the ultraviolet rays in sunlight. Fortunately, skin cancers (if diagnosed soon enough) are usually mild and often can be removed in a doctor's office. X-rays are also regarded as a cancer hazard. Both skin cancer and leukemia (cancer involving white blood cells) have been linked to x-rays. Highly radioactive substances are considered to be carcinogenic agents. Small amounts of radiation (minor exposure to the sun or medical x-rays) are not believed to be harmful.

✦ *Viruses.* In humans, viruses are linked to some cancers—the human papillomavirus (HPV) is closely linked to the development of cervical cancer in women. Researchers also believe that the Epstein-Barr virus (the cause of mononucleosis) is related to the development of some types of lymphoma, stomach cancer, and Hodgkin's disease.

Once a cell has become potentially cancerous, environmental factors such as the diet and general health of the person as well as his genetic makeup help promote the conversion of initiated cells into cancer-producing cells.

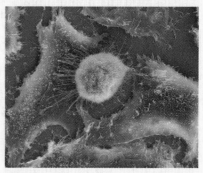

In addition to diseases and disorders, humans face the problem of aging and the prospect of death. Scripture indicates that both aging and death are a part of the Curse, which God pronounced in Genesis 3. Aging is the continual degeneration (wearing out) of our bodies. Stooped posture, thinning and graying hair, and wrinkling and drying skin are just some of the outward signs of the deterioration that is taking place throughout the body. Organs gradually wear out; the heart and blood vessels become less elastic; bones become more brittle; joints stiffen; brain cells die; the eyes become less capable of focusing; dental problems make eating difficult; injuries heal more slowly; and the resistance to disease diminishes.

Aging

Much research is being conducted in the field of **gerontology** (JEHR un TAHL uh jee), the science of aging. Researchers have determined that degeneration begins when a person is in his twenties, shortly after his growth has been completed. Not all scientists agree about the physical causes of aging. One theory holds that it is the cumulative effect of wear and tear, the "battle scars" of the continual fight against disease, injury, and the rigors of life itself.

Aging has also been attributed to waste accumulation in the cells. Wastes are continually being removed from cells, but some types of wastes are not completely removed. Therefore,

as cells grow older, the concentration of certain wastes increases, and presumably the function of the cell is impaired. Some researchers believe that as a waste product called *lipofuscin* (LIH poh FUS in) accumulates in older nerve cells, the rate of cell division and other metabolic processes is decreased. If the lipofuscin is removed from the cells, scientists have seen the cell division and metabolic rates increase in some cell types. Although this research is very interesting, much more needs to be done in this area of study.

Another theory is that aging and death are genetically programmed into the body. From the moment the zygote is formed, the individual follows a prescribed program through the different stages of life: embryo, infancy, childhood, adolescence, young adulthood, middle age, old age, and death. Thus, aging and death are a natural part of the life cycle, directed by genetic machinery. Some gerontologists believe there are genes that trigger degenerative processes as life progresses. Research has shown that certain human cells have only about one hundred generations built into them. After their appointed number of cell divisions, they degenerate and die.

There also seems to be a correlation between aging and the length of the telomeres on the ends of the chromosomes. Each time a cell divides, the telomere gets shorter, changing the expression of the genes on that chromosome, and eventually the cell stops dividing altogether.

Christians view aging as a divinely ordained modification of the physical being as a result of the Fall. Christians should be willing to accept aging, knowing that ultimately those who know Christ will receive a glorified body that is incorruptible (unaging) (1 Cor. 15:51–57). Unregenerate people, not having this hope, would like to find some way to block the sequence of aging and thus "beat the system."

Death

Someone made a statement once that the surest way to live to a ripe old age is to choose ancestors who lived to ripe old ages. This statement contains a basic element of truth: genes do set limits upon a person's life. But within these limits there is considerable room for individual variation. In addition to having "good genes," a person must avoid accidents and diseases as well as practice good living habits (proper diet, rest, exercise, and a lifestyle conducive to good mental and emotional health).

Yet, barring the Lord's return for His saints, death is inevitable for every person. Death appears to be a part of the universal degenerative trend and is in keeping with Scripture, which says, "It is appointed unto men once to die" (Heb. 9:27). It appears that only God Himself could change this plan.

But what exactly is death? In former times men relied on two main indicators—absence of heartbeat and absence of breathing. But the cells of the body do not die immediately when these functions cease. Victims of cardiac arrest, drowning, and electrical shock can often be successfully revived if action is taken quickly. Even brain cells, the most fragile type of cell, can live for a couple of minutes after their blood supply is cut off before suffering irreversible damage. A person resuscitated within this time may suffer little, if any, permanent damage. If the body temperature is lowered (as may happen when a person drowns in cold water), he may

be revived after half an hour or longer without permanent brain damage.

Therefore, a definition of death based only on the presence of a heartbeat and breathing is unsatisfactory. Now scientists usually consider brain activity as the indicator of the state of the physical organism. A functioning brain produces minute electrical impulses called *brain waves*. An electroencephalogram (ih LEK troh en SEF uh luh GRAM), or EEG, measures these waves. If there is no electrical activity in the brain, the EEG is "flat." Medical authorities say that if there are no brain waves for twenty-four to forty-eight hours, there is no hope for the patient. A person in this condition is **clinically dead**.

What if the heart of a clinically dead person (no brain activity) is still beating and the person is still breathing without the aid of any supporting equipment? Would it be proper to bury such a per-son? Even though the brain may be "dead," it would seem strange indeed to bury a body that was still breathing. To avoid such a dilemma, many authorities insist that all three criteria exist before burial—no spontaneous heartbeat or breathing and no brain-wave activity. Dilemmas often arise when "life-support" equipment is being used to maintain the heartbeat or breathing while there is little or no brain activity.

The Christian recognizes death to be the departure of the spirit from the body. For the Christian, to be absent from the body is to be present with the Lord (2 Cor. 5:8). Medical instruments cannot measure the nonphysical aspects of the body to determine when it has degenerated too far to be inhabitable by the spirit. However, caution is in order: God does work miracles!

Most Christians doubt that man will ever be able to conquer physical death through scientific endeavors. The biological problems are numerous, and Scripture says that the power of life and death is in the Lord's hands (Rev. 1:18). Although man may not be able to conquer death, there is One Who promises victory over it. Christ's death on the cross conquered physical death; He rose from the grave having a new, incorruptible (unaging), glorified body.

The just and eternal fate of every human being is hell, a place that God prepared for Satan and his hosts. However, Christ also conquered spiritual death so that those who believe on His name will live eternally with Him in their glorified bodies. With these promises in the Word of God, it is appropriate for the Christian to say, "O death, where is thy sting? O grave, where is thy victory?" (1 Cor. 15:55).

Cancer can develop almost anywhere in the body and can arise from almost any type of cell. Since tissues are often made of more than one type of cell, multiple kinds of cancer can develop in one type of tissue. What a person does to prevent one form of cancer may have no effect on another form. For example, reducing exposure to ultraviolet radiation will decrease the risk of developing skin cancer but would have no effect on the risk of developing lung cancer.

Who develops cancer also depends a great deal on genetic makeup. Certain cells in some people are naturally resistant to some forms of cancer. Another person, however, may have inherited a weaker cell and might develop that form of cancer even if he is careful to avoid factors believed to initiate or promote that cancer.

It is, of course, wise to avoid those things that are known to induce or promote cancers, but you must also be careful not to be caught up in every new fad or advertising gimmick designed to cash in on people's natural fear of cancer.

Steps to Reduce Cancer Risk

Although cancer is probably not something most high-school students usually worry about, there are some steps you can take now to reduce your risk of developing cancer in the future.

1. Do not start smoking, and if you do smoke—STOP.
2. Avoid areas where people smoke. Request nonsmoking areas in restaurants and other public places. Breathing secondhand smoke is almost as bad as smoking itself.
3. Do not start using oral tobacco, such as chewing tobacco or snuff. If you do chew—STOP.
4. Reduce your exposure to UV radiation in sunlight AND tanning lamps.
5. Increase the amount of fiber and fresh vegetables in your diet. (Read the list of chemicals on the packaging of processed foods and sodas.)
6. Eat a balanced diet, including minerals and vitamins.
7. Reduce the amount of fat in your diet.
8. Avoid exposure to strong chemicals such as pesticides, insecticides, herbicides, and solvents.
9. Avoid sexual relations until you are married and then remain faithful to your spouse.
10. Test your home (especially those with basements) for radioactive radon.

Of these ten steps, numbers one and four are the most important—and they are free. Even if you follow each of these steps, you are not guaranteed to remain cancer free; however, you will certainly reduce your risk and will develop a healthy lifestyle. A healthy lifestyle will also reduce the risk of developing other diseases and ailments as you grow older.

Cancer Treatments

There are three primary methods used to treat cancer—surgery, radiation, and chemotherapy. Physicians may use just one or a combination of these methods, depending on the type and extent of the cancer.

The leading method of cancer treatment is surgical removal of the affected tissues. Many lives have been saved in this way; however, the surgeon is successful only if he completely removes all the cancerous cells. Thus, it is important to diagnose cancer early, while the affected area is still small. If the disorder has metastasized (traveled) to the lymph nodes or to vital organs, the chances for survival are reduced. Nonetheless, surgery may still help relieve pain, restore lost bodily functions, and slow the spread of the disease even if it is impossible to remove all the cancer.

Another treatment for certain cancers is *radiation*. Physicians use x-rays or emissions from radioactive isotopes to destroy cancer cells with the hope of sparing most of the normal cells. This method can sometimes completely eradicate localized cancers. In more widespread malignancies, radiation therapy may provide relief from pain and prolong the individual's life, but complete recovery is rare. Additionally, radiation treatment can have extremely harsh side effects and may even cause new tumors.

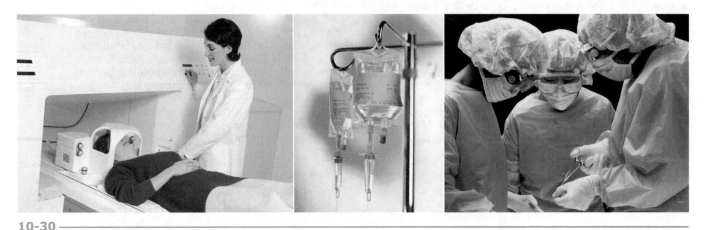

10-30
Methods of cancer treatment include radiation (left), IV chemotherapy (center), and surgery (right).

Physicians sometimes treat cancer by *chemotherapy*. In this case the chemical is directed not at an invading pathogen but at cancer cells. Often the chemical chosen interferes with cell division. Since cancer cells divide more frequently than normal cells, the chemical will affect these cells more than the rest of the body. Physicians may use two or three different types of chemotherapy, each one attacking the cancer cell at a different point in its metabolism to increase the overall effectiveness of the treatment.

Regrettably, cancer chemotherapy also affects other cells in the body. Normally, the most rapidly growing cells of the body are affected first. This explains anemia (lack of red blood cells), loss of hair, sensitive skin, and general deterioration of the patient's health. Once the chemotherapy stops, the symptoms are usually reversed and, it is hoped, only the cancer cells have been permanently harmed. Repeated chemotherapy treatments can destroy several forms of cancer. Sometimes chemotherapy is used to kill those cancer cells that were not removed in the surgical procedure, thus increasing the probability that the cancer will not start to grow again.

Review Questions 10.7

1. List and describe three major types of disorders. Give examples of each.
2. In what ways can benign tumors cause physical difficulties?
3. What are some of the differences between normal body cells and cancer cells?
4. List and describe three possible causes of cancer.
5. List and describe several present cancer treatments.
6. Why is it important to detect cancer early?
7. List several things you can do now to reduce your risk of developing cancer in the future.
⊙8. List some possible causes of aging.
⊙9. Discuss several criteria that have been used to determine when death has occurred.
⊙10. Does a human soul ever die? What is spiritual death? Support your answers from Scripture.

11

The Kingdom Protista

Microbiology Part II

Facet

Until now, our study of microbiology has included only prokaryotic organisms and viruses. The study of the kingdom Protista begins an investigation of those organisms that are composed of eukaryotic cells. The kingdom Protista has often been called a "catchall" kingdom, the group in which taxonomists place eukaryotic organisms that do not seem to fit into any of the other kingdoms of eukaryotic organisms. The organisms in this kingdom are called **protists**, and many taxonomists define *protist* as any eukaryotic organism that is not animal, plant, or fungus. Nonetheless, protists are often described as being animal-like, plantlike, or funguslike. The kingdom is also diverse in the structure of its organisms. While many are unicellular organisms, some are multicellular. Some protists also form colonies.

Like all kingdoms, Protista is divided into a number of phyla. For convenience, the phyla will be covered in three sections in this chapter: the protozoans, which are animal-like; the algae, which carry on photosynthesis; and the funguslike protists. Although many of the organisms presented in this chapter may be unfamiliar to you, they are important in nature because they play key roles in food chains and in decomposition. Many also have more direct connections to humans because they may cause illness, and some are even sources of food.

11A—The Protozoans

Protozoan means "first animals," and organisms in this group are considered the most animal-like. In fact, some early microscopists believed that these "animalcules," as Anton van Leeuwenhoek called them, gave rise to larger animals, which in turn gave rise to even larger animals. From a Christian worldview, it is unreasonable to conclude that these organisms are the source of all animal life. God made organisms to reproduce after their own kinds (Gen. 1:21, 25). God made these tiny creatures, much as they are now, perhaps on Days 5 and 6 of the Creation week.

This evolutionary paradigm also reveals an ignorance of the complexity of these tiny organisms. In the space of a single cell, protozoans perform all the functions necessary to maintain their living condition. In that respect, they are even more amazing than large animals.

11.1 General Characteristics

Protozoans are microscopic unicellular (or, occasionally, colonial multicellular) organisms which are usually *motile* (MOH tuhl), able to move from place to place. Their movement, often quite rapid, prompted the early microscopists to classify them as members of the animal kingdom.

Most protozoans are aquatic and are found in freshwater lakes, streams, and ponds, as well as in oceans. Many others are found in soil where water is readily available. Protozoans are heterotrophic, and many live inside other organisms. Some of these are parasitic, while others actually benefit their host organism.

Their small size helps protozoans in certain life functions. The protozoans, inhabiting moist environments, merely exchange dissolved gases between their cytoplasm and their surroundings. Soluble wastes diffuse from protozoan cytoplasm through their membranes into the environment. Protozoans do not have cell walls.

Most protozoans reproduce asexually by binary fission. A few species can exchange genetic material, thereby increasing genetic diversity.

Many species of protozoans can respond to changes in their environment. Some have areas of pigment called **eyespots** that can detect light intensity and either move toward or away from the light; others can detect and respond to chemical changes. During times of harsh environmental conditions, many protozoans survive by forming a cyst. While in the cyst form, the metabolic rate is slowed and a hard covering is formed around the protozoan.

As the different kinds of protozoans are presented, the text explores these characteristics as well as others in further detail. Exploring and observing what protozoans do and how they do it are some of the most interesting activities in introductory biological studies.

11.1

Objectives

- Describe some general characteristics of protozoans
- Summarize how some protozoans react to adverse conditions

Key Terms

protist
protozoan
eyespot
cyst

11-1

Giardia is a flagellated protozoan that can cause severe diarrhea in humans.

motile: (L. *movere*—to move)
cyst: (Gk. *kustis*—bladder or pouch)

Review Questions 11.1

1. Why is the kingdom Protista considered a catchall kingdom?
2. In what types of environments would you find protists?
3. How do some protozoans respond to adverse environmental conditions?

11-1 Protozoan Phyla and Their Characteristics

Phylum	Common name	Means of locomotion	Means of nutrition	Examples
Sarcodina	sarcodines	amoeboid movement	heterotrophic; some parasitic	amoebas, radiolarians, foraminifers
Ciliophora	ciliates	cilia	heterotrophic; some parasitic	*Paramecium, Stentor, Vorticella*
Sporozoa	sporozoans	none in adults	heterotrophic; all parasitic	*Plasmodium*
Zoomastigina	zooflagellates	flagella	heterotrophic; some parasitic	*Trypanosoma, Giardia*

11.2 Protozoan Classification

Just as it caused an upheaval in bacterial classification, current molecular biological research is causing reevaluation of the classification of the different protozoans within the kingdom Protista. Historically, protozoans are considered the most animal-like protists. Since the classification scheme is somewhat unsettled, the protozoans will be divided into four phyla based on their means of locomotion.

Phylum Sarcodina: The Sarcodines

Members of the phylum Sarcodina (sar koh DYE nuh) are characterized by their lack of a standard body shape. These single-celled protists are enclosed in a flexible cell membrane that allows them to constantly change their shape. When dormant, they may be nearly spherical. When moving or feeding, they will form numerous extensions called pseudopodia that function as "false feet." Some species of sarcodines construct and inhabit shells. Sarcodines live in a variety of habitats, including fresh water, the sea floor, and the human mouth and intestines.

Amoeba: A Typical Sarcodine

Amoeba proteus, the common amoeba, looks like little more than a blotch of gray jelly. Usually amoebas are not swimmers but are part of the slimy covering of submerged rocks or plants. Here they ingest organic debris and small microorganisms. Like other sarcodines, amoebas do not form colonies.

The cytoplasm in the amoeba is divided into two types: the endoplasm, located in the interior of the organism, and the clear ectoplasm, which is the outer portion next to the cell membrane. A single disk-shaped nucleus controls all the cell's metabolic activities. **Contractile vacuoles** collect and eliminate water, regulating homeostasis between the amoeba and its environment.

The most outstanding feature of the amoeba is its **amoeboid movement**. When an amoeba moves, its endoplasm streams toward one area of the cell membrane, causing a bulge to form. This bulge is gradually extended into a long **pseudopodium** (soo duh POE dee um). The cytoplasm of the amoeba flows into a pseudopodium and draws the cell membrane with it, causing the organism to move.

The amoeba can respond to a variety of stimuli, generally by a change in the speed or direction of movement. Such actions in response to stimuli are called **taxes** (TAK seez) (sing., taxis). For example, if a floating amoeba touches a solid object, it will move toward the object and adhere to it. But the response to touch varies under certain circumstances. If an attached amoeba is touched with a glass rod, it will retreat from the stimulus. Amoebas will advance toward an area containing substances diffused from foods and recoil from an area of high saltiness.

Once the amoeba has located and selected food (by taxes, amoebas do show "food preferences"), it uses pseudopodia to engulf it. This is the process of

11.2

Objectives

- Describe the movement and feeding method of a typical sarcodine
- Describe the structure and use of cilia in protozoans
- Explain the complex life cycle of *Plasmodium*, a typical sporozoan
- Describe the movement of a typical zooflagellate
- Recognize one key organism for the sarcodines, ciliates, sporozoans, and zooflagellates

Key Terms

contractile vacuole
amoeboid movement
pseudopodium
taxis
ciliate
pellicle
macronucleus
micronucleus
gullet
trichocyst
conjugation
zooflagellate

11-2

Amoeba proteus, one of the most familiar sarcodines

Sarcodina: (Gk. *sarkodes*—fleshy)
ectoplasm: ecto- (Gk. *ectos*—outside) + -plasm (to mold)
endoplasm: endo- (within) + -plasm (to mold)
taxes: (Gk. *tassein*—to arrange)

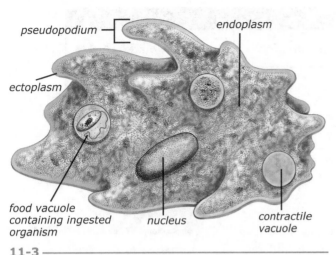

pseudopodium

endoplasm

ectoplasm

food vacuole containing ingested organism

nucleus

contractile vacuole

11-3
Typical structures of an amoeba

11-4
An amoeba phagocytizing food

Other Sarcodines

Some sarcodine species live as parasites of other organisms. *Entamoeba coli* live in human intestines and *E. gingivalis* live in human mouths, but neither is pathogenic.

Many sarcodines form tests, or shells, around themselves and project their pseudopodia through holes in the test. The *foraminifers* (fohr uh MIN uh furz) are tested ocean-bottom dwellers. Most foraminifers form tests of calcium carbonate. When they die, their shells add to the ooze on the ocean bottom.

The *radiolarians* (RAY dee oh LEHR ee uhnz) form silicon tests and also contribute to the ocean bottom ooze. About a third of the ocean floor is covered with this ooze, which may be up to 4000 m thick. This is impressive considering that 1 g may have 50 000 sarcodine tests.

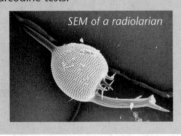

SEM of a radiolarian

Ciliophora: Cilio- (eyelid) + -phora, -phore, -phera, or -fera (Gk. *phoros*— bearing)

phagocytosis described in Chapter 3. The food material, such as algae or other protozoans, is thus sealed into a food vacuole within the amoeba. Lysosomes in the cytoplasm fuse with the food vacuole and deposit their enzymatic contents into it, and digestion proceeds within the vacuole. The soluble foods then diffuse into the cytoplasm, and the insoluble materials that remain are released from the cell or egested.

When an amoeba reaches maximum size, it undergoes binary fission, dividing into two daughter cells (under normal conditions, about every three days). The process begins with the replication of the genetic material of the amoeba. Mitosis occurs, and the cytoplasm divides, resulting in two complete, functional daughter amoebas in about thirty minutes. Amoebas are not known to reproduce sexually.

Amoebas are one of the species that will respond to life-threatening environmental conditions, such as dryness or lack of food, by becoming a cyst. These dormant organisms will become active upon the return of favorable conditions.

Although most sarcodines are nonpathogenic to humans, *Entamoeba histolytica* can cause a severe intestinal infection called amoebic dysentery. In some areas of the world, large portions of the population are carriers, although they lack symptoms. Once ingested, the *E. histolytica* destroys the cells of the intestinal walls, causing ulceration, which can result in severe diarrhea and lead to death. In times past the disease was often fatal, but today various drugs can kill the parasite.

Phylum Ciliophora: The Ciliates

The **ciliates** (SIL ee its), members of the phylum Ciliophora (SIL ee AWF uh ruh), are among the most intricate and fascinating organisms in the kingdom Protista. These organisms may be up to 3 mm long, quite large for a protist, and exist in a variety of shapes.

The characteristic that distinguishes this group is the possession of multiple hairlike projections called cilia. The cilia of these protozoans beat rhythmically to either propel the organism toward its food or move the food toward the organism. The arrangement of the cilia varies. Some ciliates are completely covered with cilia; others bear rings or patches of cilia. Most ciliates are free-swimming, but a few can attach to submerged objects.

Paramecium: A Typical Ciliate

Protozoans of the genus *Paramecium* (PEHR uh MEE see um) are common free-swimming inhabitants of stagnant lakes and ponds. Paramecia have a distinctive slipper shape that is maintained by the pellicle, a firm yet flexible protein-rich covering that is outside the cell membrane. Cilia completely cover the paramecium and can beat either forward or backward, enabling the organism to turn, rotate, and travel in any direction.

The kidney-shaped macronucleus is the most conspicuous feature of the cytoplasm. The macronucleus appears to be multiple copies of the genetic material of the cell. These copies of genes probably aid the high metabolism typical of most ciliates and are important in asexual reproduction. The smaller micronucleus functions during the exchange of genetic information during sexual reproduction.

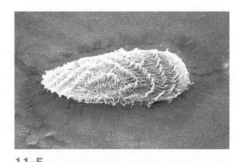

11-5

The number of cilia varies on different kinds of paramecia.

The body of the paramecium has a funnel-shaped indentation called the *oral groove*. Cilia that line the oral groove sweep food material through the *mouth pore* into a short, blind tube called the gullet. As food is directed into the gullet, an enlargement forms at the end. This enlargement is eventually pinched off, becoming a food vacuole. The vacuole is then circulated throughout the cytoplasm, where enzymes from lysosomes digest the food. Any indigestible material is expelled from the paramecium through the anal pore, a tiny opening in the cell.

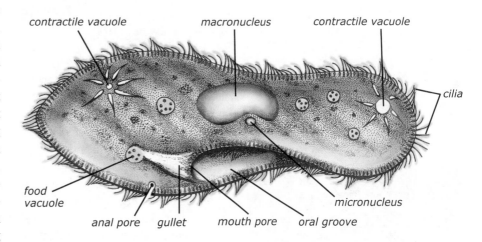

11-6

Typical structures of a paramecium

The paramecium has two star-shaped contractile vacuoles. The rays of the stars are canals that collect excess cell water and empty it into the vacuole at the center. When the vacuole is full, the water is expelled from the cell. The swelling and squeezing of these vacuoles is often visible under the microscope. The concentration of salts in the paramecium's environment determines the rate at which these organelles operate.

Paramecia tend to swim in a forward spiral motion. When the organism collides with an obstacle, it will stop, back up, turn slightly, and then move forward again. Paramecia will also avoid temperature extremes and irritating chemicals in the same way. This *avoidance reaction* is an example of taxis in paramecia. However, acidity will attract these protozoans. Since bacteria tend to thrive in acidic environments, this positive taxis to acidity helps paramecia locate bacteria that serve as their food. Paramecia also eat other smaller protists and organic debris.

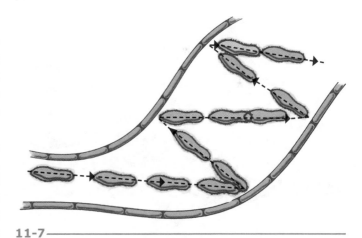

11-7

A paramecium demonstrating avoidance reaction

In response to certain stimuli, trichocysts (TRIK uh SISTS), which are tiny organelles under the pellicle, discharge stiff filaments into the water. Most researchers think that these filaments function as a defense mechanism.

Reproduction in the Paramecium

Paramecia reproduce by two methods. The first is asexual binary fission. During this process the micronucleus divides by mitosis, and the macronucleus enlarges and divides in half. Next, the paramecium elongates, and a second

pellicle: (L. *pellis*—skin)

macronucleus: macro- (Gk. *makros*—large) + -nucleus (central part)

micronucleus: micro- (small) + -nucleus (central part)

Other Ciliates

The *Stentor* is a giant among protozoans, reaching a size of 2.5 mm. This trumpet-shaped organism has a ring of cilia about its gullet. The cilia generate a current that draws food into the gullet. A hungry *Stentor* can consume one hundred smaller protozoans per minute.

Vorticella, a ciliate that lives attached by a corkscrew-shaped stalk to a submerged object, has cilia only on the top.

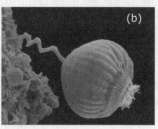

Stentor *(a) and* Vorticella *(b)*

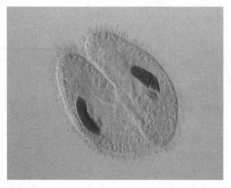

11-8
Paramecia undergoing conjugation

gullet forms. Finally, a furrow forms across the middle of the organism, and it divides into two complete daughter paramecia.

Paramecia also reproduce by a type of sexual reproduction called **conjugation** (different from the bacterial conjugation discussed in Chapter 10). Two paramecia attach to each other by their oral surfaces. Each organism undergoes a variety of nuclear changes. At one point the cells exchange a portion of their genetic material through the cytoplasmic bridge that joins them. The cells separate and undergo more nuclear changes. Finally, each paramecium divides to form four paramecia. Conjugation allows a mixing of genetic material that is impossible in asexual reproduction.

Phylum Sporozoa: The Sporozoans

The members of the phylum Sporozoa (SPOHR uh ZOH uh) are unique among protozoans since, as adults, they do not have pseudopodia, cilia, flagella, or any other special structures for locomotion. As the name implies, sporozoans form spores at some stage of their life cycle. First, the nucleus of the sporozoan divides several times, and a small amount of cytoplasm gathers around each nucleus. Then the organism breaks apart, and each of the nuclei becomes a spore that may be surrounded by protective coats.

All sporozoans are parasitic, often having complex life cycles involving a number of vectors and animal or human hosts. They feed by absorbing dissolved materials from the host's cells and body fluids. Asexual reproduction may occur by spore formation or by cell division. Many sporozoans also have some means of sexual reproduction.

Plasmodium: A Typical Sporozoan

A devastating disease of the tropics is *malaria*, an illness that causes fatigue, thirst, and high fever alternating with chills. If the disease progresses, victims usually die from kidney failure, anemia, or brain damage. Malaria, meaning "bad air," was once believed to be caused by tropical swamp air. Some infectious-disease researchers claim that malaria is the most significant disease of man, directly or indirectly causing more deaths than any other disease and significantly influencing human events.

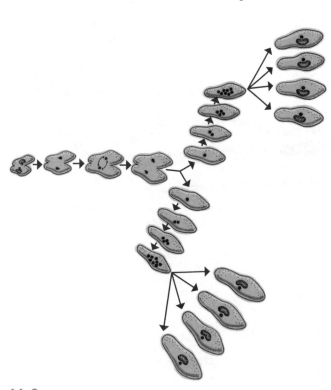

11-9
A simplified diagram of conjugation in paramecia

Disease and the Panama Canal

In the late 1800s, the French attempted to dig a canal across Central America that would join the Atlantic Ocean and the Pacific Ocean. The workmen sent to accomplish this task soon fell ill; many died either from yellow fever (caused by a virus) or malaria. This epidemic was one crucial reason why the French project was canceled. This opened the door for the United States to build a canal. In 1880, Dr. Charles Laveran discovered the malarial protozoan, a member of the genus *Plasmodium* (plaz MOH dee um). Later research unlocked the complete life cycle of this sporozoan.

Understanding the *Plasmodium* life cycle has helped scientists discover ways to control malaria in many areas of the world. When building the Panama Canal, for example, Colonel William Gorgas of the United States Army ordered swamps drained and brush cleared in the area where Americans were working. Destroying the mosquito breeding areas eliminated the organisms that transmitted malaria as well as yellow fever. This action enabled the Americans to finish the canal.

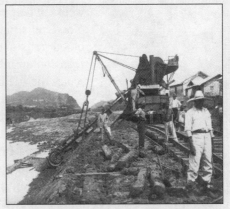

Construction of the Panama Canal was halted until outbreaks of yellow fever and malaria were controlled. Although malaria is caused by a species of Plasmodium *and yellow fever by a virus, both are spread by mosquitoes.*

Plasmodium, the genus that causes malaria, is probably the best-known representative of this phylum. The female *Anopheles* mosquito is the vector that spreads the disease from one person to another. The female *Anopheles* mosquito becomes infected by ingesting the blood of a human who has malaria. The *Plasmodium* cells mature and develop in the mosquito and migrate to the insect's salivary gland. Before feeding on the blood of another human, the infected mosquito injects saliva into the puncture wound of its new victim and thereby puts the parasite back into the human bloodstream.

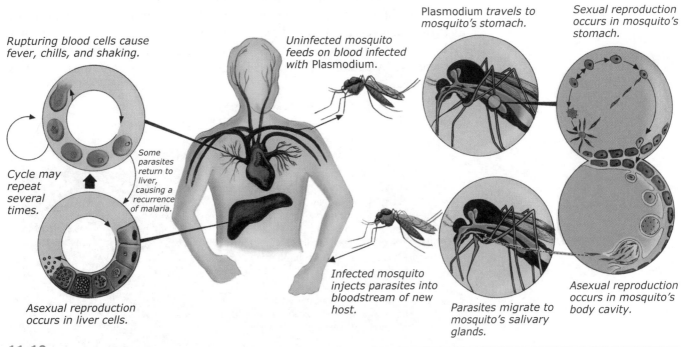

Rupturing blood cells cause fever, chills, and shaking.

Uninfected mosquito feeds on blood infected with Plasmodium.

Plasmodium *travels to mosquito's stomach.*

Sexual reproduction occurs in mosquito's stomach.

Cycle may repeat several times.

Some parasites return to liver, causing a recurrence of malaria.

Asexual reproduction occurs in liver cells.

Infected mosquito injects parasites into bloodstream of new host.

Parasites migrate to mosquito's salivary glands.

Asexual reproduction occurs in mosquito's body cavity.

11-10
Life cycle of a *Plasmodium*

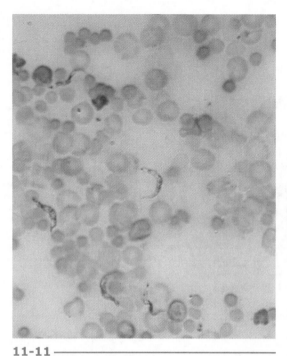

11-11

PMG of the trypanosome (purple crescent-shaped organism) that causes African sleeping sickness

Once inside the human, the spindle-shaped *Plasmodium* protozoans penetrate liver cells, where they grow and reproduce in about two weeks. The parasite then enters red blood cells and reproduces. The blood cells burst, releasing *Plasmodium* cells that may invade other blood cells and repeat this reproductive process. When a mosquito feeds on an infected individual, it ingests the parasite and thus completes the life cycle.

The periodic rupturing of red blood cells releases cell contents and *Plasmodium* cells that are associated with violent episodes of chills and fever due to the person's immune response. The person also suffers anemia (inability of the blood to carry oxygen) and an enlarged spleen. The most successful method of controlling the disease has been to limit human exposure to *Anopheles* mosquitoes. Today a series of drugs can be used to prevent and cure the disease.

Phylum Zoomastigina: The Zooflagellates

The phylum Zoomastigina (ZOH uh MAS tuh JIE nuh) consists of the **zooflagellates**, protozoans that propel themselves by means of one or more flagella. Most of these organisms are unicellular and live in ponds and lakes, where they feed on smaller organisms by phagocytosis or absorb nutrients from decaying organic matter directly through their cell membranes. Other zooflagellates live inside other organisms and absorb nutrients from them.

Most of these organisms reproduce asexually by binary fission, but some do have a sexual life cycle also. Several human diseases that are caused by zooflagellates involve an insect vector. One such disease, African sleeping sickness, or *trypanosomiasis* (tri PAN uh soh MIE uh sis), is caused by members of the genus *Trypanosoma*.

African sleeping sickness is passed from person to person by the blood-sucking tsetse fly, which is found only in Africa. The *Trypanosoma* grows and divides in the insect's intestine and then migrates to the salivary glands. When the fly bites a person, the protozoan may enter the victim's bloodstream, where it reproduces.

After the protozoan has been in the blood, it may invade the central nervous system and cause inflammation of the brain, resulting in weakness, mental lethargy, and sleepiness that can lead to a coma and possibly death. Controlling the tsetse fly controls the disease. Drugs can now be given to those suffering from the disease.

Another Zooflagellate

Zooflagellates of the genus *Trichonympha* actually keep many termite species alive. Although termites ingest wood, they are unable to digest the cellulose in it. These zooflagellates live in the gut of the termite and produce enzymes that digest the cellulose for themselves and for the termite host. The protozoan has a safe place to live, and the termite can "eat" wood; therefore, both thrive. If they are separated, both will perish. This close relationship presents a real challenge for evolutionists. How did the termite or the zooflagellate survive without the presence of the other?

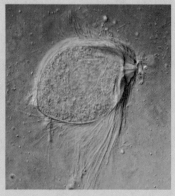

⮞ **Review Questions 11.2**

1. Describe the typical movement of a sarcodine.
2. How do amoebas obtain food?
3. What is a taxis, and what taxes do amoebas have?
4. Describe the reproduction of an amoeba.
5. Describe the typical movement of ciliates.
6. How do paramecia obtain food?
7. What taxes do paramecia have?
8. Describe both sexual and asexual reproduction of paramecia.
9. Why is Sporozoa a significant protozoan phylum?

10. Describe the *Plasmodium* life cycle.
11. List five protozoans (other than the amoeba, the paramecium, and the plasmodium) and tell their significance.
12. When visiting certain countries, people are often warned not to drink the water in smaller towns. Why is this good advice?
13. Why do people in the United States not have to worry about catching African sleeping sickness from a person with the disease who visits in their home?
⊙ 14. Why is it inaccurate to call a protozoan a simple organism?

11B—The Algae

Algae have been called the "grass of many waters." In fact, every natural water supply contains some type of alga. Many of these organisms are microscopic and either unicellular or colonial. The larger multicellular organisms are commonly called **seaweed**. Some algae may appear to be floating mats of green slime or velvety coatings on submerged rocks, while other species make underwater forests. Their presence may color the water various shades of red, brown, blue, or green. Algae can be found growing on soil, rocks, and trees and inside small animals.

Since algae are generally photosynthetic, at one time taxonomists placed them in kingdom Plantae. When the presence of true tissues became a distinguishing characteristic of plants, the seaweed and all other algae were moved out of that kingdom. Today most taxonomists recognize seven algal phyla in the kingdom Protista.

11.3 General Characteristics

To most people, algae are a nuisance, either clouding their swimming pool water or making a slimy green mat that floats on their favorite fishing pond. But algae are the source of daily sustenance for fish and thus indirectly for most other aquatic organisms as well.

Algae and the Environment

Algae are the photosynthetic organisms that capture the sun's energy in aquatic environments. In almost all bodies of water, there are thousands of tiny floating organisms called **plankton**. There are two types of plankton:

- **zooplankton** (ZOH uh PLANGK tun), tiny floating animals or protozoans
- **phytoplankton** (FYE toh PLANGK tun), tiny floating photosynthetic organisms, predominantly algae

Phytoplankton manufacture their own food, and in turn, they are the food that zooplankton and many larger organisms consume. Thus, they are the *primary food-producing organisms* in aquatic environments. Many fish have a specialized feeding apparatus that permits them to strain plankton from the water. Even the largest known living animal, the great blue whale, weighing over 107 metric tons, exists by straining the ocean's "plankton soup."

Although algae are an important food source, they benefit other organisms in another way as well. The algae in the vast expanses of the open ocean

11.3

⮞ **Objectives**

- Describe some of algae's ecological and economic significance
- Describe the basic algal body structures
- Summarize the methods of reproduction used by algae

⮞ **Key Terms**

algae	sessile
seaweed	zoospore
plankton	zygospore
zooplankton	oogonium
phytoplankton	antheridium
thallus	fragmentation

algae: (L. *alga*—seaweed)

plankton: (Gk. *planktos*—wandering)

zooplankton: zoo- (Gk. *zoion*—animal or living being) + -plankton (wandering)

phytoplankton: phyto- (plant) + -plankton (wandering)

11-12
Algae uses—wrapping sushi and thickening ice cream

perform about 70% of the oxygen-producing photosynthesis that takes place on this planet. Thus, the oxygen–carbon dioxide cycle appears to depend more on algae than on the rooted green plants to which oxygen production is normally ascribed.

Uses for Algae

In man's attempt to subdue the earth and make it useful, humans have found various uses for algae. In some areas of the world, sea algae are used as natural fertilizer. In recent times algae have been processed to obtain compounds used to make fertilizers, salt, and other products. Algae are an inexpensive source of certain vitamins and minerals in the feed of domestic livestock.

In various areas of the world, people eat algae. The Japanese cultivate algae as a human food crop. In New England a marine alga called sea kale is eaten as a vegetable. Many algal products are used in food processing. Irish moss, an alga that grows along the Atlantic coast, supplies a substance used as a thickener in puddings, jellies, and ice cream.

Some people have begun to look to the sea as a future major source of food for humans. The earth has a limited amount of land suitable for farming, and there is a limit to how much food man can produce in any given location. *Aquaculture* (the farming of ponds, lakes, and the sea) can produce considerably more organic materials than conventional farming of the same land area. Algae are an important part of aquaculture as food for fish and for humans. Although several algal food substances are edible and wholesome, many people consider them undesirable.

Structure

The basic unit of an alga is called the **thallus** (pl., thalli). The form of the thallus varies from species to species. In some the thallus is a single cell; in others it is a multicellular organism that has structures that look like leaves, stems, or even roots. However, algal cells do not form true tissues or organs. Under a microscope a large algal thallus appears as many intertwined filaments. Each of the algal cells composing such an organism can live independently in the environment where the thallus grows and (at least theoretically) can become a completely new thallus.

Many algae are free-living, unicellular organisms. Others form colonies of globular masses held together by a common slime coat. Some algae, however, form highly complex colonies that look like true plants. One of the simpler colonies is a filament—a slender, chainlike thread of cells. Some algae form branched filaments; others form broad plates or sheets of cells. Those algal colonies that are not free-floating form special cells called *holdfasts* that anchor them to submerged objects. In larger algal colonies the holdfasts may be a group of cells. Algae that grow attached to something are called **sessile** algae.

Other algae form air bladders—small air-filled spaces—that cause the thallus to float. Often a large algal thallus will have a holdfast on one end and air bladders on the other end, causing it to appear to stand upright in the water.

Reproduction

Many species of algae can reproduce both asexually and sexually, while others reproduce only asexually. In the discussion that follows, the text presents examples of asexual and sexual reproduction in a unicellular and a multicellular algal form. It should be remembered that these are general discussions and that there may be variations between different species.

thallus: (Gk. *thallein*—to sprout)

Reproduction in a Unicellular Alga

The *Chlamydomonas* (kluh MID uh MOH nas) is an example of a unicellular alga that can reproduce both asexually and sexually. The adult, or mature form, of this species exists as a haploid organism. During asexual reproduction, the flagellum is absorbed and the cell undergoes up to three mitotic divisions. Within the parent cell, the newly formed daughter cells develop into flagellated **zoospores** (ZOH uh SPORZ). A zoospore is any spore capable of moving on its own. The zoospores break out of the parent cell and grow to full size.

Sexual reproduction in *Chlamydomonas* is often triggered by adverse environmental conditions. It begins with the parent cell undergoing several mitotic divisions that produce gametes that are designated as either "plus" or "minus" gametes. There are special differences in the chemical makeup that determine whether a gamete is plus or minus. The gametes break out of the parent cell, and when a plus and a minus gamete come into contact, they fuse to form a *diploid* zygote. The diploid zygote forms a thick, protective covering and then enters a resting stage—the **zygospore**. While in the zygospore stage, the diploid cells undergo meiosis, and when they emerge, the *Chlamydomonas* is once again haploid.

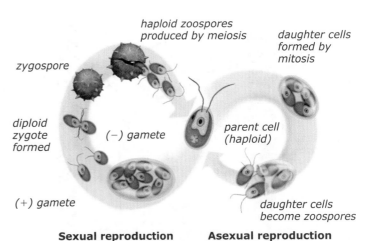

Sexual reproduction   **Asexual reproduction**

haploid zoospores produced by meiosis

daughter cells formed by mitosis

zygospore

diploid zygote formed

(−) gamete

parent cell (haploid)

(+) gamete

daughter cells become zoospores

11-13 —
Chlamydomonas life cycle

Reproduction in a Multicellular Alga

There are three basic forms of sexual reproduction carried on by algae—conjugation, isogamete formation, and heterogamete formation.

Under proper conditions, *Spirogyra* filaments line up parallel to each other and a conjugation tube forms between two cells. One cell has a plus gamete. The plus gamete migrates through the conjugation tube and unites with the minus gamete in the other cell, forming a zygospore. While in this stage, meiosis will occur to produce zoospores. The zoospore will later produce a new *Spirogyra* filament.

Specialized gamete cells that are not obviously different from one another are called *isogametes*. *Ulothrix* (YOO luh thrix), a freshwater alga, reproduces using isogametes. Each gamete has two flagella, and when released, they swim, unite, and form a diploid zygote that forms a zygospore. The diploid cells in the zygospore undergo meiosis, forming zoospores. The zygospore wall ruptures, and the zoospores begin to form a new *Ulothrix* filament.

Gametes that differ in size, structure, or both are *heterogametes*. In some algae, such as *Oedogonium* (ED uh GOH nee um), a nonmotile gamete called an ovum (egg) is produced in a specialized cell called the **oogonium** (OH uh GOH nee um). Sperm—the motile gametes—are produced in a different cell called the **antheridium** (AN thuh RIHD ee um). A sperm unites with an ovum to form a zygote. The zygote forms a zygospore that may remain inactive. The zygote undergoes meiosis to form zoospores, which break out of the protective covering and begin to form a new *Oedogonium* filament.

Many of the species listed as reproducing sexually may also reproduce asexually. These sessile algae may form zoospores rather than gametes. Once released the zoospores use their flagella to swim away and establish new colonies. The illustrations of *Ulothrix* and *Oedogonium* show their reproductive cycles.

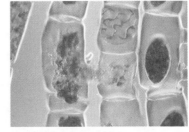

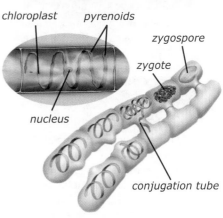

chloroplast   pyrenoids

zygospore

zygote

nucleus

conjugation tube

11-14 —
PMG and illustration of *Spirogyra* undergoing conjugation

zoospore: zoo- (animal) + -spore (seed)
zygospore: zygo- (yoke) + -spore (seed)
oogonium: oo- (egg) + -gonium (Gk. *gonos*—seed)

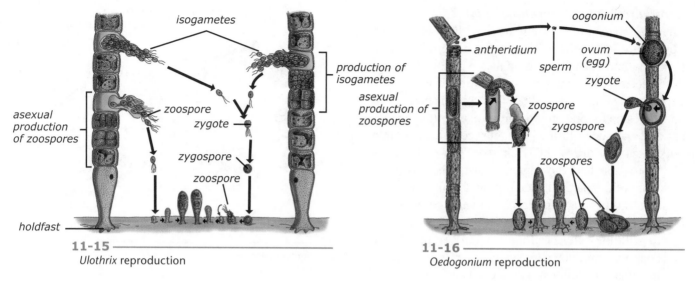

11-15 ────────────────────────────
Ulothrix reproduction

11-16 ────────────────────────────
Oedogonium reproduction

Algal colonies can also form new colonies asexually by **fragmentation**. If the colony is broken by a physical disturbance (currents or fish), each fragment can grow into a complete colony.

Review Questions 11.3

1. What characteristics separate algae from (a) protozoans and (b) plants?
2. Describe several types of algal colonies.
3. List two types of asexual and four types of sexual reproduction common in algae.
4. What is the difference between a zygospore and a zoospore?
5. Compare and contrast sexual reproduction of *Ulothrix*, *Oedogonium*, and *Spirogyra*.

11.4 Algal Classification

The classification of algae is open for debate. Because of the great differences—and similarities—in various kinds of algae, there is some disagreement among taxonomists. As discussed in Chapter 10, the prokaryotic blue-green algae—once considered members of the kingdom Protista—have been moved to the kingdom Eubacteria as the group cyanobacteria. For a time algae were placed in the kingdom Plantae; however, algae lack tissue differentiation and have no true leaves, stems, or roots.

Today most taxonomists agree that algae should remain members of the kingdom Protista and can be classified into seven phyla based on their thallus form, type of chlorophyll they contain, cell wall composition, and food storage. The remainder of this section lists the seven algal phyla and highlights some of the major characteristics of their members.

Phylum Euglenophyta: The Euglenoids

Members of the phylum Euglenophyta (YOO gluh NAWF uh tuh) present scientists with some of the greatest taxonomic difficulties. Why? They have structures and characteristics that could just as easily classify them as protozoans; in fact, for many years euglenoids were considered protozoans.

The phylum takes its name from the genus *Euglena*, probably its best-known member. Each of these spindle-shaped organisms has two anterior flagella for locomotion—one quite long and the other very short. The euglena also has chlorophylls *a* and *b*, as well as carotenoids (kuh RAHT uhn OYDZ) (yellow orange pigments), in its chloroplasts. Under ideal conditions the euglena survives primarily on the products of photosynthesis. None of these

11.4

Objectives

- List the characteristics used to classify algae
- Identify the seven algal phyla and their major characteristics
- Explain why *Euglena* might be considered both an alga and a protozoan
- Describe an algal bloom and its potential ecological and economic significance

Key Terms

euglenoid
movement
pyrenoid

diatom
bloom condition
dinoflagellate

carotenoid: caroten- (L. *carota*—carrot) + -oid (form or shape)

11-2 Summary of Algal Phyla

Phylum	Thallus form	Pigment types	Food storage	Cell wall composition
Euglenophyta (euglenoids)	unicellular	chlorophylls *a* and *b*, carotenoids	starch	no cell wall; protein-rich pellicle
Chlorophyta (green algae)	unicellular, colonial, multicellular	chlorophylls *a* and *b*, carotenoids	starch	cellulose
Chrysophyta (golden algae)	mostly unicellular, some colonial	chlorophylls *a* and *b*, carotenoids	oils	cellulose
Bacillariophyta (diatoms)	mostly unicellular, some colonial	chlorophylls *a* and *c*, carotenoids, xanthophylls	oils	silicon
Phaeophyta (brown algae)	multicellular	chlorophylls *a* and *c*, fucoxanthin	laminarin (a complex carbohydrate)	cellulose, algin
Rhodophyta (red algae)	multicellular	phycobilins, carotenoids	starch	calcium carbonate, cellulose, pectin
Dinoflagellata (dinoflagellates)	unicellular	chlorophylls *a* and *c*, carotenoids	starch	cellulose

characteristics present classification problems. However, the euglena is also saprophytic, absorbing dissolved food from the surrounding environment. In low light or darkness, its photosynthetic apparatus shuts down and may even degenerate. The euglena can sustain itself on dissolved nutrients.

The body of the euglena is covered with a shape-sustaining pellicle. At its anterior end, a small gullet enlarges into a reservoir. Though the reservoir looks much like a mouth, euglenas probably do not ingest food through the gullet and reservoir. Near the reservoir is a tiny red light-sensitive eyespot. The euglena has a single nucleus containing a large nucleolus. Like most protozoans, the euglena has a contractile vacuole that maintains the organism's water balance by expelling excess water from the cytoplasm into the reservoir.

Euglenas usually move by whirling their anterior flagella to pull themselves through the water. They can also move by a modified amoeboid movement. The euglena draws its cytoplasm in, making itself almost completely round, and then extends itself forward. Employing this **euglenoid movement**, euglenas can propel themselves with a wormlike motion.

The euglena reproduces by binary fission along the length of its body. Under ideal conditions the euglena will reproduce about once a day. Sexual reproduction in the euglena is unknown. For the present, taxonomists have decided that the euglena is more plantlike and should be considered an alga rather than a protozoan.

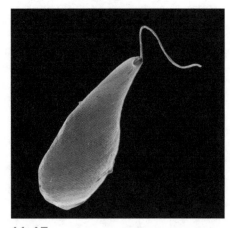

11-17
SEM of a euglena

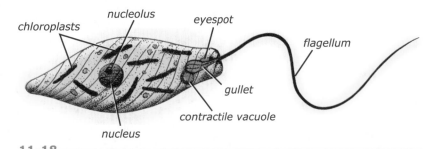

11-18
Typical structures of a euglena

Phylum Chlorophyta: The Green Algae

Phylum Chlorophyta (kloh RAWF uh tuh), the *green algae*, is one of the largest algal phyla; it contains over seven thousand species. Most green algae are freshwater organisms; however, many terrestrial and a few marine species exist. Usually, green algae are unicellular or form simple colonies, but some marine species form extensive thalli. One

11-19
Euglenoid movement

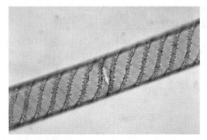

11-20

The name *Spirogyra* comes from the spiral-shaped chloroplast in this algae.

green algal species that grows off the coast of Mexico produces leaflike blades that reach 25 cm by 7.5 m (10 in. by 24 ft). Many green algae produce holdfasts and grow attached to submerged substances, but some unicellular species move by flagella. Many kinds merely float.

Green algae contain chlorophylls *a* and *b* and carotenoids; therefore, they often appear in various shades of yellow green. The chloroplasts in green algae are often unusual. Many algal species have cells containing only one large chloroplast with an unusual, but specific, shape. For example, the chloroplasts of the genus *Spirogyra* are spiral shaped. The cell walls of these organisms contain cellulose.

Green algae typically store their food as starch. Large starch granules can often be found around a protein-containing organelle, the **pyrenoid** (pye REE NOYD). Pyrenoids contain the enzymes necessary for the manufacture of starch from the simple sugars produced by photosynthesis. These organelles are also found in some members of other phyla.

Other Members of Chlorophyta

Unicellular Green Algae

✦ *Chlamydomonas.* Members of this genus are common inhabitants of stagnant freshwater pools. They possess two flagella and can reproduce both sexually and asexually (see p. 265).

✦ *Desmids.* Desmids are free-floating algae. Usually they are unicellular, but occasionally cells are joined end to end, forming a filament-like colony. Desmids often appear pinched in the middle, forming two symmetrical halves. The cell walls often have unusual patterns, making desmids some of the most interesting freshwater algae to observe.

✦ *Protococcus.* This genus contains terrestrial members, many of which grow as a green film on damp rocks or tree bark. *Protococcus* exists unicellularly but may form clumps of cells.

Filament-Forming Green Algae

✦ *Ulothrix. Ulothrix* filaments are usually attached to submerged structures by a holdfast but can be found free-floating.

✦ *Oedogonium.* This genus of unbranched filamentous algae grows in fresh water. It reproduces both sexually and asexually.

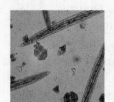

✦ *Spirogyra.* Named after its unusual spiral-shaped chloroplasts, *Spirogyra* is typical of many filamentous green algae. It, along with other similar species, is usually found as floating green mats in ponds or the backwaters of streams.

Desmids   *Terrestrial* Protococcus

Phylum Chrysophyta: The Golden Algae

Many algae of the phylum Chrysophyta (kruh SAWF uh tuh) are similar to many green algae. There are approximately 850 species in this phylum. The golden algae contain large amounts of carotenoids that give them their golden color, as well as chlorophylls *a* and *b*. Unlike Chlorophyta, members of phylum Chrysophyta store their food as oils. All are photosynthetic and most live in fresh water, although some species live in a marine environment.

Some species are colonial, although most are unicellular. Members of the genus *Dinobryon* exist as single flagellated cells, but they can also join to form a colony. *Vaucheria* is a common genus characterized by long filaments. Unlike the other algal filaments you have studied, the filaments lack interior divisions into separate cells. As a result, they are *multinucleate*.

Phylum Bacillariophyta: The Diatoms

Diatoms are a group of organisms in the phylum Bacillariophyta (BAS uh LEHR ee AWF uh tuh), which contains 11 500 species. Diatoms, in some respects the most important group of algae, are found in abundance in almost every environment on earth. Among the plankton of the open ocean, diatoms are important in the food chain and are responsible for more oxygen-supplying photosynthesis than any other group of organisms. Usually diatoms are unicellular, but some species exist in chains or other groupings. Diatoms store their food as oil that often appears as large drops in the cell.

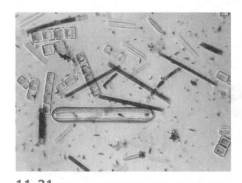

11-21

Diatoms are abundant in both marine and freshwater environments.

The cell walls of diatoms are in two separate halves, one of which fits inside the other. These walls contain silica, an exceptionally hard substance and one of the primary components of glass. These silica walls remain long after the diatom has died.

Large quantities of diatom shells can be found on the ocean floor. *Diatomaceous earth*, a crumbly substance made of diatom shells, is found in many parts of the world. In California some beds of this substance are 900 m thick. Diatomaceous earth is probably the result of diatoms that were deposited during the Flood. It often serves as a filter in industry and in home aquariums. Since it is a poor conductor of heat, it is used to insulate boilers. The shells of diatoms are often the abrasive agent in silver polish and toothpaste.

Although highly protective, the hard cell wall of diatoms presents some problems when the cells are reproducing. Since half of the shell goes with each new daughter cell, some cells become increasingly smaller as cell divisions continue. After several divisions some diatoms shed their cell walls, grow, and then produce new shells. Some diatoms reproduce sexually following several asexual cell divisions.

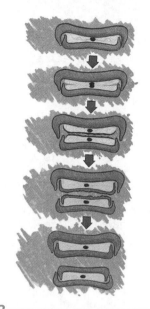

11-22
Asexual reproduction in diatoms

Phylum Phaeophyta: The Brown Algae

Species of the phylum Phaeophyta (fee AWF uh tuh) are all multicellular and almost all marine. The brown algae, found in colder waters, usually grow attached to rocks or to the ocean floor. They contain chlorophylls *a* and *c* and a large amount of the pigment fucoxanthin (FYOO koh ZAN thin). It is the fucoxanthin that gives the phylum its brown color. Larger species live

Other Members of Phaeophyta

✦ *Fucus. Fucus* (often called rockweed) is a common shoreline alga. The thick 30–90 cm (1–3 ft) leatherlike thallus is not affected by periods of dryness between tides. The swellings at the ends of the thallus are called receptacles, and they contain the multicellular reproductive structures.

✦ *Kelp.* The kelps normally grow attached to rocks in water up to 23 m (75 ft) deep. Some of these large algae may reach 30 m (98 ft) long.

✦ *Sargassum.* This genus is similar to kelp but grows in temperate to tropical areas. One species forms floating mats several hundred meters wide in an

area of the North Atlantic called the Sargasso Sea.

DANGER! RED TIDE! AREA CLOSED!

You and your family have just arrived at your favorite vacation spot to enjoy a fun day at the beach when you gasp—the water is red! Intimidating skull-and-crossbones signs posted on the beach inform you that you will have to change your plans. But what IS that stuff in the water?

Historical records inform us that you are far from the first person to be startled by this phenomenon. In 1729, Captain W. Dampier, sailing off the coast of Peru, wrote the following:

"[When] our ship [was] about ten leagues off shore, . . . we were suddenly surprised with the change of the colour of the water, which looked as red as blood to as great a distance as we could see. . . . At first we were mighty surprised. . . . We then drew some water up in buckets and poured some in a glass. It continued to look very red, till about a quarter of an hour after it had been in the glass; when all of the red substance floated to the top,

and the water underneath was as clear as usual. The red stuff which floated on top was of a slimy substance, with little knobs, and we all concluded it could be nothing but the spawn of fish."†

It is impossible to know exactly what caused the red sea condition described by Captain Dampier. The separating of the water and the "red stuff" and the slimy feel of the substance indicate that he and his crew were probably sailing through an algal bloom that today is most often called a "red tide."

First officially recorded in Florida in 1844, a red tide is an abnormally high concentration of microscopic algae that occurs naturally in most coastal areas.

Most algae, or phytoplankton, are beneficial and serve as food for the smallest of aquatic life. However, under the right conditions—warm water and the presence of oxygen, sufficient nutrients, and increased sunlight—the organisms multiply rapidly, resulting in a **bloom condition**. In many cases, this is part of a natural population cycle found in both marine and freshwater environments that causes little or no harm to other organisms. When enough algal organisms accumulate, the water may appear discolored—red, brown, blue-green, purple, or even orange! Ill effects of an algal bloom condition can occur when the algae die; decomposition and oxygen depletion then pollute the water.

Since red tides are not dependent on the tidal movement of water and may not even be red, toxic blooms,

colorful or even invisible, are more broadly and accurately referred to as *harmful algal blooms (HABs)*.

In the Gulf of Mexico, along the North Carolina coast, and on the east coast of Florida, most HABs are caused by an overgrowth of the dinoflagellate algal species *Karenia brevis*.

Swimmers exposed to sea spray carrying the toxins released into the water by *K. brevis* may develop coughing, sneezing, irritated skin, or watery eyes. Eating shellfish polluted with *K. brevis* can be deadly since cooking does not kill the toxin. Blooms can last for months, causing the tourism and fishing industries to lose millions of dollars because of ugly beaches, massive fish kills, and tainted shellfish.

Authorities in coastal areas of the United States and other industrialized nations keep track of algal blooms and make HAB information available to tourists and fishermen. Laboratory tests are the only way to determine if the sea and seafood are dangerous due to toxic algae, so we must be sure to heed "no swimming or fishing" warnings posted at beaches. Scientists are very interested in studying HABs to learn how to predict and prevent them to protect people, sea life, and the environment.

Several other species of algae also pose a serious danger to humans, causing neurological, respiratory, or gastrointestinal problems; paralysis; and even death. Some colorful algal blooms make swimming or fishing unappealing but are not harmful.

† *The Distribution of Discolored Water*, U.S. Navy Hydrographic Office Pilot Chart No. 1401 (Washington, DC: Hydrographic Office, January 1955).

The First Plague: Algal Bloom or Miracle of God?

Some critics of the Bible have suggested that the first plague of Egypt described in Exodus 7 was a natural algal bloom. Verses 20–21 state that "all the waters that were in the river were turned to blood. And the fish that was in the river died; and the river stank, and the Egyptians could not drink of the water of the river."

At first glance it may appear that an algal bloom could have caused this phenomenon. Other sections of this passage, however, demonstrate that this miracle was not a bloom of microorganisms. The miracle happened "in the sight of Pharaoh, and in the sight of his servants" (v. 20). An algal bloom does not happen instantly; it takes days to build up. Blood was also in all the pools, ponds, and Egyptian vessels (v. 19). Conditions for a bloom in the flowing Nile River would not have caused a simultaneous bloom in the standing waters.

Christians must be careful not to attribute miracles to "natural" causes. When the Bible states that God has done a particu-

lar thing, His statement must be accepted. If the water were turned the color of blood, would not God have stated this? The Egyptians, as well as Moses, the human writer of this passage, would have known the difference between blood and algal-polluted, colored water. The repetition of the word "blood" in this passage, as well as in other passages referring to the incident (Pss. 78:44; 105:29), is further evidence that it was blood and not the bloom of some microorganism.

God can cause certain "natural" circumstances to work together for His purpose, and He often does so even today. God can also, as recorded in Scripture, act supernaturally to effect His purpose. These supernatural acts are miracles. To try to explain God's miracles as merely natural circumstances is to deny the power of God as well as the clear meaning of His Word.

in deeper waters, while smaller species thrive along the coast and are often exposed during low tides. Food is stored as a complex carbohydrate called laminarin. *Algin*, a gelatinous coating found on many brown algae, serves as a thickener in commercially produced ice cream and other foods.

The thallus of a brown alga is usually composed of a holdfast (often an extensive network of cells), a stemlike structure, and leaflike blades. The stemlike structure usually contains air bladders, causing the alga to float near the surface. Brown algae produce sperm and ova in multicellular antheridia and oogonia.

Phylum Rhodophyta: The Red Algae

The algae of phylum Rhodophyta (roh DAHF uh tuh) are almost all marine, multicellular, and red. Their red color comes from pigments called *phycobilins* (FYE koh BYE linz). Most red algae grow in shallow, warmer waters, but sometimes they are found at a depth of over 150 m (490 ft). Some red algae produce ribbonlike thalli, but many are feathery, and others are slender filaments. Almost all red algae are less than 30 cm (1 ft) long. Their reproduction is sexual because they produce nonmotile gametes and spores.

Some members of this phylum absorb calcium from seawater and deposit it as calcium carbonate in the gelatinous coat that surrounds them. This calcium carbonate is important in building some reefs and islands. Other red algae products, such as carrageenan (KAR uh GEE nun), are used in foods, cosmetics, and the gelatin shells of drug capsules. Probably the best known of these is *agar*, a gelatinous substance used in growing bacteria.

11-23
Red coralline algae

Other Members of Rhodophyta

✦ *Chondrus*. Chondrus (Irish moss) is the marine red alga that is the source of agar.
✦ *Corallina*. This genus contains species that grow in short clumps sometimes called coral moss.

Chondrus crispus

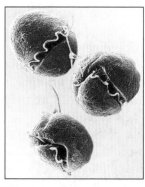

11-24
Karenia brevis (left) is a dinoflagellate that caused the red tide that shows up as dark areas just north of the Florida Keys in this satellite image (right).

Phylum Dinoflagellata: The Dinoflagellates

The phylum Dinoflagellata includes the **dinoflagellates** (DYE noh FLAJ uh lits). These unicellular algae normally possess two flagella of unequal length. One flagellum is normally kept in a groove around the algal cell; the other is used for movement.

The cell walls of the dinoflagellates are made of multi-sided cellulose plates. Yellowish brown pigments color these algae. Most dinoflagellates are marine, but some species live in fresh water. Some dinoflagellates are bioluminescent (light producing) and, when the water is disturbed, can emit enough light to cause the sea to glow.

11-25
Some algae luminesce when disturbed.

Dinoflagellata: Dino- (Gk. *dinein*—to whirl) + -flagellata (small whip)

Review Questions 11.4

1. List the characteristics used to determine algal classification.
2. List the seven algal phyla. For each phylum, list the major characteristics and give an example organism.
3. What has made classifying euglenoids difficult for taxonomists?
4. Describe diatoms and list several uses of diatoms.
⊙5. List several characteristics of an algal bloom that help make it an unacceptable explanation for the changing of the water into blood in Exodus 7.
⊙6. List several ways in which algae are (a) harmful and (b) beneficial. (You may need to refer to Subsection 11.3 to answer this question completely.)
⊙7. "Resolved: Algae are the most significant organisms in the world." Prepare a paragraph for the negative position and a paragraph for the affirmative position. Which do you think is the better position?

11.5–6

Objectives

- Summarize the differences between the funguslike protists and fungi
- Describe the differences between water molds and slime molds
- Explain the life cycles of the two types of slime molds

Key Terms

water mold
sporangium
slime mold
fruiting body
cellular slime mold
plasmodial slime mold

11-26
Dead insect colonized by water mold (left) and potato plant infested by *Phytophthora infestans*

11C—The Funguslike Protists

The funguslike protists were once considered to be members of the kingdom Fungi because of their similarity in appearance and habitat. Like fungi, the funguslike protists live in moist environments. Some are aquatic. They are heterotrophic, have minimal tissue specialization, and are important decomposers. However, as researchers continued to study these organisms, they found differences significant enough to justify moving them from the kingdom Fungi to kingdom Protista. The features that distinguish them from fungi include the following: their cell walls contain cellulose rather than chitin, and they are motile. There are two basic groups within the funguslike protists—the water molds and the slime molds.

11.5 The Water Molds

Anyone who has an aquarium has probably seen samples of water molds. **Water molds** are often seen on dead fish or other floating dead organic matter, and they appear as a branching filamentous growth of cells. Most are aquatic, although some do live in the soil.

The aquatic water molds live on dead or decaying organic material, while those that live on land parasitize plants. They reproduce both sexually and asexually. During asexual reproduction, special cells called **sporangia** (sing., sporangium) produce flagellated zoospores that swim away in search of food. Once food is located, the zoospores develop into new organisms.

During sexual reproduction, the water mold has cells that develop sperm and ova. Fertilization tubes grow between these cells, allowing the haploid gametes to form diploid zygotes. The zygotes develop into new organisms.

Members of this phylum are the cause of many kinds of *blight*, a disease of plants that causes rapid destruction of the leaves and stems, resulting in the death of the plant. Blights are easily spread and can wipe out crops very quickly.

Other Water Molds

Although it may not seem that water molds could be very important, one water mold had a huge economic and historical impact on two countries—Ireland and America. The water mold *Phytophthora infestans* causes a disease in potatoes called blight. Blight causes the plants to quickly rot, completely destroying the crop. From 1845 to 1847, there was a famine in Ireland as a result of the potato blight. Over a million people starved. Thousands of Irish men, women, and children immigrated to the United States to escape the hardships of their homeland and start a new life. Ireland's population dropped from eight million to four million during that time. Many of these Irishmen became laborers in America and were important in building the infrastructure of canals and railways of this country. Thus, one water mold played a major role in the economies of two nations.

Another water mold, *Plasmopara viticola*, infects many different kinds of fruits and vegetables and is sometimes called downy mildew. In the mid-1870s, downy mildew was inadvertently introduced into French vineyards from American grapevines when growers from both countries tried to produce a hybrid of the French and American varieties that would be more resistant to aphids. The French vineyards were almost completely destroyed before it was discovered that a mixture of copper sulfate and lime would kill the *Plasmopara*.

Thousands of Irish were left destitute when the potato crops failed. Cut surface of potato spoiled by the blight (right)

11.6 The Slime Molds

Slime molds live in cool, moist environments that are rich in organic matter. They play a large role as decomposer organisms. Slime molds are often seen in the woods or in a compost pile. The slime molds are characterized by their unusual two-phased life cycle. In the feeding phase, the cells look like amoebas, moving about using pseudopodia and eating bacteria and organic matter. During this phase, they appear as slimy masses that can be white, yellow, or red. The second phase is primarily a reproductive phase in which funguslike structures called **fruiting bodies** are formed. These fruiting bodies produce spores.

One phylum contains organisms called **cellular slime molds**. When food is plentiful, they live as amoeba-like unicellular organisms that crawl around moist areas decomposing organic matter. If food becomes scarce, they gather together into a single colony called a *pseudoplasmodium* (pl., pseudoplasmodia) that looks like a slimy slug. Although the pseudoplasmodium appears and moves as a single organism, the cells retain their cell membranes and individuality. Eventually, the pseudoplasmodium forms a fruiting body. When the fruiting body bursts open, the wind carries the spores to new locations. Once a spore has landed, it develops into a new amoeba-like cell and the life cycle repeats itself.

Acellular or **plasmodial slime molds** fall within a separate phylum. Like the cellular slime molds, plasmodial slime molds start their life as amoeba-looking cells. During the feeding phase, members of this phylum form an amorphous

11-27
Yellow "scrambled egg" slime mold (top) and "raspberry" slime mold (bottom)

11-28 ────────────────────────────

Physarum life cycle: plasmodium (left) and fruiting bodies (right)

structure called a *plasmodium* (pl., plasmodia) that may cover several square meters. What is the difference between a pseudoplasmodium and a plasmodium? A plasmodium is *multinucleate*—all the cell walls have dissolved and the thousands of nuclei are floating around in a mass of cytoplasm. The plasmodium creeps along the ground like a giant blob as it phagocytizes organic material.

When food and water become scarce, the plasmodium begins to form fruiting bodies that form spores. The spores are released and carried by the wind. Once environmental conditions become more favorable, the spores open and haploid reproductive cells emerge. These cells fuse and undergo mitosis without cytokinesis to form a new plasmodium.

Review Questions 11.5–11.6

1. What characteristics of funguslike protists caused taxonomists to move them out of the fungi kingdom?

2. How are water molds different from slime molds?

3. What are fruiting bodies? At what point do they appear in the life cycle of slime molds?

4. How do pseudoplasmodia differ from plasmodia?

12

The Kingdom Fungi

The Kingdom Fungi

Have you ever started to make yourself a sandwich, only to lose your appetite when you find that the bread has black spots all over it? Or have you ever wondered what the black film coating your shower walls is? Have you ever contemplated why bread rises? Do you enjoy mushrooms on your pizza? Have you ever noticed orange blotches on grass blades in your lawn? If you answered yes to any of these questions, then you have some experience with fungi.

The study of fungi is called **mycology**. Like the protists discussed in the previous chapter and the organisms in the kingdoms that are presented in later chapters, fungi are made of eukaryotic cells. Although the text does show certain shared characteristics, most of the material explores the differences that set the fungi apart from the other five kingdoms.

12A—Fungi and Man

Ergot of rye is a fungus that causes a purplish black swelling in rye grain. In the Middle Ages, the disease caused by eating this infected rye was termed

12A

Objectives
- Describe some beneficial and some destructive fungi
- Describe the ecological role that fungi perform

Key Terms
mycology
mycorrhiza

12-1

Head of rye infected with *Claviceps purpurea*, the causative agent of ergot of rye. It can infect rye in the field and later grow if the rye is stored in a cool, damp area.

(a)

(b)

(c)

12-2

Elm tree infected with Dutch elm disease (a), elm bark beetle, a carrier of Dutch elm disease (b), and the fungus that causes Dutch elm disease—*Ophiostoma ulmi* (c)

St. Anthony's fire. The reference to fire comes from the burning sensation in the arms and legs that is one of the hallmark symptoms of the disease. Ergot is also a hallucinogen (causing false perceptions or visions) and causes victims to imagine that their skin truly is on fire. "St. Anthony" comes from the Catholic Order of St. Anthony, which was founded to treat people suffering from the disease.

In August 1722, the men and horses of Peter the Great, poised for an invasion of Turkey, consumed rye brought to them by the local serfs. By the following morning, over a hundred horses were paralyzed. About 20 000 people in the area died, and the invasion was effectively stopped. In 1951, people in a small village in France ate flour ground from infected rye. More than two hundred people became severely ill from the poison; thirty people went temporarily insane, imagining that demons and snakes were chasing them; and four people died.

Interestingly, ergot of rye was the first source for lysergic acid diethylamide (LSD), a hallucinogenic drug. Drugs derived from ergot, which causes muscles to contract and blood vessels to constrict, have in the past been used in medicine, such as in treating migraines and stimulating contractions during childbirth. Ergot of rye appears to be both a bane and a blessing.

12.1 Destructive Fungi

The average American family spends several hundred dollars annually to control and pay for damage caused by fungi. One kind of damage that must be controlled is damage to food crops. With the possible exception of bacteria, the molds of the kingdom Fungi are man's biggest competitors for food. One estimate indicates that plant diseases cost America over $3 billion annually. Thankfully, in America food production far exceeds need.

Other plants are also affected by fungi. Some destructive fungi have caused permanent damage by destroying their host species. One of the most important nut and hardwood lumber trees was the American chestnut. Prior to 1900, these trees grew in forests from Maine to Florida. It was estimated that chestnut trees made up 25% of the tree population. However, the chestnut blight fungus has nearly eliminated this tree from American forests.

Another fungus, one spread by the elm bark beetle, causes Dutch elm disease. This fungus causes the breakdown of the water conduction tissues of the tree, and the tree inevitably dies. Many towns that once had stately elms lining their streets have been forced, in the space of a few years, to remove them.

Some fungi cause problems for homeowners when there is water damage from natural disasters or even leaky plumbing. Species of *Stachybotrys*, *Aspergillus*, and others use the cellulose in building materials as nutrients. This is a particular problem in homes that have been flooded since they will need to undergo treatment to prevent structural damage and possible problems with human health.

There are also fungi that are pathogenic to humans. The most common fungal diseases are superficial skin infections, such as ringworm, which grows as a small ring of filaments called hyphae under the skin. Other skin infections include athlete's foot and thrush, which are discussed later in the chapter. Fungal skin infections are usually easy to treat, either with over-the-counter medications or sometimes with a prescription. Certain other fungal diseases cause deep infections in the internal organs and can be very dangerous. An example is *histoplasmosis*, caused by the fungus *Histoplasma capsulatum* (see Figure 12-5). The fungus grows in soil contaminated with bird or bat droppings. If a person disturbs the soil and inhales the spores of this fungus, it can cause damage to lung tissue and sometimes even invade other organs. Antifungal medications can be used to treat the disease.

12.2 Beneficial Fungi

Many people think that the edible mushrooms on pizza are the only beneficial fungi. However, there are many edible fungi. Many cheeses are the result of fungal growth. For example, enzymes created by certain fungi act on milk products to form natural Swiss and cheddar cheeses. The fungi that form Limburger and blue cheese are actually eaten as part of the cheese.

One of the most important fungi is a group of microscopic organisms called *yeasts* (see p. 283). Yeasts leaven many baked goods. Leavening is a substance that produces bubbles of carbon dioxide in dough or batter. The carbon dioxide is produced as the yeast go through cellular fermentation. The resulting bubbles cause the dough to "rise."

As discussed in Chapter 10, Sir Alexander Fleming discovered another important use of fungi. He noted that a ring of dead bacteria formed around a particular fungus growing on his bacterial culture. The chemical that killed the bacteria was isolated and named penicillin after the common mold *Penicillium notatum* from which it is derived. Penicillin became a common antibiotic used in treating certain bacterial diseases. Many fungi are used in the biotechnology industry to produce enzymes and biochemicals.

However, in their most important and most common function, fungi under natural conditions serve as *decomposer organisms*. Living in the soil and other dark, damp places, these fungi break down complex organic substances into simple, soluble forms that plants can use.

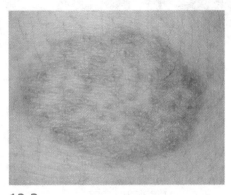

12-3
Ringworm is a skin infection caused by fungi that live on the surface of the skin. The medical term for ringworm is *tinea*.

12-4
Fungi are used in the manufacture of many types of cheeses, such as Swiss (left) and Roquefort (center).

Mycorrhizae

Many fungi live with plant roots in relationships called **mycorrhizae** (MYE kuh RYE zee) (sing., mycorrhiza). Over 90% of plants have fungi associated with their roots. The fungi effectively increase the surface area of the plant's roots to enhance the absorption of minerals from the soil. In return, the fungi receive carbohydrates from the plant. This symbiotic relationship (two organisms living together for mutual benefit) allows plants and fungi to live in environments that would otherwise be unable to support them.

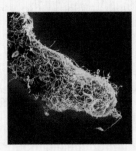

A mycorrhiza is a symbiotic relationship between a fungus and plant roots.

Leavening in the Bible

In the Bible, leavening is often compared to sin or false doctrine (Matt. 16:6–12; 1 Cor. 5:6–8). Just as a little leaven will affect the entire lump of dough, so a little error will corrupt the whole man or even groups of men. Most Bible scholars agree that in the Old Testament references to sacrifices, leaven often represents sin permeating all men.

There are many edible fungi. Yeast is used in many types of bread, and a mold makes the enzymes that form blue cheese. The morel is highly prized for its delicate flavor. Young puffballs fried in seasoned butter are considered a delicacy.

In Europe, specially trained dogs or pigs seek the truffle, a tasty underground fungus. Dogs, while not as good at finding the fungi, are preferred truffle hunters. The pig handlers must be alert and always ready to keep the pigs from eating the truffles. Dogs sniff out the fungi but refuse to eat it.

When someone speaks of eating fungi, most people think of mushrooms. Technically, all mushrooms are edible. The nonedible ones (either because they are tasteless, taste bad, or are poisonous) are toadstools. This, however, is not a good method of classification, since some of the best-tasting mushrooms and some of the most deadly toadstools are in the same genus.

For example, the *Amanita caesarea*—said to be a favorite of Julius Caesar—is a delicious mushroom, while *Amanita verna* (known as the destroying angel) is a highly poisonous toadstool. This deadly pure white toadstool causes no symptoms for several hours while the poison enters the tissues. Six to fifteen hours after eating, the person has abdominal pains, vomiting, and diarrhea. Another species, *A. muscaria*, the fly amanita, is fatal in large doses and causes psychological effects in smaller doses. Most mushroom collectors usually avoid all species of *Amanita* to reduce the possibility of accidentally picking a poisonous one.

Most people limit their mushroom picking to their grocer's shelves, where they usually find *Agaricus bisporus* (or a similar species) available in any season. They were not collected in a field or forest but were grown in buildings where the temperature and humidity are controlled. Pure cultures of the mushroom fungus are grown in laboratories and are then spread on sterile soil mixed with a rich organic material (horse manure is often used).

After several weeks of growth, another layer of soil is spread on the mushroom bed, and the mushrooms begin to form. When the buttons emerge, they must be picked quickly because, as a mushroom matures, it becomes tough and loses its flavor. The bed may be used for a few additional crops of mushrooms, but as the organic material is depleted, the whole process must be repeated.

Many people add mushrooms simply for flavor; however, they may also be as healthful as many vegetables. They are low in calories, high in fiber, and are virtually fat free, cholesterol free, and sodium free. Unlike many vegetables, mushrooms contain a small amount of all the essential amino acids. While they do not contain a wide variety of minerals and vitamins, many are good sources of phosphorus, potassium, and copper as well as the vitamins niacin and riboflavin.

A. verna A. muscaria

A. caesarea

Cato Edvardsen

Amanita caesarea *(edible)*, Amanita verna *(poisonous)*, and Amanita muscaria *(poisonous)*

Mushroom farm

Truffles are considered a delicacy by some.

➡Review Questions 12A

1. List several ways in which fungi are (a) harmful and (b) beneficial.
2. What is the most important function of fungi?

⊙3. List several fungi people commonly eat.
⊙4. Why is it safe to assume that the mushrooms in grocery stores are not of a poisonous variety that might look similar?

12B—Characteristics of Fungi

12B

Early in taxonomic history, fungi were considered plants because they are nonmotile, have cell walls, and have rootlike structures (rhizoids). However, they have several unique features that set them apart, not only from plants but also from all other organisms.

- Fungi do not have chlorophyll and cannot produce their own food. They are nonphotosynthetic heterotrophs.
- The bodies of fungi are made of masses of filaments woven together; there is no tissue differentiation.
- The fungal cell walls contain **chitin**, a complex polysaccharide. They do not contain cellulose as plants do.

12.3 Nutrition and Respiration of Fungi

One of the primary characteristics that separate the true fungi from both the plant kingdom and the algae is that fungi lack chlorophyll and therefore are heterotrophic. Some fungi are saprophytes, and others are parasites. A few fungi can be either, such as the fungus that causes histoplasmosis. In the soil, this fungus is saprophytic and appears as long strands called filaments. However, if a human or animal inhales its spores, the fungus can become a parasite; it no longer appears filamentous but exists as individual clumps of cells. This ability to change form in reaction to a different environment is called *dimorphism*.

Fungal cells lack specialized organelles for digestion. Some parasitic fungi obtain nutrition directly from the cytoplasm of the host. Most fungi, however, carry on *external digestion*. External digestion requires the secretion of enzymes that digest the food into a soluble form outside the organism. The soluble nutrients are then absorbed. The enzymes of external digestion are often the harmful substances of parasitic or poisonous fungi.

Most fungi require oxygen for their metabolism, but a few can grow without an abundant supply of free oxygen. All fungi require abundant moisture for active growth, but many species can withstand periods of dryness by forming spores.

➡Objectives

- Describe the distinguishing characteristics of fungi
- Explain how fungi obtain nutrients
- List and describe fungal colonial structures
- Summarize the methods of sexual and asexual reproduction in fungi

➡Key Terms

chitin   sporangiophore
hypha   conidiophore
septate   fruiting body
mycelium

(a)

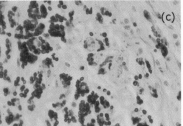
(b)

(c)

Images (b) and (c) Courtesy of www.doctorfungus.org © 2004

12-5

Dimorphism in *Histoplasma capsulatum*: a colony growing on soil (a), filamentous hyphae of soil colony (b), and parasitic form (purple oval structures) in human lung tissue (c)

World's Largest Organism?

Although some fungi are unicellular, most are multicellular colonial organisms. They range in size from microscopic unicellular yeast to the fungus *Armillaria*—the largest organism by area in the world. One exceptionally large *Armillaria ostoyae*, found in a forest in Oregon, covers a space larger than 1665 football fields and is estimated to be 2400 years old. How can it cover this much ground? Actually, only a small part of the fungus appears above ground. Most of the fungus is an extensive underground network of filaments.

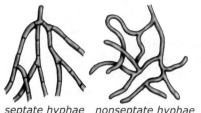

septate hyphae   nonseptate hyphae

12-6

Kinds of hyphae

hyphae: (Gk. *hyphe*—web)

septate: (L. *saepes*—hedge)

rhizoid: rhiz- (Gk. *rhiza*—root) + -oid (shape or form)

haustoria: (L. *haustus*—absorption)

sporophore: sporo- (seed) + -phore (bearing)

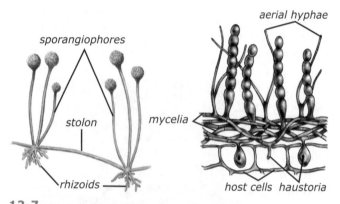

sporangiophores

stolon

rhizoids

mycelia

aerial hyphae

host cells   haustoria

12-7

Various fungal structures

12.4 Colonial Structure of Fungi

With few exceptions fungal colonies are composed of slender filaments called **hyphae** (HYE fee) (sing., hypha). Even large fungi such as mushrooms are merely organized masses of interwoven hyphae.

In some fungal groups the hyphae are **septate**; that is, they are divided into individual cells by cell walls called *septa* (sing., septum). The septa usually have a hole or pore in them, permitting cytoplasm to pass between the cells of the filament. Some fungi have hyphae that lack septa and are called *nonseptate*. Nonseptate hyphae are typically multinucleate. Although hyphae are only 5–10 µm wide, some may be almost 1 cm long.

Fungi are made of masses of intertwined hyphae called **mycelia** (mye SEE lee uh) (sing., mycelium). Mycelia are visible without magnification. A mycelium may be as simple as a clump of bread mold or as highly organized and specialized as a mushroom.

Hyphae are named according to their shape or function.

✦ *Rhizoids* (RYE ZOYDZ) are hyphae that are embedded in the material on which the fungus is growing. They support the fungus and secrete enzymes to digest food.

✦ *Haustoria* (haw STOHR ee uh) are hyphae of parasitic fungi that perforate a host's cells and obtain nutrition directly from the cytoplasm.

✦ *Aerial hyphae* are not embedded in the medium on which the fungus is growing. They absorb oxygen, produce spores, and spread the fungus.

✦ *Sporophores* are aerial hyphae that produce spores.

✦ *Stolons* (STOH lunz) are aerial hyphae that connect groups of hyphae together.

12.5 Reproduction of Fungi

Asexual reproduction occurs when environmental conditions are favorable; it is the most common type of reproduction in fungi. Yeast cells reproduce by *budding*—a part

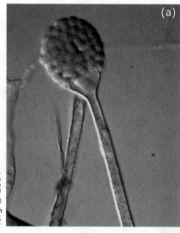

12-8

(a) The common bread mold *Rhizopus* produces sporangiophores and sporangia. The hyphae of bread mold are clear or white. The black appearance of this mold on bread is the result of the colored sporangia and spores. (b) *Aspergillus*, a common mold, produces blue-green conidia.

of the cell pinches off to produce more yeast cells. Fungi can also reproduce asexually by fragmentation of the hyphae or mycelia. However, the most common method of asexual reproduction is by spore production.

Sporophores function in the asexual production of spores. If the sporophore forms spores within an enclosure or sac, it is called a **sporangiophore** (spuh RAN jee uh FOHR), and the structure in which the spores are formed is a *sporangium* (spuh RAN jee um). If the spores are not in an enclosure, the sporophore is called a **conidiophore** (kuh NID ee uh FOHR), and the spores are *conidia* (kuh NID ee uh).

Sexual reproduction usually occurs during times of unfavorable environmental conditions, and it does not occur in all species of fungi. There are no male or female fungi; they have mating types that are denoted "plus" (+) or "minus" (−). Sexual reproduction occurs when the hyphae of different mating types come into contact. The fused hyphae produce a specialized structure called the **fruiting body** that forms and releases the spores.

sporangiophore: spor- (seed) + -angio- (Gk. *angeion*—container) + -phore (bearing)

conidiophore: conidio- (Gk. *konis*— dust) + -phore (bearing)

➡ Review Questions 12B

1. What characteristics separate fungi from (a) plants and (b) algae?
2. What is the difference between a septate fungus and a nonseptate fungus?
3. Describe five different types of hyphae based on their functions.
4. Describe asexual reproduction in fungi.
5. Describe sexual reproduction in fungi.

12C—Classification of Fungi

The kingdom Fungi contains over 100 000 species of colonial and unicellular heterotrophic organisms and is usually divided into three phyla— Zygomycota, Ascomycota, and Basidiomycota. Fungi are traditionally grouped into phyla based on their colonial structure and method of sexual reproduction.

12C

➡ Objectives
- Summarize the characteristics that distinguish the fungal phyla
- Describe the life cycles of the three phyla of fungi
- Explain the symbiotic relationships of lichens

➡ Key Terms
zygosporangium	stipe
ascus	gill
yeast	imperfect fungi
budding	symbiosis
basidium	lichen
cap	

12-1 Fungal Phyla Characteristics

Phylum	Hyphal structure	Sexual reproductive structures	Asexual reproductive structures	Examples
Zygomycota (common molds)	nonseptate	zygospores produced by zygosporangia	spores produced by sporangia	species of genus *Rhizopus*
Ascomycota (sac fungi)	septate or unicellular	ascospores produced by asci	budding, conidia produced by conidiophores	yeast, powdery mildews, species of genus *Penicillium*
Basidiomycota (club fungi)	septate	basidiospores produced by basidia	various types of spores	mushrooms, toadstools, rusts, smuts, earthstars

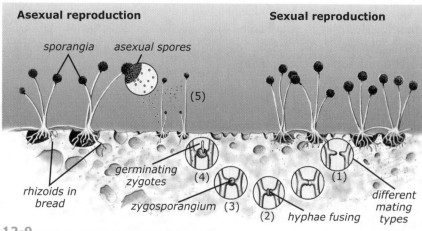

Asexual reproduction

sporangia   asexual spores

(5)

rhizoids in bread

germinating zygotes

(4)

zygosporangium (3)

(2) hyphae fusing

Sexual reproduction

(1)

different mating types

12-9

Reproduction in a typical zygomycete—black bread mold (*Rhizopus*)

1. Sexual reproduction: Two hyphae from different mating types grow until they touch each other. Nuclei in the tip of each hypha divide several times.
2. The hyphae fuse at the point where they touch.
3. A zygosporangium forms at the point where the hyphae fused. The nuclei of opposite mating types pair, forming diploid zygotes (2*n*). The zygosporangium forms a thick outer covering and enters a dormant state.
4. When environmental conditions are favorable, the zygotes in the zygosporangium undergo meiosis and germinate, forming new hyphae.
5. Asexual reproduction: As the hyphae grow, sporangiophores are formed and new spores are released.

12.6 Phylum Zygomycota

Many of the common molds are in the phylum Zygomycota (ZYE goh my KOH tuh). They are primarily land dwellers. The name of this phylum is derived from the thick-walled sexual structures called **zygosporangia** (sing., zygosporangium) that characterize the members of this phylum.

The genus *Rhizopus* contains a food mold that is found almost everywhere and is familiar to us all—the black bread mold (either *R. nigricans* or *R. stolonifer*). During asexual reproduction, the airborne spores, produced by the sporangia, land in a favorable environment and germinate, forming a small hypha. This hypha grows into a mycelium with rhizoids, stolons, and finally, sporangiophores and sporangia. Depending on the species, the spores are either black or shades of brown or green. It is best not to eat moldy foods. Some species may cause allergic reactions, and others produce digestive enzymes or other substances that are toxic to humans.

Members of Zygomycota reproduce sexually when the hyphae of different mating types touch and form a zygosporangium. When environmental conditions are favorable, the zygotes in the zygosporangia undergo meiosis and germinate. The newly formed hyphae grow into stolons and sporangia, completing the life cycle.

12.7 Phylum Ascomycota

The phylum Ascomycota (AS koh my KOH tuh) is named for its members' characteristic reproductive structure—the **ascus** (AS kuhs). The ascus (pl., asci) is a microscopic saclike structure in which the spores (called ascospores) are formed.

12-10

The powdery mildews are named for the cobwebby spores they form over the leaves of roses, apples, and other plants.

Zygomycota: Zygo- (yoke) + -mycota (Gk. *mukes*—fungus)

Ascomycota: Asco- (Gk. *askos*—bag or bladder) + -mycota (fungus)

12-11

The cup fungi are ascomycetes that form bowl-shaped fruiting bodies. Spores are dispersed primarily as falling drops of water splash them out of the cups.

Ascomycetes are also known as sac fungi based on this characteristic reproductive structure.

Asexually, conidia (spores) are produced at the tip of aerial hyphae in structures called conidiophores. Air currents carry the spores away, where they germinate and form new hyphae and mycelia.

Molds of the genus *Penicillium* are typical ascomycetes. These molds form green spores with surrounding white rings that are visible on the rinds of oranges and on other fruits and foodstuffs. Other species in this genus produce the flavors of Roquefort (blue cheese), Camembert, and other cheeses. Other members of Ascomycota include powdery mildews and cup fungi.

Asexual reproduction

conidia

conidiophore

(2)

(1)

germination

germination

released ascospores

Sexual reproduction

fruiting body

dikaryotic hyphae

(3)

asci

(4)

different mating types

ascospores

12-12

Reproduction in a typical ascomycete

1. Asexual reproduction: Asexual spores (conidia) are formed by conidiophores. Once they reach a favorable environment, they germinate.
2. Sexual reproduction: When two hyphae of different mating types meet, a tube joins them. Nuclei from one hypha cross over into the other. The nuclei pair but do not fuse.
3. The resulting dikaryotic (having two nuclei) hyphae intertwine with the hyphae of the parent fungus to form the fruiting body.
4. The asci form at the tips of the dikaryotic hyphae, where the haploid nuclei now fuse, undergo meiosis, and form eight haploid ascospores. The ascospores are released, and when they germinate, new hyphae are formed.

a Closer Look

The Yeasts

Yeasts are unicellular, predominantly saprophytic fungi found in soil or water. A few yeasts are parasites, and a few cause human diseases. Most of the over 500 yeast species are in phylum Ascomycota.

Yeasts are typically egg-shaped cells slightly larger than bacteria. Some yeasts produce vitamins B and D. Often yeast or ground yeast products are added to vitamin tablets as the source of vitamins B and D. Various yeasts are also grown commercially to produce enzymes used in the manufacture of syrup, cheese, soft-center candies, and medicine.

Yeasts reproduce asexually by **budding**. First, the nucleus divides by mitosis. Then a small pouch forms on the side of the yeast cell, and one of the new nuclei moves into it, forming the bud. After a period of growth, the bud usually separates from the parent. During periods of unfavorable conditions, yeast cells can form special spores, which can remain dormant for long periods.

Active dry yeasts contain the living spores of baker's yeast for use in home baking. These yeasts go through alcoholic fermentation instead of aerobic cellular respiration, producing alcohol and carbon dioxide as byproducts. The resulting "bubbles" of CO_2 gas cause the bread to rise. The alcohol produced evaporates as the bread is baked.

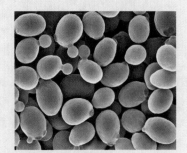

Budding in yeasts

12.8 Phylum Basidiomycota

This phylum contains the most familiar fungi—mushrooms, puffballs, earthstars, shelf fungi, rusts, and smuts. Members of the phylum Basidiomycota (buh SID ee oh my KOH tuh) are called the club fungi because they form four sexually produced *basidiospores* (buh SID ee uh SPORZ) on a club-shaped cell called a **basidium** (buh SID ee um) (pl., basidia).

basidiospore: basi- or basidio- (L. *basis*—pedestal) + -spore (seed)

12-13

Puffballs and earthstars produce their spores within protective membranes. Puffballs (left) release their dust-fine spores when they are disturbed, such as by rainfall. Earthstars release their spores through an opening in the center of the "star."

Mushrooms are usually saprophytic fungi. The well-known but short-lived **cap** and **stipe** (stalk) are only the fruiting body of an extensive underground network of septate hyphae, which may cover several square meters of soil and may be several decades old. When the humidity and temperature are right, the hyphae stop growing and "regroup" stored substances into one or several tiny knobs near the surface of the soil. These knobs are actually twisted networks of mycelia that take on the shape of a compressed miniature mushroom. Once the structures form within their thin covering membranes, the knobs are called the button stage of the mushroom. When the proper temperature, moisture, and other conditions are met, the cells of the button fill with water, and the mushroom forcibly reaches its full size.

Underneath the cap are flat structures called **gills** that radiate outward from the stipe. The gills have thousands of basidia, each of which produces four basidiospores. The basidiospores are released from the basidium and then fall between the gills onto the soil. Wind and water carry the spores to other areas.

The *shelf fungi* are either saprophytes of deadwood or parasites of living trees. Some parasitic species add a new layer to the shelf each year, resulting in shelves almost as big in diameter as the trees on which they grow. Shelf fungi produce spores in pores on the underside of the shelf. When released, they fall out of the pore and are carried away by the wind.

12-14

Shelf fungi

The Imperfect Fungi

Most of the fungal organisms in the phyla Zygomycota, Ascomycota, and Basidiomycota are sometimes grouped together and called the *perfect fungi*, or true fungi. A perfect fungus is one that has a known form of sexual reproduction and can thus be classified into one of the three fungal phyla. There are, however, several fungi that are not known to reproduce sexually. These fungi are called the **imperfect fungi**.

The list of imperfect fungi was once quite long. As scientists have cultured and observed them, they have been able to identify their sexual reproductive structures and classify many of them into an appropriate phylum. Today most scientists believe the remaining imperfect fungi do have methods of sexual reproduction that simply have not yet been discovered. Some believe they will likely be placed in the phylum Ascomycota since many of them resemble ascomycetes in other ways.

Although some of the imperfect fungi are saprophytes, many are plant parasites, and a few are parasites of man. Athlete's foot is a familiar one. This itchy, uncomfortable condition is caused by a fungus that commonly infects the skin between the toes—a warm, damp environment.

Thrush, or candidiasis, is caused by the fungus *Candida albicans*, which is one of the normal inhabitants of mucous membranes lining the mouth, nose, and throat. Normally the population of *Candida* is kept in check by competition from bacteria that grow in the same area. If something happens to upset the balance, such as prolonged use of an antibiotic that kills off the competing bacteria, the *Candida* can quickly cause an infection.

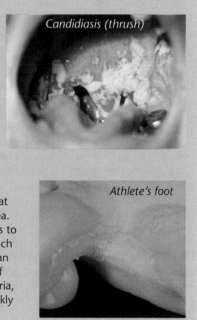

Candidiasis (thrush)

Athlete's foot

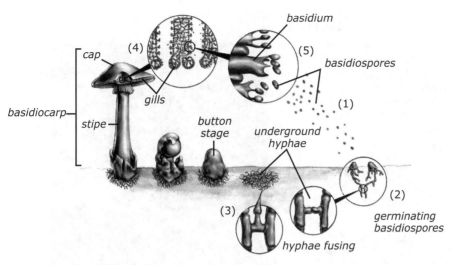

12-15

Sexual reproduction in mushrooms
1. Basidiospores are released into the air from the gills underneath the cap. Under favorable conditions, they germinate.
2. Once the basidiospores germinate, they form either plus or minus mating-type hyphae that grow underground.
3. When compatible mating types meet, the hyphae fuse, but the nuclei do not. This forms a dikaryotic hypha. These dikaryotic hyphae begin to form a fruiting body called the basidiocarp, composed of the stipe, cap, and gills.
4. Basidia are formed on the gills where the haploid nuclei fuse, forming a zygote.
5. The zygote undergoes meiosis to form the basidiospores.

Rusts and Smuts: Plant Parasites

In the phylum Basidiomycota are several very harmful plant pathogens, including *rusts* and *smuts*. Smuts produce several different spores in their life cycle and cause hundreds of millions of dollars worth of damage to crops each year. The rusts usually have complicated life cycles, producing several different kinds of spores and alternating between two different plant hosts.

Wheat rust, for example, causes dark, rust-colored patches on the stems of wheat and other grains. These patches produce red urediospores (yoo REE dee uh SPORZ), which can be carried by wind and water to infect other wheat plants generally during the same growing season. At the end of the season, the plant begins to yellow, and the hyphae produce teliospores (TEE lee uh SPORZ). Teliospores can live through the winter and then germinate to form basidia and basidiospores. The basidiospores, however, cannot directly infect

Corn smut

wheat. They first infect the underside of the leaves of the barberry bush. There the fungus forms tiny cups in which aeciospores (EE see uh SPORZ) are produced. The aeciospores then infect wheat, causing the rust-colored urediospores. An alternate host (in this case, the barberry) is essential for a rust to complete its life cycle.

In colonial Massachusetts a law was passed ordering the destruction of all barberry bushes. It was believed that eliminating the barberry would eliminate the wheat rust. Although this action did achieve a measure of control, the fungus on a single infected barberry bush can produce enough aeciospores to infect a wheat field hundreds of miles away. Today wheat rust is controlled by breeding resistant varieties of wheat, as well as by controlling the number of barberry bushes. Other significant rusts in the United States include the apple cedar rust and the white pine blister rust.

Wheat rust

Lichens

The relationship of two organisms living together for mutual benefit is called **symbiosis** (SIM bee OH sis). Many fungi live with other organisms in symbiotic relationships. Some of the most common examples of these pairings are called lichens.

A **lichen** (LYE kun) consists of a fungus and an alga living together. In lichen symbiosis the alga captures energy from the sunlight and manufactures sugars for itself and the fungus; the fungus provides support and protection for its partner. Lichens are common on tree bark, fence posts, and brick walls.

Scientists place the approximately 16 000 lichen species into three categories, based on their appearance and distribution.

Lichen morphologies: (a) crustose, (b) foliose, and (c) fruticose

Some lichens are very unusual. Reindeer moss, an important food source for caribou and reindeer in the tundra, exists in mats several centimeters thick and several kilometers wide.

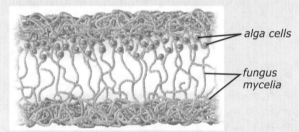

Lichens are a symbiotic relationship between algae and fungi.

Most lichens reproduce by releasing dustlike pieces of lichen called *soredia* (suh REE dee uh) (sing., soredium). A soredium contains both the alga and fungus and must be immediately transferred to an environment suitable for lichen growth. Birds and crawling animals help spread lichen soredia.

Although lichens are able to withstand some of the harshest of natural conditions, they are extremely sensitive to air pollution. The growth of lichens can indicate air quality, and they are sometimes used as air pollution monitors. In many industrial cities, lichens have died, but farther away from such places lichens grow abundantly.

- Crustose (KRUS TOSE) lichens appear as flat smears of dried paint. They often grow as a layer on rocks and trees.
- Foliose (FOH lee OSE) lichens look like small crinkled leaves.
- Fruticose (FROO tih KOSE) lichens usually have small stalks or branches with conspicuous fruiting bodies on their ends.

symbiosis: sym- (Gk. *sun*—together) + -biosis (life)

crustose: crust- (L. *crusta*—crust) + -ose (*osus*—having characteristics of)

foliose: foli- (L. *folium*—leaf) + -ose (having characteristics of)

fruticose: frutic- (L. *frutex*—shrub or bush) + -ose (having characteristics of)

⟿ Review Questions 12C

1. Give characteristics and list examples of each of these phyla: (a) Zygomycota, (b) Ascomycota, and (c) Basidiomycota.
2. Describe the structure of yeast. How do they perform asexual reproduction?
3. What is an imperfect fungus, and why is the list of imperfect fungi growing shorter?
4. Describe the structure and the life cycle of a mushroom.
5. Distinguish between lichens and mycorrhizae. Give the function of each member in these symbiotic relationships.
6. List the three basic forms of lichens and describe each.
⊙ 7. Many fungi grow in a ring shape. Why is this typical of fungi?
⊙ 8. Lichens are often called "pioneer organisms." What characteristics of lichens permit them to serve as pioneer organisms?

The Plant Kingdom

Botany Part I

The Kingdom Plantae

God created plants on the third day of Creation, and except for a few areas on the earth, they dominate the landscape. The abundance of plants is a testimony to God's loving care for humans and the animals. He made plants to meet the physical needs of His creatures—especially to provide them with food (Gen. 1:29–30). Man also depends on plants for shelter and for medicines, energy, and many other uses. This section begins the study of plants, or botany.

While most people initially think of green grass and trees or beautiful flowers when they think about plants, the kingdom Plantae includes much more than this. For starters, not all plants are green. Some have other pigments that mask the green chlorophyll, while others have little or no chlorophyll. Although most plants are autotrophic—they carry on photosynthesis using chlorophyll—there are thousands of heterotrophic plants. Conversely, not all green organisms are plants. Some fungi, protozoans, and even some bacteria contain enough chlorophyll to be green.

The value of plants to man cannot be overestimated. The plant kingdom provides almost all of man's food. Cereals, such as corn, wheat, and rice,

botany: (Gk. *botane*—herb or plant)

13-1 General Characteristics of the Kingdom Plantae

Cellular structure	All are eukaryotic.
	All are multicellular with tissues; some have organs.
	Most have chlorophyll and other pigments in plastids.
	All have cell walls with cellulose.
Nutrition	Most are autotrophic; some are heterotrophic or parasitic.
Reproduction	All reproduce sexually.
	Many have asexual forms of reproduction.
Motility	Most are sessile; some float.
	Some produce motile gametes.

along with legumes, such as peas, beans, and soybeans, are the primary food sources. People also consume many other fruits and vegetables directly. Even spices, honey, coffee, tea, cocoa, and juices are derived from plants.

Although scientists believe that algae produce the majority of photosynthesis-derived atmospheric oxygen, plants do contribute a significant portion of this life-essential gas. Plant products include paper, gum, wax, alcohol, turpentine, cork, lumber, cloth fibers, coal, petroleum, medicines ranging from castor oil to codeine, and cellulose, which is made into plastics.

Finally, plants are a source of inestimable beauty. The quiet stateliness of a lush forest and the delicate simplicity of a pale flower are not merely afterthoughts of God's creation. They are as much an integral part of His plan as the "useful" aspects of the kingdom Plantae.

Heterotrophic Plants

Mistletoe is a partially parasitic plant. It contains chlorophyll, and although it does get water and minerals from the host tree, it obtains little food from the tree.

Dodder, a parasite, has so little chlorophyll that the entire plant appears yellowish orange. Since few of its pigments are capable of photosynthesis, the dodder obtains almost all its food by rootlike structures that penetrate the stems of a host plant, similar to the haustoria of certain fungi.

Indian pipes take parasitism one step further. Containing no chlorophyll or other pigments, these plants appear ghostly white. They are believed to absorb food made soluble by certain fungi growing in association with their roots (mycorrhizae). Why are Indian pipes and similar plants not classified as fungi? Simply put, they contain tissues and produce flowers, fruits, and seeds.

Mistletoe   Dodder   Indian pipes

13.1

Objectives

- List and characterize the three groups of nonvascular plants
- Describe the life cycle of a typical moss
- Explain the concept of alternation of generations

Key Terms

botany
vascular tissue
seed
annual
biennial
perennial
leafy shoot

rhizoid
gametophyte
antheridium
archegonium
sporophyte
alternation of
generations

vascular: (L. *vasculum*—vessel)

13A—Plant Classification

Botanists have divided the kingdom Plantae into twelve phyla. This text divides these twelve phyla into three groups based on the presence or absence of **vascular** **tissues** and **seeds**. Vascular tissues are specialized structures that conduct water and dissolved materials in a plant. A seed is a structure that contains a young embryonic plant and stored food inside a protective seed coat.

13.1 Nonvascular Plants

Nonvascular plants lack vascular tissues for conducting water and nutrients and thus are generally quite small, most being under an inch (2.5 cm) in height. They also produce spores rather than seeds during their life cycle. Today

Annuals, Biennials, and Perennials

Another method of classifying plants is based on the length of time that they grow. Plants can be placed into three groups based on the number of seasons they grow (or regrow after dying back).

✦ *Annual plants.* Most herbaceous plants are **annual** plants. Annuals sprout, grow, flower, and produce seeds all in one growing season. Many showy flower beds contain herbaceous annuals like zinnias, pansies, and marigolds.

✦ *Biennial plants.* Plants like foxglove and sweet William that sprout and develop in one growing season but do not flower and produce seeds until the following growing season are called **biennial** plants. After the second year most biennials die.

✦ *Perennial plants.* **Perennial** plants grow year after year. Woody plants are usually perennials, but some herbaceous

plants have thick underground stems that live many years, even though the aboveground leaves and stems die each year. Tulips, irises, peonies, and gladioli are common examples of herbaceous perennials.

Annual—pansy   Biennial—foxglove   Perennial—tulip

biennial: bi- (L. *bi*—two) + -ennial (L. *annus*—year)

---

botanists classify nonvascular plants into three phyla, the largest of which is Bryophyta (brye AWF uh tuh), which contains the true mosses. (Many things that are called mosses, like Irish moss, Spanish moss, and reindeer moss, are actually in other phyla or even other kingdoms.)

Phylum Bryophyta: The True Mosses

Phylum Bryophyta contains the mosses. Mosses exist in many different climates. In tropical areas the abundant mosses can give a tree trunk the appearance of being several times its actual diameter. Mosses often are the dominant vegetation in vast areas of tundra, where they must live for months under ice and snow. Most mosses growing in temperate regions appear as velvety clumps in shaded areas or as delicate green carpets covering rocks or logs near streams, waterfalls, or other sources of moisture.

A clump of moss is actually many individual plants densely packed together. The most obvious part of a moss is called the **leafy shoot**. This structure is generally less than 3 cm long and transmits water in small spaces between the cells in much the same way that a paper towel absorbs water. Each of the leaflike structures of a moss is one cell-layer thick, except near the center where additional support is needed. Leafy shoots of various mosses may appear vastly different.

On the bottom of each leafy shoot is a tangled mass of rhizoids. **Rhizoids** may appear to be rootlike, but because they lack vascular tissues, they are not true roots. Moss rhizoids are usually filaments of cells used only for anchorage, not absorption. Most mosses can absorb water and minerals directly through the leafy shoot. Unfortunately, this also allows them to dry out more readily than most plants. As expected, mosses thrive in moist environments.

The Bryophyte Life Cycle

The leafy shoot is only one stage in the life cycle of a moss. There are two different types of leafy shoots: the tip of one has the male reproductive structures; the other has the female structures. Since the leafy shoot produces the gametes, this stage in the life cycle is called the **gametophyte** (guh MEE tuh FITE). The top of a male gametophyte bears saclike **antheridia** (sing., antheridium), which produce sperm. The top of a female gametophyte has one or more vase-shaped **archegonia** (AR kih GOH nee uh) (sing., archegonium). Each archegonium contains an ovum (egg cell).

13-1

(a) Moss showing leafy shoots of the gametophyte generation, and (b) moss showing the stalk and capsule of the sporophyte generation

Bryophyta: Bryo- (Gk. *bruon*—moss) + -phyta (plant)

gametophyte: gameto- (spouse) + -phyte (plant); hence, the plant that produces the next generation of its kind

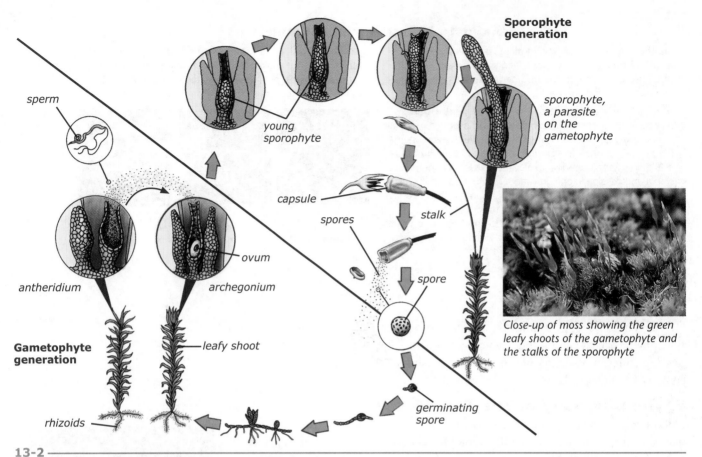

Sporophyte generation

sperm

young sporophyte

sporophyte, a parasite on the gametophyte

capsule

spores

stalk

ovum

antheridium

archegonium

spore

Gametophyte generation

leafy shoot

germinating spore

rhizoids

Close-up of moss showing the green leafy shoots of the gametophyte and the stalks of the sporophyte

13-2

Life cycle of a typical moss showing the alternation of generations

When water touches the top of the antheridium, the sperm are released. When the water also touches the top of a female plant, the archegonium releases a chemical that stimulates the flagellated sperm to swim toward the ovum. Water and clustering of the gametophytes are essential for fertilization of bryophytes, since the sperm must swim from the antheridium to the ovum.

The fertilized ovum (zygote) grows out of the female gametophyte into a stalk with a capsule on the top. Since this structure produces spores in the capsule, it is called the **sporophyte** stage. The sporophyte is the result of the union of two gametes and is therefore diploid. The capsules and stalks are often brown, red, or orange. When the capsule is mature, the cap comes off and the wind distributes the released spores. When environmental conditions are suitable, a spore begins to grow, eventually producing the leafy shoots and rhizoids of the gametophyte stage.

Botanists call the type of life cycle illustrated by the bryophytes **alternation of generations**. The haploid gametophyte gives rise to the diploid sporophyte generation, and the spores give rise to the next gametophyte. Spores cannot produce sporophyte stalks and capsules, nor can zygotes produce gametophyte leafy shoots. The two generations must alternate. In bryophytes, because the gametophyte generation (leafy shoots) is the stage more often seen, it is called the *dominant generation*. The alternation-of-generations type of life cycle is seen throughout the plant kingdom.

sporophyte: sporo- (seed) + -phyte (plant)

Other Nonvascular Plants

Remember from the discussion of the Doctrine of Signatures in Chapter 1 that liverworts were used to treat diseases of the liver. The term *wort* is from the old English word *wyrt* that means herb or plant. Liverworts are nonvascular plants, and their life cycle is similar to that of the mosses. The liverwort gametophyte is a narrow, flattened, leathery structure, whose shape reminded the early botanists of a liver. It grows along the ground, anchored to the soil by rhizoids growing from the underside. Umbrella-shaped archegonia grow out of the upper surface of the gametophyte, and the spore-producing sporophyte generation is found attached to the underside of these structures.

Hornworts are also nonvascular plants. Hornworts resemble liverworts in that the gametophyte grows along the surface of moist, shaded soil. However, the sporophyte that grows from out of the gametophyte looks like a thin, green horn.

The Marchantia, a common liverwort (left), shows the umbrella-shaped archegonia growing up out of the gametophyte. The sporophyte capsules are found on the underside of the "umbrella." The hornwort's sporophyte (right) distinguishes it from a liverwort.

Review Questions 13.1

1. In what ways are plants valuable to man?
2. List the major characteristics of the plant kingdom.
3. What are the three general groupings of plant phyla? What two characteristics are used to put plants into these groups?
4. Describe a characteristic of nonvascular plants that limits their size.
5. What two factors are essential for bryophyte reproduction? Why?
6. Describe the gametophyte and the sporophyte of a typical moss. Explain their relationship to each other.
⊙7. If you examine a clump of moss, you will find that only some of the moss has sporophyte stalks and capsules. Why?
⊙8. Why would mosses be most abundant in areas such as the tundra and tropical rainforests, where extreme environmental conditions exist?

13.2 Vascular Plants Without Seeds

All plants other than true mosses, liverworts, and hornworts have vascular tissues that conduct water and dissolved minerals throughout the plant. Thus the vascular plants are not limited to the small size range of mosses. These next four phyla, however, are unusual in that they do not produce seeds. They produce spores, single cells with a protective coat, which are used to spread the species. The best-known representatives of this group are the ferns.

Phylum Pteridophyta: The Ferns

The phylum Pteridophyta (TEHR uh DAWF uh tuh) contains an interesting and diverse group of plants. Most **ferns** are like the typical forest-floor fern or the Boston fern—a favorite potted plant. Some, however, are epiphytes. Epiphytes are plants that grow on other plants but are not parasitic. Many other plant phyla contain epiphytic members as well.

13.2

Objectives
- Summarize the characteristics of the phyla in this group
- Describe the life cycle of a typical fern

Key Terms
fern
epiphyte
frond

sorus
sporangium

epiphyte: epi- (Gk. *epi*—upon or over) + -phyte (plant)

Some ferns grow as vines; others grow floating on water. Although they are most prominent in shaded, cooler areas of the tropics, some species grow in desertlike conditions, and others grow next to glaciers. Others are very delicate, among the rarest plants on earth, and quickly perish with the slightest environmental change.

The ferns common to most of North America are typical of the phylum Pteridophyta. Each fern leaf, commonly called a **frond**, grows from an underground stem called a *rhizome* that produces roots. Most fern fronds are long and delicate looking. Ferns usually grow in clumps, produced by a single rhizome. The rhizome can produce new clumps asexually.

The Fern Life Cycle

Occasionally a person will notice what he thinks are insect eggs or some fungus on the underside of fern fronds. Actually, these are **sori** (SOR eye) (sing., sorus), collections of spore-producing **sporangia** (sing., sporangium). The location and types of sori are two characteristics botanists use to name and classify ferns.

When released, the powdery fern spores can be carried by the wind. Under proper conditions fern spores germinate and form a tiny heart-shaped *prothallus* (proh THAL us), which is one cell-layer thick. The underside of the prothallus develops rhizoids (which absorb water and minerals), archegonia, and antheridia. Flagellated sperm are released from the antheridium and swim to the ovum at the bottom of the archegonium. Thus, the prothallus is the gametophyte generation of the fern.

13-3
Typical Boston fern (top), tropical tree fern (middle), and an epiphytic tropical fern (bottom)

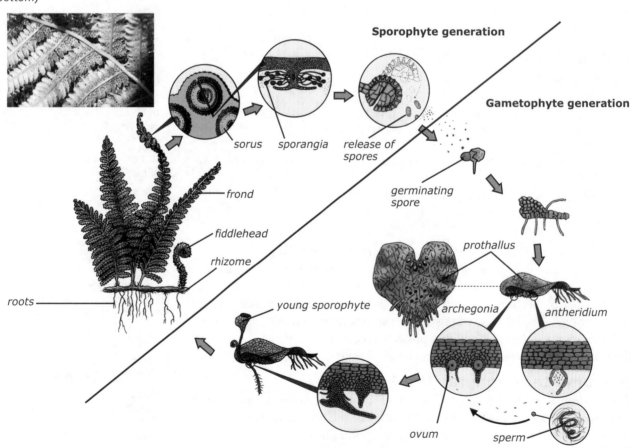

13-4
Life cycle of a typical fern showing alternation of generations. The sori are located on the underside of the fern frond (upper left).

After fertilization, the zygote that forms in the bottom of the archegonium matures and sends the first leaf up and the first root down, beginning the sporophyte generation. The young sporophyte is a parasite on the gametophyte until it is large enough to manufacture its own chlorophyll and begin carrying on photosynthesis. Soon after this, the prothallus dies. In ferns, the more prominent generation is the sporophyte. The inconspicuous gametophyte prothallus is seldom noticed.

The first leaf of a fern is often a fan-shaped blade. The second is usually a *fiddlehead*, a coiled young leaf. The fiddlehead has tissue in its coiled end that is able to produce new cells even after the base of the leaf has reached its full size. This permits some ferns to have unusually large leaves. Young fiddleheads of some ferns are edible and can be purchased for use in salads.

Other Seedless Vascular Plant Phyla

✦ *Horsetails* (phylum Sphenophyta). Most horsetails grow in wet environments. They have a thick underground stem that continues to grow year after year. The green, erect annual stems and the roots grow from this main stem. The vertical stems are ridged, hollow, and segmented by nodes, which may produce whorls of branches or of thin, needlelike leaves. Spores form in small conelike structures at the tops of stems. The stems contain silica deposits. American colonists used the tough, hard nodes to clean pans, hence the common name "scouring rushes."

✦ *Club mosses* (phylum Lycophyta). Club mosses, or ground pines, look like large moss plants. They usually have a creeping stem that may be under the surface of the soil and that occasionally sends up erect stems. These erect stems have spore cases collected into conelike structures either at the tip or at the base of each leaf. Most species are evergreens and are less than 30 cm high, but one tropical species may reach about 60 cm. Fossils of club moss trees have been uncovered that measure over 30 m high and over 1 m in diameter.

✦ *Whisk ferns* (phylum Psilotophyta). Whisk ferns are rare and unusual plants that lack true roots and leaves and are not actually ferns. A thick stem covered with rhizoids may creep underground or in horizontal cracks in the bark of tropical trees. From this stem may grow smaller green stems which produce small scales instead of leaves.

Together these three phyla contain approximately 1750 living species, all relatively small plants. The fossil record, however, reveals that before the Flood many members of these phyla were quite large and widespread, possibly the dominant foliage. Fossil horsetail specimens, for example, show that they were large, unusually shaped trees. Most coal and oil deposits are thought to be remains of these and similar plants. To have these large plants in abundance would require extensive areas of special environments that are in only small isolated areas today. This fact provides further support for an antediluvian environment that was much different from ours today.

Horsetails (left) and whisk ferns (right)

➠ *Review Questions 13.2*

1. Describe the gametophyte and sporophyte of a typical fern. How is their relationship to each other the same as in the mosses?

2. Where are the sporangia of a fern located?

3. How are the life cycles of a fern and a moss similar? How are they different?

4. Why is water necessary for the sexual reproduction of bryophytes and ferns?

⊙5. What is meant by alternation of generations in the plant kingdom, and how do the moss and fern life cycles illustrate alternation of generations?

➡ **Objectives**

- Distinguish between gymnosperms and angiosperms
- Apply the alternation of generations to a typical conifer life cycle
- Summarize the differences between a monocot and a dicot plant

➡ **Key Terms**

gymnosperm	fruit
angiosperm	monocot
pollen cone	dicot
seed cone	cotyledon

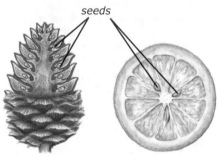

seeds

13-5

A pinecone's (gymnosperm) seeds are on scales and are not enclosed within an ovary; an orange's (angiosperm) seeds are enclosed in an ovary.

gymnosperm: gymno- (Gk. *gumnos*— naked) + -sperm (seed)

angiosperm: angio- (container) + -sperm (seed)

13.3 Vascular Plants with Seeds

The vascular plants that produce seeds are divided into two groups: the **gymnosperms** (JIM nuh SPURMZ) containing four phyla and the **angiosperms** (AN jee uh SPURMZ) with one phylum. Gymnosperms are nonflowering plants that produce seeds which are not enclosed in an ovary when mature. The pinecone, for example, has seeds that lie on tiny shelves when they are ripe. Angiosperms, on the other hand, do produce flowers and have enclosed seeds like the seeds of an apple or an orange.

The Cedar of Lebanon and Solomon's Temple

One of the best-known plants in the Bible is the cedar of Lebanon, a gymnosperm of the pine family. These trees reach a height of 30 m (98 ft) and a circumference of 8 m (26 ft). After the Flood they were the tallest, most massive trees known in the Middle East. As with most cedars, the wood of these trees is very durable since it resists rotting and insect infestation. The wood is also delightfully fragrant and exceptionally beautiful.

Solomon built the temple and his own palace using this wood; his palace is sometimes called "the house of the forest of Lebanon" (1 Kings 7:2). Scripture states that Solomon made a pact with King Hiram, the ruler of the Lebanon area, to supply the lumber for this project. Solomon was to send three shifts of ten thousand men to help Hiram's slaves and lumbermen cut the trees.

Later Israelite kings as well as rulers of other nations used cedar lumber for everything from chariots to ships. Unfortunately, no reforestation program was practiced, and today the once extensive forests of Lebanon are mere patches of 30 to 150 trees each. However, the cedar of Lebanon is not becoming extinct. It is now protected by the government and is easily cultivated wherever proper soil and climatic conditions exist.

These Lebanon cedars are similar to those that Solomon used in the temple. Even today cedar is used in producing beautiful furniture.

Useful Gymnosperms

Sometimes gymnosperms are called *evergreens*. More than 20 million of them are cut each year to supply American homes with Christmas trees. Some gymnosperms, however, are not truly evergreen because they lose their foliage in the fall. And some plants, such as holly, keep their green leaves all year even though they are not gymnosperms.

The primary economic importance of gymnosperms is found in their use as lumber and pulp. America was built from the gymnosperm forests that covered extensive areas of the continent. But Americans rashly used almost all the available virgin timber. Over one hundred years ago nearly every accessible stand of the eastern white pine, a highly prized lumber tree, had already been cut. Today most of the lumber and pulp

The bristlecone pines—Pinus longaeva—live in very harsh mountain environments and grow at a very slow rate—as little as 0.25 mm (0.01 in.) per year. This is not an actual photograph of Methuselah. Its exact location is kept secret so that souvenir seekers do not harm the tree.

used in America comes from cultivated gymnosperms, specially selected for rapid growth and quality lumber. These trees can be harvested every twenty to thirty years.

Gymnosperms range from ornamental varieties, which creep along the ground, to giant redwoods measuring over 114 m (375 ft) tall. The bristlecone pine is a species of gymnosperm growing in the White Mountain region of California. By making special drillings, scientists think that one of the bristlecone pines, named Methuselah, is the oldest known living organism. According to the number of its growth rings, it is more than 4700 years old. In other words, this particular bristlecone pine was a seedling while the Egyptians were building the pyramids.

Phylum Coniferophyta: The Cone-Bearing Plants

Among the gymnosperms the phylum Coniferophyta (kuh NIF uh RAWF uh tuh) is the largest. All conifers produce seeds in cones. Although not all cones are like the familiar pinecone and not all conifers look like pine trees, the life cycle of a pine tree is typical of this phylum.

In the spring, pine trees produce two types of cones: pollen cones and seed cones. **Pollen cones**, usually numerous, small, and short-lived, are found near the tips of the branches. The abundant *pollen* produced by these cones contains the male reproductive gametes. Since pine pollen is carried to the seed cones by the wind, vast amounts of pollen must be produced; otherwise, the likelihood of fertilization would be very small.

(a)　(b)　(c)

13-6

(a) Pollen released from pollen cones lands on (b) immature seed cones. (c) When mature, the seed cone opens to disperse the seeds.

Pollen lands on the open scales of the small, green, immature **seed cones**, usually found on other branches of the same tree. The scales then close tightly. The ova on the scales of the seed cones may not be fertilized until months later, and in some species they may not develop into seeds for several years. When the seeds are mature and environmental conditions are right, the scales of the woody seed cones open to release the seeds.

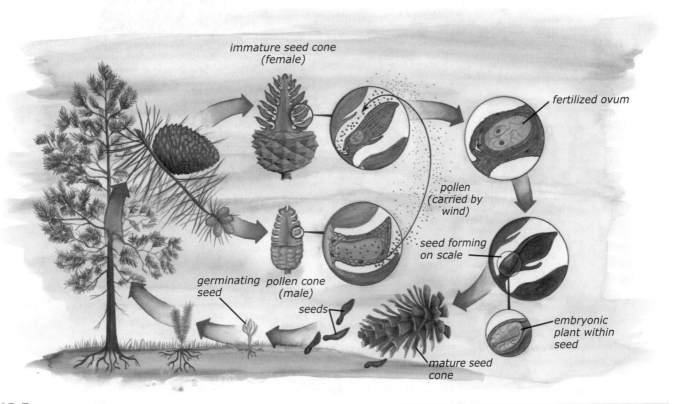

13-7

Life cycle of a pine tree

13-8

(a) A hemlock from the pine family, (b) a yew from the yew family, (c) a juniper from the cypress family, and (d) giant sequoias from the redwood family

Other Gymnosperm Phyla

There are three other phyla of living gymnosperms other than Coniferophyta.

✦ *Phylum Cycadophyta.* Only about a hundred species of cycads (SIE KADZ) remain. They are native to tropical areas but have been transplanted to the warmer areas of the United States as ornamental plants. The cycads look like palm trees, and some of them have fiddleheads. Sometimes called the sago palms, cycads bear seeds in conelike structures.

Cycad

✦ *Phylum Ginkgophyta.* Botanists once thought that all members of the phylum Ginkgophyta were extinct. However, in the 1800s, several large specimens were discovered being cultivated in monasteries in China. Today ginkgo (GING koh) trees (also called maidenhair trees) can be found in parks all over the world. The ginkgo sheds its broad, leathery, fan-shaped leaves during late autumn. Ginkgo trees are either male, producing pollen, or female, producing yellowish cherry-sized, foul-smelling fruit.

Ginkgo

✦ *Phylum Gnetophyta.* The gnetophytes (NEE tuh FITES) are an unusual group of gymnosperms that produce pollen and seeds in cones that look like flowers. Members of the genus *Ephedra* grow in the southwestern United States and are a natural source for the drug ephedra (the active ingredient in many allergy and nasal decongestant medications). A very different-appearing gnetophyte—the *Welwitschia mirabilis*—grows in the desert of southwestern Africa. This plant grows to be only a few centimeters tall, but the diameter of the stem may reach up to 1 m with two very long leaves.

Welwitschia mirabilis

There are several distinct conifer families within the classification system; many may be recognizable from their descriptions in the following paragraphs. Although most have cones similar to the pine tree, some families have cones that are very different.

The *pine family* is the largest and most economically important conifer family. The pines have the needles and cones normally associated with gymnosperms. Pine needles vary in length, roundness, number per cluster, and presence or absence of stripes. These characteristics, along with the type of cone produced, are used to classify pines, firs, hemlocks, and other members of this group.

The *yew family*, prized in gardens for its slow growth and waxy green needles, produces fruit that looks like red open-ended berries.

The *cypress family* includes the *junipers* and the *arborvitae*, both widely cultivated shrubs. Some members of this family bear small dry cones, but others have cones that remain fleshy and look like blue green peas. Most junipers and arborvitae have overlapping evergreen scales instead of needles.

The *redwood family* contains some of the largest living organisms in the world: the giant sequoias. The living organism with the greatest volume is the General Sherman giant sequoia (sih KWOI uh) tree in California. It is 31 m (102 ft) in circumference, 83 m (272 ft) tall, and has bark about 30 cm (12 in.) thick. It has been estimated to be about 2200 years old.

Phylum Anthophyta: The Flowering Plants

Plants in the phylum Anthophyta (an THAWF uh tuh), often called angiosperms, are the dominant vegetation on the earth today. There is so much diversity among the 250 000 species that they are grouped into nearly four hundred different families, based primarily on floral parts. All angiosperms have seeds enclosed in an ovary and produce flowers. Unlike what might be expected, many angiosperms do not produce colorful blossoms. For example, corn tassels and the catkins of oak trees are flowers.

An *ovary* of a plant is the structure that encloses the seeds, and a mature ovary is called a **fruit**. Apples, cherries, and tomatoes are easily identified as fruits. But pods of peas and kernels of corn and wheat are also fruits (see Ch. 14, p. 338).

Monocots and Dicots

Angiosperms are divided into two classes, *Monocotyledoneae* (MAHN uh KAHT uh LEE duh nee ee) and *Dicotyledoneae* (dye KAHT uh LEE duh nee ee), commonly called **monocots** and **dicots**. The basic distinction between these two is the number of cotyledons in the seed. A **cotyledon** (KAHT uh LEE dun) contains stored food to nourish the embryonic plant while it is in the seed and to supply the young sprout with energy until it can carry on photosynthesis.

Leaf venation can often be used for classification of monocots and dicots. Monocots, such as grass, generally have *parallel leaf venation*. The veins in their leaves start at the stem and go to the tip of the leaf. The veins are roughly parallel. Dicots, such as maple trees, generally have *netted leaf venation*. Netted veins continually branch within the leaf blade. Of course, some plants that have leaves are not even angiosperms. For example, ferns have leaves.

Monocots usually have floral parts in threes or sixes. Dicots usually have floral parts in fours, fives, or multiples of four or five. Classification by floral parts can be misleading, and there are exceptions. Other characteristics used to determine whether a plant is a monocot or a dicot are even less reliable.

13-9
A corn tassel (top) and oak catkins (bottom) contain many small flowers.

13-10
The peanut (top), a dicot, easily splits in half to reveal its two cotyledons. Corn, on the other hand, is a monocot. A corn seed has only one cotyledon and does not split apart.

Monocotyledoneae: Mono- (single) + -cotyledon- (Gk. *kotuledon*—hollow-shaped cup)

Dicotyledoneae: Di- (two) + -cotyledon- (hollow-shaped cup)

Plant type	Seeds	Leaves	Roots	Stems	Flower parts
Monocot tulips, daylilies, irises, corn, bananas	one cotyledon	parallel venation	usually fibrous	Young stems have scattered vascular bundles; mature stems may be hollow.	usually in threes or sixes
Dicot columbines, cacti, oaks, roses, carnations	two cotyledons	netted venation	usually a taproot	Young stems have vascular bundles arranged in a ring; stems are solid.	usually in fours, fives, or multiples of these numbers

Review Questions 13.3

1. Name and describe the two types of cones produced by conifers. Tell their relationship to each other.
2. How is the life cycle of a pine different from the life cycle of a fern or moss?
3. List several well-known families of conifers and describe each.
4. What characteristics do gymnosperms and angiosperms have in common? In what ways are they different?
5. What are the two classes of the phylum Anthophyta, and what are the primary characteristics that separate them?
⊙6. Seed plants are called the dominant vegetation of the earth today. In what ways are seed plants dominant over other types of plants?

13B—Plant Anatomy

One of the defining characteristics of the kingdom Plantae is the presence of tissues, which are groups of cells that work together to perform a particular function. In many plants, these tissues form organs that have a high degree of specialization. Plants have three organs that are not involved in reproduction—the leaves, roots, and stems. In addition, they have three reproductive organs—the flowers, fruits, and seeds. Plant reproductive organs will be discussed in the next chapter. In this section, the text presents the different kinds of tissues that are found in plants and shows how they form the various plant organs.

13.4 Plant Tissues

Although there is great diversity in plant organs, they are made of three basic tissue types—dermal tissue, vascular tissue, and ground tissue. In general the dermal tissue forms the "skin" of the plant, the vascular tissue functions as the "delivery system" for the plant, and the ground tissue makes up the remainder of the plant's "insides." A fourth tissue—the meristem—carries on mitosis, allowing for continued growth of the plant.

Dermal Tissue

The outside covering of the plant is made of the **dermal tissue**. In younger plants it is composed of an outer layer of cells called the **epidermis**. The epidermis covers the leaves and, in young plants, the roots and stems. In some plants a thick, waxy layer called the **cuticle** covers the leaves and stem. The

13.4

Objectives
- Summarize the four plant tissue types
- Distinguish between xylem and phloem
- Describe the function of the meristem

Key Terms

dermal tissue	sieve tube cell
epidermis	companion cell
cuticle	ground tissue
cork	parenchyma
vascular tissue	collenchyma
xylem	sclerenchyma
tracheid	meristematic
vessel cell	tissue
phloem	

cuticle: (L. *cutis*—skin)

cuticle helps prevent water loss. Hairlike growths from the epidermis covering the leaves and stem also help reduce water loss, while similar extensions on the root ends help the plant absorb water and nutrients.

As woody plants mature, the epidermis covering the stems and roots is replaced by a thick layer of **cork** cells. These cork cells soon die and provide a nonliving protective covering for the stem or root. The cork waterproofs the plant, taking over the function of the cuticle. In addition to protecting the plant from physical damage and disease-causing organisms, the dermal tissues function in gas exchange and in the absorption of minerals.

Vascular Tissue

Vascular tissue transports water and nutrients throughout the plant. Plants have two different types of vascular tissue—xylem and phloem. Each is made of columns of specialized cells stacked one on top of another, forming a complex system of pipes that extend from the root ends to the tip of the uppermost stem and laterally from the trunk to the leaves.

Xylem

The xylem (ZYE lum) carries water and dissolved minerals within a plant, primarily from the roots upward, and is made of two types of cells. Long thick-walled cells called **tracheids** (TRAY kee idz) are narrow and tapered at each end. Water can diffuse easily from one tracheid to another through indentations in the cell walls called pits. The other xylem cell type is called the **vessel cell**. Vessel cells are also stacked one on top of the other, forming long continuous tubes. However, vessel cells have actual holes (perforations) in their cell walls through which water can pass.

When xylem tissues mature, the cells die, leaving long conducting tubes. In addition to their transportation function, the thick walls of the xylem cells provide strength and structural support for the plant.

Phloem

Mature phloem (FLOH EM) is composed of living cells called **sieve tube cells**. The sieve tube cells form continuous conduits that carry water and dissolved foods (usually sugars produced by photosynthesis) from the leaves throughout the plant. Sieve tubes also have holes in their cell walls that allow the water and other nutrients to freely flow between cells. As the sieve tube cells mature, they lose their nucleus and most of their organelles. However, adjacent to each sieve tube cell is a **companion cell**, which provides "life support" for the sieve tube cell. Companion cells perform cellular respiration, protein synthesis, and other metabolic functions for the sieve tube cells.

Generally, materials in xylem move upward and materials in phloem move downward. However, in some parts of the plant or at different times of the year, the opposite can be true. For example, during the summer, the sugars produced in the leaves are carried by the phloem to the roots. But in the spring, phloem may carry water and food upward. The rising sap aids in the formation of new leaves, stems, and flowers.

Ground Tissue

The **ground tissue** makes up the remainder of the inside of the plant. Ground tissue cells remain alive throughout the life of the plant. The cells that compose the ground tissue function to help support the plant; provide storage for water, sugar, and starch; and perform the metabolic processes of the plant, such as photosynthesis. Nonwoody stems and roots as well as leaves are made of mostly ground tissue.

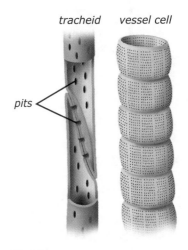

tracheid   vessel cell

pits

13-11
The xylem carries primarily water and dissolved minerals.

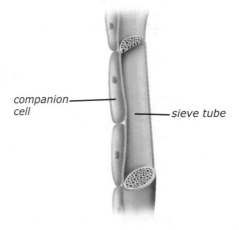

companion cell

sieve tube

13-12
Water and dissolved organic substances such as sugar are transported by phloem.

xylem: (Gk. *xulon*—wood)
phloem: (Gk. *phloios*—bark)

a Closer Look

Plant Cell Types

Ground tissues are made of three types of cells.

Parenchyma (puh RENG kuh muh) cells have thin cell walls with a large central vacuole and a small amount of cytoplasm. Much of the ground tissue is composed of parenchyma cells, which include the palisade mesophyll and the spongy mesophyll (see p. 302). These cells are important in many of the metabolic processes of the plant and in storage of water and starch. For example, in leaves, where most of the photosynthesis takes place, the parenchyma cells are packed with chloroplasts.

Collenchyma (kuh LENG kuh muh) cells have thicker cell walls than parenchyma cells. These thicker cell walls provide much of the structural support for the plant, especially in areas where the plant is still growing. Although they are strong, they are still flexible—for example, the "strings" of celery contain a high proportion of collenchyma cells.

The cell walls of **sclerenchyma** (skluh RENG kuh muh) cells are very thick and rigid. They provide support and strength to the plant in nongrowing areas. When these cells mature, most die, leaving behind their thick cell walls to maintain the framework of the plant.

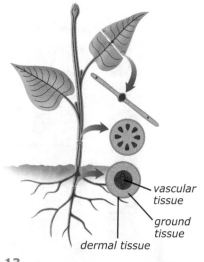

vascular tissue

ground tissue

dermal tissue

13-13

Three tissue types are found throughout a plant: dark green—dermal, light green—ground, blue—vascular.

Meristematic Tissue

Meristematic (MEHR ih stuh MAT ik) **tissues** are composed of plant cells that are capable of continuous mitosis and are responsible for the ongoing growth of the plant. Meristematic cells are small, thin-walled, and undifferentiated, but they can develop into any tissue found in that particular plant. Meristematic tissues are found in the growing areas of plants, such as buds and the tips of roots and stems.

Review Questions 13.4

1. Describe and give the functions of the three major categories of plant tissues.

2. Compare and contrast the structure and function of xylem and phloem.

3. What is the function of the meristematic tissue, and where is it located?

13.5

Objectives

- Describe the different types of leaf venation
- Draw and label a cross section of a typical leaf
- Summarize the functions of leaf structures

Key Terms

blade
petiole
parallel
  venation
netted venation
stoma

guard cell
palisade
  mesophyll
abscission layer
spongy
  mesophyll

petiole: (L. *petiolus*—fruit stalk)

13.5 The Leaf

Although it may seem somewhat odd, a leaf is an organ. Its primary function is to absorb light energy from the sun. Without this efficient energy-gathering ability, photosynthesis sufficient to support the plant would not take place. Ranging in size from a fraction of a centimeter to over 4 m long, leaves may be thick and heavy or light and delicate. While most leaves are a deep, rich green, some are pale green, yellow, red, pink, or even white. Although modified leaves may help protect the plant, attract and catch insects, store water, or even hold the roots of the plant to a tree trunk, the basic function of a typical leaf is photosynthesis.

Structures of a Typical Leaf

A typical dicot leaf has a large flattened area called the **blade**, which is connected to the stem by a stalk called a **petiole** (PET ee OL). Many monocot leaves, like those of corn, lack a petiole, and the base of the blade encircles, or sheaths, the stem. Leaves that lack a petiole are termed *sessile*.

Some leaves have *stipules*—structures attached at the base of the petiole that can exhibit many different shapes. Stipules may be thin tissues that covered the leaf as it was forming, winglike structures attached to the petiole, or leaflike structures at the base of the petiole.

One way botanists identify plants is by studying the basic shapes of the leaves and leaf parts, including the edges of the leaves, called *margins*. Identifying plants by their leaves, however, is not always foolproof. One type of oak may have over twenty different leaf shapes on a single tree.

Leaf Venation

There are two basic patterns of leaf venation: parallel and netted. In **parallel venation** a series of veins originates at the stem and proceeds to the tip of the leaf in a roughly parallel fashion. This occurs in monocots like corn, grass, irises, and orchids.

In **netted venation** large veins branch to form a network of smaller veins throughout the leaf. There are two types of netted venation. If the veins branch off one large central vein, the venation is *pinnate*. African violets, oaks, and apple trees have simple pinnate leaves. If there are two or more main veins originating from a single point, the venation is called *palmate*. Maples, ivies, and geraniums have simple palmate leaves.

If there is only one blade on one petiole, the leaf is a *simple leaf*. If a leaf on a single petiole is divided, the leaf is a *compound leaf*. Each of these blade divisions is called a *leaflet*.

If a leaf is basically pinnately veined and the leaflets are arranged down the midrib, the leaf is *pinnately compound*. If the leaf is palmately veined and all the leaflets originate from a single point, the leaf is *palmately compound*. Occasionally there is a *bipinnately compound* leaf, in which the venation is pinnate with the leaflets borne on the secondary veins, not just on the midrib. Some plants even have tripinnately compound leaves.

Some leaflets are larger than many simple leaves, and a petiole may look like a stem rather than like part of the leaf. One of the easiest ways to tell the difference between a leaf and a leaflet is to look for a stipule. Leaflets usually do not have stipules, but most leaves do. Another distinguishing characteristic is the presence of buds. A bud may be found at the base of a petiole, but not at the base of a leaflet. Leaves are found in different planes on a stem, but leaflets are always found on the same plane.

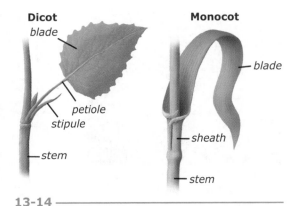

Dicot — blade, petiole, stipule, stem

Monocot — blade, sheath, stem

13-14 —
Comparison of typical dicot and monocot foliages

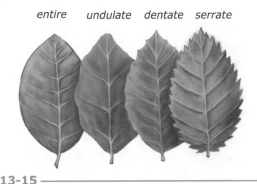

entire   undulate   dentate   serrate

13-15 —
Types of leaf margins

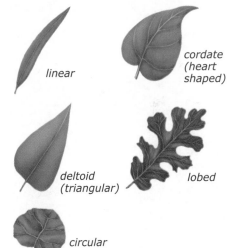

linear

cordate (heart shaped)

deltoid (triangular)

lobed

circular

13-16 —
Basic leaf shapes

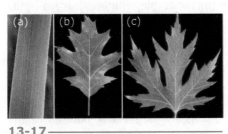

13-17 —
(a) Parallel venation in grass, (b) pinnate netted venation in an oak leaf, and (c) palmate netted venation in a maple leaf

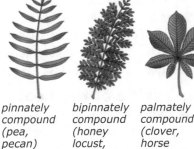

pinnately compound (pea, pecan)

bipinnately compound (honey locust, mimosa)

palmately compound (clover, horse chestnut)

13-18 —
Leaf venations

13-19 —
(a) Simple leaf (oak) and (b) compound leaf (clover)

13-20
Epidermal hairs on an African violet leaf

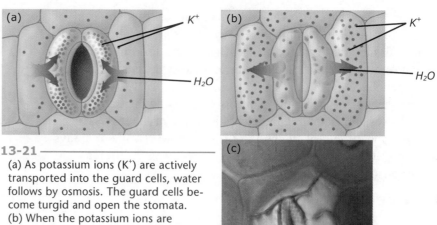

13-21

(a) As potassium ions (K⁺) are actively transported into the guard cells, water follows by osmosis. The guard cells become turgid and open the stomata. (b) When the potassium ions are pumped out of the guard cells at night, they become less turgid as the water also exits, closing the stomata. (c) A photomicrograph of a closed stoma on a leaf

The Covering of a Leaf

A cross section of a typical leaf seen under a microscope will appear to have layers of tissue. The top and bottom layers—only one cell thick—compose the *epidermis*. The epidermal cells lack chlorophyll and serve primarily as protection.

Epidermal cells often secrete a waxy substance that forms a cuticle to prevent water loss, giving the leaf a shiny appearance on one or both surfaces. The epidermal cells of some leaves produce *epidermal hairs*. Sometimes these hairs give the leaf a velvety appearance, as in African violets. Some epidermal hairs secrete sticky substances or chemicals with specific odors. For example, when a geranium leaf is touched, a distinctive scent is released by the broken epidermal hairs. Some epidermal hairs are thin and sharp enough to penetrate a person's skin without his feeling the prick. The epidermal hairs of stinging nettles can not only puncture the skin but also leave an irritating mixture of formic acid and histamines that may cause redness, itching, or a burning sensation of the skin.

On the underside of most leaves are little openings called stomata (sing., stoma) that permit the exchange of gases between the atmosphere and the spaces inside the leaf. Around each stoma are two **guard cells**. These specialized epidermal cells are shaped as opposing crescents and function to open and close the stomata. During the day, the adjacent epidermal cells pump potassium ions into the guard cells, decreasing the water concentration inside the guard cells. To equalize the concentration, water diffuses into the guard cells by osmosis. The influx of water causes the guard cells to become turgid and open the stomata. At night, the potassium ions are pumped out of the guard cells, and the water flows out by osmosis, causing the guard cells to shrink and the stomata to close.

Other factors such as enzymes and temperature play important roles in the opening and closing of the stomata. Generally, when water is in abundance and photosynthesis is taking place, the stomata are open. At night or when the plant lacks water, the stomata close to reduce loss of water into the atmosphere.

The Inside of the Leaf

Between the upper and lower epidermis are ground tissues, made up primarily of parenchyma cells. The parenchyma cells are the primary photosynthetic areas of the leaf. The upper layer or layers are called **palisade** mesophyll (MEZ uh FIL). These cells are column shaped and are tightly lined up side by side. This arrangement permits a large number of cells to be present in a small

stomata: (Gk. *stoma*—mouth)
mesophyll: meso- or mes- (Gk. *mesos*—middle) + -phyll (leaf)

Falling Leaves

In the tropics and subtropics, where the growing season lasts almost all year, most plants lose their leaves a few at a time and are never completely bare. Most houseplants are tropical plants and therefore have leaves all year.

In more temperate regions many woody plants are *deciduous*; they lose their leaves before winter to conserve water. A deciduous leaf has a narrow, light green layer at the base of the petiole. This **abscission layer** is formed even before the leaf is completely developed. Part of this layer is made of cells that begin to die as the days shorten, regardless of the temperature. For this reason deciduous plants that are kept indoors will still lose their leaves in the fall. The abscission layer also forms a layer of cork cells that seals the vascular tissues and leaves a leaf scar on the stem. Approximately two weeks after the separation starts, enough cells have died that a crack has formed between the base of the petiole and the stem. In a breeze, or under its own weight, the leaf falls.

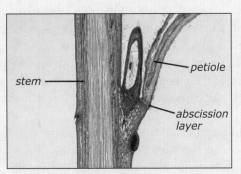

The abscission layer produces several layers of cork, which separate the petiole from the stem, leaving a leaf scar.

Most gymnosperms and a few other plants have *persistent foliage*. The leaves do not fall until the following spring when new leaves have formed. Since many persistent leaves retain their chlorophyll all year, the plant has an evergreen appearance.

The beautiful autumn colors seen in North America are evidence that green leaves contain other pigments besides chlorophyll. When the abscission layer begins to block the water supply, the leaf can no longer produce chlorophyll. As chlorophyll wears out and is not replaced, the other pigments begin revealing their magnificent yellows, oranges, reds, and purples.

In many leaves, certain pigments are formed only in cool temperatures as the sun reacts with sugars that remain after the abscission layer has formed. The brightest leaf colors, therefore, appear when the fall has sunny days with temperatures that dip to about 4 °C (39 °F). Cloudy, warm fall seasons in some parts of the world do not allow leaf colors to be as brilliant.

Some leaves do not produce these other pigments. When these leaves die, they appear a brownish color, as even the most beautifully colored leaves do eventually. This coloring is due to tannic acid, the product of the chemical breakdown of plant cell contents. The large concentration of tannic acid found in some leaves is readily dissolved in water to form tea, the most widely consumed beverage in the world. Most other plant leaves also release tannic acid when they are placed in water, but many of them release poisonous or distasteful chemicals as well.

Plant Pigments

Carotenoids are fat-soluble pigments found in the plastids of some plants. Carotenes and xanthophylls, two types of carotenoids, account for the yellow or orange as well as some reddish colors in certain plants. Carotenoids are responsible for the pale yellow green areas of leaves that are grown without sufficient light as well as for the colors of such plant parts as daffodil flowers, carrots, pumpkins, and corn kernels. When found in conjunction with chlorophyll, some carotenes help capture light necessary for photosynthesis and somehow help protect chlorophyll from intense light. Since carotenoids are the basic substances from which the body forms vitamin A, a balanced diet contains yellow or red fruits or vegetables.

Anthocyanins (AN tho SY uh ninz) are red, blue, and violet water-soluble pigments found in the vacuoles of plant cells. Leaves of the scarlet maple and some variegated-leaf plants like the coleus have large quantities of anthocyanins, which also color blueberries, grapes, plums, cherries, geraniums, roses, orchids, beets, and radishes.

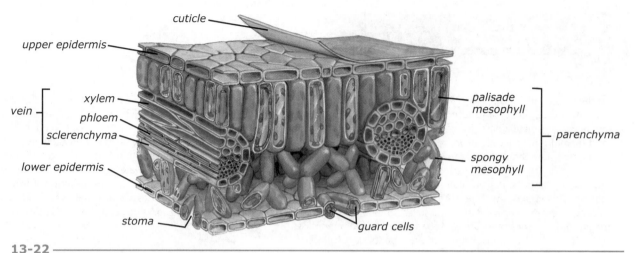

cuticle

upper epidermis

vein
xylem
phloem
sclerenchyma

lower epidermis

stoma

guard cells

palisade
mesophyll

parenchyma

spongy
mesophyll

13-22
Structures of a typical leaf

surface area. These cells have abundant chloroplasts that circulate around the central vacuole by *cytoplasmic streaming*. Each chloroplast gets its turn at the top of the cell where it can absorb the most direct sunlight.

The **spongy mesophyll** is made of irregularly shaped cells with many air spaces between them. These air spaces do not permit this layer to have as many chlorophyll-containing cells as the palisade layer. For this reason the underside of many leaves is not as dark a green as the upper surface.

Plants do not inhale and exhale air as people do, but in a sense plants do breathe. Gas exchange through the stomata and diffusion in the air spaces of the spongy mesophyll supply the carbon dioxide and oxygen exchange necessary for photosynthesis.

The xylem and phloem in the veins of the leaf continue through the petiole to the xylem and the phloem of the stem. The large veins of a leaf contain thick-walled strengthening sclerenchyma cells around the vascular tissues. As the veins branch and become smaller, the amount of supporting tissue decreases. The smallest veins contain only a single xylem vessel. Xylem is necessary to conduct the water and dissolved minerals from the roots to the cells of the leaf. Each cell in the leaf is just a few cells away from the end of a xylem vessel. The sugars and starches made in the leaf pass by diffusion from one leaf cell to another until they come to a phloem cell. The phloem then carries these leaf products to other areas of the plant for use or storage.

Review Questions 13.5

1. Differentiate between (a) parallel and netted venation, (b) pinnate and palmate venation, and (c) simple and compound leaves.
2. Describe the stomata of a leaf and give their function.
3. Draw and label a cross section of a typical dicot leaf.
4. Describe the process whereby leaves change color and fall in the autumn.
5. What accounts for the brown color of old leaves?
⊙6. Why is it essential that every cell in a leaf be near a xylem vessel, while it is not essential that every cell be near a phloem tube?

13.6 The Root

Most of us think of **roots** as the underground parts of plants. However, not all roots are below ground. The roots of *epiphytic* plants, such as orchids, creep along tropical tree branches and obtain water and minerals from substances that collect in cracks in the bark. The roots of parasitic plants such as mistletoe grow into the tissues of their host. Some aquatic plants have short roots dangling into the water from their floating leaves. Other roots, such as those in the ivy plant, hold the plant onto a rough surface like a brick wall.

Most roots serve to anchor the plant, even though they may not be in soil. They absorb water and the dissolved minerals necessary for plant growth. Roots also transport these absorbed substances to the place where they are needed in the plant. Roots may also function in food storage, as they do in carrots, radishes, and beets.

Root Systems

Plant roots differ greatly, depending not only on the plant itself but also on the conditions in which the plant is grown.

If the original root that sprouts from the seed—called the *primary root*—continues to grow as the predominant root, the plant has a **taproot system**. Not all taproots are fleshy, like beets or carrots; they may be long and thin instead. Taproots produce small, branching secondary roots. If the plant lacks a taproot but has many secondary roots, it has a **fibrous root system**.

Unless a plant is very slow growing or lives in very moist soil, its roots will need much more surface area than the leaves. Yet most root systems do not go deeper than 1–2 m (3–6 ft) into the soil. The roots of some large trees, like the pecan, are often less than 1.5 m (5 ft) deep but may spread out in a circle 30 m (100 ft) in diameter.

Primary Growth of a Root

If the root of a germinating seed is marked at millimeter intervals and then permitted to grow for twenty-four hours, the marks closest to the seed will still be a millimeter apart, but the second or third sections marked from the root tip will be several millimeters apart. If the experiment is continued, those sections that expanded the most during the first twenty-four hours will remain about the same length, but the section nearest the tip will continue to get longer. This growth in length is called **primary growth**. Why is primary growth found only in the tip of the root? Examining a longitudinal section of a young root under a microscope will provide the answer.

The tip of the root is covered by the **root cap** and is made of dead thick-walled cells that protect the delicate tissues of the root tip as it pushes through the soil. Just above the root cap is the **meristematic region**, where tiny undifferentiated cells carry on mitosis. The cells formed in the meristematic region begin to grow and establish large vacuoles in the **elongation region**, located just above the meristematic region. As the cells complete the elongation phase, they begin to differentiate and become the various tissues of a young root. The area where most differentiation takes place is called the **maturation region**.

While the primary growth of a root results from the division of cells in the meristematic region, the actual lengthening of the root tip occurs because of cell growth in the elongation region. Once the cells have elongated, they will not grow any more in length. This explains why the marks closest to the seed in Figure 13-24 did not move farther apart, but those in the meristematic and elongation regions did.

➡ Objectives
- Describe the different types of root systems
- Compare and contrast primary and secondary root growth

➡ Key Terms

root	root hair
taproot system	cortex
fibrous root system	endodermis
	vascular
primary growth	cylinder
root cap	vascular
meristematic region	cambium
	pericycle
elongation region	secondary growth
maturation region	

13-23 —————
Basic root systems—taproots (top) and fibrous roots

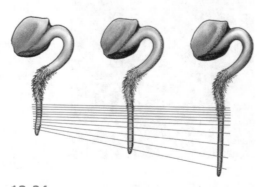

13-24 —————
Demonstration of the primary growth of a root

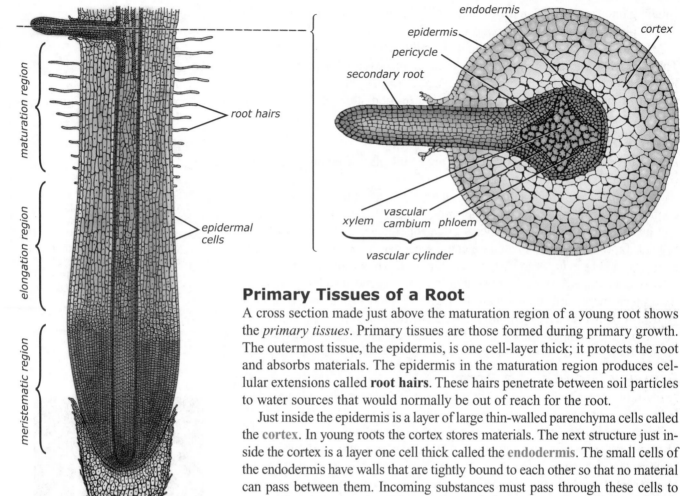

root hairs

epidermal
cells

maturation region

elongation region

meristematic region

root cap

endodermis

epidermis

pericycle

secondary root

cortex

xylem

vascular
cambium

phloem

vascular cylinder

13-25
Longitudinal section of a root tip (left)
and a cross section through a secondary
root (right)

Primary Tissues of a Root

A cross section made just above the maturation region of a young root shows the *primary tissues*. Primary tissues are those formed during primary growth. The outermost tissue, the epidermis, is one cell-layer thick; it protects the root and absorbs materials. The epidermis in the maturation region produces cellular extensions called **root hairs**. These hairs penetrate between soil particles to water sources that would normally be out of reach for the root.

Just inside the epidermis is a layer of large thin-walled parenchyma cells called the **cortex**. In young roots the cortex stores materials. The next structure just inside the cortex is a layer one cell thick called the **endodermis**. The small cells of the endodermis have walls that are tightly bound to each other so that no material can pass between them. Incoming substances must pass through these cells to enter the vascular tissues and pass to the rest of the plant. The endodermal cells ensure that only certain materials are permitted into the plant's transport system.

The **vascular cylinder** is the central area of the young root. The center of the vascular cylinder in a dicot is composed of the xylem vessels. In a cross section of a dicot root, the xylem forms a pattern that often has four arms. Sometimes the pattern may have only two or three arms.

The phloem tissues are located between the arms of the xylem. The xylem and phloem are separated by the **vascular cambium**—a layer of meristematic tissue that can produce additional xylem and phloem. Between the vascular tissues and the endodermis is the **pericycle** (PEHR ih SYE kul), which is also a meristematic tissue. When a secondary root is formed, it will originate in the pericycle and vascular cambium.

In monocots, pith is surrounded by vascular tissues. The vascular cambium is between the xylem and phloem. The endodermis separates the vascular cylinder from the cortex and epidermis.

cortex: (L. *cortex*—bark or rind)
endodermis: endo- (within) + -dermis, -derm, derma-, or dermi- (Gk. *derma*—skin)
pericycle: peri- (Gk. *peri*—around) + -cycle (Gk. *kuklos*—circle)

13-26
Root hairs greatly increase the surface area of a root.

Secondary Growth of a Root

When the primary tissue cells have enlarged as much as they can, growth in diameter stops. Typically, roots of monocot and herbaceous annuals will grow no more in diameter. However, other roots, especially those of woody plants, do increase in diameter by means of **secondary growth**. Any tissues manufactured after the primary growth has ceased are called *secondary tissues*. The vascular cambium produces secondary xylem around the core of primary xylem. Secondary phloem is also produced by the vascular cambium.

If you look at the cross section in Figure 13-25, you will see that there is no meristematic tissue to manufacture more cortex and epidermis. As the vascular cylinder enlarges by adding secondary xylem and phloem, these new tissues push outward and crush the cortex and epidermis. To prevent the vascular cylinder from being exposed directly to the soil, the pericycle produces cork cells. The thick-walled dead cork cells seal the vascular cylinder so that it cannot absorb water. Since this portion of the root is now sealed, absorption can occur only through the epidermis near the tip of the root. The root must continue to grow in length if it is to continue to absorb water. Thus, large perennial plants must have extensive root systems to overcome this restriction in absorption of water.

Review Questions 13.6

1. List the four primary functions of a root.
2. Describe the primary growth of a root. List the primary tissues of a root.
3. What is the function of root hairs?
4. Describe the secondary growth of a root, using the names of a root's secondary tissues in your answer.
⊙ 5. Roots must continue primary growth even though the plant is well anchored. Why?

13.7 The Stem

There are almost as many types of stems as there are types of plants. Aboveground stems may be erect (trees and many flowers) or creeping (watermelons and cucumbers). Stems may be thin and herbaceous (most annuals) or thick and woody. Stems may also be subterranean (the cattail, Irish potato, tulip, and onion).

Plant stems are classified as either *woody* or *herbaceous* (hur BAY shus). The trunk and branches of a tree are woody. Woody parts generally are strong because their cells have thick walls. In most woody plant parts some tissues remain undifferentiated (do not become specialized). Additional plant tissues can form from this undifferentiated tissue in future years, as is apparent in the discussion of secondary root growth. This growth may continue for centuries and produce tall trees with thick trunks.

Herbaceous plant parts, on the other hand, usually live for only one growing season, usually less than a year. Because leaves, flowers, and nonwoody roots and stems lack the thick cell walls of woody structures, they sometimes rely on turgor pressure for support. The herbaceous parts of a plant usually remain green until they die. When dead, many herbaceous plant parts, such as corn stalks, are strong and leave stems that appear woody.

Herbaceous plants lack woody structures; those that have both woody and herbaceous structures are called *woody plants*.

Most stems perform two major functions: (1) they manufacture, support, and display leaves, and (2) they conduct to and from the leaves many of the materials needed for and manufactured by photosynthesis. Most stems carry on photosynthesis when young, and in a few plants, like the cactus, even mature stems are the primary photosynthetic organs.

13.7

Branching Patterns

In stems there are three main branching patterns:

✦ *Excurrent.* In plants such as pines and hollies, the apical bud of the main stem has dominance over the lateral buds. This **apical dominance** is exhibited in varying degrees, depending on the species. Apical dominance results in an excurrent branching pattern, usually forming cone-shaped bushes or trees with one main stem.

✦ *Deliquescent.* Most angiosperms (like the apple, oak, maple, and many shrubs) have a deliquescent (DEL ih KWEH sent) branching pattern. In these plants apical dominance exists when they are young, but later the lateral buds become more active. When these plants are mature, it may be difficult to distinguish the main stem.

✦ *Columnar.* Woody monocots and some dicots have columnar growth, typified by a crown of leaves atop an unbranched stem. Palm trees and many tropical plants exhibit columnar growth.

excurrent　　　　*deliquescent*　　　　*columnar*

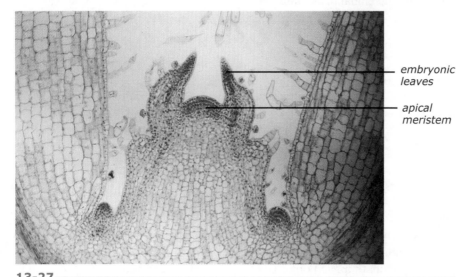

embryonic leaves

apical meristem

13-27
PMG of apical meristem

Woody Stems

A longitudinal section of the tip of a woody stem shows a small meristematic area called the **apical meristem**. Cells formed in this area of active cell division differentiate into leaves, stem tissues, and flowers.

The bud at the end of the twig is called the *apical bud.* Buds that form along the twig are called *lateral buds.* Lateral buds usually develop only if the apical bud is damaged. Those woody stems that live through a winter period form *dormant buds* in the fall.

A longitudinal section of a dormant apical bud reveals *bud scales* protecting tiny leaves and, if the tree is to bloom in early spring, flower parts. These leaves and flowers are formed in the preceding autumn. In the spring, water from the roots causes these structures to expand to almost full size overnight. When the bud scales fall, they leave rings of *bud scale scars* around the twig. Since dormant buds are formed only in the fall, counting the terminal bud scale scar areas that completely encircle the twig reveals its age.

The places where leaves are produced on a stem are **nodes**, and the areas between nodes are called *internodes.* Examination of the number of nodes and the length of internodes produced in a given year shows how productive the plant was that year.

Internal Structure and Growth of Woody Stems

A cross section of a young stem reveals an exterior covering of *epidermis.* Just under the epidermis is a layer of meristematic tissue called the **cork cambium** that will produce cork cells to protect the stem after it has grown too large for the original epidermal cells.

Inside the cork cambium is a layer of cortex, which stores materials and is usually photosynthetic. In older stems the cortex serves for storage, but as secondary growth continues, it eventually disappears completely. The phloem, with its strengthening sclerenchyma fibers, is inside the cortex. A thin layer of meristematic cells called the vascular cambium separates the phloem from the xylem. Inside the xylem is the **pith**, the largest area of a young stem. The pith, composed primarily of parenchyma cells, stores food and carries materials to different parts of the plant.

In the first year little secondary growth takes place, but in the following growing seasons secondary growth in woody stems noticeably increases the diameter. Pith continues to be the most central tissue, but its diameter does not increase. **Vascular rays**, which permit the horizontal movement of water and dissolved substances, extend from the central pith region to the outer areas of the stem.

The vascular cambium produces secondary xylem and phloem cells. If an undifferentiated cell is on the inside of the vascular cambium, it will form xylem; if on the outer surface, it will form phloem. The secondary xylem tissues are called **wood** and make up the majority of the tissue produced by secondary growth. During a growing season the

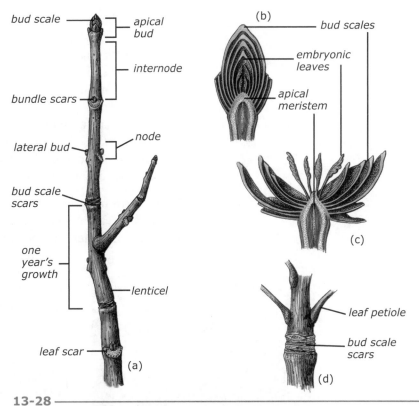

13-28

Young woody stems: (a) dormant twig, (b) longitudinal section of a dormant apical bud, (c) opening of a bud in spring, (d) young stem showing bud scale scars

Kinds of Wood

The central xylem of a mature woody stem is composed of dead cells that have been sealed off with tannins, gums, and other materials. This heartwood is often darker. The lighter sapwood is used in storage and is able to conduct water and dissolved minerals. Wood varies in density, hardness, durability, strength, grain, and texture.

Hardwood trees are usually angiosperms such as oaks, maples, walnuts, hickories, and cherries. Because of their strength and grain, hardwoods are used primarily for cabinets, furniture, veneers, tool handles, and musical instruments.

Softwood usually comes from gymnosperms. Softwoods are easier to work with than hardwoods, but they lack the strength, density, and beauty of the hardwoods. Softwoods, including firs, pines, spruces, and cedars, are used for construction, pulp, plywood, and pencils.

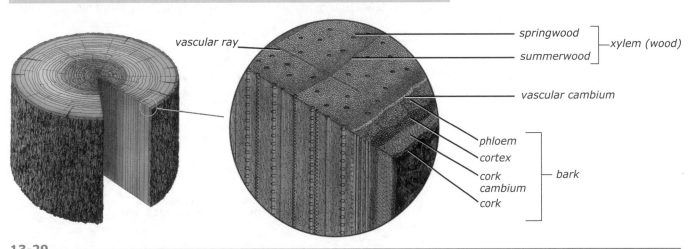

13-29

Section of a mature woody stem

vascular cambium constantly produces new xylem. In the spring when there is abundant water, xylem vessels with large diameters and thin walls are produced. Drier summers usually result in the formation of xylem vessels of smaller diameter with thicker cell walls. The xylem of larger diameter is called *springwood* and is lighter in color than the denser *summerwood*.

A cross section of a woody stem will show these layers of springwood and summerwood as **annual rings**. By studying annual rings, scientists can tell not only how old a tree is but also how the weather and other conditions have changed. Some trees produce more than one ring per year; in moist tropics, growth rings may be poorly defined and often do not represent a year's growth.

The Growth of Bark

The vascular cambium also produces secondary phloem but in much smaller quantities than xylem. Secondary phloem in woody stems is usually limited to an area just outside the vascular cambium. Sclerenchymal tissue is often found between the phloem and adds strength to the phloem tissues. All the tissues outside the vascular cambium make up the **bark**. The bark is the outer protective covering of a woody plant. The phloem and the cortex compose the *inner bark*.

The cork cambium produces flattened thin-walled cells that fill with a fatty substance called *suberin*. These cork cells then die but are impenetrable to water, gases, and most parasites. As the diameter of the stem increases, the cork cambium produces more cork; however, the cork cannot expand along with the cork cambium. As secondary growth continues, the outermost layers of cork split, forming the textured *outer bark*. *Lenticels*, tiny openings in the cork layer of a mature woody stem, allow the stem to receive the oxygen necessary for respiration.

Different types of trees produce different types of outer barks. The white birch trees of Canada and the northern United States produce thin layers of white cork that peel off the trees. Oaks have thick, rough bark; maple bark is relatively smooth. Experts can often determine the family, genus, and even species of a tree merely by examining the outer bark.

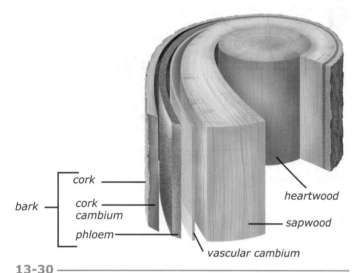

13-30
Anatomy of a log

cork
bark
cork cambium
phloem
heartwood
sapwood
vascular cambium

13-31
Cross section of a woody twig through a lenticel

Herbaceous Dicot and Monocot Stems

A typical herbaceous dicot, such as clover, marigold, or daisy, does not have a cork cambium and therefore will retain its epidermis throughout its entire

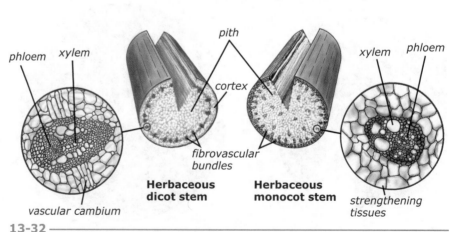

phloem   xylem
pith
cortex
fibrovascular bundles
Herbaceous dicot stem
Herbaceous monocot stem
xylem   phloem
strengthening tissues
vascular cambium

13-32
Structure of herbaceous stems

Modified Plant Parts

Not all plants have only typical leaves, roots, and stems. Many plants have modified structures that permit them to carry on unusual functions or thrive in areas where it would otherwise be impossible to grow.

Modified Leaves

Leaves often perform some very interesting functions in addition to, and sometimes in place of, photosynthesis. Spanish moss (a member of the pineapple family) has no roots but drapes gracefully on everything from tree branches to telephone poles by "holding on" with its leaves. Its scaly silver leaves also absorb water like a sponge when it rains. Some common leaf modifications are discussed in the following paragraphs.

✦ *Tendrils.* A plant part that wraps itself around something to help support the plant is called a *tendril*. Tendrils are often extensions of the midribs of compound leaves, as in the pea. They may be special leaves that lack blades, or outgrowths of the petiole, as in grapes.

Grape tendril

✦ *Spines. Spines* are hard, sharp, usually nongreen plant protectors. A spine may be a tough petiole, as in the black locust, or a hard, sharp end of a vein, as in holly. Bladeless leaves of a cactus become dead spines that protect the thick photosynthetic stems.

Holly leaf spines

Cactus spines

✦ *Succulent leaves.* Some plants store water in thick *succulent leaves*. These usually have a tough cuticle and epidermal coating and, as in the aloe, are often edged with protective spines.

Succulent leaf (aloe)

✦ *Aquatic leaves.* Plants that live in the water often have *aquatic leaves* with enlarged spongy parenchyma, which hold large quantities of air. This causes the leaves to float and supplies oxygen and carbon dioxide for respiration and photosynthesis.

Aquatic leaves (water lily)

✦ *Bracts.* Some showy flowers lack petals but have brightly colored leaves called *bracts*. Often there are small flowers in the center of a group of bracts, as in poinsettias and flowering dogwoods.

Poinsettia bracts

Modified Roots

✦ *Storage roots.* Thick and fleshy roots, usually containing starches and oils with pigments and other chemicals, are called *storage roots*. Taproots are often storage roots. Storage roots generally have secondary growth of xylem and phloem; and the pericycle, rather than producing cork, produces a thin layer of protective cells. When grown in temperate regions, most cultivated root crops such as carrots, beets, and radishes are annuals; however, they most often are hybrids of perennials with fleshy taproots similar to dandelion roots. Sweet potatoes illustrate another type of storage root. This plant produces fibrous roots that occasionally enlarge into fleshy sections.

Carrots

✦ *Adventitious roots.* Roots that grow from a stem, a petiole, or a leaf are called *adventitious roots*. They usually help anchor the plant. Adventitious roots that grow from the nodes of corn are called prop roots. Many vines (like ivy) have adventitious roots that sprout

Adventitious roots: climbing roots in ivy (left) and prop roots in corn (right)

from the stems. These roots may support the plant by growing into the cracks of a structure and enlarging to fill the cracks. These are called climbing roots.

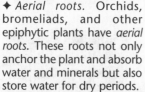
Aerial roots (orchid)

✦ *Aerial roots.* Orchids, bromeliads, and other epiphytic plants have *aerial roots.* These roots not only anchor the plant and absorb water and minerals but also store water for dry periods.

✦ *Aquatic roots.* Roots growing under water are called *aquatic roots.* They often lack root hairs, but many of them have special tubes containing vascular tissues that permit gas exchange in the growing parts of the roots.

Aquatic roots (water hyacinth)

Emily Earp and Josh Hillman—FloridaNature.org

✦ *Parasitic roots.* Parasitic roots grow into the vascular tissues of the host plants and absorb water and dissolved minerals from the xylem. Some also absorb sugars from the phloem.

Modified Stems

✦ *Succulent stems.* Succulent stems contain large quantities of water

Succulent stem (milkweed)

under thick cuticles and usually remain photosynthetic.

✦ *Stolons.* Horizontal extensions of the stem that grow along the surface of the ground are called **stolons**, or "runners." They produce secondary roots and aerial branches where the nodes touch the soil. Strawberries and grasses such as Bermuda and St. Augustine produce stolons.

Strawberry stolon

A number of different types of modified stems can be found underground and are often thought of as part of the roots. They are involved in producing leaves, keeping herbaceous plants alive over a period of years, and producing new plants.

✦ *Rhizomes.* Thick, fleshy, horizontal underground stems that produce leaves or leaf-bearing branches are called **rhizomes** (RYE ᴢᴏʜᴍᴢ).

Rhizomes of many plants, such as cattails, are edible. Rhizomes are found in many ferns, orchids, peonies, irises, and water lilies.

✦ *Bulbs.* The underground structures that are found on plants such as onions are actually collections of storage leaves, or *bulbs,* that branch from small discs of stems. Roots grow from the bottom of these stems. Other plants producing bulbs include the hyacinth, lily, daffodil, and tulip.

✦ *Tubers.* Irish or white potatoes are actually storage stems called **tubers**. Tubers produce roots and have "eyes," which are the nodes on these underground stems.

✦ *Corms.* Corms are thick underground stems that produce aerial leaves. Several underground leaves often cover a corm. The solid fleshy tissue is a storage area that enlarges each year. Lateral buds form at the nodes of the corm, and additional corms are formed in successive years.

rhizome: (Gk. *rhiza*—root)
tuber: (L. *tuber*—lump or swelling)

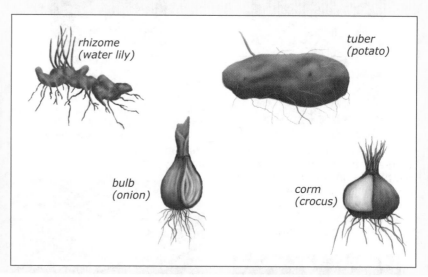
rhizome (water lily)
tuber (potato)
bulb (onion)
corm (crocus)

life span. Its stem cortex is usually photosynthetic. The vascular tissues do not form continuous concentric circles, as in woody stems, but are arranged in **fibrovascular** (ғʏᴇ broh ᴠᴀѕ kyuh lur) **bundles**. These bundles are usually located around the edge of the stem. Inside the fibrovascular bundle ring is the pith, which may contain chlorophyll and be photosynthetic, especially in young stems.

Secondary growth in herbaceous dicot stems is limited. Between the xylem and phloem of the fibrovascular bundles there usually is a layer of

fibrovascular: fibro- (L. *fibra*—fiber) + -vascular (vessel)

meristematic tissue that can produce secondary xylem and phloem. As the fibrovascular bundles grow, however, there can be little increase in the size of the epidermis, since there is no meristematic tissue available to add to that made during primary growth. Some large herbaceous dicots split their epidermis and rely on secondary strengthening tissues around the fibrovascular bundles to protect their stems.

Except at the nodes, monocot stems generally lack meristematic tissue. Secondary growth, even in "woody" monocots, is limited. The slender columnar growth of palm tree trunks is an example of this. Herbaceous monocots, such as corn and grass, have fibrovascular bundles, which when cut in cross section, usually reveal a "monkey face" of xylem, phloem, and strengthening tissues. These fibrovascular bundles are scattered throughout the pith of the stem. In some monocots, like bamboo, the central pith disappears, leaving a hollow, mature stem.

Girdling

The functions of the xylem and the phloem in a woody plant can be illustrated by girdling. If a strip of bark, including the vascular cambium, is removed from around the trunk of a tree, the tree has been girdled. Sometimes gnawing animals girdle a tree. A chain pulled back and forth in a sawing motion can also remove enough tissue to girdle a tree. Nutrients are still sent up the sapwood of a girdled tree through the xylem; therefore, the top of the tree will continue to produce leaves and carry on photosynthesis. The trunk above the ring will continue secondary growth.

However, below the girdled area the trunk will not grow in diameter. Why? Since the girdling disrupted the phloem, it is unable to transport sugars produced in the leaves down to the roots. The roots are forced to use stored food in order to continue growing. As food reserves are depleted, the roots will be unable to produce new roots and root hairs. Without new roots and root hairs, water and dissolved minerals cannot be absorbed and transported to the leaves. The lack of water will cause the leaves to wither and shut down photosynthesis. Without photosynthesis the tree finally dies because it cannot make its own food.

Review Questions 13.7

1. What are the major functions of a stem?
2. Describe the external anatomy of a twig. Tell the functions of each structure.
3. Describe secondary growth in a stem.
4. What is the difference between (a) heartwood and sapwood, (b) springwood and summerwood, and (c) hardwood and softwood?
5. Describe the formation of bark.
6. List two differences and two similarities between herbaceous monocot and herbaceous dicot stems.
⊙⊙7. What types of plants have fewer than the average number of root hairs?
⊙8. What can and cannot be determined about a tree by examining its annual rings?
⊙9. Compare and contrast (a) the primary growth of a stem and a root and (b) the secondary growth of a stem and a root.

14

The Life Processes of Plants

Botany Part II

14A—Plant Physiology

Although most plants require basically the same materials, different plants must have them in different amounts to grow properly. A corn plant requires 135–180 kg (150–190 L, 40–50 gal) of water during a growing season; however, a healthy desert cactus may not get that much in ten years. If the corn were transplanted to the desert and the cactus to the field, one would quickly wither, and the roots of the other would rot. Cypress trees grow best in flooded areas, but palm trees could not grow there. A forest-floor fern would get too much sun and die if it were transplanted into a meadow. Similarly, a meadow fern transplanted to the forest would also die, but for the opposite reason.

Our world has many vastly different habitats. God designed various kinds of plants to thrive in many of these environments. But because the conditions for one plant's healthy growth are so different from the needs of another plant, most plants grow best only in certain areas, survive in others, but cannot live in others.

Someone who plants a banana grove in northern Canada would be thought foolish. He would be ignoring either the Canadian weather or the requirements of banana plants. A farmer must consider all the environmental conditions of his farm before he plants a crop, or he may experience a crop failure. Even when a person selects a plant for his yard or his room, he should consider the conditions available to the plant.

14.1 Plants and Water

Plants contain large amounts of water. If 10 g of grass is dried, it weighs only 3 g. Most herbaceous parts of plants are over 80% water. Plants also require large amounts of water. An acre of rapidly growing corn gives off 1 million liters of water per day. A single large oak tree may lift 1000 L (264 gal) in twenty-four hours.

Plants use water for several purposes:

♦ *Photosynthesis*. For each molecule of glucose made by plants, six or more molecules of water must be broken down into hydrogen ions and electrons, which are necessary for photosynthesis.

♦ *Turgor pressure*. The abundant presence of water in the cells stiffens the herbaceous parts of plants. The water is contained in the central vacuole of the plant cell. As the central vacuole fills with water, it exerts outward pressure against the rigid cell wall.

♦ *Hydrolysis*. Plants often break large organic molecules apart by combining them with water molecules. Complex organic molecules can be broken into simple monomers by hydrolysis reactions.

♦ *Circulation*. The organic molecules and minerals that a plant needs for metabolism and growth must be dissolved in water to be moved from one area of the plant to another.

Water in the Soil

Most plants cannot absorb the water that falls on or collects on their leaves. However, the leaves often are arranged so that the water falling on them will drop to the soil in the area of the roots of that plant. The root tips, with their abundant root hairs, then absorb the water.

Some people think that plants will grow well if the roots are kept in just water. However, just like other living tissues, the roots of a plant must have oxygen to carry on cellular respiration. Although many plants can use the oxygen dissolved in water, a plant will die if the oxygen is not replaced. Other types of plants cannot extract dissolved oxygen from water. Without the oxygen contained in the spaces of the soil, these plants will die.

Plant Circulation

Chapter 13 discusses the various vascular tissues in a plant that are responsible for transporting water, nutrients, and minerals throughout the plant. But just how does this occur? How are water and minerals transported from the underground roots all the way to the top of a 300 ft tall giant sequoia? In humans, the heart circulates the blood throughout the body, but plants have no such organ. God designed plant vascular structures and ordained the physical properties of matter to accomplish circulation.

Transportation of Water

When there is a good supply of water in the soil, water is absorbed by the root hairs. The concentration of water molecules is higher outside the root cells, and the concentration of solutes is higher inside the cells. This concentration gradient permits the movement of water molecules across the root cell membranes by osmosis. The root cells must expend energy to maintain a higher concentration of solutes in their cytoplasm. This concentration gradient allows for the continued

14.1

Objectives

- Summarize the transpiration-cohesion theory
- Describe the pressure-flow model of translocation

Key Terms

root pressure
capillarity
transpiration
cohesion
transpiration-cohesion theory

turgor pressure
nastic movement
translocation
pressure-flow model

net movement of water from the soil into the root cells. The absorbed water molecules pass into the cells of the cortex and then into the vascular cylinder.

As more water enters the roots and collects in the vascular cylinder, pressure builds up. This is **root pressure**. It causes water (and dissolved minerals) to move up the xylem of the stem. If the stem of a well-watered plant is cut, drops of water will ooze from the vascular tissues as a result of root pressure.

Initially scientists thought that root pressure was the only mechanism that provided the force needed for plant circulation. However, root pressure rarely exerts more than a fraction of 1 kilogram per square centimeter of pressure (14 pounds per square inch). Research determined that dozens of kilograms per square centimeter of pressure would be necessary to force water up plants that are over 1 m tall. Another explanation had to be found.

Capillarity (KAP uh LEHR ih tee) is a property of water that, at one time, was considered a possible explanation for water movement in a plant. Water rises slightly on the surfaces of a glass container due to the attraction, or *adhesion*, between the water molecules and the glass crystals. Operating by the same principle in which water is absorbed into a paper towel, water rises in a thin-diameter tube. The thinner the diameter, the higher the water rises. But even tubes as thin as xylem vessels cannot produce capillary action to the height of tree leaves.

Although the forces of root pressure and capillarity help move water in a plant, most biologists now support another theory of water transportation: instead of being pushed up from the roots, water is pulled up from the roots. This theory is based on two considerations—transpiration and cohesion.

During the day, water is constantly evaporating from the leaves of a plant. This release of water vapor into the atmosphere is called **transpiration**. For example, a typical maple tree absorbs 361 kg (796 lb) of water in a day, but only 0.36 kg (13 oz) of this water is used by the tree. The rest (360.64 kg) passes out the stomata as water vapor.

Cohesion is the property of water molecules that causes them to "stick together." Cohesion is a result of the hydrogen bonds between water molecules. These bonds explain why water forms beads on a smooth surface and why there is enough surface tension for a needle to float on water. The molecules in the columns of water in the xylem are attracted to each other by hydrogen bonds. The water molecules also demonstrate adhesion to the xylem tube walls, aiding in the upward pull of the water column.

As water molecules evaporate from the leaves in the upper portion of the plant (transpiration), the cohesion between the water molecules pulls the entire column of water up toward the leaves. As

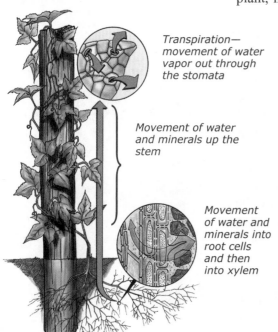

Transpiration—movement of water vapor out through the stomata

Movement of water and minerals up the stem

Movement of water and minerals into root cells and then into xylem

14-1

Transpiration-cohesion theory

Guttation

Most of the stomata of a leaf are closed at night. In some leaves the stomata at the ends of veins or along the margin of the leaf do not close. When soil moisture, humidity, and temperature are right, these leaves may guttate. In *guttation* (guh TAY shun), drops of water are forced through the stomata at the margin of the leaf. Guttation is not dew; dew comes from condensation of water vapor in the air. Guttation water originates in the roots and therefore, unlike dew, contains dissolved minerals.

transpiration: trans- (across) + -spiration (L. *spirare*—to breathe)

the water is pulled up from the vascular cylinder of the roots, additional water molecules move into the roots. Minerals are carried in with the water and are distributed throughout the plant. This explanation of water movement in plants is called the **transpiration-cohesion theory**.

Turgor Pressure and Wilting

Most of the water that comes up a stem is given off as water vapor in transpiration. The second major function of water in plants is maintaining turgidity. **Turgor pressure** is the presence of water inside a plant cell in sufficient quantity to give the cell stiffness (see Ch. 3, pp. 67, 69). As long as there is sufficient water available to living cells, they will remain turgid, and the plant will be stiff.

During hot conditions a lack of water in the soil can cause *temporary wilting*. The closing of the stomata stops as much transpiration as possible, but there is still a loss of water into the environment. The loss of water reduces turgor pressure, which causes the herbaceous parts of the plant to droop. Normally, temporary wilting is corrected during cooler conditions when transpiration slows down or when the plant is watered.

Occasionally, insects, fungi, bacteria, or other factors damage vascular tissues and cause *permanent wilting*. Permanent wilting can also occur when the available water in the soil is exhausted and not replaced before the tissues die. While being transplanted, a plant's root tips can be destroyed, causing permanent wilting because water can no longer be absorbed.

14-2 ————————
The plant on the top is well watered, and the leaves and stems are turgid. On the bottom is the same plant after water has been withheld for several days. The loss of turgidity has caused the leaves and stems to wilt.

Nastic Movements

Most people notice that tulips, morning glories, buttercups, and similar flowers open during the day, close in the evening, and reopen the following morning. These are examples of nastic movements. **Nastic movements** are the result of the loss or gain of turgor pressure in certain cells—most often at the base of the petal or leaf—and do not depend on the direction of the stimulus.

Nastic movements occur as potassium ions are pumped into or out of the cells at the base of each petal or leaf. If potassium ions are pumped out, water will also leave the cells by osmosis. This decreases the turgor pressure in these cells, causing the leaves to droop. If potassium ions are pumped into the cells, the leaves rise as the turgor pressure increases. Since nastic movements depend on the presence or absence of water, they are temporary, reversible changes.

There are two basic categories of nastic movements—thigmonastic and nyctinastic. *Thigmonastic* movements are characterized by a rapid loss of turgor pressure in response to touch. If the leaves of the sensitive plant (*Mimosa pudica*) are touched, they fold up. A few moments after the leaves have folded, they reopen. The rapid closure of the Venus flytrap leaves is also a thigmonastic response.

Nyctinastic movements— sometimes called "sleep movements"—are more gradual

Mimosa pudica *showing thigmonastic movement after being touched*

movements that appear to be in response to the cycle of light and darkness. Plants that show this type of movement include morning glories, wood sorrel, and clover. The prayer plant, a common houseplant, also folds its leaves at night (after sunset); however, putting a prayer plant into a dark room at midday will not cause it to fold its leaves. The leaves will not move until the usual time (after sunset). If the plant is placed in an area of constant temperature, humidity, and dim light, it will still open and fold its leaves by its usual schedule, not in response to the man-made environment. Therefore, some nastic movements that appear to be triggered by external factors are actually controlled by some unexplained factor within the plant.

Prayer plant: day (a) and night (b)

Morning glory: morning (a) and afternoon (b)

nastic: (Gk. *nastos*—pressed closed)

thigmonastic: thigmo- (Gk. *thigma*—to touch) + -nastic (pressed closed)

nyctinastic: nycti- (Gk. *nux*—night) + -nastic (pressed closed)

The Bible refers to the Word of God as the water whereby a Christian grows. If a Christian does not continually seek the water of the Word, his spiritual life will wilt. Psalm 1:3 says that a person who delights in God's Word will "be like a tree planted by the rivers of water, that bringeth forth his fruit in his season; his leaf also shall not wither; and whatsoever he doeth shall prosper." As a tree needs water to grow, so a Christian needs the water of God's Word for spiritual growth. And as leaves of a tree deprived of water wilt, so the spiritual life wilts unless a Christian continually seeks God's Word. That water that keeps a person's spiritual life full is available to all who have accepted Christ as their Savior.

Translocation of Carbohydrates

Carbohydrates are produced in the leaves and green stems by photosynthesis. The phloem transports these molecules throughout the plant: to meristematic tissue as an energy source for growth and to various storage areas in the roots, stems, and fruits. Botanists call the movement of carbohydrates throughout the plant **translocation**.

The model currently used to explain translocation is called the **pressure-flow model**, an entirely different process from the transpiration-cohesion theory. In the pressure-flow model, the areas where carbohydrates are stored or manufactured are called *sources*, and the places where the carbohydrates are used or stored are called *sinks*.

In the pressure-flow model, carbohydrates are actively transported from the source (e.g., leaves) into the phloem tubes. As the concentration of carbohydrate molecules in the phloem increases, water from the adjacent xylem moves into the phloem by osmosis. This movement increases the pressure within the phloem, forcing the carbohydrate-rich fluid in the phloem to move away from the source to an area where the carbohydrates are needed.

When the fluid arrives at a sink (storage area), the carbohydrate molecules are actively pumped out of the phloem and into the sink. The water will then move back into the xylem. This movement decreases the pressure in the phloem, causing the water to flow away from the sink.

The flow from source to sink can occur either up or down in the plant. For example, when carbohydrates are being produced in the leaves (the source), they can be transported to one of several sinks—to meristematic tissue for plant growth or to fruits and roots for storage. When the leaves are not producing carbohydrates, the roots, where the carbohydrates are stored, then become the source, and areas in other parts of the plant are sinks.

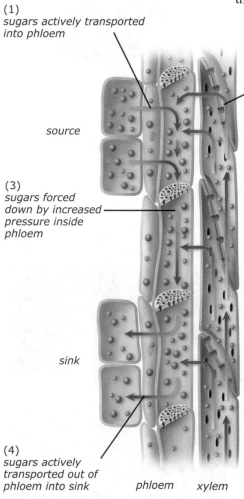

(1)
sugars actively transported into phloem

source

(2)
water moved from xylem to phloem by osmosis

(3)
sugars forced down by increased pressure inside phloem

sink

(4)
sugars actively transported out of phloem into sink

phloem   *xylem*

14-3

The pressure-flow model of translocation

From Sap to Syrup

The fluid carried in xylem and phloem vessels is sometimes called *sap* and can be used in the production of syrup. Although sap can be obtained from other trees, maple trees are most common. To collect sap, holes are cut into the maple tree, and tubes are inserted. The sap can then flow through the tubes and into collecting buckets.

After the sap is collected, it is transported to the "sugar house," where it is boiled down. Amazingly, ten gallons of sap boil down to only about one quart of syrup!

Vermont is the leading producer of maple syrup in the United States.

⮕ **Review Questions 14.1**

1. Why will all plants not grow equally well in the same conditions?
2. What are the four main uses of water in plants?
3. Why are root pressure and capillarity inadequate to explain the movement of water in a stem? Name and explain the method of water movement that most botanists agree occurs in plants over a few centimeters tall.
4. Compare and contrast temporary and permanent wilting.
5. What causes nastic movements?
6. Explain how the pressure-flow model can result in "two-way" movement of organic molecules in a plant.

14.2 Plants and Minerals

For centuries man has known that plants absorb materials from the soil. If a particular type of plant is grown and harvested repeatedly in the same field, the crop yield decreases year by year as the necessary materials are depleted. This observation led to the belief that plants eat soil, much as animals eat food. Even today people speak of "plant food," although it is an incorrect phrase.

Except for a few parasitic and saprophytic plants, plants do not "eat food." The sun is the energy source for food manufactured by plants during photosynthesis. Plants do, however, absorb soluble minerals from the soil and use them to manufacture various substances.

Minerals in the Soil

Minerals are inorganic substances found in the soil. Many different minerals are necessary for proper plant functioning and survival. The known essential nutrient elements for plants are listed in Table 14-1. Different plants require varying amounts of these elements. For example, nitrogen may be 1%–3% of the dry weight of a plant, potassium 0.3%–3%, and calcium as low as 0.1% or as high as 3.5%. Most minor essential elements and those found in only a few plants are mere traces of a plant's dry weight.

Plants are not always able to use an element in its free state. Nitrogen, for example, is a major component of the air. However, plants cannot use this atmospheric nitrogen. Plant roots absorb nitrogen in the form of nitrates or ammonium ions.

In the American colonial period and the period of westward expansion, farmers did not replenish the minerals removed from the soil by crops. This was partly because most farmers did not understand how to care for the land properly and also because they did not see a need—there was still plenty of unused land available when an area "wore out."

In the southeastern United States, cotton was the major crop in the nineteenth and early twentieth centuries. Cotton, however, rapidly depletes the soil of nitrogen. For many years, farmers would periodically allow the land to "lie fallow," not planting any crops and allowing minerals to be replenished to the soil. In the early 1900s, however, farmers realized that planting legumes (such as peas, beans, and peanuts) would also replenish the soil with nitrogen. Not only do these plants grow well in nitrogen-poor soil, but they also produce nodes on their roots that contain nitrogen-fixing bacteria. These bacteria provide nitrates for the plant; when the plant dies, the nitrates are added to the soil. These nitrates are then available for future crops. Farmers discovered that after a few years of planting legumes, they could again grow cotton successfully. This *crop rotation* is one method of keeping a mineral complement in the soil.

Today a farmer or homeowner can have his soil tested to find out the elements it contains. He can also learn what elements the various plants he wishes to grow need. If his soil lacks a particular material, he can add chemical *fertilizers* to the soil. The contents of a container of fertilizer are often expressed by three

14.2

⮕ *Objectives*

- Discuss the importance of crop rotation
- Summarize methods of replenishing mineral content in soil
- Explain how plants absorb minerals

⮕ *Key Term*

insectivorous plant

14-4 ⸻

Nitrogen can be replaced in the soil naturally by nitrogen-fixing bacteria in peanut roots or artificially by using fertilizer.

14-1 Elements Needed by Plants

Nutrient elements	Deficiency symptoms	Some functions
Major essential		

nitrogen
  nitrates $(NO_3)^-$
  ammonium $(NH_4)^+$
	light green to yellowish lower leaves; little growth	production of amino acids, proteins, nucleic acids, chlorophyll, coenzymes

phosphorus phosphates $(H_2PO_4)^-$ or $(HPO_4)^{-2}$	dark green to purplish leaves; stunted growth	formation of ATP, nucleic acids, some fats, coenzymes
potassium K^+	yellowish leaves, turning brown at the margin; weak stems	protein synthesis, cell membranes, nucleic acids
sulfur sulfate $(SO_4)^{-2}$	yellowing of young leaves	some proteins, amino acids, coenzymes
calcium Ca^{+2} (lime)	disintegration of young shoots and root tips	cell walls, aids in regulation of the uptake of other elements
magnesium Mg^{+2}	death of leaves from the stem up	chlorophyll, needed for some enzyme actions
iron Fe^{+2} or Fe^{+3}	gradual yellowing of leaves between small veins, then between larger veins	chlorophyll formation, part of many enzymes
Minor essential		
boron chlorine copper manganese molybdenum zinc	stunted growth or poorly formed plant parts, especially flowers and fruits	primarily serve as activators for enzymes
Found in some plants		
aluminum cobalt selenium silicon sodium	stunted growth or poorly formed plant parts	various functions in plants specialized for certain environments

Plant "food" is really fertilizer.

numbers, such as 10-10-10. The first number is the percentage of nitrogen; the second, phosphorus; and the third, potassium. Other elements may also be listed. In this way, a farmer adds only the needed minerals to his soil.

A third method of replenishing the minerals in the soil is to add decomposing organic matter, often called *mulch*. This method builds and maintains the soil naturally. Cultivation often removes the leaves and other materials that would normally become mulch; organic debris can be added after cultivation to replace depleted substances.

Absorption of Minerals by Roots

Nutrient minerals reach roots in a soluble form. Since the concentration of solutes is greater inside the cell cytoplasm than in the soil water outside the root cells, the cells must expend energy to move these substances into the cells against the concentration gradient. This *active transport* can be demonstrated by stopping the supply of oxygen necessary to keep root cells alive. The absorption of minerals stops as the oxygen supply decreases. Why does stopping the oxygen supply halt the absorption of minerals?

If a plant is grown in a soil that has too many soluble materials (such as might happen when an area is overfertilized), there is a higher concentration of ions outside the cell. Then, rather than being moved by active transport, the ions enter the

Plants That Catch Insects

There are over 500 species of **insectivorous** (IN sek TIV ur us) **plants**, plants that have leaves designed to catch and digest insects. These plants do not obtain energy from the insects they digest; they all contain chlorophyll and produce their own sugars. Most of them, however, live in soils lacking usable nitrogen. Insectivorous plants obtain nitrogen from digested insects. Some of these plants, if grown in soil with ample nitrogen, will not properly form their insect-catching traps.

✦ *Venus flytrap.* The Venus flytrap is found in moist soils in coastal regions of North and South Carolina. The leaf blade is an effective insect-catching trap. The red color of the trap and the sweet fluids it secretes attract flies and other insects. When an insect disturbs a set of epidermal hairs near the midrib of the leaf, the two sides quickly shut (a thigmonastic movement) and squeeze the victim. Inside the closed trap, secreted enzymes digest the insect in a week to ten days. Then the blade reopens, ready to receive another victim. Usually the leaf dies after it has closed for the third time.

Venus flytrap

✦ *Sundew.* A small ground-hugging insectivorous plant with red-tinged leaves is the sundew. Each leaf has about one hundred tiny tentacles equipped with a drop of sticky, sweet, enzyme-containing material that attracts insects. When an insect touches a tentacle, the others converge on it. Enzymes digest trapped insects.

Sundew

✦ *Butterwort.* On the inside of the narrow, curled leaves of the butterwort are hairs that secrete a sweet, sticky substance containing enzymes. When an insect touches these hairs, the leaf curls shut until the insect is digested.

Butterwort

✦ *Pitcher plants.* There are many different types of pitcher plants, some of which are tropical epiphytes. A pitcher plant produces hollow leaves that usually have a lip over the top to prevent excess rainwater from entering the pitcher. A sweet fluid that attracts insects and contains digestive

Pitcher plant (dissected to show interior)

enzymes is located in the bottom of the hollow leaf. Often there are hairs in the neck of the pitcher plant, all of which point downward. When an insect begins to move toward the sweet fluids, the hairs prevent it from crawling out. In time it falls into the pool of enzymes.

✦ *Bladderwort.* A floating plant, the bladderwort has tiny sacs that serve as insect traps. Each small sac has several hairs near its opening, which snap shut over the opening when touched, pushing the aquatic insect inside. Digestive enzymes are then secreted.

Bladderwort

---

root cells in large concentrations by diffusion, and water molecules leave the cytoplasm to go into the soil. This loss of water causes *plasmolysis* and death of the root epidermal cells. If this cellular destruction is extensive, the plant will wilt, turn brown, and die. Because of their appearance, plants that have been damaged by overfertilization are said to be "burned" by the fertilizer.

Salt-Loving Plants

Some soils having high concentrations of salt—even higher than seawater—can still support plants. Plants that grow in these salty conditions are called *halophytes* (HAL uh FITES). They have higher concentrations of salts inside their cells than in the soil around them. Salt crystals often form on halophyte leaves as water evaporates from them.

halophyte: halo- (salt) + -phyte (plant)

14-6 ——

The beach grass *Ammophila arenaria* (right) is common on both East and West Coast beaches of the United States. Halophytic turf grasses (left) have been specially developed for coastal golf courses and can tolerate salt spray and flooding by sea surge without any damage to the greens.

1. List three ways of replenishing the mineral content of the soil.
2. Explain how plants absorb minerals from the soil.
⊙ 3. What value does catching insects have for insectivorous plants?
⊙ 4. Why is it inaccurate to speak of "plant food"?

⮞ **Objectives**

- Define *hormone*
- Explain how plant hormones affect plant growth
- Describe the major categories of plant tropisms

⮞ **Key Terms**

hormone	tropism
auxin	phototropism
gibberellin	gravitropism
cytokinin	thigmotropism
ethylene	chemotropism
abscisic acid	

14-7
Fritz W. Went

14.3 Plant Hormones

In the mid-1920s, Fritz W. Went, a Dutch plant physiologist, performed a series of experiments on oat seedling coleoptiles (KOH lee AHP tilz). A coleoptile is the protective sheath of a growing embryonic stem. It protects the tissues as the young stem pushes up through the soil.

When the coleoptile tip containing the apical meristem was removed, cell elongation stopped. When Went replaced the tip, elongation continued.

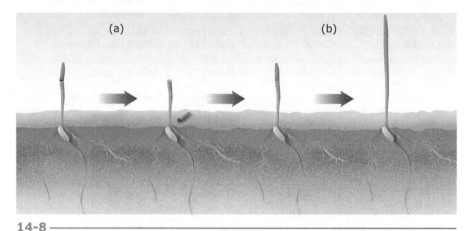

14-8
Went's first coleoptile experiment: (a) coleoptile tip removed ➤ elongation stopped; (b) coleoptile tip replaced ➤ elongation continued

Next, Went treated a thin piece of agar by placing a coleoptile tip on it and leaving it undisturbed for a period of time. After he removed the tip, he placed the agar on the coleoptile, and elongation resumed. However, if a piece of untreated agar was put on the seedling, no elongation occurred. The untreated agar showed that the elongation was not caused by the weight of the tip but by some unknown substance produced by the coleoptile tip that entered the agar and then passed to the seedling.

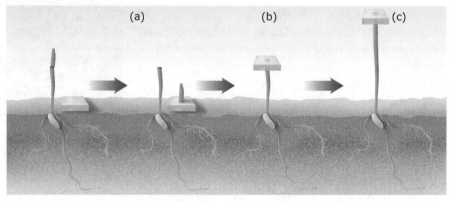

14-9
Went's second coleoptile experiment, showing the result with treated agar:
(a) coleoptile tip removed and placed on agar; (b) treated agar placed on seedling in place of coleoptile tip; (c) elongation continued

In another experiment a section of treated agar was put on only one side of the seedling. Elongation of the coleoptile occurred only on that side, causing the coleoptile to bend toward the opposite side. The chemical that Dr. Went presumed to cause elongation was later isolated and named *auxin* (AHK sin).

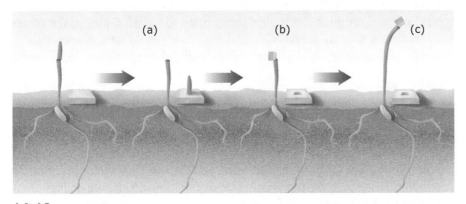

14-10

Coleoptile experiment in which the treated agar is placed on only one side of the stem: (a) coleoptile tip allowed to rest on agar; (b) treated agar placed on only one side of seedling; (c) elongation occurred only on treated side

Types of Hormones

The chemical that Went discovered is a hormone. Although hormones produced by the human body are more familiar to most people, plants also produce and use hormones. A **hormone** is any chemical produced by one area of an organism and transported to another part where it causes a response. Plant hormones control the growth and development of plants, as well as their response to different environmental conditions. Many botanists refer to plant hormones as *growth regulators*. There are five major groups of plant hormones—auxins, gibberellins, cytokinins, ethylene, and abscisic acid.

Auxins

Auxins are commonly found in stems, seeds, leaves, fruits, and in smaller quantities in roots. Auxins vary greatly in their effects on different plants and on different areas of the same plants.

The elongation of the coleoptile tips observed by Went was caused by the most common auxin, *indoleacetic acid* (IAA). IAA is necessary for the elongation of cells in the maturation region of stems. The rate and amount of elongation varies with the amount of the hormone. If the quantity of IAA in a leaf or a fruit drops below a certain level, the abscission layer at the base of the petiole forms cork cells, and the fruit or leaf eventually drops off. Fruit growers often spray IAA on their orchards a few weeks before harvesting so that all the ripe fruit remains on the tree and can be harvested at one time.

Plant chemists have produced several synthetic auxins, some of which are used to force unfertilized flowers to develop seedless fruits. Potatoes are sometimes sprayed with synthetic auxins to inhibit sprouting of the eyes (lateral buds). Another synthetic auxin, when applied in low concentrations, kills many dicots but does not affect most monocots like grasses. It is often used as a lawn weed killer.

Gibberellins

At the same time Went was doing his work, Japanese scientists were studying "foolish seedling disease" in rice. The rice plants grew very quickly but were too spindly to stand. It was discovered that *Gibberella fujikuroi*, a fungus that

14-11

The rice seedlings on the left have "foolish seedling disease."

auxin: (Gk. *auxein*—to grow)

14-12
Gibberellins are often used to increase the size of celery stalks.

Control 17 days GA-Treated 17 days

14-13
Green tomato and one treated with ethylene

14-14
The leaves of this African violet are growing toward the light.

tropism: (Gk. *tropos*—turn or change)

phototropism: photo- (light) + -tropism (turn)

can infect rice seeds, produced a chemical that caused sprouts to grow rapidly. The chemical was named **gibberellin** (JIB uh REL in). At first it was thought that this chemical was foreign to plants; however, over eighty different gibberellins have since been isolated from various plant tissues.

Gibberellins stimulate both cell division and cell elongation in leaves and stems. When these plant hormones are applied to dwarfed plants, the plants will grow to a normal height. Gibberellins are involved with flower, pollen, and fruit formation in some plants, and they stimulate growth of some seeds.

Cytokinins

Cytokinins stimulate cell division in plants and promote lateral bud growth. They can also retard the aging of flowers, fruits, and leaves. This class of hormones is produced in actively growing tissues such as roots and developing fruits and seeds. Note that cytokinins have some effects that are opposite of auxins. For example, auxins inhibit lateral bud growth, but cytokinins promote it. Auxins cause elongation of existing cells; cytokinins stimulate cellular division. Growth hormones often function together, and the ratio of the hormones present in the plant will determine what type of growth is observed.

Ethylene

Unlike the previous hormones, ethylene is a gas at room temperature. **Ethylene** is produced by the fruit and stimulates ripening. It can easily diffuse from the plant into the air and affect adjacent fruit. Fruit growers and distributors can take advantage of ethylene's effect. If the fruit is picked green, it can be shipped over greater distances with less chance of spoilage or bruising (green fruit is less likely to be bruised than ripe fruit). Once the green fruit arrives at the distribution warehouse, it is usually treated with ethephon, a synthetic chemical that releases ethylene gas as it breaks down, ripening the fruit.

Ethylene can also promote the formation of the abscission layer of leaves, flowers, and fruits. Many nuts are harvested by shaking the trees. If the trees are sprayed with this hormone just before harvesting, more nuts can be shaken out of the trees.

Abscisic Acid

Although originally thought to induce abscission (hence its name), **abscisic** (ab SIS ik) **acid** is now considered an inhibitor of many other hormones. For example, it appears to inhibit many of the growth responses of auxins and gibberellins. Abscisic acid induces dormancy in seeds and buds, causing buds to produce bud scales. When applied to growing plant tips, it forces the formation of dormant buds.

Tropisms: Plant Growth Responses

A growth response in plants as a result of a stimulus is called a **tropism**. Auxin concentrations are responsible for producing tropisms. Movement toward the stimulus is called a positive tropism, and movement away from the stimulus is called a negative tropism.

There are several types of tropisms, named according to the environmental condition that causes them.

♦ **Phototropism**, a growth response to light, is not the same in all plants or in all plant parts. Plant stems respond with growth toward the light—*positive phototropism*. Light causes auxins to migrate to the side of the stem opposite the light. The stem side with the highest concentrations of auxins elongates the most, causing the stem to grow away from the dark, hormone-containing side and toward the lighted, hormone-deficient side. On the other hand, since roots grow away from the light, they exhibit *negative phototropism*.

✦ **Gravitropism** is a response to gravity. Normally roots have positive gravitropism, while stems have negative gravitropism. This explains why a seed does not need to be "upright" when it is planted. Gravity seems to affect the root, causing more auxins to be on the lower side of the root. Because roots are very sensitive to auxins, this actually inhibits growth on the lower side, causing the root to grow more on the upper side and bend downward.

✦ **Thigmotropism** is the tropic response to contact with an object. The elongation of a stem on the side opposite to the side touched is a positive

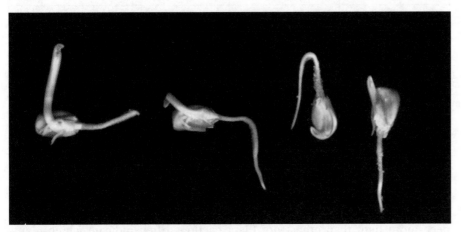

14-15

Gravitropism of corn seedlings. The stems exhibit negative gravitropism, and the roots exhibit positive gravitropism.

Hydrotropism and Heliotropism: Are They Really Tropisms?

Hydrotropism, the growth of roots toward water, was once held as an explanation for the full growth of tree roots in water-rich soil while drier areas around the tree remain rootless. There is, however, a much simpler explanation for this growth. When plant roots enter watered areas, they grow in abundance, while roots in less favorable conditions do not grow as well.

In experimental conditions, plant roots were observed to grow at the same rate until some of the roots reached an area of abundant water. While the watered roots grew faster, the other roots continued at the same rate and in the same direction they were growing rather than turning toward the water. Plants cannot sense where water is or grow toward it. Evidences of hydrotropism are usually only evidences of good and bad growing conditions; it is not really a tropism at all.

Heliotropism, or solar tracking, is a term used to describe the movement of a plant as it "follows" the sun across the sky. A good example is the movement of sunflowers; but is it a tropism? A tropism is a growth response to a stimulus, but heliotropism does not result in cellular growth. Is heliotropism a nastic movement? No, nastic movements are independent of the direction of the stimulus. Heliotropism is its own unique type of movement and is similar to the internal cause of some nyctinastic movements. Researchers are still searching for its exact mechanism of action.

hydrotropism: hydro- (water) + -tropism (turn)

heliotropism: helio- (sun) + -tropism (turn)

gravitropism: gravi- (L. *gravis*—heavy) + -tropism (turn)

thigmotropism: thigmo- (touch) + -tropism (turn)

thigmotropism. The tendrils of grapes and peas, as well as the stems of some vines like the morning glory, exhibit positive thigmotropism.

◆ **Chemotropism** is the growth response toward or away from certain chemicals. For example, the pollen tube grows in response to certain chemicals produced by the cells in the ovary (see Section 14B).

14-16
The growth of grape tendrils around a wire is an example of thigmotropism.

chemotropism: chemo- (chemical) + -tropism (turn)

14-2 Plant Tropisms

Tropism	Stimulus	Example
phototropism	light	growth toward light
gravitropism	gravity	roots growing down
thigmotropism	touch/contact	tendrils growing around wire
chemotropism	chemical	pollen tube growing toward ovary

Review Questions 14.3

1. Describe Went's coleoptile experiment and tell its significance.
2. List five plant hormones. In what ways do these hormones affect plants?
3. How do tropisms affect plants? List several examples of positive and negative tropisms.
⊙4. How do tropisms differ from nastic movements?

14.4

Objectives

* Summarize the importance of photoperiodism
* Discuss the role of phytochromes in regulating photoperiodism

Key Terms

etiolated
photoperiodism
critical dark period
short-day plant
long-day plant
day-neutral plant
phytochrome
dormancy

14-17
The plants on the left were grown in darkness and are etiolated. The plants on the right were grown in normal light.

14.4 Plants and Light

Although all plants require light for photosynthesis and other processes, some need more light than others do. Plant catalogs often indicate which plants grow best in shady areas and which require full sun. Sun-loving zinnias will be small and produce pale flowers if planted in the shade, and shade-loving impatiens will be scorched if planted in a sunny spot. As most window gardeners know, even the direction of light affects plants. Phototropism will cause geranium leaves to hug the window glass if the pot is not faithfully rotated. Until the early 1900s, however, most people were not aware that even the length of time that light is available to a plant is important.

Intensity of Light

A plant that does not receive adequate light becomes **etiolated** (EE tee uh LATE ud): the stem growth is rapid, but both the size and number of leaves are greatly reduced. The plant appears to be stretching for light. Since light is necessary for chlorophyll production, etiolated plants are pale green, yellow, or even white.

Some plants, like the dogwood tree, have two growth patterns, one for full sun and another for shade. Most trees produce different types of leaves in the shade and in the sun with differing amounts of palisade mesophyll, air spaces, stomata, and chlorophyll to accommodate the different conditions. In some trees, like the maple, sun leaves and shade leaves can be produced on various parts of the same plant, but the difference between the two types can be found only by careful observation.

Photoperiodism

A variety of tobacco plant called Maryland Mammoth grows to be 3–5 m (10–16 ft) high and produces many leaves, but when grown in Maryland, it does not begin to flower until the arrival of cold weather. The plants are then killed by frost before the seeds can form.

In the early twentieth century, this presented a problem for tobacco farmers in the area. They wanted their tobacco plants to grow taller, but they first had to understand why this variety of tobacco flowers so late. Research was begun in the early 1900s to find out why. Researchers found that seeds planted in Maryland greenhouses in the late summer flowered at the same time as those planted outside, but they were not nearly as tall as their outdoor counterparts. After more than fifteen years of research, studies showed that the flowering of the Maryland Mammoth, as well as many other plants, is affected by the length of day and night, a phenomenon now called photoperiodism.

14-18

Interrupting the critical dark period will cause a long-day plant to flower.

Initially the early researchers thought that day length was the controlling factor of photoperiodism. However, through further experimentation, researchers found that if flashes of light interrupted the dark periods, the plants would not flower. Researchers now know that a period of uninterrupted darkness rather than uninterrupted light is required for flowering. This is called the **critical dark period**. Knowledge of a plant's photoperiod enables scientists to predict when it will flower. This is particularly important to plant producers. By artificially creating the proper critical night length, they can force plants to flower—plants that normally flower in the summer or spring can have blossoms in midwinter.

Most plants grown in temperate areas are divided into three groups. **Short-day plants** flower when the period of light is less than twelve hours—sometimes less than ten hours is required for flowering. Some examples are poinsettias, goldenrods, asters, soybeans, corn, and strawberries.

Long-day plants flower when the period of light is more than twelve hours. Some examples are sunflowers, hollyhocks, sweet clover, irises, and radishes.

Day-neutral plants flower independently of the photoperiod. These plants usually flower continuously if temperature, moisture, and various other conditions are favorable. Hybrid roses, beans, tomatoes, carnations, snapdragons, peas, marigolds, and zinnias are common day-neutral plants.

As soon as many plants mature, they produce flowers if the photoperiod is right. Some plants will not flower if a photoperiod is off by ten minutes.

The physiological processes of photoperiodism are regulated by plant pigments called phytochromes (FYE tuh KROHMZ). Two forms of the phytochrome responsible for photoperiodism are found in plants. Light changes one form to the other, and darkness changes it back. Plants flower when the ratio of these two phytochromes is right for that particular variety. In many plants a single exposure to the proper photoperiod induces flowering. In most plants, if the dark period is broken even by a short exposure to dim light, the proper ratio of the two types of phytochrome will not occur, and the plants will not flower.

In combination with the photoperiod, plants often flower in response to temperature, available moisture, and other factors. Many spring-flowering woody plants (such as fruit trees, azaleas, and dogwoods, along with many

photoperiodism: photo- (light) + -periodism (Gk. *periodos*—circuit)

phytochrome: phyto- (plant) + -chrome (color)

Most plants encounter conditions unfavorable for growth at some time during the year. Annuals die during this period, leaving seeds to continue the species. Some herbaceous plants leave roots or underground stems that sprout the following year. Some woody plants lose their foliage and form dormant buds, which sprout when favorable conditions return. Where the conditions do not get too severe, some plants merely stop growth and wait for better times. For all these plants the period of inactivity is called **dormancy**. Plants have been designed by the Creator to have their dormant periods terminated by precise conditions so that unusual quirks in the season do not cause early breaking of dormancy, thus destroying the plant.

After a tulip blooms in early spring, it forms a floral bud and by all out-ward signs is ready to sprout again by midsummer. The summer sun, however, would scorch and kill the leaves of the plant. The tulip is designed to enter a period of dormancy after the bud is formed. A period of near freezing temperatures, which destroys certain chemicals in the bulb, breaks this dormancy. These chemicals prevent the tulip from sprouting in the summer or fall.

Some seeds have chemicals that inhibit sprouting. Only when these chemicals are washed away will the seed sprout. This is important for plants that need a rainy season in order to grow. On the other hand, maple tree seeds sprout as soon as they fall from the tree in early spring. It is essential that the maple seedling grow for a full season before winter comes; therefore, maple seeds will sprout even on a block of ice.

Corn seeds and other seeds that are formed in the fall require about 13 °C (55 °F) to sprout. If a corn seed sprouted as soon as it formed, winter weather would kill the young plant. A period of dryness and the spring temperatures are necessary to break the dormancy of corn seeds.

Some areas have two growing seasons in a year. These areas usually have dry, hot summers that force plants to grow only in the spring and fall. Some plants are designed to conquer this problem by rapidly growing and producing seeds in the few months of spring or fall alone. Some plants in these areas have leaves that can become dry, often curling up and becoming brittle; but when moisture returns, the leaves fill with water and growth resumes. Without these God-designed mechanisms, the plants in these areas would have been killed long ago.

flowering bulbs like tulips, crocuses, and hyacinths) form their flower buds one year (often in response to photoperiod changes and to decreased temperatures) and flower the next year following a dormant period, regardless of the photoperiod.

➡️ Review Questions 14.4

1. List several characteristics of light that are significant for plant growth.
2. Describe short-day, long-day, and day-neutral plants, and give an example of each.
3. Summarize how a period of dormancy may help certain plants.
4. List several conditions that break dormancy for different plants.
5. The plant that thrived in the living room window all summer slowly turned yellow and lost its leaves when Mom moved it to a window in the back bedroom. Mom said she had been giving it the same amount of water and "plant food." List as many factors as you can that could account for the change in the health of the plant.

14B—Plant Reproduction

Chapter 13 discusses the life cycles of various plant groups. This section concentrates on the reproduction of the flowering plants—phylum Anthophyta. Sexual reproduction of flowering plants involves the formation of flowers, fruits, and seeds; the germination of those seeds; and the growth of the plant to maturity. Although some anthophytes complete this life cycle in the space of a few weeks, others often require many years.

In addition to reproducing sexually, many plants can reproduce asexually. Both methods of reproduction have advantages and disadvantages. Since there is genetic variation each time there is sexual reproduction, a plant with the exact desired traits (such as large fruit, certain flower color, or disease resistance) may not be produced by sexual reproduction. For example, a high-quality red rose may produce hundreds or thousands of inferior rose plants by sexual reproduction before it has a high-quality offspring. However, once a plant breeder has developed the desired traits, he can reproduce those exact traits by vegetative reproduction.

14.5 Vegetative Reproduction

Vegetative reproduction is asexual; therefore, the offspring is a clone of the parent plant, having the same genetic makeup. Although all plants are capable of sexual reproduction, many ornamental and food plants are reproduced vegetatively. Since growers already know the characteristics of the variety, they want to maintain those characteristics throughout successive generations. If the plant were allowed to reproduce sexually, some of the desired characteristics might be altered or might disappear altogether.

Vegetative reproduction of plants can occur naturally or be induced. *Natural* vegetative reproduction sometimes results because a portion of a plant forms the structures to make another complete plant. In some cases, the parent plant forms small complete plants called *plantlets* on special stems or leaves that can grow independently. Man uses *induced* methods of vegetative reproduction to force the meristematic tissues to grow the missing parts of the plant.

Vegetative reproduction often produces a plant capable of maturing years sooner than a seedling. In plants such as fruit trees, berry bushes, and many ornamentals, sexual reproduction may be used to develop new varieties; and while the seeds of the new variety could be gathered and planted, it might be years before the plant matures and becomes productive. By using certain methods of asexual reproduction, plant breeders can obtain productive plants more quickly. However, if all plants were reproduced vegetatively, there would not be any new varieties of plants or plant products.

14.5

Objectives
- Describe the advantages of vegetative reproduction
- Distinguish between natural and induced vegetative reproduction
- Explain the differences between grafting and budding

Key Terms

vegetative reproduction
grafting
budding

14-19
Each of these plantlets is a natural clone of the parent.

Commercially Important Plants

Vegetatively Reproduced
   almond, apple, banana, blackberry, cherry, chrysanthemum, fig, grape, holly, iris, Irish potato, lily, olive, orange, orchid, peach, pear, pecan, pineapple, poinsettia, rose, strawberry, sweet potato, tulip

Sexually Reproduced
   barley, bean, broccoli, cabbage, carrot, celery, coconut, corn, cotton, cucumber, lettuce, maple, marigold, melon, oak, pea, peanut, pine, rice, soybean, tomato, turnip, wheat

Vegetative Reproduction by Underground Parts

The common cattail, if started from a seed, can produce almost one hundred shoots from its spreading rhizomes in a single year. The seeds establish cattail rhizomes in new areas. Vegetative reproduction, on the other hand, produces new plants around the parent. Plants like iris, bamboo, lilac, elderberry, grass, and many herbaceous weeds reproduce by rhizomes or shoots produced from spreading roots.

Corms and bulbs are underground stems that can reproduce vegetatively. Once a corm or bulb matures, tiny cormels or bulblets develop from lateral buds at the base of the parent plant. Tulips, hyacinths, and daffodils reproduce this way.

14-20
The cattail can reproduce asexually by rhizomes (inset).

Although not technically forms of vegetative reproduction because no new plant is produced, grafting and budding are used to obtain many of the genetically identical plants needed for commercial purposes. Woody plants such as apples, peaches, and roses are reproduced this way.

Grafting is the process in which a stem, called the *scion* (SYE uhn), is cut off from one plant and placed in contact with the stem of a rooted plant, called the *stock*. Thus the scion begins to grow on the stock. There are several ways in which a graft can be cut and joined to the stock; each type of plant, however, has a way that works best. The idea is to get the vascular cambiums of each plant as close together as possible. The vascular cambiums produce cells which in time become attached to the xylem and phloem and form strengthening and covering tissues. Once this happens, the graft has "taken," and the scion will grow on the stock.

Budding is similar to grafting, but rather than using a stem, a bud with a sliver of bark is placed under a slit in the bark of the stock. The top of the stock is cut off, and the grafted bud continues apical growth.

Usually grafting and budding are done while the plants are dormant. If the plants were growing, the water demands of the leaves on the scion would not be met while the new cells at the graft site were forming. The wounds of the plant must be supported and sealed, usually with elastic tape and a waxlike or plastic substance, to prevent the stock and scion from wiggling loose and drying out. Usually, successful grafting or budding involves members of the same genus or at least the same family. Most often grafting is done with woody plants, but it may be done in some herbaceous plants.

Since the meristematic tissues of the stem produce the fruits and since there is no exchange of genetic material between the meristematic tissues of the roots and the stems of a graft, the stock usually does not affect the scion and vice versa. Roots supply water and dissolved minerals. As long as there is an adequate supply of both, the grafted-in portion of the plant will grow and ex-

"Starkrimson Delicious" apple trees are the result of grafting.

press its genes normally. Grafting and budding are widely practiced today, for they permit man to obtain many genetically identical copies of a desirable plant part. They also allow production of a single plant with the desired traits from two different plants, such as a tree with two different colors of flowers.

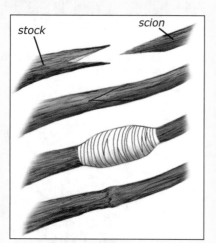

stock

scion

Biblical Grafting

The practice of grafting is centuries old and was a well-established agricultural practice in biblical times. The apostle Paul was thoroughly familiar with the principle and procedures of grafting. In Romans 11 he compares Gentile Christians to a wild olive tree branch that is grafted into the stock of a cultivated olive tree, the house of Israel. The scion is warned not to boast that the original branches were pruned for its sake. The Jews were pruned for the Gentiles' sake, but they were also pruned because of their unbelief. God says that He will prune any rotten branches and will graft in old branches if they repent.

Tubers are underground stems that often produce abundant lateral buds, each of which may develop into a new plant. Cutting the tuber apart and planting its "eyes" (lateral buds) will produce new plants. A familiar example is the white or Irish potato.

Vegetative Reproduction by Stems

Stolons are stems that produce leaves and roots—plantlets—at their nodes. When the plantlets grow too heavy to be supported, the stolon bends to the soil where the adventitious roots set in, and a new plant develops. This method of natural asexual reproduction is seen in strawberry plants, spider plants, and eelgrass.

Stems are often used in induced vegetative reproduction. A common practice among plant nurseries is *layering*. In layering, the stem of a parent plant is usually bent over and buried in moist soil. Often the stem is wounded (cut) to expose the meristematic tissues and then treated with rooting hormones. In time most plants produce roots from nodes or at the wound. Many plants that produce spindly stems or low-lying branches often bend to the soil surface and layer themselves naturally.

A *stem cutting*, or slip, is usually a section of a stem that is placed in water or in moist sand, soil, or similar medium. Exposing the meristematic tissues in lateral buds or nodes to moist conditions promotes the growth of roots in some plants, whereas others will rot under these conditions. Several plants commonly reproduced by cuttings are hollies, yews, pineapples, cacti, camellias, and roses.

The stems of many vines reproduce asexually. Vine nodes often have adventitious roots for climbing. If these roots contact the soil, they can begin to grow and thereby produce new plants.

Under proper conditions, leaves can be forced into vegetative reproduction. Most leaves, however, wilt and die long before their tissues develop new plants. A couple of notable exceptions are members of the African violet family and succulent-leaved tropical and semitropical plants (often seen as houseplants). A leaf rooting in water or moist soil is called a *leaf cutting*.

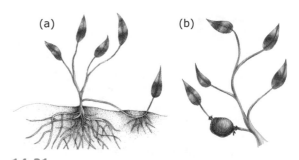

14-21
Simple layering (a) and air layering (b) are methods commonly used to propagate plants asexually. In air layering, the stem is wrapped in moist sphagnum moss rather than being buried in soil.

Vegetative Reproduction by Tissue Culturing

In the 1950s, scientists learned that some plant cells are *totipotent*; that is, they are capable of differentiating into all the cell types of the plant. Using special culturing media, scientists can force a single cell to differentiate into all the kinds of tissues needed for an entire plant, called *tissue culturing*. Once researchers have established the desired traits of a plant by breeding or genetic engineering, thousands of genetically identical plants can be produced from a very small amount of tissue. Tissue culturing has been used extensively in the flower industry to produce identical flower blossoms and disease-free plants.

14-22
Leaf rooting (left) and stem rooting are simple methods to reproduce certain species.

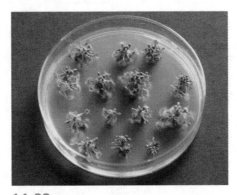

14-23
Tissue culturing can produce hundreds of genetically identical plants.

Review Questions 14.5

1. Why are some plants reproduced asexually for commercial purposes? What is the value of sexual reproduction of commercial plants that are usually asexually reproduced?
2. Differentiate between natural and induced methods of vegetative reproduction.
3. List several plants that are asexually reproduced by (a) underground roots and (b) underground stems.
4. Describe the process of layering, and name some methods of layering.
5. What advantages does reproducing plants by tissue culturing offer consumers?
6. Describe the grafting process. How does this differ from budding?
7. Why is it important to match up the vascular cambiums of the stock and scion?
8. Why does the stock of a graft not affect the type of fruit and flowers the scion produces?
9. Why are some commercially valuable plants like fruit trees and Irish potatoes reproduced asexually, while others like corn and wheat are not?

- Describe the functions of the structures of a complete flower
- Summarize the steps of gamete formation and fertilization of a flower
- Compare and contrast typical dicot and monocot seed structures
- Discuss the factors required for seed germination
- Describe several advantages and methods of seed dispersal

→ **Key Terms**

pedicel	style
receptacle	ovary
sepal	ovule
petal	pollination
stamen	pollen tube
filament	endosperm
anther	double
pollen grain	fertilization
carpel	germination
stigma	

14-24 ————————————
A tulip cut to show flower parts

14.6 Sexual Reproduction

In the plant kingdom sexual reproduction is carried on in a variety of ways. The anthophytes (angiosperms, phylum Anthophyta), which are discussed in this section, have a basic similarity in their sexual reproductive processes but almost limitless variety in the structure of their flowers, fruits, and seeds. The goal of sexual reproduction, no matter how it is accomplished, is to produce new plants and to recombine genes into different groupings, thereby increasing the genetic variability.

The Flower

Although flowers vary greatly, there are only six basic floral parts. The lower and outermost structures (pedicel, receptacle, sepals, and petals) are actually accessory flower parts. The inner two (the anther and carpel) are the reproductive parts of a flower.

Usually each flower has only one each of the first two flower parts.

✦ *Pedicel.* The **pedicel** is the stalk that supports the flower.

✦ *Receptacle.* The enlarged end of the pedicel that bears the remainder of the flower parts is the **receptacle**.

The remaining four flower parts are arranged roughly in concentric rings, and flowers often have multiples of each.

✦ *Sepals.* The outermost ring is composed of the sepals. Sepals are often green and protect the other floral parts as they form within the bud. In other flowers such as the tulip, sepals may be colored and even indistinguishable from the petals.

✦ *Petals.* **Petals**, often large and brightly colored, are just inside the sepals.

✦ *Stamen.* The **stamens**, the male reproductive structures, form the next ring. The stamen is made of the **filament**, or the stalk that bears the **anther**. An anther is a sac in which **pollen grains** are formed. Pollen grains contain the male gametes of the plant and vary greatly in size and surface structure.

14-26 ————————————
Pollen, which contains the male gametes, is formed in the anther.

✦ *Carpel.* The female reproductive structure, the **carpel** (pistil), is the innermost floral part. The uppermost tip of the carpel is the **stigma**, which when mature, has a sticky surface to receive the pollen. The **style** is a stalklike structure that supports the stigma. Located at the base of the carpel is the ovary, which contains the ovules. If an *ovum* (egg) in the ovule is fertilized, the **ovule** will develop into a seed.

petal
stamen
carpel
receptacle
pedicel
stamen { anther
filament
pollen grain
stigma
pollen tube
style
ovary
ovule
petal
sepal

14-25 ————————————
Structures of a typical flower

sepal: (blend of Gk. *skepe*—covering + L. *petalum*—petal)

ovary: (L. *ovum*—egg)

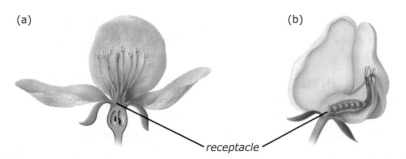

14-27
An apple blossom has an inferiorly positioned ovary (a), and the pea flower has a superiorly positioned ovary (b).

14-28
The oak catkin (a male flower) lacks petals but produces abundant pollen for wind pollination.

Plant ovaries may contain a single ovule, as peaches and pecans do, or they may contain many ovules, as beans, apples, and tomatoes do. Some flowers have multiple carpels. The strawberry and magnolia have many individual carpels on a single receptacle. Ovaries of flowers may be superior, above the receptacle, or inferior, within the receptacle.

If a flower has sepals, petals, and at least one stamen and one carpel, it is called a *complete* flower. Flowers lacking any of these structures are termed *incomplete* flowers.

Many flowers, usually incomplete ones, are pollinated by wind. They have no need of showy petals and most often do not have them. Some plants produce male flowers, containing only the stamen, and female flowers, containing only the carpel. Squash and cucumbers develop from female flowers that contain the carpel. They are insect-pollinated plants with separate male and female flowers located on the same plant. Many trees, like the oak and pecan, produce catkins, which are collections of stamens. Catkin pollen is carried by the wind to inconspicuous female flowers on other parts of the plant, which will mature into the acorn or nut. Plants such as spinach, date palms, willows, hollies, ashes, and some maples produce male flowers on one plant and female flowers on another.

The sunflower has a large receptacle with hundreds of complete, individual flowers arranged on it. On the outside rim there are sterile flowers that produce large petals but lack most other floral parts. A flower composed of many small flowers is called a *composite* flower. Daisies, chrysanthemums, dandelions, and clover are other composite flowers.

14-29
Each of the petals of a mum (chrysanthemum) actually has male and female parts. Since they are all on a common receptacle, the mum is a composite flower.

Pollination

The process whereby pollen is transferred from the anther to the stigma is called **pollination**. Many complete flowers have showy petals that aid in the transfer of pollen by insects or birds. *Nectar*, a sweet fluid secreted at the base of the petals, is the lure which brings these animals to the flowers. The petals are often designed as a sort of bull's eye to guide the pollinators to the nectar.

The anthers and the stigma are usually positioned so that anything attempting to reach the nectar must pass them, thereby aiding pollination. Although some flowers, like the members of the pea family, have their petals wrapped around the stamen and carpel, ensuring self-pollination, most flowers have mechanisms that increase the probability of cross-pollination. Often the stamen and the carpel mature at different times. Some carpels are arranged so that an insect entering the flower will place pollen from another flower on the stigma, but as it leaves with pollen from the stamen of this flower, it will not

14-30
As the bee flies from flower to flower, pollen is deposited onto the stigma of the carpel.

pass the sticky part of the stigma. Some species will not produce seeds if their carpel receives pollen from a stamen of the same flower, or in some cases, from the same plant.

Plant Fertilization—the Union of the Gametes

Each pollen grain contains several nuclei. When a pollen grain lands on the stigma of the right species, chemicals manufactured by the stigma stimulate one of these nuclei to form the **pollen tube**. The pollen tube grows down the soft tissues of the style, enters the ovary, and then enters the ovule through a small opening.

Two other nuclei in the pollen are called sperm nuclei. One sperm nucleus fuses with the egg cell to form a diploid zygote that will eventually develop into an embryo. The other sperm nucleus fuses with two nuclei from the ovule, resulting in a triploid ($3n$) nucleus. This triploid nucleus forms the endosperm, which will provide food for the embryonic plant while in the seed and when first sprouting. Because seed formation requires the fertilization of two nuclei in the ovary by two sperm nuclei, it is termed **double fertilization**. Double fertilization occurs only in angiosperms.

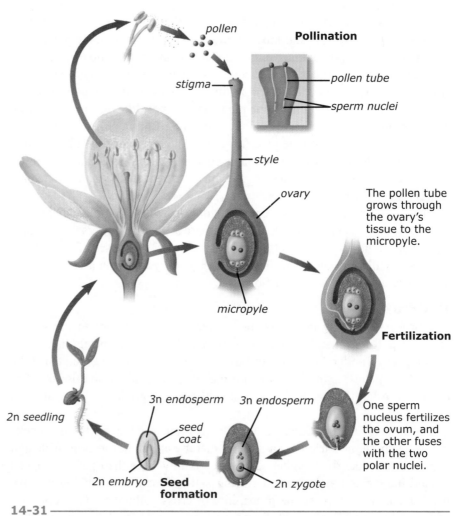

Pollination

pollen

stigma

pollen tube

sperm nuclei

style

ovary

The pollen tube grows through the ovary's tissue to the micropyle.

micropyle

Fertilization

One sperm nucleus fertilizes the ovum, and the other fuses with the two polar nuclei.

2n seedling

3n endosperm

3n endosperm

seed coat

2n embryo   **Seed formation**

2n zygote

endosperm: endo- (within) + -sperm (seed)

14-31
Pollination and fertilization

Formation of Gametes

Ovules begin as a swelling along the ovary wall, where a large megaspore mother cell undergoes meiosis. The embryo sac forms around the four haploid cells produced by this division. Three of these haploid cells degenerate; the remaining cell is called the megaspore. Its nucleus divides by mitosis three times, resulting in eight identical haploid nuclei. These nuclei organize into two groups of four and move to opposite ends of the embryo sac. One nucleus from each group migrates back to the center of the embryo sac; together they are called *polar nuclei*. During this time the embryo sac has been growing and has become covered by two *integuments* (in TEHG yoo munts). The integuments attach the ovule to the ovary wall by a stalk called the placenta. The *micropyle* is a small gap between the integuments that forms an opening into the ovule. One of the nuclei located near the micropyle enlarges to become the ovum.

In the flowering plants, the pollen (male gametes) forms inside the anther. Large microspore mother cells undergo meiosis to form four haploid microspores. Each microspore undergoes mitosis but does not divide its cytoplasm. Within each microspore, one of the haploid nuclei becomes the *tube nucleus*; the other becomes the *generative nucleus*. Later, the generative nucleus divides to form two *sperm nuclei*. Each of these nuclei performs different functions during the process of fertilization. A protective covering forms around the cell to form the pollen grain.

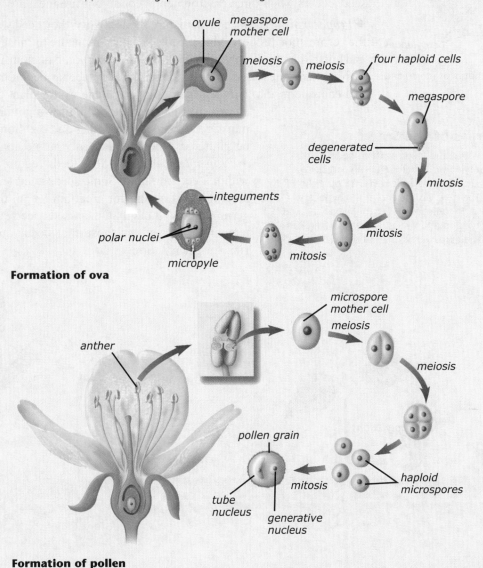

Formation of ova

Formation of pollen

integument: in- (in) + -tegu- (L. *tegere*—to cover) + -ment (E. -*ment*—action or process)

micropyle: micro- (small) + -pyle (Gk. *pule*—gate)

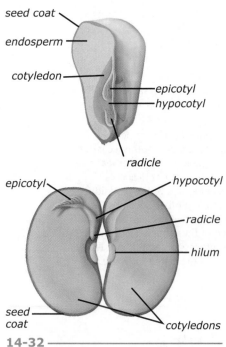

14-32

Structures of typical monocot (corn) and dicot (bean) seeds

Seeds

Seeds are made of a tiny embryonic plant, stored food, and a seed coat. The seed coat (sometimes called the testa) may be smooth and cover the content of the seed tightly, or it may be relatively loose fitting. Some seeds, such as the milkweed, have fluffy extensions of the seed coat to aid the seed in dispersal. On many seeds, a small scar called the *hilum* can be seen; this is the point where the seed was attached to the ovary wall.

The endosperm, or stored food, contained in a mature seed may be in several forms. The bean, for example, has an endosperm, but a portion is used during the formation of the embryonic bean plant. The remainder of the endosperm is contained in its two cotyledons. Corn, on the other hand, has a single cotyledon *and* an endosperm within its seed coat. Remember that anthophytes are classified into two classes based on the number of cotyledons contained in the seed.

Seed Germination

Germination, the beginning of the growth of an embryonic plant within a seed, occurs when three environmental conditions are met.

✦ *Proper moisture.* Since most mature seeds are dry, they need water to initiate the germination process. Many seed coats need moisture to cause them to swell and burst, thereby supplying moisture to the embryonic plant inside. The water activates enzymes used to break down the polysaccharides in the cotyledons and endosperm into simple sugars to provide energy for plant growth. If the embryonic plant were to begin growing without adequate moisture, it would soon wither and die. Some seeds must be saturated; others require only humid conditions.

✦ *Proper temperature.* Most seeds require temperatures within a certain range. Some seeds need temperatures to remain near freezing for a certain length of time before germination will begin. Other seeds need extremely high temperatures, such as those that occur during a forest fire, for germination.

Areas of the Seed

The embryonic plant within the seed has three general areas, based on where the cotyledons attach to it. The part below where the cotyledons attach is the *hypocotyl* (HYE puh KAHT ul), the embryonic stem. The *radicle* is the end of the hypocotyl, which will develop into the primary root of the plant. The end above the point of cotyledon attachment is the *epicotyl* (EP ih KAHT ul), or plumule. The epicotyl has one or two tiny, completely formed leaves.

radicle: (L. *radix*—root)

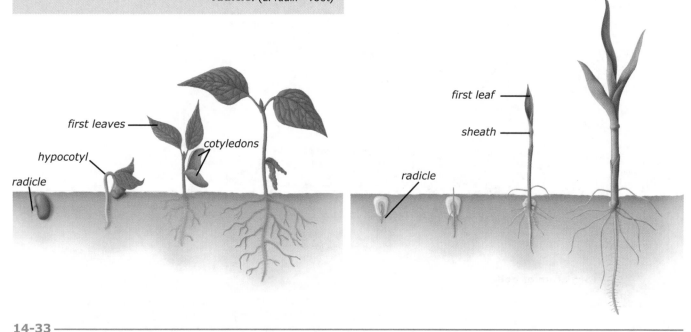

14-33

Germination of a dicot (bean) and monocot (corn)

♦ *Proper oxygen.* Oxygen is necessary for normal cellular respiration. The embryonic plant also needs oxygen to use the food stored in the endosperm and cotyledons.

Exact amounts of moisture, heat, and oxygen necessary for germination vary considerably among species. Most seeds do not require light for germination. A seed contains enough stored energy for the embryonic plant to get its roots established, absorb water, and force its first leaves above the soil. Once this occurs the stored food in the seed is usually depleted, and if the plant is unable to perform photosynthesis, it soon dies.

Once started, the enlargement of a seedling is usually rapid. All the cells necessary to establish the root and first leaves were formed before the seed matured. These cells merely need to be filled with water to enlarge and develop.

Seeds produced in the fall usually must go through a period of dormancy before they can germinate. For some seeds this dormancy may be broken by a period of cold; others require scraping of the seed coat, and some require water to wash away inhibiting chemicals from the seed coat. If kept cool and dry, seeds may remain *viable* (able to germinate) for many years. The embryonic plants in seeds use stored food to maintain life during dormancy. However, if the seed does not get oxygen, the stored food cannot be used and the embryo will die. The seeds will become nonviable. Some seeds, especially those produced in the spring, germinate without periods of dormancy.

Seed Dispersal

To continue the species, plant seeds must be dispersed, or spread out. If all the seeds of a plant fell around the parent, they would compete with one another and the parent plant for light, minerals, and water. The fruits of many plants aid seed dissemination. Animals and man often carry fleshy fruits away. When the fleshy portion is eaten and the seeds are discarded, dissemination has taken place. Many seeds have heavy protective coats that allow them to pass through the digestive system of an animal undamaged. The seeds then sprout in the droppings of the animal that consumed the fruit.

Some fruits have sticky coatings or hooks that attach to passing animals. Winged fruits (called samaras) like those of the maple and seeds like those of the dandelion are carried by the wind. The seeds of many plants, such as the poppy, are small and easily carried away from the parent plant by wind or water. Some mature pods and capsules snap open, throwing the seeds away from the parent. Other plants, such as the coconut palm, produce fruits that have waterproof walls and air chambers that allow them to float for long distances.

Fruits

A *fruit* is a mature ovary with seeds (matured ovules) inside. Various plant hormones cause the ovary to mature into the fruit. Many fruits, however, consist of accessory parts of the flower that mature along with the ovary. Fruits protect seeds, aid in dispersal, and delay sprouting. Classification is usually based on the number of carpels or flowers that form the fruit and on whether it is dry or fleshy at maturity.

Table 14-3 shows the classification for fruits as well as some examples of each type. Notice that some fruits that are commonly called nuts or berries may not be true nuts or berries.

14-34
Each of these seeds has structures designed to enhance its dissemination.

viable: (life)

Description	Examples
I. Simple fruit is produced from a single ovary.	
A. A dry fruit has a hardened or leathery exterior at maturity.	
1. A legume usually splits down the middle.	bean, pea, peanut
2. A capsule has no specific direction of split.	poppy
3. A follicle usually splits down one side.	milkweed
4. A grain has a thin seed coat fused to the ovary wall.	corn, wheat
5. A samara has one or two thin wings.	maple, elm
6. A nut has a thick, tough ovary wall.	acorn, chestnut
7. An achene has a thin ovary wall.	sunflower, dandelion
B. A fleshy fruit has thickened fleshy parts† at maturity.	
1. Some have a single seed and are called drupes. They have a hard layer around the seed.	olive, peach
2. Some have multiple seeds.	
a. A true berry has a thin skin.	cranberry, tomato
b. A modified berry has different kinds of skins.	
(1) A pepo has a thick, hard rind that is hard to remove.	cantaloupe, cucumber
(2) A hesperidium has a thick, leathery skin that is easy to remove.	orange, lemon
c. A pome has seeds contained in a core surrounded by a thin ovary wall; the remainder of the fruit is a fleshy layer† made from the receptacle.	apple, pear
II. Compound fruit‡ is produced from several separate ovaries.	
A. An aggregate fruit results when several separate ovaries from one flower fuse.	raspberry, blackberry
B. A multiple fruit results when ovaries of more than one flower fuse, forming one fruit.	pineapple, fig

† The fleshy parts of many fruits are composed of some additional part of the flower (often the receptacle) that enlarges along with the ovary. These fruits are also called accessory fruits.

‡ Many compound fruits can be considered fleshy and accessory fruits.

Review Questions 14.6

1. Name the accessory flower parts.
2. Name and describe the male reproductive structures of a flower.
3. Name and describe the female reproductive structures of a flower.
4. Explain the difference between pollination and fertilization.
5. Describe the process of fertilization in flowering plants.
6. List the basic parts of a seed and tell their functions. List the basic parts of the embryonic plant within the seed.
7. What three basic conditions must be met for a seed to germinate?
8. Compare and contrast (a) simple and compound fruits, (b) aggregate and multiple fruits, and (c) fleshy and dry fruits.
9. Describe and give an example of each of the following fruits: drupe, pome, true berry, modified berry, legume, capsule, samara, nut, grain, and achene.

The Invertebrates

Zoology Part I

15A—Introduction to the Animal Kingdom

The study of the kingdom Animalia is called zoology (ZOH AHL uh jee). Zoology is not limited to scientists in a back room of a zoo, to researchers in a laboratory, or to naturalists in a forest looking at the fauna. Almost everyone is, to some degree, a zoologist.

Animals have always fascinated mankind. The first recorded human act is Adam's naming the animals (Gen. 2:19–20). This was a monumental task since scientists now estimate that there are more than 1 267 000 living species of animals, and there are thousands that have become extinct. Furthermore, at least 15 000 new species of creatures are described and named each year. It is quite likely that the "kinds" in the original creation have given rise to many more species that are known today, since species are artificial designations made by man. Even if Adam did not name the animals by species, just recognizing all the different "kinds" would have been a major undertaking.

Animals are an important, inescapable part of man's life. Animals have served man well. They have been his beasts of burden, his food, his transportation, his sport, and even his companions. The Fall, however, affected animals

zoology: zoo- (animals) + -logy (the study of)

as well, causing their death and many other negative effects. Animals have eaten man's crops, destroyed his homes, killed his domestic animals, and have bitten, stung, attacked, and even killed man himself. Humans should exercise wise dominion over animals to gain the good they have to offer and to limit the bad that they can do.

15.1 Characteristics of the Kingdom Animalia

Animals are so much a part of daily experience that it seems unnecessary to ask, "What is an animal?" However, the answer is not as simple as it might seem. Kingdom Animalia encompasses a wide assortment of life, from jellyfish to jaguars, from weevils to whales. Despite the differences among animals, this group shares certain common characteristics. The basic characteristics of the kingdom Animalia appear in Appendix B. For almost every characteristic, however, there are some animals that are glaring exceptions. It should be remembered that this is man's classification system, and it is far from perfect.

Organisms in the kingdom Animalia not only share certain common characteristics but also engage in the following processes:

✦ *Movement.* Most animals engage in some type of locomotion (movement through the environment) to obtain food. Although most animals are motile during all or part of their life cycle, a few animals are *sessile*, fixed permanently to some object. These animals draw food to themselves by moving their environment.

✦ *Support.* Some animals are small and require little or no structure for support, but many animals achieve tremendous sizes and weights. Large animals must have a means of support, or their own weight would crush their internal structures. Some animals, such as sponges, are supported by a network of interlocking needlelike structures. Others, such as lobsters and insects, are supported by an **exoskeleton**—a system of tough plates covering the outside of the animal. Other animals have a lighter, more flexible internal system of bones or cartilage termed an **endoskeleton**.

✦ *Protective body covering.* An animal's body may be covered with only a single layer of cells or with such complex structures as feathers, shells, fur, skin, or scales. Even though body coverings differ widely, they all serve to protect the animal from predators, harsh environments, microbes, or even chemicals. Sharp spines or quills, poisonous or caustic skin secretions, shells, tough hides, or camouflaging coloration can be effective protection for the animal.

✦ *Nutrition.* Animals are heterotrophic—they must find and consume food to supply the building materials and energy necessary for life. Nutrition includes several processes: ingestion (in JEHS chun), the intake of food; digestion, the breaking down of food into substances the animal can use; and assimilation, the absorption of the food for later use as a source of energy or building material. Because nearly every living thing serves as food for some animal, the ways that animals obtain their food are quite diverse. Unicellular animals and sponges are limited to small food items because they must digest the food within each cell. All other animals and humans have a digestive cavity where larger items can be broken down before being absorbed into the cells. Many aquatic animals have special structures for filtering small food items from the water.

✦ *Respiration.* Some animals have respiration only at the cellular level; each cell of the organism exchanges oxygen and carbon dioxide directly with the environment. Some animals exchange gases through their body coverings. Other animals possess organs and systems that transport the necessary oxygen

15-1 ——————————
The mud crab has a very tough exoskeleton.

15-2 ——————————
Animals must consume food for energy since they cannot produce it.

exoskeleton: exo- (outside) + -skeleton (Gk. *skeletos*—dried body)

endoskeleton: endo- (within) + -skeleton (dried body)

ingestion: in- (in) + -gestion (to bear or to carry)

respiration: re- (again) + -spiration (L. *spirare*—to breathe)

to their cells. Those that obtain oxygen from the air usually have lungs, and those that obtain oxygen from water usually have gills.

✦ *Circulation.* Circulation is the transport of materials throughout the animal. This process helps feed and maintain those specialized tissues that cannot provide for themselves. The blood, or a similar fluid, may be contained within vessels (closed system), or it may exit vessels to bathe the surrounding tissues (open system). Some organisms simply circulate the environment through their bodies to transport substances.

✦ *Excretion.* Excretion is the elimination of waste materials. Without a means of excretion, the animal's cells would become clogged or poisoned by accumulated waste. Ammonia is the primary waste product released by the cells of animals.

✦ *Response.* The ability to perceive and respond to stimuli in the environment is known as irritability. Irritability involves such processes as the reception of stimuli and the conduction of a nerve impulse. Except for sponges, all multicellular organisms have nerve cells that make this possible. Some animals have brains that can sort stimuli and coordinate responses. A few animals possess large brains that add complex reactions such as memory, disposition, and emotion.

✦ *Reproduction.* Animal reproduction preserves the life of the species. Asexual reproduction, which does not involve the union of male and female gametes, includes budding and regeneration. While some animals can reproduce asexually, sexual reproduction is characteristic of the animal kingdom. Sexual reproduction involves the union of two gametes to form a zygote. An animal's fertilized zygote goes through **embryonic** (EM bree AHN ik) **stages** as it develops into an independent organism. Some animals have **larval** (not like the adult) **stages** as they mature.

embryonic: em- or en- (in) + -bryonic (Gk. *bruein*—to grow)

▶ *Review Questions 15.1*

1. List and describe the main characteristics of the kingdom Animalia. You may consult Appendix B.

2. List the nine main life processes of animals.

15.2

▶ *Objectives*

- Distinguish between the three types of symmetry seen in animals
- Apply relative terms to familiar examples of animal symmetry and anatomy
- Recognize the relative abundance of species among the major phyla and classes of animals

▶ *Key Terms*

symmetry
spherical symmetry
radial symmetry
bilateral symmetry
invertebrate
vertebrate

asymmetrical: a- (without) + -sym- (same) + -metrical (measure)

radial: (L. *radius*—spoke of a wheel)

15.2 Animal Anatomy and Classification

Basic Anatomy of Animals

Animals have vastly different shapes and structures. This variety, however, can be easily described with several general terms that identify parts of the animal's anatomy.

One feature of an animal's body structure is its symmetry. A symmetrical organism can be cut into equal halves. Some organisms, however, are *asymmetrical* (AY sih MET rih kul) in that their body patterns cannot be divided into equal halves. Asymmetry is characteristic of few adult animals. Different types of **symmetry** are defined by the different ways organisms can be divided into equal halves. Symmetry is a consideration of the animal's exterior only. Not all interior structures are symmetrical. The basic types of symmetry in the animal kingdom include the following:

✦ **Spherical symmetry**—a body pattern that can be divided into equal halves by a cut in any direction as long as the cut passes through the center of the body. A spherically symmetrical animal has no top, bottom, or sides.

✦ **Radial symmetry**—a body pattern that can be divided into equal halves by a cut made through the center of the animal and along its length (not a

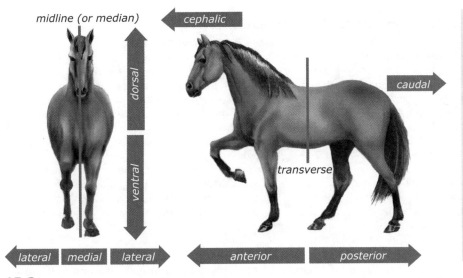

15-3
Some terms of anatomy

**Common Terms of
Animal Anatomy**

Cephalic: concerning the head
Caudal: concerning the tail
Anterior: toward the front; forward
Posterior: toward the rear; farther back
Dorsal: on or near the upper surface;
  back
Ventral: on or near the lower surface;
  front
Lateral: on or toward the side
Medial: on or toward the middle
Midline (median): divides into right
  and left
Transverse: crosses perpendicular to
  midline

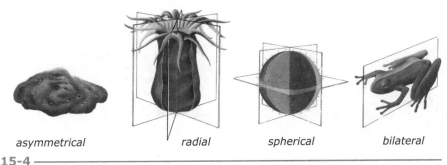

asymmetrical   radial   spherical   bilateral

15-4
Kinds of symmetry

transverse cut). A radially symmetrical animal has a top and bottom but no right and left sides. All radially symmetrical animals are aquatic.

✦ **Bilateral** (bye LAT ur ul) **symmetry**—a body pattern that can be divided into equal halves only by a cut that passes longitudinally (top to bottom) and divides the animal into right and left sides. Most often such an animal has a front (head) and a hind (tail) region.

Classification in the Animal Kingdom

Although the following division is not always used as a formal classification, the kingdom Animalia is often divided into two groups:

- **Invertebrates** (in VUR tuh BRAYTS)—animals without backbones
- **Vertebrates** (VUR tuh BRAYTS)—animals with backbones

Vertebrates include most of the familiar animals: fish, birds, snakes, lions, mice, and elephants. There are, however, not only more kinds of invertebrates but also a much larger total number of invertebrates. At least 95% of the more than 1 267 000 species of animals are invertebrates. Figure 15-5 shows the relative numbers of species in some of the various phyla and classes of the animal kingdom.

Though they lack backbones, not all invertebrates lack support. The jelly-fish and the octopus are soft bodied. Some soft-bodied animals are kept stiff by a fluid-filled internal cavity. Clams, lobsters, and insects have rigid outer structures to support their boneless bodies.

cephalic: (Gk. *kephale*—head)
caudal: (L. *cauda*—tail)
anterior: (L. *ante*—before)
posterior: (L. *posterus*—coming after)
dorsal: (L. *dorsum*—back)
ventral: (L. *ventralis*—belly)
lateral: (L. *lateralis*—side)
medial: (L. *medialis*—middle)
transverse: trans- (across) + -verse or
-vert (L. *vertere*—to turn)
bilateral: bi- (two) + -lateral (side)
invertebrate: in- (L. *in-* —not) +
-vertebrate (*vertebra*—joint)

Classification, Cladistics, and Evolution

Descriptions of the different ways animals accomplish the life processes from Subsection 15.1 are used as tools in the modern classification system. Classification of organisms in the kingdom Animalia was originally based on appearance. This caused problems. The bat, for example, has the wings, size, and flight abilities of a bird. The bat's fur coat, reproductive process, and bone structure, however, clearly identify it as a flying mammal. Thus, the structures that animals use to accomplish their bodily functions and that help zoologists to classify animals may cause problems rather than solve them.

The over twenty phyla in the animal kingdom can be arranged in a branching diagram to represent a supposed evolutionary history. If a member from each animal phylum were lined up, a person would see, after a few rearrangements, a duplicate model of proposed animal evolution. That this can be done is no accident. Evolutionists have arranged and rearranged the animal groups to make them fit their pattern of evolution. This rearrangement, though, has not altered the simplicity and usefulness of the levels of the classification system. (Nor does this man-made reclassification prove evolution. The changes have added only apparent evolutionary significance to the system.) The field of study that arranges organisms into their assumed evolutionary pathways is known as cladistics.

The use of the present levels of classification is scientifically sound. This system is the most widely used, and it is easily understood. The classification system used in this text is only an organizational tool. Any evolutionary meaning attached to it is artificial.

The remainder of Chapter 15 presents some common examples of the invertebrate phyla. Chapter 16 surveys the largest invertebrate phylum, the arthropods. Chapters 17 and 18 discuss the various classes of phylum Chordata, the only phylum containing vertebrates.

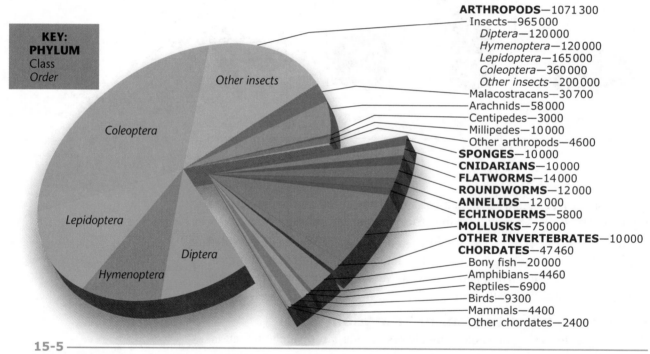

KEY:
PHYLUM
Class
Order

ARTHROPODS—1071300
Insects—965000
Diptera—120000
Hymenoptera—120000
Lepidoptera—165000
Coleoptera—360000
Other insects—200000
Malacostracans—30700
Arachnids—58000
Centipedes—3000
Millipedes—10000
Other arthropods—4600
SPONGES—10000
CNIDARIANS—10000
FLATWORMS—14000
ROUNDWORMS—12000
ANNELIDS—12000
ECHINODERMS—5800
MOLLUSKS—75000
OTHER INVERTEBRATES—10000
CHORDATES—47460
Bony fish—20000
Amphibians—4460
Reptiles—6900
Birds—9300
Mammals—4400
Other chordates—2400

15-5
A comparison of the number of species within groups in the animal kingdom

Review Questions 15.2

1. Describe the types of symmetry common in animals.

2. In a turtle, what is (a) the most dorsal structure, (b) the most ventral structure, (c) the most anterior structure, and (d) the most posterior structure?

3. Imagine a fish cut into halves along its midline. List three of its external structures that would be cut in half.

4. Imagine a transverse cut through the dorsal hump of a camel. (a) Name three structures that would be in the cephalic portion of the camel. (b) Name three structures that would be in the caudal portion of the camel.

5. Which is the larger group in each of these sets: (a) vertebrates or invertebrates, (b) mammals or insects, (c) insects or mollusks, and (d) sponges or vertebrates?

⊙6. Why would scientists occasionally want to rearrange the classification of organisms within the kingdom Animalia? Are these reasons justifiable?

15B—Poriferans and Cnidarians

Many people assume that animals are similar to humans. Indeed, in many respects they are. Many animals have systems (circulatory system, nervous system, etc.) and organs (heart, brain, liver, etc.) similar to those of humans. Some animals that are quite unlike humans even have systems and organs that function very much like similar structures in humans.

There are some animals, however, that have no organs or systems. These animals are often called the "tissue animals" because the only level of cellular organization they have is that of tissues—a group of cells working together to accomplish a function.

This section discusses two phyla of "tissue animals": Porifera (the sponges) and Cnidaria (the hydra, jellyfish, coral, and similar organisms). All of these organisms are aquatic. It will become apparent from this study why they must live in water.

15.3 Phylum Porifera: The Sponges

Sponges are unobtrusive organisms found mainly in marine habitats, though there are freshwater species. The adult sponge is sessile. As an organism, the sponge is unresponsive to most environmental changes. Because the sponge lacks certain animal characteristics, there was much scientific disagreement about it for many years. Aristotle called it a plant. Others classified it as an animal. Some said it was neither. The debate was not resolved until the mid-1800s, when stronger microscopes were available.

Where should sponges be classified? Their cellular structure and their development from zygote to mature organism clearly make them animals. Sponges are in the phylum Porifera (puh RIF uh ruh). **Poriferans** ("pore bearers") act as living pumps, drawing water into their bodies through tiny **ostia** (AHS tee uh) (sing., ostium) and expelling it through a larger opening called an **osculum** (AHS kyuh lum). This pumping of its environment through its body establishes the sponge as an animal.

Sponges range in size from 1 cm (0.5 in.) to more than 2 m (6.5 ft) across. Worldwide, there are more than 10 000 species of sponges. The freshwater sponges are in the minority with only about 150 species; these are often dull in color, carpeting rocks, sticks, and underwater plants.

15.3

Objectives
- Trace the path of water and food through a typical sponge
- Describe how a sponge's body is supported
- Contrast asexual and sexual reproduction in sponges

Key Terms

poriferan	spicule
ostium	collar cell
osculum	budding
epidermis	gemmule
amoebocyte	

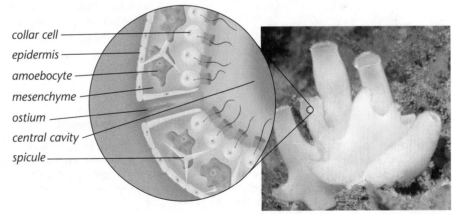

collar cell
epidermis
amoebocyte
mesenchyme
ostium
central cavity
spicule

15-6
Grantia, a typical sponge

Structure of Sponges

The body of a typical sponge is like a sac with walls composed generally of two cell layers separated by a thin, jellylike layer. The outermost cell layer is the **epidermis**, which protects the exterior of the sponge's body. The inner layer of cells lines the cavities inside the sponge.

These two layers are separated by a jellylike matrix (noncellular material) called the *mesenchyme* (MEZ un KIME). Although the mesenchyme itself is

Porifera: Pori-, pora-, pore-, or poro- (Gk. *poros*—opening or passage) + -fera (bearing)

ostium: (L. *ostium*—door, opening)

osculum: (L. *osculum*—little mouth)

epidermis: epi- (over) + -dermis (skin)

mesenchyme: mes- (middle) + -en- (in) + -chyme (Gk. *khein*—to pour)

15-7

The spicules in the wall of a sponge add both support and protection.

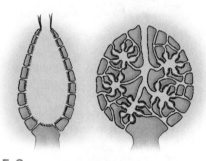

15-8

Cross section of a simple sponge with one central chamber (left) and cross section of a complex sponge with a network of canals and chambers (right)

amoebocyte: amoebo- or amoebi- (Gk. *amoibe*—change) + -cyte (cell)

noncellular, it contains a variety of structures and cells vital to the existence of a sponge. Crawling cells called amoebocytes move freely within the mesenchyme. They transport food throughout the sponge and carry wastes to the sponge surface where they can be expelled. Amoebocytes also operate in digestion.

The sponge is supported by a network of **spicules** found in the mesenchyme. Amoebocytes produce these spicules. Spicules may assume a variety of shapes, from needle and multipronged "jack" shapes to hook and barb shapes. These hard, mineral structures, distributed and intertwined throughout the body of the sponge, provide a supporting framework for the animal. The prongs or barbs of the spicules may protrude through the outer cell layer of the sponge to give it a spiny or velvety appearance. Some sponges have spicules composed of calcium carbonate (lime); others, of silica—the major component of glass. A third, softer group of sponges is supported by a branching network of tough, proteinlike fibers called *spongin*. Some sponge species contain spicules embedded in a network of spongin. Scientists use the arrangement and shape of the spicules to classify sponges.

Sponges that have thin-walled, saclike bodies are simple sponges. Complex sponges are usually larger and have a system of canals and chambers that empties into a central chamber. Some complex sponges grow to be over 2 m in diameter.

Life Processes of Sponges

The sponge consumes microscopic algae, bacteria, and organic debris. Since the sponge cannot move about to seek this food, it must draw food particles into itself. **Collar cells**, which line the cavities of sponges, have flagella that beat vigorously to generate the water current of the sponge. The water enters through the numerous ostia and leaves through the osculum. The collar cells engulf food particles carried in by the current, partially digest the food, and transfer it to the amoebocytes in the mesenchyme. The amoebocytes complete the digestion and transport the food throughout the sponge.

Sponges perform excretion and respiration at the cellular level by diffusion. Each cell can perform these functions because the sponge's thin-walled structure keeps every cell either in direct contact with or very near the environment.

Uses of Sponges

The sponge has few natural enemies. Its hard, sharp spicules and unpleasant taste and odor generally make it an undesirable mouthful for potential predators. Animals have used sponges for a variety of purposes. Many animals use the hollow sponge as a ready-made home. Certain crabs stick pieces of sponge to their shells as a living and growing camouflage. Not all sponges are harmless homes or decoration for animals. One type of sponge, for instance, attaches to mollusk shells and destroys both the shell and the animal.

Throughout history man has used the soft, durable, and remarkably absorbent spongin skeletons of some sponges. These skeletons were used in the past as padding for Greek armor, as Roman paintbrushes, and in place of a drinking cup at the Lord's crucifixion (Matt. 27:48).

Most commercial sponges today are man-made from polymers and are formed with many absorbent pores. Natural sponges, after they are cleaned and processed, can be durable cleaning tools or, as in the case of the elephant ear sponge, a fine swab for surgical or artistic purposes. The demand for these sponges has made sponge harvesting a small but profitable enterprise.

A processed natural sponge (left) and a synthetic polymer sponge (right)

Reproduction in Sponges

In sponges, asexual reproduction assumes a variety of forms. In favorable conditions a group of cells from the sponge's body may enlarge and separate from the parent to form a new individual in a process called **budding**. This is the most common type of asexual reproduction in sponges.

During periods of freezing temperatures or drought, many freshwater sponges form **gemmules** (JEM yoolz)—clusters of cells encased in a tough, spicule-reinforced coat. When harsh conditions kill the parent sponge, the gemmules survive in a dormant state. When favorable conditions return, the gemmules open and new sponges form.

The third means of asexual reproduction is regeneration. A small piece of a sponge can regenerate into a new, complete adult. Commercial sponge growers exploit this remarkable ability. By breaking a living sponge into several pieces and spreading them in beds, they cultivate a large number of sponges in the same area. This type of cultivation makes sponge harvesting much easier and more productive than traditional sponge harvesting.

Sexual reproduction in sponges occurs primarily in the spring. Sperm released by one sponge can swim and are carried by water to another sponge. A sperm enters a collar cell, and an amoebocyte transfers it to an ovum in the mesenchyme. The zygote develops into a flagellated larva, which leaves the parent sponge and swims until it finds a suitable place to attach itself and grow into a new sponge.

regeneration: re- (again) + -generation (L. *generare*—to beget)

Review Questions 15.3

1. What characteristics of sponges would cause Aristotle to classify them as plants?
2. Briefly tell how a sponge accomplishes each of the nine life processes. Where possible, include structure names.
3. How is the regenerative process of a sponge used commercially?

15.4 Phylum Cnidaria

Anyone who has spent much time at the ocean is probably familiar with members of the phylum Cnidaria (nye DEHR ee uh)—also called cnidarians. The jellyfish with its bell shape and painful sting is a common seaside annoyance. Like the jellyfish, all cnidarians are aquatic, and most are marine. Not all cnidarians, however, are flimsy sacs of jelly.

Cnidarians assume two basic forms, both of which have radial symmetry. The polyp (PAHL ip) is a cup-shaped, tubular cnidarian with a mouth and tentacles at one end and a sticky **basal disc** for attachment at the other. The medusa (mih DOO suh) form has an expanded bell-shaped body and swims freely. By contracting and relaxing the margin of its body, the medusa glides through the water in a jerky upward motion. Although some cnidarians may exist only as a polyp or medusa, many have both polyp and medusa stages in their life cycles.

Scientists have divided the cnidarians into three different classes: Hydrozoa, Scyphozoa, and Anthozoa. The remainder of this subsection examines a representative species for each class.

Hydras (Class Hydrozoa)

The characteristics of the phylum Cnidaria are clearly illustrated by the hydra, a small freshwater cnidarian commonly found in quiet lakes or ponds. Hydras are often white, green, or brown. The basic polyp body plan of the hydra is a

15.4

Objectives

- Describe the two different cnidarian body forms
- Summarize how the cnidocytes function in capturing prey
- Compare reproduction in the hydra and *Aurelia*

Key Terms

cnidarian
polyp
basal disc
medusa
gastrovascular cavity

cnidocyte
nematocyst
nerve net
hydrostatic skeleton

cnidarian: cnid- or cnido- (Gk. *knide*—nettle) + -arian or -aria (E. *-ary*—related to or pertaining to)

polyp: poly- (much or many) + -p or -pod (Gk. *pod*—foot)

medusa: (Gk. mythological woman with snakes growing out of her head)

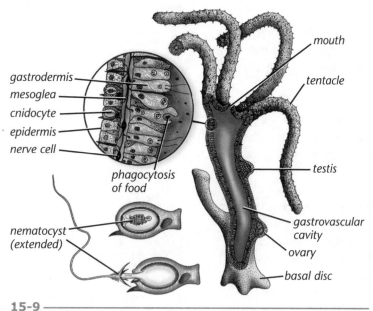

gastrodermis
mesoglea
cnidocyte
epidermis
nerve cell

phagocytosis
of food

nematocyst
(extended)

mouth

tentacle

testis

gastrovascular
cavity

ovary

basal disc

15-9
The structures of a hydra

hollow tube with a single opening or mouth. All cnidarians have only one opening to their interior cavity. That means the mouth functions in both ingestion and egestion. Long, movable tentacles, which the hydra uses to catch food, surround the mouth. The internal cavity of the animal is the **gastrovascular cavity** where digestion and food circulation occur.

The hydra's body is composed of two cell layers. The outer layer, the epidermis, is mostly protective epithelium. The cells of the *gastrodermis*, the layer lining the interior of the animal, primarily perform digestion. A jellylike layer called the *mesoglea* (mehz uh GLEE uh) separates the two cell layers.

The cells of both the epidermis and gastrodermis have long contractile fibers at their bases. Because these fibers can contract like muscle cells, the animal is able to bend its body, move its tentacles, or contract into a tiny ball when disturbed. Although the hydra can move about by a strange somersaulting motion and by various other methods, it usually remains attached by its sticky basal disc to some underwater base, like a rock or plant. Polyps are typically sessile, but some may release themselves to float or tumble to a new site.

The Feeding Process of Hydras

When feeding, the hydra hangs limply from its underwater base and allows its tentacles to dangle in the water. Batteries of stinging cells called **cnidocytes** (NYE duh SITES) line the tentacles. These stinging cells produce **nematocysts** (nih MAT uh SISTS), which are capsules containing poisonous barbs, long coiled threads, or a sticky substance. When an unwary worm or other small invertebrate brushes against the tentacles, the nematocysts discharge explosively, piercing the prey with dozens of poisonous harpoons, entangling it in a mesh of threads, or sticking to it.

The hydra's tentacles then draw the paralyzed, helpless prey into its mouth. The food is forced into the gastrovascular cavity, where digestive enzymes secreted by gland cells in the gastrodermis begin the *extracellular* phase of digestion. The cells of the gastrodermis then engulf the partially digested food particles, and digestion is completed in an *intracellular* phase. The hydra egests indigestible matter through its mouth by contracting its body.

The hydra has no special structures for excretion or respiration. The thinness of its body allows all its cells to exchange gases with and excrete waste into the environment either directly or indirectly through the gastrovascular cavity.

The Responses and Reproduction of Hydras

The hydra, unlike the sponge, is capable of dramatic responses to stimuli. The hydra automatically begins drawing in its tentacles and opening its mouth when food touches its tentacles. If the hydra is touched with a pin, however, it will contract into a tiny ball. The hydra's nerve net makes these different reactions possible. The **nerve net** is a network of nerve cells and fibers extending throughout the hydra's body that allows it to coordinate its feeding movements for greater efficiency. Once the nerve net receives a stimulus, impulses travel throughout the animal's body, and the entire hydra may respond.

gastrovascular: gastro-, gastri-, or gastr- (Gk. *gaster*—belly) + -vascular (vessel)

gastrodermis: gastro- (belly) + -dermis (skin)

mesoglea: meso- (middle) + -glea or -glia (Gk. *gloia*—glue)

cnidocyte: cnido- (nettle) + -cyte (cell)

nematocyst: nemato- (Gk. *nema*—thread) + -cyst (pouch)

Portuguese Man-of-War

The Portuguese man-of-war lives on the surface of the water. Although often thought of as a large floating medusa-type jellyfish, this highly poisonous animal is actually a colony of many hydrozoan polyps. One large gas-filled polyp acts as the float for the rest of the colony. A variety of special polyps, each with a particular job or function, hang from the float. Some sting and paralyze prey; others feed, consuming the prey; others produce gametes. The entire colony is joined and nourished by a common gastrovascular cavity. The stinging polyps can trail out as much as 15 m (50 ft) in the water. Although not usually fatal to humans, the stings from their tentacles are extremely painful.

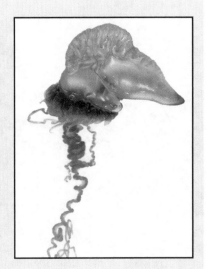

When the seasonal conditions are favorable, hydras reproduce asexually by *budding*. The bud starts as a small bump on the side of the adult. It grows tentacles and elongates until it looks like a miniature hydra; it then separates from the parent and becomes a new hydra. Hydras can also reproduce by *regeneration*: if a hydra is cut into small pieces, many of these pieces will regenerate into new, complete animals.

The sexual reproduction of hydras occurs usually in the fall or winter. The ova form in ovaries, which are small swellings of the animal's body wall. Free-swimming sperm form in similar swellings called testes (TES teez) (sing., testis). Most hydras have only one type of reproductive organ, although occasionally the same animal may have both ovaries and testes. As the ovum develops, it breaks through the body wall to become exposed to the water. At this time, normally, sperm released from the testis of another hydra fertilize the ovum. While still attached to the parent, the zygote undergoes several divisions to become a mass of cells with a tough protective coat. The mass of cells then separates from the parent and, after a dormant period during the winter months, forms a new hydra.

Jellyfish (Class Scyphozoa)

The *Aurelia* (aw REEL yuh), a common jellyfish, is a scyphozoan that represents the two-stage life cycle. As a free-swimming medusa (the dominant form), the *Aurelia* spurts through the water and feeds upon the organisms that bump into its *oral arms*, four long flaps of tissue hanging from its mouth. The male *Aurelia* produces sperm that are released into the sea. Some of these sperm may enter the gastrovascular cavity of a female, fertilizing the ovum deposited there.

The zygote then leaves the cavity of the female through the mouth and clings to the oral arms. After a period of development, the zygote becomes a ciliated larva, which leaves the female. The larva swims away and eventually attaches itself to some underwater base. It then becomes a small polyp about 12 mm in length.

The polyp stage of *Aurelia* can survive for months, store food, and produce new polyps by budding. Under certain seasonal conditions, the polyp changes dramatically: its body develops a number of horizontal constrictions, causing

15-10

Nomura's jellyfish (*Nemopilema nomurai*) is one of the largest in the class Scyphozoa, measuring up to 2 m (6.5 ft) in diameter.

oral: (L. *os*—mouth)

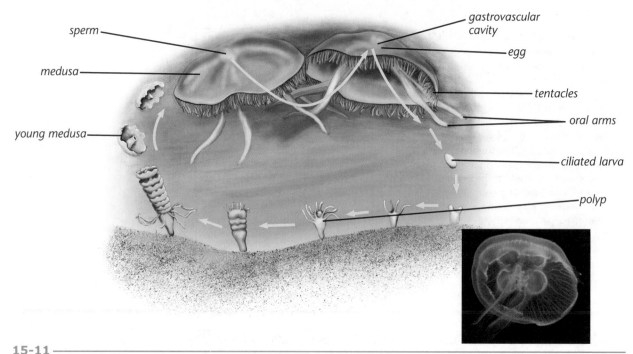

15-11
Typical jellyfish life cycle with *Aurelia* (inset)

Dangers to Coral Reefs

Known as the rainforests of the ocean, coral reefs are valuable to humanity. Reefs provide food, medicine, protection against storm surge, recreation, and beautiful aquarium fish. These resources, however, are in danger from both natural and human threats.

One natural threat to coral is the crown-of-thorns starfish, which can sometimes explode in population and eat away at coral. Also, natural ocean warming caused by El Niño may be a cause of coral bleaching. When the ocean becomes too warm, special algae (zooxanthellae) that symbiotically support coral die, killing coral polyps and leaving the coral looking bleached.

Humans also contribute to reef damage, mainly by overexploiting the reef's resources. One of the methods of capturing fish for aquariums is cyanide fishing. In cyanide fishing, divers go down to the reef and squirt cyanide at fish, which stuns them and makes them easier to catch. The cyanide, however, damages the coral. Also, mostly in poor communities, coral is mined to make building material. There is debate whether global warming adds to coral bleaching.

Despite all these natural and human threats, reefs are amazingly resilient. Humans should, however, be good stewards over these magnificent gifts from God.

it to look like a stack of saucers. These "saucers" separate, swim free, and eventually develop into large medusas, completing the life cycle.

Corals and Sea Anemones (Class Anthozoa)

The class Anthozoa, a name meaning "flower animals," includes some of the most beautiful cnidarians. *Corals* are common and important inhabitants of the sea. The actual coral cnidarian is a minute polyp that lives in a self-made stone skeleton or cup. When not feeding, the polyp can contract into the cup for protection. These fragile animals are important because they frequently cement themselves one upon another and thus form colonies. A colony may consist of

15-12
This coral reef (bottom right) was built up by thousands of individual soft-bodied polyps (top right). The larger sea anemones (left) belong to the same class but do not secrete a rocky skeleton.

thousands of polyps living in a common limestone (calcium carbonate) skeleton. Only the top layer contains living coral. These skeletons may grow so large that they form underwater ridges called reefs, which may stretch for miles and encircle an island. Corals might even form the foundation for an island.

Sea anemones (uh NEM uh neez), the "flowers" of the sea, live on the sea floor in coastal areas attached to rocks and other submerged objects. These cnidarians are supported by a hydrostatic skeleton. **Hydrostatic skeletons** depend on two layers of muscles in the body wall and a fluid-filled interior. As the muscles contract, the water pressure in the anemone increases, redistributing the water and causing its body to flex and twist in response. These large, stout-bodied polyps possess rows of tentacles that can inflict paralyzing nematocyst stings upon careless fish that get too close. While most have weak nematocysts and are unable to penetrate human skin with their barbs, a few types are very painful and dangerous to careless divers.

➧ Review Questions 15.4

1. List the nine life processes of animals and briefly tell how the hydra accomplishes each of these processes. Where possible, include structure names.

2. How do cnidocytes and nematocysts help in the food-getting and protection of a cnidarian?

3. Compare and contrast the polyp and medusa stages of a typical jellyfish.

4. List and describe several unusual cnidarians.

⊙ 5. Compare and contrast the asexual and the sexual reproduction of the hydra and sponge.

⊙ 6. If a male hydra is cut into pieces, and each piece regenerates to become a new hydra, will the new hydras be male or female? Explain your answer.

⊙ 7. Poriferans and cnidarians can be called "tissue animals." In what ways is this a good descriptive term for these animals?

15C—The Worms

Worms are soft-bodied, long, legless organisms, although some have appendages. An *appendage* is a structure or organ that is attached to the main trunk of the body. All worms have a cephalic, or head, region where most of the sensory organs and nerve cells are concentrated. This characteristic is called cephalization (SEF uh lih ZAYE shun). Worms also exhibit bilateral symmetry.

Another characteristic shared by worms is that their bodies develop from three cell layers—the epidermis, the mesoderm, and the gastroderm. These three cell layers are examined in detail as the various types of worms are studied. This section discusses Platyhelminthes, Nematoda, and Annelida, the three most common worm phyla.

15.5 Phylum Platyhelminthes: The Flatworms

Body shape is the most obvious characteristic that separates the phylum Platyhelminthes (PLAT ih hel MIN theez) from the other worm phyla. Most worms are cylindrical, but the platyhelminth has a thin, flat body, hence the common name **flatworm**. Flatworms are among the least familiar of animals. Although some flatworms dwell within human bodies, most people have never seen them. Those that are free-living shun the daylight to live under rocks or under the leaves of water plants. Most flatworms, however, are parasitic and live on or within other animals.

15.5

➧ Objectives

- List and describe the purposes of the three planarian body layers
- Summarize the feeding and digestion processes of planarians
- Contrast asexual and sexual reproduction in planarians
- List several major structural differences between parasitic and free-living flatworms
- Compare the life cycles of the sheep liver fluke and the pork tapeworm

➧ Key Terms

cephalization	parasite
flatworm	tegument
planarian	fluke
mesoderm	tapeworm
pharynx	scolex
flame cell	proglottid
hermaphroditic	

appendage: ap-, ad-, or ac- (L. *ad*—to or toward) + -pend- (to hang) + -age (E. *-age*—related to)

cephalization: cephal- (head)

Platyhelminthes: Platy- (Gk. *plat*—flat) + -helmin- (*helmis*—parasitic worms)

Free-Living Flatworms: The Planarians

The class Turbellaria, the free-living flatworms, is illustrated by the planarians (pluh NEHR ee unz). These small flatworms commonly inhabit freshwater lakes and streams and may vary in color from black or brown to white. The body of the planarian is essentially a strip of flat tissues about a centimeter long that ends in a triangular point. This point—the planarian's "head"—is marked on the dorsal side by a pair of *eyespots* with a cross-eyed appearance. These do not focus on objects but are sensitive to the presence and direction of light.

The planarian's body is composed of three cell layers. The outermost layer is the *epidermis*. It not only protects but also provides the animal with a means of locomotion. The innermost layer—the *gastroderm*—lines the digestive tract. The third layer is between the other layers and is called the mesoderm. Many different organs and systems develop from the cells of the mesoderm.

Life Processes of Planarians

The ability to move is important because the planarian must seek and capture its food. Special cells on its ventral surface enable it to move; these cells secrete a layer of slime under the planarian, and ciliated cells propel the flatworm over the slime in a smooth, gliding fashion. Contractions of the muscle layers beneath the epidermis contribute to even larger movements.

Planarians usually feed by scavenging pieces of decayed animal or plant matter and eating small organisms, such as protozoans, that they capture. The mouth is located at the center of the animal's ventral surface. As with cnidarians, the mouth is the single opening for the entrance of food and the elimination of indigestible waste products. The planarian extends a muscular, tubelike pharynx (FAR ingks) through the mouth and sucks up the food particles. The food is then drawn into the branched *intestine*, where it is broken down into small pieces and where enzymes partially digest the food. Cells lining the intestine then engulf the food and digest it completely. Once digested, the food diffuses throughout the body. Any indigestible material remaining in the intestine is egested through the pharynx.

15-13
Planarians

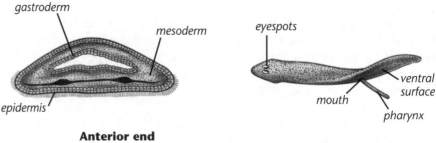

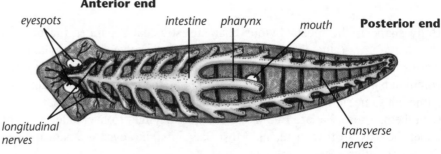

15-14
Cross section of a planarian (top left); planarian showing exterior structures (top right); structures of the planarian (bottom)

planarian: plan- or plano- (flat) + -arian (related to or pertaining to)

mesoderm: meso- (middle) + -derm (skin)

pharynx: (Gk. *pharunx*—throat)

The thin body and extensive digestive tract of the planarian keep most of the animal's cells near food and the environment. Diffusion on the cellular level can thus accomplish respiration and circulation efficiently.

The cells excrete wastes mostly by diffusion, but this excretion is also assisted by a system of tubules that extends throughout the planarian's mesoderm, connected to **flame cells**. These cells are hollow bulbs containing a tuft of cilia that beat vigorously to maintain a current in the tubules. The current carries wastes to the excretory pores to be released to the environment. This system also helps maintain the animal's water balance by excreting excess water.

The planarian's feeding activities require an elaborate nervous system. The most obvious portion of the system is a mass of nerve tissue that coordinates responses to the environment. Many nerves branch directly from this mass to sensory structures such as the eyespots and sensory organs for taste, smell, and touch located in the anterior portion of the animal. Two *longitudinal* (LAHN jih TOOD nul) *nerves* extend from the brain down the length of the animal. A number of *transverse nerves* unite the longitudinal nerves in a ladder-like pattern, enabling the planarian to coordinate its responses.

15-15 ———

Many marine, free-living flatworms are much more colorful and ornate than the familiar planarian.

Reproductive Processes of Planarians

Planarians can reproduce both asexually and sexually. To reproduce asexually, the planarian simply pulls itself apart in the middle. Each half of the animal regenerates into a complete adult. Regeneration will produce complete adults even when the animal is cut into many small pieces.

The planarian possesses an elaborate system of organs for sexual reproduction. Every planarian is **hermaphroditic** (hur MAF ruh DIT ik), meaning it has both male and female reproductive organs. Even though it possesses both reproductive systems, a planarian mates only by cross-fertilization with another individual. After planarians mate, each will release several fertilized eggs enclosed in a capsule; the capsule then attaches to a rock or plant in the water. In less than a month, small planarians hatch and emerge from the capsule.

15-16 ———

Planarians have remarkable powers of regeneration. If one is cut, almost any size piece can regenerate into a new planarian. In time, the planarian will pull itself apart, becoming two animals.

Parasitic Flatworms

A **parasite** is an organism that fastens itself to a *host* (another live organism), depends upon the host (usually for nourishment), and often harms the host. Parasitic flatworms, unlike the free-living flatworms, do not have a variety of sensory organs or efficient means of locomotion to find and capture food. These systems are less important because the host feeds the parasitic adult. Instead, it needs protection against the host's digestive juices and an effective means of remaining attached to its source of food. Characteristics of parasitic flatworms include the following:

- Few sensory organs as adults
- No external cilia in adults
- A thick **tegument** (TEG yuh munt) (a protective body covering)
- *Suckers* or *hooks* or both for attachment to a host

Scientists divide the parasitic flatworms into two major groups: the flukes and the tapeworms.

The Flukes (Class Trematoda)

The *sheep liver fluke* is a typical member of class Trematoda. As an adult, this flatworm inhabits the liver of sheep or other grazing animals, causing a disease known as "liver rot." The anatomy of these leaf-shaped parasites is quite similar to that of free-living planarians. The adult **fluke** attaches to the host

hermaphroditic: (Gk. mythological male Hermaphroditus who became united in one body with a female [hence, an organism with organs of both sexes])

tegument: (L. *tegere*—to cover)

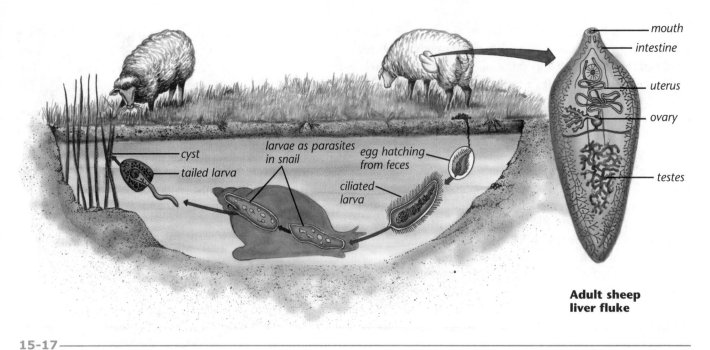

mouth
intestine
uterus
ovary
testes

**Adult sheep
liver fluke**

cyst
tailed larva
larvae as parasites
in snail
egg hatching
from feces
ciliated
larva

15-17

Life cycle and structures of the sheep liver fluke

using powerful suckers and feeds upon the host's tissues and fluids by drawing them into its anterior mouth using a muscular pharynx.

The sheep liver fluke reproduces sexually in a life cycle typical of flukes. The adult produces eggs, which pass through the sheep's bile ducts into its intestine; the eggs are then egested in *feces* (solid waste material). The eggs hatch in a body of water, releasing ciliated larvae, which enter a particular species of snail. The snail becomes an *intermediate host* (an animal that temporarily harbors the immature form of a parasite).

Within the snail, the immature flukes undergo a series of changes while reproducing asexually. Eventually the young flukes leave the snail as tailed larvae, attach to vegetation, and form cysts. If a sheep eats these cysts, its stomach digests the cyst wall, releasing an immature fluke. When developed, the fluke migrates to the sheep's liver and starts the cycle again.

The Tapeworms (Class Cestoda)

Tapeworms are also parasitic flatworms, but they bear little resemblance to the flukes. The pork tapeworm is a human parasite that, as an adult, makes its home within human intestines. This tapeworm has a small bulb-shaped "head" called the **scolex** (SKO leks) and a ribbonlike body that can grow to more than 6 m (20 ft) long. Suckers and hooks on the scolex attach the worm to the intestinal lining of its host.

A short distance from the head, the body of the tapeworm is divided into segments called **proglottids** (pro GLAHT idz). These segments originate by budding just below the scolex, and they grow larger as they mature. The small, immature proglottids are immediately beneath the scolex, while the mature, larger ones are near the posterior end of the worm.

The pork tapeworm has no mouth or digestive organs. It can absorb and use only food that the host's body has already digested. For this reason, the adult tapeworm lives in the small intestine. Here the host has completed most of the digestive processes, and the parasite can simply absorb the predigested food. The tapeworm's tegument protects it from the harsh digestive juices in

feces: (L. *faex*—dregs)
scolex: (Gk. *skolex*—worm)

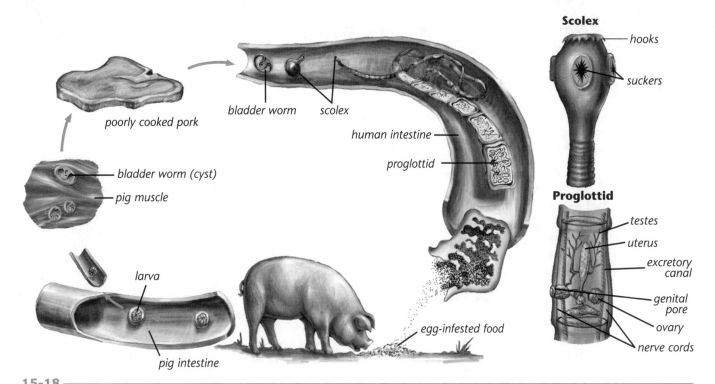

Scolex
— hooks
— suckers

poorly cooked pork

bladder worm   scolex

human intestine

proglottid

bladder worm (cyst)
pig muscle

larva

pig intestine

egg-infested food

Proglottid
— testes
— uterus
— excretory canal
— genital pore
— ovary
— nerve cords

15-18 ————————————————————————————
Life cycle and structures of the pork tapeworm

the intestine while allowing the digested food to pass unobstructed into the worm's body.

Most of the tapeworm's metabolism is dedicated to reproduction. Containing a complete set of male and female reproductive organs, each proglottid produces, fertilizes, and stores eggs. When mature, the proglottid (which now contains thousands of eggs) breaks off and is egested in the host's feces.

If a pig (the pork tapeworm's intermediate host) eats the proglottid, the eggs are released and hatch in the intestine. The microscopic larvae burrow through the pig's intestinal lining to blood vessels and are carried through the bloodstream to muscle tissues throughout the animal. Within the pig's muscles these larvae form cysts, or bladder worms. Each bladder worm contains a completely formed tapeworm scolex. If a person eats muscle tissue of an infected pig that has not been cooked enough to kill the cyst, the scolex emerges, attaches to the human's intestinal wall, and begins to produce proglottids.

Though tapeworms deprive the host of some of his food and release their toxic wastes for the host to absorb, these parasites usually pose little physical danger to the host. Serious danger arises if the worm becomes so long and tangled that it clogs the intestine or if a human accidentally consumes some of its proglottids. In the second case, the human can become the intermediate host, with the worms forming cysts in his muscles, eyes, brain, or other organs. Man can easily avoid tapeworm infection by thoroughly cooking meats that might contain the bladder worms and by practicing sanitary precautions. With the modern sanitation practices in the United States, there are few cases of human tapeworm infestations; but in developing countries where pigs may be eating food contaminated by human feces, this disease is more common.

In each of these three worm phyla there are worms that are human parasites. Some are more of a nuisance than a major threat, but others have changed the course of history and could cause major problems for anyone who does not take proper precautions.

Parasitic Flatworms: Phylum Platyhelminthes

✦ *Human blood fluke.* The human blood fluke lives as an adult in human blood vessels and produces eggs, each possessing a spine. (This spine helps the eggs penetrate the intestine or urinary bladder.) The eggs are expelled from the host with the wastes. After a period of development within a snail, the young flukes become swimming larvae that invade a human by penetrating his skin. This fluke's life cycle lacks a cyst stage.

✦ *Tapeworms.* In humans, the beef tapeworm (which has a life cycle similar to the pork tapeworm's discussed in the text) is most common. There are other tapeworms people can get by eating poorly cooked game such as bear, deer, and raccoons. Of the human tapeworms, the largest is the fish tapeworm. This worm exists in a larval form within a microscopic invertebrate and encysts in the muscle of fish. When a person eats infected fish that is raw or poorly cooked, the tapeworm is released into the person's intestinal tract, where it matures. This tapeworm may grow up to 18 m (60 ft) long and dwell in its host for many years.

Parasitic Roundworms: Phylum Nematoda

✦ *Ascaris.* (See pp. 357–59.)

✦ *Pinworms.* Pinworms commonly afflict children. These small worms live in the large intestine and migrate to the anus to lay their eggs during the night. This causes irritation and invites scratching,

but scratching contaminates hands, bed linens, and even the air with eggs. If a human ingests the eggs, reinfection occurs. The best prevention of pinworm infection is cleanliness. Drugs are available to cure pinworm infections.

✦ *Trichina worm.* The trichina (trih KYE nuh) worm is a parasite passed from one meat-eating host to another. If an animal eats meat containing encysted trichina worms, those cysts open in the animal's digestive tract, and the freed worms grow, reproduce, and burrow into the animal's intestinal walls. There the female trichina worms produce larvae that migrate to the muscles of the host. Once inside the animal's muscles, the larvae form cysts. If this infected animal is eaten, the process will repeat itself.

Pigs often ingest trichina worms by eating infected rats or food scraps containing meat with cysts in it. Humans, in turn, may contract trichinosis (TRIH kuh NOH sis), or trichina worm infection, by eating insufficiently cooked pork that contains the encysted trichina worm. The extent and effects of trichinosis make the disease an important health consideration. Some studies indicate that nearly two out of ten people in the United States will have trichinosis during their lifetimes. Mild cases of the disease cause little discomfort. Serious trichina worm infection, however, can cause extreme muscle pain, fever, body swelling, and even death.

Parasitic Segmented Worms: Phylum Annelida

✦ *Leeches.* The common 2.5 cm (1 in.) freshwater leech found in the United States feeds upon the blood of vertebrates like fish and turtles. With powerful suckers, the leech attaches to its temporary host. Once attached, it

punctures the skin of the host with its jaws, fills the wound with an antibiotic and with a chemical that prevents clotting, and sucks blood from the host. It even releases an anesthetic to deaden the pain. In one feeding, the leech will consume three to four times its own body weight in blood. This single feeding will sustain the leech for months.

A related species, found in Europe, is known as the medicinal leech. For centuries doctors exploited the medicinal leech's appetite for blood. It was once believed that "bleeding" a person could cure many illnesses; applying a 4 in. leech to a sick person became a common medical practice. Although this practice has been discredited, doctors have learned from their study of leeches and are beginning to use them again. For example, chemicals that the leeches use to prevent blood clotting have been used to prevent blood clotting during surgery. Live leeches are also used to drain off excess blood that sometimes pools in certain tissues after the reattachment of amputated limbs or the transplant of skin or muscle flaps.

Jewish Dietary Laws and Trichinosis

Some people have suggested that the Jewish dietary law forbidding the eating of pork (Lev. 11:7) was to prevent diseases like trichinosis. The law is logical because the trichina worm is an abundant, dangerous parasite, and the Israelites' cooking facilities were not ideal.

The health advantages of the ban on pork are undeniable since even today there are no completely effective drugs for treatment of trichinosis.

It is impossible to say whether health was God's reason for issuing the law. God did not ban all foods that may have been harmful, and the Israelites could have had a safe source of energy and protein in their diets if God had permitted them to eat only well-cooked pork. Many of the dietary laws appear to be simply a test of obedience for Israel.

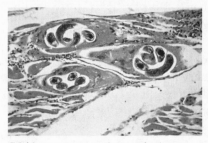

Trichina worm cysts in muscle tissue

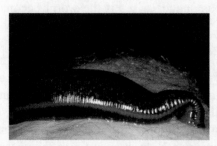

Medicinal leeches are still used in modern hospitals for some treatments.

1. What characteristics do worms from the various worm phyla have in common?

2. Name the three cell layers in flatworms and give the function of each one. In relation to each other, where are these three cell layers? Are flatworms only three cells thick? Explain.

3. List the nine life processes of animals and briefly tell how a planarian accomplishes each of them. Where possible, include structure names.

4. What are the four primary structural differences between a free-living flatworm and a parasitic flatworm?

5. Describe the life cycle of the sheep liver fluke. Why would removing snails from an area help control these parasites?

6. Describe an adult tapeworm.

7. Outline the life cycle of the pork tapeworm.

⊙ 8. Most evolutionists claim that parasitic flatworms evolved from earlier free-living forms sometime after their vertebrate hosts. Explain why this would be devolution and not evolution.

⊙ 9. Describe two flatworms, other than the pork tapeworm, that infect humans.

⊙10. Describe the life cycle of the trichina worm.

⊙11. Describe a leech.

15.6 Phylum Nematoda: The Roundworms

Members of the phylum Nematoda (NEM uh TOH duh), the **roundworms** or nematodes, are tiny cylindrical worms usually less than 2.5 cm (1 in.) long. But what the roundworms lack in size, they more than make up for in sheer numbers. Although often not visible, these tiny animals inhabit virtually every environment and every living thing. Roundworms may be found in a shovelful of dirt, an oak tree, or a human.

Not only do these worms exist in great numbers, but they also live in places where other animals would not. Roundworms thrive in the frozen arctic tundra, the heat of hot springs, the heights of mountaintops, and even the pressurized depths of the ocean floor. To survive in these environments, roundworms endure extreme environmental conditions. For example, the vinegar eel—a roundworm found in vinegar—thrives in an acidic environment that would kill most other forms of life. Other roundworms can survive in areas of drought or famine. The range and durability of these animals are virtually unmatched elsewhere in the animal kingdom.

Ironically, although roundworm species can be found in a wide range of habitats, most survive only in very specific environments. They are even highly selective about where they live in their chosen host. A parasitic roundworm living in one part of a certain animal may not survive in another part of the same animal or even in the same part of another animal. For this reason humans need not fear most parasitic roundworms, for many simply have no taste for man.

A Typical Roundworm: *Ascaris*

Ascaris (AS kuh ris), an intestinal roundworm, is a good representative of the phylum Nematoda. *Ascaris* is one of the larger parasitic roundworms, reaching 30 cm (1 ft) in length. It has an elongated, cylindrical body that tapers to a point at both ends. The body's structure is basically a tube within a tube. The outer tube consists of the epidermis and a thick, noncellular *cuticle* secreted by the epidermis. The cuticle contains fibers that give the roundworm support and protection from the host's digestive enzymes.

15.6
Objectives

- List some of the diverse habitats that harbor roundworms
- Explain why the *Ascaris* has been described as a "tube within a tube"
- Compare *Ascaris* reproduction with that of other worms already studied

Key Terms
roundworm
anus

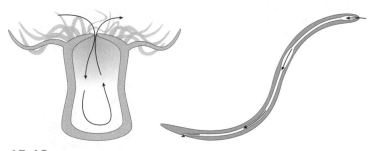

15-19 — The sea anemone (left) is a cnidarian with a single opening serving as both mouth and anus. The roundworms (right) have a complete digestive tract with two separate openings.

Nematoda: Nemat- (thread) + -oda or -ode (Gk. *odes*—of the nature of)

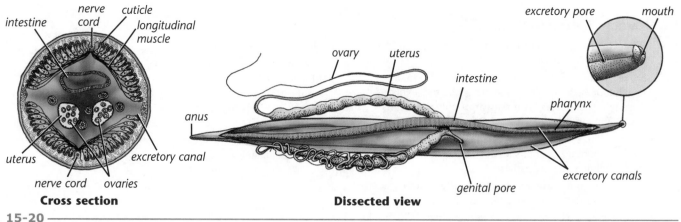

intestine · nerve cord · cuticle · longitudinal muscle

uterus · nerve cord · ovaries · excretory canal · anus

Cross section

ovary · uterus · intestine · excretory pore · mouth · pharynx · excretory canals · genital pore

Dissected view

15-20

Ascaris anatomy (female)

Within the outer tube is another tube, the digestive canal, that extends the length of the worm and has an anterior and a posterior opening. This phylum is the first in this chapter to have a digestive canal with two openings. The posterior opening, the **anus**, permits the egestion of indigestible materials (wastes) without interrupting food intake at the mouth. *Ascaris* is supported primarily by the organs and the fluid that fill the space between the two tubes.

In contrast to the smooth gliding movement of the flatworm, *Ascaris* moves by a frantic thrashing motion. *Ascaris* possesses only longitudinal muscles arranged in four bands along the animal's length. The muscle contraction in only one direction and the stiffness of its cuticle determine this roundworm's type of movement.

Ascaris feeds upon the digested food matter found in its host's intestine. The animal sucks food into its digestive tube through a long nonmuscular intestine. The intestine absorbs digested materials and eliminates wastes through the anus.

Having no organs for respiration or circulation, this roundworm accomplishes these processes on the cellular level, aided by the fluid that fills the space between the worm's digestive tube and outer body wall. Digested food from the intestine and respiratory gases from the epidermis diffuse into this fluid, which is kept circulating by the worm's movements. Two lateral excretory canals run the length of the animal and empty wastes through a single excretory pore just below the mouth.

Worms in the Word

Although the word *worms* appears numerous times in the Scriptures, a closer study of the original languages reveals that few of these belonged to any of the worm phyla in this chapter. The worms that ate the hoarded manna (Exod. 16:20) and that ate egotistical Herod (Acts 12:23) were probably wormlike maggots of some flying insects, while the worm that destroyed Jonah's gourd (Jon. 4:7) was most likely a caterpillar form of a butterfly or moth.

The best possibility of true worms would be those that Job said had "clothed" his body (Job 7:5), since these were probably some type of internal parasite.

C. elegans: A Model Worm

Although few nonscientists have heard of it, *Caenorhabditis elegans* is the most studied nematode in the world. Barely 1 mm long, this transparent worm is common in soil where it dines on bacteria and decaying organisms. Its popularity is due to the ease with which it can be maintained in the lab, its simple chromosomes, and the fact that it has many tissues and organs in common with larger and more complex animals. As a result, it serves as a model for many studies in genetics, embryology, cell biology, pharmacology, biochemistry, molecular biology, and much more.

In 1998, *C. elegans* was the subject of the first complete genome sequence for a multicellular organism ever completed. Approximately 100 000 base pairs were identified. Armed with this information, thousands of researchers worldwide are advancing our understanding of animal development and gene expression.

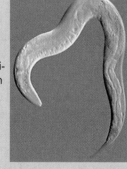

The nervous system of *Ascaris* consists of a ring of nervous tissue around the pharynx and two long nerve cords that extend from the ring down the animal's dorsal and ventral sides. The roundworm's sensory apparatus is concentrated around the mouth. Roundworms may also have chemical receptors along their sides.

Ascaris, like most roundworms, reproduces only sexually; but unlike the previously discussed worms, it is not hermaphroditic. The female *Ascaris* is larger than the male. The primary male reproductive organ is a single long coiled tubule, the testis. The female possesses a pair of coiled tubules called the ovaries.

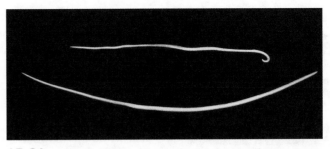

15-21
The *Ascaris* male has the hook; the female is larger.

After mating, the female *Ascaris* within the human host's intestine begins producing eggs at a rate of up to 200 000 eggs per day. The eggs are egested daily with the host's feces. If a human swallows these eggs, they hatch into tiny larvae that burrow through the intestinal wall. The blood carries these young, extremely small roundworms to the lungs. They penetrate the lungs and migrate up the air passage to the throat, where they are swallowed. Once back in the digestive tract, the worms mature and settle in the intestines, mate, and begin producing more eggs.

A single adult intestinal roundworm usually poses little threat to an otherwise healthy host. However, when the *Ascaris* population in a host gets too large, the worms may completely clog the intestine or migrate to other organs such as the liver, where they can cause serious damage.

> **Review Questions 15.6**

1. Give a general description of the organisms in the phylum Nematoda.

2. List the nine life processes of animals and briefly tell how *Ascaris* accomplishes each of these processes. Where possible, include structure names.

15.7 Phylum Annelida: The Segmented Worms

"Little rings" is a literal translation of the word *Annelida*, and the name reveals the most outstanding characteristic of phylum Annelida—the bodies of **annelids** are divided into similar rings, or segments. This segmentation is not limited to external appearance but extends into the internal organs and systems of the worm. Internally, a thin membrane, or **septum** (pl., septa), separates each segment of an annelid's body. Most annelids inhabit salt water, but freshwater and terrestrial species are not uncommon.

Although there are many members in this phylum, the earthworm is probably the most familiar. Obvious features of the earthworm include its segments and the **clitellum** (klye TEL um), a barrel-shaped swelling usually covering segments 32 through 37 (segments are numbered from the anterior to the posterior end). The clitellum is used in reproduction and helps identify the worm's anterior end. From outward appearances, the earthworm has no distinct head, although its anterior end is more pointed and is usually darker. A small mouth beneath a liplike *prostomium* (proh STOH mee um) is at the anterior tip of the earthworm. The anus is at the posterior tip.

15.7

> **Objectives**

- Describe the major life processes of the earthworm
- Compare and contrast the leech and earthworm in regard to structures and habits

> **Key Terms**

annelid
septum
clitellum
seta
esophagus
crop

gizzard
closed circulatory system
nephridium
ganglion

clitellum: (L. *clitellae*—packsaddle or saddlebag)

prostomium: pro- (before) + -stomium (mouth)

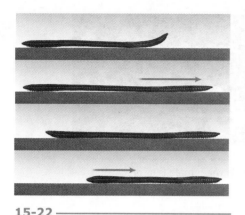

15-22
Stages of earthworm movement

Along the worm's ventral and lateral surfaces are setae (SEE tee) (sing., seta), tiny bristles, four pairs to each segment, that can be retracted into the worm's body. The earthworm moves by means of its setae and two muscle layers—a circular layer that makes the worm longer and thinner when contracted and a longitudinal layer that contracts to make the worm shorter and thicker. To move forward, the worm contracts the circular muscles, lengthening the anterior portion of its body. Next, the setae anchor the anterior end while the longitudinal muscle layer contracts, drawing the posterior end forward. The setae then anchor the posterior end of the worm as it reextends its anterior end. When on the surface of the ground or tunneling through loose earth, the earthworm uses this method to push itself along. If the soil is firmer, the worm tunnels by literally eating its way through the soil.

Earthworms and the Soil

The worms' eating soil not only sustains the worms themselves but also enriches and improves the soil. The burrows dug by earthworms permit the passage of oxygen and water deep into the ground and enable plants to send down roots much more easily. Their burrowing also helps mix the organically rich surface soil with mineral-rich subsoil, forming a fertile topsoil mixture. Earthworm casts (solid excreted wastes) are rich both in minerals that plants need for growth and in bacteria that are important in the decomposition of organic matter in the soil.

These "intestines of the soil," as Aristotle called them, have an enormous impact on soil quality. The feeding process of a single earthworm produces about 230 g (0.5 lb) of rich, fertile humus in one year. Considering that 50 000 earthworms easily inhabit 1 acre of land, earthworms can produce over 11 500 kg (24 000 lb or 12 tn) of topsoil per acre per year.

Nutrition of the Earthworm

The earthworm feeds upon vegetation, refuse, and decayed animal matter in the soil. It draws soil containing these substances into its mouth with its muscular sucking pharynx. The food-laden soil is passed down the esophagus (ih SAHF uh gus) to the **crop**, where it is stored temporarily, and then it is passed to the **gizzard**. The gizzard's muscular contractions grind food against the ingested soil and break the food up into small pieces.

The actual chemical process of digestion occurs in the intestine, where the food is broken down by digestive enzymes. Earthworms, like most other annelids, have a straight, tubular digestive tract. The indigestible materials that the worm has consumed are passed down the intestine and expelled through the anus. These solid wastes are called casts. The digested food is absorbed by the blood circulating through the walls of the intestinal tract and transported to all parts of the worm.

Circulation, Respiration, and Excretion of the Earthworm

The earthworm's size and complexity demand that it have some means of transporting vital substances such as food and oxygen to all the cells of its body. This need is met by a **closed circulatory system**. The earthworm's system involves a series of vessels containing blood. The circulatory system is closed; that is, the blood remains in the blood vessels throughout the entire cycle. The blood absorbs nutrients and oxygen from the digestive and respiratory structures and then discharges these substances into the tissues of the body.

The major vessels of the circulatory system are the *dorsal blood vessel*, which carries the blood toward the anterior end of the animal, and the *ventral blood vessel*, which transports blood toward the posterior end. Many small vessels connect these major vessels to the body wall, the internal organs, and to each other. In the region of the esophagus, five pairs of thickly muscled vessels called *aortic* (ay OR tik) *arches* connect the dorsal and ventral blood vessels. These arches help regulate blood pressure.

seta: (L. *seta*—bristle)
esophagus: (Gk. *oisophagos*—gullet)

Respiration is closely linked to circulation. The earthworm's thin epidermis is the site for the exchange of gases. Tiny vessels bring the blood close to the body surface. Here carbon dioxide diffuses out of the blood, and oxygen is absorbed into the blood. The blood then carries the oxygen to all the animal's tissues. The epidermis is covered with a thin, protective cuticle that must be kept moist to permit the passage of oxygen through it.

During heavy rains, an earthworm is in danger of suffocating. Although the worm can obtain dissolved oxygen from water, the rainwater that filters down through the soil into the burrow has lost most of its oxygen. To escape suffocation, the earthworm often surfaces during a rain and then faces the dangers of hungry birds and man's feet.

To excrete metabolic wastes, the worm has a pair of tubelike **nephridia** (nuh FRID ee uh) (sing., nephridium) in each body segment except the first three and the last. These organs filter wastes from the blood and expel them through small openings on the animal's side called *nephridiopores*.

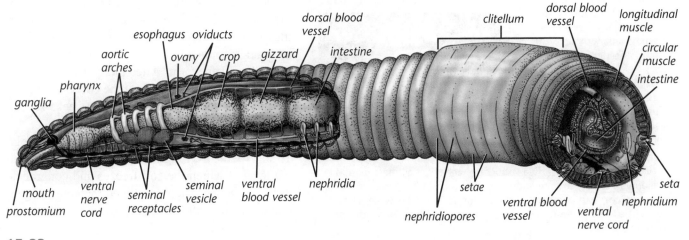

15-23
Structures of an earthworm

Nervous and Reproductive Systems of the Earthworm

The major structure of the earthworm's nervous system is its "brain," which is actually a pair of **ganglia** (GANG glee uh) (sing., ganglion), a mass of nervous tissue. These ganglia above the pharynx are joined by two nerves to another ganglion below the pharynx. A double *ventral nerve cord* extends from the large ganglion to the end of the worm. Along this cord are smaller ganglia, one for each segment of the earthworm. Nerves leading to the muscles and sensory structures of each segment branch from these ganglia.

Earthworms are sensitive to light, touch, and certain chemicals. Their sensory cells are not grouped into large sensory organs but are distributed over the animal's entire body. The anterior and posterior ends of the worm have high concentrations of sensory cells.

The normal mode of earthworm reproduction is sexual, although certain species of earthworms, when cut in two, can produce complete adults by regeneration. However, not all segmented worms have the ability to regenerate.

The earthworm is hermaphroditic, with male and female sex organs located in separate segments. The earthworm's male organs are two pairs of testes located in segments 10 and 11. Sperm from its own testes are stored in *seminal vesicles* until mating; seminal receptacles in segments 9 and 10 store sperm

nephridium: nephr- or nephro- (Gk. *nephros*—kidney)
seminal: semin- or semen- (L. *semen*—seed)

Other Segmented Worms

Not all segmented worms live timidly in holes in the ground, feeding on soil, as does the earthworm. Segmented worms live in a variety of habitats and feed in many unique ways. Some are parasites that feed only every few months; others are voracious predators; still others gather tiny particles from the water.

Clam worms

The clam worm is often found along the seashore at low tide and preys upon small animals and other worms. Unlike the earthworm, the clam worm has a definite head with sensory organs and two rows of fleshy appendages called parapodia along its body. It resembles a centipede because of its large setae, but it does not have an exoskeleton or jointed legs. To catch its prey, the worm extends its pharynx and retracts it back into its mouth, drawing the prey with it.

Feather worms

The feather worm, another marine species, constructs a sturdy tube to live in and feeds by sliding its head out of the tube. The worm's head has several rings of long tentacles covered with mucus to entrap food

particles. When food clings to the tentacles, cilia direct it to the animal's mouth. The spray of feeding tentacles at the end of the body tube gives the animal its feather-duster appearance. When threatened, it quickly pulls its elaborate feeding apparatus into the safety of the tube.

Feather worm

Leeches

While blood-sucking parasitic species of leeches get all the attention, a majority of species are predators or scavengers. Most species are small (under 5 cm), but one tropical species may grow to be 30 cm (1 ft) long. The facet "Parasitic Worms in Humans" contains more information on leeches.

received from its mating partner. The worm's female organs are a pair of ovaries in segment 13. The eggs discharged by the ovaries are collected in the tubular oviducts in preparation for mating.

In sexual reproduction two worms must exchange sperm. While pointing in opposite directions, the worms join the ventral surfaces of their anterior ends. A slime tube secreted by both worms holds them together. The sperm released from the seminal vesicles of one worm leave through the sperm ducts on segment 15, travel along a groove on the worm's body, and enter the seminal receptacles of the other worm. After exchanging sperm, they separate.

A few days later a cocoon forms around the clitellum of each worm. The worm then backs out of the cocoon. As the cocoon passes over the openings of the oviducts, eggs are released. Then the seminal receptacles release the sperm obtained from the other worm. After the worm has pulled out of the cocoon, the ends of the cocoon seal, and fertilization occurs. In two to three weeks several small earthworms emerge from the cocoon.

Some earthworms live to be several years old. To survive the freezing winter, they burrow beneath the level at which the soil freezes and congregate in large masses. Their combined metabolism generates the warmth necessary to keep them alive during dormancy.

Review Questions 15.7

1. Describe segmentation.
2. Describe the movement, nutrition, and excretion of the earthworm. Where possible, include structure names.
3. Describe two annelids other than the earthworm.

⊙ 4. Compare the following life processes in the hydra, sponge, planarian, tapeworm, *Ascaris*, and earthworm: (a) nervous systems and responses; (b) circulation; (c) asexual reproduction; (d) sexual reproduction.

15D—Other Invertebrates

Based on numbers, the largest animal phylum is Arthropoda, with more than a million species of insects, crabs, lobsters, and spiders. Arthropoda is covered in Chapter 16. The next largest phylum, Mollusca (muh LUS kuh), has 75 000 species of snails, slugs, clams, octopuses, and many others. Mollusks may seem like strange, secretive animals with no direct relation to humans, but they are the largest source of invertebrate food, and their sheer numbers make them significant.

Echinodermata (uh KYE noh DUR muh tuh), the phylum at the end of this section, contains 5800 species and includes starfish, sea cucumbers, and sea urchins. The members of this small phylum are common inhabitants of the seashore, and their body structure is unique in the animal kingdom.

15.8 Phylum Mollusca

Centuries ago, mariners lived in fear of enormous sea monsters with snakelike arms that could crush the hull of a ship. Though there is no reliable documentation of ships being mangled by such creatures, these tentacled monsters were no myth. Very large mollusks, commonly called monsters and devilfish, existed at one time (and probably still do); they contrast sharply with the small, quiet mollusks that are commonly known.

The phylum Mollusca includes octopuses, oysters, snails, slugs, squids, and clams. Although they are an extremely varied group, **mollusks** share most of the following characteristics:

✦ **Mantle**—a sheath of tissue that encloses the vital organs of the mollusk, secretes its shell, and forms its respiratory apparatus

✦ **Shell**—the tough multilayered structure secreted by the mantle as a means of protection or body support

✦ **Visceral** (VIS ur ul) **mass**—the part of the mollusk's body that contains its heart and its digestive and excretory organs, often covered by the mantle

✦ **Foot**—a fleshy, muscular organ that is used for locomotion and assumes a variety of forms, depending on the animal

✦ **Radula** (RAJ oo luh)—a small organ covered with many tiny teeth that scrapes up food particles and draws them into the mollusk's mouth

The phylum Mollusca is divided into six classes based on the shape and type of shell (if any) and the kind of foot. The three classes presented in this section are Bivalvia (the two-shelled mollusks), Gastropoda (the stomach-footed mollusks), and Cephalopoda (the head-footed mollusks).

Class Bivalvia: Two-Shelled Mollusks

The class Bivalvia contains over 10 000 species, including clams, oysters, mussels, and scallops. The clam, a typical animal of this class, has a soft, flat, oval-shaped body encased in a two-piece shell connected by a hinge. Each half of the clam's shell is called a valve, hence the common name **bivalve**, meaning "two-shelled."

The shell is secreted by the mantle and is composed of the following three layers:

- the thin, protective, horny outer layer
- the thick, prismatic middle layer, composed of calcium carbonate
- the smooth, pearly innermost layer

The two-piece shell, which enlarges as the animal grows, protects it not only from exterior attack by predators but also from interior irritation. If a grain of sand or a parasite slips between the valves and is caught between the mantle and the shell, the mantle will cover the foreign object with a secreted material identical to that in the pearly layer. The result is a pearl.

The *foot* is the clam's organ for locomotion and burrowing. The animal extends the foot between its valves into the sand. The end of the foot is enlarged into a hatchet-shaped anchor. Then the clam contracts its foot muscles, and the shell is dragged toward the foot and deeper into the sand.

Because the clam is a typical bivalve, it displays the most common life processes of this class. Respiration is accomplished with two pairs of **gills**. The gills are thin-walled structures richly supplied with blood vessels. Cilia on the gill surfaces sweep water containing food and oxygen into the animal.

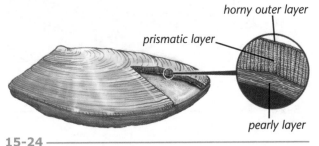

15-24
Shell of a bivalve

horny outer layer
prismatic layer
pearly layer

The Galloping Scallop

The scallop is an edible bivalve that is not limited to movement using only its muscular foot. When startled, the scallop rapidly opens and closes its shell, bouncing along the sea floor like a clacking false teeth toy. It is alerted to danger by as many as one hundred small eyes located along the margins of the valves. These simple eyes are very sensitive to light and movement.

Mollusca: (L. *mollis*—soft)
visceral: (L. *viscus*—body organ)
radula: (L. *radere*—to scrape)

15-25
A scallop has tiny eyes on individual stalks. These alert it to danger so that it can flee.

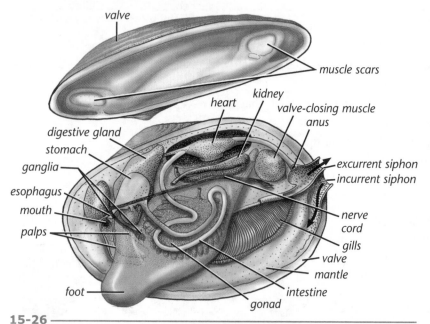

valve

muscle scars

heart
kidney
valve-closing muscle
anus

digestive gland
stomach
ganglia
esophagus
mouth
palps

excurrent siphon
incurrent siphon

nerve cord
gills
valve
mantle
intestine

foot
gonad

15-26
Structures of a clam

The water flows across the gills, where oxygen and carbon dioxide are exchanged within the bloodstream. An open circulatory system consisting of colorless blood, a pumping heart, and a few blood vessels distributes oxygen and digested food. A kidney filters wastes from the blood.

Clams are called *filter feeders* because they feed on the organic materials they strain from the water. The clam circulates water from its environment through two tubes, or **siphons**. The *incurrent siphon* brings water into the animal; the *excurrent siphon* carries water out of the animal. Bivalves lack a radula. As food particles in the water enter the clam, they are trapped in stringy, sticky mucus secreted by the gills and by the palps, a pair of lobelike organs that surround the clam's mouth. Cilia on these structures move the mucus and food into the animal's mouth. The food passes through the stomach and is digested in the digestive gland that is attached to the stomach. Indigestible materials pass into a looped intestine and are expelled through the anus near the excurrent siphon.

The clam's nervous system consists of three major ganglia joined by nerves and various sensory organs, such as balance receptors in the foot, sensors around the incurrent siphon to detect various substances in the water, and light and touch receptors around the mantle's edge.

The life cycle of the clam involves sexual reproduction and larval stages. Many mollusks have a ciliated larval form called the **trochophore** (TROHK uh FOHR) in their life cycles. The trochophore larva is a key characteristic used to identify members of the phylum Mollusca.

Class Gastropoda: Stomach-Footed Mollusks

The class Gastropoda (ga STRAHP uh duh) includes snails, slugs, and nudibranchs. These animals, the **gastropods**, can be found on the land and in fresh or salt water. The snail best illustrates the major characteristic of this class. Its foot is located immediately below the visceral mass, the portion of the mollusk's body containing the stomach. The snail is therefore "stomach-footed." The snail moves by laying down a thin layer of slime on which it glides by rhythmical contractions of its muscular foot. This activity enables the snail to achieve speeds of about 3 m (10 ft) per hour.

The snail is a *univalve*, meaning it has a single shell (usually coiled in a spiral) that serves as its protective home. When the environment or predators threaten, the head and foot of the snail can contract into its shell that, in many snails, is then closed with a special platelike "door."

The snail grazes upon plant material, using its tonguelike *radula* to grate and ingest it. The food is broken down in a coiled digestive tract, and indigestible material is expelled at the anus located above the mouth. The strange position of the anus is the result of *torsion*, the twisted, asymmetrical body arrangement unique to gastropods, produced by uneven growth in certain muscles of the snail.

The *mantle cavity*, the space between the mantle lining, the shell, and the animal's soft body, is thickly supplied with blood vessels and functions as a lung in respiration. A series of ganglia around the head region constitutes this mollusk's "brain." The snail's sensory organs include a pair of eyes set at the ends of two flexible, telescoping *tentacles*. Other receptors for smell, touch, and some chemicals are located in the head and foot.

The size of snails ranges from the small but destructive plant-eating garden snail to the giant Australian trumpet, a sea snail with a shell almost 0.6 m (2 ft) long. Although most snails are completely harmless to man, one group of Pacific sea snails forms a poison and injects it into worms or small animals that serve as its food. This poison immobilizes and kills the prey and is powerful enough to kill a human.

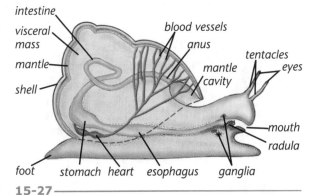

15-27
Structures of a snail

Class Cephalopoda: Head-Footed Mollusks

Class Cephalopoda (SEF uh LAHP uh duh) includes the "sea monsters," such as the squid and octopus, as well as the less fearsome nautilus and cuttlefish. The foot of the **cephalopod** extends from the animal's head region (thus its class name). The foot is usually divided into a number of sucker-bearing arms that are used to catch food.

The cephalopods live active lives in the open sea. They possess special defense mechanisms, enabling them to survive in this dangerous environment. For example, most cephalopods have pigment cells that enable them to change colors to blend with the environment. Also, when threatened by predators, most cephalopods can squirt a dense black, inky fluid into the water to confuse their attackers.

15-28
Like many other gastropods, the wavy volute of Australia is a predator.

15-1 Comparison of Major Mollusk Classes				
Class	**Environment**	**Shells**	**Cephalization**	**Locomotion**
Bivalvia	freshwater/marine	2, external	indefinite	predominantly sessile
Gastropoda	terrestrial/freshwater/marine	1, external—most species; none—slugs; nudibranchs	definite	crawling; some aquatic species swim
Cephalopoda	marine	none except nautilus	definite	swimming

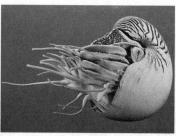

15-29
The octopus (top left), squid (bottom left), and nautilus (right) are all cephalopods, and like the other members of their class, they live in the sea.

The squid is a torpedo-shaped cephalopod with a rod-like internal shell that gives it shape and support. The large eyes of this mollusk are much like mammalian eyes and are capable of very accurate vision. The largest known invertebrate, reaching an estimated length of 13 m (42 ft), is the giant squid. Another huge invertebrate, the colossal squid, may weigh even more than the giant squid but most likely is not as long.

The octopus is a familiar mollusk "monster." This cephalopod has eight heavily suckered arms, which it can use to drag itself along the sea floor. When it needs to move rapidly, the octopus can "jet" backward through the water like a squid by forcing water out of its siphon. Unlike the fictional man-killer of adventure stories, the octopus is extremely timid and usually avoids man.

octopus: octo- (Gk. *okto*—eight) + -pus (foot)

nautilus: (Gk. *nautilos*—sailor)

The nautilus lives in a large external shell. This shell is a series of chambers, each of which the mollusk once occupied but outgrew. The animal simply adds a larger chamber to its shell when it needs more living space.

Review Questions 15.8

1. What characteristics do all mollusks share that separate this phylum from other phyla in the kingdom Animalia?
2. Describe the valve of a clam.
3. Briefly tell how a clam accomplishes respiration, response, and reproduction. Where appropriate, include structure names.
4. Describe the locomotion of bivalves.
5. How do gastropods differ from bivalves?
6. How do cephalopods differ from (a) bivalves and (b) gastropods?

15.9

Objectives

- Describe the operation of the water-vascular system
- Compare sexual and asexual reproduction in the starfish
- Name and briefly describe one example for each of the five echinoderm classes

Key Terms

echinoderm
water-vascular system
tube feet
madreporite
coelom

Echinodermata: Echino- (Gk. *ekhinos*—spiny) + -dermata (skin)

Asteroidea: Aster- (Gk. *aster*—star) + -oidea (shape or form)

15.9 Phylum Echinodermata

The phylum Echinodermata includes marine animals ranging from the prickly sea urchin to the most familiar **echinoderm**, the starfish. This diverse group contains more than 5800 species divided into six classes, five of which are discussed here. Most members of this phylum exhibit radial symmetry and have an endoskeleton of plates called *ossicles*. While the skeleton may appear to be external, there is usually a thin layer of skin covering these plates and holding them together. The spines that give this phylum its name often project through the epidermis. Echinoderms are also characterized by a unique water-vascular system, a marvel of hydraulic engineering that is best seen in the starfish.

Starfish (Class Asteroidea)

Starfish are probably one of the most familiar members of the phylum Echinodermata. They inhabit coastal waters worldwide and come in a number of colors and shapes. This radially symmetrical animal has five or more arms or *rays* tapering gradually from a *central disc*. Its shape and support are maintained by a system of small, hard plates joined by connective tissue beneath the epidermis. These plates have spines that give the starfish its characteristic rough, spiny appearance.

The starfish shares with all the echinoderms a unique **water-vascular system**—a series of canals and tubules for locomotion and food capture. The most obvious parts of the system are the hundreds of **tube feet** in the deep grooves along the lower surface of the echinoderm's rays. These hollow feet are joined by a water canal that extends along the starfish ray to the ring canal within the central disc. The water-vascular system opens at the **madreporite**

(MAH druh POOR ite), a sievelike structure on the dorsal surface of the animal. This system is used to vary the water pressure in the tube feet. Thus, each foot can, by suction, grip an object or release it. The animal can then slide along the ocean bottom or cling securely to rocks.

Some starfish also use their tube feet to pry open the shells of their favorite food—clams and oysters. By applying steady pressure with these feet upon the mollusk's shell and by working the tube feet in shifts, the starfish can fatigue the mollusk's muscles and open its shell. Once it has opened the shell, the starfish protrudes its stomach through its mouth (located in the center of its lower surface) and into the shell of the prey. The everted stomach secretes enzymes to reduce the soft parts of the clam or oyster to a soupy broth and then absorbs this material and transports it to highly branched digestive glands for digestion to be completed.

Respiration and excretion in the starfish occur between the coelom (SEE lum), or body cavity, and the environment. The fluids of this internal cavity are brought close to the environment through a number of small fingerlike projections on the surface of the animal called skin gills. Gases easily exchange through these thin-walled extensions of the coelom.

The skin gills are also important excretory organs because many wastes diffuse out of the body directly through them. Also, cells called amoebocytes move through the fluid of the coelom and gather other waste materials. Once laden with this metabolic waste, the cells penetrate the walls of the gills and escape into the environment, taking the wastes with them.

In starfish, as in most echinoderms, the sexes are separate and fertilization is external. To reproduce sexually, they expel eggs and sperm into the water, where these gametes unite. After fertilization, the zygote develops into a ciliated, bilaterally symmetrical larva. The larva eventually settles to the sea floor and becomes a radially symmetrical adult.

Even with no head or brain, the starfish efficiently senses and responds to its environment. A nerve ring circles the mouth area and ties to radial nerves extending into each ray. These radial nerves bring back sensory input from the extremities and also coordinate the movement of thousands of individual tube feet. A secondary nerve net, just beneath the skin, controls the movement of the spines and the skin gills. A light-sensitive eyespot at the tip of each ray does not produce real images but does provide some protection.

everted: e- (out) + -verted (to turn)
coelom: (Gk. *koiloma*—cavity)

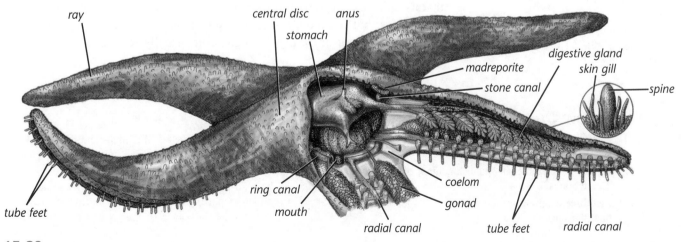

15-30
Structures of a starfish

15-31

Although they seem quite different, the sea urchin (top) and the sea cucumber (bottom) belong to the same phylum.

Echinoidea: Echin- (spiny) + -oidea (shape or form)

Holothuroidea: Holothur- (Gk. *holothourion*—water polyp) + -oidea (shape or form)

Ophiuroidea: Ophiur- (Gk. *ophioukhos*—serpent) + -oidea (shape or form)

Crinoidea: Crin- (Gk. *krinon*—lily) + -oidea (shape or form)

Starfish, like all echinoderms, have remarkable powers of regeneration. If a starfish is dismembered, a single ray can grow into a complete adult if it has even a small portion of the central disc attached to it.

Though the slow-moving starfish may appear to be a relatively harmless creature, it often is not. One starfish can eat a dozen clams and oysters a day. Indeed, a few starfish can practically destroy a commercial shellfish bed in a short time. Starfish can also severely damage other forms of sea life. The crown-of-thorns starfish, a spiny, poisonous echinoderm, has done tremendous damage to many coral reefs by eating large quantities of the young coral polyps. Biologists have not been very successful in controlling these animals.

Sea Urchins and Sand Dollars (Class Echinoidea)

The sea urchin has been called the "porcupine of the sea." This globular animal has long pointed spines and equally long tube feet in rows over its body. Sea urchins are scavengers, eating almost any organic material. They often graze on algae. Their spiny body protects them from most predators but not from the damselfish. This fish patiently bites off the urchin's spines and then eats the helpless animal. Sand dollars are also members of class Echinoidea. When alive, they also are covered with movable spines, but the spines are very tiny.

Sea Cucumbers (Class Holothuroidea)

The sea cucumber is a slow-moving, sac-shaped echinoderm that feeds upon the organic materials it traps in its sticky, feathery tentacles. Because it lacks the bony endoskeleton found in most other echinoderms, it is vulnerable to predators. Its only defense is self-mutilation: when attacked, this echinoderm jettisons some of its internal organs. This meal often appeases the attacker while the sea cucumber retreats and regenerates a new set of organs. When found on the beach, it can be identified by its dark, leathery, picklelike appearance.

Brittle Stars and Basket Stars (Class Ophiuroidea)

The brittle star resembles the starfish but has thin, tentacle-like arms. The star moves by the snakelike action of its arms and is faster than other echinoderms. This echinoderm is called "brittle" because it readily discards its arms when attacked or disturbed. The basket star feeds at night by raising its rays into the form of a basket to trap particles and small organisms suspended in the water.

Sea Lilies and Feather Stars (Class Crinoidea)

The sea lily, a plantlike echinoderm, spends its adult life fixed to the ocean floor by a long stalk. This sessile animal feeds on the minute food particles that its ciliated arms direct to its mouth. Feather stars look a little like fuzzy brittle stars, but the mouth is on the dorsal surface. They are able to crawl along the sea floor or swim, although adults usually remain in one place for long periods.

➡ *Review Questions 15.9*

1. Describe the water-vascular system. How does it help the starfish obtain food?
2. What characteristics distinguish echinoderms from other invertebrates?
3. List three examples of the phylum Echinodermata, other than the starfish.
4. List the nine life processes of animals and briefly tell how a starfish accomplishes each. Where possible, include structure names.
5. Trace a particle of food through the digestive system of (a) a planarian, (b) an *Ascaris*, (c) an earthworm, (d) a clam, and (e) a starfish.

The Arthropods
Zoology Part II

16A—Introduction to Arthropods

The phylum Arthropoda (ar THRAHP uh duh), which includes lobsters, spiders, scorpions, millipedes, and insects, encompasses over one million species. Four out of five animal species are **arthropods**, and they are easily the most abundant of all the visible animals. Thousands of collected specimens have yet to be named, and new ones are being collected every day, especially in tropical habitats. Some scientists believe that possibly only one-tenth of all the arthropods on the earth have been discovered and named.

Arthropods live virtually everywhere, and their influence upon man and the environment cannot be ignored. They can destroy crops, compete with other animals for food, and transmit deadly diseases. They also produce such valuable things as wax, honey, drugs, and silk. Perhaps more importantly, they help maintain plants and crops by assisting cross-fertilization, and some even help control the numbers of harmful arthropods. In addition, engineers have studied their designs and have discovered ways to improve machines ranging from robots to aircraft.

Arthropoda: Arthro- (Gk. *arthron*— joint) + -poda, -pod, or -ped (foot)

Objectives

- Describe the structure of the exoskeleton and the process of molting
- Describe the typical arthropod nervous system, including all connected structures or organs
- List and give examples for the four subphyla of the phylum Arthropoda

Key Terms

arthropod   cephalothorax
molt   abdomen
appendage   compound eye
thorax

16-1

The cicada emerges from its old exoskeleton and flies away when its wings have hardened.

Classification of Arthropods

The phylum Arthropoda contains four subphyla divided into as many as twenty different classes. The primary characteristics that separate the groups are body divisions, number and kinds of appendages, and means of respiration.

- Subphylum Trilobita—trilobites (extinct)
- Subphylum Crustacea—pill bugs, crabs, shrimp, crayfish, and krill
- Subphylum Chelicerata—spiders, scorpions, ticks, and mites
- Subphylum Uniramia—centipedes, millipedes, and insects

cephalothorax: cephalo- (head) + -thorax (Gk. *thorax*—breastplate)

16.1 Characteristics of Arthropods

Despite the diversity in this phylum, arthropods possess many common characteristics.

✦*An exoskeleton.* The most obvious characteristic of an arthropod is its exoskeleton. The exoskeleton is a nonliving body covering—a triple layer "suit of armor"—secreted by the epidermis. An outer coating of proteins and lipids repels water and prevents water loss in species on land. The middle, supporting layer is primarily protein and *chitin*, a polysaccharide with strong hydrogen bonds. The inner layer is similar to the middle layer but provides more flexibility at the joints.

Chitin gives the body covering toughness and flexibility. Mineral salts in the exoskeleton protect the animal from harsh chemicals in its environment and harden the exoskeleton to withstand attacks from predators. The exoskeleton also supports the arthropod's body. Unlike a human's internal skeleton, an arthropod's skeleton is worn externally.

An exoskeleton provides good protection, but it comes at a cost. The weight of this armored body covering limits to some extent the size of the animal. As long as the arthropod is small, it has sufficient muscle power to move the exoskeleton easily; but if it were to get very large, the total weight of an exoskeleton necessary to support it on land would make it difficult for the animal to move. Most large arthropods—such as the Japanese crab, which can have a span of 4 m (13 ft) between its claws—are aquatic. No known living insect has a measurement greater than 28 cm (11 in.).

The exoskeleton also poses an obstacle to body growth. Because it is nonliving and cannot grow, the arthropod must periodically **molt**, or shed its covering, in order to grow. To molt, the arthropod epidermis produces enzymes that eat away at the inside of the old exoskeleton while a new exoskeleton is produced beneath it. Some arthropods then take in water and air to swell and rupture the old shell; others merely wiggle out. Even the legs and antennae pull out of the old shell. Once free of its former exoskeleton, an arthropod may go into seclusion until its new, expanded exoskeleton hardens.

✦*Jointed appendages.* The term *Arthropoda* means "joint-footed" and describes the **appendages**, or limbs, of this animal group. Although arthropods, vertebrates, and man all have jointed limbs, the structure of arthropod limbs is unique. The joints of a man's arm or leg lie between internal bones that are covered with muscles. The muscles of the arthropod, however, move the limbs from within its exoskeleton.

Appendages on arthropods come in a variety of forms and functions. The delicate limbs of the spider and the powerful jumping legs of the grasshopper are used for locomotion. The "fangs" of the centipede enable it to capture its food. Appendages may also be a means of defense: crabs and lobsters bear formidable claws for that purpose. Some jointed appendages (like antennae) are for sensory reception, some for chewing food, and others for sexual reproduction. The scorpion's stinging tail is used for defense as well as predation.

✦ *Body segmentation.* The body of the typical arthropod is divided into three major segments: the head, the **thorax** (the head and thorax are often united into a cephalothorax [SEF uh luh THOHR AKS]), and the **abdomen**. Further segmentation may occur within these major divisions. Certain members of this phylum have body segments designed for specific functions. The muscular abdominal segments of the crayfish work with the flaps or tail fin to move it rapidly backward through the water as it escapes danger.

✦ *An open circulatory system with a dorsal heart.* The circulatory system of the arthropod does not limit its blood to blood vessels. Instead, the dorsal heart pumps blood through short vessels that empty into cavities within its body and bathe its organs. The *open circulatory system* is not as efficient as the closed circulatory system because it depends in part on gravity for its operation.

✦ *A ventral nervous system.* The arthropod nervous system consists of a pair of ganglia centered over the esophagus and joined by two major nerves to a ganglion below the esophagus, and a ventral nerve cord. The ventral position of the nervous system places the vital nervous pathways of the arthropod in a place of maximum safety: the ventral nerve cord is protected not only by the hard exoskeleton but also by the bulk of the body of the animal. This ventral position also places the nervous system near the appendages, the structures most often requiring nervous commands and coordination.

The nervous system receives sensory information through a variety of organs. *Antennae*, appendages in the head region, provide taste, smell, and touch sensation to most arthropods. Many arthropods bear sensory bristles scattered over their bodies.

Virtually all arthropods have some type of eye. **Compound eyes** (in most insects and crustaceans) contain thousands of individual lenses. These lenses are each set at a slightly different angle, giving the arthropod a mosaic image of the world. Some compound arthropod eyes see size, shape, and movement, and a few can sense color. *Simple eyes*, like those of the spider, contain only one lens and present a very limited view. Some simple eyes produce a poor image, and some respond only to the presence or absence of light. Some arthropods seem to be able to sense ultraviolet light with their eyes. This helps many of them find flowers.

16-2 ———————
Simulation of what scientists believe a compound eye would see

---

➤ Review Questions 16.1

1. List the major characteristics that distinguish Arthropoda from the other phyla in kingdom Animalia.

2. List some advantages and disadvantages of an exoskeleton.

⊙3. Why do you think the largest arthropods are found in water and not on land?

16.2 Subphylum Crustacea

Lobsters, crabs, crayfish, shrimp, pill bugs, and barnacles are all members of subphylum Crustacea. Most **crustaceans** are free-living and aquatic; however, parasitic and terrestrial species exist. Because of its large size and availability, the crustacean most often used for classroom study is the *crayfish*, or "crawdad." The crayfish is common in lakes and streams around the world. It resembles the lobster, a sea dweller, although most lobster species are larger.

Like all arthropods, the crayfish is covered with a tough exoskeleton composed of chitin. The exoskeleton is thick and immovable over most of the animal's body but thin and flexible at the joints. The body of the crayfish is divided into two major sections: the cephalothorax, which is covered by the **carapace** (KEHR uh PASE), a single exoskeletal plate, and the abdomen, which is composed of six segments, each covered by a set of plates. Beneath the abdomen are a number of paired, small, flipperlike appendages called

16-3
This shrimp is one of the many marine crustaceans.

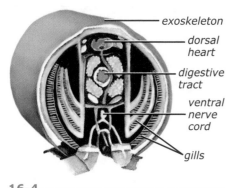

- exoskeleton
- dorsal heart
- digestive tract
- ventral nerve cord
- gills

16-4
Cross section of an arthropod

pericardial: peri- (around) + -cardia- (Gk. *kardia*—heart)

sternal: stern- or sterno- (Gk. *sternon*—breast or breastbone)

swimmerets, used for swimming and reproduction. Four pairs of *walking legs* and a pair of prominent *chelipeds* are attached to the cephalothorax. The chelipeds bear pincers that are used for capturing food and for defense.

Life Processes of Crayfish

Crayfish are scavengers, eating virtually any edible material. To feed, the mouthparts of the crayfish first reduce the food to swallowable size. The food passes through the mouth and a short esophagus into the anterior portion of the stomach. There it is ground into fine particles by the muscular action and chitinous teeth of the *gastric mill*.

In the posterior portion of the stomach, the food is sorted: fine particles are directed to digestive glands where digestion is completed, and coarse particles are moved to the intestine. These coarse particles, along with indigestible residue from the digestive glands, are passed through the intestine and eliminated through the anus, which is in the last abdominal segment.

The respiratory apparatus of the crayfish consists of two sets of feathery *gills* found in two lateral gill chambers along the thorax region. The gills are covered by the lateral portions of the carapace, leaving openings only along the animal's ventral surface. Some gills are attached to various appendages. Appendage movement and feathery mouthparts help to keep oxygenated water flowing over the gills. Blood traveling through the thin-walled gills releases carbon dioxide and absorbs oxygen. Many crustaceans can store water in their gill chambers in order to walk on land. In other words, on land they can "hold their water" like people "hold their breath" in the water.

The crayfish has an open circulatory system. Blood collects within the pericardial (pehr ih KAHR dee ul) sinus, the cavity surrounding the heart. The blood enters the dorsal heart through tiny openings. As the heart contracts, valves close the openings, preventing the blood from passing back into the sinus.

The blood is forced through a series of arteries that empty into spaces within the body cavity. Once the blood has drained through these spaces and bathed the organs, it collects in the large ventral *sternal sinus*. The blood then passes through the gills, becomes oxygenated, and returns to the pericardial sinus. As the blood circulates, the *green glands*—structures near the base of the antennae—filter out waste materials. The fluid waste is then excreted through a pore just anterior to the mouth.

The ventral nervous system receives information about the environment from a number of sensory sources. The most obvious sensory organs of the

---

Other Crustaceans

Some crustaceans are not as familiar as shrimp, lobsters, and crayfish. Many children have played with the *pill bug*, or "roly-poly," the tiny animal that rolls into a ball when disturbed. The pill bug is a terrestrial crustacean that keeps its gills moist by living in moist soil under rocks or logs.

The *barnacles* that encrust wharves and ship bottoms are another unusual member of this group. Once thought to be mollusks, these sessile animals live in shell-like exoskeletons containing calcium and feed by sweeping food into their "shells" with jointed appendages. These animals will attach to virtually any underwater object, including live lobsters and whales' teeth. Their accumulation on ship hulls can become so great that the vessel must be dry-docked to remove them.

Some crustaceans are much smaller. The Sea-Monkeys sold in toy and museum stores are a type of brine shrimp that can survive for years as a dry egg until salt water triggers their development. *Daphnia* (the water flea), another tiny freshwater crustacean, can be found in almost any freshwater pond or stream.

These barnacle colonies get a free ride on the head of a whale.

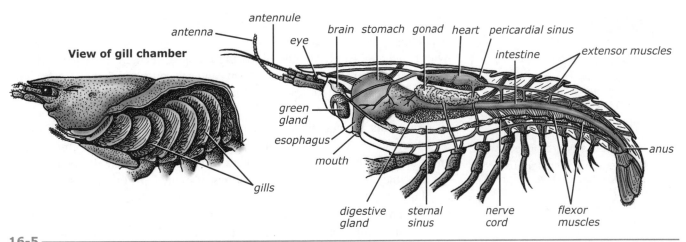

View of gill chamber

antenna — antennule — brain stomach gonad heart pericardial sinus
eye — intestine — extensor muscles
green gland
esophagus
mouth
gills
digestive gland — sternal sinus — nerve cord — flexor muscles
anus

16-5

Internal structures of a crayfish

crayfish are its compound eyes (which sit atop a pair of movable stalks), antennae, and *antennules*, shorter sensory appendages. Tiny bristles for touch are found on many of the appendages.

At the base of each antennule is a *statocyst* (STAT uh SIST), the organ of balance. The statocysts are sacs, lined with tiny sensory hairs, that contain several grains of sand. As the animal's body tilts, the grains shift within the statocysts, stimulating particular sensory hairs. The impulses from the organs travel to the brain, and the crayfish then rights itself. When the crayfish molts, it loses the grains in its statocysts and must find replacements.

Regeneration and Reproduction in Crayfish

Although the crayfish cannot reproduce asexually, it is capable of *regeneration*. In arthropods, regeneration is a method of repair. When battling with predators, the crayfish may lose or deliberately discard a limb. This can be done without excessive blood loss, thanks to a self-sealing double membrane in each limb. After the loss of a limb, the animal will gradually grow a new appendage to replace it.

Crayfish usually mate in the fall. During this season, the male transfers its sperm to special receptacles in the body of the female, using its reproductive swimmerets. The female stores the sperm until she lays her eggs in the spring. The eggs are fertilized as they pass out of the oviduct in slimy bunches. When the egg clusters are attached to the swimmerets to develop, the female is "in berry." Within five to six weeks the eggs hatch, releasing young that are miniature replicas of the adult. These young crayfish continue to cling to the mother for several weeks.

As the young crayfish grows, it seems trapped within its exoskeleton. The crayfish then begins molting. During its first year the crayfish will molt seven times. After that the animal molts about twice a year for the rest of its life (3–8 yr).

16-6

The female crayfish carries the eggs and newly hatched crayfish under her abdomen.

statocyst: stato- (Gk. *statos*—standing) + -cyst (bladder or pouch)

➡️ **Review Questions 16.2**

1. What characteristics distinguish crustaceans from other arthropods?

2. List the nine life processes of animals and briefly tell how a crayfish accomplishes each of these processes. Where possible, include structure names.

3. List and describe three crustaceans other than the crayfish.

- Name and describe the special mouthparts that give Chelicerata its name
- Compare and contrast the arachnid's body plan and structures with the typical insect's
- List several ways spiders use their silk other than for capturing prey
- Name three arachnids other than spiders

→ Key Terms

chelicera	spinneret
arachnid	book lung
pedipalp	trachea

16.3 Subphylum Chelicerata

The subphylum Chelicerata (kuh LIS uh RAHT uh) includes three classes, with Arachnida (uh RAK nud uh) being the most recognized. The other, less familiar classes include horseshoe crabs and sea spiders. All chelicerates have mouthparts called **chelicerae** (kih LIS uh REE) (sing., chelicera), which may appear as claws or fangs.

Arachnids include such animals as spiders, scorpions, ticks, and mites. Although some arachnids do have painful stings and a few transmit disease, most are completely harmless to man. Most arachnids actually benefit man because they help control the populations of many harmful insects.

Many arachnids, with their small, compact bodies and long, delicate limbs, are often mistaken for insects. The arachnids, however, have several characteristics that distinguish them from the insects and other classes of this phylum:

- Four pairs of walking legs
- Body divided into two major segments: cephalothorax and abdomen
- No antennae or mandibles
- Respiration by book lungs
- Usually four pairs of simple eyes

Horseshoe Crabs

In yet another case of confusing names, the horseshoe crab is not classified as a crustacean with the real crabs but is more closely aligned with spiders and scorpions. The four horseshoe crab species have their own class under Chelicerata. These dangerous-looking arthropods are harmless despite the swordlike tail that is used as a rudder in the water and to right themselves if they flip over on the beach.

Arachnida: (Gk. *arakhne*—spider)
pedipalp: ped- or pedi- (foot) + -palp (L. *palpus*—a touching)

Spiders

The spiders are the largest, most familiar group within the class Arachnida. They are cunning hunters and skillful architects. Spiders are also efficient predators, attacking and eating insects, small crustaceans, and even birds and fish.

In contrast to other arthropods, spiders have no antennae. The spider does possess six pairs of appendages:

- One pair of chelicerae armed with poisonous fangs used to paralyze prey
- One pair of pedipalps (PED uh palps) used for sensory reception and, in the male, for the transfer of sperm
- Four pairs of *walking legs* used for locomotion

All spiders spin silk that may be used for containing eggs, restraining prey, or building a web. The silk originates as a liquid protein in silk glands and is released through organs called **spinnerets** at the rear of the abdomen. The liquid silk solidifies quickly after being released. Not all spiders spin webs to catch prey. Some, such as tarantulas and wolf spiders, prowl on the ground and ambush unsuspecting creatures.

Although the means of capturing food vary, the method spiders use to eat the food does not. When the prey has been immobilized by a bite from the chelicerae, the spider injects digestive juices into the victim. The spider later sucks up the partially digested tissues of the prey, using its muscular stomach and pharynx. This food is then transported to the digestive gland in the abdomen of the spider, where it is stored for future use. A few wolf spiders and tarantulas can consume the body of their victims as well as the fluids.

16-7
Internal structures of a spider

Labels: heart, intestine, ovary, sucking stomach, brain, eyes, digestive gland, poison gland, pedipalp, chelicera, anus, book lungs, branches of stomach, pharynx, mouth, spinnerets, silk glands, genital pore, oviduct, abdomen, cephalothorax

Only arachnids have and use book lungs for respiration. Air enters the spider through a slit in the abdomen and flows between the pagelike folds of the **book lungs**, where oxygen and carbon dioxide are exchanged. Some spiders have **tracheae** (TRAY kee ee) (sing., trachea), minute tubules that transport oxygen directly to tissues.

The sexes are separate among spiders, with the female often larger than the male. The male places his sperm in a tiny web sac that he stores in special cavities of his pedipalps. The male will then transfer the sperm sac to the seminal receptacles on the ventral surface of the female's abdomen. This action is usually preceded by some courtship ritual such as the male's offering the female a gift of food.

In a few species the female will eat the male after mating. This provides food for her at a time when she needs lots of extra input for egg production. As the female lays her eggs, they are fertilized by the stored sperm. The eggs are usually placed within a silk cocoon that she attaches either to her body or to a web. When the young spiders hatch, they remain in the cocoon for several weeks.

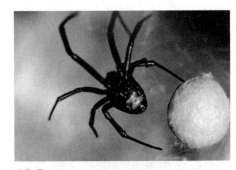

16-8

A black widow spider with a cocoon containing eggs

Dangerous Spiders

Most spiders produce poisonous venom, but this venom is rarely harmful to man. Although some spiders in various parts of the world can inflict bites harmful to humans, in the United States only two spiders have dangerous bites.

◆ *The black widow.* The female black widow has a shiny black body with a berry-shaped abdomen bearing the characteristic red "hourglass" marking on its underside. The male spider is small and harmless. The female's name "widow" implies that she eats the male after mating; however, this does not often occur. The black widow usually constructs its irregular web in dark, secluded places such as in cellars or under objects. The venom of the spider is a neurotoxin, affecting the nervous system and causing such symptoms as intense pain, muscle spasms, and vomiting. The bite is serious but rarely fatal.

◆ *The brown recluse.* A small brown spider with a dark violin-shaped marking on the dorsal surface of its cephalothorax, the brown recluse produces a venom that kills tissue around the bite, often leaving a large sunken scar. This spider is native to about fourteen states from the Gulf Coast up through the Midwest. It is most often found in houses, where it can be abundant, and people are often bitten while they are sleeping or cleaning.

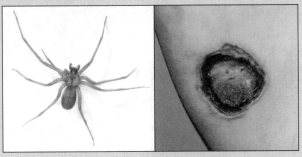

The brown recluse (left) and the black widow are the two poisonous spiders frequently encountered in the United States. The bite of a brown recluse can cause extensive tissue destruction.

Tarantulas, the Not-So-Dangerous Spiders

The over eight hundred species of *tarantulas* are among the largest of all spiders. Although they look extremely frightening, most are good-natured and rarely bite. However, some tropical species can be deadly. None of the species native to the United States have a bite any more dangerous than a wasp's sting.

This tarantula's pedipalps shield its poisonous chelicerae from view.

The Arthropods   **375**

16-9
This scorpion has assumed a defensive posture.

Other Arachnids

The *scorpion*, an arachnid found in tropical countries and throughout much of the United States, has a long segmented body bearing two outstanding features: a pair of pedipalps (large pincers used to capture food) and a segmented abdomen with a poisonous stinger in its last segment. The scorpion keeps its tail-like weapon coiled, ready to strike and immobilize an attacker or its prey. The strong chelicerae of scorpions enable them to chew their prey. Like spiders, scorpions are feared by most people, although only a few species found in the United States can cause man any significant harm. Large scorpions can inflict very painful stings, and a few tropical varieties have killed men.

Mites and *ticks* have bodies fused into unsegmented ovoid shapes. Both groups can cause varying amounts of damage to man and animals. Ticks, blood-sucking parasites of vertebrates, transmit diseases such as Rocky Mountain spotted fever and Lyme disease. Mites are much smaller than ticks but can also cause considerable discomfort. Some mites can cause mange, a skin disease of animals, while other harmless ones are abundant on our skin, even living in our pores, where they feed on oils and dead skin cells. Chiggers, or red bugs, are larval-stage mites that may infest the skin of man, causing itching and redness. The itchy sores of chigger bites are caused by the body's reaction to the salivary digestive enzymes injected during the bite of these tiny creatures.

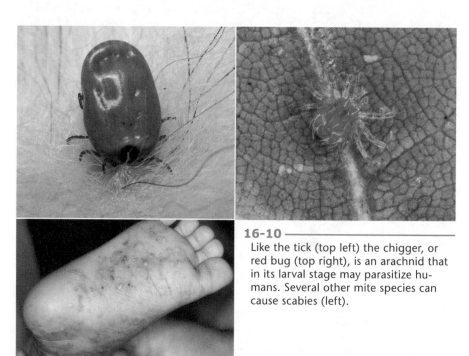

16-10
Like the tick (top left) the chigger, or red bug (top right), is an arachnid that in its larval stage may parasitize humans. Several other mite species can cause scabies (left).

Review Questions 16.3

1. List several characteristics that distinguish the class Arachnida from the other classes in phylum Arthropoda.
2. List the spider's six pairs of appendages and tell the functions of each.
3. Describe (a) the food-ingesting process, (b) the respiration process, and (c) the reproductive process of spiders.
4. List and briefly describe two arachnids other than the spider.

16B—Class Insecta

Class Insecta encompasses nearly one million species, almost 80% of all kinds of animals. Not all **insects** are as familiar as grasshoppers, flies, bees, and butterflies. Many of God's most unusual creatures belong to this diverse yet surprisingly homogeneous group.

Certain characteristics of insects help explain why they are so successful. One of the most important is that most insects can fly. This ability enables them to escape from danger, find a mate, look for food, and search for a suitable environment. Insects also have tremendous reproductive capabilities; some species can produce thousands of eggs in a single day. Finally, insects vary so much that they rarely compete with one another for food or living space. Some insects eat leaves or fruit; others suck plant juices, drink human blood, or prey upon animals. Many are specialized to survive only on one particular plant or animal host. Some even parasitize larger insects or other arthropods.

The sheer number of insects makes them a major influence in nature and an important part of man's life. Insects serve as food for many larger organisms and are crucial in accelerating decomposition of dead plants and animals. Some insects produce honey, silk, or wax and perform beneficial services. Others destroy crops, transmit deadly diseases, and are just plain pests.

16.4 Characteristics of Insects

Basic Insect Structure

The members of the class Insecta share several characteristics that distinguish them from the other arthropods:

- Three pairs of walking legs
- Wings usually present
- Body divided into three segments: head, thorax, abdomen
- One pair of sensory antennae

Legs of Insects

Insects have three pairs of jointed legs located along the thorax. The structure of the limb varies with the type of insect and is suited to the movement it performs. The legs of the fly have tiny claws and sticky pads, enabling it to climb smooth surfaces. The grasshopper uses heavily muscled hind legs to jump. The closely set bristles on the long legs of the water strider allow it to skitter across the surface film of water.

The legs of some insects were designed for functions other than locomotion. For example, the front limbs of the praying mantis are powerful "claws" for seizing prey. The fuzzy legs of the bee have special combs and hairs that it uses to groom its body and carry pollen. Some insects use their legs to make sounds to attract a mate, and certain male beetles have exaggerated front legs used for courtship.

Wings of Insects

Although most insects have two pairs of chitinous wings, some species have one pair, and wingless species do exist. Insects are the only invertebrates that can fly. Their flying styles and speeds vary. The butterfly flaps its wings five to six times per second. The bee, on the other hand, whips the air with up to two hundred beats per second. Flight speeds also vary among insects. The fly averages 8 km/h (5 mph), while the dragonfly cruises at 40 km/h (25 mph).

16-11 —————
Membranous wings of a dragonfly (top) and outer leatherlike wings of a grasshopper in flight (bottom)

Insecta: (L. *insectum*—segmented)

Legs and More Legs: Centipedes and Millipedes

Centipedes and millipedes share the same outstanding characteristic: a large number of jointed legs. Both groups are similar to insects in internal anatomy. Occasionally these arthropods have been mistaken for the larval forms of insects. Although the common names *centipede* and *millipede* are often used interchangeably, these names identify two distinct classes of animals. Together with class Insecta, these three classes form the subphylum Uniramia.

Class Chilopoda: Centipedes

The centipede, whose species make up the class Chilopoda (kye LAHP uh duh), is a carnivore having a flat body divided into a number of similar segments. Its head bears a pair of antennae and several pairs of mouthparts. Each body segment bears one pair of limbs. The body is flattened dorso-ventrally.

The centipede moves with amazing speed and agility, considering the problem of coordinating all its appendages. The limbs

Centipede

on the first body segment are poisonous claws used to immobilize and capture the insects and small animals the centipede eats. The common house centipede is a small member of the class, reaching only about 4 cm (1.6 in.) in length. Some tropical centipedes may be 30 cm (12 in.) long and may be capable of subduing worms and even snakes. The bite of most centipedes, though occasionally painful, is rarely fatal to humans.

Class Diplopoda: Millipedes

Millipedes do not really have the thousand legs suggested by their name. Millipedes make up the class Diplopoda (duh PLAH puh duh), which literally means "double feet." The millipede does indeed have "double feet" in that it has two pairs of feet per body segment. These dark-colored animals vary in size, most being about 5 cm (2 in.) long. Unlike that of the centipede, the segmented body of the millipede is rounded and cylindrical. The millipede moves slowly by the graceful wavelike motion of its legs and feeds mostly on vegetation and organic debris. It has no poisonous fangs, and when disturbed, it often curls up into a ball. If coiling fails as a defense, some millipede species can emit a foul-smelling, somewhat acidic secretion from pores along their bodies. While it may deter many predators, it is not considered harmful to humans.

Millipede

centipede: centi- (hundred) + -pede (foot)

Chilopoda: Chilo- (Gk. *kheilos*—lip) + -poda (foot)

millipede: milli- (thousand) + -pede (foot)

Diplopoda: Diplo- (double) + -poda (foot)

Insects' wings, like their legs, differ in structure and function.

✦ *Membranous wings.* Membranous wings are thin, transparent, and criss-crossed with supporting veins. These are the flying wings of most insects.

✦ *Scale-covered wings.* The flight wings of butterflies and moths are covered with delicate, beautifully colored scales that rub off easily.

Two other wing types are protective:

✦ *Leatherlike wings.* The grasshopper has its membranous wings covered by a front pair of leatherlike wings. When the insect is not in flight, these wings protect the flying wings.

✦ *Horny wings.* The front wings of beetles, such as the ladybug, are horny wings—thick shields that cover not only the membranous wings but also most of the dorsal surface of the insect.

Mouthparts of Insects

The mouthparts of an insect suit its particular food. The typical insect mouthparts are for chewing. They consist of the upper lip, or *labrum* (LAY brum), a pair of chewing **mandibles**, a pair of feeding appendages called *maxillae*, and a lower lip, or *labium* (LAY bee um), bearing another small pair of appendages called labial palps. Insects such as grasshoppers and beetles use these mouthparts to consume plant material, often with disastrous effects on crops.

labrum: (L. *labrum*—lip)

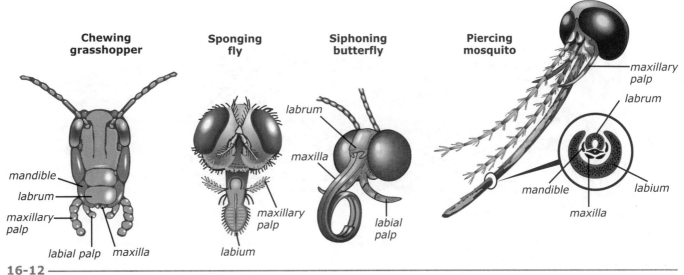

Chewing grasshopper

mandible
labrum
maxillary palp
labial palp   maxilla

Sponging fly

labrum
maxillary palp
labium

Siphoning butterfly

maxilla
labial palp

Piercing mosquito

maxillary palp
labrum
mandible
labium
maxilla

16-12

Types of insect mouthparts

Insects like the mosquito have mouthparts designed for piercing the surface of a plant or animal and feeding on its fluids. Moths and butterflies have a third type of feeding apparatus, a long, flexible siphoning tube for drawing nectar from flowers. When not in use, the tube is coiled under the head like a rolled-up party blower. Other insects have mouthparts that consist of an elongated labium with two enlarged lobes at its tip designed for sponging or lapping up liquid food. Grooves along the lower surface of these lobes soak up liquid food. The common housefly has sponging mouthparts. Insects such as giant water bugs are large enough to kill tadpoles and fish with a long piercing beak used to inject venom and digestive enzymes into their prey.

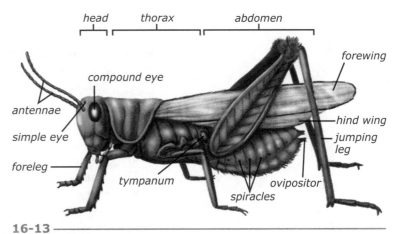

head   thorax   abdomen
forewing
compound eye
antennae
simple eye
foreleg
hind wing
jumping leg
tympanum
ovipositor
spiracles

16-13

The external anatomy of a grasshopper shows many of the typical insect characteristics. However, due to the great diversity of insects, there is no typical insect.

Nutrition of Insects

Beyond the differences in mouthparts, the digestive system is virtually the same for all insects and is divided into three main parts:

✦ *Foregut.* Food enters the foregut through the mouth, where it is moistened with secretions from the *salivary glands*. The food passes through a short esophagus to the crop, a thin-walled sac in the thorax used primarily for storage. The food continues to the gizzard, a muscular organ lined with chitinous plates, where the food is thoroughly ground.

✦ *Midgut.* The gizzard opens by a valve into the midgut, or stomach, located in the abdomen. Pouchlike organs called *gastric ceca* (SEE kuh) surround the stomach, supplying it with digestive juices. The midgut is the major site of digestion and absorption since tough chitin plates line the rest of the digestive tract.

✦ *Hindgut.* Any solid residue left in the midgut after digestion passes through the hindgut, or intestine, where water and salts are absorbed, and out of the body through the rectum and anus.

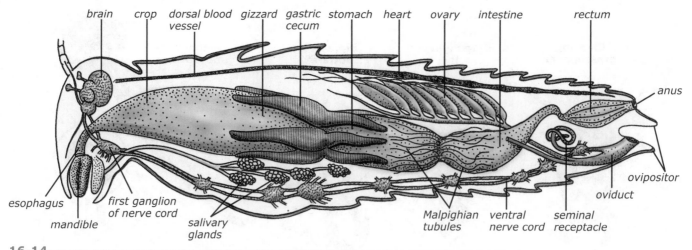

brain crop dorsal blood vessel gizzard gastric cecum stomach heart ovary intestine rectum

anus

esophagus
mandible
first ganglion of nerve cord
salivary glands
Malpighian tubules
ventral nerve cord
seminal receptacle
oviduct
ovipositor

16-14
Internal structures of a grasshopper

Some insects need a little help in digestion. Termites carry a healthy crop of protozoans in their digestive tract to break down the cellulose from wood into smaller molecules the termite can absorb. This amazing design is testimony to God's care for His creatures even on a tiny scale.

Respiration, Circulation, and Excretion in Insects

The respiratory system of insects is an elaborate system of tubules called *tracheae* that branch throughout the animal. This system is so complete that oxygen is transported through the animal without using the circulatory system. These tubules open to the outside through a series of **spiracles**, small pores that run along each side of the animal. The insect "breathes" as abdominal contractions pump air into and out of its tracheae.

The primary circulatory organ of the insect is a tubular heart dorsal to the digestive tract. A clear or yellowish fluid blood is pumped through the heart toward the head and then pours through the thorax and back over the abdominal organs. After the blood has passed through the body cavities, it returns to the heart through small pores along its length. This open circulatory system is involved only in the transport of nourishment and the collection of waste. As the nourishing fluid washes over the organs, it passes over a ring of thin tubules encircling the juncture of the stomach and intestine. These **Malpighian (mal PIG ee un) tubules** extract nitrogenous wastes from the blood and pour them into the intestine for elimination. In the rectum, most of the water in the waste is reabsorbed in special glands.

Irritability in Insects

In keeping with the arthropod pattern, the insect's nervous system consists of an anterior "brain" connected to an anterior ganglion and a ventral nerve cord. The system receives information from a variety of sensory organs. The sense of smell is usually centered in the antennae. Smell is important because many insects produce and release chemicals called **pheromones** to find a mate. Sensors in the antennae may also be used to measure humidity and flight speed. Taste receptors are abundant on the mouthparts. Numerous tactile hairs on the antennae, limbs, and body make the sensation of touch possible. The compound eyes are often so large that they bulge, dominating the head and giving the animal tremendous visual range. Many insects also have several simple eyes that respond only to the presence of light.

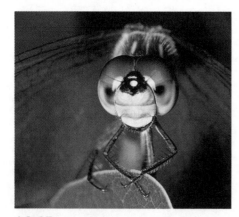

16-15
The eyes of a dragonfly wrap around the head so far that they provide almost 360° of peripheral vision.

spiracle: (to breathe)

Insects outnumber other animals not only in the number of species but also in the sheer number of individuals. Insects survive so abundantly because they are equipped with a large number of varied defenses. These defenses may serve to keep the individual insect or the entire species alive. Some of the common insect defenses are discussed here.

Physical Appearance as Defense

The first line of insect defense is the insect itself. The body color and shape are often simple but effective camouflage, making some insects almost invisible to potential predators. When they rest open-winged on the bark of a tree, some mottled-colored moths are virtually invisible. Some insects blend into the scenery by resembling objects like leaves, twigs, flower petals, and even thorns. The body of the walking stick, for example, resembles a twig.

The leaf insect resembles dead leaves.

Many insects have brilliant colors that make them conspicuous. In many cases this is warning coloration, informing potential predators that the insect tastes bad or can defend itself. The *monarch butterfly*, for example, tastes bad to birds and causes them to vomit. A bird that has had an unpleasant experience with a monarch will remember the distinctive appearance of the insect and not try to eat another one.

Other insects "outwit" their predators by resembling insects that predators avoid. The *hornet moth*, with its furry yellow and black body and thin transparent wings, looks like a hornet. An organism that has been stung by a real hornet will avoid this harmless imposter.

Stinkbugs release chemicals when disturbed.

Weapons of Insects

Some insects depend on special weapons to repel attackers. The most familiar weapons are the poisonous stingers of bees and wasps. Many insects from the order Hemiptera can use their piercing beak designed for feeding to inject a painful mixture of chemicals and digestive juices.

The insect world has its share of "skunks": insects that secrete foul odors to repel their enemies. *Stinkbugs* are perhaps the most infamous, but they are not the only ones. The *bombardier beetle* has a chemical laboratory in its body that can produce an explosive chemical mixture.

Social Defenses of Insects

The timeworn phrase "There is safety in numbers" assumes great meaning in the insect world. Several insect species, mostly within the orders Hymenoptera and Isoptera, ensure their survival by forming large colonies.

A striking example of the benefits of an insect society is the *termite colony*. A single termite is wingless, blind, and has an exoskeleton so thin that it offers little protection. Left to itself, the termite would die within hours. Within the organization of the

Some termite mounds are as strong as concrete.

termite colony, however, termites thrive. A major task of the colony is to build shelter. In some species the home of the colony may be a rotten log. Others manufacture mounds of a stonelike material that may reach 6 m in height. Within this home the members of the colony maintain an environment suitable for their survival.

Reproductive Defenses of Insects

Many insects produce enormous numbers of young. In its lifetime the common *housefly* will produce 1000 eggs. Some queen termites can produce 10 000 eggs a day, and they can live more than thirty years. Even if disease or predators claim many of these young, the sheer number of eggs guarantees that some young will survive, mature, and reproduce.

Many insects improve the chances of survival for their young by providing special care to the eggs. The *mud dauber wasp*, for example, builds a set of clay tubes for its young. The nest, therefore, is divided into separate cells in which the eggs are placed. Each cell also contains a spider paralyzed by the female wasp to be food for the larvae.

Insect Behavior as a Defense

Insects' behavior can also serve them in times of danger. The most common response to danger is escape. The ability to fly is a great asset because many predators cannot follow the insect into the air. Some insects will "play dead" when faced with danger. This usually involves folding the legs tightly under the body and remaining motionless until the threat passes. If attacked or provoked, some insects will turn and fight back. For example, when even a large animal breaks into a wasp nest, the wasps will not flee but will sting the attacker.

A swarm of hornets charging a person who disturbed their nest may appear to be mounting a well-organized attack. Actually, each charging individual is responding to the same stimulus—often triggered by the release of an attack hormone.

16-16

Some ichneumon wasps can drill several inches through solid wood with their slender ovipositor to deposit an egg into a living host.

Several insect species possess organs for hearing, often in unexpected places. The grasshopper hears with its *tympanum* (TIM puh num), a membrane-covered chamber in the first abdominal segment. The katydid listens with an organ in its foreleg. Hearing is important to these insects, not for escaping their enemies or finding food, but for finding mates.

Reproduction in Insects

In insects the sexes are separate. The female stores the sperm in a seminal receptacle. As the female lays the eggs, they are fertilized. The last segment of a female's abdomen forms a pointed extension called the *ovipositor* (OH vuh PAHZ ih tur) that is used to deposit the fertilized eggs.

The exact structure of the ovipositor varies. The grasshopper uses its short, stout ovipositor to dig a burrow for its eggs. The ichneumon wasp, with a slender ovipositor reaching 15 cm (6 in.) in length, inserts this probe deep into the bark of a tree to place its eggs in wood-burrowing beetle larvae. The wasp larvae will feed upon the beetle larvae, eventually killing the host, but not before they complete their own development. Some female insects cover their eggs with a protective secretion. The praying mantis lays its eggs in a gummy mass that hardens into a protective case. Other insects "shellac" their eggs to leaves or bark to hold them in place and waterproof them.

Metamorphosis in Insects

During its development from egg to adult, an insect may assume a variety of forms that differ vastly in appearance, structure, locomotion, diet, and habitat. This series of developmental changes is called **metamorphosis** (MET uh MOHR fuh sis). For many insects these changes permit their various stages to accomplish different functions. For example, some insects eat only during their immature forms and become adults only to reproduce and die in a few hours.

A few insect species do not exhibit metamorphosis. For the majority that do, the process of metamorphosis takes one of two major routes.

Incomplete metamorphosis—the line of development followed by such land insects as grasshoppers, cicadas, and true bugs—is a process involving three basic stages: *egg, nymph,* and *adult.* The **nymph** that hatches from the egg usually looks like a miniature, oddly proportioned adult. The nymph lacks wings and external reproductive structures. As the nymph grows and molts, it becomes more like the adult. When it finally develops wings and mature reproductive structures, the insect is an adult. The diet and habitat of the nymph are identical to those of the adult.

Complete metamorphosis, the process of development used by almost 90% of insects, involves four stages: *egg, larva, pupa,* and *adult.* The egg hatches into the larva, a segmented wormlike stage. The larvae of various insects have common names: *maggots* for flies, *grubs* for beetles, *wigglers* for mosquitoes, and *caterpillars* for moths and butterflies. The most important activity of any larva is to feed.

After a period of eating and molting, the larva enters the pupal stage. In some insects the **pupa** is as active as the larva, but in many insects the pupa is a resting stage. During this stage the pupa forms a case around itself or weaves a silken *cocoon.* Though outwardly quiet, the pupal stage is a period of considerable activity. Within the protective case the body structures and organs of the larva are dismantled and completely reshaped. Eventually the pupa case opens, releasing a fully grown, fully developed adult insect. The process of complete metamorphosis is controlled by *hormones*, chemicals secreted into the blood by glands.

tympanum: (Gk. *tumpanon*—drum)

ovipositor: ovi- (egg) + -positor (L. *positio*—placed)

metamorphosis: meta- (involving change) + -morphosis (shape)

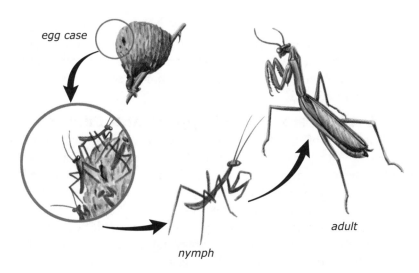

Incomplete metamorphosis—praying mantis

egg case

nymph

adult

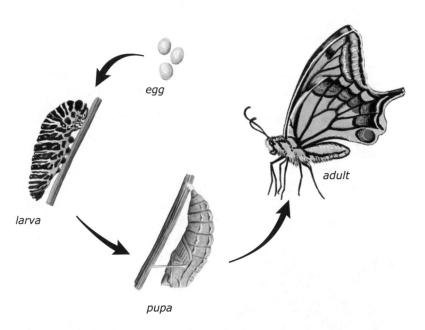

egg

larva

pupa

adult

Complete metamorphosis—European swallowtail butterfly

16-17
Insect metamorphosis

> **Review Questions 16.4**

1. What characteristics separate the class Insecta from other classes of the phylum Arthropoda?

2. List three different functional types of insect (a) legs, (b) wings, and (c) mouthparts and give an example of each.

3. Briefly describe and distinguish between the centipedes and the millipedes.

4. List the three major parts of the insect's digestive system. List the structures within each part and give their functions.

5. Describe the respiration of insects.

6. Describe insects' sense of (a) touch, (b) sight, and (c) hearing.

7. List the steps in incomplete metamorphosis and give several examples of insects that have incomplete metamorphosis. Do the same for complete metamorphosis.

8. What controls the molting process?

⊙9. List and describe five factors that help insects survive and thrive in our world.

16.5 Major Orders of Insects

Taxonomists divide the class Insecta into thirty-one different orders. Fifteen of the most common orders and some of the characteristics used to classify them are given in Table 16-1. A few of the more common insect orders are discussed in the following sections.

Objectives

- Know common insect names from each of the six orders covered in the text
- Compare metamorphosis of the dragonfly with that of the butterfly
- Describe at least three types of biological controls used against insects

Key Terms

chrysalis
caste
quarantine
insecticide
biological control

Order Orthoptera: Grasshoppers and Cockroaches

Although it is one of the smaller insect orders with fewer than 20 000 species, the order Orthoptera contains many significant and well-known species. Incomplete metamorphosis, chewing mouthparts, and one pair of membranous wings covered by a pair of leathery ones set this group apart. Several families, however, contain wingless species.

Of the many grasshopper species only a few are economic pests. The migratory species, called locusts, that live in grasslands are usually the most harmful. Although all grasshoppers have hind legs that they can use for jumping, not all of them are good flyers. Some, however, have larger wings and can fly for long distances. The cricket, another orthopteran known for its jumping ability, can rub its legs together to chirp. Only male crickets chirp, and they do so to attract females and ward off other males.

Few insects are more despised than the cockroach. Of the over two thousand species of roaches, only a few have invaded human dwellings. Known for their long, slender antennae and shiny black or brown leathery body-covering wings, these insects prefer dark, warm, humid environments.

16-18
The cockroach, an orthopteran

Order Odonata: Dragonflies and Damselflies

Nearly every body of fresh water hosts a population of aerial acrobats from the order Odonata. With 5300 species, they come in nearly every size and color. The largest odonates have wingspans up to 20 cm (8 in.). Some tiny species have wingspans of less than 2 cm (0.8 in.). Both dragonflies and damselflies are characterized by long, slender abdomens; paired, netted wings; and large eyes. Most dragonflies have eyes that cover nearly the entire head surface, while damselfly eyes are separated. Dragonflies also keep their wings extended after landing—like an airplane—while damselflies fold theirs together above their thorax.

Both of these groups of insects lay their eggs in water, where the egg hatches into a nymph called a *naiad*. Naiads are nymphs that do not closely resemble their adult forms. They are wingless and have gills to accommodate their aquatic lifestyle. These larvae are eating machines, and dragonfly naiads are aided by a retractable labium that is rapidly extended to grab any moving prey, some nearly as large as the larva itself. After emerging from water and molting, they complete their metamorphosis and become winged wonders, capable of flying forward and backward and hovering while snatching insects as small as mosquitoes from the air.

16-19
Can you tell the difference between a damselfly and a dragonfly?

Tiny Servants

God has used insects many times to manifest His power. Flies (Exod. 8:21), locusts (Exod. 10:4–6), and hornets (Exod. 23:28) are but a few examples. Even the tiniest of His creatures are at His bidding.

Orthoptera: Ortho- (Gk. *orthos*—straight) + -ptera (*pteron*—wing)

Order Coleoptera: Beetles

Beetles are the largest order of insects, including 360 000 different species. With their compact bodies covered by heavy armor, most beetles look like miniature war machines. Some bear large chitinous horns or mandibles

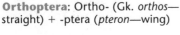

16-1 Major Insect Orders

Order (literal meaning)	Type of wings	Type of adult mouthparts	Type of metamorphosis	Example	Economic impact
Thysanura ("tassel tail")	none	chewing	none	silverfish	eats cellulose in books and clothing
Phthiraptera ("lice without wings")	none	piercing or sucking	none	sucking lice	parasitizes humans, mammals, and birds; carries disease
Orthoptera ("straight wing")	1 pair membranous, 1 pair leatherlike	chewing	incomplete, land	grasshoppers, crickets, cockroaches	destroys crops and stored foods
Isoptera ("equal wing")	2 pairs membranous	chewing	incomplete, land	termites	feeds on wood in buildings; recycles forest wood
Hemiptera ("half wing")	1 pair (anterior half leatherlike; posterior half membranous)	piercing and sucking	incomplete, land	true bugs— stinkbugs, bedbugs, water striders	damages crops; annoys man
Homoptera ("same wing")	2 pairs membranous	piercing and sucking	incomplete, land	cicadas, aphids, leafhoppers	destroys plants and crops
Dermaptera ("skin wing")	1 pair leatherlike	chewing	incomplete, land	earwigs	destroys plants and crops
Odonata ("toothed")	2 pairs membranous	chewing	incomplete, water	dragonflies, damselflies	destroys harmful insects; naiads serve as food for fish
Ephemeroptera ("lasting-only-a-day wing")	2 pairs membranous	do not feed; nonfunctional	incomplete, water	mayflies	naiads serve as food for freshwater fish
Coleoptera ("sheath wing")	1 pair horny, 1 pair membranous	chewing	complete	beetles, weevils	destroys crops and annoys man; destroys harmful insects
Lepidoptera ("scale wing")	2 pairs scaly	siphoning	complete	butterflies, moths	damages crops and clothing; produces silk; pollinates flowers
Hymenoptera ("membranous wing")	2 pairs membranous	chewing or siphoning	complete	bees, ants, wasps	pollinates flowers; produces honey; annoys man; parasitizes pests
Diptera ("two wings")	1 pair membranous, 1 pair balancers	piercing or sponging	complete	flies, mosquitoes, gnats	annoys man; transmits disease
Siphonaptera ("siphon without wings")	wingless adult	sucking	complete	fleas	parasitizes birds, mammals; carries disease
Neuroptera ("nerve wing")	2 pairs membranous	chewing	complete	dobsonflies, ant lions, lacewings	destroys harmful insects

16-20 —————————————

Beetles come in all colors, shapes, and sizes.

enlarged into fearsome-looking pincers. The exoskeleton of these insects is remarkably strong; in some instances it can sustain over 1500 times its own weight.

The most obvious parts of a beetle's body are the horny wings, a pair of thick, often colorful, protective sheaths that cover most of the body. These armored wings give the order its name: Coleoptera (KOH lee AHP tuh ruh), which means "sheath wing." Carefully folded under the horny wings is a pair of small, thin membranous wings used for flying. Although all beetles fly, many are poor fliers since the heavy outer wings must be held in an open position, creating drag. These same heavy horny wings that impair flight also help protect the beetles if they cannot escape in time. Some beetles are also remarkably fast runners.

The appetite of some beetles makes them both a curse and blessing to man. Armed with chewing mouthparts, these beetles can destroy a variety of crops and plants. Many beetles are most destructive during their feeding larval stage. Adult Japanese beetles plague gardeners in the United States. The rice weevil makes its home where man stores his food, destroying products such as wheat, barley, oats, corn, rye, and rice. Some beetles, however, have appetites that benefit man. The familiar red and black ladybug feeds almost exclusively upon aphids and scale insects—animals that ravage crops. A single ladybug can eat ninety adult scale insects and three thousand of their larvae during its own larval lifetime.

Order Lepidoptera: Butterflies and Moths

At one stage in the life of a lepidopteran (LEP ih DAHP tur un), it is a slow-moving worm, perhaps bristling with spines. At another stage it is a graceful fluttering insect with large powdery wings. Of the 165 000 species in order Lepidoptera, more than 85% are moths.

The complete metamorphosis from caterpillar to butterfly or moth is one of the most noticeable characteristics of this order. The monarch butterfly lays delicately sculptured eggs that hatch and release caterpillars. The monarch caterpillar has sixteen legs and, after several molts, bears brilliant zebra-patterned coloration. The caterpillar has chewing mouthparts designed for its vegetarian diet. During its brief existence, the caterpillar does little more than feed, grow, and molt. When signaled by hormonal changes, the caterpillar

Coleoptera: Coleo- (Gk. *koleon*—sheath) + -ptera (wing)

Lepidoptera: Lepid- or Lepido- (Gk. *lepis*—scale) + -ptera (wing)

attaches itself to a leaf and enters the pupal stage. Within a protective case called a **chrysalis** (KRIS uh lis), the insect makes its most dramatic change. After two to three weeks, the adult monarch butterfly emerges. The mouthparts of the insect are now a coiled tube for sucking nectar from flowers. The adult has the characteristic six legs.

Although the main activity of the adult is to reproduce (in fact, some lepidopteran adults do not eat but live only long enough to reproduce), many monarch butterflies migrate. The adult monarchs alive in the fall migrate from most of the United States and Canada to special areas of Mexico and Florida where they land on certain trees and "rest" through the winter. In the spring they migrate back through North America.

Farmers, however, do not always like the beautiful butterflies and moths. The larvae of certain lepidopteran species—army worms and cutworms, for example—cause tremendous damage to crops. Some moth larvae also damage stored goods and clothing.

16-2 Summary of Differences Between Moths and Butterflies	
Moths	**Butterflies**
pupa in a spun silk cocoon	pupa in a chrysalis
adult antennae feathery	adult antennae thin with knobs on the ends
adult body thick and often appears furry	adult body slender
wings usually held horizontally while at rest	wings usually held vertically while at rest

Order Hymenoptera: Bees, Wasps, and Ants

Hymenoptera means "membrane winged" and is an apt description of the bees and wasps found in this order. However, not all of the 120 000 species in this group are considered winged, because ants also belong here. The female workers that are the most conspicuous members of most ant colonies lack wings, although other ants may have wings.

Many members of the order Hymenoptera are social insects; they live in large groups and have **castes** (classes) that perform different functions for the group. In a bee colony, for example, there are a large number of female workers, several drones (males), and one queen. The queen and drones serve primarily to reproduce, while the workers build and maintain the hive and make honey.

One of the interesting behaviors of the worker bees is their form of dancing as communication. God designed these bees to perform different kinds of dances to let the rest of the hive know where to look for food. Bees within the hive interpret the style of dance, the angle of the bee, and the speed of the movements to know what type of food is available and how to find it.

Order Hymenoptera also includes the ants. In most ant species, however, the male and female reproductive castes are the only ones that have wings. Since males usually die after their mating flight and the queen usually loses her wings after mating and workers never have wings, most ants that people see are wingless. Many ant species lack stingers but can nonetheless inflict painful bites with their powerful jaws. A few species possess a stinger similar to that of a bee.

16-21 —
Hornets are hymenopterans that chew up wood and plant fibers to manufacture an enclosed papier-mâché-type nest. The layered cells inside are brood chambers for the production of the young.

16-22 —
An ant's mandibles are used for defense, carrying, and eating (SEM 7×).

Hymenoptera: Hymen- or Hymeno- (Gk. *humen*—membrane) + -ptera (wing)

Suicide Missions

Although many insects can inflict painful bites, only hymenopterans can sting. The ovipositor of the workers in some social hymenopteran groups is not used to lay eggs but to inject venom. When a disturbed bee implants its stinger into its victim, it then rips off its last abdominal segment, which continues to pump more venom. Soon after the bee stings, it dies. Hymenopteran stings cause soreness and pain. However, a person who is allergic to the venom may die within several minutes without proper medical attention.

Some insect activities are essential to life on the earth as modern man knows it. Many plants are pollinated only by insects. From honeybees, man obtains 320 million kg of honey and 2 million kg of wax annually; from silkworms, over 30 million kg of silk per year; and from other insects, valuable drugs and dyes. However, insects also eat grain, fruit, and vegetables; injure and destroy ornamental trees and shrubs; destroy wood and buildings; damage fabrics; carry disease; and parasitize plants, animals, and man. In the United States alone the losses from insects and the amount spent to control harmful insects total over $7.5 billion annually.

Only 0.01% of all the world's insects are harmful to crops, but those that are can be devastating. Scientists attempt to control harmful pests while preserving beneficial insects. Many methods of control have been attempted with varying amounts of success.

Quarantine

The U.S. Department of Agriculture oversees the entry of plant materials and insects into the United States. Widespread damage has been caused by more than seventy-five harmful insect species that have been introduced into the United States.

Problems with Insecticides

Insecticides have several drawbacks. Often the quantity of insecticide necessary to kill a harmful insect also kills beneficial insects and other organisms. Some insecticides are very stable; that is, they do not easily break down into smaller compounds in natural conditions. If a stable insecticide is applied repeatedly in an area, the concentration may build up. Although the quantities applied may not be dangerous to organisms other than insects, the buildup may become dangerous to other organisms and man.

Occasionally it is said that an insect has "developed" a resistance to an insecticide. Evolutionists would say that insects that are resistant to the insecticide have evolved because of their being exposed to the insecticide. Actually, by killing those insects that are most sensitive to the insecticide, only those with the strongest resistance to the insecticide are left. The insects having the resistance were already in the population; otherwise they would have been killed by the insecticide.

Although it may be possible for insects to have a mutation with a variation that enables them to tolerate certain insecticides, this is not evolution. There would be no new insect, just a variation of an existing insect.

Fire Ants

Angry Alien Invaders

The well-named fire ants are feared and despised in the fifteen states where they now occur. Hundreds of millions of dollars are spent each year to try to control them, yet they continue to advance. They inhabit the area from Virginia to southern California and many points south of that area. If scientists are correct, they have the potential to spread to more than half the country. The most significant threat is their aggressive stinging behavior. A single ant will clamp to the flesh with its tenacious jaws, arch its abdomen, and inject venom from a stinger on its tail end. It then pivots its head, rapidly repeating the painful process as many as seven or eight times before finally releasing. When a nest is disturbed, the first responders release an attack hormone that incites the rest of the ants to aggressive action. Consequently, many ants are usually stinging simultaneously. Some large colonies may contain more than 300 000 ants! Wildlife and livestock are frequently killed, and humans with allergic reactions have also died from ant attacks. Farmers dislike them because they can kill crops like corn, soybeans, and even young citrus trees. Their hills also damage farming equipment in the field.

These ants are actually not part of North America's native insect population. They arrived in Mobile, Alabama, sometime in the 1930s, most likely in soil used as ballast. Ships from either Brazil or Argentina probably introduced them to American shores. In the years since then, they have fanned out from that area, sometimes advancing by winged flights, sometimes being carried by flood waters, but moving more effectively by human transport in sod and plant shipments. This is how they hopscotch to new, disconnected areas.

Chemists have developed dozens of products to kill the ants with varying degrees of success. Some must be put directly at the hill, while others can be sprayed onto fields from tractors or airplanes. Close inspection of shipments is also a crucial measure in halting the importation and further spread of the ants. Recent efforts at biological control involve the release of tiny phorid flies that lay a single egg in an ant. When the egg hatches, the larva migrates to the head of the ant, causing it to fall off. The "head-hunter" larva then completes its development inside this protective structure. Microscopic internal parasites called microsporidia are also being tested since they are found in the South American fire ant populations and seem to limit their numbers. In the end, a combination of chemicals, caution, and biological controls will probably be necessary if these alien ants are to be halted.

Many insecticides are applied by aircraft.

The gypsy moth was accidentally introduced into America in 1869. The larva (inset) consumes leaves. In large groups they can strip land bare.

The gypsy moth, the Japanese beetle, the Mediterranean fruit fly, and the fire ant are not native to North America, but they are among the most serious insect threats in the United States. In regard to most plant and animal material, therefore, the United States is now under **quarantine**. Materials are carefully inspected before they are permitted into the country to be sure that an unwanted organism is not entering also.

Insecticides

Chemicals that man uses to poison insects are **insecticides**. Usually insecticides affect insects in one of two ways:

✦ *Stomach poisons* are applied to the leaves of plants being attacked by a foliage-eating insect. When the insecticide is eaten in sufficient quantity, the insect dies. Some insecticides are systemic, meaning they enter the plant and remain inside the plant tissues. When an insect sucks the plant juices or eats part of the plant, it consumes the poison.

✦ *Contact poisons* are applied near insects that for some reason would not eat a stomach poison. Contact poisons can either affect the tracheae of the insect and cause suffocation or dissolve into the blood of the insect, usually poisoning the nervous system.

Environmental Changes

The egg, larval, and pupal stages of the mosquito are aquatic. If a pond or swamp where mosquitoes breed is periodically drained, the mosquito population can be controlled.

Knowing insects' life cycles can halt their threat to a particular crop by temporarily switching to a different crop. The practice of crop rotation changes the insects' environment and thereby helps control their population. Crop rotation also helps control plant diseases and prevents depletion of soil minerals.

Biological Controls

Biological control of insects involves using one of several techniques to destroy an insect but not harm the rest of the environment. One popular method is the use of natural predators and parasites. Today, many gardeners encourage frogs, toads, snakes, spiders, birds, and predatory

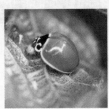

The ladybug or ladybird beetle is a voracious predator of aphids.

insects to live near their gardens. The ladybug is a favorite among farmers for its tremendous appetite for destructive aphids and scale insects. Some seed companies sell ladybug and praying mantis eggs to be placed in gardens.

One of the most promising insect control methods involves *pheromones*. Pheromones are chemicals that insects secrete to communicate—often to attract a mate. The female gypsy moth, for example, produces a pheromone that attracts males from great distances. By using that same chemical substance as a lure, the male moths can be drawn into a trap where they can be killed or where their populations can be estimated in order to time some other treatment.

Farmers can also release pheromones into fields or orchards to create a confusing blend of chemical signals that hinder the male insects from ever finding a mate. The advantage of using pheromones is that they are extremely specific, attracting only the target insect, and seldom harm other species.

Another biological control of insects involves either capturing or growing males of the species and then bombarding them with radiation to make them sterile. The sterilized males are then released and mate with females. The eggs of those females will not be fertilized. If this genetic sabotage is continued, the unwanted insect's population will decrease. Scientists have used this technique effectively against the screwworm fly, a menace to livestock, and against the Mediterranean fruit fly.

Bacteria and viruses are other potential weapons against insects. One such bacterium known commercially as Bt produces a substance toxic to the larvae of butterflies and moths. This microscopic killer is very specific, causing no harm to vertebrates, other insects, or even the parasites within the body of the caterpillar.

Biological control techniques are not a perfect answer to insect control. Natural predators, which are safe to the environment, are often difficult to find and use effectively. The development and safe use of man-made biological controls are time-consuming, difficult, and expensive. Many are still in the experimental stages and not ready for general use. Nevertheless, biological control techniques show the greatest promise in combating unwanted insects without the harmful side effects of chemical insecticides.

This trap contains a pheromone, undetectable to humans, that attracts gypsy moths so that their numbers can be monitored.

insecticide: insecti- (insect) + -cide (L. *caedere*—to kill)

Most ants live in a caste society with divisions of labor and make their living in many fascinating ways. The leaf-cutter ants make expeditions from their underground home to cut and gather bits of foliage for the colony. In special underground chambers, ants chew the leaves, mixing them with salivary secretions to produce a suitable material on which to grow fungus. These ants then use the cultivated fungus, not the leaves, as food.

African or South American army ants have no permanent homes. During their hunting raids, a temporary shelter is built for the queen and the young by using the bodies of the ants themselves. Other types of ants capture smaller ants and make them work as slaves.

Order Diptera: Flies, Mosquitoes, and Gnats

Insects of the order Diptera (DIP tuh ruh) have caused man much pain and discomfort. Although most insects eat plant parts, many dipterans use man or other mammals as their targets. More than 120 000 species have been identified.

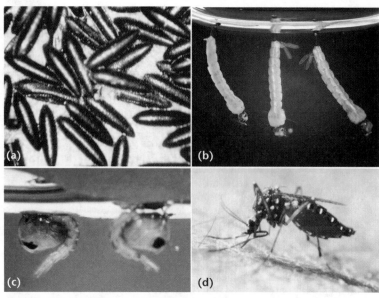

(a) (b) (c) (d)

16-23

Life cycle of the mosquito: the eggs form a small floating raft (a) and hatch into larvae (b) that swim about feeding, returning to the surface to breathe. The pupal stage (c) is also aquatic, feeds actively, and must breathe at the surface. The adult female (d) must have a blood meal before she can lay her eggs.

Diptera: Di- (two) + -ptera (wing)

The mosquito is an annoying and potentially dangerous member of this order. It lights so softly that the victim is usually not aware of its presence. Only the female feeds on blood; she needs it to nourish the eggs developing in her body. The mosquito's mandibles and maxillae taper into a set of four slender needles that easily pierce the skin. Its needlelike mouthparts then fill the wound with a chemical to prevent blood clotting. Finally, a slender feeding tube sucks up the blood. When the meal is over, the victim is left with only a little blood loss but with a painful, itchy swelling caused by the anticlotting chemical. A mosquito can be more than a pest since it may harbor yellow-fever virus, malarial organisms, and the larvae of parasitic worms.

The common housefly, another dipteran, is potentially dangerous because of what it can carry. A single housefly may transport 33 million microorganisms in its intestinal tract and 500 million on its body. When the fly lights on human food, some of the microorganisms from its body may be left. When a fly feeds, it expels digestive fluids onto its food and then laps the fluids and digested food. In the process many microorganisms from the fly's intestinal tract, some of which may be pathogenic, remain.

➡ Review Questions 16.5

1. List the characteristics and give several examples of these insect orders: (a) Orthoptera, (b) Odonata, (c) Coleoptera, (d) Lepidoptera, (e) Hymenoptera, (f) Diptera.

2. What are the three castes seen in the social insects? What are their functions within the insect colony?

3. What characteristics separate the insects into various orders?

⊙ 4. List and describe four methods man uses to control insects.

⊙ 5. Describe the two main types of insecticides.

⊙ 6. Give three methods of biological control of insects.

⊙ 7. What value are pheromones to (a) insects and (b) farmers?

⊙ 8. Is it possible for a small butterfly to grow and become a larger butterfly? Explain.

⊙ 9. If you were a biologist considering the importation of a tiny wasp to kill caterpillars that were eating valuable crops, what kinds of things would you have to know before you could safely release the wasps?

17

The Ectothermic Vertebrates

Zoology Part III

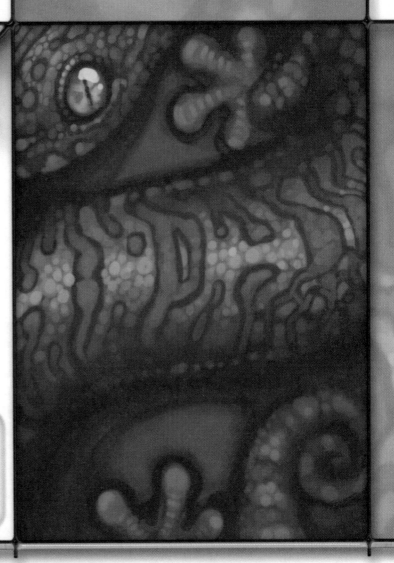

17A—Introduction to Phylum Chordata

The shark swimming soundlessly in the sea, the squirrel chattering from a treetop, the swallow swooping across the sky, the snake slithering in the grass, and the small sea squirt attached like a plant near the shore—all are members of the phylum Chordata (kohr DAH tuh). As diverse as they seem, **chordates** share many common characteristics.

17.1 The Chordates

Characteristics

✦ *Dorsal notochord.* The **notochord** (NOH tuh KORD) is a rod of tough, flexible tissue running the length of the animal's body and serving as its primary support. For some chordates the notochord remains throughout the animal's life. In one group, the notochord is present only in the larva, disappearing by the adult stage. But in most chordates, before birth or hatching, the notochord is replaced by **vertebrae** (VUR tuh bray) (sing., vertebra). Vertebrae, which serve as the primary support for the animal, are either a tough, flexible material

17.1

➡ Objectives

- List and describe the characteristics of the phylum Chordata
- Explain the major differences between the three subphyla of Chordata
- Discuss the limitations of being ectothermic versus endothermic

➡ Key Terms

chordate	pharyngeal
notochord	pouches
vertebra	vertebrate
nerve	ectothermic
cord	endothermic

Chordata: (L. *chorda*—cord)

notochord: noto- (Gk. *noton*—back) + -chord (cord)

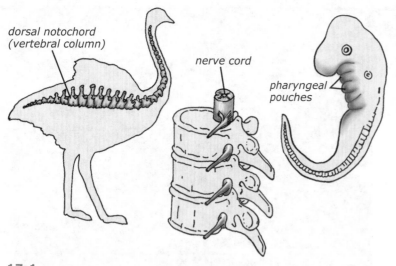

dorsal notochord
(vertebral column)

nerve cord

pharyngeal
pouches

17-1
Chordate characteristics

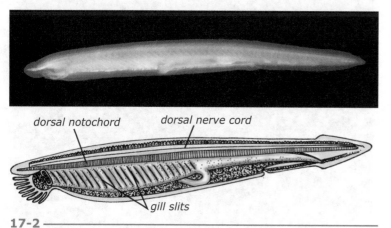

dorsal notochord

dorsal nerve cord

gill slits

17-2
Amphioxus

17-3
Not all tunicates are this colorful, but they all pump water through their bodies.

pharyngeal: (throat)

Cephalochordata: Cephalo- (head) + -chordata (cord)

Urochordata: Uro- or -ura (Gk. *oura*—tail) + -chordata (cord)

or bone. Together, all the vertebrae are called the *vertebral column*, or backbone.

✦ *Dorsal tubular nerve cord.* The **nerve cord** is dorsal to the notochord or vertebral column. It and the brain—connected at the anterior end of the cord—compose the main part of the animal's nervous system. In many species the vertebrae encase the nerve cord to protect it.

✦ *Pharyngeal pouches.* During embryonic development, all chordates have folds of skin along the neck (pharyngeal region) called **pharyngeal** (fuh RIN jee uhl) **pouches**. In most aquatic chordates, openings called pharyngeal slits, or gill slits, develop at these pouches. These openings permit water to flow over the gills, which develop inside the pouches. In nonaquatic chordates, the pouches never open and are never associated with lungs or respiration. The pharyngeal pouches of terrestrial animals and of man develop into various structures of the lower face, neck, and upper chest.

Classification Within Chordata

Phylum Chordata is divided into three subphyla: Cephalochordata, Urochordata, and Vertebrata. The organisms in the first two subphyla are aquatic, relatively small, few in number, and of little economic importance. However, the members of Vertebrata are the most familiar and make up 95% of Chordata.

✦ Members of subphylum Cephalochordata (SEF uh luh kohr DAH tuh) retain their notochords throughout their entire lives. A typical notochord-retaining organism is the *amphioxus*, or lancelet, a slender marine animal about 5 cm (2 in.) long. This tiny eel-like creature exists in tropical and temperate coastal waters. The amphioxus lives half-buried in the sand with its head exposed and feeds by filtering plankton from the water. Cilia around the mouth help circulate the water through its pharyngeal slits as it feeds.

✦ Members of subphylum Urochordata (YOOR uh kohr DAH tuh) have notochords at the larval stage. The sea squirts, or *tunicates*, perhaps best represent the urochordates. True to their name, the sessile sea squirts pump water through small pores on their bodies. They reproduce sexually by producing swimming, tadpole-shaped larvae that have all three of the chordate characteristics. Of the three, they lose all but the pharyngeal slits, which they keep into adulthood.

✦ Members of subphylum Vertebrata (vurt uh BRAHT uh) develop vertebral columns, usually before they are born or hatched. The members of this subphylum are studied in depth in this chapter and in Chapter 18.

Vertebrate Variety

Within phylum Chordata the subphylum Vertebrata contains the **vertebrates** (animals with backbones) and is unmatched in its variety. It includes fish, amphibians, reptiles, birds, and mammals. Vertebrates live in virtually every habitat, from the ocean floor to the fringes of the atmosphere. The bar-headed goose has been sighted flying over the Himalayas at a height of 9.1 km (5.6 mi). At that altitude the temperature is well below freezing. In contrast, the brotulid fish thrives under nearly 8 km (5 mi) of ocean water, a depth where light never penetrates. Environments hostile to most forms of life will support certain chordates. The polar bear, a large carnivore, dwells on the seemingly lifeless ice floes of the Arctic Circle. Equally forbidding, the deserts support a variety of vertebrates, including snakes and lizards.

Vertebrate Classifications

Traditionally, vertebrates have been put into two groups: cold-blooded and warm-blooded. Although not part of the modern classification system, the groupings are still useful. An animal is **ectothermic** (EK tuh THUR mik), or cold-blooded, if it does not have the ability to generate its own body heat and thus maintain a consistent body temperature above that of its surroundings. Ectothermic animals depend upon heat from sources outside themselves to stay alive. These animals are usually sluggish when cold and active when warm. To warm themselves, ectothermic animals such as lizards and snakes often lie in the morning sun. If they get too warm, they seek cooler, shaded places. Amphibians, reptiles, and fish are all ectothermic.

17-4 —————
Is this creature ectothermic or endothermic?

Endothermic (EN doh THUR mik), or warm-blooded, animals generate their own body heat through physiological changes and often have elaborate mechanisms for maintaining body temperature within a narrow range. Endothermic animals can be active regardless of the external temperature, but most often they adjust their activity to permit their bodies to generate more heat or lose excess heat. This increased convenience comes at a considerable cost because an endothermic animal may burn thirty times as much energy as an ectothermic animal of the same mass. Birds, mammals, and humans are endothermic organisms.

The subphylum Vertebrata has seven classes—five are ectothermic and are discussed in this chapter. This includes three classes of fish and fishlike animals, amphibians, and reptiles. The two endothermic classes (birds and mammals) are discussed in the next chapter.

ectothermic: ecto- (outside) + -thermic (heat)

endothermic: endo- (inside) + -thermic (heat)

17-1 Classes of Subphylum Vertebrata			
Class	**Key characteristics**	**Examples**	**Species number (approx.)**
Agnatha	jawless fish, lack bone or paired fins, ectothermic	lamprey, hagfish	80
Chondrichthyes	cartilaginous skeleton, paired fins, ectothermic	sharks, rays	800
Osteichthyes	bony skeleton, ectothermic	perch, bass, salmon	20 000
Amphibia	lay eggs in water, aquatic larval stage, ectothermic	frogs, toads	4500
Reptilia	dry scaly skin, amniotic egg, internal fertilization, ectothermic	turtles, snakes, alligators	6900
Aves	hollow bones, flight, feathers, endothermic	birds	9300
Mammalia	hair, nurse young with milk, endothermic	dogs, lions	4400

⇨ Review Questions 17.1

1. List and describe three characteristics of organisms in phylum Chordata.
2. Describe the amphioxus and tell why it is in the same phylum as the vertebrates.
3. Describe what it means to be ectothermic and endothermic. List five organisms that are ectothermic and five that are endothermic.

17.2

➥ **Objectives**

• Contrast the endoskeleton with the exoskeleton, giving advantages of each

• List the major divisions of the vertebrate skeleton

• Name the major structures and functions of the typical vertebrate circulatory system

• Distinguish between the three vertebrate feeding strategies, giving examples of each

• Compare the three methods of vertebrate reproduction

• Name the five major lobes of the vertebrate brain with the major function of each

➥ **Key Terms**

axial skeleton	omnivorous
appendicular skeleton	oviparous
	viviparous
artery	ovoviviparous
capillary	olfactory lobe
vein	cerebrum
hemoglobin	optic lobe
kidney	cerebellum
herbivorous	medulla
carnivorous	oblongata

Vertebrate Support and Movement

The exoskeletons of the arthropods and other invertebrates greatly restrict their growth and movements. Vertebrates, however, are characteristically supported by an internal skeleton, or endoskeleton, composed of both bone and cartilage. The structures of the endoskeleton often surround, protect, and support delicate organs such as the brain, spinal cord, heart, and lungs. Movements are accomplished by muscles that cover the bones. This structural difference permits greater freedom of movement than the exoskeleton armor of the insects, crabs, and other invertebrates permits. Providing the same amount of support, the endoskeleton is considerably lighter than an exoskeleton. Therefore, a vertebrate moves more freely and can also be much larger than an invertebrate with an exoskeleton.

The skeleton of the vertebrate has the following two major divisions:

✦ *Axial skeleton.* The **axial skeleton** consists of the vertebral column, the skull, and the ribs.

✦ *Appendicular skeleton.* Most vertebrates have two girdles suspended from the axial skeleton for the attachment of limbs. The *pectoral* (PEK tur uhl) *girdle* is in the anterior region, and the *pelvic girdle* is in the posterior. Attached to each girdle are the bones of the limbs, which may be in the form of fins, flippers, legs, or wings. The girdles and limb bones form the **appendicular skeleton**.

Vertebrate Circulation and Excretion

Vertebrates have a closed circulatory system consisting of a heart, located ventral to the vertebral column, and blood vessels. The heart may have two, three, or four chambers, depending on the species. There are three basic types of blood vessels in vertebrates.

✦ **Arteries** carry blood away from the heart to body tissues.

✦ **Capillaries** (KAP uh LAIR eez) are the thinnest branches of the arteries. They are tiny thin-walled vessels that pass through the body tissues. They supply the tissues with nutrients and oxygen and remove wastes and carbon dioxide from the same tissues.

✦ **Veins** begin in capillaries and carry blood from body tissues back to the heart. Normally, red blood cells do not leave the blood vessels.

All vertebrates have red blood because certain of their blood cells contain **hemoglobin** (HEE muh GLOH bin), a red oxygen-carrying pigment. In most vertebrates the blood passes through a pair of **kidneys**, where wastes are filtered out.

skull   pectoral girdle   vertebral column   pelvic girdle

ribs

☐ axial skeleton
☐ appendicular skeleton

17-5

Divisions of the vertebrate skeleton, using a cat skeleton as an example

axial: (L. *axis*—hub)

capillary: (L. *capillus*—hair)

hemoglobin: hemo- (Gk. *haima*—blood) + -globin (L. *globus*—globe)

herbivorous: herbi- (L. *herba*—herb) + -vorous (*vorare*—to devour)

Vertebrate Nutrition

Although there is a great deal of similarity between members of the subphylum Vertebrata, there is also considerable variety in feeding methods and foods. These feeding strategies, however, fit into three categories:

✦ **Herbivorous** (hur BIHV ur us) animals eat plants. Grazing animals such as cows and horses are familiar herbivores. A plant diet presents some problems.

The cellulose walls that encase plant cells are difficult to digest. To break up these walls, most herbivores use grinding teeth to thoroughly chew the plants. Bacteria in the digestive tract often supply the enzymes necessary for breaking down plant materials. Some nonmammal herbivores include certain tortoises, lizards, fish, and birds.

✦ **Carnivorous** (kar NIHV ur us) animals feed upon other animals. Sharks, lions, and eagles, for example, capture and tear their prey with sharp teeth, beaks, or claws. Frogs and chameleons snag their insect victims with rapid-fire sticky tongues. It should be remembered, however, that all animals were originally herbivores, and that the Fall caused animals to eat each other.

✦ **Omnivorous** (ahm NIHV ur us) animals, such as pigs, bears, and rats, eat both plants and animals. These animals often have varied teeth; some are designed to tear meat, others to chew plants.

In vertebrates, the alimentary canal (food tube) is composed of an esophagus, stomach, and intestines (often with a liver, gallbladder, and pancreas as accessory organs). The digestive system is ventral to the vertebral column.

Vertebrate Reproduction

In most vertebrates the sexes are separate. The males have paired testes that produce sperm, and the females have paired ovaries that produce ova. There are two basic methods by which the sperm come in contact with the ova.

✦ In *external fertilization* the male fertilizes the ova by releasing sperm onto them after the female lays them. External fertilization takes place in water, and there cannot be a shell on the ova.

✦ In *internal fertilization* the male places the sperm inside the female's body, where the ova are fertilized. Certain aquatic animals and most land animals breed by internal fertilization. Among egg-laying vertebrates, internal fertilization permits a shell to be placed on the egg as it is formed. Eggs usually have some form of protective coat (shell) that permits them to be laid and to develop outside the mother's body.

Although exceptions exist, there are three basic methods of development for young vertebrates:

✦ **Oviparous** (oh VIHP ur us) organisms, such as birds and many reptiles, produce offspring from eggs that hatch outside the body.

✦ **Viviparous** (vye VIHP ur us) organisms produce live offspring that have been nurtured to birth inside the uterus or similar structure in the mother. In mammals, the placenta, an attachment between the mother and the fetus, supplies nutrients to the offspring. Most mammals are viviparous.

✦ **Ovoviviparous** (OH voh vye VIHP ur us) organisms produce live offspring that were not nurtured through a direct connection to the mother. The fertilized eggs, often with a shell, remain in the mother's body and hatch there, and then the young emerge. This is a benefit for the eggs since they are better protected and kept at a more constant temperature. The garter snake is a good example, along with some other reptiles and sharks.

Vertebrate Nervous System

The nervous system of vertebrates contains the brain, spinal cord, cranial nerves (which branch from the brain), spinal nerves (which branch from the spinal cord), and sensory organs (such as eyes, ears, and taste buds).

17-6
Like almost all mammals, cows are viviparous.

carnivorous: carni- (L. *carn-*—flesh) + -vorous (to devour)

omnivorous: omni- (L. *omnis*—all) + -vorous (to devour)

oviparous: ovi- (egg) + -parous (L. *parere*—to give birth)

viviparous: vivi- (alive) + -parous (to give birth)

ovoviviparous: ovo- (egg) + -vivi- (alive) + -parous (to give birth)

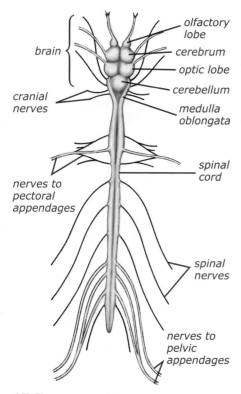

brain
- olfactory lobe
- cerebrum
- optic lobe
- cerebellum
- medulla oblongata

cranial nerves

nerves to pectoral appendages

spinal cord

spinal nerves

nerves to pelvic appendages

17-7
Typical vertebrate nervous system

a Closer Look

Which Came First—the Baby or the Egg?

Some evolutionists try to trace evolution through means of reproduction. Supposedly, oviparous reproduction developed first, followed by ovoviviparous and then viviparous reproduction. When examined superficially, this may appear to follow an evolutionary path. Fish and amphibians reproduce oviparously with external fertilization, reptiles and birds are oviparous with internal fertilization, and mammals are viviparous with internal fertilization.

However, when examined carefully, the theory breaks down. Sharks, which supposedly developed early in vertebrate history, reproduce with internal fertilization both ovoviviparously and viviparously. Though fish and reptiles are predominantly oviparous, some, like the mosquito fish and the rattlesnake, reproduce ovoviviparously. Mammals are predominantly viviparous, but the echidna, or spiny anteater, a mammal found in Australia, is oviparous. Creation offers a far simpler explanation of the various forms of vertebrate reproduction. God created animals with diverse capacities for reproduction to show His glory.

The tiny echidna hatches from an egg incubated by the mother in a temporary pouch.

The primary functions of the five major lobes (divisions) of the typical vertebrate brain are listed below from anterior to posterior.

- The **olfactory lobes** receive impulses from the smell receptors of the nostrils.
- The **cerebrum** controls voluntary muscle activity.
- The **optic lobes** receive impulses from the eyes.
- The **cerebellum** coordinates muscle activity and some involuntary activities.
- The **medulla oblongata** (mih DOOL uh • AHB long GAH tuh) transports impulses to and from the spinal cord, including some reflexes.

In most vertebrates, the larger the portion of the brain given to a certain function, the more skilled the animal is at performing that function.

olfactory: (L. *olfacere*—to smell)
cerebrum: (L. *cerebrum*—brain)
optic: (Gk. *optos*—visible)
cerebellum: (L. *cerebellum*—small brain)

➡ *Review Questions 17.2*

1. What are the two major divisions of the vertebrate endoskeleton? What are the major groups of bones in each of these divisions?
2. List distinguishing features of the three major types of blood vessels.
3. Why is the blood of vertebrates red?
4. Based on feeding methods, what are the three major groups of vertebrates and what does each group eat?
5. What are the two basic types of fertilization found in vertebrates? Which requires mating in water, and which may be accomplished in water or on land?
6. List and describe the three basic methods of development of vertebrate embryos.
7. What are the five major parts of a vertebrate nervous system?
8. List the five typical vertebrate brain lobes and tell the primary function of each.
⊙ 9. Is there a method of embryonic development found in vertebrates that is more efficient than the other methods? Explain. Can efficiency of embryonic development be used to support a belief in evolution? Why or why not?

17.3 Vertebrate Behavior

The way an animal responds to its environment is its **behavior**. Although it can be said that the responses of a worm or a hydra are behaviors, the most noticeable and interesting animal behaviors are found in the vertebrates.

An animal's behaviors are perhaps even more important than its physical characteristics in ensuring its survival. God's design in the animal kingdom extends even to His equipping them with the proper behaviors. Most scientists agree that different vertebrates have various levels of three types of behavior: inborn, conditioned, and intelligent.

Inborn Behavior

Inborn behavior (innate behavior) is the behavior that the organism has from birth and does not need to develop. It predominates in the activities of many animals. Inborn behavior can be divided into two groups:

✦ Reflex behavior is an automatic, involuntary response to a stimulus. Blinking an eye when it is touched and recoiling from pain are *reflexes*. The sucking reflex in young mammals is a critical reflex behavior.

✦ Instinct behavior is a fascinating and little-understood form of behavior. *Instincts* are elaborate behaviors, apparently the result of a stimulus or series of stimuli. For example, the mating of salmon often involves migrating hundreds of miles inland. From the ocean the salmon swims upstream to its birthplace. The female then digs a shallow "nest." At last the salmon releases her eggs. This complex mating cycle is an unlearned instinct apparently stimulated by temperature, length of days, available foods, and even odors in the water.

Often an animal's instincts are highly specific. Birds raised in captivity that have never seen the mating ritual of their species can perform that ritual perfectly. After finding a mate, a bird also relies on instincts to build a suitable nest. Some instincts, such as self-preservation (often called the "flight or fight" instinct), are found in almost all animals. Most animals, for example, will instinctively flee from a large adversary. If cornered or trapped, however, they will turn and fight the intruder, even if they stand little chance of winning the fight.

Conditioned Behavior

Conditioned behavior (learned behavior) is a response learned by experience. Many vertebrates are capable of this second type of behavior. The antics of a trained circus elephant or seal are behaviors they would not perform instinctively. They are behaviors carefully encouraged through a system of rewards. The process is familiar to anyone who has attempted to train the family dog to sit up or roll over. Even fish and pet turtles can be trained to come to certain areas for feeding.

Not all learned behaviors, however, are induced by man. After being sprayed by a skunk, most animals learn to leave the black animal with white stripes alone. A dog that gulps down a toad and vomits it after it releases strong toxins will probably never bother another toad. Some animals learn important behaviors by watching others of their species in a process known as modeling. Many of the infant-care behaviors of chimpanzees are developed in this way.

Intelligent Behavior

Some vertebrates are capable of the third type of behavior: intelligent behavior. **Intelligence**, although it is difficult to define, can be expressed as the ability to use tools to manipulate the environment, reason out the solution to a

Objectives

- Define *behavior* as it applies to animals
- Describe the three main types of vertebrate behavior, giving an example of each

Key Terms

behavior
inborn behavior

conditioned behavior
intelligence

17-8
A reflex controls the size of the openings of a cat's eyes in different light intensities.

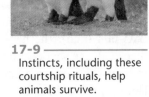

17-9
Instincts, including these courtship rituals, help animals survive.

17-10
Marine mammals such as this killer whale (orca) are among the most trainable creatures.

problem, or communicate with symbols. Such behavior is seen in some birds, most mammals, and humans.

Some chimpanzees, for example, have figured out how to use large stones to crack open tasty nuts, while others fashion twigs to "fish" in termite mounds. The woodpecker finch, lacking the long, probing tongue of the real woodpeckers, uses a pointy cactus spine held in its beak to dig grubs out of tree branches. The first finch to accomplish this improved feeding method is described as having *insight*, a flash of intelligence allowing it to solve a problem with a novel response. Once an individual chimpanzee or woodpecker finch becomes successful with the new method, others may benefit by learning through modeling.

Many animals communicate with each other. Much of this communication is instinctive or learned, but in a few species, the level of communication indicates a level of intelligence above what most scientists expected of animals. Some apes and other mammals have even been trained in the use of specialized languages (sign language or a language of symbols) and have even taught some of the language to other members of their species. Nevertheless, no animal comes close to the thinking and speaking abilities of even young humans. This distinction is expected in a Christian worldview. Humans alone are made in God's image.

Review Questions 17.3

1. Outline and define the basic types of animal behavior.
2. Give examples of each of the types of animal behavior.
3. It appears that animals with more intelligence have fewer instincts. Why would this be logical?

17B—The Fish

In Genesis 1:26 man is told to "have dominion over the fish of the sea," and they have been a main food source for people since early times. Fish are also important in other areas of human life. They are a good form of fertilizer, and they can be used to make glue and other industrial products. Also, aquariums provide people the joy of seeing these wonderful creatures at home.

Nearly three-quarters of the earth's surface is covered by water, and in almost every body of water—fresh or salt—there are fish. In fact, nearly half of all vertebrate species are fish.

Many groups of vertebrates spend at least some of their lives in the water. The fish, however, are most recognized as being designed for aquatic life. Although fish may superficially appear to be a homogeneous group, there are three separate classes in subphylum Vertebrata that are considered fish. The two that have cartilaginous skeletons throughout their entire lives are Agnatha (AG nuh thuh) (the hagfish and lamprey) and Chondrichthyes (kahn DRIK thee eez) (the sharks and rays). The vast majority of fish have some bony parts in their skeletons and belong to the class Osteichthyes (AH stee IK thee eez). Because these three groups are separated at the class level, they are in some ways as different from each other as robins are from rhinos. At the same time, their aquatic habitat requires them to have many things in common.

17.4 Class Osteichthyes

Normally when one thinks of fish, members of the vertebrate class Osteichthyes (bony fish) come to mind. Some members of the class, such as the sea horse, are unusual in their appearance; and some nonbony fish, such as the shark, may appear similar to the bony fish. However, there are major differences that separate the true bony fish from the other classes.

17.4

Objectives

- List and describe the characteristics of Osteichthyes
- Trace the circulation of blood through a typical fish
- Describe the structure and function of the gills in fish
- Identify and describe the function of four structures of the fish nervous system
- Explain the steps of oviparous reproduction in fish

Key Terms

swim bladder	gills
scale	atrium
mucus	ventricle
chromatophore	lateral line
operculum	spawn

Osteichthyes: Oste- or Osteo- (Gk. *osteon*—bone) + -ichthyes (*ikhthus*—fish)

Most fish have the typical flat, spindle-shaped body ideally designed for movement in the water. Some fish, however, are flat top to bottom. Some are round like a pencil; others resemble boxes, pyramids, balls, or almost any other shape imaginable. Most fish are less than 1 m (3.3 ft) in length, but the Philippine goby is only 1 cm (0.4 in.) long, while certain river sturgeons are over 4 m (13 ft) long and weigh over 900 kg (2000 lb).

Support and Movement of Fish

Members of the class Osteichthyes have bony skeletons. Bone is basically a special type of cartilage hardened by mineral deposits. It is considerably harder and offers much more support than cartilage. The bony fish, however, usually have only a vertebral column and skull of bone. The ribs and the pectoral and pelvic girdles (if present) are often of cartilage, but some fish have both cartilage and bone in these structures.

A fish moves through the water primarily by the whipping motion of its body. The power behind this action lies in the wavy, muscular bands found in the fish's trunk and tail. The effective use of these muscles, combined with the friction-reducing shape and smooth body coating, helps the fish move with a minimum of wasted effort.

Because fish are denser than water, they face the problem of staying afloat. Many bony fish have a **swim bladder**, a thin-walled sac in the body cavity that enables the fish to control its depth and to maintain that depth without swimming. Gases diffuse into and out of the swim bladder through the bloodstream or by contact with the digestive system. By increasing the gas volume in the bladder, the fish will rise in the water; by decreasing it, the fish will sink. Bottom-dwelling fish such as the flounder and sculpin have no swim bladder.

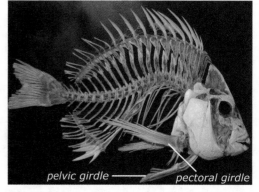

pelvic girdle —— pectoral girdle

17-11 —————
Fish skeleton

Fins of Fish

Fins project from the fish's body and are supported by either bone or cartilage. Fish generally have two sets of *paired fins*:

- The *pectoral fins* are close to the head on either side of the body.
- The *pelvic fins* are below and slightly behind the pectoral fins.

The *unpaired fins* of fish are often greatly modified for special functions:

- The *anterior dorsal fin*, on the back of the fish, is often supported by sharp spines, forming a convenient weapon against predators and careless fishermen.
- The *posterior dorsal fin*, behind the anterior dorsal fin, is generally smaller and softer than the anterior fin. Both dorsal fins are stabilizers, keeping the fish upright while swimming.
- The *anal fin* is on the ventral surface of the fish behind the anal opening.
- The *caudal fin*, at the extreme posterior end of the fish, is the tail fin, used to propel the animal through the water.

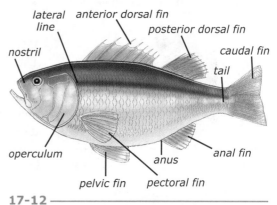

17-12 —————
Exterior structures of a bony fish

The fins of a fish are most often used for guidance or slow swimming movements. In some fish, however, the fins may serve a number of unusual functions. The sea robin, for example, uses its pectoral fins to "walk" along the sea bottom. The walking catfish can, by holding water in its gills, jump along on land for long distances if the pond it is in begins to dry up. The pectoral fins of the flying fish serve as gliding wings that enable some species to leap out of the water to escape from predators and glide at speeds of 55 km/h (34 mph) and distances of 46 m (150 ft).

17-13 —————
Sea robin

Facets of Biology

Lobe-Finned Fish

Most of the familiar fish belong to a group called the ray-finned fish. Their fins are made of stiff rays or spines connected by skin. Eight species of fish, composing the lungfish and the coelacanth groups, have very different fins that are fleshier and that contain small bones or cartilage. These are often called the lobe-finned fish. Evolutionists believe that these fins were precursors to amphibian legs as fish moved from water to land and became tetrapods. This is the same reasoning used when discussing the structure of whale fins, except the whales are supposed to have evolved from a land mammal that moved to water.

The six lungfish species have an unusual swim bladder that actually functions more as a lung. In fact, some species have no gills at all as adults but instead gulp air from the surface. Although lungfish have four small fins positioned where amphibian legs would be, they are very thin and too weak to function as legs. Lungfish species can be found in Africa, South America, and Australia.

It's Alive! Oops!

Evolutionists dated coelacanth fossils as 400 million years old and presumed the fish had been extinct for at least 65 million years. They are abundant in the fossil record, with 120 species found representing every continent except Antarctica. In 1938, a strange, metallic blue 2 m (6.5 ft) long fish was caught in the Indian Ocean along the east African coast. Biologists were astonished at the "living fossil." Since that time, many other living coelacanths have been caught, some as far away as Indonesia,

Coelacanth

and submersibles have even observed them in their deep-sea caves. The thickened lobes supporting the coelacanth's fins were thought to have helped the fish crawl across muddy areas on land. The coelacanth, however, is a deep-sea fish, and although its fleshy fins are moved in the same sequence of crawling salamanders, it has never been observed using them to walk on the bottom. The same scientists who had considered coelacanths the missing link between fish and amphibians struggled to explain why this particular type of fish came so close to changing and then seemingly stopped.

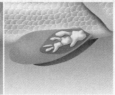
Ray fin (left) and lobe fin (right)   *Lungfish*

17-14
Fish scale

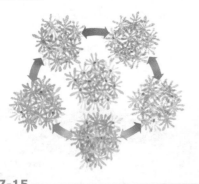

17-15
The operation of chromatophores explains the ability of certain fish (and other animals) to change color.

chromatophore: chroma- or chromato- (color) + -phore (bearing)

Body Covering of Fish

The body covering of most fish consists of overlapping **scales**. The number of scales a fish has usually does not change; as the fish grows, its scales merely grow in proportion. Thus, you can estimate the age of some fish by counting their scale rings. Usually, fish that live in waters that remain the same temperature throughout the year do not develop scale rings because they have steady growth rather than growing and dormant periods. Since scale rings obscure the colors of the skin, these fish are often more colorful. Not all fish possess typical scales. Some, like certain catfish, are scaleless. Other catfish are covered with bony plates. The sturgeon has rows of large enamel-like plates.

Special glands beneath the scales secrete slimy **mucus** that covers the exterior of the fish. The mucus coats the scales with a nearly waterproof covering, protects the fish from parasites, and lubricates the fish for smoother movement through the water. Studies have shown that the slimy covering reduces water friction by 66%.

Fish exist in a variety of colors. Those that live among colorful coral reefs are generally brilliant red, yellow, blue, and purple. Fish in other environments are subdued browns and greens. **Chromatophores** (kroh MAT uh FORZ) are branched cells responsible for producing some of the pigments that color fish. In some fish all the chromatophores produce the same color pigment. In other fish, however, there are different colored chromatophores that, when viewed together, produce a color.

Some fish have the ability to alter color and color pattern by rearranging the pigment in various chromatophores. When stimulated by certain nerves, the pigment is either concentrated in the center of the cell or dispersed throughout it. When the pigment is dispersed, that color is apparent on the body of the fish.

Fish often change color in response to temperature, diet, states of excitement (aquarium fish chased by a net often show a change in color or markings), and physical condition. Normally the fish's colors are brightest during the mating season. The flounder is probably the champion "chameleon" of the fish world. In minutes it can change its pigmentation from dark to light and from solid to spotted. It has colors ranging from yellow to black to green to red, which it uses to match the color of its background.

Many fish possess a form of camouflage known as *countershading*. The upper half of the fish is a dark color so that when viewed from above, it blends with the bottom of the body of water. The lower part of the fish is light colored, which makes it blend with the brightly lit water surface when viewed from below. Countershading helps protect fish from predators above and below it.

17-16
The flounder has a remarkable ability to change its color and pattern to camouflage itself.

Digestion in Fish

Fish feed on a variety of foods ranging from plankton, worms, insects, and plants to other fish and even mammals. Food is ingested through the mouth, which often contains teeth for biting or holding the prey. The mouth and teeth vary according to the diet of the animal. Predatory fish such as the barracuda have large anterior mouths armed with sharp teeth. The kissing gourami has large lips and a sucking mouth for scraping food from submerged objects. The parrotfish uses its tough, beaklike mouth to break chunks of coral reef and dislodge the tiny animals that it eats. The direction the mouth points reveals whether the fish feeds from the surface, the bottom, or somewhere in between.

Food passes whole through a flexible throat, or pharynx, and a short esophagus into the saclike stomach for storage. Digestion occurs in a short intestine that extends from the bottom of the stomach. Short tubes called pyloric ceca,

17-17
Countershading is common in many aquatic animals.

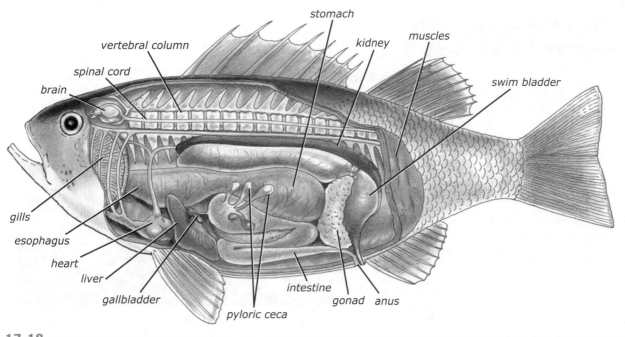

17-18
Internal structures of a fish

The sharp teeth of a piranha are designed and positioned for quickly removing flesh.

located at the junction of the stomach and intestine, are thought to secrete digestive enzymes. A large liver found near the stomach secretes bile that aids in the digestion of fats. Most fish also have a gallbladder that stores excess bile and a pancreas that secretes other digestive enzymes. The indigestible substances are eliminated via the anus.

Fish use various methods to capture or obtain their food. The razor-toothed piranhas of South America travel in schools, and some species viciously attack animals. A school of one such species reduced a 45 kg (99 lb) mammal to bone in less than a minute.

For some fish, brute strength gives way to stealth. The goosefish waits motionless on the sea floor, camouflaged by its speckled coloration. A portion of its dorsal fin forms a spine topped with a fleshy "worm" that it wiggles before its mouth as bait. When the prey comes close, the goosefish opens its huge mouth, creating a powerful suction that pulls the prey inside. The archer fish accurately "shoots down" insects up to 1.2 m (4 ft) away from the surface by firing a spurt of water out of its mouth to knock the insect into the water. The fish then quickly eats the water-bound victim. This fish has a remarkable aim considering that it must correct the firing angle to take into account the refraction of light waves through water.

Respiration in Fish

Almost every fish has an **operculum** (oh PUR kyuh lum) (pl., opercula), a conspicuous plate behind the eye on each side of the head. The opercula pulsate regularly. Beneath each operculum is a series of **gills**, the major organs of respiration. The gill consists of two rows of thin gill filaments on a band of cartilage called the gill arch. Close examination of the gill filaments reveals that they are stacks of thin plates covered by a thin epithelium and richly supplied with blood vessels. The *gill rakers*, a number of cartilage projections on the inner margin of the gill arch, prevent food and debris from passing over and clogging the gills.

operculum: (L. *operculum*—cover)

Saltwater and Freshwater Fish

Most fish dwell in either salt water or fresh water; both environments pose special problems for the fish. In fresh water, the concentration of salt in the water is much lower than that of the fish's body fluids. By osmosis, water enters the fish while salt escapes from it.

Although the scales and slime coat are waterproof, water gain and salt loss do occur through the gills of freshwater fish. To counteract this flow and maintain body-fluid balance, the kidney of a freshwater fish excretes excess water with wastes, while special salt-absorbing chloride cells in the gills actively transport salt from the water into the fish's blood.

Ocean water has a salt concentration higher than the body fluid of the fish, causing the fish to gain excess salt and lose body water. To obtain the water necessary for life, the marine fish must drink seawater, which also brings excess salt into the fish. The excess salt is either passed from the fish by the chloride cells in the gills, left in the intestine to be ejected with solid wastes, or excreted by the kidney.

The process by which this homeostasis is maintained is called osmoregulation, and it is under the control of hormones. Some fish, such as the salmon, eel, and steelhead trout, travel regularly from salt water to fresh water as part of their breeding cycles. These fish have highly adaptable fluid regulation devices to deal with the different salinities (salt levels).

Salmon before breeding (left) and after breeding (right). Salmon go through both internal and external changes.

The Path of Blood in Fish

1. Blood is pumped from the ventricle through the ventral aorta, which branches into a series of afferent brachial arteries.
2. These arteries supply the two sets of gills. As the blood passes through the gills, it accepts oxygen and eliminates carbon dioxide.
3. Oxygenated blood in the efferent brachial arteries enters the dorsal aorta.
4. Branches from the dorsal aorta supply various parts of the fish. One branch carries blood to the digestive organs. Digested food is absorbed into the blood to be distributed throughout the fish. The blood that passes through the kidneys is filtered of its waste materials.
5. The sinus venosus empties into the atrium, which in turn empties into the ventricle, and the process begins again.

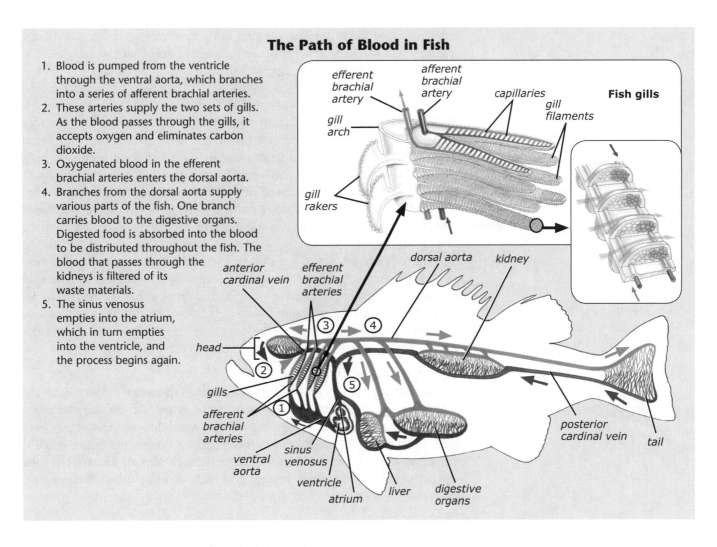

The fish obtains oxygen for cellular respiration by utilizing the oxygen dissolved in water (not the oxygen of the water molecule). In respiration the fish closes the opercula and expands the gill arches laterally, drawing water into its mouth. The fish then closes its mouth, contracts the gill arches, and opens its opercula, forcing the water over the gills.

As the water passes over the gills, oxygen diffuses into the numerous tiny blood vessels in the gills. Waste carbon dioxide diffuses out of the blood and into the water. This gas exchange process is so efficient in some fish that 95% of the oxygen dissolved in the water is absorbed in one pass over the gills.

Circulatory and Nervous Systems in Fish

The circulation of blood enables oxygen and nutrients to reach every cell of the fish. The moving force behind circulation is a two-chambered heart, located posterior and ventral to the gills. The **atrium** (AY tree um) (pl., atria) is the chamber in the heart designed to receive the blood from the body tissues. The atrium empties the blood into the **ventricle** (VEN trih kul), a muscular chamber designed to pump the blood forward through the arteries. This is a single-loop system in which the heart pumps blood only to the gills.

The major organs of the fish's nervous system are the brain and spinal cord. Ten pairs of cranial nerves branch from the brain, and many pairs of spinal nerves come from the spinal cord. The brain contains the major lobes (divisions) that are typical of most

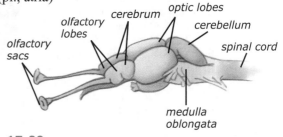

17-20

The brain of a fish

vertebrates. You might assume, judging from their large optic lobes and large eyes, that fish would be able to see well. Fish have excellent vision, but due to light diffraction and suspended particles in the water, their range of vision is probably limited to 6 m (20 ft).

The sense of smell is one of the fish's keenest senses. Although people speak of tasting water, fish actually smell the water that enters the olfactory sacs (small pouches behind the nostrils on the fish's snout). Fish can detect incredibly minute amounts of substances dissolved in water. Some experts believe that certain migratory fish—such as salmon—are guided to their home rivers by the precise combination of smells produced by the plants in that river. The tongue is used for both taste (a rather dull sense in the fish) and touch.

Fish have no external ears; however, they can detect sound vibrations through their skulls. A bony fish also has a sensory canal, or **lateral line**, on each side of its body. Actually a string of sensory structures, this line of sensors branches over the head for improved reception. The sensory hair cells in the pits along this line respond to changes in current, enabling the fish to detect underwater movements and pressure changes. The system is so sensitive that blind fish are still able to locate and catch prey. It also enables the close schooling behavior of fish that seem to swim as one organism.

Reproduction in Fish

Although some fish are ovoviviparous and even viviparous, most follow the same basic oviparous reproductive plan: the eggs are laid and then fertilized externally. The female fish has paired ovaries, which in some species can produce hundreds of thousands of eggs. The male sex organs, the testes, produce sperm. When environmental conditions are right, the female **spawns** (lays the eggs). The male then covers the eggs with *milt* (a milky fluid containing the sperm).

Unusual Forms of Fish Reproduction

Different species of fish practice elaborate migration, courtship, and even nest-building rituals. One of the most familiar reproductive behaviors is the migration of salmon to their spawning grounds. The salmon hatched in freshwater streams migrate to the open waters of the ocean. They remain at sea for years. At the proper season when these fish are mature, they return to the very stream where they were hatched. In the process, the salmon leap waterfalls, battle fierce currents, and navigate complicated waterways to reach their spawning grounds.

In many species of fish, the female scatters the eggs in the spawning area as the male darts behind, fertilizing as many as possible. This completes the mating instincts of both parents, and they then turn and eat as many eggs as they can find. Other fish lay eggs in the open or in shallow depressions in the sand and then leave them to whatever predator cares to find and eat them. Fortunately, such fish usually produce huge numbers of eggs so that even if only a fraction of them hatch and develop, the species will survive.

The male stickleback guarding its nest

The male stickleback, however, constructs an elaborate nest of aquatic vegetation, using a sticky body secretion to glue the nest together. The male then entices the female into the nest to spawn. Once she has laid the eggs, the female leaves. The male fertilizes the eggs and then guards them till they hatch. Even after they hatch, the male stickleback remains a dedicated parent.

Deep sea anglerfish employ perhaps the most unusual form of fish reproduction. The male anglerfish is far smaller than the female and is equipped with an amazing sense of smell to help him find the female in the deep dark sea. After the male finds the female, he locks onto her with his teeth, and over time his body becomes fused with her body, remaining attached to her for the rest of their lives. The male's digestive system becomes nonfunctional and his blood vessels connect to hers, making him a parasite. The female benefits by having a male around to fertilize her eggs whenever she releases them.

Proper conditions for spawning often include slight changes in temperature, types and amount of food, pH of water, and other factors. Any home aquarist who has tried to raise certain types of fish will attest to their finickiness. God designed fish to spawn only when environmental conditions will allow the greatest chance of survival for the young fish.

The fertilized eggs develop into embryos. In the egg, the embryo usually forms atop a ball of food material called *yolk*. As the embryo grows, the yolk shrinks. Eventually a tiny larval fish—usually called a fry—escapes the egg and fends for itself. The process from egg to larva may take only a few hours for some tropical freshwater species or as long as ninety days for the brook trout.

Unique Fish

To survive in the highly predatory aquatic world, many fish use unusual behaviors or defense mechanisms. Camouflage is a common method of self-defense. The long, slender body of the pipefish looks like a strand of the seaweed in which it hides. The leafy sea dragon, resembling a sea horse plastered in seaweed, is one of the most fascinating and best camouflaged of all fish.

Some fish are also protected by other markings that confuse predators. Eyespots on the tail end and dark bands through the eyes give safety to many fish. Many predators mistake the head for the tail and are too confused to give chase when the fish seems to retreat backward.

Some fish bear frightening weapons. The dazzling, striped coloration of the tropical lionfish makes it a conspicuous target, but its long, elegant fins have extremely poisonous spines.

Fish differ in their relationships and interactions with other animals. Some travel with other members of their species in huge groups called schools. Others are loners, confronting other fish only when about to eat, be eaten, or mate.

Some fish have beneficial partnerships with other animals. For example, certain clownfish have found a partner in the sea anemone. When most fish swim within range of the anemone's tentacles, the nematocysts, or stinging cells, in the tentacles immediately paralyze them. The clownfish, however, can swim among the tentacles without being harmed because of a special layer of mucus that protects it by keeping the anemone from recognizing it as food. Thus, whenever a predator is near, the clownfish can enter a refuge where few enemies can follow. The anemone benefits by feeding on scraps of food dropped by the fish.

leafy sea dragon

lionfish

clownfish

Review Questions 17.4

1. List several characteristics of the class Osteichthyes that separate its members from other vertebrate classes.
2. List and describe the six major types of fins found on a typical fish.
3. Describe several unusual fish fins and their uses.
4. How do fish maintain their depth in the water?
5. Describe the body surface and coloring of a fish.
6. How do chromatophores cause color changes in fish? What can trigger a color change?
7. What are the advantages of countershading?
8. Describe the digestive system of a fish.
9. List several methods specific fish have for obtaining food.
10. Where do fish obtain their oxygen, and how do they use the oxygen for respiration?
11. What methods do fish use to maintain the water and salt balance between their body fluids and their environment?
12. Describe the heart of a fish.
13. Describe the senses of sight, smell, taste, hearing, and touch in fish.
14. Describe the reproductive process used by most fish.
⊙15. A fish out of water soon dies from a lack of oxygen even though it is in an environment that has more oxygen than its normal environment. Explain why.
⊙16. Why is it important for fish to mate only when certain environmental conditions are met? In what way is this timed mating anti-evolutionary?

Objectives

- Contrast the habits of the hagfish and the sea lamprey
- Name three ways cartilaginous fish are different from bony fish
- Compare the skates and rays with the sharks, giving three differences

Key Term

gill slit

Agnatha: The Jawless Fish

The lamprey and hagfish are in the class Agnatha. These animals are identified by a lack of scales, an eel-like shape, unpaired fins, a skeleton of cartilage, and a jawless, sucking mouth. Although this class has few species, its members have caused tremendous damage in certain areas.

The adult sea lamprey may reach 60 cm (24 in.) in length. It has two single dorsal fins and a tail fin. A single nostril separates its two eyes. The nostril leads to a sac supplied with olfactory nerves. Seven oval **gill slits** appear on each side of the head of the organism. These openings lead to a chamber containing gills.

The most conspicuous feature of this animal is its circular, funnel-shaped mouth lined with rows of teeth. The mouth is used to cling to the body of a fish. Once attached by suction, the lamprey bores a hole into the fish and sucks its blood and body fluids. Sometimes the fish hosts survive the attacks, but often they do not.

Lampreys leave the ocean to spawn in freshwater streams. The male and female lampreys dig a shallow depression in the stream bottom into which the female deposits 25 000 to 100 000 eggs. The eggs are fertilized externally. In about two weeks the eggs hatch, releasing tiny wormlike larvae. The larvae float downstream until they reach an area with a quiet, muddy bottom. The larvae burrow into the mud and remain there, feeding on organic debris, for three to seven years. After this period of growth, the larvae change into adults, swim downstream to the ocean, and begin their lives as parasites.

The hagfish is a jawless fish that spends its entire life in the sea. It is a bottom dweller that usually stays in burrows on the ocean floor, scavenging or feeding on small organisms. It often invades bodies of dying or dead animals, eating their internal organs. Like the lamprey, it lacks jaws but has small teeth for tearing and a muscular, sucking mouth. It does not have true eyes but depends on light-sensitive structures to keep itself hidden. An acute sense of

17-21
Hagfish

Agnatha: A- or an- (Gk. *a-* —without) + -gnatha (*gnathos*—jaw)

A Great Lakes Invader

Not all sea lampreys spend their adult lives in the ocean. By the 1830s, sea lampreys had made their way up the St. Lawrence River and appeared in Lake Ontario. Canals completed in the early 1900s made it possible for them to bypass the natural barrier of Niagara Falls and reach Lake Erie, and by the 1940s they were devastating commercial fisheries in all five of the Great Lakes. In some areas, populations of important fish like lake trout dropped by 90%.

Early attempts to control them centered on mechanical structures that limited their access to spawning streams. These structures used mesh barriers and electrical fields generated by metal rods hanging in the water. While these had some success, floods and floating debris limited their effectiveness. After testing more than six thousand chemicals, scientists settled on a toxin known as TFM that killed the lamprey larvae but did not seem to be a threat to the environment or man. While lampreys are still present in all of the lakes, chemicals are now controlling their numbers. A poison that kills them in their larval stage is applied to about 250 of the tributary streams where they spawn. Future goals include decreased dependence on chemicals and improved mechanical barriers for some streams. The release of sterilized males is also being tested in some areas.

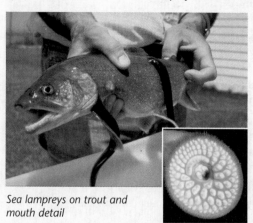

Sea lampreys on trout and mouth detail

smell is more important in finding food. To protect itself, this creature secretes large amounts of foul-tasting slime and can also tie itself into a knot to keep from being pulled from crevices.

Chondrichthyes: The Cartilaginous Fish

Some of the most misunderstood animals in the ocean belong to the class Chondrichthyes, a class including sharks, rays, and skates. The term *Chondrichthyes* reveals one of their major characteristics—an elaborate skeleton composed of strong yet flexible cartilage. Although they do not have the calcified bones found in members of Osteichthyes, they do not lack support. The cartilage is surrounded by tough protein to increase the strength of the skeleton. While the hagfish and lampreys also have a cartilaginous skeleton, members of Chondrichthyes differ by having jaws and paired fins.

Tales of the crushing jaws of the shark and the deadly sting of the stingray have given this group of animals a sinister reputation. Yet not all cartilaginous fish are fearsome. Even most sharks are timid and harmless. For example, the whale shark is the largest shark (18 m, or 59 ft), but it feeds only on microscopic plankton.

Sharks

Sharks are easily recognized by their sleek torpedo shape, which enables them to glide through the water with a minimum of friction. Their skin appears smooth, but close examination reveals that it is actually covered by many tiny hook-shaped scales, making it as rough as sandpaper. The shark has the same basic fins as the fish in Osteichthyes. In the sharks, however, the dorsal part of the caudal fin is always larger than the ventral part, which is not true of bony fish.

Behind and below the anterior point, or rostrum, of the shark is the mouth. The mouth has rows of razor-sharp triangular teeth that are arranged in rows that point inward. This makes it hard for prey to escape once it is grasped. These teeth can be straightened for biting or tearing and folded into the mouth when not in use. A shark's teeth are not anchored into bone like a human's but are anchored just into the skin. If a tooth is lost, another moves forward to replace it. It is not unusual for a shark to go through 20 000 teeth in its lifetime.

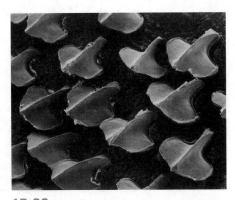

17-22
Shark scales (magnified 35×) are sharp and give the animal the feel of sandpaper.

Whether they feed on large fish and other animals like seals or on microscopic plankton, all sharks are carnivorous. Sharks will generally seize their prey with their powerful jaws and sharp teeth. Sharks usually swallow their food whole, passing it to the stomach for storage. Sharks have been known to store undigested whole dolphins for as long as a month. Digestion occurs in a short intestine.

Sharks obtain their oxygen by passing water over their gills. These gills are exposed to the environment by a series of gill slits on each side of the shark's body. Most sharks have five slits per side, but a few species have six or seven. When motionless, a shark can be observed pumping water through its gills. The circulatory system is similar to that of bony fish. The fish of Osteichthyes are able to move up and down in the water column thanks to the buoyancy provided by the air-filled swim bladder. Because sharks lack this structure,

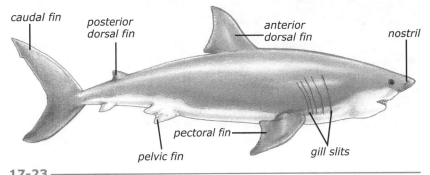

17-23
External structures of a shark

Chondrichthyes: Chondr- or chondro- (Gk. *khondros*—cartilage) + -ichthyes (fish)

they need another mechanism. A large liver filled with oils less dense than water helps keep them afloat. Combined with the lift provided by the forward motion of swimming, sharks are able to maintain their level without sinking to the bottom.

An elaborate set of sensory structures assists the shark in its search for prey. Sharks have very sensitive hearing, tuned to the low frequencies produced by injured or struggling animals. Hearing may be the first sense sharks use when locating prey. The lateral line, a sensory canal, runs the length of the shark's body. The cells in this canal respond to minute water vibrations caused by distant moving objects. Many sharks investigate any unusual vibrations as a potential food source. Paired nostrils on the ventral side of its head lead to olfactory pouches. These organs of smell enable the shark to detect extremely small amounts of blood in the water. Prey, however, cannot be detected at great distances. Tiny pores scattered across the head of a shark are connected to chambers that can sense minute electrical fields from muscle activity of nearby animals. These same organs may also be sensitive to changes in temperature and salinity. The shark's two large eyes function efficiently only at close range.

In sharks the sperm are transmitted from the male to the female with the aid of claspers, a specially shaped portion of the male's pelvic fins. Fertilization is internal. Oviparous sharks release the fertilized eggs in a capsule that attaches to a submerged object. Some sharks are viviparous, a rarity outside the class Mammalia, and some are ovoviviparous.

17-24
Manta ray

17-25
Stingray

Rays and Skates

Rays and skates are flat and thin, not torpedo-shaped like sharks. The pectoral fins of the ray form flat wings that propel the fish through the water. These animals swim like underwater birds flying in slow motion.

Skates and rays have a ventral mouth, ideally positioned for their feeding style. They dine mainly on mollusks and small crustaceans that live in or on the bottom sediments. Some use their mouths like a vacuum, sucking up invertebrates.

Though graceful and beautiful, some rays can be quite dangerous. The stingray has one or more sharp, venom-filled spines along its tail that can inflict painful, slow-healing wounds. Not all rays live in the sea. The freshwater stingray inhabits rivers in the Amazon basin and cannot tolerate salt water at all. Most rays and skates are less than 1 m (3.3 ft) long, but one of the largest rays, the manta ray or "devilfish," sometimes measures more than 6 m (20 ft) across. Despite its fearsome size, the manta ray eats plankton.

Review Questions 17.5

1. Give an example of class Agnatha and list five characteristics of the class.
2. List several characteristics of class Chondrichthyes that separate it from class Agnatha and class Osteichthyes.
3. List and describe two members of the class Chondrichthyes.
4. Why is it important for a shark's teeth to be replaced whenever the shark loses them?

17C—Class Amphibia: The Amphibians

The salamander and toad seem drastically different. The salamander has a smooth, almost tubular body with fragile limbs. The toad, on the other hand, is a fat, lumpy-skinned blob with bulging eyes, a wide grin, and limbs that seem comically disproportionate to its body. Although very different, these two animals are both members of the class Amphibia (am FIB ee uh) and share many characteristics.

An **amphibian** typically has a two-part life cycle, beginning as an unshelled egg laid in or near water. After hatching, it spends time in a swimming stage, using gills for respiration. Most develop into adults that live on land and breathe with lungs. The skin lacks scales and may be moist. There are no claws on the feet, and webbing may be present for those adults who spend much time in water.

17.6 Introduction to Amphibians

Classification of Amphibians

The class Amphibia is divided into three orders on the basis of body shape and type of limbs.

✦ *Order Apoda* (AP uh duh). This order contains a small group of strange, wormlike animals called caecilians. Although some resemble giant earthworms, they differ by having the backbone, jaws, and closed circulatory system of vertebrates. These amphibians lack limbs and limb girdles. Most species live underground and are totally blind as adults. They burrow to find the worms and invertebrates that are their food.

The 160 species of caecilians are found in warm, moist habitats, and some species are totally aquatic. While the largest species can grow to 1.5 m (59 in.), most are less than one-third that long.

✦ *Order Caudata* (kaw DAH tuh). Salamanders are often mistaken for lizards. With their shape and movement, these animals do resemble reptiles; however, their life cycle and body covering clearly label them as amphibians. Like other amphibians, they hatch in the water and lack the scales and claws that reptiles possess. Salamanders are distinguished from other amphibians by the presence of a slender body with a definite head, trunk, and tail. Most salamanders have two pairs of small identical limbs at nearly right angles to the trunk.

There are almost four hundred species of salamanders, and while most are less than 16 cm (6 in.) in length, a few species grow to a length of 1.5 m (59 in.). Salamanders are found in cooler, more temperate habitats than those that support the caecilians.

✦ *Order Anura* (uh NOO ruh). The frogs and toads have wide heads that fuse with short, pudgy bodies. Their large eyes bulge on the top of their heads. They have wide mouths that appear to be locked into permanent

17.6

Objectives
- Identify one example for each order in the class Amphibia
- List and describe the characteristics of the class Amphibia
- Discuss metamorphosis in frogs, using terms related to diet, habitat, and respiration
- Compare the amphibian circulatory system with that of the fish

Key Terms
amphibian
metamorphosis
tadpole
lung

17-26 —
Caecilian

17-2 Amphibian Orders			
Order	Key characteristics	Example	Species number (approx.)
Apoda	worm- or snakelike body; tropical; no appendages	caecilians	160
Caudata	slender body with four identical limbs on trunk	salamanders	400
Anura	large head; tailless; short front limbs with large muscular hind limbs	frogs	3900

Amphibia: Amphi- (Gk. *amphi*—on both sides) + -bia (life)

Apoda: A- (without) + -poda (foot)

Caudata: (L.—tail)

Anura: An- (without) + -ura (Gk.—tail)

17-27
Red-eyed tree frog: eggs, tadpole, and adult

17-28
Northwestern salamander: eggs and adult

17-29
The skin secretions of some poison dart frogs are used to poison the dart and arrow tips of certain Latin American natives.

smirks. The major characteristics that distinguish frogs and toads from other amphibians are their lack of tails and their uniquely designed limbs. The front legs are small and function primarily as props for sitting and shock absorbers for landing from jumps. The hind legs are larger and heavily muscled. These limbs are designed for jumping on land and swimming in the water. The toes of the hind legs are webbed for swimming. Unlike caecilians and most salamanders, frogs have the ability to flick out their tongues to catch insects.

Anurans (members of the order Anura) are by far the most prominent amphibians, with nearly 3900 species, and they are found worldwide except in polar climates. Despite their need for water, some species are designed to survive even in deserts, burrowing underground during the long dry periods between rains.

Metamorphosis

The term *amphibian*, meaning "double life," aptly describes the remarkable dual existence of most members of this vertebrate class. They undergo **metamorphosis**, characterized by drastic changes in habits and body structures. Their life cycle begins in the water. The female amphibian lays jellylike eggs in the water. Toads lay strings of eggs; frogs produce egg masses. Usually many eggs are produced, but the number varies dramatically. A small Cuban frog produces only a single egg. A land salamander may produce thirty; a large toad, 30 000. Those species that lay low numbers of eggs usually provide care for the young before and even after hatching, in contrast to the typical strategy of laying huge numbers of eggs and abandoning them. In most amphibians fertilization is external: the male fertilizes the eggs by discharging milt over the eggs. Usually, after a period of development, an aquatic larva escapes the egg.

The larval salamander resembles the adult but bears feathery external gills about its head. The larval frog is the familiar **tadpole** with its oval body tapering to a slender tail. A flap of skin covers the gills of the tadpole. A tadpole is an eating machine that undergoes major physical changes to become an adult: the gills degenerate as the lungs develop, the characteristic frog limbs appear, the eyes grow and bulge, the mouth widens, and the tail is absorbed. At the same time, the digestive system is changing to switch from a menu of algae and plant material to a strict carnivorous diet. The resulting land-dwelling, air-breathing, bug-eating adult bears little resemblance to the tadpole.

The basic scheme of amphibian metamorphosis has a number of variations. The mudpuppy, a common salamander, remains in an aquatic gill-breathing form throughout its life. Certain salamanders remain in aquatic forms as long as they have water to live in. When their watery homes dry up, certain internal mechanisms are triggered, and these larvae change into air-breathing adults. Some salamanders lay their eggs in moist areas on land. The young hatch as miniature adults, skipping the aquatic larval stage.

Skin

Amphibians are covered with a smooth, scaleless skin that is richly supplied with blood vessels and moistened by secretions from glands in the skin. This skin is more than a body covering. Some amphibians secrete poisonous substances over their skins to discourage potential predators.

These poisons may only taste bad, or they may be strong enough to burn the animal that touches the amphibian. About sixty-five species of Central and South American poison dart frogs secrete a poison that makes them deadly to eat. It is hypothesized that the frogs do not synthesize the poison themselves but instead absorb it from poisonous centipedes and ants they feed on. Some native hunters have used the secretions of poisonous frogs to poison the tips of their darts and arrows, speeding up the kill when the prey is hit.

Chromatophores color the skin of amphibians. These cells may provide camouflage coloration for some, but many amphibians are brilliant shades of red, yellow, gold, and blue. In most cases these are warning colorations; the bright color labels the animal as dangerous or inedible.

The skin can also function as a respiratory organ. The tiny blood vessels in the thin, moist skin bring the blood into close contact with the environment, where the exchange of oxygen and carbon dioxide can occur.

Respiratory System

Amphibians may use one or more of four mechanisms to obtain the oxygen they need. The first method, used by every amphibian during some phase of its life, is respiration by gills. Since most amphibians spend the early part of their lives as aquatic animals, the necessity of such an apparatus is obvious. The gills degenerate when the amphibian begins its terrestrial existence.

Second, many amphibians have **lungs**, internal organs for the exchange of gases between the atmosphere and blood. One such amphibian is the adult frog. The frog does not depend only on lungs to obtain oxygen—they are not the primary means of respiration. In fact, most amphibians can spend long periods without refilling their lungs.

Third, the lining of an amphibian's mouth and throat is abundantly supplied with blood vessels and may serve as a means of respiration. Finally, the most unusual amphibian respiratory organ is its skin. The skin accounts for 90%–95% of the respiration of certain lungless salamanders. Frogs depend on their skin for respiration while underwater for long periods of time—especially when they are hibernating at the bottom of a pond.

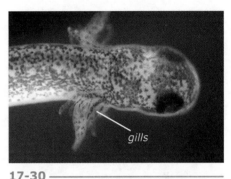

17-30
Most larval salamanders respire in the water using external gills.

Circulatory System

The major differences between the circulatory system of amphibians and that of fish are (1) the amphibian's three-chambered heart and (2) an additional "loop" in the circulatory cycle. The upper part of the amphibian's heart consists of two atria that receive blood from different parts of the body. The atria empty into a single muscular ventricle, the chamber that pumps the blood to the organs and tissues of the animal.

The added loop is a connection between the heart and the respiratory apparatus. Oxygenated blood enters the left atrium. The right atrium collects deoxygenated blood from the body. Both atria empty simultaneously into the ventricle, where the oxygenated and deoxygenated blood mix.

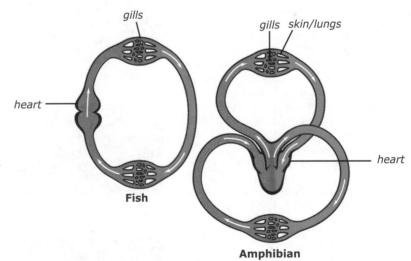

17-31
Fish and amphibian circulation compared

The ventricle then pumps the blood through a single large artery. A portion of that blood goes by one branch of this artery to the lungs and skin. This blood is oxygenated and returned to the heart (left atrium), completing the respiratory loop. The remaining blood goes by other branches to the body, supplying the tissues with oxygen, and returns to the heart (right atrium).

Amphibians are ectothermic; their body temperatures and levels of activity vary with the surrounding temperature. When the temperature drops, the amphibian's body processes slow down. During this time the amphibian usually burrows into the mud of a stream or lake to wait out the cold weather in a "lifeless" state called *hibernation*. The animal becomes inert; its heartbeat and circulation almost cease. In some very cold climates, amphibians may even form a compound similar to antifreeze in their blood to prevent tissue damage.

In this slowed state of metabolism, the animal needs little food. Stored food is slowly used to meet the animal's nutritional needs. The small amounts of oxygen needed usually diffuse through the skin, and at cooler temperatures, water is capable of holding more dissolved oxygen. When warm weather returns, the amphibian "awakes" and resumes its normal activities.

Extreme summer heat can also create problems for cold-blooded animals. With no mechanisms to keep cool, some amphibians must escape the heat by staying at the bottom of a pond, burrowing into the mud, or hiding in a cool place. This period of inactivity is *estivation* (ES tuh VAY shun). When the temperatures are suitable again, the animal becomes active.

American Frogs and Toads

The terms *frog* and *toad* are often used, but they do not have any true scientific distinction. That is, there is not a "frog" family and a "toad" family. Instead, there are at least seventeen different families of anurans. Most of the families fit into one of three informal groups: true frogs, toads, and tree frogs.

The most common true frogs in America are in the genus *Rana*. They generally have a smooth, shiny skin that easily dries. For that reason, frogs usually stay in or near the water. Common American frogs include the spotted leopard frog (*R. pipiens*) and the green frog (*R. clamitans*). The largest American frog is the bullfrog (*R. catesbeiana*), a common inhabitant of the East Coast, reaching a length of 23 cm (9 in.). This amphibian has a deep voice, and its hind legs often appear on the dinner table as an entree.

Leopard frog

Many toads belong to the genus *Bufo*. Unlike the frog, the toad has a dry, rough, warty skin. Despite superstitions to the contrary, this warty skin does not produce warts on humans. The toad is strictly a land dweller, entering the water only to mate. Often these amphibians can be found in the leaves and moist soil of the forest floor far from a body of water. *B. terrestris* is a common toad of the eastern United States. As a rule, toads have shorter hind limbs and do not leap as far as frogs do.

Toad

Tree frogs (genus *Hyla* and others) are generally smaller than frogs or toads. These frogs have enlarged sticky discs at the ends of their toes that enable them to cling to tree branches. The spring peeper (*H. crucifer*) is a tiny tree frog whose peeping call is a familiar sound on spring evenings.

Tree frog

American Salamanders

Most salamanders in the United States are small, rarely more than 15–25 cm (6–10 in.) long and are also subdued in color. Of course, several notable exceptions exist. Some species are brilliant oranges, reds, and yellows.

Larger American salamanders include the hellbender, a camouflaged stream dweller of the eastern United States that may reach 60 cm (24 in.) in length, and the mudpuppy, an inhabitant of muddy stream bottoms and lakes in the Midwest.

The eastern newt is a common inhabitant of the United States. After an aquatic larval stage, this salamander spends a part of its life on land as a subadult called an eft before returning to the water to become an aquatic adult. The term *newt* is sometimes used to refer to any salamander but most properly applies to these salamanders.

A typical salamander

1. List several characteristics of the class Amphibia that separate its members from other vertebrate classes.
2. What characteristics set salamanders apart from other amphibians?
3. What characteristics set the anurans (frogs and toads) apart from other amphibians?
4. Identify at least three functions of an amphibian's skin.
5. List four structures amphibians use for respiration. Tell which are normally used during each stage of life.
6. Describe the heart of an adult amphibian.
7. What are the primary differences and similarities between hibernation and estivation?
8. What are the three main groupings of frogs? What characteristics separate them?

17.7 Structure and Function in the Frog

The frog is a common laboratory animal. Frogs are inexpensive and easy to obtain, but more importantly, their anatomy and physiology partially resemble those of mammals and man.

External Structures of the Frog

The most conspicuous feature of the frog's head is its large bulging eyes. A colored **iris** surrounds the dark **pupil**, the opening for light. The iris can enlarge or reduce the size of the pupil in response to varying amounts of light. At the bottom of the eye is the transparent folded **nictitating** (NIK tih TAY ting) **membrane**. This "third eyelid" moves across the frog's eye, keeping it moist and protected.

The two small nostrils on the frog's snout open into its mouth. These structures allow the animal to breathe with just the top of its head above the water. A circular membranous structure situated behind each eye is the **tympanic** (tim PAN ik) **membrane**. This membrane serves as an eardrum, transmitting sound vibrations to the ear cavity underneath.

Mouth of the Frog

The frog has a large mouth designed to capture the insects it eats. A large sticky tongue is attached at the front of the floor of the mouth. When an insect comes within range, the frog flips out its tongue, snagging the prey. The frog then flips the tongue and insect back into its mouth. Frogs "blink" when they swallow. This is not because the food tastes bad; the eyes are actually being pressed toward the mouth cavity to help crush the food and force it down the gullet.

The inside of the frog's mouth contains several openings. At the front of the roof of the mouth are the internal nostril openings. The openings of the **eustachian** (yoo STAY shun) **tubes** uniting the ear cavity with the mouth are at the back corners of the mouth. The passage to the digestive tract is the *gullet* at the extreme back of the mouth.

Immediately in front of the gullet is a small swelling parted by a slender opening. This is the *glottis* leading to the lungs. Male frogs have a pair of openings in the back of the lower jaw leading to the **vocal sacs** in the floor of the mouth. The frog passes air through these openings into the sacs when it "croaks," increasing the volume of the sound.

The frog possesses two sets of inconspicuous teeth: a ridge of tiny maxillary (MAK suh LEHR ee) teeth in the upper jaw and two sets of vomerine (VOH muh RINE) teeth between the internal nostril openings in the roof of the mouth. These two sets of teeth assist the frog in grasping its prey, not in chewing. There are no teeth in the lower jaw.

17.7

➡️ **Objectives**

- Trace the path of food as it passes through the frog's digestive tract
- Name and describe the four possible means of respiration for the frog
- Describe frog reproduction, including all relevant structures

➡️ **Key Terms**

iris	eustachian tube
pupil	vocal sac
nictitating membrane	mesentery
tympanic membrane	trachea

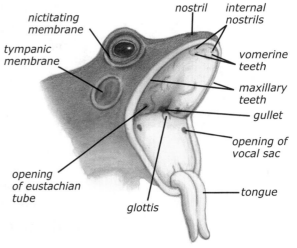

17-32
Structures of the frog's head

glottis: (Gk. *glotta*—tongue)
maxillary: (L. *maxilla*—upper jaw)

Digestion in the Frog

Although tadpoles are herbivores, feeding on plants, algae, or plankton, the adult frog is a carnivore, ingesting insects, spiders, slugs, worms, or any other moving object it can fit into its mouth. Some larger frogs have been known to eat rats. The food, once ingested, passes through a short esophagus to the stomach, an enlarged portion of the digestive tract that serves as a storage sac and site of preliminary digestion. The stomach narrows at the region called the pylorus (pye LOR us), where the muscular pyloric valve controls the further passage of the food into the small intestine.

The stomach joins the small intestine, a thin tube consisting of the enlarged duodenum (DOO uh DEE num) joined to the stomach and the coiled ileum (IL ee um). Here, the dissolved food is absorbed into the bloodstream. The small intestine empties into the short, stubby colon, or large intestine. From here all indigestible material is passed through the short, tubular cloaca (kloh AY kuh) and expelled from the body. The cloaca is a common passageway for both liquid and solid wastes, as well as eggs and sperm during reproduction.

Two glands outside the digestive tract contribute to the digestive process. The liver is the large maroon-colored, three-lobed organ that dominates the body cavity in Figure 17-33. It produces bile, a substance that assists digestion. The bile collects in the small saclike gallbladder and flows to the upper portion of the small intestine through the common bile duct. The liver also performs a variety of other functions, including the storage of digested food.

The pancreas, another digestive gland, is a small strip of tissue near the stomach. Pancreatic secretions also pass into the small intestine through the common bile duct. All the digestive organs are enclosed in transparent membranes called **mesenteries** (MES un TEHR eez) that bind the organs to the dorsal body wall. Blood vessels within the mesenteries carry blood to and from the internal organs.

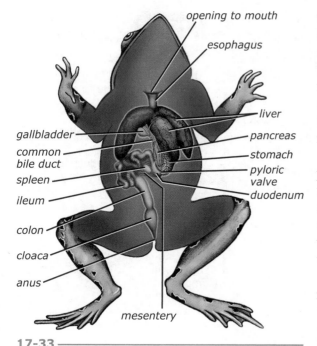

17-33
Digestive system of the frog

Respiration in the Frog

The frog obtains oxygen by using its skin, mouth lining, lungs, and, during its larval stage, gills. When its oxygen requirements are low (for example, during hibernation), the frog can obtain all the oxygen it needs through its skin. The frog may also use its capillary-rich mouth lining to supplement its respiration. When active, however, its increased oxygen needs must be supplied by its pair of small lungs.

The lungs join to a single tube called the **trachea**, which opens into the mouth cavity through the slitlike glottis. The frog lacks the ribs and muscles needed to inhale and exhale as humans do. To breathe, the frog draws air into the mouth through the nostrils by lowering the floor of its mouth. Then with nostrils closed and glottis open, the frog swallows the air, forcing it into the lungs. Contraction of the body wall expels the air.

Circulation in the Frog

Blood circulating through capillaries in the lungs becomes oxygenated and flows through the pulmonary vein to the left atrium of the heart. At the same time, deoxygenated blood from the body organs and muscles flows through three large veins called the venae cavae (sing., vena cava) to a thin sac at the back of the heart, the sinus venosus, which empties into the right atrium. Once filled, both atria contract simultaneously, emptying the blood into the muscular ventricle.

cloaca: (L. *cloaca*—canal)
mesentery: mes- (middle) + -entery
(Gk. *enteron*—intestines)

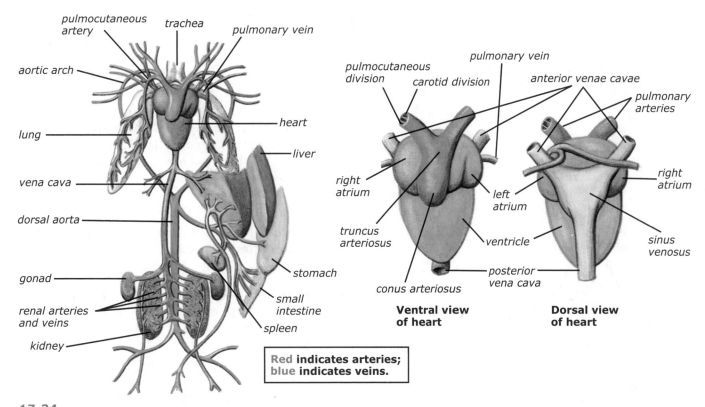

17-34

Circulatory system of the frog

The ventricle contracts, forcing oxygenated and deoxygenated blood through a single large vessel, the conus arteriosus, which branches into a right and left truncus arteriosus. From each truncus arteriosus branches a carotid (kuh RAHT id) arch (blood to the head), a pulmocutaneous (PUHL moh kyoo TAY nee us) arch (blood to the lungs and skin), and an aortic arch. The right and left aortic arches fuse into a single dorsal aorta (blood to the body organs and muscles).

pulmocutaneous: pulmo- (L. *pulmo*—lung) + -cut- (skin)

Frog and Toad Reproduction

During the mating season, which is most often triggered by temperature changes and rainfall, the male frog croaks to attract a mate. Each kind of frog has a distinct call to prevent the wrong frogs from getting together. The female will respond to the call when her eggs are ripe. After they have found each other in the water, the male, who is frequently smaller than the female, clasps her from behind—a process called *amplexus*. This action stimulates the female to release her eggs. Each egg is coated in a jelly produced by an oviduct. As the eggs are released, the male covers them with milt containing sperm.

After fertilization, the parents usually abandon the eggs, making them easy prey for various insect larvae, leeches, and flatworms. Fish, and even adult frogs, feed on tadpoles. As a result, it is typical for only about 5% of the fertilized eggs to survive to adulthood.

The process of egg laying varies dramatically in certain frog species. During mating, the male Surinam toad presses the fertilized eggs into the spongy tissue on the female's back. The young hatch and develop within the skin of the mother. After several months, the young toads emerge.

With lakes and ponds scarce in Jamaica, a species of tree frog on this island lays its eggs in the pool of water found in the leaves of a native plant. The plant maintains this pool even in the driest seasons, providing an ideal home for the developing tadpoles. Some male frogs carry the eggs and tadpoles in their vocal sacs until the young are fully developed.

Frog amplexus

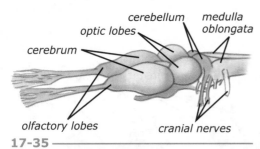

17-35
Structures of a frog's brain

Nervous Response of the Frog

The nervous system of the frog can be divided into two sections. The central nervous system consists of the brain and spinal cord encased in the bony skull and spinal column. The amphibian brain has the five basic lobes common to most vertebrates. The lobes of the amphibian cerebrum occupy a much larger part of the brain than do those of the fish. The optic lobes, however, are proportionally smaller in the amphibian. The amphibian cerebellum is unusually small, being only a thin strip of tissue behind the optic lobes. Since the cerebellum controls muscle coordination, the small size of this brain region may explain why frogs are not very graceful or coordinated.

The second division of the frog's nervous system is the peripheral (puh RIF ur ul) nervous system. Included in the peripheral nervous system are all the nerves that transmit impulses between the central nervous system and the frog's muscles and sensory organs.

Excretion and Reproduction in the Frog

Fixed against the dorsal body wall is a pair of long, red brown structures, the kidneys. Blood passes into these organs through a series of renal arteries and out through renal veins. The kidneys filter wastes and excess water from the blood, concentrating them in the form of urine. This liquid waste empties by a pair of thin, tubular ureters into the cloaca. It may then be excreted from the body or passed through a small passage into the urinary bladder, a sac that can store the urine for future disposal.

The female frog has large lobed ovaries above the kidneys. The ovaries fill with eggs until, during the breeding season, the thin walls of the ovaries burst, spilling the eggs into the female's body cavity. Abdominal contractions direct the eggs into tubular oviducts to be transported out of the body. Cilia in the oviducts propel the eggs through the long coiled tubes and into the saclike uterus (YOO tur us), where the eggs are stored until they are laid. To lay the eggs, the frog passes the eggs from the uterus into the cloaca, from which they are ejected during mating.

The male frog possesses a pair of oval testes, found on the ventral side of the kidneys. The sperm produced by these organs travel to the kidneys in thin tubes, the vasa efferentia. These tubes empty through the kidneys into the cloaca.

17-36
Nervous system

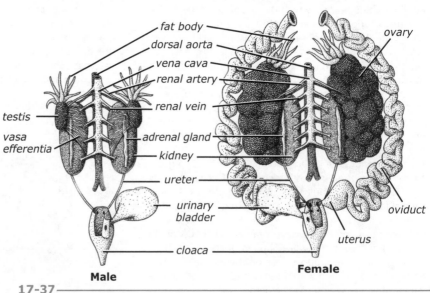

17-37
Urogenital system of the frog

peripheral: peri- (around) + -phera- (bearing)

uterus: (L. *uterus*—womb)

1. Describe the methods frogs use to capture and swallow food.
2. List and tell the function of each of the openings within the mouth of a frog.
3. List the structures a morsel of food passes through from the time it is swallowed by a frog until it is egested.
4. Describe how a frog takes in and releases air.
5. What are the two major divisions of the nervous system of the frog, and what structures belong in each division?

6. List and give the functions of the organs in the frog's excretory system.
7. Describe (a) the male reproductive system of a frog, (b) the female reproductive system of a frog, and (c) the mating process of frogs.
⊙8. Describe the major differences and similarities between the circulatory system of a fish and that of a frog.
⊙9. Does a larger brain lobe indicate a more acute function of that brain lobe in frogs? Explain.

17D—Class Reptilia: The Reptiles

To many people, **reptiles** have an unredeemable reputation. The appearance and behavior of these animals have become symbols of the power and subtlety of evil. More than a few students of biology would prefer to ignore them. However, in order to obey God's command to exercise wise and good dominion "over every living thing that moveth upon the earth" (Gen. 1:28), man must study the domain given to him by God. And, of course, part of that domain is the world of reptiles. The reptiles, like all animals, reveal something about God. Specifically, they show that even the less appreciated creatures play a role in God's plan. While the serpent is often aligned with temptation or punishment, it also serves as a symbol of God's power (Exod. 4:3) and redemption (Num. 21:8–9). All of creation, which He called "very good," performs His bidding.

17.8 Characteristics of Reptiles

Although many amphibians are terrestrial, most of them must stay in moist environments. The amphibian's thin skin must be periodically moistened, and amphibian eggs must be deposited in water or a wet area to develop. The members of the class Reptilia, however, have a special protective skin, possess lungs from birth, and lay eggs encased in shells. These features make many reptiles well suited for life on land. All reptiles must breathe air; therefore, even marine reptiles, such as certain turtles and sea snakes, can drown.

One characteristic most reptiles share with amphibians is a three-chambered heart. Three-chambered reptile hearts have a partially divided ventricle that helps keep oxygenated and deoxygenated blood from mixing. Some reptiles (alligators and crocodiles) have a four-chambered heart similar to that of mammals and birds. Reptiles, if they have limbs, have claws on their toes, a characteristic not shared by the amphibians.

Skin

Snakes and lizards are often called "slimy," an inaccurate description. Their skin is better described as cool, dry, or leathery. The scales, which give the reptile's body this texture, are part of a thick skin with few glands. Scales are composed mainly of **keratin**, a fibrous protein that is waterproof and quite durable.

17.8

➡️ Objectives

- List and describe major characteristics of the class Reptilia
- Describe and give the functions of each part of the amniotic egg
- Compare the different styles of reptilian hearts with the amphibian heart
- Give evidence to support the argument that man and dinosaurs coexisted

➡️ Key Terms

reptile	chorion
keratin	allantois
amniotic egg	septum
shell	Jacobson's
amnion	organ
yolk sac	

17-38
Some snakes shed their skin many times each year.

17-39
Reptiles' scales are specially designed to help absorb solar energy.

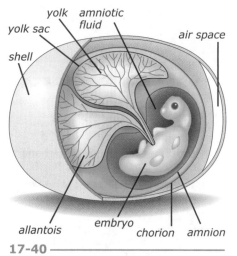

17-40
The amniotic egg

17-41
The crocodile shown here is carrying one of her babies to water—not eating it.

chorion: (Gk. *khorion*—afterbirth)

Generally, reptilian scales are nonliving and cannot grow with the animal. The reptile must periodically shed, or molt, its outgrown scales and replace them with a larger set, which the reptile grows before it sheds the old. Unlike amphibian skin, which is thin and moist to permit respiration, reptilian skin and scales are thick and dry to prevent water loss. Since many reptiles live where water is scarce, this water-preserving function is essential to the animal's survival.

The thick skin of a reptile helps it conserve some heat in its body. Reptiles are predominantly ectothermic. Many reptiles do "sun themselves" to warm their bodies, but too much heat can kill them. Most reptiles therefore hide in cool spots during the hot part of the day. In cool weather many reptiles hide and enter an inactive state similar to hibernation.

Amniotic Egg

Reptilian eggs differ from those of water-breeders such as fish and amphibians. In water, small eggs covered with a gelatinous material are fertilized and develop without difficulty; on land, such eggs would quickly dry up and die. The reptile must, therefore, have an egg that maintains a suitably moist environment for the developing animal. The **amniotic** (AM nee AHT ik) **egg**, which is produced by all egg-laying reptiles, meets this condition.

The amniotic egg must be fertilized within the body of the female. Once fertilized, the egg is encased in a protective **shell** and is passed from the female's body. The shell is porous, permitting the exchange of gases between the egg and the environment but preventing excessive water loss. The tiny embryo grows atop a ball of stored food material called the *yolk*.

Four embryonic membranes form during development within the amniotic egg. The first membrane, the **amnion** (AM nee ahn), grows around the embryo, protecting it in a fluid-filled sac. The second membrane is the **yolk sac**, which surrounds the yolk. As the yolk is consumed, this sac is slowly drawn into the embryo until it disappears altogether. The chorion (KOHR ee AHN) is a membrane that lines the inner surface of the shell. The fourth membrane, the **allantois** (uh LAN toh iss), is a sac richly supplied with blood vessels from the embryo. Respiration and excretion occur through the vessels of this membrane. The embryo in an amniotic egg is an independent, self-sustaining unit provided with a protective home, a moist environment, and a built-in food supply.

Although all reptiles practice internal fertilization, not all are oviparous. Some lizards and snakes are viviparous, with the young developing in membranes inside their mother, with whom they have a direct link for nourishment. Others are ovoviviparous, retaining the eggs inside the body cavity until the young hatch and crawl out. The empty eggs are then expelled. All turtles, crocodiles, and alligators are oviparous as described above.

Unlike birds, reptiles provide little or no care for their offspring beyond laying the eggs in a suitable spot for development. In fact, most reptiles probably never even see their parents. Exceptions include most crocodiles and alligators, which may carry the new hatchlings to the water and protect them for a limited period of time. A few types of lizards also exhibit this protective behavior.

Respiration

Reptiles do not have gills for respiration but rely on lungs from the time of birth. Their lungs include many thin air-filled sacs that give them more surface area for gas exchange than is found in amphibian lungs. Even aquatic reptiles

Dinosaurs and Man

What is a dinosaur? The term *dinosaur* (literally, "monstrous lizard") was coined by Sir Richard Owen in 1842, about the time of the first fossil finds of large reptiles. Although normally thought of as large beasts, most dinosaurs were about the size of horses and some were as small as chickens. Normally, however, the term is used to describe the larger of the presumably extinct reptiles. Technically, it applies only to the land-dwelling reptiles that walked erect, that is, with their legs beneath them. They are roughly divided into two groups—Ornithischia (bird-hipped) dinosaurs that fed on plants and Saurischia (lizard-hipped) dinosaurs that included both carnivores and herbivores. This definition means that flying reptiles like *Pteranodon* and swimming reptiles like *Plesiosaurus* are not true dinosaurs, although they are often discussed in the same context.

Biblical and scientific evidence tends to support the idea that men and dinosaurs existed at the same time. Two references to animals believed to be dinosaurs are found in the final chapters of Job. God, calling Job's attention to His mighty works of cre-

One artist's conception of an Apatosaurus

ation, describes behemoth, a large, powerful, swamp-dwelling vegetarian whose bones were like bars of iron and whose tail resembled a cedar tree (Job 40:15–24). The description fits the *Apatosaurus* or *Brachiosaurus* (both formerly called brontosaurus) far better than the hippopotamus or elephant, as proposed by some scholars. Elephant and hippopotamus tails are very *unlike* cedar trees. In verse 19 behemoth is called "the chief of the ways of God," implying that it was the greatest animal God ever made. An elephant, hippopotamus, or similar creature simply does not fit that description.

In Job 41, another mighty creature, the leviathan, fits the description of a large marine reptile such as the *Plesiosaurus* or *Mosasaurus*—not the crocodile as some scholars suggest.

Not all Bible scholars agree on the time that Job lived since there is little within the Bible to pinpoint a date. In any case it would be illogical for God to ask Job to think of something with which he was not familiar. Although Job may never have seen these ani-

Unearthing fossils

mals, it is safe to assume he was at least familiar with oral traditions about them. This is a good indication that man was alive at the same time as dinosaurs.

Most people assume that dinosaurs are extinct. It is true that these large reptiles are not in zoos or regular sights on wild animal safaris. Many people believe that changing conditions drastically reduced the areas in which dinosaurs could live and the dinosaurs simply died out.

Evolutionists claim that dinosaurs died out 65 million years ago and assume their extinction. Creationists place the change as recently as 4000 years ago and should not be surprised if extinction is not complete. In fact, there is some evidence that what are known as dinosaurs may not have disappeared entirely.

Stories are told of dinosaur-like animals. In central Africa some stories describe a land monster, the *mokele-mbembe*, as a water elephant with a head on the end of its trunk, a possible description of an *Apatosaurus*-like creature. In Zambia some people tell of a meat-eating, long-necked *chipekwe*, which attacks rhinoceroses, elephants, and hippopotamuses. They describe it as having a single horn and other features similar to a *Ceratosaurus*.

Artwork left in caves and on rock faces throughout the world also reveals figures that seem to depict large types of dinosaurs, but this evidence is open to speculation.

These stories may be fanciful since investigation has not produced any actual proof of such animals living today. Perhaps, however, scientists have not yet looked hard enough for evidence because most of them do not believe it exists. This is yet another example of how presuppositions affect research.

like sea turtles respire with lungs, but they are very efficient at holding their breath for long periods. Most reptiles draw air into their lungs by expanding the rib cage.

Circulation

Like the amphibians, most reptiles have a three-chambered heart with two atria and one ventricle. The major difference is that the ventricle is partially divided by a **septum** (wall) that keeps the oxygenated and deoxygenated blood

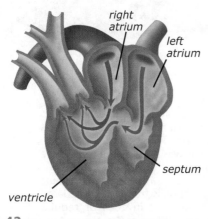

17-42 ——————
Reptilian heart

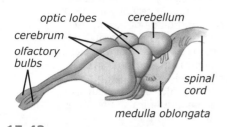

17-43 ——————
Reptilian brain

somewhat separate. In fact, 60%–90% of the blood stays separate as it leaves the heart to go out either to the lungs or to the other tissues that need oxygen. During periods of activity, the movement of this septum may even completely isolate the two circulations, keeping the blood unmixed.

A Change of Heart

Crocodiles and alligators have a four-chambered heart with a structure similar to the heart of birds, mammals, and humans. To conserve energy, crocodiles have a special cog-style valve in their heart that can prevent blood from being pumped to the lungs, pumping it instead to the rest of the body. This may explain how they can stay submerged for such long periods. Crocodiles also have the ability to slow their heartbeat to one or two beats per minute.

Nervous System

Reptilian brains usually amount to less than 1% of their total body mass, and a crocodile's is less than 0.5%. The brain does not fill the cranium (brain case) as it does in birds, mammals, and humans, and it is sometimes not even totally surrounded by bone. Despite these seeming disadvantages, reptiles have many highly acute senses.

The cerebrum, the lobe associated with controlling and coordinating behavior, is much larger than in fish or amphibians. Because most reptiles have fairly large eyes and vision is important, they have large optic lobes. Most reptiles, with the exception of snakes, have a tympanic membrane, or eardrum, often inside a canal. It is connected to an inner ear that processes sound waves.

One of the most important sensory organs for all snakes and many lizards is a pair of **Jacobson's organs**. These pits in the roof of the mouth contain nerve endings sensitive to chemicals captured from the air by the animal's tongue. This sense can help these reptiles find prey, locate mates, and even avoid predators.

➡ *Review Questions 17.8*

1. What characteristics separate the class Reptilia from other vertebrate classes?
2. List and describe the structures that compose an amniotic egg.
3. What type of fertilization is associated with animals that develop from amniotic eggs?
4. Why is the reptilian heart more efficient than the fish or amphibian heart at providing the increased energy needed for life on land?
⊙5. Tell why the behemoth described in Job 40:15–24 is not a hippopotamus or an elephant.
⊙6. List evidences to support the belief that man and dinosaurs lived at the same time.
⊙7. What does *ectothermic* mean? How would humans have to change their lifestyles if they were ectothermic? List things that ectothermic animals do to compensate for this condition.
⊙8. What characteristics do reptiles have that permit us to call them land animals?
⊙9. Considering the respiratory mechanisms, number of offspring, amount of parental care, and types of reproduction of fish, amphibians, and reptiles, some people infer an "evolutionary trend" from water to land. What would be your answer to an evolutionist who used this "trend" to support his theory of evolution?

17.9 Classification of Living Reptiles

The class Reptilia encompasses nearly seven thousand living species. The smallest reptile is a lizard about as thick as a pencil and 5 cm (2 in.) long. Crocodiles have been measured at 7 m (23 ft); however, most are less than 4.5 m (15 ft) long. The anaconda, a snake of the tropics, may reach 11 m (36 ft) in length. In the United States most lizards are under 60 cm (24 in.), and a snake over 1.5 m (5 ft) long is considered large.

Reptiles are classified in four orders:

- Squamata—snakes and lizards
- Testudinata—turtles and tortoises
- Crocodilia—alligators and crocodiles
- Rhynchocephalia—tuatara (*Sphenodon*)

17-3 Reptilian Orders

Order	Key characteristics	Examples	Species number (approx.)
Squamata	disarticulating lower jaw; some venomous species	snakes, lizards	6600
Testudinata	bodies encased in protective shell	turtles, tortoises	300
Crocodilia	four-chambered heart; some parental care	alligators, crocodiles	21
Rhynchocephalia	spiny dorsal crest	tuatara	2

17.9

Objectives

- Distinguish between the four surviving reptilian orders, giving an example of each
- Describe several ways that snakes move
- List three ways that snakes can kill their prey
- Contrast the structure of lizards and snakes

Key Terms

scute	hemotoxin
constriction	quadrate bone
fang	carapace
neurotoxin	plastron

Order Squamata: Snakes

While snakes and lizards have some obvious differences, they share many similarities, including long scaly bodies and the presence of teeth. These features, as well as skull similarities, led taxonomists to put them in the same order—Squamata. Indeed, a lizard, if you can imagine it without legs (see "Unique Lizards" box, p. 424), has much in common with a snake.

Many people, when they see a snake, flee in terror or attempt to kill it. It is true that there are some poisonous snakes that are a real danger to humans; however, most snakes are harmless to man, and many actually benefit him. Without snakes feeding on rats, mice, and insects that plague man, the uncontrolled populations of vermin could become serious problems.

Movement of Snakes

Snakes lack appendages and, depending on the snake and the terrain, use one of four locomotion techniques.

✦*Serpentine movement*, used by the snake both in crawling and swimming, involves the winding of the snake across the ground in a series of S curves. Twigs and stones on the ground aid traction.

✦*Concertina movement* involves the snake's drawing itself into a tight S shape and then extending itself forward. This is normally used when the snake is between two structures.

✦*Rectilinear* (REK tuh LIN ee ur) *movement* involves **scutes**, which are broad scales on the ventral surface of the snake. With wavelike muscular action, the snake forces the scutes forward. As the scutes grip the earth, the animal pulls itself forward in a straight line.

✦*Sidewinding* is used by snakes in deserts or sandy areas where traction is poor. The snake shuffles sideways through the sand by continually looping its body forward. Except at two or three points, it keeps its body raised above the sand.

Most snakes are good swimmers and some are excellent climbers. A few can even spread their sides to help them glide on air when leaping from one branch to another, and some even "swim" under the sand, exposing only their nostrils for breathing. Although some snakes seem to move very quickly, none are faster than man, and most will retreat in the opposite direction unless they are cornered.

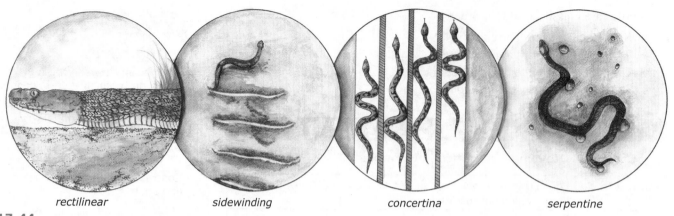

| rectilinear | sidewinding | concertina | serpentine |

17-44

Types of snake locomotion

Feeding of Snakes

Snakes are strictly carnivorous and feed on living prey. A snake's diet includes rodents, insects, lizards, small mammals, eggs, and even other snakes. To locate its prey, the snake uses a variety of sensory organs. The snake's lidless, staring eyes are not exceptionally sensitive (except in certain tree-dwelling species), but several acute senses compensate for weak eyesight.

Nostrils lead to nasal cavities that are provided with olfactory nerve endings. In the roof of the snake's mouth, a pair of sensory pits, called Jacobson's organs, aids the sense of smell by gathering chemical samples undetectable to the nose. As the snake's forked tongue flits out of its mouth, it gathers traces of chemicals and transfers them to the Jacobson's organs. These pits, combined with sensations from the nasal cavities, enable the snake to detect very slight traces of odors.

A group of snakes called the pit vipers has heat-sensing pits between the nostrils and the eyes. Certain pythons have a whole row of these pits along the upper jaw. They enable the snake to track and strike accurately at warm-blooded animals, even in total darkness. Snakes have no external ears and are

The Original Serpent

The world's most notorious reptile is introduced in the early pages of the biblical record: "Now the serpent was more subtil than any beast of the field which the Lord God had made" (Gen. 3:1). Other passages indicate that mankind encountered more than a snake that day. Revelation refers to the Devil as "that old serpent," the one who "deceiveth the whole world" (Rev. 12:9; 20:2). It seems logical, then, to conclude that Satan took control of a snake's body for the purpose of tempting Eve.

Since both the serpent and Satan were involved in the temptation of Eve, God's curse fell on both: "Because thou hast done this, thou art cursed above all cattle, and above every beast of the field; upon thy belly shalt thou go, and dust shalt thou eat all the days of thy life: and I will put enmity between thee and the woman, and between thy seed and her seed; it shall bruise thy head, and thou shalt bruise his heel" (Gen. 3:14–15). Verse 14 focuses on the snake's curse, and verse 15 focuses on Satan's. It is hard not to speculate from verse 14 what the snake may have looked like before this curse. Some have supposed that prior to this snakes could talk or that they had limbs or wings and were marvelously beautiful. But this is only speculation. All

that is known for certain is that the present anatomy of snakes is the result of a curse from God.

Verse 15 is more significant for Christians. It reveals in very brief language God's program for human history. From this point on, God says, the human race will be divided into two groups: the seed of the woman and the seed of the serpent. The former are the people of God, who are led to victory through the triumph of their Leader, the Messiah (John 1:12–13; 1 John 3:9–12). The latter are the enemies of God and His people, who are led to defeat by their "father," the one who was a serpent that day in Eden (John 8:44).

With this understanding, Christians are able to reflect on the significance of snakes with greater profit. God's double curse expressed in Genesis 3:14–15 demonstrates that sometimes zoological study can be theologically instructive. As you study snakes, their bodily structure and behavior, you are studying a metaphor of God's working in the world. The enemy of God and His people is cunning and subtle, but he is also humiliated and certain to be destroyed. Although the heels of God's people may be struck, the head of their enemy will most certainly be crushed.

totally deaf to airborne vibrations, but they can detect some vibrations in the ground.

After locating the prey, snakes most often use one of three methods to capture it. The simplest method is to swallow the prey alive, a process often used by insect-eating snakes but occasionally seen in snakes that eat animals even larger than they are.

Many nonvenomous snakes kill their prey by **constriction**. When the snake spots its prey, it seizes it in its mouth. It uses its teeth on the upper and lower jaws to keep prey from slipping away—not to tear or chew. The teeth of the snake curve inward and provide a firm grip. The snake then coils its body around the prey and squeezes it to death. The prey is not crushed by the pressure (usually the ribs are not broken), but it is quickly suffocated because it is unable to inhale.

The final method of prey capture is with *venom*. A few snakes have glands that supply poison to a pair of needlelike **fangs** on the upper jaw. To attack, the snake opens its mouth wide and thrusts its fangs into its victim. The poison flows from the venom glands in the top of the head, through grooves or tubes in the fangs, and into the prey. Usually the snake then releases the prey and waits for it to die before swallowing it. The venom not only kills the animal; it also begins internally digesting it. Venom is also used as a defense when the snake is threatened.

The snake's venom (poison) takes one of two forms (although some snakes produce both):

✦ **Neurotoxin** (NOOR oh TAHK sin) attacks the nervous system and often quickly paralyzes the prey.

✦ **Hemotoxin** (HEE muh TAHK sin) is more deadly for large animals and humans. Slower than neurotoxin, it destroys blood vessels and red blood cells.

How can a snake that is unable to tear its food swallow something that may be four to five times bigger around than the snake itself? The secret is in the skull architecture. The snake's lower jaw fastens to the **quadrate bone** rather than directly to the skull. This bone acts as a hinge, enabling the snake to open its mouth wide. In addition, the lower jaw bones join at the front by an elastic ligament, permitting flexibility. When swallowing, the snake can move its lower jaw bones independently; one side of the jaw stretches forward to gain a new grip on the prey while the other pulls back to draw the prey into the snake. Since the teeth point inward, even an animal that recovers from the constriction or the poison is not able to wiggle free. The skin of the mouth and neck region is extremely elastic. When the snake is swallowing or digesting large prey, its ribs may even temporarily dislocate.

Order Squamata: Lizards

Though both lizards and snakes belong to the reptilian order Squamata, lizards differ from snakes in several ways.

- Lizards usually have two pairs of limbs while snakes have none.
- Lizards have external ear openings, enabling them to sense airborne sounds to which snakes are deaf.
- Most lizards have eyelids and can close their eyes; snakes' eyes are permanently open and covered by a protective membrane.
- The lizard's belly is covered with scales similar to those on the rest of its body; the large scutes across the snake's belly assist in crawling.

17-45 —
A snake's head is specially designed to allow the snake to swallow huge meals, like this egg.

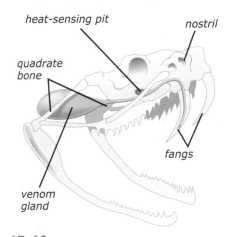
17-46 —
Structures of a snake's head

17-47 —
The colorful chameleon has a prehensile tail and toes for life in the trees.

neurotoxin: neuro- (Gk. *neuron*—nerve) + -toxin (L. *toxicum*—poison)

hemotoxin: hemo- (blood) + -toxin (poison)

quadrate: (L. *quadrus*—a square)

Unique Lizards

Taxonomists struggle to classify the nearly 150 species of worm lizards because of their blend of snake and lizard features. They have snake-shaped bodies with small square scales covering all surfaces. They do have tiny eyes and ears, but they are covered with skin. None have back limbs, but a few species have paddle-shaped limbs very close to the front. They spend most of their time burrowing beneath the ground where they feed on worms and other soft invertebrates. Sometimes called amphisbaenians, the worm lizards are either seen as a family under the lizards or as a third distinct suborder of Squamata, separate from the lizards and the snakes.

Nearly eighty species of lizards are found in the family Anguidae. Many of these lizards are legless and superficially resemble snakes. Unlike snakes, however, they have ear openings and movable eyelids. They also have stiff scales containing thin plates of bone, with rows of scales going across the belly; snakes, on the other hand, have long scales called belly scutes.

These fascinating reptiles are also known as glass snakes. They earn this name by their ability to sacrifice a portion of their tail as a means of protection. In this process, known as *autotomy*, a significant portion of the tail falls off and thrashes around to distract the would-be predator. The lizard can often escape during the confusion and will eventually regrow the missing portion.

This glass "snake" is actually a legless lizard.

Lizards are the largest group of reptiles, and species can be found in a range of habitats from deserts to rainforests. Tiny geckoes cling to tree leaves with their special clinging toes, while large green iguanas munch on fruit high overhead in the canopy. In the Australian outback the tan scales of the bearded dragon camouflage it against the rocky terrain.

True chameleons—tree dwellers of Africa, Madagascar, and India—are able to change color in response to changes in light, temperature, and even excitement level. The chameleon has a number of unique structures that enable it to hunt and thrive in trees. Its toes can grasp branches securely, and its prehensile tail can wrap around branches for extra support. Its keen eyes can act in unison or move independently of each other. When the chameleon spots an insect, it "fires" its long tongue and snags the prey with the sticky end. The chameleon's extended tongue may be longer than the chameleon itself.

The two known venomous lizards belong to the same genus. The Gila monster of the southwestern United States and the Mexican beaded lizard have venom glands in their lower jaws. Once these lizards clamp onto their prey, they hang on tightly, release venom into their saliva, and inject it into their victims by their chewing. Fortunately, the chance of being bitten by one of these slow-moving lizards is slight.

In the fourth century, some members of the monitor lizard family were described as "land crocodiles." There are about sixty species of monitors, and this group contains many of the largest lizards. These reptiles usually live on land, but many of them swim and climb with ease. In 1912, a visitor to the island of Komodo in the East Indies discovered the largest of the lizards, the Komodo dragon, which reaches a length of 3 m (10 ft) and can weigh 135 kg (297 lb). The Komodo dragon has a long forked tongue that is often a pale orange color. From a distance, it might look like the "dragon" is breathing fire.

17-48

The Komodo dragon is the world's largest surviving lizard.

Order Testudinata: Turtles

The term *turtle* actually refers to any of the shelled reptiles that are in the order Testudinata (tes TOOD ih NAY tuh), although it is sometimes used just for those types that are aquatic. The land dwellers of this group are usually called *tortoises*.

A hard shell consisting of a dorsal **carapace** and a ventral **plastron** covers these organisms. Like all reptiles, turtles are vertebrates, and most of their backbone is fused into the inside of the carapace. From within the shell, the head, thick limbs, and tail protrude. Some species, including the box turtles, can pull

Testudinata: (L. *testudo*—tortoise)

these appendages completely into the shell and close the front opening with a "trapdoor" on the plastron, becoming impenetrable fortresses when attacked.

Each limb of most turtles and tortoises is short and thick and has five claws. Some turtles have paddlelike flippers. An upper eyelid, a lower eyelid, and a transparent nictitating membrane inside the other lids protect the eyes. The animals lack teeth but have jaws formed into tough horny beaks that can cut and tear food. The nostrils are placed high on the snout, enabling the animal to hide underwater with only a small portion of its head above the surface. Females usually bury their eggs in a burrow and then abandon them. No parental care is supplied to the young.

Order Crocodilia: Crocodiles and Alligators

Alligators, caimans, crocodiles, and gharials belong to the order Crocodilia and are among the largest living reptiles. Caimans and gharials are among the less familiar crocodilians. Six different species of caiman are found in Central and South America. They are placed in the alligator family, and most are less than 2.5 m (8 ft) in length. Another crocodilian is the gharial, with its long skinny snout lined with sharp teeth for snaring fish. Found only on the Indian subcontinent, the single gharial species is unique enough to be placed into its own family.

Often crocodiles and alligators are mistaken for one another. Examination of each animal's head reveals that the alligator has a blunt, thick snout and teeth that fit within the jaws; the crocodile has a thinner, more pointed snout with some teeth protruding from the jaws when the mouth is closed. Most people, however, do not get close enough to these animals to evaluate their snouts or teeth.

Crocodiles are stealthy carnivores. They glide smoothly just below the surface of the water when they search for prey. The crocodile's eyes and nostrils are on the top of its head so that the animal can see and breathe while keeping most of its head underwater. When it comes upon a potential victim, the crocodile seizes it in its powerful jaws and thrashes it about, tearing the victim with its teeth. While

17-49
Gharial

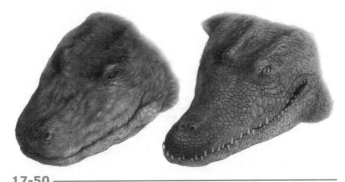

17-50
Heads of an alligator (left) and a crocodile (right)

it does this, its nostrils can be closed, and valves keep water from flowing into its ears. These reptiles also have a special flap that can cover the windpipe to prevent drowning during underwater feeding.

All crocodilians are oviparous. American alligators deposit eggs in a mound of decaying vegetation. The heat generated by the rotting vegetation warms the eggs. After eleven to fourteen weeks, the young alligators make audible "peeps," signaling the mother to dig them out. Once they are unearthed, the mother gently carries her young to the water in her mouth.

Order Rhynchocephalia: Tuataras

The *tuatara* (TOO uh TAR uh) is a 60 cm (24 in.) reptile found only on a few islands near New Zealand and the only living member of the order Rhynchocephalia (RING koe suh FAYL yuh). Although it resembles a lizard, it has several internal structural differences, including holes in the skull and a unique tooth arrangement, that warrant its classification in its own order. The tuatara is also unusual in that it has traces of a light-sensitive structure on the top of its head. This third "eye" is nonfunctional and is covered by scales in the adult.

Evolutionary biologists consider the two surviving tuatara species to be living fossils, virtually unchanged from the age of the dinosaurs, 220 million years ago. Most Creationists view these reptiles as small survivors of a group of reptiles that became extinct several thousand years ago because of their inability to survive changes that occurred on the earth.

17-51
Tuatara

Rhynchocephalia: Rhyncho- (Gk. *rhunkhos*—beak) + -cephalia (head)

Review Questions 17.9

1. List the four major groups of living reptiles. (Use common names.)

2. Based on Genesis 3, what three characteristics do some people claim the snake had before the Curse? Which, if any, of these characteristics are necessary according to Scripture?

3. What senses do snakes use to locate prey?

4. What are the three methods snakes use to kill their prey?

5. Name the two main types of venom produced by snakes and tell what part of the human body each type of venom affects.

6. What structures of the snake permit it to swallow large prey whole?

7. What are the four major differences between snakes and lizards?

8. List and describe four lizards.

9. List several characteristics of turtles that separate them from other reptiles.

10. What are the observable differences between alligators and crocodiles?

11. Describe the tuatara.

18

The Endothermic Vertebrates

Zoology Part IV

Facets

18A—Class Aves: The Birds

In a Greek myth, Daedalus and his son Icarus sought escape from an island prison. Seeing the birds fly to and from the island, they decided to attempt flying to freedom. Wearing wings made of wax and feathers, they leaped from a cliff and soared over the ocean. The father reached land, but Icarus, enjoying flight, went higher and higher. He flew too near the sun; his wings melted, and he was lost in the sea.

Through the ages, men have envied the apparently effortless flight of birds and have dreamed of flying. Leonardo da Vinci's sketchbooks contain drawings of winged machines. Many odd contraptions were designed by would-be fliers over the years; a few got off the ground, but none stayed there. In 1903, Orville and Wilbur Wright designed an awkward-looking biplane and took a twelve-second flight at Kitty Hawk, North Carolina. The dream became a reality, and man was flying.

Since then man has learned that flight requires more than just a pair of wings. Anything that flies, be it an airplane or a bird, must meet specific design requirements if it is to get off the ground and stay aloft. Flight requires a structure that is streamlined and lightweight yet extremely strong. Airplanes

→ **Objectives**

- List characteristics of class Aves that distinguish its members from most other vertebrates
- Name the features and functions of birds that enable flight
- Describe the structure of the typical flight feather, using appropriate terms

→ **Key Term**

feather

18-1 ——————————————

Bird bones are hollow but still strong due to internal supports.

Aves: (L. *avis*—bird)

require intricate designs and large quantities of fuel to take off, direct their flight, and land. Birds, however, can fly with considerably more control and proportionally far less energy than the best airplane. The bird's body is a clear illustration of the completeness and perfection of God's design.

18.1 Bird Characteristics

God created an abundance of animals with great variety in their characteristics and appearance. This variety sometimes makes it difficult to classify some animals. For example, some sharks look like bony fish and some amphibians look like reptiles. Birds, however, are simple to identify since many of their characteristics are obvious even to the untrained eye and are unique to the class Aves (AYE veez).

Flight of Birds

The more than nine thousand species of birds share many common characteristics, nearly all of them promoting their ability to fly. Even those few kinds that are too heavy to fly or that "fly" through water instead of air exhibit these features. Foremost, all birds have **feathers**, the protein-based structures that provide covering, insulation, and shape. No other group of animals possesses anything like them. Birds have their front limbs configured as wings with adjustable flight surfaces. Wings alone would not ensure flight if the bird were too heavy, so bird bones are thin walled and hollow, supported by tiny struts that span the hollow cavities and lend more support.

Flight requires enormous amounts of energy, which demands lots of oxygen for respiration. Birds have just two lungs, but they are connected to many air sacs, effectively multiplying their volume. The empty spaces in some bones are even tied in with the respiratory system, allowing more surface area for gas exchange. Birds must have rapid access to energy, and so of necessity they must be endothermic. With their high metabolism, they maintain an internal temperature averaging 41 °C (106 °F).

Several other characteristics keep birds light enough to get off the ground. Unlike fish, amphibians, and most reptiles, birds have no teeth. The horny

---

Backyard Dinosaurs?

Many evolutionists point to the archaeopteryx as the "missing link" between reptiles and birds. In fact, some even say that the birds found in our backyards are actually feathered dinosaurs that survived. This idea began with the discovery of a remarkable crow-sized fossil animal in 1861, just two years after Charles Darwin published *On the Origin of Species by Means of Natural Selection*. Seven other archaeopteryx fossils have been uncovered, but even evolutionists are divided over their significance.

Those supporting the "reptile-to-birds" path of evolution point out the nonbird characteristics such as teeth, unfused vertebrae, and wing claws as evidence that these fossil animals were not birds. Some even question whether they could fly at all. They suggest that these creatures merely ran along the ground, sweeping up prey with their feathered appendages, which they also used as insulation in a cooling climate. (Many evolutionists think bird feathers evolved from

One of the eight archaeopteryx fossils

elongated reptile scales.) Those that believe bird flight evolved in feathered dinosaurs are divided over whether the earliest fliers climbed trees and glided from there or ran along the ground and eventually became airborne. Because some recent fossil finds that evolutionists date as much older than the archaeopteryx are clearly flying birds, some have concluded that perhaps the archaeopteryx was a dead end.

Creationists view the archaeopteryx as a fascinating animal but very much a bird or birdlike creature. It had feet for perching on limbs as well as fully formed feathers. Feathers are very different from reptile scales; they grow, like hair on humans, from follicles deep in the skin, while scales are merely folds of the outer epidermis. The skeleton of the archaeopteryx also shows adequate attachment sites for flight muscles. There is nothing that would prevent the Creator from putting teeth on some birds or feathers on some reptiles, but neither of these would prove that one gave rise to the other.

beak material, made of durable keratin, is just as effective but lighter. All birds are oviparous, keeping their developing young in a nest rather than carrying them in their bodies. These features, and many more, point to vertebrates that are designed for flying.

Feathers of Birds

While feathers are critical for moving air during flight, birds also depend on them for other functions. Layers of feathers shape the body to reduce drag and make the bird aerodynamic. Because birds are endothermic, they must have some means of retaining heat. Feathers are ideal for this. They also serve as a cushion—protecting the fragile skeleton during collisions. The colors and patterns provided by feathers may help camouflage the bird or even attract and identify a mate.

The strong yet virtually weightless feather grows from a tiny structure of the skin called a papilla (puh PIL uh). The feather is composed of a flat *vane* and shaft. The shaft within the vane area is the *rachis* (RAY kis), and the shaft from the papilla to the vane is the hollow *quill*.

The vane consists of parallel rows of thin barbs that originate in the rachis. Parallel rows of tiny barbules extend from each barb, making each barb almost like a miniature feather. The barbules of one barb firmly interlock with the barbules of another by means of microscopic hooks. If adjacent barbs are forced apart, the bird can reattach them by using its beak to stroke the edge of the vane.

Many birds oil their feathers to keep them from becoming brittle. A gland at the base of the tail produces the oil. In preening, the bird applies the oil with its bill while it adjusts and smoothes its feathers. The oil also provides a waterproof shield valuable to swimming birds such as ducks. Without such protection, the feathers of these birds would become soaked, making it difficult for the animal to remain afloat and stay warm.

The following are the most common types of feathers:

✦ *Down feathers* have barbules that do not interlock with one another. As a result, the barbs of those feathers do not form an orderly vane but a "feather duster" tuft instead. These feathers are an underlayer, providing insulation as well as a cushion.

✦ *Contour feathers* are the common vaned feathers. They cover most of the animal's body, giving it its external shape and color. The large *flight feathers*, which extend from the wings and tail, are also contour feathers.

Fully developed feathers, like scales and hairs, are dead. Periodically, the bird sheds or molts old feathers and replaces them. In many birds this process occurs in late summer so that the new feathers are ready for the fall migration.

Unlike the skin shedding of reptiles, the molting of birds is a gradual, highly ordered process. A bird missing many flight feathers on one wing would be crippled; therefore, birds shed these feathers in corresponding pairs (one from one wing and the same one from the other wing). Once one pair of flight feathers has been lost, another pair will molt only when the replacements for the original pair have appeared. Some birds molt certain feathers just prior to mating and have new, colorful plumage in order to attract a mate.

Appendages of Birds

Except for some flightless birds, such as the penguin, which uses its wings for swimming, most birds use their wings to propel them through the air. No bird uses its wings to manipulate items, but many birds use their long flexible necks as substitutes for arms and hands. The lower parts of the hind limbs of

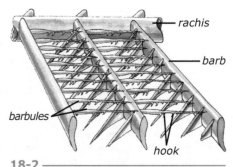

18-2
Structures of a feather

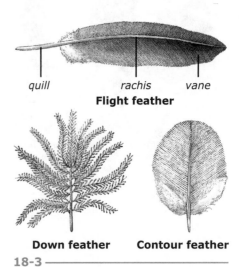

Flight feather

Down feather   **Contour feather**

18-3
Feather types

rachis: (Gk. *rhakhis*—spine)

Wings of Birds

Based on the kind of flying that birds need to do, the designs of their wings differ. Most wings fit into one of four design categories.

◆ *Elliptical wings* are short and wide and provide for quick takeoffs and landings, low-speed flight, and maximum maneuverability. The elliptical wings of sparrows and woodpeckers permit them to change direction quickly in their forest habitats.

◆ *High-speed wings* have a long, thin, tapered shape that generates little drag in the air. Terns, swifts, and sandpipers have this wing type and are among the fastest fliers.

◆ *Soaring wings* are long and thin and resemble the wings of a glider. These wings enable water birds such as gulls to expend a minimum of energy in staying aloft. This increased flight efficiency, however, is obtained at the expense of maneuverability.

◆ *High-lift wings* are large and convex and provide a tremendous amount of lift even at low speeds. These wings allow birds of prey such as hawks, owls, and eagles to carry large prey. Many of these birds are skilled at soaring, riding thermal air currents, and changing course with subtle movements of their wing tips.

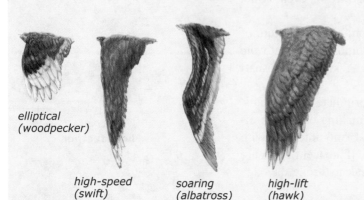

elliptical
(woodpecker)

high-speed
(swift)

soaring
(albatross)

high-lift
(hawk)

Common Types of Birds' Feet

◆ *Wading.* Wading feet have long, thin, widely spread toes, which distribute the weight of the bird so that it does not easily sink into a muddy bank. The thin toes and legs can move through the water without creating turbulence that would scare away the water animals the bird eats.

◆ *Swimming.* Many swimming birds, such as the duck, have webbed feet, which serve as effective paddles.

◆ *Climbing.* Birds that climb, such as the woodpecker, spend much of their time clinging on vertical surfaces, sometimes even upside down. Their climbing feet with two front toes and two back toes provide grip and balance.

◆ *Running.* Many flightless birds have long, heavily muscled legs for running and strong thick toes that function almost like cattle hooves.

◆ *Grasping.* The foot of a bird of prey is designed to grasp and kill prey; the toes bear *talons* (long curved claws). Birds of prey, such as the osprey, that must catch and control slippery fish even have spiked scales on the bottoms of their toes to maintain a grip.

◆ *Perching.* The perching feet of most songbirds, such as the robin and the wren, have three front toes and one back toe. A "locking device" closes the toes as the bird lands on a branch and remains tightly closed until the bird releases it. This is how a bird can sleep without falling out of a tree.

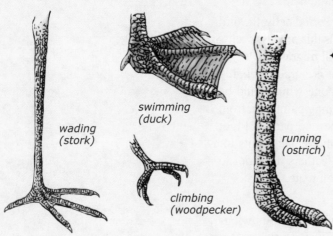

wading
(stork)

swimming
(duck)

climbing
(woodpecker)

running
(ostrich)

grasping
(falcon)

perching
(wren)

the bird are usually thin and covered with scales, and most birds have three or four clawed toes.

The design of the legs and feet suits the needs of the particular bird. The hummingbird, which spends most of its time in flight, has tiny legs and feet. The flightless ostrich, however, has well-muscled legs, permitting it to run overland at 65 km/h (40 mph).

Skeleton of Birds

The most outstanding feature of the bird's skeleton is the numerous air-filled cavities within the bones, a feature that results in extreme lightness. As light as feathers are, the average bird has a skeleton that weighs less than its feathers. The reduction in weight, however, does not reduce the strength of the bones.

Several other characteristics make the bird's skeleton unusual:

- The upper jaw (maxilla) and the lower jaw (mandible) are elongated and form the bill.
- Many neck vertebrae grant free movement of the head, an especially useful feature since birds lack the ability to manipulate items with their forelimbs and since most do not have movable eyes.
- The vertebrae of the tail are free moving and help in guiding flight.
- The trunk vertebrae, the flat ribs, and the sternum (breastbone) are fused to make the trunk a rigid framework.
- The large sternum has a central ridge called the *keel*. The keel provides attachment for flight muscles.
- The clavicles (collarbones) are enlarged and fused, forming the "wishbone." This also provides attachment for flight muscles.

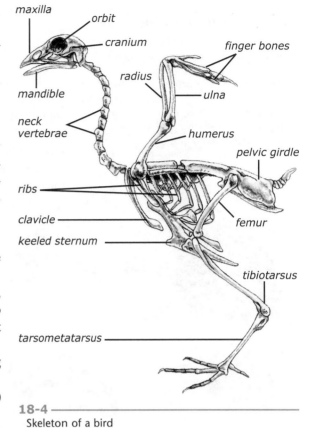

18-4
Skeleton of a bird

Review Questions 18.1

1. Describe a feather. How do birds care for their feathers?
2. Describe the two major types of feathers and tell their functions.
3. List and describe four types of birds' wings and give an example of a bird for each type.
4. List and describe four types of birds' feet.
5. List six characteristics of the bird's skeleton.
⊙ 6. Which characteristics of class Aves separate birds from other classes of vertebrates?

18.2 Bird Systems

Birds must accomplish the same internal processes as all other vertebrates—digesting and absorbing food, bringing oxygen and nutrients to the cells, eliminating carbon dioxide and other wastes, and responding to their environment. The challenge is accommodating these diverse tasks in a lightweight flying package. Each of the familiar life processes discussed in this section is uniquely modified by the Creator to serve a bird's high-energy lifestyle.

Digestion in Birds

The high body temperatures and flying activities of birds create an enormous energy demand, which birds satisfy by eating large quantities of food. To accuse someone of eating like a bird may mean the opposite of what was intended. Young birds may eat more than their own weight in a day. Some adults of small species eat over a quarter of their weight a day. Once they ingest the food, it is quickly digested by an efficient digestive system. Rapid digestion is

18.2

Objectives

- Trace the movement of food through a bird's digestive tract
- Compare a bird's respiratory and circulatory systems with those of the ectothermic vertebrates
- Relate the sizes of the lobes in a bird's brain with the significance of its various senses

Key Terms

crop   air sac
gizzard   syrinx
cloaca

tearing (hawk)

probing (ibis)

cracking (bunting)

filtering (flamingo)

18-5

Types of bird bills

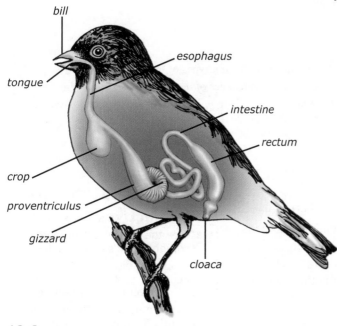

bill

esophagus

tongue

intestine

rectum

crop

proventriculus

gizzard

cloaca

18-6

Digestive system of a bird

syrinx: (Gk. *surinx*—shepherd's pipe)

essential because birds cannot afford to carry heavy quantities of food in their stomachs for long periods.

The bird's unique structure for food gathering is its bill. The appearance and use of the bill vary with the bird's diet. Birds that spear fish or pluck insects from cracks in tree bark usually have long, slender bills. Short, stout bills are used for cracking and crushing seeds. Predatory birds often have hooked, pointed bills for tearing prey apart. The flamingo and other birds that feed on tiny plants and animals have sieves in their bills for filtering mud from their meal. The ibis and other birds that probe underground for food have many touch sensors in the tips of their bills for locating prey. Many hummingbirds have slender bills and long tongues uniquely designed to lap nectar from only certain types of flowers.

Digestion begins as the food, usually swallowed whole, passes down the elastic esophagus and enters the **crop**. The crop is an enlargement of the esophagus that stores the food until it can pass into the two-part stomach. The first part of the stomach, the *proventriculus* (PROH ven TRIK yuh luhs), produces digestive juices. The muscular and thick-walled **gizzard** is the second section; it contains particles of sand and small stones swallowed by the bird. Muscular contractions of the gizzard grind the food against the sand and stones and mix it with the digestive juices.

The food then passes into the intestine, where final digestion and absorption occur. Undigested wastes are passed into the **cloaca** and then out of the body. The cloaca in birds provides a common opening for the intestine, kidney ducts, and reproductive organs.

Respiration in Birds

The bird's small, inelastic lungs are unable to supply all the oxygen the bird needs to maintain its high metabolism; therefore, a series of **air sacs**, structures unique to birds, aids the lungs. About 25% of the air drawn through nostrils in the bill enters the gas exchange areas of the lungs. The other 75% is routed into the air sacs. As the bird exhales, the air in the lungs passes out, and the fresh air in the air sacs passes through the lungs. Thus, oxygen-rich air passes through the gas exchange areas of the lungs during both inhalation and exhalation. Air sacs also fill spaces that would be filled by fluids and fat in other animals. Thus, air sacs help make the bird light.

Respiration not only provides oxygen; it is also the bird's primary means of cooling. High metabolism creates much heat, yet birds lack many cooling devices that some mammals possess. They do lose some heat through sweat under the feathers, but this loss is limited. When very hot, birds can "pant"—rapidly moving air in and out of their air sacs, thus expelling excess heat with the expired air.

Just as humans rely on the respiratory system for speech, birds use theirs to vocalize. One of the most familiar features of birds is their song. Bird songs are produced by the **syrinx** (SIHR ingks), or song box, an enlargement of the trachea just above where it divides to enter the lungs. The syrinx, with the surrounding complex arrangement of muscles, can produce a variety of sounds. Remarkably, most songbirds must perfect their simple, innate calls by imitating others of their own species.

Circulation and Excretion in Birds

The bird's circulatory system has complete separation of the oxygenated and deoxygenated blood, a characteristic also found in mammals. This separation is possible because of a four-chambered heart. The right side of the heart receives deoxygenated blood from the body and pumps it to the lungs. Once the blood is oxygenated in the lungs, it flows into the left side of the heart and then is pumped to the body. Comparatively speaking, birds have the largest hearts of all vertebrates, up to twice as large as those of mammals of a similar size.

To maintain the high metabolic rate, the heart of a bird must beat rapidly. Beats per minute (bpm) range from 135 to 570, and the heartbeats of some birds, like the chickadee, can range up to 1000 bpm. It has been said that birds do not die of old age; they "burn themselves out"—not really a scientific description, but nonetheless appropriate. Because their hearts pump so efficiently, they have much higher blood pressure than mammals, sometimes leading to heart failure during periods of high stress.

Metabolic wastes are filtered from the blood by a pair of kidneys that lie along the back of the body cavity. The wastes empty directly from the kidneys into the cloaca. Birds have no bladder for storing urine and also do not store feces. To reduce their weight and make flight easier, most birds empty their cloaca just before takeoff and at other frequent intervals.

Responses in Birds

God's design becomes more obvious from a study of the various animals. Reptiles, amphibians, and fish rely heavily on the senses of smell and taste. In contrast, their vision and hearing are weak. Many of these animals live in a watery environment that muffles sound and reduces light, making hearing and vision of little value. Even those reptiles and amphibians living on land have little use for strong eyes; they live in or close to the ground, limiting their field of vision. God provided the specific senses these animals need for their activities and environment.

A bird's habits differ vastly from those of fish and reptiles, and it requires different sensory abilities. A bird in flight must be able to spot obstacles in its path soon enough to avoid them. Locating food from the air requires a keen eye. Hearing is important because birds communicate and seek mates by various sounds. In addition, some nocturnal birds such as owls rely on their ears to locate small prey. To match these needs, God has given birds good hearing and the keenest sight in the animal kingdom.

The bird's eyes are large and set deep in the skull. Because they are so large in relationship to the size of the skull, they are usually immovable. In order for the bird to look about, it must move its flexible neck. Bird eyes are usually set on either side of the head, permitting a wide field of vision. Some birds have exceptionally keen sight; the vision of a hawk is eight times more acute than that of man. A hawk can spot a crouching rabbit at a distance of more than 1.5 km (about 1 mi). An owl can spot prey in light $^{1}/_{10}$ to $^{1}/_{100}$ the brightness required by man to see.

The ear openings of a bird are often covered with feathers. Ear canals open behind the eyes and lead to a complex sensory mechanism within the skull. Hearing is sensitive in the upper sound range, where most birdcalls are made.

The slur "bird brain" notwithstanding, a bird's brain is comparatively larger than that of all other vertebrates besides mammals. The structure and size of the major areas of the brain

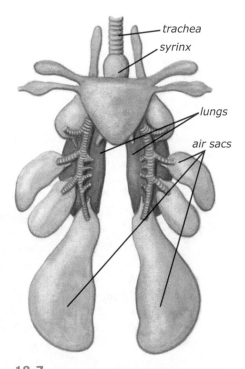

18-7
Respiratory system of a bird

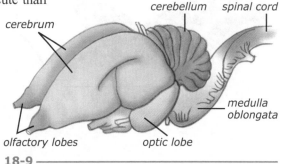

18-8
While most songbirds have a standard, limited repertoire, some do imitate others.

18-9
The brain of a bird

indicate the sensory, physical, and mental strengths and weaknesses of the animal. The olfactory lobes are small in the avian brain. Smell is not important to many birds, except for vultures and others that must locate dead animals (carrion) to scavenge. The cerebrum, which probably serves in the bird's many complex instincts, is large. The optic lobes beneath the lobes of the cerebrum are large, as would be expected for such visual animals. The cerebellum, which coordinates the muscular activities, is large because flight requires such a complex combination of movements.

Review Questions 18.2

1. Which structures of the bird's digestive system compensate for its lack of teeth?
2. What can be learned by studying birds' bills?
3. Give two functions of the air sacs in a bird's respiratory system.
4. Give a method for conserving heat and a method for eliminating heat in a bird's body.
5. Describe a bird's heart. What advantage does this design provide for birds?
6. What are the most acute senses in birds? How does this illustrate design?
⊙7. What types of bird bills usually go with what types of bird legs and feet?
⊙8. Birds are the first animals we have studied that are endothermic. What advantages and disadvantages does this characteristic present for birds?
⊙9. List all of the characteristics of birds that help them fly. Be sure to include those characteristics that make birds lighter than other animals of the same size.

Objectives

- Describe the structure of a typical bird egg
- Distinguish between and give examples of precocial and altricial chicks
- Name some advantages of migration for birds

Key Terms

courtship
albumen

18.3 Bird Family Life

For birds, the process of reproduction and raising young is quite complicated, usually involving courtship and nest construction. **Courtship** is the male bird's attempt to attract a mate. Some species attempt to catch the eye of a mate with their dazzling appearance; thus, male birds are generally more colorful than the females. The female peacock (or peahen), for example, is a dull-brown bird. The peacock (male), however, has an iridescent blue head and neck and has extravagant tail feathers. A more common courtship tool is the bird's song, which enables the animal to call for a mate of its own species.

Birds are oviparous, producing amniotic eggs. The male reproductive organs, a pair of testes, produce sperm. During mating, the sperm are transferred to the cloaca of the female. The female reproductive system usually consists of a single ovary for the production of ova and an oviduct that transports the ova to the cloaca.

Fertilization occurs within the oviduct. The fertilized ovum is mostly yolk, the food source for the developing embryo. The embryo appears as a small nucleus in the yolk. The yolk is then coated with a protein-containing substance called **albumen** (al BYOO mun), the egg white, and finally is encased in a hard shell by special glands in the oviduct. Thick spiral fibers of the albumen, called chalazae, keep the yolk with its attached embryo suspended securely in the center of the egg. The shell may range in color from white to almost black, and it may have markings like spots, splotches, or streaks. While a bee hummingbird egg is the size of

18-10
Often, the showier bird is the male.

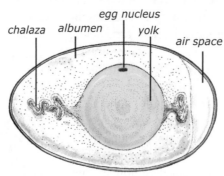

18-11
A chicken egg

chalaza   albumen   egg nucleus   yolk   air space

albumen: (white)

Bird Nests

Most birds construct nests to serve as nurseries for their developing young. The bird accomplishes this remarkable feat using only its bill and feet. The design, location, and materials of the nest vary with each species. The flamingo nests on the ground by building a large volcano-shaped mud mound into which it lays its single egg. The grebe builds a floating, raftlike nest on the water. Many woodpeckers use their chisel-like bills to excavate nest cavities deep inside trees, while the burrowing owl makes its nest in the burrow of a squirrel or prairie dog.

Weaverbird   *Emperor penguin*

Often the nest design serves to protect the eggs from predators. Weaverbirds of Africa construct intricate ball-shaped nests. Long strands of plant fiber are woven like a basket, and the birds even tie knots to hold it together. The entrance to the nest is usually at the bottom to prevent rainwater and predators from entering. As added protection, the nest hangs from a branch too thin for most predators to climb.

The male and female hornbills select a suitable hollow tree; the female enters the hollow, and both birds wall up the opening with clay and debris, sealing the female inside. The hornbill lays,

Typical cavity nest

incubates, and hatches the eggs within the closed chamber. The male passes food to the female through a tiny hole in the clay "door." After a while the female chisels her way out of the "nursery chamber." The young birds promptly reseal themselves into the chamber, where they will remain for weeks.

Not all birds follow typical nest-building, young-raising patterns. In Antarctica, the domain of the flightless emperor penguin, the scarcity of materials makes building a nest impossible. The male penguin, therefore, must balance the egg on his feet, covering it with a fold of skin. He stays in this position for two months without eating or swimming, sometimes losing as much as half of his body weight before the egg finally hatches and the female returns to regurgitate food for the chick. The female brown-headed cowbird lays its eggs in the nests of other birds. The baby cowbird hatches early and shoves all the other chicks out of the nest. The foster parent bird ends up feeding only the cowbird chick, which may quickly grow to be much larger than the "parent." This reproductive strategy enables the cowbird to live in almost any habitat of North America where it can find enough insects for food.

---

a pea, an ostrich egg may weigh 1.5 kg (3.3 lb) and could hold 4700 of the small hummingbird eggs.

Bird Families

Eggs and young birds require parental care. Most bird eggs must be incubated (kept warm) during the development of the embryo. Usually one or both parents accomplish this by sitting on the eggs until they hatch. Depending upon the species, the chick emerging from the egg fits into one of two categories.

✦ *Altricial* (al TRISH ul) *chicks* usually hatch in less than two weeks and emerge from the eggs naked, blind, and helpless. When hatched, all they can do is open their large mouths and consume food. The parents supply food until the chicks are feathered and able to fly. Altricial birds, such as robins, thrushes, and sparrows, usually produce fewer than six eggs at a time.

✦ *Precocial* (prih KOH shul) *chicks* have longer incubation periods, sometimes as long as a month. When hatched, the young are well developed, alert, and able to move and feed themselves. Their bodies are covered with soft down. Though independent compared to altricial birds, these chicks usually stay close to their parents for protection. Precocial birds, which include ducks, quail, and most water birds, often have families numbering fifteen to twenty. Their nests are more likely to be found on or close to the ground.

Most birds have a reproduction cycle of one year, with a different mate each year. Some birds, however, mate for life. Eagles, for example, usually choose a mate as they reach maturity and remain with that partner until one of them dies, which for some may be ten to twenty years. Eagles, unlike most birds, usually build only one nest, which they refurbish throughout their lives. For some eagle families, the young birds are still around watching the parents hunt when the next set of eggs is laid a year later.

18-12
Robin chicks are altricial, while geese have precocial chicks.

With the exception of the fish, the class Aves is the largest vertebrate group. Over thirty-five birds are singled out in the Bible, making Aves the group with the largest number of individuals represented in Scripture. The list of unclean birds in the Jewish dietary laws (Lev. 11:13–19) is the most complete list of any group of animals in the Bible.†

The living members of the class Aves have been organized into as many as thirty orders. For convenience, this text categorizes birds into general groups according to appearance, activities, and habitats.

Flightless Birds

Though all birds have wings, there are many flightless birds. The ostrich is the tallest living bird, reaching a height of 2.5 m (8 ft). It lives in the dry African grassland. Its slender, flexible neck and long, powerful legs are naked, but the body is covered with long, luxuriant feathers, often used to adorn hats. A passage in Job 39 describes the fact that the female will often cover her eggs with sand during the day when she goes feeding. Ostriches can kick powerfully and can cut off fingers with their bills. Swift

Ostrich female (left) and male (right)

runners and fierce fighters, they do not hide their heads in the sand when threatened; however, when surprised, they will often place their heads near the ground, looking carefully at whatever disturbed them. Other flightless birds include emus and kiwis. The elephant bird of Madagascar, a creature that became extinct in the mid-1600s, stood 3 m (10 ft) tall and probably weighed 455 kg (1000 lb).

Penguins are fitted in a "tuxedo" of tiny, dense feathers and a layer of fat under their skin, enabling them to thrive in the icy waters and on the ice floes of the Southern Hemisphere. Using paddle-shaped wings, penguins swim gracefully. Being social birds, they usually travel in large flocks.

† This list also includes the bat, a "fowl" that is classified not as a bird but as a flying mammal.

Birds of Prey

The birds of prey are certainly the "kings of birds." The power, grace, and deadly efficiency of these aerial hunters have won them the admiration of mankind. Birds of prey can be placed into three groups: daytime hunters, carrion feeders, and night hunters.

Eagles, hawks, and falcons are daytime hunters. Their basic method of hunting involves sitting upon a high perch or floating in the air and scanning the surroundings with keen eyes. When food is sighted, they begin a high-speed dive, and then, with their razor-sharp talons, they grip the prey and snatch it from the ground or the water. The force of the dive may stun or instantly kill the victim. The hunter may feed where it is or carry its meal to some inaccessible spot before eating.

The eagle is often mentioned in the Bible. The imperial eagle, now scarce in Palestine, and the golden eagle were probably the most common during Bible times. In the Bible are references to eagles as majestic birds (Ezek. 17:3), as builders of high solitary nests (Jer. 49:16), and as strong, fast fliers (Exod. 19:4; Deut. 28:49; Prov. 23:5; Jer. 4:13; Rev. 12:14). Daytime birds of prey often seen in North America include the large golden eagle, having a wingspan of 2 m (6.5 ft), and the red-tailed hawk.

Vultures generally have a bad reputation because they feed on carrion. The griffin vulture, the largest raptor in Israel and probably the most common vulture in Palestine, is the one most often referred to in Scripture. Until a few years ago the vulture population in Palestine was quite large, but the modern practices of hunting and proper disposal of dead bodies have considerably reduced the vulture population. These vultures are also poisoned by eating rats that have died from poison

Turkey vulture

Great horned owl

put out by farmers. The turkey vulture, or "buzzard," is a bareheaded scavenger often seen circling the kills of other predators in North America. All vultures serve an important role in helping clean up dead animals.

As the eagles and hawks dominate the day, so the owls rule the night. The owl has special equipment for night hunting: soft flight and body feathers enable the bird to fly and dive upon its prey without making a sound; large eyes in the front of the head give it keen vision and depth perception; amazing night vision and keen ears can detect the slightest movement in the grass. Common owls include the barn owl, with its white heart-shaped face, and the great horned owl, whose "horns" are actually tufts of feathers.

Game Birds

Game birds, such as the turkey, quail, pheasant, partridge, dove, and chicken, frequently appear on the dinner table. Their tender breast meat, or "white meat," is actually their flight muscles. The white meat lacks the abundant blood supply (with its supply of food and oxygen) that the other muscles have. Thus the flight muscles of these birds fatigue quickly, allowing them to fly only in quick bursts before tiring and gliding to a landing.

A domesticated game bird often mentioned in the Bible is the dove. The word *dove* is used loosely for all the small members of the family Columbidae, and the word *pigeon* is used for all the larger members. *Turtledove* refers to members of a particular genus of doves. The

Rock dove

writers of Scripture used the dove as a symbol of beauty (Song of Sol. 1:15; 5:12) and gentleness (Matt. 10:16), as an illustration of panic (Hosea 7:11), and as an example of lamenting—because the cooing of a dove sounds like human sighs (Isa. 38:14; 59:11; Ezek. 7:16).

Doves and pigeons are easily raised and were a common food source for the poor in Bible times. They were also sacrificial animals. When Mary and Joseph presented Jesus at the temple, they offered two turtledoves in accordance with their social status and the Law, which states, "If she be not able to bring a lamb, then she shall bring two turtles [turtledoves], or two young pigeons; the one for the burnt offering, and the other for a sin offering" (Lev. 12:8).

Water Birds

The water birds fit into three groups: the swimming birds, the diving birds, and the wading birds.

Ducks and geese, familiar swimming birds, have oval-shaped bodies supported by a pair of short legs with webbed feet. This group includes the mallard duck with its distinctive green head; the wood duck with its "hood" of feathers; and the beautiful mute swan, a pure white bird with a long, elegant neck.

The gannet, a diving bird, can soar to a height of 90 m (295 ft), arch over, and plummet headlong toward the water. Open wings perform subtle course adjustments until impact, when the wings fold against the body. The bird pierces the water and snaps up a fish. The pelican is a comical diving bird with a pouch on its lower bill for holding its catch. It plunges into the water from much lower heights and scoops up its scaly meal, allowing the excess water to drain out.

Gannet diving sequence

The final group contains the wading birds such as the heron, flamingo, crane, and egret. With long necks and long, stilt-like legs, these birds can see above the plants along the water's edge while they carefully look for fish and other organisms that may be snatched, speared, or filtered from the water.

Songbirds

Sparrows, wrens, orioles, and robins belong to the largest group of birds—the songbirds. Their common names, such as chickadee, pewee, cuckoo, and towhee, are actually imitations of the birds' songs. Many songbirds have short bills for eating seeds and berries. Others consume insects, and a few songbirds are predators. Male songbirds attract females with their vocal abilities.

The male cardinal is a dazzling red songbird with a bold crest on its head. The female of the species is drably colored. The mockingbird has a gray and uninteresting appearance, but its song and mimicry are brilliant. A menace to any cornfield, the crow (a black bird with a long beak) feeds upon the crop, supplementing its diet of insects, berries, young birds, and even small mammals. Unfortunately for farmers, crows are intelligent birds unintimidated by scarecrows.

Most songbirds are small, but the raven sometimes measures 69 cm (27 in.) long. Other members of the same family include the crow, rook, magpie, and jay. Ravens are frequently referred to in the Bible. Noah first sent the raven, a strong flier, from the ark. The raven did not return because it was able to sustain itself, a sign that the waters were receding (Gen. 8:6–7). God chose ravens to feed Elijah by the brook Cherith (1 Kings 17:2–6). While the raven is thought to be one of the most intelligent of all birds, it is also one of the least likely to give up food since it hoards food for later meals. That is why ravens can often be seen carrying large morsels in their beaks. This makes the miracle of God's provision through the ravens' actions even more remarkable.

Domesticated Game Birds

Benjamin Franklin described the turkey as "a bird of courage, [that] would not hesitate to attack a grenadier of the British Guards who should presume to invade his farm yard with a red coat on." Had Franklin had his way, the turkey rather than the bald eagle would have become America's national emblem. He disliked the eagle due to its tendency to occasionally steal food from other birds and to scavenge on dead animals. Although the turkey is a fine bird, it would seem a little unpatriotic to eat the national symbol for Thanksgiving.

The red jungle fowl is a ground-dwelling bird of Southeast Asia. Its head and bill are decorated with red flaps of tissue. Its body feathers range from gold to black. Many centuries ago these birds were domesticated or selectively bred by man and trained to fight one another, perhaps as the first spectator sport. Through careful breeding, they became a major source of meat and eggs. By 1500 BC, the bird had spread into Central Europe. Today's descendants of the red jungle fowl are common barnyard chickens. By New Testament times, the chicken was a common domestic bird in Israel. Probably the best-known scriptural reference to a rooster is the one in Matthew 26, which records that Peter denied his Lord three times before the cock crowed.

Red jungle fowl   *Domestic rooster*

18-13
Some birds migrate in formation.

Bird Migrations

The ideal place for raising offspring may not provide suitable year-round conditions. Some birds overcome this dilemma by migrating. Many animals *migrate* from summer to winter feeding grounds, but few animals make journeys as long and dramatic as those of birds. Nearly half of all bird species in the Northern Hemisphere travel south in winter and north in summer. Migration differs from other movements of birds because it is seasonal, predictable, and repeated on an annual basis. The purpose of the migration is twofold.

- Migration enables birds to live year-round in warm climates where food remains abundant enough to supply their enormous needs.
- Migration provides the best possible environment for raising young. Most birds nest in northern ranges where the summer days are long and predators are few.

It is well documented that birds monitor the length of days to determine when to migrate. The time, route, destination, and other aspects of bird migration are inborn. These God-given instincts are necessary for the survival of many bird species.

The mechanisms birds use for navigation during their migrations are not completely understood. Birds may use landmarks to guide them, but some species travel over open water with no such guides. Experiments show that birds have an instinctive sense of direction. Some species appear to be able to detect the magnetic field of the earth, and some use the sun and stars as guides. These guidance systems, though highly efficient, are not perfect. Occasionally, birds can become lost in fog or be blown off course by high winds.

Many birds cover great distances in their seasonal journeys. The arctic tern covers the greatest distance, traveling 17 700 km (11 000 mi) from its breeding grounds north of the Arctic Circle to Antarctica in winter and returning the next summer. The bobolink breeds in the upper regions of North America. During winter migration it covers over 11 000 km (6820 mi), finally coming to rest near Argentina. Such long flights require a tremendous reserve of energy. For this purpose, some birds store fat composing 50% of their body weight. Fat not only releases a high amount of energy, but its breakdown also produces water, protecting the bird from dehydration.

Jeremiah, when complaining that the Israelites were disobeying God, wrote, "Yea, the stork in the heaven knoweth her appointed times; and the turtle [turtledove] and the crane and the swallow observe the time of their coming; but my people know not the judgment of the Lord" (Jer. 8:7). Today these migratory birds are common passersby in Israel. The arrival of the stork is anticipated as a harbinger of spring in the Holy Land, much as the robin's arrival is in the northern United States.

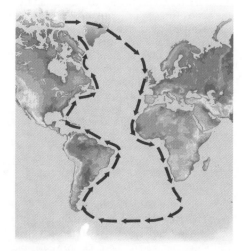

18-14
The arctic tern is the current champion of long-distance migration.

➡ *Review Questions 18.3*

1. What are the two most common methods male birds use to attract mates?
2. What is the yolk of a bird's egg? What is the proper name of the egg white, and what is it made of?
3. Describe and give examples of the two major types of chicks in bird families.
4. What are the two major reasons a bird migrates?
5. What triggers the migration instinct in birds?
6. List three flightless birds, three birds of prey, three game birds, three water birds, and three songbirds.
7. What are the three groups within the birds of prey? Give an example of each.
8. What are the three groups within the water birds? Give an example of each.

18B—Class Mammalia: The Mammals

Most people think that the class Mammalia (muh MALE yuh) contains the "real" animals. Dogs, cats, lions, horses, cows, bears, giraffes, elephants, and monkeys are all mammals. Mammals fill zoos, serve as beasts of burden, are eaten as food, can be trained to do tricks, are loved as pets, and have been written about, photographed, and painted. No other single group of organisms affects humans in such diverse ways.

18.4 Characteristics of Mammals

The **mammals** make up one of the smallest vertebrate classes; however, the class encompasses animals of great variety that inhabit virtually every habitat imaginable. The elephant and rhinoceros are giants of the land; whales and porpoises are fishlike mammals. Bats are the mammals of the air. In the trees are the squirrels, on the land are deer and rodents, and underground are the moles. These strikingly different animals share many characteristics.

Hair of Mammals

All mammals have hair, even though it may be hard to find on some of them. Hair is sparse on elephants and nearly absent on whales and some other aquatic mammals. Hair, a collection of nonliving cells filled with filaments of the protein *keratin*, is produced by structures located in the skin called hair follicles. Two types of hair usually compose a mammal's fur coat.

- The *underhair* is a soft, insulating fur layer next to the animal's skin.
- The *guard hair* is coarser, longer, and found over the underhair. This layer gives the animal its color.

Hair or fur can serve a number of functions. The insulating ability of fur is best illustrated by the beaver's fur, which has underhair so dense that water does not penetrate it. When the beaver enters an icy stream, it never gets soaked to the skin. On cold days many land mammals fluff their fur to trap more insulating air in the underhair.

The hair may also provide camouflage. Some weasels have a dark coat in summer, but they shed that hair in the fall and grow a new white coat for the winter. This change enables the animal to remain inconspicuous against its background.

The "whiskers" of animals such as cats, dogs, and seals are enlarged hairs that have follicles supplied with sensory nerves. These "whiskers" permit the animal to detect objects in its path or around its head in total darkness.

Hair may also be a means of repelling attackers, as the porcupine aptly demonstrates. Some of this rodent's hair is in the form of long, sharp quills that separate easily from its skin. When approached by a predator, the porcupine erects and rattles its quills ominously. Any wise would-be attacker will seek food elsewhere. A predator that ignores the porcupine's warning will probably find itself pierced by the barbed quills. These quills are difficult to remove and tend to work themselves deeper into the victim as he moves.

Limbs of Mammals

Mammals usually possess two pairs of limbs used for locomotion. Their forms, however, are as varied as the animals themselves. The forelimbs of the

Objectives
- List and describe the defining characteristics of the class Mammalia
- Name the three main types of mammal teeth and their functions
- Distinguish between the three main types of mammalian reproduction
- Give examples of both marsupials and monotremes

Key Terms

mammal	marsupial
rumen	monotreme
cud	uterus
cecum	estrus
diaphragm	gestation
larynx	mammary
placenta	gland

18-15 Mammal hair may be used as (a) insulation—grizzly bear, (b) sensors—tiger (whiskers), or (c) weapons—porcupine.

Mammalia: (L. *mamma*—breast)

18-16 ———————
Kangaroos use their powerful hind legs and tail to travel rapidly across the outback of Australia.

bat, for example, have elongated fingers that are connected by a thin membrane to form a wing. Short, stout limbs, wide paws, and large claws serve the mole as shovels. The mole can burrow underground at a rate of 3.5–4.5 m (11–15 ft) in one hour. The whale has forelimbs fused into massive swimming paddles to compensate for a complete lack of hind limbs. For high-speed travel, the large, heavily muscled hind limbs of the kangaroo enable it to leap up to 9 m (30 ft).

The cheetah holds the overland speed record, attaining speeds over 95 km/h (60 mph). The speed is fast when compared to the 69 km/h (43 mph) of a racehorse or the 25 km/h (15 mph) of a sprinting human.

Limbs can also be feeding tools. The paws of the lion, a major predator of the African grasslands, have needle-sharp retractable claws that tear into prey. The gorilla has heavily muscled arms that attain a combined span of almost 2.5 m (8 ft) and that are powerful enough to bend a 5 cm (2 in.) diameter tempered steel bar. The gorilla, however, is a vegetarian and uses its awesome strength to climb trees and tear down foliage.

Digestion in Mammals

If something is edible, there is probably some mammal that eats it. Mammals eat grass, leaves, fruits, seeds, bark, tree sap, microscopic organisms, blood, honey, invertebrates such as insects and snails, and vertebrates, including other mammals.

Unlike most animals studied to this point, most mammals have a variety of specialized teeth. While a few mammalian species have no teeth and survive by crushing food with their gums, nearly all mammals have several types of teeth arranged in jaws. In many cases, the types and numbers of teeth are used to classify them into certain orders and families. A scientist can learn a great deal about the diet of a mammal by examining its teeth. There are three basic types of mammalian teeth:

- *Incisors* (in SYE zurz) are flat, thin teeth in the front of the mouth, used in gnawing or biting.
- *Canines* (KAY nines) are rounded, pointed teeth toward the front of the mouth, used for tearing.
- *Molars* (and premolars) are usually thick, squat teeth in the back of the mouth, used for grinding and chewing.

Carnivores such as lions and dogs have enlarged canine teeth. These animals bite into their prey and rip off large pieces of meat to be swallowed whole. Gnawing animals like rats and beavers have large incisors and no canines. Herbivores cut foliage and grind it until it can be easily digested. Thus, the cow, horse, giraffe, and camel have sharp incisors and large flat molars. Omnivorous animals such as monkeys and bears have well-formed teeth of each type.

Digestive organs also vary among mammals with different diets. Some herbivorous mammals, such as cattle, obtain nourishment from cellulose. No vertebrate, however, can digest cellulose without help. Plant material ingested by the cow passes down the esophagus to the **rumen**, a special section of its multichambered stomach. Animals that have a rumen are called *ruminants*. Bacteria maintained within the rumen produce enzymes that digest cellulose.

incisor: in- (in) + -cisor (to cut)
canine: (L. *canis*—dog)

For improved digestion, the ruminants regurgitate **cud** (partially digested food) into their mouths and chew it a second time, mixing the enzymes from the rumen with the tough plant materials. They then swallow the food again. Ruminants have long digestive tracts, allowing the digestive system more time to digest cellulose.

While cattle, deer, camels, and giraffes all ruminate, some herbivores do not have such a complex digestive system and do not chew cud. They do, however, have the ability to break down large carbohydrate molecules into forms that can be absorbed by keeping the food inside for a longer period. The manatee, one of the few herbivores in coastal waters, has 45 m (150 ft) of intestines. Manatees, horses, elephants, and many other nonruminant herbivores also have a **cecum**—a pouchlike extension between the small intestine and large intestine—that contains bacteria and acts as a fermentation tank for the digestion of the cellulose.

Carnivores, such as the seal and wolf, eat primarily protein. Protein is easier to digest than cellulose and does not require special stomachs or cultures of digestive bacteria. The carnivore's digestive tract is considerably shorter than the ruminant's and has an unsegmented stomach similar to that of a human.

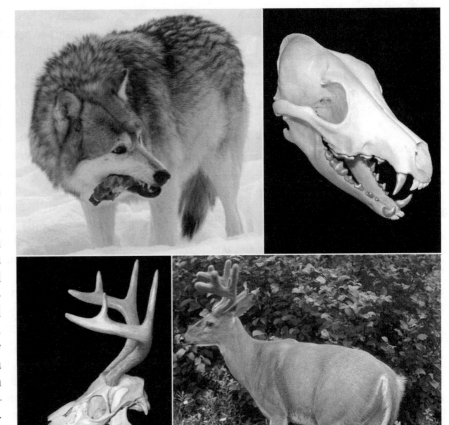

18-17
A wolf has teeth for killing and tearing, while a deer's teeth are ideal for browsing and chewing.

- [ ] incisors
- [ ] canines
- [ ] molars

Respiration and Circulation in Mammals

The major organs of mammalian respiration are a pair of oblong spongy lungs in the upper chest cavity. The lungs are separated from the abdominal organs by the muscular **diaphragm**. Air is drawn into the lungs by the contraction of the dome-shaped diaphragm. As air passes down the throat, it goes through the **larynx** (LAHR ingks), or voice box, a cartilaginous organ containing vocal cords. Controlled vibration of these cords enables various mammals to bark, squeal, meow, grunt, or make other sounds.

Like the avian heart, the mammalian heart has four chambers. Mammals have a circulatory system similar to man's (see Sec. 22A).

Mammals expend energy to maintain a constant body temperature, enabling them to function steadily at a high metabolic rate, even in varying climates. This constant temperature is made possible by various heat-regulating structures in the mammal's body. Mammals in frigid habitats possess mechanisms and behaviors to conserve warmth. The polar bear, for example, is designed to conserve heat. Its small head, tiny ears, and compact body shape expose a minimum amount of body surface to the cold. A coat of dense, transparent, hollow hairs provides heat-conserving insulation.

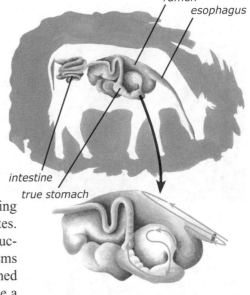

rumen
esophagus
intestine
true stomach

18-18
The ruminant digestive system showing the path of food

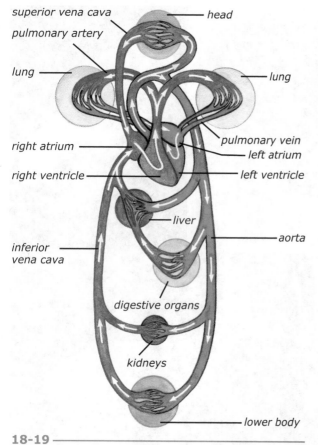

superior vena cava
pulmonary artery
lung
right atrium
right ventricle
inferior vena cava
head
lung
pulmonary vein
left atrium
left ventricle
liver
aorta
digestive organs
kidneys
lower body

18-19
The mammalian circulatory system

18-20
An elephant loses excess heat through its oversized ears, while a dog loses heat through its tongue.

Mammals in hot environments have mechanisms and behaviors to eliminate excess heat. Many mammals lose excess body heat through their skin, but occasionally, certain mammals must use other methods. A dog pants to cool the warm blood that is directed to the tongue and mouth lining when the animal is overheated. The African elephant has a thick skin that not only serves as protection but also holds heat. A network of vessels in the elephant's huge, thin ears, however, brings the blood near to the body surface and can cool the blood as much as 5 °C (9 °F) in a single passage. Bathing and wallowing in mud also cool the elephant.

Many small mammals escape the low temperatures and reduced food supply of winter by entering *hibernation*. Ground squirrels and woodchucks enter a state of true hibernation. After filling itself and its den with food, the mammal gradually reduces its body temperature, sometimes as much as 30 °C (54 °F). Its metabolism slows to a point where the animal becomes unconscious. The mammal's heart and respiration rates drop drastically. Occasionally during this period, the animal awakes to pass its wastes and then quickly returns to sleep. An animal in true hibernation needs hours to raise its metabolism back to its normal, active state.

Some large mammals like the bear and badger sleep or become *dormant* through the winter but do not truly hibernate. Their heart rate slows, but their body temperature remains close to normal. Unlike true hibernators, these animals waken if bothered.

Response in Mammals

The *cerebrum*—the center of intellect and instinct—dominates the mammalian brain, making mammals the most intelligent animals. The cerebrum is also the seat of the memory, which accounts for the trainable nature of most mammals. The tussling of lion cubs or the play of a baby chimpanzee is similar to the activity of young children. The care, protection, and training most mammalian parents give their young have counterparts in human society. Many animals also appear to express emotions such as fear, anger, contentment, excitement, and happiness.

Though these mammal-human comparisons are interesting and in many cases valid, care must be taken not to read too much into them. A mammal, though intelligent, is not capable of the complex emotions and elaborate thoughts that characterize human mental activity. (These contrasts are discussed in Section 20A.)

The mammalian brain is much larger in proportion to its body than the brain of any other land animal. While mammals rely to some extent on all of their senses, most have a particular dependence on one or two for their daily survival.

The sensory organs of mammals are, for the most part, similar to those of man; their sensitivity, however, may be quite different. The bat's hearing, for example, is considerably more sensitive than human hearing. Its large cup-shaped ears are the receiving devices for its sonar system. The bat produces high-frequency sounds that echo off objects in its flight path. By listening to the echoes (echolocation), the bat can avoid obstacles without using its eyes.

A familiar example of keen mammalian smell is the legendary bloodhound. These dogs can follow the scant traces left by an animal or person walking through the woods. The police also use trained dogs to sniff out and locate hidden explosives and drugs. Because most mammals are keenly attuned to smells, many of them possess glands that produce substances that mark their territory.

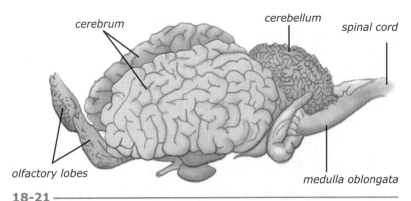

18-21 —
The brain of a mammal (horse)

Reproduction in Mammals

The most familiar mammals have young that develop internally, nurtured by a **placenta**. The placenta is the interface between the mother and the developing offspring through which gases, nutrients, and wastes are exchanged. Some mammals, however, may produce young without the use of a placenta. These include the marsupials (mar SOO pee ulz) (pouched mammals) and monotremes (MAHN uh TREEMZ) (egg layers). Their unique variations on reproduction are discussed in the next section.

Placental Mammals

Although there is a great range in size and body form for placental mammals, they are quite similar in their reproduction. Periodically the female mammal's paired ovaries produce and release ova, which are guided through a tubular oviduct toward the **uterus**. The uterus is a muscular chamber where the young develop. When the female's ova are ready to be fertilized, the animal enters a period called estrus, or "heat." Various odors released by the female at this time attract males of her species.

Among mammals, the male's testes produce sperm. The sperm are carried in a fluid that allows them to swim up to the egg, where fertilization takes place. All mammals practice internal fertilization. In most mammals the sperm fertilizes the egg in the oviduct. Even before the fertilized egg, or zygote, reaches the uterus, it has divided several times. These first few cells of the new mammal are implanted in the uterine wall. A placenta, composed of a portion of the uterine wall and tissues of the embryo, forms. The placenta has a rich blood supply from both the mother and the embryo. The two blood supplies exchange nutrients, gases, and wastes, even though they do not actually mix.

The blood vessels from the placenta to the ventral surface of the embryo are coated in protective membranes. These blood vessels and the membranes form the *umbilical* (um BIL ih kul) *cord*.

While an embryo is in the mother's womb (the uterus), it receives nourishment from her. This period of pregnancy is gestation (jeh STAY shun). The length of the gestation period varies greatly for different mammals. Generally speaking, the longer the gestation, the more developed the young are when they are born. Most mice have a gestation period of only 21 days; the young are born hairless, blind, and weak. The horse has a gestation period of 335 days; a foal is born with a full coat of hair and can stand, walk, and even run within hours of its birth.

Parental care continues after the young are born. The first order of parental business is to feed the young. The female supplies milk from **mammary** (MAM uh ree) **glands**, usually through a series of nipples along her ventral

18-22 —
A dog's sense of smell aids law enforcement.

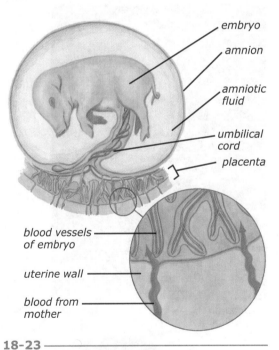

18-23 —
Diagram of an embryonic pig

marsupial: (L. *marsupium*—pouch)

monotreme: mono- (single) + -treme (Gk. *trema*—perforation)

estrus: (L. *oestrus*—frenzy)

umbilical: (L. *umbilicus*—navel)

gestation: (to carry or to bear)

koala

Tasmanian devil

kangaroo with joey

18-24
Marsupials are a diverse group.

surface. Milk is a substance rich in nutrients and antibodies to help the off-spring grow strong and healthy.

Many mammals train their young to hunt, select fruit, build shelters, and perform various other actions before they leave the "family." While some small mammals such as rodents may mature and leave "home" in a few weeks, orangutans and some other large primates mature slowly and may stay with their mother for several years.

Marsupials: The Pouched Mammals

Kangaroos, koalas, and other marsupials produce young without a placenta. The fertilized egg begins growing and dividing within the uterus but does not implant itself or form a true placenta. Instead, it is nourished by a small yolk sac. After a short time the yolk sac is depleted and the immature offspring crawls out of the uterus and onto the mother's fur. Instinctively this tiny creature moves toward the mother's pouch, where the nipples are. Often the mother will lick a groove in its fur for the young to follow. This tiny creature is hairless and bears little resemblance to the adult. If it survives the perilous journey and makes it to the pouch, the young animal locks its mouth over a nipple. Sustained by the mother's milk, the young marsupial completes its development within the pouch. Most young marsupials begin to leave the pouch and explore the world before they are totally *weaned* (stop drinking the mother's milk), and for a period they enter and exit their "nursery" until they become more independent.

The male red kangaroo, the largest-known living marsupial, stands about 1.5 m (5 ft) tall and may weigh 82 kg (180 lb) when full grown, but it is smaller than a man's little finger when it is born. After only thirty-three days of gestation, it spends up to seven months in the pouch before the mother evicts it to make room for the next joey. Because the new sibling is small and there are multiple nipples for feeding, the older joey may continue to nurse until its first birthday.

The koala is a tree-dwelling, tailless animal with a large head, furry ears, and handlike paws. Its resemblance to a teddy bear has endeared it to many people. The koala feeds only on the leaves of eucalyptus trees. Unlike the kangaroo, the koala has a pouch that opens to the posterior.

America's Marsupial

One of the few marsupials living outside Australia is the opossum—a common inhabitant of American woodlands. This mammal, looking much like an overgrown rat, is known for its ability to "play dead." When approached by a potential predator, the animal often flops onto its side and appears to be dead—its eyes glassy, its mouth locked in a hideous grimace. When the predator departs, the opossum ambles away. The opossum, however, is a dangerous animal; its sharp teeth can inflict a painful bite. The opossum searches for its mammal, bird, egg, or insect prey at night, but it also eats carrion and fruit. The mother opossum produces from six to twenty young that leave the womb after about two weeks. They stay in the mother's pouch for about two months.

Young opossums ride "possum-back" after they mature enough to leave the pouch.

It appears that many placental mammals have marsupial counterparts, because there are also some marsupials whose habits and habitats mirror those of moles, rabbits, monkeys, rats, anteaters, and various others. Some speculate that marsupials have survived in and around Australia because they are not in competition with placental mammals, which dominate the rest of the world.

Monotremes: The Egg-Laying Mammals

The monotremes are the exception to many mammalian characteristics. The very name *monotreme* indicates that they have a single perforation or hole for the digestive, excretory, and reproductive systems. Members of this single order are also the only mammals to lay eggs and incubate them like birds. Once hatched, the young feed on their mother's milk. Female monotremes, however, do not have nipples. The milk empties onto the skin, and the young lap the milk off the mother's fur.

The best-known monotreme is the duck-billed platypus. The platypus reaches about 46 cm (18 in.), including its broad, flat tail, which it uses in swimming. The platypus uses its toothless ducklike bill to dig for invertebrates in the mud and its webbed, clawed feet to dig burrows up to 30 m (100 ft) long in stream banks. The male's hind legs have sharp spurs linked to poison glands, making the platypus one of the few known venomous mammals. In a moss-lined "nest" hidden in the burrow system, one or two tiny hairless platypus young hatch from eggs. Here they stay for three to four months until they begin to swim and feed on their own. They live in streams, rivers, and some lakes along the east coast of Australia.

The two species of echidnas, or spiny anteaters, also lay eggs, but with a twist. The single egg is laid directly into a shallow pouch on the lower abdomen while the mother lies on her back. Here, the young hatches about ten days later. It laps milk from patches near the opening of the pouch. The sharp spines do not develop until the period when the youngster is leaving the pouch.

18-25
The duck-billed platypus (top) is found only in Australia, while the echidna (bottom) is native to Australia and New Guinea.

Review Questions 18.4

1. What characteristics separate the class Mammalia from the other vertebrate classes?
2. What are the two primary types of hair found on most mammals? Of what value is each to an animal?
3. List, describe, and give the functions of the three basic types of teeth found in animals.
4. How do ruminants digest cellulose?
5. Name and describe two structures of the mammal's respiratory system that differ from those of the bird's respiratory system.
6. List and describe three structures important during the period of gestation.
7. Tell how the development of a kangaroo and the development of a platypus differ from the development of a typical placental mammal.
⊙8. Why are monotremes considered mammals instead of birds?
⊙9. Compare altricial and precocial mammal young.
⊙10. Differentiate between true and false hibernation in mammals.

18.5 Classification of Placental Mammals

Placental mammals living today are divided into as many as twenty orders based primarily on the type of reproduction, the type of teeth, and the design of their limbs. Some of the best-known orders are presented here.

Order Rodentia: Gnawing Mammals

Order *Rodentia* is the largest mammalian order, including such animals as mice, squirrels, porcupines, and beavers. Approximately 1800 species are

Rodentia: (L. *rodere*—to gnaw)

▶ Objectives

- List and describe traits that distinguish Primates from other mammalian orders
- Compare the ten other mammalian orders covered in the text, giving several distinguishing traits for each

▶ Key Terms

blowhole
opposable thumb
bipedalism
prehensile tail
ungulate

18-26

The beaver is the largest North American rodent.

spread across every continent except Antarctica. Large chisel-like incisors in both the upper and lower jaws characterize the animals in this order. The rodent uses these teeth for biting and gnawing. A rodent's incisors continue to grow throughout its life, allowing the animal to gnaw almost incessantly. In fact, the rodent must use its incisors, or they grow so long that the animal cannot close its mouth. Hard enamel covers the front of the rodent tooth; the back consists of softer material. As the animal gnaws, the back of each incisor wears away faster than the front, keeping the teeth sharp.

Most rodents (with the notable exception of porcupines) are defenseless and frequently fall prey to bird, reptile, and mammal predators. How then do these animals keep from being wiped out? The secret lies in reproduction: rodents become sexually mature sooner, can produce larger litters, and can reproduce more frequently than almost any other mammal. Field mice, for example, are weaned at three weeks, are independent at four weeks, and can bear litters at six weeks. Captive meadow mice can produce thirteen to seventeen litters in one year, yielding a total of over seventy-five offspring. This reproductive ability has allowed the house mouse, *Mus musculus*, to spread from Asia to become the most widespread mammal in the world.

Order Carnivora: Meat-Eating Mammals

The order *Carnivora* (kar NIV uh ruh) includes most of the meat-eating mammals. In a way, they are all "kings" of their domains. Many carnivores are the largest predators of their areas and need not fear being eaten by other beasts. Not all carnivores are large, however, nor are they monarchs living in ease and safety. Carnivores must often fight with one another over kills or territory. Often the carnivore's prey puts up a vicious fight, resulting in the predator's death.

The structure of the carnivore is very different from that of other animals. Its enlarged canine fangs and sharp molars are used for tearing flesh. The carnivore's limbs are usually supple and powerful, providing maximum speed and agility during pursuit. Some have sharp claws for gripping their prey. The most important feature of the mammalian carnivores is their ability to learn behaviors and use intelligence in hunting. To survive, they must know how to catch and kill prey that may be much larger and faster than they are.

The cat family (Felidae), a familiar group of carnivores, ranges in size from the tiny domesticated cat to the Siberian tiger, which averages over 3 m (10 ft) from the nose to the tip of the tail. Members of the cat family are known for their quiet stalking of prey, followed by a high-speed dash and a leap for the kill. Big cats of the Americas include the jaguar, a stout-bodied jungle cat, and the puma, or mountain lion, found in the western United States and in South America.

Bears (family Ursidae) are the largest land carnivores. The largest is the Kodiak bear, a type of brown bear that attains a length of 2.8 m (9 ft) nose to tail and that can weigh over 545 kg (1200 lb). Bears are actually omnivorous, eating whatever is available. When they hunt, bears depend on raw strength and sharp claws.

Cats use stealth, bears use strength, but dogs (family Canidae) use endurance and teamwork. Canines such as wolves usually hunt in packs, surrounding their victim, harassing it, tiring it, and finally killing it. These large

raccoon

jaguar

18-27

Though they come in all sizes, all carnivores have prominent canine teeth.

The Lion in the Bible

The lion, one of the best-known carnivores, is described in Proverbs as having a stately tread and courage, for it "turneth not away for any" (Prov. 30:30). The Persian lion, a subspecies of the familiar, slightly larger African lion, is the lion to which the Bible refers. A Persian lion weighs about 180 kg (397 lb) and is about 1.5 m (5 ft) long with a 1 m (3 ft) tail. Since lions are a threat to domesticated animals, shepherds were often called upon to defend their sheep from these large carnivores, as did David (1 Sam. 17:36). As often happens, these predators of livestock were killed by man and became almost extinct. Today special animal preserves are keeping many biblical animals, including the lion, alive in Palestine.

Several times in the Bible the Lord used the lion, which normally avoids humans, to kill men. The disobedient prophet in 1 Kings 13, the man in 1 Kings 20:36, and the foreign settlers in Samaria in 2 Kings 17:25 were killed by lions.

Although the Bible does not call the lion the "king of beasts," it is often compared to strong or majestic things. Judah is described as "a lion's whelp [cub]: . . . who shall rouse him up?" (Gen. 49:9). Proverbs 28:1 says, "The righteous are bold as a lion," implying that they need fear nothing. Even God Himself is described in terms of lion characteristics, such as "Behold, the Lion of the tribe of Juda, the Root of David, hath prevailed" (Rev. 5:5). On the other hand, Satan in his vicious attack on Christians is also described as a "roaring lion . . . seeking whom he may devour" (1 Pet. 5:8).

carnivores often pick old, weak, or young animals for their prey and thus help prevent overpopulation. An old Eskimoan saying aptly describes this relationship: "The wolf and the caribou are one; the caribou feeds the wolf, but the wolf keeps the caribou strong."

Fearing the loss of his livestock, man has hunted and poisoned many of the larger canines. Wolves were once common throughout North America, Europe, the Middle East, and most of Asia. As man moved in, wolves moved out. Now there are only a few wolves in northern regions. The coyote, however, is still common in the American Great Plains.

Many smaller animals such as raccoons, mink, otters, weasels, badgers, and mongooses are also carnivores.

Order Cetacea: Aquatic Mammals

Aquatic mammals include the fishlike whales, dolphins, and porpoises and are often called *cetaceans* (sih TAY shuns). Unlike seals and walruses, which are also carnivores, these mammals do not come ashore; they even bear their young in the water. The ninety species of cetaceans are divided into two groups—the toothed whales, ranging from dolphins to killer whales, and the baleen whales, which filter invertebrates from the water with a net of baleen in their mouth. The great blue whale, the largest creature on earth, is a baleen whale. The toothed whales feed on squid, fish, and even other whales.

cetacean: (L. *cetus*—whale)

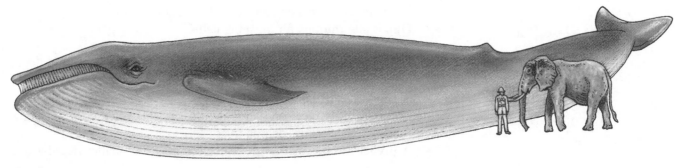

18-28
Giants of sea and land compared to a human

18-29
Unlike fish, marine mammals swim by flexing their tails vertically.

18-30
Most cetaceans breathe through a blowhole rather than nostrils.

Cetaceans steer with a pair of pectoral flippers supported by bones. They may have a small dorsal fin to help stabilize them. The power for movement is provided by a pair of tail flukes that are horizontal rather than vertical like the caudal fin of a fish. Powerful muscles in the tail section of the body pump the boneless flukes up and down, propelling the creature up to 40 km/h (25 mph).

Whales and dolphins must surface to breathe air. Some have a single nostril on the top of their head. This **blowhole** is for breathing, although in colder areas the release of warm breath gives the appearance of water being forced out. The toothed cetaceans find prey and navigate by echolocation. They emit clicks and "songs" that are focused by a special chamber in the forehead called the melon. The returning sound waves are detected elsewhere in the head, perhaps in the lower jaw.

Some researchers believe that dolphins and porpoises are among the most intelligent animals since they are easily trained and appear to have an intricate method of communicating by underwater sounds.

Dolphin vs. Porpoise

While the terms *dolphin* and *porpoise* are often used interchangeably, they are two different groups of cetaceans, each with its own family. In addition to the differences in the table below, dolphins are much more social, often congregating in large groups called pods, and they seem less frightened by man.

To further confuse matters, there is a bony fish from the perch family sometimes called dolphin that appears on many seafood menus. This tasty fish is not the dolphin mammal and is often referred to by one of its other names, mahi-mahi or dorado.

Dolphin vs. Porpoise	Teeth	Dorsal fin	Head shape
Dolphin	conical	taller and curved	beak on snout
Porpoise	spade-shaped	smaller and triangular	more rounded

Order Primates: Erect Mammals

The 235 species of the order Primates (prih MAY teez), the erect mammals, include the great apes, the monkeys, and several smaller animals. Primates share most of the following characteristics:

- Freely moving arms and legs with large hands and feet
- **Opposable thumbs** for grasping items
- Nails on most fingers and toes
- Forward-facing eyes, permitting *binocular vision* with depth perception
- A poor sense of smell
- Ability to walk erect on two legs (**bipedalism**) when on land, though most of their time is spent in trees
- A **prehensile tail** that is useful for gripping for smaller tree-dwelling primates
- A strong tendency to be social, often living in large groups

bipedalism: bi- (two) + -pedal- (foot) + -ism (characteristic)

- Less specialized teeth than those of most other mammals
- A mostly vegetarian diet, often supplemented with eggs, insects, and occasionally other small animals

The gorilla, the largest ape, lives in the dense forests of Africa. Early reports of these apes described terrifying creatures. Indeed, the broad shouldered gorilla is fearsome in appearance. When confronted, the gorilla rears up on its hind legs and beats its barrel chest in an awesome display. For all this frightening show, however, the gorilla is a timid creature.

The chimpanzee is a smaller ape with arms longer in proportion to its body, a more dome-shaped head, and large ears. Many researchers consider chimps the most intelligent of animals. In the wild, chimpanzees are one of the few groups of animals that use tools. They have been observed to use sticks as weapons and blades of grass to "fish" termites out of their mounds.

In experiments, chimpanzees have demonstrated the ability to reason in order to solve minor problems. Additional research has been done in training chimps to communicate. Using sign language, chimps have learned large vocabularies and appear to be able to communicate effectively. Just how far this learning and communication can go remains to be seen.

18-31
Primates come in various sizes and colors.

Lemurs, bush babies (galagos), and tamarins are a few of the many smaller primates, and like most other members of their order, they live in tropical forests. Because tropical forests are being destroyed and primates have such low reproduction rates, the primates are perhaps the most endangered of all the mammalian orders.

Order Perissodactyla: Odd-Toed Ungulates

At one time, all hoofed mammals were in the order Ungulata (UNG gyuh LAY tuh). While that order has now been divided into two distinct orders, the term **ungulate** is still used to describe them. *Perissodactyla* (puh RISS oh DAK tih luh) includes the horse, ass, zebra, tapir, and rhinoceros. Ungulates of this order have one long-hoofed toe or three enlarged toes, as in the case of the tapir and rhinoceros. While all of these mammals are herbivores, they rely on a cecum rather than a multichambered stomach for digestion.

18-32
While this Malayan tapir may resemble a pig, its three-toed feet put it in the order with horses and rhinos.

Perissodactyla: Perisso- (Gk. *perissos*—uneven) + -dactyla (*daktulos*—finger)

18-33
The order Artiodactyla includes herbivores even smaller than these 60 cm (2 ft) Thomson's gazelles and as large as the giraffe at 5.3 m (17 ft).

18-34
Sloths are sometimes ridiculed as being the slowest-moving mammal, but in the rainforest canopy, speed is of little use.

18-35
The thirst for blood of vampire bats is more fiction than truth. Contrary to stereotype, they do not suck the blood but rather make a tiny incision and lap the blood as it oozes out.

Artiodactyla: Artio- (Gk. *artios*—even) + -dactyla (finger)

Chiroptera: Chiro- (L. *chiro*—hand) + -ptera (wing)

Order Artiodactyla: Even-Toed Ungulates

The mammals in the order *Artiodactyla* (AR tee oh DAK tih luh) have long legs with two or four functional hoofed toes. The even-toed hoofed animals are divided into ten different families and about 210 species. Pigs, hippopotamuses, camels, and giraffes are among the smaller families. The deer family, Cervidae (SUR vuh dee), and the cow family, Bovidae (BOH vuh dee), are the dominant groups. Far more than cows, Bovidae includes antelope, gazelles, buffalo, and wild sheep and goats. Cattle, deer, camels, giraffes, antelope, goats, buffalo, and sheep have rumens and, using a biblical expression, "chew the cud" (Lev. 11:4). Some artiodactyls, like pigs, have simple stomachs and do not chew cud.

Order Edentata: Toothless Mammals

Anteaters, armadillos, and sloths are in the order *Edentata* (EE den TAH tuh), which literally means "no teeth." In reality, only the anteaters are completely toothless. The others do have peglike teeth, but the teeth lack enamel. The twenty-nine species of edentates range from the pure insect diet of anteaters to the leafy diet of sloths.

Order Chiroptera: Flying Mammals

The only mammals capable of true flight are in the order *Chiroptera* (kye RAHP tuh ruh). The more than nine hundred species of bats are divided between the large herbivorous ones and the more abundant small bats. The large bats are sometimes called flying foxes and feed mainly on fruit and nectar. They live in the tropics and subtropics of Africa, Asia, and Australia, but not in the Western Hemisphere. The largest flying fox has a wingspan of 1.7 m (5.6 ft). They navigate by sight, not by sound.

Typically, the small bats feed on insects, which are caught during flight. These bats have an unusual sense called *echolocation*, similar to the ability shown by cetaceans. Those that do not catch insects may feed on fruit, nectar, blood, or even fish and amphibians.

Horns or Antlers?

Horns and antlers are characteristic of many even-toed ungulates. In some species, only the males grow antlers or horns, but in others, both sexes have these head ornaments. While they are sometimes used for fighting or protection, among males they probably play a more important role in attracting and keeping mating partners.

Horns are made of bone covered with keratin, are unbranched, and grow larger as the years pass. They are found on members of Bovidae. The bony branched *antlers* of the family Cervidae are shed each year and replaced.

Cervidae   Bovidae

Facets of Biology

Domestication and Dominion

The Formative Years

A survey of farm animals reveals many creatures that, although useful to man, would probably not fare well in the wild. Although some types of horses, swine, and goats have successfully reverted to living in the wild after escaping or being intentionally released, most breeds of domesticated animals depend on humans for their survival. Did Adam and Eve ride horses through the Garden of Eden or collect eggs from chickens? What exactly is the relationship between farm animals and their wild forebears? Did God create some animals just to be used by humans? Although we do not know the answers to all of these questions, Scripture, history, and a comparison of living animals can give us some direction.

The history of domestication is as old as the human race, and for those accepting biblical authority, that started with the very first family. We read in Genesis 4 that Abel was a keeper of sheep (v. 2) and that he brought offspring from his flock (v. 4) to present as a gift to God. The Hebrew text is more general than the word *sheep* implies and probably includes goats as well. Clearly, at this early stage in history, at least one (and possibly more) type of animal had been selected and kept for the benefit of man. By the eighth generation of man, Jabal is described as "the father of such as dwell in tents, and of such as have cattle" (Gen. 4:20).

The Bible does not reveal any details on how the first animals were selected to be raised by humans, but it is possible that God created some kinds of animals with dispositions that made them, through repeated association with humans, more willing to be confined and cared for by people. Soon, most of the common farm animals were being selectively bred to make them more manageable and valuable to humans.

This artificial selection was most likely targeted at enhancing characteristics such as thicker wool coats on sheep, richer milk from goats, and more pulling strength from cows. Through this selective breeding and management of species, man was fulfilling his mission of having dominion over the earth as defined in the Creation Mandate of Genesis 1:28.

Archaeologists, using radiocarbon dating, claim that sheep and goats

Many believe that the sheep was first domesticated from the wild mouflon sheep.

were the first mammals to be domesticated and that this happened less than 10 000 years ago. Pigs, cattle, horses, donkeys, and dogs followed this within a few thousand years. These figures proposed by secular scientists are figures that creationary scientists already know and base their models on—domestication began 6000 to 7000 years ago.

Today, 15 different species of mammals have been domesticated to provide food and labor. Not all are typical farm animals, though, as yaks, llamas, and even Asian elephants have also been bred and raised to serve man.

The Flood and the Dispersion

When God decreed that He would destroy the earth by a global flood, He also graciously put in place a plan of preservation for His creatures. One pair of each land mammal, bird, and reptile was to be saved in the ark. In fact, those animals that were considered clean were to have a population of seven. Most livestock probably fell into this clean category.

Over the centuries since the Flood, man has been propelled by war, natural disasters, the confusion of languages God caused at the Tower of Babel (Gen. 11:6–8), and even curiosity to disperse and populate the far reaches of the planet. As he moved into new areas, domesticated animals certainly accompanied him. Further selective breeding in diverse environments produced breeds that could better tolerate the temperatures and conditions of their new homes.

For instance, the dense shaggy coats of Scotch Highland cattle protect them from extreme weather in their country of origin, but the same coat would be a detriment on the plains of Central Africa. Some African cattle breeds have been selected for heat tolerance and resistance to tick-borne diseases.

Animal breeders are constantly trying to improve their productivity, and this can often be accomplished through crossing separate breeds to see what new characteristics might occur. If a population with im-

Scotch Highland cow

proved traits can be established and is able to consistently pass on those desired traits, it may be accepted as a new breed. Over time, the number of breeds has mushroomed to more than 400 recognized breeds of sheep and over 800 breeds of cattle.

The Future

Because domesticated animals exist to serve the needs of man, breed popularities can rise and fall. In fact, we don't really know what the wild or early forms of many domesticated animals looked like because those traits have been blended into hundreds of generations of breeding. Those breeds of cows, sheep, horses, or pigs that have traits desirable to the greatest number of people will be preserved as their populations expand. Conversely, breeds that become less popular may be allowed to die out— the farm equivalent of extinction.

Today, there is increasing recognition that certain rare breeds of livestock are worth preserving, even if they do not have immediate widespread appeal or commercial value. Many have characteristics that make them ideal for use in regions where more popular breeds cannot compete. Some are suited for small farms or less intensive management. Furthermore, they may carry genes that might be very valuable in future breeding efforts. If popular breeds were to be devastated by disease or they themselves no longer meet market demands, some of the rare breeds might suddenly become valuable. Even if they are not useful in their present form, they might be crossbred to confer new traits.

There are now organizations that encourage the preservation of certain rare breeds. Through becoming aware of the genetic resources in rare breeds and ensuring their survival, man is again serving as an agent of dominion in fulfilling the Creation Mandate.

Order Insectivora: Insect-Eating Mammals

Shrews, moles, hedgehogs, and some lesser-known animals make up the order *Insectivora* (IN sek TIV uh ruh). Although this group contains about four hundred species, it is one of the least studied orders of mammals. Insects are the chief food of many, but some insectivores eat worms, slugs, eggs, carrion, and even fruit. With their small sharp teeth, insectivore skulls resemble miniature carnivore skulls.

Members of this order are characterized by very high metabolism, and some have an almost constant need for food. The sense of smell is acute in this group, and pointy, flexible noses are common. As might be expected in a group that is often nocturnal and that contains many burrowing species, their eyes are usually small and, in some cases, almost nonexistent.

Order Sirenia: Sea Cows

Manatees and dugongs make up the small order *Sirenia* (si REE nee uh). The four species live primarily in coastal tropical and subtropical areas, sometimes moving up into estuaries and rivers. When mariners first discovered them, their slow movements in the water reminded them of mermaids or sirens, hence the order name. With their flat tail fluke, they superficially resemble small whales, but they have many differences. Sirenians are all herbivores, feeding on underwater and floating foliage, and they all breathe through nostrils rather than a blowhole.

Order Proboscidea: Trunked Mammals

Two species of elephants, the African and Asian types, are all that remain of the order *Proboscidea* (PROH buh SIHD ee uh). The mammoth, an extinct member of this order, at one time flourished across Africa, Eurasia, and even North America, perhaps becoming extinct as recently as two thousand years ago. There is strong evidence that Native Americans killed them for food. Other pressures that may have led to their extinction include the climate change that may have occurred following the Great Flood and predators like the saber-toothed tiger.

Elephants today are distinguished by a flexible and useful trunk, which is essentially an elongation of the nose and upper lip. The ivory tusks that attract great attention from poachers are the two upper incisor teeth that continually grow. The only other teeth in an elephant's mouth are four brick-sized molars. Weighing up to 6000 kg (13 200 lb), the bull African elephant is the largest land animal.

18-36

The West Indian manatee, the species found along the Atlantic and Gulf coasts, is often injured or killed by powerboats. This may be due to its slow movements and inability to hear low sound frequencies.

Insectivora: Insecti- (insect) + -vora (to devour)

Proboscidea: Pro- (in front) + -bosc- (*boskein*—to feed)

▶Review Questions 18.5

1. List several characteristics and examples of (a) the order Rodentia, (b) the order Carnivora, (c) the order Cetacea, and (d) the order Primates.

2. What are the two major groups of hoofed mammals? Which contains the ruminants?

3. List several key differences between the large bats and the small bats.

4. Name the largest land animal and the largest animal on the planet.

⊙ 5. How is the domestication of animals an example of mankind obeying the Creation Mandate?

⊙ 6. Which orders other than Carnivora contain meat eaters? Why are these animals not part of the meat-eating order?

19

Ecology

19A—The Ecosystem

In the mid-1800s (about the time Mendel published his findings about inheritance factors in peas), the term *ecology* was first used by the German zoologist Ernst Haeckel. His definition of *Ökologie* included "the investigation of the total relations of the animal both to its inorganic and its organic environment."†

Today, ecology encompasses the study of limited portions of the environment, as well as the impact of man upon those ecosystems. In fact, some dictionaries include the study of man's detrimental effects on the environment as an acceptable definition for ecology.

19.1 The Nature of Ecology

Though often thought of in terms of environmental hazards, **ecology** is not just the study of pollution. It is rather the study of what most people would call "nature." It does not deal with the individual parts (that is, the individual organisms) but with the whole. How living and nonliving things affect one another is ecology.

---

† Haeckel, *General Morphology of Organisms*, 1866.

19.1

⇒ Objectives
- Define *ecology* and relate it to the term *pollution*
- Describe the typical responsibilities of an ecologist

⇒ Key Term
ecology

ecology: eco- (Gk. *oikos*—house) + -logy (the study of)

19-1

An ecologist inspects the paw of an anesthetized polar bear.

19-2

How has the advancement of technology affected man's ability to alter his environment?

In its earliest form, ecology was purely descriptive. A scientist would enter a geographic area and describe it. He would find certain organisms in deserts and others in forests. A listing of what lives where is one of the first steps in any ecological study. But as scientists began to ask "Why?" ecology really became interesting and, as the answer unfolded, difficult.

For example, a scientist might ask, "Why are polar bears found in the Arctic and not in the Antarctic?" To answer this question, he must study the polar bear as well as its environment. Polar bears eat seals and fish. They live on ice floes, swim in ice water, and sleep in caves they dig in the snow. Although they can live in temperate regions (as they do in zoos), they appear to be uncomfortable in the summer heat.

After studying polar bear ecology, the scientist concludes that the conditions for polar bear life exist in the Antarctic. There are no polar bears on ice floes around the southern ice cap only because the polar bear has not migrated there, and the perils of the trip through the temperate and tropical regions appear to prevent them from doing so.

The polar bear example is a simple one because there are few factors involved. Ecologists today often explore complex issues involving environmental problems such as the disappearance of species, the spread of unwanted species, or other long-term impacts. In these situations, a multitude of different factors must be considered.

The Role of Ecology

Most frequently, ecologists are called upon to

- study the relationships between existing organisms and their environment (example: study how deer populations affect the trees in an area);
- predict what would happen if some factor were changed (example: predict what effects an increase in the number of deer-hunting permits might have on the deer and, ultimately, the trees);
- recommend steps to change an environment or the organisms in it (example: suggest steps to control the deer population, including the introduction of a deer predator).

The ecologist is not always called upon to correct existing problems. Often the ecologist recommends ways to use a natural resource wisely, without destroying it and others. For example, determining which crop should be planted on what land and what land should be used as pasture or to grow trees is also the work of ecologists.

Making such recommendations may sound simple, but it is actually quite difficult. Consider how many organisms and conditions would have to be understood to make a valid prediction. If the studies are haphazard or if they favor the interests of a particular group, then there may be misuse of the natural resources and disastrous results.

Three hundred years ago man's decisions about how to use a natural resource and other ecological concerns were minor in comparison to current decisions. The wastes of a small town were few and easily dealt with in the earlier environment. When man had to use his own muscle (or that of his animals) to alter the environment, the effect was minor even if he was not wise in his actions. Decisions made today (as with nuclear wastes) can quickly affect the environment on a wide scale for hundreds or thousands of years. Today man can use technology such as strip mining to alter the environment on a major scale. If the decisions he makes are unwise, the results of his actions may be permanent scars. Man's decisions about the use of natural resources and other ecological concerns are, therefore, of major consequence.

Everyone needs to make wise decisions regarding the use of natural resources. It is advisable to know something about how ecologists determine possible outcomes and about what realistic Christian values should be set on the use of natural resources.

➡ *Review Questions 19.1*

1. Differentiate between ecology and a study of man's pollution of the environment.

2. What functions do ecologists most frequently perform?

19.2 The Ecosystem Defined

An **ecosystem** is all the living things and all the nonliving factors and their interactions within a limited area. In general terms an ecosystem is everything in any area being studied or discussed. Although some global issues such as climate change necessitate ecologists to take a very broad view, most focus on a much smaller scale. Normally the area spoken of as an ecosystem would be more the size of a pond, a field, or a section of a forest, coastline, or stream.

Even when an ecosystem is small, scientists may divide it into areas that could be considered ecosystems in themselves. In a 100 m (328 ft) stretch of river, for example, conditions and organisms found near the shore will be very different from those found in the open current. In the quiet backwaters certain organisms flourish that are found only as "passersby" in the rapids.

A major problem with trying to limit an ecosystem is that all ecosystems depend upon the ecosystems around them. To study completely the ecosystem of a river, scientists must study the ecosystems of the banks and even the entire land area that drains into the river. If the surrounding land is forest, the river will have one set of conditions. If the forest burns, is cut down, is changed to farmland, or is converted into a housing subdivision, the water in the river will change; thus, the river ecosystem changes.

Ecosystems are made of two distinct but interacting parts: the **abiotic** (ay bye AHT ik), or physical, **factors** and the **biotic** (bye AHT ik) **factors**. The abiotic factors (the physical environment) are all the nonliving aspects of the ecosystem. The biotic factors, or *biotic community*, are all the living things within the ecosystem. These two components affect each other. The biotic factors affect each other and the abiotic factors, which also affect the biotic factors.

The Abiotic Factors

Bananas do not grow in Canada, because it gets too cold in the winter. Neither do they grow in the Sahara Desert; it is too dry. Anyone who wants to grow

---

19.2

➡ *Objectives*

- Define *ecosystem* and give examples
- List common abiotic factors and tell how they work together to create different living conditions
- Trace the steps of the water cycle
- Clarify the relationships between the terms *population*, *biotic factors*, and *ecosystem*
- Discuss the three types of survival strategies
- Contrast consumers and producers and give several examples
- Construct a food web for a meadow with at least six organisms, including both plants and animals
- Define the seven potential relationships between any two organisms in terms of their impact on one another

➡ *Key Terms*

ecosystem	ecological
abiotic	pyramid
factor	trophic level
biotic factor	biomass
population	biodiversity
density	food web
producer	neutralism
consumer	competition
primary	predation
productivity	amensalism
food chain	parasitism
detritus	commensalism
decomposer	mutualism

The Ecosystem

Abiotic Factors

- Radiation
  - Light
  - Heat (temperature)
- Atmosphere
- Rotation of the earth
- Wind (currents in aquatic environments)
- Water
- Topography (depth of water in aquatic environments)
- Soil and geologic substrate (mineral concentration in aquatic environments)
- Gravity
- Fire

Biotic Factors

- Producers (green plants, algae)
- Consumers
  - Feeders (animals, protists)
  - Decomposers (fungi, bacteria)

---

abiotic: a- (without) + -biotic (life)

bananas should buy land in the wet tropics. But bananas do not grow everywhere in the wet tropics. Certain areas are too wet or lack certain minerals. In some areas physical conditions may encourage the development of plant diseases. Someone wanting to grow bananas might wonder what abiotic factors contribute to healthy banana tree growth. To answer that question, one must carefully study the requirements of the organism.

Radiation, Winds, and Atmosphere

The sun supplies *radiation* as *heat* and *light* to the earth. (Other forms of radiation are also sent to the earth by the sun. Most of them are either filtered out by the atmosphere or are relatively insignificant.) Plants and algae use light in photosynthesis to produce sugar, which serves as food for the plants themselves and for all other living things.

If the amount or quality of light in an ecosystem changes, then the ecosystem will change. The heat of the sun also affects ecosystems. The heating of large air masses and the rotation of the earth cause *winds*. Winds strongly affect the biotic factors. Most living things both give off and require certain atmospheric gases. Wind mixes atmospheric gases, keeping them within livable concentrations.

In aquatic environments *currents* take the place of winds. The speed at which water moves determines what organisms live in it. Certain aquatic organisms are found only in rapidly moving streams; others, only in the slower-moving areas; and others, only in standing water.

Currents, or the lack of them, help determine the concentration and distribution of substances such as dissolved minerals and plankton. A rapidly moving stream contains large amounts of dissolved oxygen, a necessity for some aquatic animals, while standing water may have dissolved oxygen only to a depth of about a meter. Without currents, most bodies of water would become stagnant and uninhabitable for most organisms. Currents are also critical in determining the water's temperature at different levels.

Water

Water is one of the most crucial abiotic factors in any ecosystem. Often the main difference between a lush tropical forest and a desert is the quantity of water, as irrigation of deserts has repeatedly shown.

One estimate indicates that if all the water in the atmosphere were to fall as rain, the entire earth would get about 2.5 cm (1 in.) of rainfall, which is about a ten-day supply. Most areas average about 89 cm (35 in.) of rainfall per year, but some areas of the world have experienced over 1200 cm (472 in.). Obviously the water that forms clouds and rain originates from sources other than the atmosphere.

Most free water on the earth is *cyclic*; it is in a constantly moving cycle. This movement is called the *water cycle*. Through *transpiration* (the loss of water from plants) and *evaporation*, water enters the atmosphere. As the warm, moisture-laden

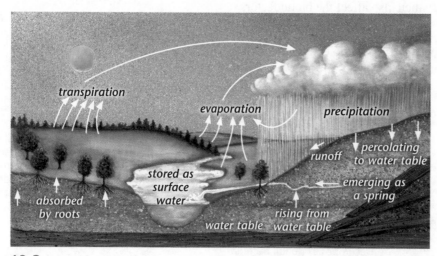

19-3

The water cycle

air rises, it cools, condensing the moisture and forming clouds. As clouds become cooler, the water falls as *precipitation*.

Some of the precipitation forms streams that flow toward larger bodies of water (lakes and eventually oceans). Water moving on the surface of the earth is called *runoff water*. Some of the water *percolates* (PUR kuh LAYTS), or drains into the soil. When this water reaches a layer it cannot penetrate, it collects and forms *groundwater*. Groundwater is water that comes out of the ground as springs or that man obtains by digging wells. In time the groundwater also reaches a lake or ocean.

The water cycle exemplifies the interaction of several abiotic factors. If the sun were not warming the atmosphere, evaporation would stop. The heat of the sun and the rotation of the earth cause the winds that move the moisture-laden air. These moist winds cool the atmosphere and move clouds over the land to prevent the rain from falling only over the oceans, where most evaporation takes place. And if gravity were not pulling at the rain, then runoff water, groundwater, lakes, and oceans would not form.

Other Abiotic Factors

Topography is the "lay of the land"—whether it is flat, hilly, or mountainous. The presence of rocks, lakes, and ocean trenches affects the topography. The slope, height, and expanse of a mountain range determine such factors as winds, temperature, and rainfall hundreds and even thousands of kilometers away. Mountains and bodies of water often serve as barriers, limiting the range not only of individuals and plants but also of whole species. Even a rock, placed in a stream, offers on its downstream side an environment suitable for certain organisms that cannot exist there if the stream bottom is smooth.

The *soil* and the *geologic substrate* (rocks and other substances underneath the soil) are closely related. Often the inorganic parts of the soil are derived from the geologic substrate. Together with topography, the soil and geologic substrate help determine the amount of precipitation that becomes runoff water and, to some extent, the organisms that are found in an area.

Most aquatic environments receive minerals either by dissolving the geologic substrate or from runoff water. The measure of the concentration of dissolved minerals in the world's oceans is *salinity*, which influences the growth of plankton. Thus the soil, topography, and substrate of the surrounding area affect a body of water and the organisms in it.

The *gravitational pull* of the earth certainly affects all organisms, but the moon also affects biotic communities. The ocean tides are caused in part by lunar gravity. Certain algae and other organisms growing along the seashore grow best when they have regular alternating periods of exposure to seawater and air. Ecosystems of thousands of organisms are found in *tidal pools*, which are regularly cut off from the ocean as the tide goes out.

Although it can be devastating, fire caused by lightning or spontaneous combustion is a normal factor in many ecosystems. In certain forests a regular clearing of small plants, fungi, and insects by a *ground fire* is beneficial. Such fires usually do not kill most vertebrates (they can find shelter) or the larger plants. One species of pine produces cones that will not release their seeds unless they are exposed to fire. A fire kills the underbrush that would compete with the pine seedling for light, water, and minerals and thus gives the seedling a better chance of survival.

The abiotic factors in an ecosystem are dynamic, just like the biotic components, changing over time. The tilt and revolution of the earth cause seasons, affecting temperature, rainfall, and light levels. Erosion, the result of wind and

19-4

These organisms are designed to survive regular periods of flooding followed by periods of exposure.

19-5

Fire is a significant, often natural, factor in many ecosystems.

In 1944, the federal government began a campaign to remind people of the importance of preventing fires in forests. A cartoon bear was created to speak for the forest, and when a partially burned black bear cub was rescued from a fire in New Mexico in 1950, it became the living Smokey Bear. (Smokey lived at the National Zoo until his death in 1976.)

Initially, the advertising campaign slogan "Only you can prevent forest fires" portrayed fire of any type as negative. Over time, many foresters and ecologists realized that periodic controlled burns, like those that had been occurring naturally, reduced the buildup of fuel and lessened the chance of dangerous, catastrophic fires. The extremely hot catastrophic fires burn through the crowns of the trees, often leveling whole forests, while the controlled burns thin out the undergrowth, actually helping the established trees.

Smokey's message was altered to remind people to prevent "wild" fires, indicating that controlled burns are a beneficial tool for managing forests. Nevertheless, there is still a vigorous debate between those supporting controlled burns (often foresters and land managers) and those advocating fire suppression (usually environmentalists).

moving water, can change the topography. Volcanoes, earthquakes, and floods can drastically rearrange the earth's surface. Some living things whose range of tolerance is broad enough may do well with these changes. Others may become dormant, migrate, or die.

The Biotic Factors

The biotic factors include all the living things in an ecosystem. Within the biotic factors are **populations**. A population includes all the members of the same type of living thing (often a species or a similar grouping) occupying the same geographic area at one time. The biotic factors within the ecosystem of a pond, for example, would include populations of lily pads, cattails, mosquitoes, frogs, and turtles. Each type of organism—for example, a fish, alga, or bacterium—constitutes a different population within the community.

Each individual population has many measurable characteristics. Ecologists may want to know the total population, its **density** (number of individuals in a defined area or volume), or even its arrangement (evenly dispersed or clumped in subgroups). They may also study the dynamics of how the population changes over time. Factors such as birthrate, death rate, and life expectancy all enter into this.

Some ecosystems may have only a few dozen populations, yet others of similar size may have hundreds. The ecosystem of a cornfield, for example, is artificially structured to favor the corn population and those other populations that aid the growth of the corn (such as beneficial insects) and to discourage populations that would hamper corn yield (such as weeds, crows, and harmful insects). The biotic factors of a similar-sized natural swamp, meadow, or forest will include a larger number of more varied populations.

Within an ecosystem the various populations affect one another in different ways. In a pasture the cattle population

biosphere

ecosystem

community

population

individual

19-6

An individual organism does not exist in isolation but is connected to a population, community, ecosystem, and ultimately, the whole biosphere.

eats the grass population. Though the fungi and bacteria of the soil may not appear to affect the cattle directly, the grass would soon die from lack of usable minerals were it not for these organisms breaking down organic materials.

Within the pasture certain birds help the cattle by eating grasshoppers and other insects that compete with the cattle for food. Some insects, however, help the cattle. For example, bees pollinate clover. The roots of clover support bacteria that help put nitrogen compounds into the soil. Without these compounds the grass would not grow, and the cattle would go hungry. Thus bees indirectly help cattle. It can be assumed that, at least indirectly, every living thing affects all the other living things within its ecosystem. This section presents some of the ways populations interact.

Producers and Consumers

Some of the most obvious relationships within biotic communities are the *nutritional relationships*: who gets food (energy and building substances) from whom or what. The populations within any biotic community can be divided into two categories based on food sources.

✦ *Producers.* Organisms that manufacture their own food, such as the green plants and algae, are the major **producers**. They carry on photosynthesis to produce sugars. (Photosynthetic and chemosynthetic bacteria are also producer organisms, but they are not discussed here because of their relatively low impact on most ecosystems.)

✦ *Consumers.* Animals, protozoans, fungi, most bacteria, and humans, which consume all or part of other organisms for food, are **consumers**.

19-7

An aquatic food chain consisting of algae, aquatic arthropods, a minnow, and a bass

The **primary productivity** of an ecosystem is the rate of photosynthesis carried on by its producers. The radiant energy of sunlight is the only significant source of energy in most ecosystems. Only about 50% of the total light energy available to plants is actually absorbed by the plants. The rest is reflected or lost as heat. Of the light that is absorbed by plants, about 2% (or approximately 1% of the total light energy available) is actually converted to sugar. Of that energy stored in sugar, the plants use about 50% for their own metabolism. This means that, of all the light energy available to plants, only about 0.5% is available to the first consumer.

Often the nutritional relationships between organisms in an ecosystem are called a **food chain**. A simple food chain might include algae, which are eaten by aquatic arthropods, which are eaten by minnows, which are eaten by bass. In this case the algae are the producers, and the other organisms are the consumers. The aquatic arthropods, since they eat producers, are the *primary consumers*. Primary consumers are normally herbivores. Minnows are *secondary consumers*. Secondary consumers are usually carnivores, often called *first-level carnivores*. Bass are the *tertiary consumers*, often called *second-level carnivores*.

At each step in the food chain there is an 80%–90% loss of energy. In other words, for each step up the food chain, a great deal of energy must be invested. Figure 19-9 shows the energy transfer from producer (algae) to primary consumer (copepods), secondary consumer (minnows), and tertiary consumer (bass). It is important to note that the energy flow is one way and does not move in a cycle as do many other natural processes.

The uneaten blades of grass and the molted grasshopper exoskeletons in Figure 19-8 contain energy that is unusable to the next food chain level; it is lost from this food chain. The respiration of the consumers, as well as the energy contained in materials they give off (excretions and secretions), is also

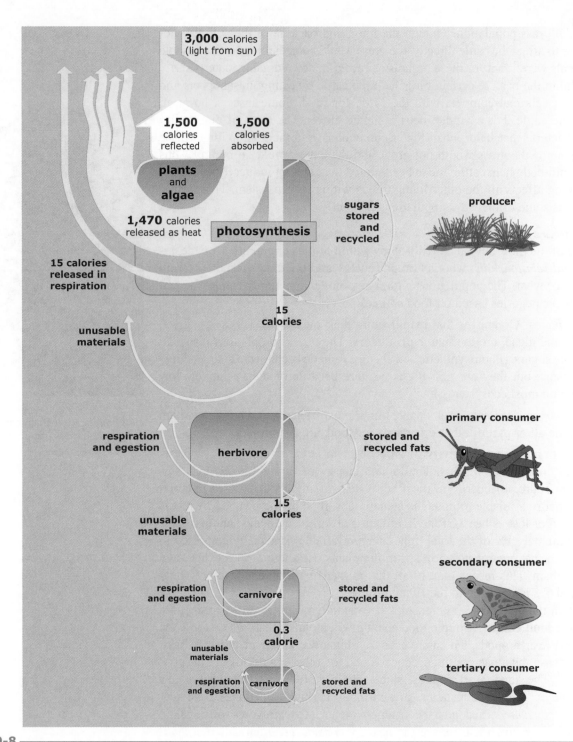

3,000 calories
(light from sun)

1,500 calories reflected

1,500 calories absorbed

plants
and
algae

1,470 calories released as heat

photosynthesis

15 calories released in respiration

sugars stored and recycled

producer

unusable materials

15 calories

respiration and egestion

herbivore

stored and recycled fats

primary consumer

unusable materials

1.5 calories

respiration and egestion

carnivore

stored and recycled fats

secondary consumer

0.3 calorie

unusable materials

respiration and egestion

carnivore

stored and recycled fats

tertiary consumer

19-8

A diagram of energy flow in a food chain

energy lost to the next level of the food chain. Much of the energy a consumer ingests is in unusable forms. For example, the exoskeletons of grasshoppers and the bones of frogs contain energy that consumers may not be able to convert to a usable form; thus, energy in these forms becomes useless to this food chain.

The stored food (fats, sugars, starches, and oils) may be consumed by the next level of the food chain, or they may be used by the organism that stored them. This use of stored energy is quite necessary because the sun does not always shine on the producers, and the consumers are not constantly consuming.

Detritus Food Chain

Dead organic matter such as fallen leaves, dead trees, dead bodies of animals, or even excrement is detritus (dih TRY tus). Detritus may be broken down by **decomposer** organisms, such as bacteria and fungi, into the nutrient molecules that plants need to grow.

Even dead animals have a purpose in God's plan for nature.

Not all detritus, though, is decomposed directly by bacteria and fungi. Before bacteria and fungi decompose detritus, it is often a part of a separate, but overlapping, *detritus food chain*. A dead leaf, for example, may become the food of a detritus feeder, or *detritivore* (dee TRY tih vore), such as an earthworm. Worms may be eaten by a bird, which, when it dies, is eaten by detritivores and broken down by decomposers. Many insects, worms, crustaceans (crabs, lobsters, shrimp, and crayfish), and a number of other invertebrates are detritus feeders. In some sense, scavenging vertebrates, such as catfish and vultures, that feed on dead and decaying material are also detritivores.

Job understood the involvement of human bodies in this process when he said of the wicked, "The womb shall forget him; the worm shall feed sweetly on him" (Job 24:20). Although souls are eternal, physical bodies return to the earth.

Many detritus feeders, of course, do not enter food chains that contain progressively larger organisms. All detritus feeders, however, help cycle materials, since the foods which pass through their digestive systems are broken into smaller pieces. For example, although most detritivores are unable to digest the cellulose of plant cell walls, their digestive systems can break down cellulose into smaller pieces. Bacteria and fungi more easily decompose these smaller pieces. Some detritus feeders eat the same material several times, thus taking advantage of the action of the digestive enzymes secreted by decomposers.

detritus: (L. *deterere*—to wear away)

Ecological Pyramids and Food Webs

Another method of illustrating the nutritional relationships of food chains is with **ecological pyramids**. The ecological pyramid in Figure 19-9 represents an aquatic food chain. Ecological pyramids are divided into **trophic levels** to show the flow of matter and energy through a food chain. The trophic levels are feeding levels through which energy exchanges take place in an ecosystem.

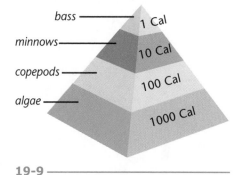

19-9

Here is an ecological pyramid demonstrating energy flow in a food chain. Notice that on average only 10% of the energy is transformed to biomass in the next level.

Since it normally takes about ten units of energy to result in one energy unit in the next higher level, ecological pyramids based on energy are always regularly shaped. If the ecological pyramid is based on the number of individuals in a population, however, the pyramid may be misshapen or reversed. For example, a single tree (producer) may support thousands of insects (primary consumers) that are the food of only one bird (secondary consumer).

Measuring the productivity of an ecosystem is difficult. Should the number of individual organisms at each trophic level be determined? This number might be misleading because some individuals, like insects, are very small and thousands could be supported by one tree, as in the preceding example. Ecologists usually measure the **biomass** of a population or system by calculating the total mass of the living tissue (often in a dried form). Healthy ecosystems are supported by greater amounts of biomass than unhealthy systems.

Although food chains and ecological pyramids may be helpful in understanding energy transfer in biotic communities, they oversimplify what actually happens in an ecosystem. For example, a bear may be a tertiary consumer (eating salmon), a secondary consumer (eating grubs), or a primary consumer

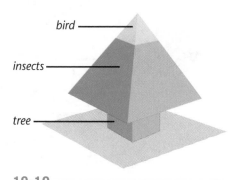

19-10

Ecological pyramid based on population size

Biodiversity

Another important measurement for ecologists is an ecosystem's **biodiversity**. This measures not the total amount of productivity, but the number of different species within a system. Generally, ecosystems with high biodiversity tend to be more productive and adaptable to change than those with lower diversity.

19-11

This brown bear is a tertiary consumer. Bears can also be primary or secondary consumers, depending on what they are eating.

(eating berries), depending on what it eats. Birds of prey may be secondary or tertiary consumers. Owls, snakes, mountain lions, and many other predators would be secondary consumers when eating rabbits, but if they ate a small carnivore they would be classified as tertiary consumers.

To describe these multiple relationships, ecologists often use **food webs**. Arrows point from a food source to its consumer. Food webs may be simple or extremely complex, depending on the number of different animals in the ecosystem and their ability to adapt to different foods.

Species Interactions

In nature there are many different ways that two species can interact with each other beyond the energy transfers of eating and being eaten. Ecologists study interactions between animals, plants, fungi, and even microscopic protozoans and bacteria. For ease of understanding, many of the illustrations that follow focus on animals, specifically the white rhinoceros and animals in its ecosystem on the African plains.

✦ **Neutralism**. Every population in an environment indirectly affects every other population; however, neutralism exists when there is no direct relationship. A large snake, the black mamba, may share space with a white rhino. Although the snake might feed on rodents who eat the same plants as the rhino, they do not have a direct connection but exhibit neutralism.

✦ **Competition**. If two populations inhibit each other because they depend on the same limited resource, they are in competition. Because the white rhino and the zebra both graze on grasses in the same area, they are said to be in competition.

✦ **Predation** (prih DAY shun). In predation, one organism (the predator) eats another organism (the prey). When a lion attacks and kills a rhino calf, it is the predator and the calf is the prey.

The broad term for any close, long-term relationship between two organisms of different species is *symbiosis*. The relationship could benefit only one or both of them. In some symbiotic relationships, one party might also be harmed. Ecologists often categorize these relationships as follows:

19-12

Can you identify the only animal in this forest-edge food web that, according to the arrows, can be a primary, secondary, or tertiary consumer?

19-13

What relationships are seen here?

✦ **Amensalism** (AH MEN suh LIZ um). When one population is inhibited or harmed by a second population and the second population is not affected by the first, they exhibit amensalism. Certain molds, for example, produce antibiotics that inhibit the growth of certain bacteria. The bacteria are harmed by the mold, but the mold is neither helped nor harmed by the bacteria. In our plains example, a rhino, while grazing, disturbs lizards that flee across the ground, only to be eaten by birds. The lizards are definitely harmed, while the herbivorous rhino does not benefit from them. Therefore, amensalism exists between the rhino and the lizards.

✦ **Parasitism**. In parasitism, the parasite depends (often for food) on a *host*. Usually a parasite is much smaller than its host. Many parasites do not kill their hosts but merely take advantage of a "free meal." Tapeworms and many fungi are parasites that usually take little from their hosts. Some parasites, like certain protozoans, flukes, and roundworms, may eventually kill their hosts. While the best-known parasites, such as tapeworms and flukes, live inside their host and are called *endoparasites*, some, such as ticks and lice, live on the outside and are *ectoparasites*. Ticks are frequent ectoparasites on the white rhino.

✦ **Commensalism** (kuh MEN suh LIZ um). If one population benefits from a second population and the second population is not helped or harmed by the first population, the relationship is commensalism. Cattle egrets, birds that follow the rhino in order to catch the disturbed insects, are commensal to the rhino.

✦ **Mutualism**. In mutualism, both populations benefit from the relationship. Some mutualism is *obligatory*; that is, neither organism would survive without the other. The protozoans that live in the digestive tract of the termite are an example. The protozoans supply the enzymes to digest the cellulose that the termite eats. The termite could not exist without these protozoans, and the protozoans do not live naturally outside the termite. Some lichens are examples of obligatory mutualism, for neither the algae nor the fungi that compose them can exist in the environment of the lichen without the other. Other mutualisms are not obligatory. Sparrow-sized birds in Africa, called tickbirds, eat ticks and other parasites that they find on grazers. The tickbird gets a free meal

amensalism: a- (without) + -mensal- (L. *mensa*—table)

commensalism: com- (together) + -mensal- (table)

mutualism: (L. *mutuus*—exchanged or reciprocal)

Animals must obtain food and avoid becoming food for something else if they are to survive. This goal may be accomplished by a number of different strategies.

Camouflage

For many organisms, hiding is almost as important as eating. In fact, were it not for their *camouflage*, many organisms would either be eaten or not be able to eat. The color and shape of the praying mantis, for example, causes it to look like a branch as it sits quietly waiting for its prey. Since the vision of insects picks up movement more than shape or color, the mantis is unnoticed as its prey walks within grabbing distance. But its camouflage also helps the mantis not be seen by insect-eating birds. Some tropical members of the mantis family are pink and look like the brightly colored flower petals where they wait for their prey and hide from their predators.

Katydid

Protective coloration can be seen in many insects. They appear as thorns, twigs, buds, flowers, leaves, or rocks. Some butterflies have wings that are brightly colored when opened flat, but when closed over the insect's back, the wings look like a dead leaf. The *countershading* of many aquatic animals is a protective coloration (see p. 401). The colors of young mammals are frequently pale to help them blend in with their natural surroundings. Many deer fawns and tapir calves have a combination of stripes and spots that camouflage them on the sun-dappled forest floor. Stripes on tigers, which make them seem so obvious when seen out of their natural habitat, actually help them blend in with the tall, dried grasses of their normal surroundings.

Animals such as the flounder even change colors to blend with their surroundings, giving them a protec-

---

Interdependence of Organisms: The Tale of the Yucca

Many symbiotic relationships, along with mimicry, protective and warning colorations, and other population relationships, pose problems for the evolutionist. Consider the example of the symbiotic relationship between the yucca plant and the yucca moth.

The yucca, a plant native to the American Southwest, produces flowers in which the anthers are isolated from the stigma. Thus, the flower cannot be self-pollinated, nor is it pollinated by wind or any insect other than the yucca moth. Both male and female yucca moths are attracted by the fragrance of the yucca flowers and mate within the flower. From the flower in which she mates, the female moth collects pollen and rolls it into a ball using special mouth appendages that only she possesses. After mating, the female flies to another yucca flower, uses her ovipositor to bore into the thinnest section of the ovary wall, deposits an egg in it, and then pollinates the flower by placing some of the pollen from the original blossom onto the stigma of the second flower. This sequence is then repeated on other yucca blooms. Without the placement of the pollen, neither the seeds nor the larvae would ever develop.

Eventually the larvae of the moth hatch and feed upon some of the seeds within the ovary. Then they burrow out and drop to the ground. The development of the moth continues in the soil. A fertilized yucca plant normally produces several hundred seeds in each ovary, and the single larva destroys only a few. The yucca moth cannot complete its life cycle without the yucca, and the yucca cannot produce seeds without the yucca moth.

Evolutionists often cite the yucca plant and the yucca moth as an example of "coevolution between two different organisms." They believe that the ancestors of both the plant and the animal were evolving at the same time and as they evolved, they became ideally suited for the complex mutualism they now exhibit.

The yucca could not develop a dependence upon the moth until the moth had developed enough of the specialized structures and behaviors to ensure the fertilization of the yucca. Otherwise the plant would have become extinct. Also, the moth could not develop a larva dependent upon yucca seeds until the yucca had developed. In addition, how did the moth figure out that she needed to collect, carry, and deposit the pollen to help her offspring survive? The probability of the evolution of two such mutually dependent organisms is virtually impossible, yet evolutionists insist that it had to happen because of their bias against a Creator Who designed these organisms this way.

Yucca moth and yucca flower

tive coloration advantage that most organisms lack.

Warning Coloration

Some animals, rather than blending with their environment, seem to clash with it. These animals have *warning coloration* that appears to tell potential predators to beware. Most of the poison dart frogs of tropical regions hunt during the day. Their brightly colored bodies are clear warnings to those who might want to eat them that by doing so they will be eating their final meal.

The skunk, with its bold black-and-white markings, sends a clear warning. If an animal comes too close, it

Skunk

quickly learns to avoid the black-and-white striped animal; in fact, some animals will avoid any animal with similar color combinations.

Some insects have "false eyes" that help startle predators. The false eyes of some caterpillars are on their back ends. The would-be predator appears to view these as snake eyes, and the caterpillar is left alone. Certain butterflies and moths have spots shaped and colored like eyes on their wings. To their predators these may appear to be owl eyes, and owls often prey on these "predators."

Mimicry

One of the best-known examples of *mimicry* is the common orange and black viceroy butterfly, which is very similar to the foul-tasting orange and black monarch butterfly. Once a bird has tried to eat a monarch butterfly, it never again eats an orange and black butterfly. This similarity—or mimicry—protects the viceroy butterfly.

Another example of mimicry is the harmless scarlet king snake. It has yellow, red, and black stripes; the highly poisonous coral snake has stripes of the same color, but in a different order. The ranges of these snakes overlap, and animals usually avoid both of them.

It should not be thought that these organisms have *become* mimics—that would be evolution. Organisms exhibit camouflage, warning coloration, and mimicry because they were designed with these protective devices.

The scarlet king snake (top) and the coral snake (bottom) exhibit mimicry.

and performs a service for the white rhino by eating its ectoparasites. This is mutualism, benefiting both of them, but it is not obligatory, because either one could survive without the other.

Some relationships between organisms cannot be easily classified, and many of them change periodically. For example, in the early spring before berries are ripe, the relationship between a berry-eating bird and a berry bush is one of neutralism. When the berries are ripe, the relationship may appear to be predation from the bird who feeds on them, but since the seeds of the berries are dispersed as they pass unharmed through the bird's digestive tract, the relationship could be considered mutualism.

19-1 Species Relationships			
Relationship		**Species A (rhino, for example)**	**Species B**
	neutralism	0	0
	competition	−	−
	predation	−	+
Forms of symbiosis	amensalism	0	−
	parasitism	−	+
	commensalism	0	+
	mutualism	+	+

In this review table, − denotes a harmful impact, + represents a benefit, and 0 indicates no impact in either direction.

1. What are the two components of an ecosystem?
2. List several abiotic factors of your school's grounds or your yard at home.
3. Describe the water cycle.
4. What is included in the biotic factors of an environment?
5. What are the two main types of populations (based on their nutritional relationships)?
6. What happens to the radiant energy available to plants? How much of it is available to the first consumer?
7. Construct a food chain diagram using an amphibian, an insect, a bird, and two other organisms.
8. Account for the energy losses between the steps in a food chain.
9. Add four organisms to the food chain established in Question 7 and make it a food web.
10. List and describe the possible species interactions or nutritional relationships between organisms.
⊙11. List and describe several strategies that animals use to survive.
⊙12. Why is it improbable for the relationships between organisms to be the result of evolution?
⊙13. What role do plants play in the water cycle?
⊙14. A large toad species is introduced to an island to control the populations of some insects that damage the sugar cane crop. Other than decreasing the populations of some insects, how might the introduction of this new species affect the ecosystem?
⊙15. Some ecologists consider the detritus food chain one of the most important in an ecosystem. Why do you suppose this could be true?

19B—The Biosphere

Despite periodic unexplained phenomena, there is no evidence to support the idea that there is life on other planets. The rocks brought back from the moon and the space probes sent to various planets have revealed no life and no set of conditions that would support life as it is found on the earth. Scientists occasionally get excited over meteorites that seem to have traces of very small organisms, but in every case to date, these have been discredited. This lack of evidence, however, is not proof that there is not life other than that on the earth.

Scripture says nothing about living beings or any form of life beyond this planet. Instead, the Bible presents the earth and mankind as God's central concern in the universe. Since the Bible gives no information on the topic, there is no way of knowing without investigation—an activity many people consider a waste of time and money.

The thin shell around the earth in which all known physically living things exist is the **biosphere** (BY uh SFEER). On land the biosphere extends as deep as 4000 m (13 123 ft) below the surface, where bacteria have been found. In the ocean most living things exist in the top 150 m (492 ft) of water, but a few species exist in marine trenches over 9.7 km (6 mi) deep. While some living things may temporarily fly or be blown a few kilometers into the sky, the lichens growing on the edges of mountain snowcaps are the highest limits of the terrestrial biosphere.

The limits of the biosphere are thus only about 16 km (10 mi) apart. The actual thickness of the biosphere at any given spot, however, is much thinner, usually less than 300 m (984 ft). Within this thin veneer exist vastly different sets of conditions, most of which support some form of life. The biosphere is composed of many ecosystems, each with its own unique assemblage of biotic and abiotic factors. Before discussing how man affects the biosphere, this text examines some of the inner workings of the biosphere and its diverse ecosystems.

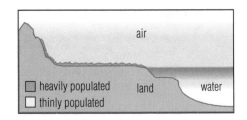

air

☐ heavily populated
☐ thinly populated

land   water

19-14
The biosphere

biosphere: bio- (life) + -sphere
(L. *sphaera*—globe)

19.3 Interactions in the Biosphere

Habitat is a general term for the area where a type of organism lives—its "address." The habitat of the leopard frog is the edges of ponds or the quiet backwaters of streams. *Habitat* applies both to the physical environment and to the biotic community in which an organism lives. The physical environment of the leopard frog must include abundant water and a certain range of temperatures. Its biotic community must include insects to serve as prey and predators to keep the frog population in check.

Some organisms have wide ranges of habitats, while others have very narrow ones. For example, the raccoon can be found in forests, in meadows, or even in suburbs and cities, while the brook trout can survive only in certain sections of streams with cold flowing water.

When the discussion turns to *how* the organism affects the ecosystem, it is no longer about habitat. What an organism does and how it fits into and affects its habitat is its **niche** (NICH). The niche concept includes both the biotic and abiotic conditions needed by an organism; it also includes how the organism uses those conditions. If the habitat is the address of an organism, the niche is its occupation. For example, the niche of the adult leopard frog is, in part, to serve as a consumer of insects and as a prey of snakes, birds, and raccoons. Its niche also includes the need for fresh water and for temperatures at which it can survive. The niche can also be thought of as the organism's lifestyle.

Two organisms may live in the same habitat, but if their niches are different, they are not in direct competition. The ladybug and the grasshopper, for example, may live in the same field or even on the same plant. The grasshopper consumes the leaves of the plants, and the ladybug consumes small insects on the plants. They occupy two separate niches in the same habitat. Even animals that have similar niches may avoid competition by feeding preferences. For instance, three different insect-eating warbler species can share the same spruce tree because one feeds near the top, one prefers the middle, and one feeds near the bottom and at the bases of branches near the middle.

Because no ecosystem exists with just one species population, the niche of each species is going to be influenced by those of others. For instance, an orchid that grows on tree branches in the rainforest cannot spread to cover every possible spot that offers the proper conditions for growth. Its niche is

19.3

➤ Objectives

- Describe and contrast the concepts of *habitat* and *niche*
- Trace the steps of the oxygen, carbon, and nitrogen cycles
- Compare the nature of matter and energy in an ecosystem
- Define and give examples of limiting factors
- Use examples to demonstrate the difference between density-dependent and density-independent factors
- Explain why logistic growth is preferable to exponential growth when modeling most real populations

➤ Key Terms

biosphere
habitat
niche
nitrogen fixation
biogeochemical cycle
limiting factor
density-dependent factor
density-independent factor
exponential growth
carrying capacity

19-15
These organisms have the same habitat, but their niches differ. How do the niches differ? What other animals with different niches would you expect in this habitat?

habitat: (L. *habitare*—to dwell)

niche: (Fr. *niche*—recess in a wall; hence, nest)

Generalists and Specialists

Some organisms have a very large niche because they can tolerate a wide range of conditions and, if they are consumers, can live on many different food sources. They are called *generalists*. The *specialists* are organisms with very little flexibility toward their habitat conditions or whose food preferences are very narrow. The English house sparrow has been introduced to many continents and has an almost worldwide distribution because of its preference to nest and feed near humans and its aggressive tendency to displace other birds. It is the ultimate generalist. The giant panda, a specialist, teeters on the brink of extinction in the mountains of China because of its almost total dependence on a single food source—bamboo. In years when stands of bamboo flower and die out, many pandas that cannot migrate to new forests starve.

The giant panda is a specialist.

limited by other species whose niches are similar. When two species in an area compete for the same resource, whether it is sunlight, food, or space, one may be more successful at the expense of the other species. This process of one species replacing another is called *competitive exclusion.*

The niches available within an ecosystem determine what organisms exist there. Both the physical environment and the other populations within an ecosystem determine what niches are available. To understand how the abiotic and biotic factors interact to determine niches, it is important to study some of the cycles found in ecosystems and the concept of limiting factors.

Matter and Energy in an Ecosystem

The quantity of available energy in a food chain becomes progressively smaller as it passes from one organism to the next (see Section 2A). Finally, the last of the energy captured from the sun by the producer organisms is released in the decomposer organisms; it is no longer available for use in the biotic community. Everything alive must constantly have usable energy, but once it has used or released some energy, that energy cannot be reused. The organism must obtain more.

The Nitrogen Cycle

The *nitrogen cycle* illustrates interdependence within the biotic community as well as between the abiotic and the biotic factors. Although nitrogen composes only about 0.5%–4% (by weight) of living organisms, those compounds that contain nitrogen are crucial. DNA and RNA contain nitrogenous bases, and amino acids contain nitrogen.

Nitrogen composes about 78% of the atmosphere. Gaseous nitrogen (N_2), however, is useless to most living things. Where do living things obtain nitrogen? Much of the nitrogen found in living things is cyclic within the biotic factors (and the soil or water). Although it frequently changes the combinations of its compounds, it remains in the *short nitrogen cycle.*

The short nitrogen cycle involves decomposer bacteria and fungi breaking down dead organic substances and converting the nitrogen contained in those compounds to ammonia (NH_3). The ammonia dissolves in the water in the soil and spontaneously converts to ammonium ions (NH_4^+). Ammonium ions can be absorbed and used by some plants.

Many ammonium ions are converted into nitrites (NO_2^-) by nitrifying bacteria in the soil; then other nitrifying bacteria convert the nitrites to nitrates (NO_3^-), which can be absorbed by plants. The ammonium ions and nitrates a plant absorbs can be used to form the various nitrogen-containing compounds essential to the plant and to those organisms that consume the plant.

Nitrogen compounds are lost from some of the steps in the short nitrogen cycle by deni-

trifying bacteria. These bacteria convert the nitrates into nitrogen gas that makes its way to the atmosphere. This shortage of usable nitrogen compounds in the soil results in reduced plant growth. To compensate for this loss, the *long nitrogen cycle* converts gaseous nitrogen into usable compounds.

For centuries farmers have known that many crops grow better in soil that has grown legumes (clover, beans, alfalfa, or peas) for several years. Early in the last century, scientists discovered that bacteria grew in the small nodules (swellings) on the roots of these plants. Later it was discovered that these bacteria were able to convert atmospheric nitrogen into nitrogen-containing compounds, a process called **nitrogen fixation**. There is a mutualistic relationship between the plant and the bacteria—the plant provides a home and food to the bacteria, and the bacteria provide usable nitrogen compounds to the plant. Any excess nitrogen compounds are released into the soil.

Various free-living soil bacteria (those not living in the root nodules) and several species of algae can also perform nitrogen fixation. The compounds made by nitrogen fixation can be used directly by plants or algae, and when they are, they enter the short nitrogen cycle. To a much smaller extent, lightning can convert atmospheric nitrogen into compounds that can enter the nitrogen cycle. Therefore, the nitrogen compounds that escape the short nitrogen cycle can be offset by the nitrogen fixation of the long nitrogen cycle.

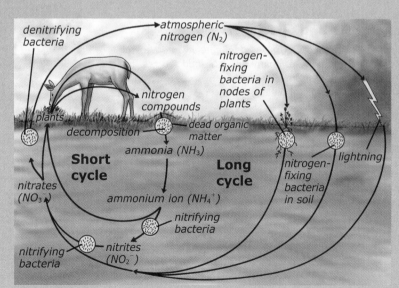

Even within the abiotic environment, energy merely passes through; it is used but not reused. For example, the radiant energy that warms a body of water and causes evaporation eventually escapes into the atmosphere. To continue evaporation, more radiant energy is necessary. The thermal or mechanical energy of an earthquake or volcano can be enormous and have far-reaching effects, but it is no longer available after its release. Thus, energy must constantly be supplied to the environment. Although it may be stored temporarily, energy in the ecosystem is *noncyclic*.

Matter, on the other hand, is cyclic. The carbon, oxygen, and nitrogen cycles and the other cycles known to scientists show that matter is repeatedly used. Because these cycles involve chemicals that pass through the biosphere but are also part of the earth's systems, they are referred to as **biogeochemical cycles**. Although some matter may escape from the cycle (for example, by falling to the bottom of the ocean or by becoming trapped in a substance like a fossil), there are reserves of most of the elements that living things need. These reserves can be used to add to the amount cycling within the ecosystem. Some of the reserves made available to the biotic community include the water deep in oceans and lakes (see the water cycle discussion, pp. 456–57), the nitrogen in the atmosphere, and elements in the soil and geologic substrates.

Reserves compensate for minor natural losses from the cycles. Occasionally man also uses them for purposes that prohibit their entering ecological cycles (at least temporarily), such as when he uses them as building materials. At other times man uses the reserves to affect the cycles directly. Nitrogen, for example, can be taken from the air, made into nitrogen compounds, and then used as fertilizer. Here man is serving as an agent of nitrogen fixation.

Limiting Factors

Whether or not an organism can exist in a particular habitat depends on various **limiting factors**—factors that in some way limit the growth or existence of an organism. Within a lake there may be sufficient dissolved minerals and adequate sunlight to support a large alga population but not enough carbon dioxide for the algae to carry on photosynthesis. Carbon dioxide is the limiting factor. If the amount of carbon dioxide in the lake were increased, the alga population would grow. But the alga population will not grow larger than the next limiting factor will allow. For example, if the supply of carbon dioxide is increased and the alga population grows, the supply of a particular dissolved mineral may become depleted. That mineral then becomes the limiting factor.

The Oxygen and Carbon Cycles

The *oxygen cycle* is a relatively simple cycle within the ecosystem. Free oxygen (O_2)—found as an atmospheric gas and dissolved in water—is essential for most living things. Both autotrophs and heterotrophs use free oxygen predominantly in cellular respiration, the breaking down of an energy source such as sugar to obtain energy. During respiration, oxygen is used and carbon dioxide is released.

Carbon dioxide (CO_2), an atmospheric gas also found dissolved in water, is the primary source of carbon in living systems and thus is a major part of the *carbon cycle*. In photosynthesis, autotrophs fix carbon dioxide and produce sugars. The carbon in these sugars is the primary source of carbon in all other living things. The respiration of sugar (glycolysis and the Krebs cycle) converts the carbon of the sugar back into carbon dioxide.

Of course, many carbon-containing compounds are either excreted or remain in the body of the organism when it dies. Decomposer organisms carry on respiration and other processes that convert most of these organic compounds into carbon dioxide and other substances. The burning of organic substances (whether "fresh" like wood and leaves or fossilized like coal and oil) also converts carbon-containing compounds to carbon dioxide and thus puts carbon back into the atmosphere.

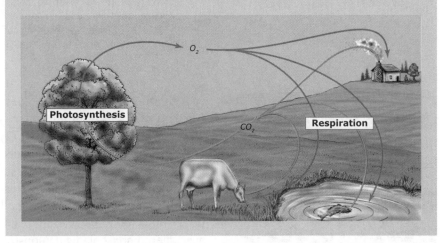

Photosynthesis O_2 CO_2 Respiration

19-16

Limiting factors in a forest include not only how much water there is but also how and when the water is supplied. If the river were the only source of water, plant growth would be limited. The quantities of snow and ice and the amount of summer rain can limit which trees grow in an area and even how well they grow.

Principles Regarding Limiting Factors

- An organism with a wide range of tolerance for all factors will have a wide distribution. The dandelion and other weed species can tolerate a wide range of environmental factors and can grow "all over."
- An organism with a wide range of tolerance for one factor may have a narrow range for a different factor. Many desert plants, for example, have a wide range of tolerance for temperature but can tolerate water only within certain narrow limits.
- When one factor is not in its optimal range (range within which the organism operates best; see pp. 70–72), the range of tolerance for another factor may be reduced. For example, certain plants cannot tolerate low quantities of water unless they are receiving sufficient quantities of certain minerals.

- The period of reproduction is the most critical time for limiting factors. Some mammals can live long, healthy lives in zoos but will not reproduce unless their diets contain the right nutrients.

Many desert plants can tolerate extreme temperatures, but too much water may kill them.

19-17

Would the development of beachfront property be a density-dependent or a density-independent factor for the nesting of sea turtles that depend on oceanfront sand dunes?

A limiting factor, however, does not always have to be one that is in short supply. Often the limiting factor is that which goes beyond an organism's *range of tolerance* (see pp. 70–72). For example, certain orchids normally grow only in shaded environments because they cannot tolerate the heat of the sun. If they are put in bright sun but kept cool, they actually grow better than when kept in the shade. In this example, high temperature is the limiting factor. Too much water is often the limiting factor in terrestrial environments. If the soil is too moist, certain plants will not grow well, or not grow at all, as overwatered houseplants demonstrate.

Ecologists often classify limiting factors as either density-dependent factors or density-independent factors. **Density-dependent factors** are those that become more limiting as a population's density increases. For instance, the lack of grazing areas for pronghorn antelope as their population grows or the lack of suitable trees for woodpecker nesting cavities would both be density-dependent factors. On the other hand, **density-independent factors** limit population growth regardless of the size of a population. Hurricanes, extreme winter temperatures, and drought are just a few examples of such factors that would affect a population equally, no matter what its density is.

Population Change and Limiting Factors

In nature, populations of organisms are rarely static. They change over time due to some obvious events. The two most significant impacts on a population's size are the *birthrate* and *death rate*. When a population's birthrate exceeds its death rate, it grows, and conversely it shrinks when the death rate exceeds the birthrate. A constant population occurs when these two rates are in balance with each other. Another, often overlooked, factor in population change is the movement of individuals into (*immigration*) or out of (*emigration*) an area.

Theoretically, the future population of an organism can be calculated if you know how many offspring are produced and how often this occurs. A basic illustration is the growth in the number of bacteria in a culture. A typical bacterium splits to form 2 cells every 20 minutes. If you continue this process, you will have 4 cells at 40 minutes, 8 cells at 60 minutes, 16 cells

19-18

Theoretical exponential growth of a bacterial colony

at 80 minutes, and so on. This might not sound too impressive, but if you do the math, you will discover that you will have several tons of bacteria in just 24 hours! This theoretical example in which a population multiplies at a constant rate at regular intervals is called **exponential growth**. Figure 19-18 shows a graph of the exponential curve representing the population change in our hypothetical flask of bacteria. Note that the larger the population becomes, the faster it grows.

Clearly, we are not slogging through mounds of bacteria as we go about our daily lives, so this exponential growth must be changing at some point. The bacteria population, like that of all organisms, will increase until one or more limiting factors halt its further growth. It has then reached its **carrying capacity**. The limiting factor for the bacteria might be a lack of chemicals for further growth, the buildup of waste materials, or other factors. At carrying capacity, a population's birth and death rates approach a balance, and the change in population size is very low. The variation that occurs in a population when it has neared carrying capacity and slowed or halted its exponential growth is sometimes called *logistic growth*.

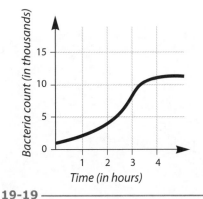

19-19
Actual logistic growth of a population

Review Questions 19.3

1. Describe the limits of the biosphere.
2. Differentiate between the habitat and the niche of an organism. Choose two organisms and tell the habitat and niche of each.
3. Briefly trace (a) the oxygen cycle, (b) the carbon cycle, (c) the short nitrogen cycle, and (d) the long nitrogen cycle.
4. Why does a closed terrarium constantly need light but does not necessarily need water? Why must a closed terrarium contain sufficient animal life?
5. List at least five possible limiting factors for a squirrel living in a city park.
6. List two factors that will directly cause population growth and two that will cause population decline.
⊙7. Why do you think secular scientists get so excited about the prospect of life occurring on Mars or other planets?
⊙8. If you could become any animal, would you rather be a generalist or a specialist? Why?

19.4 Changes in the Biosphere

Chapter 1 explains that life is a dynamic equilibrium; all living things constantly change (expend energy) to maintain the homeostasis necessary for their living conditions. In a much larger framework, the same is true of an ecosystem and even the biosphere.

Checks and balances are easily observed between organisms and in the responses of organisms to environmental factors. For example, a cold winter may kill many insect eggs. There would be fewer insects the following summer, resulting in a large crop of seeds. A well-fed rodent population would then be able to produce and raise more offspring. This would cause the hawk and owl populations to increase. Animals such as lizards that depend mainly on insects for food would probably decrease in number. A few mild winters, however, would reverse the process.

Predator-prey relationships are easy to see and understand. Some other obvious changes within an ecosystem and within the biosphere are vital to the survival of the organisms but are not always so easy to understand, perhaps because they often occur slowly. One of the most significant is known as **ecological** succession. This is the predictable changes that occur in a biotic community over a period of time. If an ecosystem develops gradually from bare rock, it is called **primary succession**. The more common progression of biotic communities in which soil and plants are already in place is known as **secondary succession**.

19.4

Objectives
• Contrast pioneer and climax stages of succession by listing several differences
• Distinguish between primary and secondary succession

Key Terms
ecological succession
primary succession
secondary succession

succession: (L. *succedere*—to go after)

Succession

A field on a North Carolina farm was purchased and became part of a national park. The first year after the farmer moved, there was no cultivation, and quick-growing, sun-loving annual weeds such as ragweed and clover covered the field. In the second and third years, perennial plants such as crabgrass, asters, and horseweeds predominated. During these *pioneer periods*, insects like grasshoppers, open-area birds like the meadowlark and the grasshopper sparrow, and small animals like the field mouse inhabited the area.

In about the third year, junipers and other sun-loving shrubs took over the field, shading out the pioneer plants but providing suitable conditions for other plants such as the morning glory and honeysuckle vines. In this *shrub period*, animal populations also changed. Rabbits were among the most noticeable animals, and the cardinal and field sparrow became the dominant birds.

The seeds of the shrubs were unable to establish themselves in the shade of the mature plants. Pine seedlings, however, are shade-resistant and were in the field starting about the twentieth year. Twenty-five to thirty years after the field had been farmed, a *pine forest* established itself and began shading out the shrubs. A whole new *understory* (plants growing in the shade of trees) arose, and changes in the animal population took place. Squirrels, raccoons, pine warblers, and towhees thrived in the pine forest.

About fifty years after the field had been farmed, hardwood deciduous trees, such as oaks and hickories, were found in the understory of the pine forest. These slow-growing trees eventually began competing for sunlight in the upper story and began to take over as the dominant plants about one hundred years after the field had been left alone, producing a hardwood deciduous forest.

The accumulation of falling leaves instead of pine needles changed the makeup and the pH of the soil, causing a considerable change in the understory. Mosses, ferns, May apples, and orchids began to fill the shaded, moist forest floor. Deciduous trees produced edible leaves, fruits, and nuts, which supported animals like squirrels, deer, raccoons, and wild turkeys. Other animals found in the hardwood forest included foxes and bears.

In many areas of North Carolina, as in many areas of the eastern United States, the hardwood deciduous forest is the *climax vegetation*. The climax veg-

19-20

Two stages in the secondary succession of an old field: pioneer stage (top) and twenty years later (bottom)

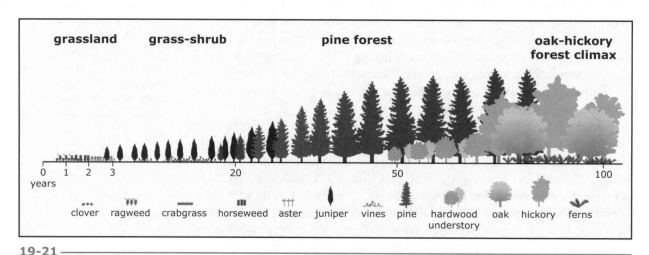

19-21

This diagram represents the succession from grassland to a climax forest in the southeastern United States.

etation is the plant community normally found in an area free of disturbances by nature (disasters such as fire and flood) or by man (farming, cutting lumber, building structures, and such). A climax vegetation is self-perpetuating. As the plants of a climax vegetation die, they are usually replaced by the same species, not new ones.

19-2 Comparison of Ecosystems in Succession

Pioneer stage	Climax stage
• few species, large populations of each • smaller organisms • organisms with higher metabolic rates • organisms with short life spans • organisms with simple forms of reproduction • large amounts of nutrients needed and large amounts lost from the ecosystem	• many species, small populations of each • larger organisms • organisms with lower metabolic rates • organisms with long life spans • organisms with complicated life cycles • nutrients cyclic within the ecosystem

Mount St. Helens: A Case Study in Succession

At 8:32 a.m. on May 18, 1980, Mount St. Helens erupted, blowing away the top 400 m (1300 ft) of the mountain and devastating 500 km² (193 mi²) of pristine forest. Miles of verdant forests, mountain streams, and crystal clear lakes were instantly converted to a sterile landscape of ash, steaming pumice fields, mudflows, and debris. It was so lifeless that then-President Jimmy Carter compared it to a moonscape. Nearly all visible ferns, mosses, and shrubs vanished, causing many scientists to predict that it could never recover.

Now fast-forward three decades. In many devastated spots, 20 ft hardwoods and scattered conifer trees are flourishing. Wildflowers and shrubs provide food for insects as well as larger animals like migrating birds, rodents, deer, and elk. Scientists have been shocked not only by the rate of the recovery but also by the sequence.

In the typical order of primary succession, mosses and tiny plants start, followed by small shrubs, and, eventually, trees. Animals such as insects usually come after sufficient plant cover has developed. In the case of Mount St. Helens, small beetles and spiders were the first pioneers, and their decaying remains served as nutrients for the first sprouting seeds. Pocket gophers returned earlier than anyone predicted, tunneling through the ash and tilling the ground for better plant growth. Their mounds trapped wind-blown seeds and accelerated recovery. Salamanders, newts, and frogs that had survived the blast because they were hibernating beneath frozen ponds found cool, safe shelter in the gopher burrows. The elk that wandered from unaffected areas brought seeds in their feces, complete with fertilizer to give them a head start.

Now, the "rules of succession" are being reevaluated as scientists observe the regrowth of the Mount St. Helens blast zone. While most models based on old field succession predict a gradual, ordered progression, in this case all major stages of forest development are occurring almost simultaneously. Within three years, 90% of the original plant species had returned to the blast zone.

That the blast zone is recovering at such a rapid rate could be applied to the landscape after the waters receded following the Flood. The animals and plants (floating seeds, organic debris, seeds from foodstuffs loaded in the ark, etc.) could have become reestablished rather quickly after the ark landed and the waters abated.

Year one

Year twenty

Pocket gopher

→ **Review Questions 19.4**

1. What is the relationship between succession and climax vegetation?
2. Describe a pioneer stage of succession and compare it to a climax stage of succession.
⊙ 3. How might a pH change that seems to kill only developing tadpoles in a pond affect other species like algae, snails, and bass?
⊙ 4. Which would have had a better chance of survival in the period right after the Flood—small animals like rats and mice or larger ones like bears and lions? Defend your answer.

19C—Man in the Biosphere

Within the last few decades, some individuals, in an attempt to stop logging projects, have set up housekeeping high in the branches of old-growth trees. Others have sabotaged car dealerships by torching brand-new SUVs to protest overconsumption. Whaling ships have been sunk by protesters opposed to the killing of whales. Those with less radical notions have stalled the building of a dam to preserve a rare minnow or blocked the development of private land to save an endangered insect.

These situations reflect the impact of a movement called *deep ecology*. Although it is more of a philosophical approach, it has had far-reaching effects in many areas of modern culture and science. In fact, many of the goals of animal rights and environmental activists are based on this movement. Deep ecology emphasizes two points:

1. All life is of equal value. That is, every species has an inherent right to life no matter how small its role in the ecosystem.
2. All life forms are interdependent on each other. This holistic view of life views man as just another species that is part of the living earth. He is not separate from it, nor does he have any special rights over any other species.

Deep ecologists teach that man has thrown off the balance of nature, and they sometimes suggest extreme changes to fix ecological problems, such as cutting the human population by 90%, abandoning modern technology, and even giving up modern medicine so that man can have a more "natural" lifespan. In their way of thinking, man is a "cancer" upon the earth. Although many who are concerned about the environment would not promote such drastic measures, the fingerprints of deep ecology can be seen in many positions and initiatives of the environmental movement.

Some in the deep ecology movement believe that the biblical Creation Mandate is the root of all present ecological problems since it gave man both the permission and the mission to completely dominate nature. Deep ecology's accusation that Christianity is the main source of present ecological problems does not hold up to scrutiny. The atheistic Soviet Union produced much more serious pollution and introduced many more ecological problems than the Christian West. Also, ancient nature-worshipping cultures such as the Roman Empire or the Easter Island natives caused severe ecological problems too.

This is not to say, however, that deep ecology is totally wrong in its criticism of Christianity. Some Christians mistreat the environment. They believe that the environment does not matter because the earth is going to be destroyed in the end anyway. This is the wrong attitude because God has given us the present responsibility of taking care of the earth (Gen. 2:15). Man's authority to

19-22 —
Some believe it would be best if the world were left in its natural state. God's creation, however, is not necessarily violated by a farmer's plow, a rancher's herd, or a family's house.

manage the world does not give him license to misuse or abuse it. There is a considerable difference between wise, conservative use and foolish squandering. God has promised to supply a Christian's needs, but not necessarily all his wants. Man can expect that God has supplied all his needs in his environment, but if man foolishly takes all he wants, someone will eventually have to pay.

In Deuteronomy 22:6–7 the children of Israel were told that if they came upon a nesting bird, they could take the young, but not the mother. "But . . . let the dam go, and take the young to thee; that it may be well with thee, and that thou mayest prolong thy days" (v. 7). This may speak of God's increasing a man's lifetime, but it may also refer to man's care of the environment by not overtaxing it and thus keeping his environment capable of supporting him longer.

19.5 Man's Ecological Niche

Organisms must either find a niche in an area or migrate to a new area, or they will die. Wherever they are, they affect their environment. Man also affects the environment in which he lives; however, he is intelligent and capable of modifying an ecosystem to supply his needs. There is nothing wrong with man's using his environment, but as with almost anything man can do, there are acceptable and unacceptable ways to do it. The question is not whether man should use the physical world, but how, when, where, and to what extent.

Man—the Consumer-Manager

Man affects his environment in two basic ways:

✦ *Consumer.* Man can be considered a consumer since he takes materials from the food chain. Man is capable of interrupting the flow of energy and materials in an ecosystem at many different points to supply his own needs. Man can be an herbivore and a carnivore, and he can even consume organisms of the detritus food chain (shrimp, lobster, catfish, and such).

✦ *Manager.* When man begins to change the environment to make it meet his needs, he becomes a **manager**. When man clears ground and plants crops; when he converts land into pasture for domestic animals; when he dams a stream to create a lake or to control flooding; when he constructs a road, a home, or a mall, man becomes a manager of the environment. When the human population, and therefore man's consumption, increases beyond what an area can support, man must either move to other areas or become a manager of his environment.

There are many ecosystems that, because of their high rates of productivity, can tolerate human "predation" without suffering. For example, the Indians of the American plains killed bison to meet their shelter, food, and clothing needs. Seasonally they moved from place to place in order to harvest the natural foods the land provided. As long as their population remained small, they were able to continue this existence as consumers without damaging the ecosystem.

Some people are willing to accept man's role as a consumer as long as he only "nibbles." If man consumes so much that he changes the environment or must become a manager of the environment to supply his needs, they feel man has gone too far. They feel that man's management spoils the environment and that he should therefore limit his habitat and be only a "light consumer."

The Bible, however, says to man, "in the sweat of thy face shalt thou eat bread" (Gen. 3:19); "be fruitful, and multiply"; and "subdue . . . and have dominion over . . . the earth" (Gen. 1:28). Not everything that man wants to do

19.5

⮞ **Objectives**
- Describe man's role in the biosphere as both consumer and manager
- Give examples of man's abuse of his role as consumer-manager

⮞ **Key Terms**

manager
invasive species
pantheism

19-23

A fish harvest. Man consumes much that comes from the sea. To a limited degree, man has even begun to manage sections of the ocean.

19-24

The American bison at one time formed extensive herds. They were wastefully killed, almost to the point of extinction. Today, small herds are managed in parks and preserves.

to his environment is acceptable; he must be a good and wise steward of what God has given him. But the problems of population growth and of subduing and having dominion over the earth force man to be a manager.

Man—the Manager

Man can carry his consumer-manager role too far and destroy some aspects of his environment. For example, when the pioneers came to the American West, they began to prey heavily on the bison, often killing hundreds of them by chasing an entire herd off the edge of a cliff and then taking only a few. The rest were left to rot. Other times men shot bison merely for sport with no intention of using the meat or hides.

Man killed so many bison that herds decreased in size, and the bison faced extinction. Only through changes in man's practices and through careful management do bison herds survive in a few parks and preserves today. If man had used wisdom in his consumption and management of bison, they would not have faced extinction. Many other ecological problems could be avoided if man were a better manager.

Unwelcome Guests

Sometimes man harms the environment not by directly destroying species but by introducing species from a different place. In many cases, these introduced species, away from the limiting factors of their old environment, get out of control. These **invasive species** often spread at the expense of species originally found in the habitat, causing the original species to decline or even become extinct. Some invasive species also cause problems for humans. The kudzu vine, European starlings, and fire ants are just a few examples of invasive species in the United States.

Kudzu is a vine that was intentionally brought to the southeastern United States but has become an invasive species, causing many problems.

Some people feel that almost anything man does to the environment is bad. If a dam is built on a river and floods a valley, some people feel that too great a sacrifice has been made. There is justification for building dams. Without dams for water storage and flood prevention, certain areas of the land would be uninhabitable and uncultivable. However, just because some dams benefit man, it cannot be concluded that dams should be built anywhere and everywhere.

Every proposed dam must be carefully considered. What would be the losses? What would be the gains? Do the gains outweigh the losses? By how much? Could the needs be met another way? With accurate answers to these and similar questions, the wise and proper decision can be made.

All too often, however, man does not seek these answers, or he ignores them. Sometimes he overrates the importance of one answer in relation to another. Some people feel that killing any animal or forcing any number of people out of their homes is too great a price to pay. On the other hand, killing animals or destroying homes unnecessarily makes little difference to some people if money is made in the process. After man gets answers to his questions, he must use his intelligence to weigh those answers.

Man—the Consumer

The Los Angeles River runs for 82 km (51 mi) through Los Angeles in a concrete-lined channel to the Pacific Ocean. It is fed by 644 km (400 mi) of concrete tributaries. During dry seasons it carries little more than a trickle of water. The city uses all the water of this once-large river for drinking, irrigation, and industry and then passes it through sewers and eventually into the ocean.

Long ago, when the water in the river began to be used up, the people of Los Angeles began to dig shallow wells to the water table. As more and more people used this water source, the water table lowered, and wells had to be dug deeper. It became evident that the water, which had seemed so abundant and easily obtained (as most natural resources are at first), was limited.

19-25

A section of the concrete-lined Los Angeles River near downtown Los Angeles

As Los Angeles continued to grow, its people needed more water. An aqueduct built to the mountains diverted the distant Owens River to Los Angeles, but in time the city needed still more water. An aqueduct 354 km (220 mi) long, which would bring water from the Colorado River, was proposed. But other southwestern states and Mexico channel water from the Colorado River. A compromise worked out among the various governments permitted dams to be built to form reservoirs during the wet seasons.

But even with rerouting the water in Los Angeles and nearby areas, there are still shortages in some seasons, and during rainy seasons there is sometimes too much. What should have been done, and what should be done now? Some possible considerations are the following:

- Although it may be loathsome to those who would have everything "natural," perhaps additional channeling of water is part of the answer.
- Perhaps areas in the Southwest should have restricted growth—prohibiting new homes, apartments, and water-using industries.
- Although it is costly, salt water can be converted to fresh in a process called *reverse osmosis*. The Pacific Ocean has no shortage of water and, if it can be desalinated economically, could supply all the water Los Angeles needs.
- Used water can now be recycled. This may be part of the answer.

There are many avenues of action that can be taken and possibly others that have not yet been developed. The questions are which solution should be pursued and when, how, and how far should it be pursued. Only through study can man know the answers needed to make wise decisions. But even then the "best" answer often upsets people who are inconvenienced by it.

Stories similar to the Los Angeles water shortage can be told of many of man's uses of natural resources such as minerals, coal, oil, timber, and even the air. For example, during the era of the early American settlers, the forests of the United States were a natural resource so abundant that they could be used without thought of depletion. With the small human population, trees that were taken were replaced naturally before others were needed. In time, as the population increased and technology improved, the rate of use exceeded the rate of growth. By the end of the nineteenth century, 30% of America's original forests had been cleared in a process now called *deforestation*. Fortunately, forests were then recognized to be a valuable resource, and measures were put in place to maintain them. Today, about 75% of the original area of America's forest reserves have been stabilized.

19-26 —

Man's need of certain materials has resulted in massive environmental manipulation such as this strip mine. Leaving a hole when a desired substance is mined out is poor environmental management. Today, laws often require companies that alter the environment to help the succession of the area reach its climax vegetation. The photo at the right shows the area after restoration.

God-Supplied Natural Resources

It is inconceivable that God would have placed man on an earth that did not have adequate natural resources to supply his needs. However, some people think that no matter what man does, the ecological damage he causes will be irreparable. This could happen, but if it does, it will occur because man has misused his environment, not because he used it.

In Europe for hundreds of years, bear-baiting was a popular spectacle. A bear, chained to a post or confined in a pen, was whipped or tortured (sometimes its nose was filled with pepper to arouse it). Trained dogs then attacked the bear, trying to seize its nose. Bearbaiting is now outlawed in most civilized countries.

Today, bullfights are popular events in Mexico, Spain, and several other countries. In a bullfight, a bull specially bred for his strength and aggressiveness is placed in a ring with trained, colorfully dressed people whose main aim is to entertain the spectators. While bloodless bullfights are sometimes staged in the United States, the real bullfights in Spain and Latin America all end with the matador stabbing and killing the bull.

Some consider spectator events in which animals are forced to fight and

Modern bullfights are colorful spectacles designed to entertain and thrill crowds.

suffer pain as cruel; those who enjoy them are considered sadistic, and those who engage in the event are thought barbaric. Others consider attending a bullfight or a cockfight an honorable part of their culture and heritage.

Some people argue that events involving animals are less offensive than football and boxing, which involve spectators watching as people inflict pain and occasionally serious injury and death on each other. Human sports are sometimes defended by the argument that the participants have chosen to be involved, whereas the animals in cruel spectator events have not chosen this as their end.

The animal that provided the meat in your last hamburger did not consent to satisfy your hunger, nor did the animal that gave his hide to supply the leather for your shoes volunteer for such a cause. What about the animals used in medical research? Do they choose to be exposed to harmful chemicals so that scientists can learn whether the chemical is safe to use in food or medicines?

What rights do animals have? What constitutes animal abuse by man? What position should a Christian take regarding the use and the abuse of animals?

In Scripture God gives man control over animals, tells him to use them for food, and gives His approval for using animals for other purposes such as labor and clothing. As with every physical blessing that God has placed under human dominion, man can choose to properly use or to abuse his authority over the animal kingdom.

If modern Americans tried to feed themselves using only wild game, there would soon be shortages. Man could wipe out every wild species considered desirable for food. This would be an abuse of dominion over animals. Today, hunting of certain animals is limited to avoid such abuse.

Man has domesticated various animals and, through farming, raises enough animals to meet his needs while not destroying wild animal populations. Although techniques differ, the concept of farming was used in ancient times and is not condemned in Scripture.

When man hunts or domesticates an animal, he must assume a portion of the responsibility for the wise use of that animal. As steward of the physical world, man is responsible for altering an animal's habitat. There is nothing wrong with man's using the physical environment, nor does Scripture directly forbid the use of animals. But that does not give man permission to abuse the environment or hunt animals to extinction.

Most people would agree that brutal or cruel treatment of animals, unnecessary scientific experimentation using animals, wasteful killing of animals, unnecessary hunting of endangered species, and similar abuses should be condemned. These are abuses of man's position of dominion over animals and cannot be justified with Scripture.

Animals in Scientific Experimentation: Diabetes and Dogs

Diabetes mellitus (muh LYE tus) is a disease that involves the inability to properly metabolize food. Before the 1920s, a person with this disease experienced gradual physical degeneration and eventual death. If a diabetic person ate foods rich in sugar, his blood sugar level could become so high that he could die in minutes. Even if he ate only foods low in sugar, he would gradually lose weight because his body could not properly utilize the food. Many adult diabetics died weighing less than 50 pounds after years of slowly wasting away.

Scientists observed deformed *islets of Langerhans* (small clusters of specialized cells in the pancreas) in the bodies of people who died from this disease, but they did not know whether this condition caused the disease or whether it was a result of the disease. Reports that a dog with its pancreas removed developed diabetic symptoms prompted Frederick Banting, a Canadian physician, to consider that diabetes might be related to a secretion from the islets of Langerhans.

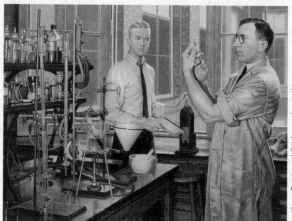

Banting, Best and Diabetes by Robert Thom, UMHS.43

Best (left) and Banting (right) discovered the link between insulin and diabetes.

In the early 1920s, Banting and Charles Best began a series of experiments at the University of Toronto. First, they needed to determine whether a lack of the materials manufactured by the pancreas caused diabetes. Removing a dog's pancreas and examining its blood confirmed that the blood sugar level increased when the pancreas was removed. They then injected extracts made from the dog's pancreas and found that the blood sugar level dropped.

Additional experiments were necessary to determine whether the islets of Langerhans were involved in diabetes. More experiments were needed to isolate the chemicals produced by the islets of Langerhans, and then additional experiments helped determine which ones were involved in the disease. Additional tests were conducted to determine how this information could be used in the treatment of human diabetics. John Macleod, who had allowed Banting and Best to use his laboratory space, joined the project at this point.

Banting and his colleagues used dogs in their experiments, in part because they were large enough to permit dealing with the pancreas. The dogs were given the best care possible, but some inevitably died. Scientists took the lives of a number of dogs to discover, isolate, and test insulin (the hormone produced by the islets of Langerhans that is responsible for proper metabolism of sugar; see pp. 592–93). There was, however, no other way that this lifesaving information could have been obtained at the time. (With today's improved medical techniques, mice or rats could probably be used.) These animals were sacrificed for medical information that saves the lives of many people.

In a war, soldiers die so that the people of their nation can live their lives the way they want to. War is not a desirable event, but in this world it is often necessary. Likewise, in the war against diabetes, dogs died so that many people could live. Using animals in this way may not be desirable to some people, but it is often necessary.

Humans are the only organisms God created with an eternal soul. God validated the worth of human souls by sending His Son to die for them. This leads to the conclusion that human life is far more valuable than animal life. The sacrifice of animals for medical knowledge is within the guidelines established in Scripture (Gen. 1:26).

The New Age and Animal Rights

The Bible teaches an absolute distinction between the Creator and the creature. Scripture teaches that God existed before He created the world and that the world was designed, created, and is sustained by Him. God will also exist after He has destroyed this present world. God's power is revealed in the creation, but God is not the world.

Many religions believe in **pantheism**—that the physical world is god. Pantheists speak of the "One" or "Ego" or "Great Spirit" of which every living thing (and for some pantheists, every nonliving thing as well) shares a part. This spiritual part that all things have in common makes all things related (mother earth, father sea, brother rock, and so on). Thus, to destroy or harm anything is to sin against or offend the pantheist's god. But to realize the "spiritual oneness" of everything is to be more aware of that god.

The *Gaia movement* of today is based on a theory that the entire biosphere functions as a single thinking organism. Humans are said to be not caretakers of the earth, but just another part of this giant life force. Gaia is an unusual amalgam of environmental science and mysticism that tries to unite various environmental organizations into a unified alliance. Adherents to this thinking may refer to Gaia as if "she" were God. Its philosophical beliefs are little more than ancient pantheism repackaged for a modern audience.

Although pantheism has long been denounced by Bible-believing Christians as a heresy, many Christians are being influenced by pantheism dressed in new clothes—the New Age movement and its environmental cousin, Gaia. By teaching doctrines that sound good and not publicizing those parts that might be offensive, the New Age movement has gained popularity even among some Christians.

One appealing pantheistic doctrine is animal rights. Pictures of suffering animals have swayed public sentiment and have caused people to believe the movement is intended only to stop animal cruelty. Certainly bringing unnecessary animal cruelty to people's attention and applying pressure to have such abuse stopped is a legitimate cause. Many animal rights activists, however, are interested not only in stopping unnecessary cruelty to animals but also in preventing the use of animals for fur or meat or even as pets.

To animal rights pantheists, no use of animals by humans is justifiable. They regard animals and humans as parts of the same god. Pantheists say that it is as wrong to hurt an animal as it is to hurt another human and that animals should have the same rights and protection under law as people have. Scripture does not support this idea.

Many claim that because the earth is becoming overpopulated, there will soon be a shortage of food. It is true that the population of the earth in 1700 was just under 1 billion, and the 3 billion mark was reached about 1970. It doubled in less than three more decades to reach the 6 billion mark by 1999. Recent studies have projected a population of 9 billion by 2050 and perhaps 11 billion by 2200. Although the rate of growth has slowed in recent decades, more children are surviving and people are living longer because of cleaner water, better nutrition, and better disease control.

World leaders question whether it will be possible to feed that many people, but history has shown that predictions of population growth are notoriously inaccurate. War and natural disasters, for instance, can stem the tide of

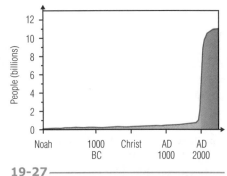

19-27

Population growth from Noah to AD 2500 (projected)

population growth. Concerns about a food shortage should also be allayed by the fact that only 8% of the earth's land surface (well under the amount that can be cultivated), if farmed using modern techniques, would be able to produce enough food to support 79 billion people.

A greater challenge than population growth is the lack of income to purchase food or even seeds. Many of the new, high-yield crops are very expensive because of the research that went into their development. Some countries are also hampered by the lack of an efficient system for storing and distributing food. In many developing countries, government corruption interferes. The financial and political issues are far from simple, but it is within man's abilities to overcome these obstacles.

Review Questions 19.5

1. What are the two main aspects of man's ecological niche?
2. In what ways is man the top consumer in the biosphere? How does man differ in his consumption from all other consumers in the biosphere?
3. What types of things does man do to manage the ecosystems?
4. What is the primary problem regarding man's use of natural resources?

19.6 Pollution

One of the major ecological problems facing man today is pollution. **Pollution** is man's placing into the environment substances that, because of either their nature or their abundance, make a significant negative change in the environment. Some of the substances that are considered pollutants (such as carbon monoxide) can occur naturally, while others (certain pesticides, for example) are man-made. When the human population was small, limited technology did not permit man to produce quantities of harmful substances that the environment was unable to deal with. With increased population and advanced technology, however, man can now leave pollutants in the environment that will alter it for many thousands of years. With such abilities comes the responsibility to use technology wisely.

Ecologists have found it useful to distinguish between two kinds of pollution. Pollutants that the environment can break down and return to the normal cycling of substances are *biodegradable*. These include sewage, paper, wood products, and many chemicals. Pollutants that stay in their original form and cannot be broken apart in the environment are *nonbiodegradable*. Glass, metal, and many chemicals used in insecticides and found in industrial wastes are nonbiodegradable pollutants.

Pollution can have an impact on the water, air, and soil. The following sections give examples of the impact that pollutants can have in these three realms.

Water Pollution

If the wastes from the sewers of a city are released into a stream (assuming that the sewage is not from industries producing nonbiodegradable materials), it is only a matter of time and distance before the natural decomposer organisms in the stream will have converted the sewage into usable materials. Depending on the amount of sewage and the size of the stream, the polluted area may be only a few feet or a few miles.

If the amount of sewage is large enough to pollute the entire stream for some distance, the pollutants may form an effective barrier to clean-water organisms, but they do supply a large niche for the decomposer organisms that normally exist in small numbers.

pollution: (L. *polluere*—to spread plague, pestilence)

biodegradable: bio- (life) + -de- (L. *de*—down or reversal) + -grad- (*gradus*—rank or step)

Problems arise when a large amount of sewage is put into a small stream or when several cities place sewage in a larger stream. The stream may be an open sewer for its entire length. If the stream is of any value for fishing and beauty or if cities rely on it for their water supply, then severely polluting a large portion of the stream is unacceptable.

Mankind has the technology to convert sewage into pure water. One common practice is to run the sewage through a series of man-made canals or tanks while bubbling air through it. This encourages the growth of the same decomposer organisms that would naturally purify the water in a stream. Rather than polluting a stream, man should clean the water before returning it to the environment.

Some substances that end up in the water are not as easily decomposed. Oil spills from tankers or storage facilities make the headlines because of the far-reaching and long-term impact and because of the enormous cleanup costs. The thick crude oil not only coats and kills birds and marine mammals such as sea otters, but it also blankets the shoreline and can completely change the ecosystem unless it is scrubbed off.

Although spills of oil and other chemicals into rivers and oceans may make the news, greater problems are caused by *nonpoint source pollution*. This occurs when rain or snowmelt runoff carries natural or man-made substances into waterways or when these substances percolate down into the groundwater. This is the kind of pollution that occurs when a farmer puts fertilizer (soluble minerals) on his crops, and then rainwater carries the chemicals into a pond. Even though there normally are soluble minerals in a pond, added minerals can easily become pollutants.

19-28

Both of these activities have the potential to pollute water.

DDT: Miracle or Menace?

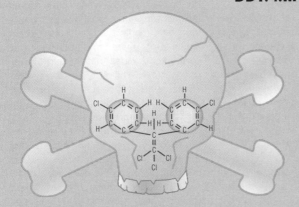

A very effective pesticide called DDT (dichlorodiphenyltrichloroethane) was developed at the end of WWII and used extensively against mosquitoes and a host of crop pests for many years. Millions of humans were spared from malaria, but by the late 1960s, some tests seemed to show that fish and birds were harmed by it. This led to a ban on its use in the United States and in many other countries.

DDT and other similar nonbiodegradable pollutants are capable of significant, widespread **biological magnification**. For example, some of the DDT put on crops eventually collects in the ocean as a result of runoff; then it collects in the oil droplets of diatoms. If DDT were a compound that could be quickly broken down and metabolized, it would not pose much

risk. Instead, it is stored in the fat of animals for many years. Because of the loss of energy from one trophic level to another and the need for each animal to feed on many smaller organisms, chemicals like DDT become more concentrated. Diatoms are eaten by minute crustaceans, which are eaten by small fish, which are then eaten by larger fish, which are finally eaten by fish-eating birds. Bald eagles, terns, herons, and pelicans were shown to have high levels of DDT in their bodies and in the eggs they produced. Some widely disputed studies indicated that this might have impaired their reproduction by causing the eggshells to be too thin for proper chick development. The chemical was also said to put humans at increased risk for cancer. It was banned for use in the United States in 1973. A closer look at the data shows little support for these initial fears.

DDT has been greatly misused in the past. However, when its use was outlawed, crop losses and insect-related human diseases increased. Since some of the newer pesticides seem to be more dangerous to the humans who apply them, DDT should not be completely ignored in the war against insects. In emergencies DDT may be the only known way to bring certain insects under control and should probably be used at such times. Meanwhile, better, nonpolluting ways of insect control are being sought. God-given intelligence can help man subdue and dominate the physical world, even mosquitoes, without needlessly destroying other life.

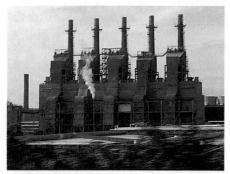

19-29
Federal regulations in the United States require industries to remove most of the harmful pollution from their exhaust or face stiff fines.

As the pond's physical chemistry and other abiotic factors change, the numbers and kinds of its populations also change. The abundance of algae caused by the added minerals causes the pond to become cloudy. When the algae die, their decomposition can kill fish. The fertilizer that was designed to help the farmer's crop may eventually destroy a pond. Such unintentional polluting of the environment has affected rivers, lakes, and even oceans. Other common sources of nonpoint source pollution include the oil and gas that wash off parking lots and roadways and the soil that erodes from fields and construction sites.

Air Pollution

Most of the pollutants that find their way into the air fall into two categories: *particulate matter* (the larger particles such as carbon soot that are suspended in the air after a fire) and *noxious gases* such as sulfur dioxide (SO_2), which is linked to acid rain. Some air pollution is the result of industry, while much of it comes from people driving cars or even spraying aerosol cans.

Good Ozone, Bad Ozone

Ozone is a triatomic form of oxygen (O_3) that is very healthful, as long as it is not in the wrong place at the wrong time. A layer of ozone in the stratosphere (the upper atmosphere) actually shields man from some of the ultraviolet (UV) rays of the sun. High levels of UV radiation have been associated with increased skin cancer, cataracts, and other health concerns. Stratospheric ozone is beneficial.

When diatomic oxygen (O_2) reacts with certain pollutants (especially those found in automobile exhaust) in the lower atmosphere, it can form ground-level ozone. This type of ozone is a component of the smog often seen in large cities, and it can be very dangerous for people with asthma or other respiratory problems. Because higher temperatures increase ozone production, summer is often the season when ozone warnings are announced.

Low-level ozone is a major component of smog.

Global Warming?

All living things release carbon dioxide (CO_2) as they respire. The level of CO_2 is normally kept in check by green plants, which utilize it for photosynthesis. The combustion of fossil fuels such as coal and petroleum also releases CO_2, which is known to be increasing in the atmosphere. The earth's temperature is kept warm enough to encourage abundant life as a result of the insulating effect of a layer of CO_2 and other gases. These gases allow sunlight to pass through but trap the radiation that reflects off the earth, keeping it from returning to space. This is known as the **greenhouse effect**. Some scientists have analyzed long-term climate data and have noted a slight increase in the earth's temperature over the past century. While the data is far from conclusive, some blame CO_2 and other "greenhouse gases" for the increase. They attribute this **global warming** to car engines, electric power plants run on fossil fuels, and other major sources of CO_2 emissions. It is important to note that just because there is a correlation between CO_2 fluctuations and temperature does not prove cause and effect. In addition, many scientists are not convinced that man's activities are making a significant contribution to these changes in CO_2 levels.

The worst-case scenario, assuming that global warming is legitimate and continues, includes the melting of the polar ice caps, an action that could raise sea levels by an unknown amount, possibly flooding many of the world's major cities (since many of them are in coastal areas) and covering a few low-lying island nations. Such warming could also change major weather patterns worldwide and shift many of the biomes, perhaps triggering mass extinction of species that cannot adapt or migrate. Drought, famines, and weather extremes would kill many people.

Critics of this doomsday outlook point out that the earth goes through regular, gradual adjustments in temperature. In the middle of the twentieth century, for example, climatologists were concerned about the effect of a global *cooling* that they were noticing. While certain ground temperatures are now showing a warming trend, readings by satellites of the actual atmosphere temperature seem to contradict the idea that the earth is getting warmer. While it is true that some studies suggest that ocean temperatures are rising, there is no clear evidence linking this to the greenhouse effect.

The global warming issue illustrates the nature of science and the influence of worldviews. While most scientists admit that there is some evidence that the earth's temperature is rising, the extent of man's impact on this variation is far from certain. A person's worldview also determines his view of the extent of the global warming threat. Christians are assured by Scripture that the world will not end as a result of global warming. Non-Christians, however, have no conviction that God will preserve the earth. To them the earth is a fragile accident of evolution and could be destroyed by human-caused climate change. It is obvious that both sides of the argument have strong feelings and interpret the data in ways that support their presuppositions and political leanings.

Improved emission control systems on cars have greatly reduced the pollution they release. Filters and other mechanical means are now used to remove much of the particulate matter from factories, power plants, and other large producers of exhaust. Scrubbers trap many dangerous gases by bubbling the exhaust through liquids.

The invisible gases society produces actually pose more risks than the particulate matter. Despite all of the antipollution laws, large amounts of potentially harmful materials are released into the air. While some dangerous gases can result only from artificial processes, others are completely natural.

Soil Pollution

Because the earth's abiotic factors are interconnected, what happens on land will almost inevitably affect the air and the water. Chemicals applied to fields can follow runoff water into streams, and evaporation or oxidation from substances on the land releases some air pollutants. Conversely, polluted water can affect soils where water meets land, and air pollution can dissolve into rain, which then reaches the soil.

Most soil pollution issues relate to man-made chemicals that are deliberately or accidentally released into the ground. It may be as seemingly insignificant as a homeowner's pouring used motor oil into a ditch or as major as the soil of a whole neighborhood being contaminated with asbestos from an area factory. Some of the most difficult challenges of soil pollution relate to the handling of hazardous wastes.

Hazardous Wastes

The problem of what to do with inert (chemically unreactive) nonbiodegradable pollutants such as plastic and glass is significant, and recycling is often the answer. In many ways, however, that challenge is not as significant as what is done with pollutants that are chemically active, especially those that are harmful to living things. Such chemicals are called *hazardous wastes*. They are classified as hazardous because they are toxic to organisms, are highly reactive or even explosive with other elements, cause corrosion, or are flammable.

Most of these wastes are chemical byproducts of making paints, plastics, textiles, and some other everyday products. Some of them are so strong that a single drop of the chemical in a swimming pool of water could kill anyone who drinks the water.

What is to be done with these substances? For a long time factories that made hazardous chemicals put them into metal drums. The drums were then dumped into the ocean, buried underground, placed in caves and old mines, or stacked on land in remote areas. In time, however, the drums rusted or the chemicals inside them ate away the protective lining, and the drums began to leak. These highly poisonous substances were then released into the environment.

For a long time the "solution to pollution is dilution" principle was applied to hazardous wastes. Then scientists discovered that many hazardous wastes that seep into the environment could still affect humans many miles away. Often these pollutants are carried by the movement of water or by the food chain.

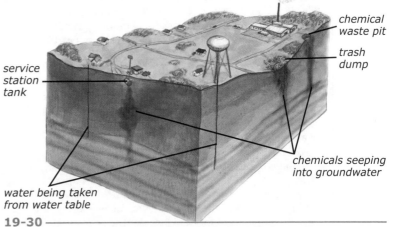

chemical waste pit

trash dump

service station tank

chemicals seeping into groundwater

water being taken from water table

19-30

The ground was thought to be a good "filter" for nonbiodegradable chemicals. But if placed in the ground, these chemicals actually pass into the water table, though the process may take many years. Some pollutants can taint water and harm organisms many miles away from their point of origin.

What a Waste!

For a long time, the saying "the solution to pollution is dilution" seemed to hold true. It was thought that if the pollutant was spread out enough, it would become unnoticeable. A little trash tucked in an unnoticeable place here or a little trash buried in an out-of-the-way place there was considered acceptable. But when the quantities began to grow and the nearby places became more crowded, the dilution solution became unacceptable. Landfills (areas where trash is placed, often with layers of dirt to speed the decomposition of biodegradable material) become full, and new areas for putting trash must be found. But no one wants to live near the trash dump. In fact, the acronym NIMBY (Not In My Back Yard) was coined to describe this attitude.

Rather than using landfills, many coastal cities resorted to putting their trash in barges and dumping it in the ocean. Some people thought that it would take extreme amounts of trash to really pollute the ocean and that as long as the trash did not interfere with swimming or fishing, there would be no problem. Biodegradable substances are easily recycled by ocean ecosystems.

But the nonbiodegradable substances continued to build up, and in recent years swimmers have found such wastes washed up on beaches. The nature of some of these materials has forced the closing of some beaches. Today, less garbage is being dumped directly into the ocean, but the ashes of incinerated trash are often disposed of this way.

Another solution to the problem of nonbiodegradable trash (such as glass and plastics) has been to burn it. But burning creates airborne chemicals and other, often more serious problems. Toxic chemicals may be concentrated in the remaining ash, and, although it is lighter and has less volume, it must still be disposed of.

Landfills eventually reach their capacity and must be closed.

Most modern landfills are lined to prevent leakage and then covered with some sort of barrier when they become full. Even with these precautions, the air and area groundwater are continually monitored to make sure pollutants do not leave the secure landfill site.

Recycling

One answer to certain pollutant problems is *recycling*. Recycling helps reduce the demands upon natural resources and keeps materials out of landfills. In the United States every man, woman, and child produces nearly 5 pounds of trash every day. This number has nearly doubled in the past 40 years. Fortunately, approximately 27% of solid waste is now being recycled.

Substances like paper, glass, plastic, and even cloth fiber can be recycled. Leftover lumber from construction/demolition sites can be chipped or ground for use in particleboard. Recycling is necessary for many metals. Since the amount of readily available metals in the earth is limited, melting them and reusing them, rather than leaving them to pollute the environment, is much to man's advantage. Some metals, such as aluminum, are much less expensive to recycle than to mine from the earth since recycling requires only 11% of the energy it takes to mine new bauxite ore and process it. That's why nearly half of all aluminum cans are turned back in for recycling.

One factor working against recycling is the difficulty of getting people to do it. If the demand for a substance is great enough, the market will pay a high enough price to make people want to turn in the materials. When aluminum prices are up, people scour the roadsides and dumpsters, picking up cans for extra income. The opposite is true when aluminum prices fall. The sometimes dirty job of sorting trash is not very appealing to consumers who get little or nothing in return or even have to pay extra to a trash company to pick up their recyclables.

Another problematic factor for a successful recycling program is "closing the loop." If consumers purchase packaging and materials that are made from recycled material, there is incentive to keep on recycling. In some cases, however, these recycled products may cost more than comparable nonrecycled products. Then consumers must decide whether the benefits to the environment are important enough to them.

The logo on the left indicates that the product can be recycled, while the one on the right shows that it contains recycled material.

Man's Ecological Future

The Bible teaches that this world will be destroyed by God because of sin. Man's rebellion is ultimately responsible for the great trumpets of judgment described in Revelation 8 and the vials of God's wrath poured out upon the earth in Revelation 16. Those events will cause massive destruction of the physical world and the cataclysmic death of most living things.

Some believe that, because the Bible says living things will exist in the end times, those living things are going to survive until then, no matter what man may do. Often these people take this belief as license to waste or misuse natural resources. It is true that the Lord will take care of tomorrow, but people who are reckless about environmental issues sin when they do not carefully use and protect what God has given. The unsaved world should not see Christians squandering the Lord's blessings. Christians should not ignore genuine environmental issues. The Lord has left believers as stewards of His creation—a responsibility not to be taken lightly.

➥ *Review Questions 19.6*

1. What are the two main types of pollutants, on the basis of their decomposition?
2. What are hazardous wastes? Give several reasons that they are difficult to deal with.
3. In what ways can a beneficial substance put in an ecosystem become a pollutant?
4. Why is biological magnification a problem with nonbiodegradable pollutants but not with biodegradable pollutants?
⊙ 5. A large swamp is to be drained by constructing a concrete-lined channel for the stream that feeds into the swamp. List some questions that should be answered before work is started.
⊙ 6. Write a paragraph either supporting or refuting the following statement: Since God has given him the know-how, whatever man wants to do to the environment is within God's will.

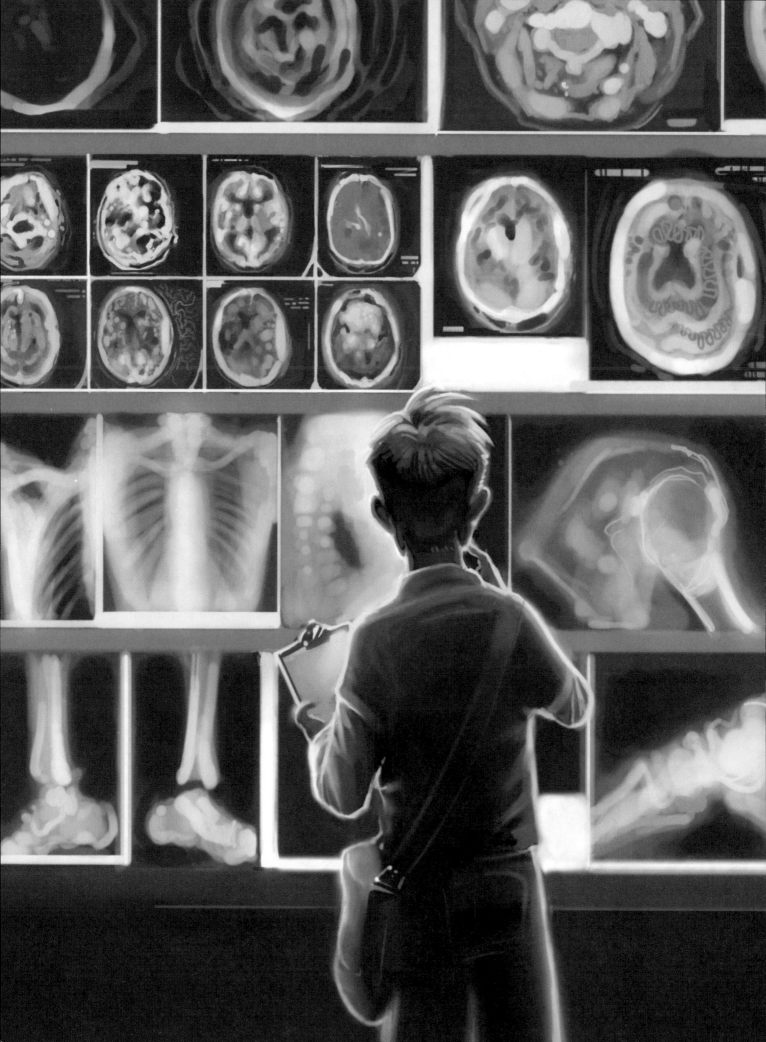

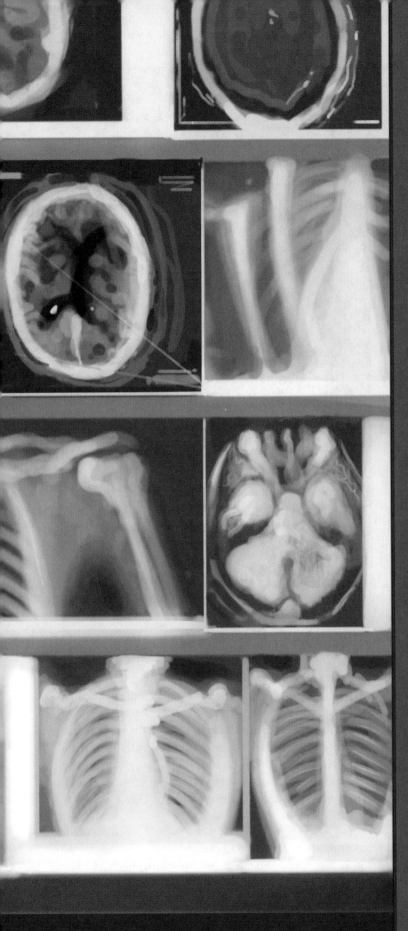

Unit 3

The Study of Human Life

Introduction to Human Anatomy and Physiology

20A—What Is a Human?

According to current evolutionary theory, modern man first appeared in the savannas of Africa about 200 000 years ago. Man had many improvements over his primate ancestors, including the ability to walk upright, a larger brain, and the use of language. But is man really just a very smart animal? The evolutionary worldview provides man with very little to distinguish himself from animals, leaving man with no more dignity or worth than any other animal. The Bible, however, paints a very different picture of man's origin and worth by asserting that he is the special creation of God.

20.1 In the Image of God

Humans are different from animals because "God created man in his own image" (Gen. 1:27). A person is more than the chemicals that form his body. He is even more than the functioning of those chemicals called life. Being created in God's image does not mean that men have a physical body like God; God is a spirit (John 4:24). Man is a special creation of God. Animals were not formed and given life and spirits in the image of God. But what does it mean to be made in the image of God? An easy answer is "that which makes

20.1

Objectives
- Describe the basic levels of human behavior
- Discuss the differences between animals and humans

man different from animals." Examining what a man is by what he does helps clarify what the image of God is.

Human Behavior: Inborn

Chapter 17 describes the three basic levels of behavior found in animals: *inborn*, *conditioned*, and *intelligent*. The lowest level of inborn behavior is reflex action, which is the only behavior for some animals. Humans also have reflexes. Focusing the eye and pulling away from a source of pain are reflex actions.

The other major type of inborn behavior is *instinct*. Instincts are complex reactions to various stimuli and account for much of the behavior of vertebrate animals. In birds, parental care is almost completely instinctive. If a bird egg is separated from other birds before it hatches, the hatched bird (when it matures) will still perform all the normal parental duties of its species as it builds a nest, incubates eggs, and cares for its nestlings.

Changing diapers and feeding and bathing babies, however, are all tasks that humans must learn or figure out by using their intelligence. Birds, lacking the intelligence to determine that a nest must be made, that eggs need incubation, and that chirping nestlings require food, must have instincts for parental care.

Is a mother's desire to care for her baby an instinct? Sadly, not all mothers want to care for their children, but most mothers find childcare rewarding. Though a well-groomed, well-behaved child may be considered a status symbol, most mothers care for their children because of love. Love, however, is an emotion and is considered part of intelligent behavior, not instinct. It is impossible to say how much of a mother's care for her child is learned from the examples she sees around her and how much is instinct. Experiments cannot be set up to test for human parental-care instincts. True human instincts are very difficult to determine. The instincts that humans may have are usually so overshadowed by learned and intelligent behavior that the instinct itself cannot be measured.

20-1
The extent of a "mothering" instinct is impossible to determine since many of its activities are learned.

Human Behavior: Conditioned

Humans have learned behaviors, as animals do. Birds have inherited the equipment necessary to fly, but they must learn to do so. Organisms learn some behaviors by themselves; other behaviors must be taught to them. Crawling and walking are behaviors that humans can learn for themselves. Spelling words, reciting Bible verses, and playing the piano are all behaviors that are learned through teaching and repetition. Some dogs have been taught to sit up, fetch, and roll over. Some dogs perform these learned behaviors better than other dogs, depending on the quality of their training. Likewise, a major factor in the success of man's learning is the quality of the teaching he has received. Because of human intelligence, most conditioned behavior does not require the extensive trial and error and repetition employed to teach a dog to sit up.

Human Behavior: Intelligent

A person can take a test to measure his intelligence. These tests, however, are only attempts to measure intelligence. The test scores may show *how much* a person has, but those who read the scores are not sure *what it is* he has. Intelligence has not been adequately defined. There are many attributes ascribed to it. Intelligent behavior is based on the ability to reason, to solve problems, to have insight, to see relationships between objects, to recognize causes and effects, to react to something that does not immediately and

20-2
Intelligence is measured with IQ (intelligence quotient) tests, but scientists disagree about what is actually being measured.

20-3

One characteristic of intelligence is problem solving.

personally affect oneself, to attach value to something that has no immediate value (such as money), and to react emotionally.

Although animals show some intelligence, we must be careful about attributing human behavior to them. Dogs, after they have chewed slippers or soiled the carpet, fear their owners, but the question is whether that fear is truly intelligent or simply learned from previous scoldings. A dog's wagging tail and frisky greeting may be more for the pats and treats he receives than for the highly complex emotion called love.

It is *not* true, however, that humans are the only organisms with intelligence. An ape, put into a cage with a banana hung from the ceiling above his reach and boxes in the corner, will usually look at the banana, stack the boxes under it, climb up on the boxes, and get the banana. A dog in a similar situation will tire himself out by jumping, and then he will lie down and eye the food, waiting for something to happen. The ape has demonstrated intelligence in solving his problem; the dog has shown a lack of intelligence.

Are humans merely highly intelligent animals? Is man a monkeylike organism that merely has enough intelligence to learn a complex communication system using word and letter symbols? Is man merely a very smart animal? Absolutely not! The ability of some animals to exhibit intelligent behavior does not provide evidence to place them on the same level as humans, nor does it support any aspect of evolutionary theory.

Although some animals do have a degree of intelligence, they do not come near the intelligence that man has. First, man has reasoning power far beyond that of any animal. Second, man has the ability to use highly advanced language. Some animals such as whales and dolphins may communicate, but this is far below human language in its complexity. Third, animals may sometimes display emotions, but they are not on par with the many-faceted emotions that man can have.

Human Behavior: Spiritual

Intelligence, emotions, and will are often cited as being part of the image of God. These undoubtedly are attributes of God and man; but to limited and varying degrees, animals also possess intelligence, emotions, and will. Some people cite man's dominion over God's creation (the Creation Mandate) as part of the image of God. Many Christian scholars also agree that the image of God includes the original knowledge, righteousness, and holiness that man lost in Adam's sin.

God originally created man in moral innocence, with true holiness and righteousness, reflecting God's holiness and righteousness. God also created man with a sense of right and wrong, and the ability to choose between them. Tragically, man chose wrong and has marred this part of the image of God. Fallen man still has a remnant of

God the Father in a Glory of Angels, Francesco Montemezzano, From the Bob Jones University Collection

20-4

Artists have often portrayed God the Father as a bearded, older man. These physical human attributes, or *anthropomorphisms*, are merely the artists' attempts to visualize God.

this moral nature in his conscience, which informs a person's sense of right and wrong. Some people can ignore their conscience and commit horrible sins, but the conscience still exists, and deep in their heart such people know they are doing wrong.

In the Christian life, the process of man becoming conformed to the image of Christ is called sanctification, and it is the daily duty of the Christian to conform more of his or her life to the commandments of Christ. One day, either at death or at the Second Coming of Christ, the believer will be conformed to the perfect moral image that God created him to have (Rom. 8:23).

God is the source of all truth and moral authority. Since God created man, he is under God's authority and is accountable to God for his actions. Animals do not have a moral nature and cannot sense right from wrong. God gave the Israelites laws that prescribe killing animals that have killed people, but many Bible scholars agree that God does not punish or reward animals eternally for their actions (Exod. 21:28–29).

Many theologians also agree that man's being created in the image of God involves the fact that both God and man are spiritual beings (John 4:24; 1 Cor. 2:11). Unlike the rest of creation, man can have a relationship with God; man can speak to Him through prayer and hear from Him through the Holy Spirit. Because man is a spiritual being, he has something more than intelligence that directs his behavior. Man is aware that he is more than just a physical body. Animals, on the other hand, are probably not aware of more than the physical realm. Significantly, animals do not make idols, but men do. Animals are also not aware of the significance of their own deaths. Man, in contrast, realizes he will die and, therefore, can prepare physically and spiritually for the inevitable. Man is an immortal spirit-being and will spend eternity somewhere. Depending on whether a person has put his faith in Christ, he will spend eternity either enjoying God's presence in the new heavens and the new earth or enduring punishment from God.

Man's ultimate problem is that he is a sinner in need of a Savior. God has provided that way of salvation through the vicarious death and resurrection of His Son, Jesus Christ, Who paid the price for man's sin. Man can place his faith in Christ for salvation from his sin, and he will spend eternity in heaven (with an everlasting glorified body) with the God Who created him. Or man can reject God (Luke 12:8–9) and spend eternity in hell. A Christian, someone who has trusted Christ for salvation, can gain rewards for good deeds done in Christ's name, or he can receive nothing for his selfish efforts (1 Cor. 3:12–15).

The Bible states that your personal relationship with Christ settles the eternal destiny of your spiritual self (John 3:16). What about you? Have you accepted Jesus as your Savior?

⇒ Review Questions 20.1

1. List the four levels of human behavior. Describe and give examples of each (when possible).

2. List several possible explanations of the idea that man is created in the image of God.

⊙ 3. Compare and contrast animals and humans in regard to the four basic levels of behavior.

⊙ 4. Compare and contrast man and angels in regard to their physical bodies, knowledge, and creation in the image of God. Use Scripture to support your comparisons.

Objectives

- Summarize the difference between anatomy and physiology
- Demonstrate proper use of anatomical terms of direction
- List and describe the four basic human tissues
- Identify the major areas and cavities of the human body

Key Terms

anatomy	matrix
physiology	epithelial
anatomical	tissue
position	muscle tissue
histology	nervous tissue
connective	
tissue	

20.2 Human Tissues, Organs, and Systems

Anatomy is the study of the shape and structure of an organism and its parts. **Physiology** is the study of the functions of a living organism and its parts. Both are important since one cannot be fully understood without the other.

Anatomical Terms

Human anatomy can be an exacting science. There is a name and description for every structure no matter how tiny or insignificant it may appear. For most high-school students, such detailed knowledge is not necessary; however, knowledge of the proper names of major areas of the body is useful. Figure 20-5 includes some terms that help in understanding the information and descriptions presented in this and the remaining chapters of the text.

To be precise, anatomists use terms of direction and locality based on the **anatomical** (AN uh TAHM ih kul) **position**. When a person stands erect, arms at his sides, with toes, palms, and face forward, he is in the anatomical position. Anatomical descriptions of the body should be given as if it were in this position. Properly speaking, the left side of a person's heart (or other body part) is that part of his heart on his left side when he is in the anatomical position, not the left side of the heart as someone else is viewing it.

Human Tissues

After the development of the light microscope, the primary study of the human body shifted from gross anatomy (the study of the parts of the body that are visible to the unaided eye) to **histology** (hih STAHL uh jee), the microscopic study of tissues. A *tissue* is a group of similar cells that work together to perform a similar function. Now, with advanced techniques, the study of individual cells and even organelles has become prominent.

Most tissues are *vascular*: that is, blood vessels run through them. Tissues that lack blood vessels are called *avascular*. Avascular tissue is usually very thin since food and wastes must enter and exit by diffusion from nearby blood vessels. Knowledge of cells and other materials that make up human tissues is important

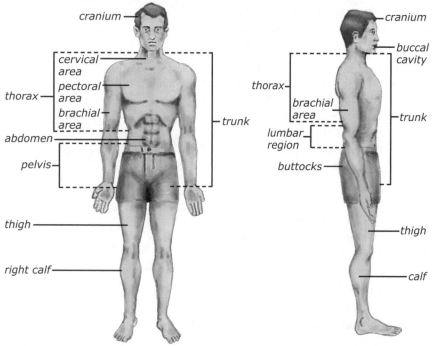

20-5
Anatomical terms

anatomy: ana- (up) + -tomy (Gk. *tomos*—a cutting)

physiology: phys-, physi-, or physio- (Gk. *phusis*—nature) + -logy (the study of)

histology: histo- (Gk. *histos*—web; hence, cells with weblike structure) + -logy (the study of)

Relative Directions on a Human Body in Anatomical Position

- *Anterior (ventral)*: toward the front (stomach side)
- *Posterior (dorsal)*: toward the back
- *Superior*: upward; toward the head
- *Inferior*: downward; toward the feet
- *Superficial*: on or near the surface
- *Deep*: toward the inside
- *Proximal*: a location closer to the trunk of the body, or another specified reference point
- *Distal*: away from the trunk; the opposite of proximal
- *Lateral*: toward the side from an imaginary line that divides the body into halves
- *Medial*: closer to the midline in relation to another part
- *Transverse*: a line that divides the body into a superior and an inferior portion
- *Sagittal*: a line that divides the body into right and left portions

Human Body Cavities

The body is divided into four main cavities, or spaces, that contain the major internal organs. The cavities protect and cushion the organs from possible injury from activities such as jumping and running and from trauma. The *cranial cavity* houses the brain, and the *spinal cavity* surrounds the spinal cord.

The trunk has two large cavities, separated by a large muscle called the *diaphragm*. The superior cavity is called the *thoracic cavity* and contains the heart, respiratory organs, and esophagus. The *abdominal cavity* is the inferior cavity and contains the digestive, reproductive, and excretory organs.

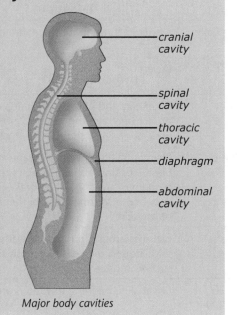

— cranial cavity
— spinal cavity
— thoracic cavity
— diaphragm
— abdominal cavity

Major body cavities

to the understanding of how the tissues function. Histologists classify human tissues into four main groups according to their structure and function.

✦ **Connective tissues**, including bone, blood, lymph, tendons, fat, and cartilage, are the most abundant and diverse of the four tissue types in the human body. They support, connect, and protect other structures in the body. The connective tissue cells are usually not adjacent to each other but have varying amounts of an intercellular substance between them called **matrix**. The matrix can be fluid, semisolid, or solid. Bone cells are embedded in a solid, crystalline matrix. The cellular components of blood and lymph are suspended in a fluid matrix that carries them throughout the body. Tendons, ligaments, and cartilage are in a semisolid, fibrous matrix.

✦ **Epithelial** (EP uh THEE lee ul) **tissue** consists of layers of cells that cover or line the external surfaces as well as internal surfaces such as the organs and the blood vessels. The epithelial cell layers vary in thickness from a single layer lining capillaries to multiple layers of cells forming the skin. The epithelial cells of the skin are bound tightly together, forming a protective barrier; those lining the capillaries allow some substances to pass between them. Some epithelial cells have specialized functions and structures. For example, the respiratory tract is lined with both ciliated and mucus-secreting epithelial cells that trap and remove foreign substances.

✦ **Muscle tissue** is composed of cells that can contract. By contracting, the muscles either move the body or move substances through the body. The human body contains three types of muscle tissue—skeletal, smooth, and cardiac; they will be discussed later in this chapter.

✦ **Nervous tissue** contains cells that receive and transmit electrochemical impulses from muscles, glands, and other nerve cells. It includes the brain, spinal cord, and nerves throughout the body. Nervous tissue coordinates muscular movements, interprets sensations received from the environment, and controls thought processes and emotions.

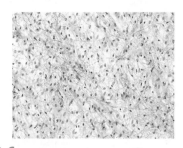

20-6 ——————
PMG of connective tissue

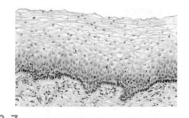

20-7 ——————
PMG of epithelial tissue

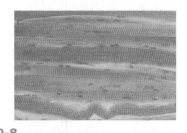

20-8 ——————
PMG of muscle tissue

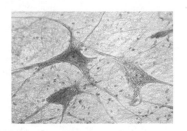

20-9 ——————
PMG of nervous tissue

Type and name	Description	Location	Function
Connective tissues	contain matrix; usually vascular	scattered throughout the entire body	connect; support; protect; store
bone	solidified matrix; few cells	skeleton	supports rigidly; protects; stores minerals
cartilage	semisolid matrix of fibers; avascular	outer ear; end of the nose; ends of the bones; between the vertebrae; between ends of ribs and sternum; in trachea	supports firmly but flexibly; cushions; reduces friction in some joints
dense fibrous connective tissue	matrix of bundles of fibers that may be arranged all in the same direction or as a membrane	ligaments; tendons; membrane around bone (continuous with ligaments and tendons); covering of the brain and spinal cord (dura mater); in the dermis of the skin	joins bone to bone or muscle to bone, permitting flexibility with strength; provides protection; supports
loose fibrous connective tissue	fibers (mainly elastic) in a soft matrix	capsules around organs; beneath facial skin; around the cells of various tissues such as muscle	holds structures flexibly in place
adipose (fat)	predominantly cells with large lipid-storing vacuoles; some fibers	under skin; padding in various areas	cushions; insulates; supports; stores fats
blood	cells in fluid matrix	in blood vessels	transports substances; protects
lymph	mostly fluid with few cells	in vessels of lymphatic system	bathes cells in fluid to supply substances and remove wastes; protects
hematopoietic (blood-forming)	cells supported by a delicate network of fibers	marrow spaces of bones; lymph nodes; spleen; tonsils; thymus gland	forms blood cells; filters bacteria; forms antibodies
Epithelial tissues	avascular; may be a single layer or multiple layers of cells that are tightly bound together	skin; covering internal organs; lining body cavities, blood vessels, heart, mouth, nose, throat, esophagus, stomach, intestines, urinary tract, and reproductive tract; found in many glands that secrete substances	cover and line to protect; secrete; absorb; filter
Muscle tissues	vascular	throughout entire body	move and support
skeletal	long (up to 3.8 cm), multinucleate, striated cells	attached to bones, other muscles, and structures	moves bones, eyes, tongue, and other structures
visceral	short, uninucleate, nonstriated cells	in walls of tubular organs such as blood vessels, intestines, and stomach; in eye; attached to hair follicles	moves substances through tubular organs; changes size of pupil; focuses lens of eye; causes goose bumps
cardiac	branching fibers; short, uninucleate, striated cells	walls of the heart	pumps blood
Nervous tissues	cells with long (sometimes several feet long) projections; often protected by other cells; avascular; most with no further mitosis once maturity is reached	brain; spinal cord; nerves; in eye; in ear; taste buds; touch receptors	have irritability; conduct nerve impulses, thoughts, and emotions

Human Organs and Systems

The tissues of the body are organized into organs. The heart, for example, is an organ that contains many different types of tissues, such as blood, cardiac muscle, nervous tissue, connective tissue, and so on. All these tissues work together to accomplish the functions of the heart.

Organs with related functions are grouped together into systems. This grouping is artificial and is designed for man's convenience. For example, most would agree that the muscular system includes all muscles. But what about the muscles that churn food? They are in the stomach—an organ in the digestive system. Can you think of other examples of systems and organs that seem to overlap?

The body was not designed to be studied easily; it was designed to carry on life. The systems approach—studying the body one organ system at a time—is probably the clearest method and is used in this text. Table 20-2 lists the eleven systems in the order in which this book discusses them.

20-2 Human Organ Systems

Organ system (section number)	Major functions	Examples
integumentary (20B)	protection	skin
skeletal (20C)	support, protection, movement	bones
muscular (20D)	movement	muscles
respiratory (21A)	exchange gases between blood and air	lungs
digestive (21B)	absorb food, eliminate wastes	stomach, intestines
circulatory (22A)	transport blood and other substances throughout body	heart, blood vessels
lymphatic (22B)	protect against disease	lymph nodes
excretory (22C)	eliminate wastes, maintain water balance	kidneys
nervous (23)	coordinate and control movement, process sensory information	brain
endocrine (24A)	regulate body functions	glands that secrete hormones (e.g., thyroid)
reproductive (25A)	produce offspring	ovaries, uterus, testes

➧ Review Questions 20.2

1. What is the difference between anatomy and physiology?
2. Write sentences using all of the following words:
   Sentence 1: anterior, posterior, calf
   Sentence 2: superior, inferior, thoracic cavity
   Sentence 3: superficial, deep, heart
   Sentence 4: proximal, distal, brachial area
   Sentence 5: lateral, medial, shoulder
   Sentence 6: transverse, cranium
   Sentence 7: cervical, thorax, lumbar
   Sentence 8: sagittal, transverse, halves
3. List the four major body cavities and their components.
4. List, describe, tell the function of, and give examples of the four basic types of tissues found in the human body.
5. List eight types of connective tissues and give their locations.
6. List three different kinds of matrixes often found in human tissues, and give an example of each.
7. List the eleven systems of the human body.
⊙8. Can human systems function independently of each other? Explain.

20B—The Integumentary System

The *integumentary system* consists of the skin, hair, and nails. Approximately 1.8 m^2 (about 19 ft^2 or 2800 in.^{2}) of skin covers the body. Although the thickness varies, it averages 3.2 mm (0.125 in.). Few people think much about their skin except for the portion that is visible in the mirror each day. However, this large but thin system and its associated structures are vital in maintaining homeostasis. If large portions of the skin are lost or damaged due to extensive burns, they must be replaced with either skin grafts from unburned areas or artificial skin. Even so, the body struggles to survive.

20.3 The Skin

Functions of the Skin

The skin has six major functions.

✦ *Protection.* The skin does not permit significant amounts of substances like water and air to enter or exit the body. Bacteria, viruses, and many common chemicals that humans constantly touch would be very harmful if they penetrated into the tissues of the body. Skin is an effective barrier to most of them.

✦ *Sensation.* The skin contains nerve receptors for touch, pressure, temperature, and pain. The various kinds of receptors are unevenly distributed throughout the skin, which accounts for the extreme sensitivity of the fingertips in comparison with the back of the elbow, which is not nearly as sensitive.

✦ *Temperature regulation.* The amount of blood being carried to the surface of the skin is regulated to control the amount of heat lost to the environment. The flush of pink after physical exercise is caused by large amounts of warm blood coming to the surface for cooling. The skin turns pale when the body is cold because blood is being withheld from the skin to conserve body heat. However, when the skin is very cold, blood is sent to it to prevent it from being damaged. Rosy cheeks and red ears after outdoor winter activities demonstrate this function quite vividly. If the body generates excessive heat, the skin secretes large amounts of sweat. Evaporation of sweat cools the skin's surface and thereby cools the blood in the skin's blood vessels.

✦ *Excretion.* Small amounts of body wastes are excreted with sweat.

✦ *Vitamin D manufacture.* The skin produces small amounts of vitamin D, which is necessary for the absorption of calcium and for proper development of teeth and bones.

✦ *Absorption.* The skin can absorb some chemicals, a few drugs, and a small amount of oxygen.

Layers of the Skin

The most superficial layer of the skin is the **epidermis**. The epidermis is made entirely of layers of epithelial cells. The deepest layer of the epidermis is called the *stratum germinativum* (GER muh nuh TIE vum) (basal layer). This single-cell layer carries on cell division, producing new cells. These new cells are constantly forced upward as more cells are produced. The older cells of the epidermis fill with a waxy protein called **keratin** and die. These dead cells form the outer layer of the epidermis, called the *stratum corneum*. This layer is constantly being sloughed off, and about every twenty-five days (fewer days

20.3

➡ **Objectives**

● Describe the functions of the integumentary system
● List and describe the layers of the skin

➡ **Key Terms**

epidermis
keratin
dermis
subcutaneous layer

for some people and some areas of the body), a completely new epidermis covers the body.

The next deeper layer of the skin is much thicker and is called the **dermis**. The dermis is composed primarily of connective tissues. The blood vessels, nerve endings, sweat glands, hair follicles, and oil (sebaceous) glands are contained within this layer.

The subcutaneous (SUB kyoo TAY nee us) **layer** is not actually a part of the skin. It is the portion of the integumentary system that attaches the dermis to the underlying muscles and is composed of connective tissues. The fat deposits found in this layer serve to insulate, cushion, and smooth the contours of the body.

Many of the fibers in the connective tissues of the dermis and subcutaneous layers are elastic. Just as a rubber band loses some of its elasticity when it is old, so the elastic fibers of the skin lose some of their elasticity with age. Hence, the skin sags and wrinkles. Surgical facelift procedures often separate the skin from the subcutaneous layer, remove some of the fat tissue, surgically

20-11

Much of the difference between young and old skin is the elasticity of subcutaneous fibers.

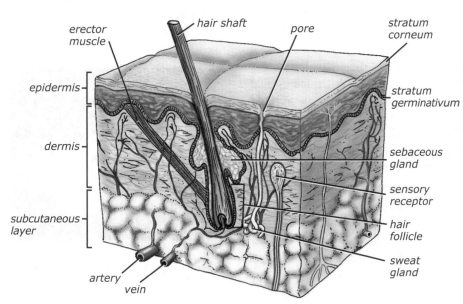

20-10

A cross section of human skin

subcutaneous: sub- (under) + -cuta- (skin)

Skin Diseases and Disorders

+ *Acne*: an inflammation of sebaceous glands caused by a blockage of the pore, resulting in the accumulation of oil and white blood cells within the gland. The causes of acne are not completely understood, but cleanliness, bacteria, diet, hormones, and heredity are involved.

+ *Blisters*: a collection of watery fluid under the epidermis. There are various causes. If they result from ill-fitting shoes, the friction causes the epidermis to separate from the dermis, and extracellular fluid collects in the space. This forms the familiar heel blister. If the skin is severely pinched, the force of the pinch can separate the two layers. Blood may fill the space and cause a "blood blister" to develop.

+ *Burns*: First-degree burns involve damage to the outer epidermal layers only. Cold water on the area, followed by an ointment, is usually the recommended first aid. Mild sunburns are first-degree burns. Second-degree burns involve damage to all the epidermal layers and some of the dermis. Cold water

and covering with gauze is usually recommended. Severe sunburns can result in second-degree burns. Third-degree burns involve tissue damage to all the dermis and often into the subcutaneous layer. The area should be covered with a moist, sterile cloth until medical attention can be obtained. Ointments and other medications should not be used on a second- or third-degree burn unless prescribed by a physician.

+ *Calluses and corns*: thickened layers of epidermis caused by pressure and friction.

+ *Psoriasis*: red, scaly patches of skin where epidermal cells have formed too rapidly. The exact causes of psoriasis are unknown, although most researchers believe that it results from a malfunction of the immune system and enzymes that regulate skin cell division.

+ *Warts*: layers of hard, dead cells resulting from a localized viral infection.

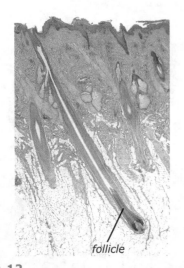

20-12
A light micrograph of a hair follicle. As the epidermal cells at the base of the follicle reproduce, they move upward, lengthening the hair shaft.

20.5

➡️ Objective
- Compare and contrast sebaceous and sweat glands

➡️ Key Terms
exocrine gland
sebaceous gland
sebum
melanin
sweat gland

follicle: (L. *folliculus*—little bag)

exocrine: exo- (out) + -crine (Gk. *krinein*—to separate)

sebum: (L. *sebum*—tallow)

tighten the muscles, and pull the skin taut. Any excess skin is removed and the patient has a less wrinkled face. Unfortunately the results are only temporary and the wrinkles eventually return. Some people opt to receive cosmetic injections to rid the face of wrinkles; however, these results are also temporary.

20.4 Hair and Nails

Hairs are located all over the body except the palms of the hands and the soles of the feet. Different body regions, however, have different amounts and types of hair. The root of a hair lies in a hair **follicle** (FAHL ih kul) located in the dermis or occasionally as deep as the subcutaneous layer. Hair cells are produced at the bottom of the hair follicle and are then filled with keratin and pigments. As the follicle produces more cells, the dead cells are pushed upward and out of the follicle. The **shaft** is the visible portion and is actually a collection of these dead protein-filled cells. Since the shaft of the hair is dead, cutting the hair does not cause it to grow faster or coarser.

Scalp hair grows about 1.3 cm (0.5 in.) per month. Diet, hormones, general health, and age can all affect growth. If the follicle opening is elliptical, the hair is kinky. Round follicles produce straight hair. There are varying degrees in between. Straightness, texture, color, and other hair characteristics are all inherited traits.

Attached to most hair follicles is an *erector muscle*. When the muscle contracts, the hair stands on end, and goose bumps form. Temperature, as well as emotions such as fear, can trigger contraction of these muscles.

The nails of the fingers and toes are very similar to hair since they develop from the skin and since the visible portion is composed of dead cells filled with keratin. Whereas fingernails are replaced about every six months, toenails may require a year.

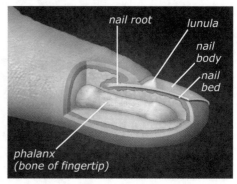

20-13
Fingernail anatomy. The whitish, moon-shaped lunula is the area of most active growth.

20.5 Skin Glands

The skin contains two kinds of **exocrine** glands. Exocrine glands release secretions through ducts, in this case to the surface of the skin.

Sebaceous (sih BAY shus) **glands** produce **sebum** (SEE bum), or oil, and are usually connected to hair follicles by ducts. There are also some sebaceous glands that do not accompany hair follicles, especially if there is only a small

---

Hair Loss

Periodically a hair follicle stops producing for a time, and the hair falls out; however, it may begin to grow again. Diet, chemicals, fever, emotions, medications, and other factors can temporarily shut down large numbers of hair follicles. Baldness, on the other hand, is an inherited trait. When the genes direct the hair to stop growing, little can be done either to prevent hair follicles from shutting down or to start them again.

Many people experiencing hair loss try various treatments. While these treatments often have high success rates (up to 90%) for slowing hair loss, fewer users (often around 12%) actually experience the growth of new hair. Hair transplants, in which functional hair follicles from one part of the body are moved to other parts, have shown higher success rates. In some cases, new growth has still been reported a year after the transplant.

Skin Color and Genetics

The pink color of a person's skin is his blood showing through; other colors are the result of pigments. *Melanocytes* are cells in the epidermis that produce **melanin**, the brown or black pigment that colors the skin. All humans have approximately the same number of melanocytes in their skin. It is the amount of melanin produced that is genetically and environmentally controlled. More melanin yields darker skin.

Ultraviolet radiation (one of the forms of radiation from the sun) stimulates melanocytes to produce melanin, causing the skin to appear darker. Melanin in the upper layers of the epidermis absorbs ultraviolet radiation before it penetrates and damages or kills skin cells. Thus a person who has inherited the genes for melanocytes that produce more dark melanin has a darker skin color and can withstand more of the sun's harmful ultraviolet rays without skin damage than a person with fair skin can.

Many people have the idea that sunbathing is good and that a deep tan is an indication of good health. Both ideas are in error. A person with a tan can be very ill, and a fair-skinned person, one with minimal ability to produce melanin and who cannot tan, can be in excellent physical condition. When sunbathing to obtain a tan, one actually forces his body to put up a screen against excessive exposure to the sun.

Excessive sunbathing damages or kills skin cells by exposing them to the mutation-inducing ultraviolet rays of the sun. These cells must then be replaced at a rate faster than normal, and the skin consequently ages prematurely. Ultraviolet radiation from the sun can also cause the mutations that cause skin cancer. Tans may be fashionable, but if a person has to work at one, he will be better off to be out of fashion.

What about skin color and the different races seen today? Actually the classification of humans into races is a device that man has created. In reality there is only one race of humans. In Acts 17:26 Paul states, "And [God] hath made of one blood all nations of men for to dwell on all the face of the earth." As discussed in the previous paragraphs, all humans have the same number of melanocytes. The Bible states that all humans are descendants of Adam and Eve. How then can all of the various skin colors be accounted for?

Chapter 5 discusses the concept that some traits are controlled by more than one gene (polygenic inheritance). Skin color is one such trait. Although the mechanism for skin color inheritance is not fully understood, scientists believe that as many as five genes, or even more, are involved. Many skin color genotypes are therefore possible from one set of parents. Does this scientifically prove that all humans came from Adam and Eve? No, Christians accept Adam and Eve as mankind's ancestors because it is recorded in the Bible—the inspired Word of God. However, this does show that the scientifically accepted knowledge of genetics fits within the biblical framework.

melanocyte: melan- or melano- (Gk. *melas*—black) + -cyte (cell)

According to a biblical worldview, all of the variations in skin color came from the gene pool of Adam and Eve.

amount of hair in that area. The sebum helps keep the hair and skin soft and reduces water loss. Sebaceous glands are concentrated on the scalp, face, and forehead; none are located on the palms or the soles of the feet.

The amount of sebum produced by sebaceous glands is inherited. Very oily skin often requires the use of special oil solvents. However, removal of too much sebum by detergents or oil-dissolving soaps can cause skin to become dry and cracked and can cause hair to become brittle and to develop split ends. Frequent washing may require the use of conditioners and lotions to keep hair and skin supple.

There are 150–300 **sweat glands** per square centimeter of skin (1 cm^2 = 0.155 in.^{2}). These tube-shaped glands produce *perspiration*, a substance that is about 99% water. Half of the other 1% is salts (primarily sodium chloride), and half is organic substances, including sugars, amino acids, and urea. Sweat production has two primary functions: elimination of waste and temperature control.

If a person's kidneys fail, the skin attempts to rid the body of wastes like urea and uric acid. This activity produces an unusual body odor. Skin is a poor excretory organ, though, and a person with kidney failure who does not receive medical treatment will soon die.

Normally, the entire body sweats constantly and excretes about 237 mL (0.5 pint) of water daily, even if a person remains in a comfortable room and does no strenuous activity. Because this amount evaporates quickly, this perspiration goes unnoticed. On a hot day with strenuous activity, a person can lose almost 7.6 L (2 gal) of water. This is one reason why it is so important to drink plenty of water; water lost through sweating must be replaced.

Sweat itself does not have an odor, but odors are released when sweat combines with substances on the skin and with the waste products of bacteria growing on the dead skin cells that collect on hair and clothing. Washing with warm water to remove the sloughed-off epidermal cells and bacteria and wearing clean clothing minimize body odor resulting from perspiration. Antiperspirants contain chemicals that inhibit the sweat glands. Deodorants are usually perfumes to cover body odor, often with a little antibacterial substance.

Review Questions 20.3–20.5

1. List and describe the functions of the human skin.
2. What are the three major divisions of the human skin?
3. Describe the epidermis and give the functions of its parts.
4. Describe a human hair. Tell how it grows.
5. What are sebaceous glands? Where are they found? What functions do they accomplish?
6. List the major functions of sweat glands.

20C—The Skeletal System

Living bones are quite different from the bones of a skeleton that hangs in a science classroom. Such a skeleton is only the nonliving matrix of bone tissue. The bones in a person's body are living organs that contain blood vessels, nerves, fat tissues, and blood cell–forming tissues.

The *skeletal system* serves as a framework for the support of the body and for the protection of delicate organs such as the brain, spinal cord, heart, and lungs. Bones serve as levers and, with the muscular system, produce movement. The bones are also storage areas for minerals, especially calcium and phosphorus. Bone marrow is involved in blood cell production.

20.6 Bone Anatomy and Physiology

The adult human skeleton consists of approximately 206 bones grouped in two principal divisions—the axial and the appendicular skeletons. The *axial skeleton* is based around the imaginary vertical centerline of the body (the axis). The axial skeleton includes the bones of the skull, the ribs, the sternum, and the vertebral column, which total 80 bones. The bones of the extremities (arms and legs) and the bones of the pectoral and pelvic girdles make up the *appendicular skeleton*. There are 126 bones in the appendicular skeleton.

The bones of an adult human can be classified according to shape. The bones of the arms and legs are long bones, while the bones of the wrists and ankles are short bones. Inside the skull are several irregularly shaped bones, and the ribs and top part of the skull are good examples of flat bones.

A long bone has two basic regions: the **diaphysis** (die AF ih sis) (pl., diaphyses), or shaft, which is the long main portion, and the **epiphyses** (ih PIF ih SEEZ) (sing., epiphysis), the ends of the bone. The surface of each epiphysis

20.6

Objectives

- Identify the major bones of the human body
- Describe the anatomy of a typical long bone
- Explain how bones grow in length and diameter

Key Terms

diaphysis
epiphysis
periosteum
endosteum
red bone
  marrow
yellow bone
  marrow

Haversian
  system
osteocyte
cartilage
ossification

diaphysis: dia- (Gk. *dia*—through) + -physis (*phusis*—growth)

epiphysis: epi- (upon) + -physis (growth)

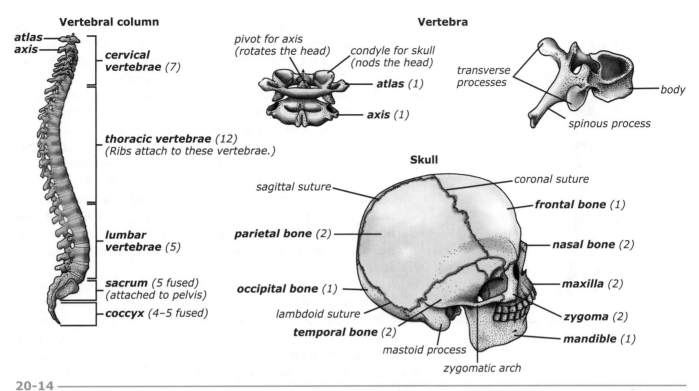

Vertebral column

atlas
axis

cervical vertebrae (7)

thoracic vertebrae (12)
(Ribs attach to these vertebrae.)

lumbar vertebrae (5)

sacrum (5 fused)
(attached to pelvis)

coccyx (4–5 fused)

Vertebra

pivot for axis
(rotates the head)

condyle for skull
(nods the head)

atlas (1)

axis (1)

transverse processes

body

spinous process

Skull

sagittal suture

coronal suture

frontal bone (1)

parietal bone (2)

nasal bone (2)

occipital bone (1)

maxilla (2)

lambdoid suture

zygoma (2)

temporal bone (2)

mandible (1)

mastoid process

zygomatic arch

20-14

Major bones and bone markings of the human skull and backbone. Bone names are in boldface. Numbers in parentheses are the total number of bones of that type found in the body.

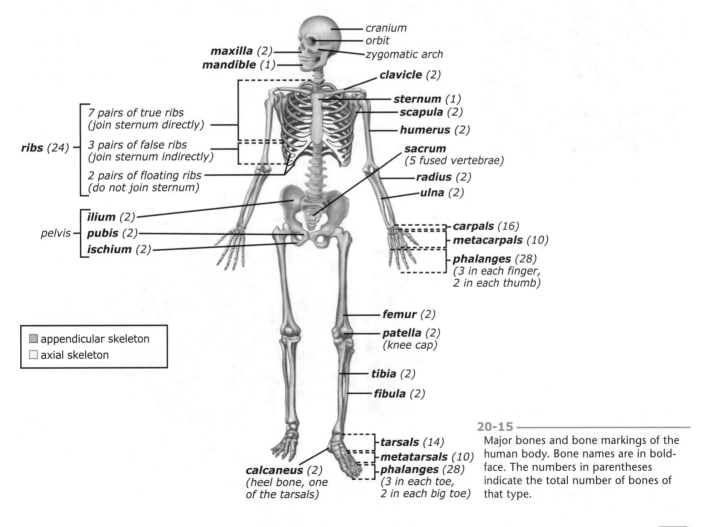

cranium
orbit
zygomatic arch

maxilla (2)
mandible (1)

clavicle (2)

sternum (1)

scapula (2)

humerus (2)

7 pairs of true ribs
(join sternum directly)

3 pairs of false ribs
(join sternum indirectly)

ribs (24)

2 pairs of floating ribs
(do not join sternum)

sacrum
(5 fused vertebrae)

radius (2)

ulna (2)

ilium (2)

pelvis

pubis (2)

ischium (2)

carpals (16)

metacarpals (10)

phalanges (28)
(3 in each finger,
2 in each thumb)

□ appendicular skeleton
□ axial skeleton

femur (2)

patella (2)
(knee cap)

tibia (2)

fibula (2)

tarsals (14)

metatarsals (10)

phalanges (28)
(3 in each toe,
2 in each big toe)

calcaneus (2)
(heel bone, one
of the tarsals)

20-15

Major bones and bone markings of the human body. Bone names are in boldface. The numbers in parentheses indicate the total number of bones of that type.

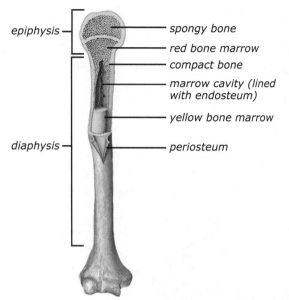

epiphysis — spongy bone
— red bone marrow
— compact bone
— marrow cavity (lined with endosteum)
— yellow bone marrow
diaphysis — periosteum

20-16
Anatomy of a typical long bone

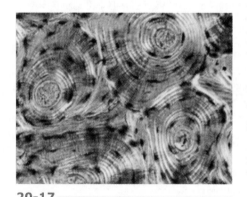

20-17
PMG of Haversian systems

is covered with a thin layer of smooth, bluish white *articular* (ar TIK yuh lur) *cartilage*. This cartilage layer provides for smooth movement at the joints during activities such as swimming and for cushioning when you return to the gym floor after a jump shot.

The diaphysis of the bone is covered with a layer of dense white fibrous tissue called the **periosteum** (PEHR ee AHS tee um), which is responsible for muscle attachment and for bone growth and repair.

The interior of the diaphysis is actually a hollow center called the *marrow cavity*. This cavity usually extends the entire length of the diaphysis. Lining the marrow cavity is the **endosteum** (en DAHS tee um) and, like the periosteum, it is involved in bone growth and repair. In young people the marrow cavity is filled mostly with **red bone marrow**, a soft tissue that produces blood cells. **Yellow bone marrow** is fatty tissue that gradually replaces the red bone marrow as people grow older.

The wall of the diaphysis is made mainly of *compact bone*—tightly packed tissue with few spaces. The epiphyses, on the other hand, contain *spongy bone*. Spongy bone is sturdy but contains many small spaces that make it look like a sponge. These spaces are filled with red bone marrow and some fat. Red bone marrow in the ends of some long bones retains its blood-producing abilities throughout an individual's lifetime.

Microstructure of a Bone

Most bones have a veneer of compact bone tissue covering spongy bone tissue. Compact bone is more regular in its microscopic arrangement than spongy bone. A common feature of all compact bone tissue is the **Haversian** (huh VUR zhun) **system**, or *osteon*. A Haversian system is a unit of bone that consists of a blood vessel in a central canal, surrounded by layers of hard matrix called lamellae (luh MEL ee).

Between the layers of matrix are fluid-filled spaces called lacunae (luh KYOO nee). These spaces contain **osteocytes** (AHS tee uh SITES), the living bone cells. Tiny canals bring nourishment from the central blood vessel to the osteocytes.

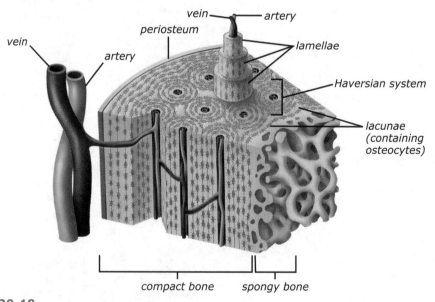

vein — artery
periosteum
lamellae
Haversian system
vein
artery
lacunae (containing osteocytes)

compact bone   spongy bone

20-18
Microstructure of a typical bone

articular: (L. *articulus*—small joint)

periosteum: peri- (around) + -oste- or -osteum (bone)

endosteum: endo- (within) + -osteum (bone)

osteocyte: osteo- (bone) + -cyte (cell)

Broken Bones and Their Treatment

A broken bone or cartilage is known as a *fracture*. Most often fractures are a result of trauma from falling, playing sports, or vehicular accidents. Sometimes fractures occur "spontaneously" when bones have become brittle or weakened by disease. This type of fracture is termed a *pathologic fracture*.

When a bone fractures, it may also injure the surrounding soft tissues. Sometimes the soft tissue injury is more serious than the fracture itself, especially when an artery is torn or a lung is punctured. If the fractured bone segments stay within the soft tissue, it is termed a *closed* fracture. Fractures in which one or both segments protrude through the skin are called *compound* or *open*.

The treatment of fractures is twofold. First, the bone must be set into proper alignment. Sometimes the emergency-room physician or orthopedic surgeon can do this manually by pulling the bones into place. Other types of fractures require surgery to properly set the fractured segments.

The second aspect of the treatment is immobilization. Once the fragments are realigned, a cast made of plaster of Paris or fiberglass holds the bone fragments in place while healing takes place. Sometimes an operation is needed to immobilize the fracture. The surgeon may use small screws and plates to properly secure the bone ends.

Often it is unwise to immobilize a mending bone completely. A limited

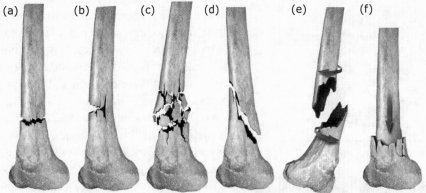

(a) A transverse fracture is complete and often at 90° to the bone surface. (b) A greenstick fracture is incomplete and usually occurs when the bone bends. (c) A comminuted fracture is complete and produces several fragments. (d) An oblique fracture occurs at a sloped angle to the bone surface. (e) A spiral fracture is the result of excessive twisting of a bone. (f) An impacted fracture is a complete fracture in which one fragment is driven into the other one.

amount of use—but not enough to dislocate the bone piece—sometimes aids healing. Therefore, "walking casts" or other limited-use devices are often prescribed.

The Healing Process

Immediately after a bone fractures, blood vessels rupture and bleed into the fracture site, forming a blood clot. Immature osteocytes called osteoblasts move from just under the periosteum into the clot (the periosteum is a source tissue for osteoblasts) and multiply near blood vessels. The osteoblasts form a layer of cartilage between the broken bone ends. At the same time, special bone cells called osteoclasts arrive to phagocytize de-

bris at the fracture site. Eventually the cartilage becomes ossified and the bone of the healed fracture becomes like other mature bone in the area.

Some bones naturally heal more rapidly than others. The humerus (upper arm), for example, may heal in three months, while the tibia (lower leg), for a similar fracture, usually requires six months. Also, fractures heal faster in a young person. A broken bone in an elderly person sometimes takes years to heal. For this reason artificial bone sections are sometimes surgically substituted for broken bones in older people. This surgery is often done for fractured femurs, a common result of falls among the elderly, and it permits the person to walk in a few weeks.

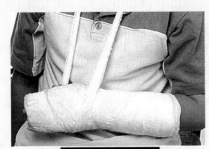

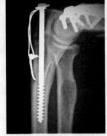

Fracture immobilization can be internal (screws and plates) or external (casts).

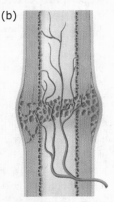

A blood clot is formed at the fracture.

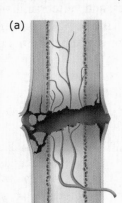

Osteoblasts replace the clot with connective tissue (cartilage).

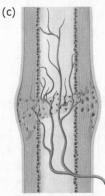

The connective tissue is replaced by bone.

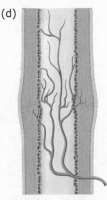

Osteoclasts remove the excess bone, making the new bone like the original.

Stages of bone healing

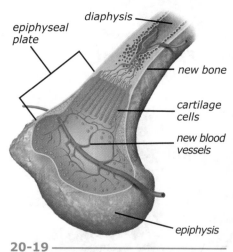

epiphyseal plate · diaphysis · new bone · cartilage cells · new blood vessels · epiphysis

20-19
Epiphyseal plate anatomy

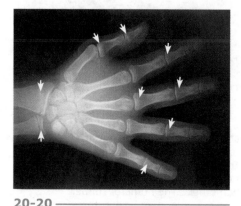

20-20
Epiphyseal plates in a 12-year-old's hand

ossification: ossi- (bone) + -fication or -fy (L. *ficus*—a making or forming into)

Bone Development

The skeletal system of an adult human consists of two main types of connective tissue: **cartilage** and bone. Bone gives rigid support, while cartilage is flexible. Cartilage tissue is composed of a soft fibrous matrix that surrounds the cartilage cells. It does not usually contain blood vessels; however, the nutrients necessary for cartilage cells can diffuse through the matrix from nearby blood vessels. In adults, cartilage is found primarily on the ends of long bones and in the nose and outer ear. It also attaches the ribs to the sternum.

An embryo's tiny skeleton is composed mostly of cartilage. During the third month of fetal development, the osteocytes in the cartilaginous skeleton begin to incorporate calcium and other minerals into the cartilage matrix; this process is called **ossification** (AHS uh fih KAY shun).

After birth, the bones continue to develop as the cartilage is replaced by mature bone. As the long bones in the legs lengthen, the person grows taller. Bones elongate by growing at the *epiphyseal plates* (growth plates), which are layers of cartilage cells lying between the epiphysis and the diaphysis.

The layer of cartilage cells nearest the epiphysis divides to produce more cartilage cells and, thus, increases the length of the bone. Meanwhile, the layers of cartilage cells nearest the shaft slowly ossify. Once the cartilage in the epiphyseal plate has been replaced by bone, the elongation stops.

X-rays of the hand and wrist show many epiphyseal plates. Illnesses delay the ossification of the epiphyseal plates. In other words, whenever a person is ill, his growth rate slows.

As a person grows, the body requires an increasingly stronger skeletal system for support. Therefore, as bones increase in length, they must also increase in diameter. This increase in diameter occurs primarily beneath the periosteum, where new bone tissue is added. Specialized osteocytes called *osteoclasts* enlarge the marrow cavity by "eating away" the internal bone tissue. This dissolved bone substance is then used for growth on the outer part of the bone. By this dual process of dissolving and building, a thin bone with a small marrow cavity becomes a thicker bone with a larger marrow cavity.

As they grow, bones continually need calcium in order to form the matrix of newly formed Haversian systems. Therefore, dietary calcium is needed for bones to properly form. Vitamin D is also needed so that the calcium in food can be absorbed into the blood.

Although most people stop growing by their late teens, ossification is not completed until the mid-twenties. The last bones to completely ossify are the sternum, clavicles, and vertebrae. Even after ossification is complete, the bone tissue is continually being "remodeled," which accounts for some of the change in facial features as a person matures physically.

➥ *Review Questions 20.6*

1. List the functions of the skeletal system.
2. What are the two main divisions of the human skeleton, and what bones compose each?
3. List the four types of bones according to their shapes.
4. Draw a longitudinal section of a long bone and label the parts.
5. Draw two Haversian systems side by side and label the parts.
6. Describe the formation of a bone.
7. Describe how a bone grows in length and how a bone grows in diameter.
⊙8. What, besides proper nutrition, is necessary for proper bone growth?
⊙9. List and describe several types of bone fractures.
⊙10. What are the two steps usually taken to treat a bone fracture?
⊙11. Describe the process of bone healing. What can be done to speed the process?

20.7 Joints

A **joint** is a connection between two or more bones or between cartilage and bone. Joints are essential in articulation, one of the most important functions of the skeletal system. *Articulation* refers to the smooth movement of one bone upon another.

Classification of Joints

Joints are designed for specific functions and are classified into three groups according to the degree of movement.

✦ *Freely movable joints.* Most joints in the skeletal system are freely movable. The ends of the bones at such joints are shaped to provide smooth articulation. The freely movable joints may be classified according to these ends of the bones that make up the joint and the kind of movement they permit.

✦ *Slightly movable joints.* In slightly movable joints a pad of cartilage between the bone ends permits limited movement. These joints connect the vertebrae. The ribs articulate with the sternum and allow for limited movement as a person inhales and exhales. The union between the pubic bones is also classified as a slightly movable joint, even though there is movement only under severe stress, as in childbirth.

✦ *Immovable joints.* In some places where bones meet or where a bone attaches to cartilage, the joints must be rigid. *Sutures*, the interlocking margins of skull bones, are also immovable joints. At birth the sutures have not formed, and the spaces between the skull

20.7

➡ **Objectives**
- List the major types of joints in the human body
- Describe the anatomy of a typical joint

➡ **Key Terms**

joint
ligament
tendon
synovial membrane

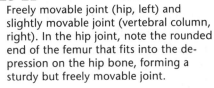

20-21
Freely movable joint (hip, left) and slightly movable joint (vertebral column, right). In the hip joint, note the rounded end of the femur that fits into the depression on the hip bone, forming a sturdy but freely movable joint.

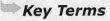

fontanel

suture

20-22
Immovable joints (sutures) in adult cranium and fontanels in a young infant

suture: (L. *suere*—to sew)

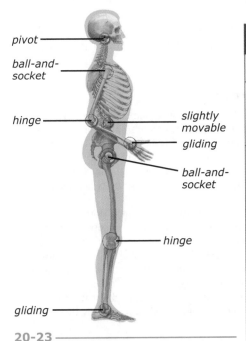

pivot

ball-and-socket

hinge

slightly movable

gliding

ball-and-socket

hinge

gliding

20-23
Major movable joints of the human body

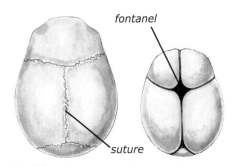

20-3 Movable Joints and Their Actions			
Type of joint	**Description**	**Action**	**Examples**
ball-and-socket	A ball-shaped head moves within a socket that is "hollowed" to receive the head.	rotating; free movement in all directions	shoulder and hip
hinge	Two cylindrical surfaces, one concave and the other convex, fit together to form the joint.	bending only in one direction	elbow, knee, and between phalanges (fingers and toes)
pivot	A ringlike formation at the end of one bone surrounds a slender projection that functions as an axle.	rotating and swiveling	between atlas and axis in neck
gliding	Opposing bone surfaces are slightly convex and concave and thereby restrict movement.	limited movement sideways and up and down	between carpals (wrist) and between tarsals (ankle)

bones are filled with fibrous membranes known as fontanels. These are the "soft spots" that allow an infant's skull to be slightly compressed during birth. More immovable joints are found at the junction of the ilium, ischium, and pubis—the three bones that make up the "hip bone."

Anatomy of a Typical Joint

The joints that allow for free movement are often complicated structures. The location of the ligaments and muscle attachments and the presence of other bones that might restrict movement govern the amount of movement at a joint.

A connective tissue sheath called the *joint capsule* covers the proximal and distal bone ends forming the joint. The space within the joint capsule is called the joint cavity (or space). The **synovial** (sih NOH vee ul) **membrane** lines the inner surface of the joint cavity and produces synovial fluid. Synovial fluid lubricates the joint and acts as a "shock absorber" between the bones.

Saclike structures called *bursas* (BUR suz), located between tendons, ligaments, and bones, are also lined with synovial membranes that produce synovial fluid. The fluid-filled bursas serve as cushions and reduce friction between moving parts of the joints.

When joints are sprained or dislocated, the fibers of the ligaments and tendons stretch and usually tear. Since these tissues heal rather slowly, it is necessary

bursa: (L. *bursa*—bag or purse)

to restrict their movement. Excessive activity could permanently stretch the ligaments or tendons, resulting in a weak joint that can easily dislocate. However, if the injured person does not move the joint at all during the healing process, the ligaments or tendons may shorten, greatly restricting the movement of the joint.

Often the doctor will advise a person to walk with the aid of crutches but still "put some weight" on the sprained ankle or knee. He may also give warnings not to engage in any strenuous physical activity that might reinjure the weakened joint since it may not be completely healed even after the pain has subsided. These warnings are a difficult, yet necessary, assignment for young athletes if they expect to be athletes at all when they are older.

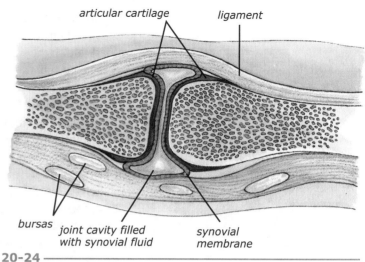

20-24

Anatomy of a typical joint

⟿**Review Questions 20.7**

1. List the three types of joints, according to their type of movement. Give examples of each.

2. Describe four types of movable joints and give an example of each.

3. Draw a typical movable joint and label the parts.

20D—The Muscular System

The two key words that describe the *muscular system* are *contraction* and *movement*. Contraction refers to the ability of a muscle tissue to shorten and thereby cause movement. Some muscles, like those used to get out of bed in the morning, are consciously controlled. Others, such as the muscles of the stomach and intestines that help digest food, operate without mental commands. Muscles that cause breathing are consciously controlled when one is talking, singing, or taking a deep breath; at other times breathing is automatic.

Muscle tissue also possesses three other important characteristics. *Irritability* is the muscle's ability to respond to impulses from the nervous system. After the muscle responds by contracting, it rebounds to its original length. This ability to return to its previous shape is *elasticity*. Therefore, the old wives' tale that you shouldn't make faces at people because your muscles will "freeze in place" is not true. Muscle tissue also possesses *extensibility*: it can be stretched when it is relaxed. Extensibility is very important because when one group of muscles contracts, often another muscle group must passively stretch to allow for movement.

20.8 Muscle Anatomy

Muscle tissue is composed of a great number of cells called **muscle fibers** and is supported by layers of connective tissues. Muscle fibers are long cells that contain many nuclei.

Types of Muscle Tissue

There are three types of muscle tissue in the human body. Each type has properties that fit it for particular functions. These muscle types are distinguished by their location, microscopic appearance, and type of nervous control.

20.8

⟿ **Objectives**

- Compare the three types of human muscle
- Describe the structures of a typical skeletal muscle fiber

⟿ **Key Terms**

muscle fiber	myofibril
skeletal muscle	myosin
visceral muscle	actin
cardiac muscle	sarcomere
fascicle	

The first type, named for its location, is **skeletal muscle** tissue. It is usually attached to bones. It is also called *striated* (STRY ay tid) *muscle* because there are dark and light stripes in its cells when viewed with a microscope. These striations are actually filaments of protein in the muscle cells. Skeletal muscle is also known as *voluntary* muscle tissue because it is primarily controlled by conscious thought. However, the term *voluntary* does not always apply (for example, when a person moves while sleeping).

The diaphragm (the primary muscle of breathing in) functions under voluntary control when a person speaks or sings, but at other times it functions involuntarily. Therefore, it is difficult to classify the diaphragm, even though it is skeletal by its location and appears striated. The muscular tissues in the walls of the pharynx (throat) are also classified as striated because of their appearance, but they are neither voluntary nor attached to a bone.

The second type is **visceral muscle** tissue. The term *visceral* refers to internal organs. The name fits well because visceral muscle is located in the walls of internal organs such as the stomach, intestines, blood vessels, and urinary bladder. Visceral muscle is also located in the iris of the eye and causes the diameter of the pupil to increase or decrease, depending on the brightness of the environment.

Most of the *sphincters* (SFINGK turz), circular bundles of muscles that regulate the diameter of various tubular organs and openings, are visceral muscles. The muscular valves at both ends of the stomach are examples of sphincters.

Visceral muscle is also called *nonstriated* or *smooth* because it does not have dark and light stripes. All visceral muscle tissue is termed *involuntary* because it is *not* directly controlled by conscious thought. In other words, visceral muscle tissue can function (and, in fact, functions most regularly) when a person is asleep.

Cardiac muscle, the third type of muscle, is located only in the heart. Cardiac muscle tissue is *striated* and *involuntary*. The striations, however, are not as regular and distinct as in skeletal muscle. Cardiac muscle fibers branch and join together. This network of interwoven fibers allows nerve impulses to spread quickly through the muscular walls of the heart and produces effective pumping of the blood.

sphincter: (Gk. *sphingein*—to bind tight)
cardiac: (heart)

20-4 Muscle Classification

Kind of muscle	Location	Function	Kind of fibers	Voluntary or involuntary
skeletal	primarily attached to bones and other movable structures	move parts of the body	striated	voluntary
visceral	walls of internal organs and blood vessels	move organ or substance within the organ	smooth	involuntary
cardiac	heart	contract heart to pump blood	striated	involuntary

Visceral and cardiac muscle tissues are discussed in greater detail in connection with other body systems (see Ch. 21–25). The remainder of this chapter deals with skeletal muscle tissue.

Muscle Structure

A skeletal muscle fiber (cell) is barely visible to the unaided eye. Each fiber is a multinucleate cylinder wrapped in a thin sheath of connective tissue. Groups of ten to a hundred fibers are bound together with other layers of connective tissue to form a **fascicle**. A thicker connective tissue layer encloses groups of fascicles to form the muscle itself. At the ends of the muscle, the connective tissue merges to form the tendons that attach the muscle to bones. The connective tissues not only hold the muscle fibers together but also support blood vessels and nerves that supply the fibers.

The cytoplasm within each muscle fiber contains numerous tiny threadlike **myofibrils** (MYE uh FYE bruhlz). These myofibrils are parallel to each other and extend the length of the fiber. There are two kinds of protein filaments in each myofibril—thick filaments made of **myosin** (MYE uh sin) and thin filaments made of **actin**. The overlapping arrangement of these filaments is repeated to produce the striated appearance. The actin filaments are anchored at their midline to a structure called the *Z line*. The distance from one Z line to another is called a **sarcomere**—the functional unit of muscle contraction.

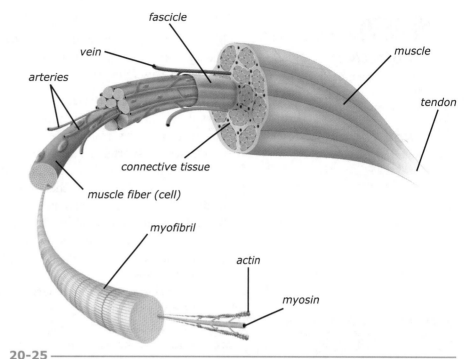

20-25

Skeletal muscle is composed of tissues. The muscle fibers are bound together by sheaths of connective tissue to form fascicles and the muscle. Connective tissue also supports nerves and blood vessels that are included in the muscle tissue.

myofibril: myo- (Gk. *mus*—muscle) + -fibril (fiber)

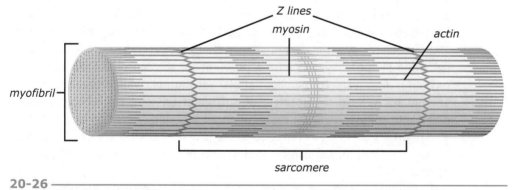

20-26

A sarcomere is the functional unit of a muscle. The horizontal green lines represent actin, and the horizontal yellow lines represent myosin. The overlapping arrangement of actin and myosin allows for muscle contraction.

Review Questions 20.8

1. What are the five characteristics of a muscle?
2. What are the three basic types of muscle tissue? Describe each and tell where it is located in the body.
3. Describe the structure of a typical skeletal muscle.

→ **Objectives**

• Describe the muscle contraction process
• Explain muscle fatigue

→ **Key Terms**

muscle fatigue
oxygen debt

20.9 Muscle Physiology

Muscles move. To move, they need ATP as an energy source. As they move, energy is released as heat. How muscles accomplish these functions is the science of *muscle physiology*.

Muscle Contractions

How does a muscle actually produce movement? The contraction process begins when impulses from nerve cells stimulate the actin and myosin filaments to interact and shorten the sarcomere (the distance between Z lines). The myosin filaments have little extensions called crossbridges that are shaped like golf club heads. Actin filaments look like a strand of twisted beads. The crossbridges of the myosin filaments attach to the actin filaments. Then the crossbridges flex and pull the actin filaments, drawing the muscle ends together and shortening the sarcomere. During a single contraction, the actin and myosin attach, disengage, and reattach multiple times, making a coordinated movement of the muscle.

Where does the energy for muscle contraction come from? Contraction requires the energy derived from ATP molecules; therefore, a continuous supply is needed. Normally, cellular ATP is derived from the breakdown of glucose to pyruvic acid and then to carbon dioxide and water (aerobic cellular respiration). The amount of ATP in a muscle cell is enough for only a few contractions; therefore, it must be continually replenished by the numerous mitochondria that are located in the muscle fibers.

Muscle contraction is an all-or-none response—the fibers either contract or remain relaxed. How does a person reduce the force of a contraction of the muscles in his hand when he is gently holding an infant's hand and then increase the force when holding on to the leash of a large dog? The number of muscle fibers stimulated determines the force of the contraction. When the number of fibers stimulated is increased, the force of the contraction is increased.

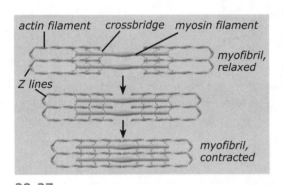

20-27

Interaction between actin and myosin during muscular contraction

Muscle Fatigue and Physical Condition

When a person is resting or moderately active, the skeletal muscles usually have enough glucose and oxygen for energy production. However, after he exercises skeletal muscles strenuously for a few minutes, his respiratory and circulatory systems cannot supply enough oxygen to the muscles. Why does this lack of oxygen cause a problem for the muscle? Remember that oxygen is necessary for aerobic cellular respiration, which produces a large amount of ATP. In the absence of sufficient oxygen, cells break glucose down into *pyruvic acid*, releasing a small amount of energy to make ATP. The pyruvic acid is then converted to lactic acid. Recall that this is anaerobic cellular respiration, not aerobic cellular respiration, and it does not produce as much ATP.

Eventually, if exercise continues, the demand exceeds the supply. The decrease of both ATP and oxygen results in the buildup of lactic acid inside the muscle fibers. Since the lactic acid cannot be used, it must be transported by the blood to the liver cells, where it can be converted into glucose through the expenditure of ATP energy. The blood, however, can transport only a fixed amount of lactic acid at a time. The excess lactic acid in muscle cells decreases the pH, resulting in an inability of the muscle fibers to respond (contract) to the nerve stimulation. This inability of the fibers to contract is called **muscle fatigue**.

The lack of oxygen that causes the accumulation of lactic acid in the muscles and liver is known as the **oxygen debt**. The amount of oxygen debt equals the amount of oxygen needed by liver cells to convert the accumulated lactic acid into glucose.

It is common to feel "out of breath" and to continue to breathe deeply following vigorous exercise. During this time of deep breathing, oxygen is supplied to the muscles. When sufficient oxygen has reached the muscles and ATP production proceeds as usual, normal breathing resumes. However, some lactic acid remains in the muscles and liver. It may take several hours or days to relieve this oxygen debt completely.

A person is in good physical condition ("in shape") when he can engage in moderately strenuous physical activity without developing fatigued muscles. The conditioned athlete can run several miles without significant oxygen debt or soreness, but the unconditioned person cannot. The conditioned athlete's muscles have more protein filaments and hold more oxygen. His circulatory and respiratory systems function efficiently to transport oxygen to the muscles.

The "out of shape" person does not have these benefits and soon begins to pant (which does not supply as much oxygen to the lungs as smooth, deep breathing) and to waste energy on excess movements. The person in poor physical condition builds up an oxygen debt much sooner and with much less physical activity than the person in good physical condition.

The body builds the muscle fibers and oxygen-holding capacity it needs for its normal activities. Daily exercise will soon prepare it for a more strenuous workout.

Big Muscles, Little Muscles

In a body-building program a person may perform numerous muscle contractions to increase the size and strength of his muscles. This increase is an enlargement of individual muscle fibers, not an increase in the number of fibers. The mitochondria within the fibers reproduce, and the number of actin and myosin filaments increases.

Because the strength of a contraction is directly related to the diameter of the muscle fibers, large muscles are strong muscles. Running and other aerobic exercises, however, usually produce only slight increases in the size and strength of the muscles involved. Running benefits the heart, blood vessels, and lungs more than it does the skeletal muscles. The runner may not have the strength or the bulging muscles of the weightlifter, but his endurance is probably greater. Today, most fitness trainers recommend an exercise program that includes both aerobic and strength training. Of course, the final results depend largely on the individual and his genetic makeup.

The proper motive for participating in any exercise program is to develop and maintain the body's health and strength in order to serve the Lord regularly and efficiently. The desire for bulging biceps to impress people is sinful pride.

A muscle that is not used or is used only to produce weak contractions usually decreases in size and strength. Mitochondria in the fibers and blood capillaries decrease in number. The amount of muscle protein (actin and myosin) also decreases. This is called muscular atrophy (AT ruh fee) and occurs when a muscle is in a cast or when an active person becomes inactive.

atrophy: a- (without) + -trophy (nourishment)

Review Questions 20.9

1. Describe oxygen debt. Identify its causes, results, and "cure."
2. When a person exercises, he becomes tired. Why does a person who exercises regularly not become tired as easily?
3. The ability to build up oxygen debt is essential for humans. Discuss some advantages you have because of this ability.

20.10 Muscles and Body Movements

There are nearly seven hundred skeletal muscles in the human body. These muscles may be tiny, like those that move the eyes, or large, like the trapezius of the upper back, which moves the spine, head, and shoulders.

Muscle Attachments

Skeletal muscles produce movements by pulling tendons, which in turn exert force on bones. There is a secure union at the tendon-bone junction, and when the muscle contracts, one bone is drawn toward the other. Usually one of the bones involved remains more or less stationary while the other bone moves.

The attachment of the muscle's tendon to the more stationary bone is the **origin** of the muscle. The attachment of the other muscle tendon to the more

20.10

Objectives
- Differentiate between muscle origin and insertion
- Describe the function of muscle groups

Key Terms
origin
insertion

Muscle	Origin	Insertion	Action
epicranius	lower part of occipital bone	skin of forehead	raises eyebrows and wrinkles forehead
orbicularis oculi	wall of orbit	skin around eyelid	closes eye
orbicularis oris	muscle fibers around mouth	skin around mouth and lips	draws lips together
buccinator	maxilla and mandible	orbicularis oris	compresses cheek as in blowing air
masseter	zygomatic arch	mandible	elevates (closes) jaw
temporal	temporal bone	mandible	elevates (closes) jaw
sternocleidomastoid	sternum and clavicle	temporal bone (mastoid process)	flexes neck; rotates head
pectoralis major	clavicle, sternum, and true ribs	proximal humerus	adducts upper arm anteriorly
pectoralis minor	ribs	scapula	pulls shoulder down and forward
rectus abdominis	pubic bone	ribs and sternum	compresses abdomen; flexes trunk
external oblique	lower eight ribs	ilium and pubic bone	compresses abdomen
internal oblique	ilium and connective tissue in lumbar region	lower ribs, pubic bone, and the opposite internal oblique	compresses abdomen
transversus abdominis	same as internal oblique	the opposite transversus abdominis	compresses abdomen
external intercostals	lower border of rib above	upper border of rib below	elevate ribs
internal intercostals	upper border of rib below	lower border of rib above	depress ribs during forceful expiration
trapezius	occipital bone and upper vertebrae	clavicle and scapula	raises or lowers shoulders
latissimus dorsi	lower vertebrae, lower ribs, and ilium	humerus	extends and adducts upper arm posteriorly
longissimus capitis	upper vertebrae	mastoid process	extends and rotates head
deltoid	clavicle and scapula	proximal humerus	abducts upper arm
biceps brachii	scapula	proximal radius	flexes forearm
triceps brachii	scapula and humerus	proximal ulna	extends forearm
flexors of hand and fingers	distal humerus	metacarpals, carpals, and palm	flex hand at wrist
extensors of hand and fingers	distal humerus	metacarpals	extend hand at wrist
rectus femoris	ilium	patella and tibia	flexes thigh and extends lower leg
gluteus maximus	ilium, sacrum, and coccyx	femur	extends thigh
adductor group	pubic bone and ischium	femur	adducts thigh
hamstring group	ischium and femur	femur	extends thigh
sartorius	iliac spine	tibia	flexes, abducts, and rotates thigh
tibialis anterior	tibia	tarsals and metatarsals	flexes and inverts foot
gastrocnemius	lower femur	calcaneus (Achilles tendon)	extends foot and flexes lower leg
soleus	tibia and fibula	calcaneus (Achilles tendon)	extends foot (plantar flexion)

Characteristics Used to Name Muscles

Size

pectoralis major—large muscle in the pectoral (chest) region
pectoralis minor—small muscle in the pectoral region
gluteus maximus—largest (maximum) muscle of the gluteus (buttocks) region

Shape

deltoid—triangular-shaped muscle
trapezius—trapezoid-shaped muscle

Location

tibialis anterior—muscle located on the anterior region of the tibia
external and *internal oblique*—muscles of the ribs that are on the outside and inside
rectus femoris—muscle near the femur
temporal—muscle near the temporal bone
orbicularis oculi—muscle near the eye (ocular organ)
epicranius—muscle that goes around (epi-) the head (-cranius)

Action

masseter (muh SEE tur)—a muscle that is involved in mastication (chewing)

adductor group—adducts (draws in) the thigh
flexors or *extensors* of the hand—flex or extend the hand

Number of Attachments

biceps brachii—a muscle that has two origins (bi-, two; -ceps, head) and is located in the arm (brachial) region
triceps brachii—three origins

Direction of Fibers

external oblique—a muscle that is located near the outside of the body and has fibers arranged obliquely (in a slanting direction)

Muscle Actions

- *Flexion*: decreases angle between bones
- *Extension*: increases angle between bones
- *Abduction*: moves bone away from midline of body
- *Adduction*: moves bone toward midline of body
- *Rotation*: produces a turning or revolving movement around an axis
- *Elevation*: produces an upward movement
- *Depression*: produces a downward movement

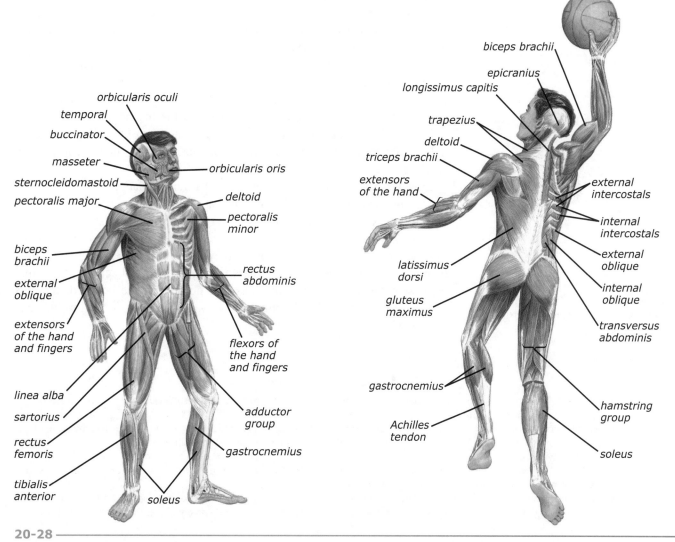

20-28

Major muscles of the human body

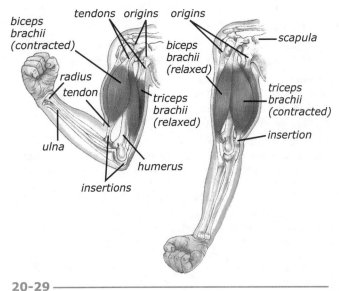

biceps brachii (contracted)
tendons origins origins
radius
tendon
ulna
insertions
humerus
triceps brachii (relaxed)
biceps brachii (relaxed)
scapula
triceps brachii (contracted)
insertion

20-29
A typical antagonistic pair

movable bone is the **insertion**. However, the origin or insertion of a muscle is not always a bone; it can be the skin, a layer of connective tissue, or even another muscle.

The movement performed by the muscle when it contracts is called its *action*. For example, the action of the biceps brachii muscle is the flexing of the arm at the elbow. When a muscle contracts, the origin is pulled toward the insertion. The fleshy portion between the origin and insertion is the *belly* of a muscle. This is the part of the muscle that is seen when someone flexes his muscles.

Muscle Groups

Body movements are not the result of only one muscle; skeletal muscles function in groups. During the flexing of the arm at the elbow, the biceps brachii muscle is the *prime mover* because it performs the main action. During this movement, the triceps brachii is relaxed, allowing the biceps to perform its action. The triceps brachii is an *antagonist* (an TAG uh nist) to the biceps because it performs the opposite action—rather than flexing the forearm, it extends the forearm. When the arm extends, the triceps brachii becomes the prime mover and the biceps brachii the antagonist. Thus, the biceps and triceps of the arm are an *antagonistic pair* because they have opposite actions.

For most body movements, there are other muscles that assist the prime mover; these are the *synergists* (SIN ur JISTS). Synergists contract at the same time as the prime mover in order to help produce the movement or stabilize a particular joint. For example, as the biceps muscle flexes the arm at the elbow, the deltoid and pectoralis major muscles also contract to hold the upper arm and shoulder in a stable position and allow the lower arm to move.

Muscles have different roles at various times, depending on the action. During one action a particular muscle may function as a prime mover, but during other actions it may serve as an antagonist or synergist.

Any movement of the body is the result of a muscle's *pulling* on its insertion point. Muscles never push; they only pull. A person may be able to push an object, such as a wheelbarrow, but only because the prime movers, especially the hips and legs, are pulling on the bones of the legs and feet. In other words, although the body is able to push, the muscles can only pull.

antagonist: ant- or anta- (against) + -agonist (Gk. *agonizesthai*—to struggle)

synergist: syn- (together) + -ergist (Gk. *ergon*—work)

Muscle Disorders

- *Atrophy*: a great reduction in muscle fibers and possible replacement by fibrous tissue
- *Convulsions*: violent, involuntary contractions of an entire group of muscles; characteristic of epileptic seizures and drug withdrawals
- *Cramps*: painful, involuntary contractions in those muscles that have been used heavily and have suffered from fatigue
- *Muscular dystrophy*: a progressively crippling genetic disease in which the muscles gradually weaken and atrophy

- *Myalgia*: muscle pain resulting from a muscular disorder or disease
- *Paralysis*: inability to move a muscle; usually because of some nervous system failure
- *Shin splints*: soreness on the front of the lower leg due to straining a muscle; often as a result of walking up and down hills
- *Spasm*: an involuntary contraction of shorter duration than a cramp and usually not as painful

Most of the well-known muscles are described and illustrated on the preceding pages. While learning the muscles, you may find it helpful to realize that various characteristics of the muscles are used to name them. The information in the boxes on page 513 may help in understanding how muscles are named.

⮕ Review Questions 20.10

1. Compare and contrast the following: prime mover, antagonistic pairs of muscles, and synergistic muscles.

2. Give an example of a muscle's name based on the muscle's (a) size, (b) shape, (c) location, (d) action, (e) number of attachments, and (f) direction of fibers.

⊙3. How can you push your pencil while the muscles performing the action are not pushing?

21

Incoming Substances

Facets

21A—The Respiratory System

If someone asked you to identify the two major substances that you take into your body and are vital to your existence, you would probably respond with food and air. We all know that without sufficient nourishment and oxygen we will die. However, it is not just the quantity of air and food we take in that affects our physical condition; it is also the quality. To make wise decisions regarding diet and other habits, you should understand what your body needs and how it uses incoming substances.

21.1 Anatomy of the Respiratory System

The *respiratory system* consists of organs that transport oxygen and carbon dioxide to and from the blood: the nose, pharynx, trachea, bronchi, bronchial tubes, and lungs. Basically the respiratory system brings air from the atmosphere into the lungs. Oxygen in the lungs, however, does most body cells little good. How is the oxygen transported throughout the body? The oxygen in the lungs goes into the blood, which then transfers it to the body cells. Carbon dioxide is removed from the cells by the blood and is taken to the lungs to be exhaled.

21.1

Objectives
- Describe the functions of the respiratory system's structures
- Explain how gases are exchanged in the lungs

Key Terms

palate	trachea
pharynx	bronchus
larynx	alveolus
epiglottis	pleura

The Nose

The term *nose* refers not only to a part of the face (external nose) but also to the *nasal cavity*. The nasal cavity is posterior to the external nose and is separated by the *nasal septum* into right and left sides. A ciliated mucous membrane lines the nasal cavities and filters bacteria, smoke, and dust particles from the air.

The nasal sinuses (air-filled cavities) open either into the nasal cavity or into the throat. Since the sinuses are also lined with ciliated mucous membranes, mucus continually drains into the nasal cavity and throat. If these membranes become infected or swollen, the sinuses become painful and congested.

Have you ever heard that it is better to breathe through your nose than your mouth on a cold day? Air drawn through the nasal cavities is not only filtered but also warmed and humidified (moistened) before it reaches the remainder of the respiratory system. This prevents the delicate linings of these structures from becoming cold and dry when air that is not particularly warm and moist is inhaled.

The upper portion of the nasal cavity has nerve endings that are involved in sensing odors. The **palate** (PAL it) serves as the "floor" of the nose and roof of the mouth. It has two parts: the anterior *hard palate* and the posterior *soft palate*. The bones of the hard palate grow together during the growth of the embryo and form a suture along the midline. If the bones do not grow together properly, the baby is born with an opening between the mouth and nasal cavity, called a cleft palate. This disorder may also include a divided upper lip, a condition called cleft lip.

The soft palate is primarily muscle tissue that ends in a structure called the *uvula* (YOO vyuh luh). When you swallow, the uvula moves posteriorly and superiorly (back and upward) to prevent food from entering the nasal cavities.

The Pharynx and Larynx

The **pharynx**, commonly called the throat, is a muscular tube about 13 cm (5 in.) long, lined with a ciliated mucous membrane. It extends from the back of the nasal cavity to the esophagus and serves as part of both the respiratory and the digestive systems.

Dennis Kunkel Microscopy, Inc./Visuals Unlimited, Inc.

21-1
SEM of the ciliated mucous membrane of the nose

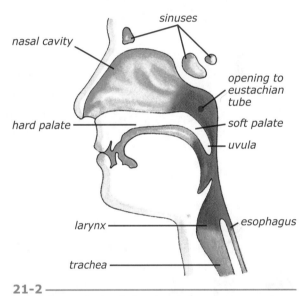

21-2
The pharynx is the dark red area.

Coughing and Sneezing

Coughing and sneezing are reflexes designed to eliminate unwanted objects from the body's air passageways. However, both the process and purpose of these reflexes differ.

Sneezing occurs when something irritates the nasal membranes. A sneeze is a deep breathing-in followed by a blast of air up through the nasal cavities and out the nose. It serves to expel the irritating substance.

An obstruction or irritation in the larynx, bronchi, or trachea stimulates the cough reflex. Coughing involves taking

a deep breath and then forcing air from the lungs against the closed glottis. When the glottis opens suddenly, a blast of air is forced through the larynx, pharynx, and mouth, usually carrying with it the substance that stimulated the cough reflex.

Sneezes expel thousands of droplets into the air, and the speed of a cough or a sneeze can reach up to 160 km/h (100 mph). The spray can travel as far as 3.7 m (12 ft). The moral of the story? Cover your mouth and nose!

The Voice

Human vocal sounds originate from the vibration of the vocal cords as air from the lungs moves over them. Decreasing the muscular tension on vocal cords produces lower-pitched sounds, and higher-pitched sounds are produced when the cords are pulled tighter. Because the vocal cords in females are usually thinner and shorter than those in males, they vibrate more rapidly, giving women a higher range of pitch than men.

Other structures convert sound produced by the vocal folds into recognizable speech. The pharynx, mouth cavity, nasal cavities, and nasal sinuses act as resonating chambers (much the same as the sound box of a violin). The narrowing and dilating of the pharynx walls, along with tongue placement, help form vowel sounds. The tongue, lips, teeth, and palate form consonant sounds.

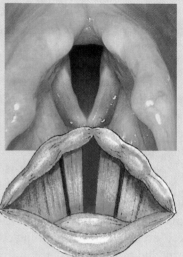

The vocal cords in the larynx as viewed from above

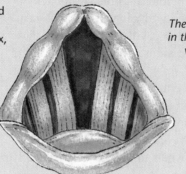

Open (normal breathing)

Partially closed (producing sound)

The pharynx has seven openings:

- Two openings from the back of the nasal cavities for airflow
- Two passageways to the middle ear called *eustachian tubes* for equalizing air pressure
- The opening to the mouth cavity for passage of food and air
- The entrance to the esophagus—the passageway to the stomach for food
- The entrance to the larynx—the passageway to the trachea for air

The **larynx**, or voice box, is a short passageway that leads from the pharynx to the trachea. The walls of the larynx consist of several cartilaginous structures held together by muscles and ligaments. The larynx contains a pair of mucous membrane folds known as the *vocal cords*, or vocal folds. You produce sounds as air from your lungs moves over the vocal cords, causing them to vibrate. The space between the vocal cords is the *glottis*. When you swallow, a flap of tissue called the **epiglottis** closes over the glottis, preventing substances from going to the lungs.

The Trachea and Bronchi

The **trachea** is a cylindrical tube extending from the larynx approximately 12 cm (4.8 in.). It is about 2.5 cm (1 in.) wide and is supported by several C-shaped cartilaginous rings. The open ends of these incomplete rings are adjacent to the esophagus. Muscular and membranous tissues between the cartilage rings provide flexibility, allowing the neck to flex in all directions without injuring the trachea.

A ciliated mucous membrane lines the trachea and helps trap foreign matter such as dust and pollen. Cilia move foreign substances up to the pharynx, where they are swallowed and destroyed by the stomach acid.

The trachea ends behind the heart, where it divides into two primary **bronchi** (BRAHNG KYE) (sing., bronchus), one leading to the right lung and one to the left lung. The bronchi resemble the trachea in structure but have a smaller diameter. Additionally, the bronchi have complete cartilaginous rings for protection rather than the C-shaped rings of the trachea.

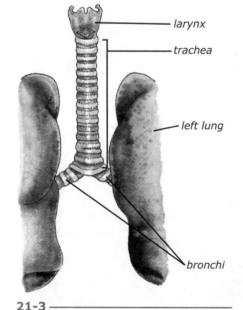

21-3
The trachea

larynx

trachea

left lung

bronchi

bronchus: (Gk. *bronkhos*—windpipe)

The Lungs

The lungs are the site of gas exchange between the external environment and the blood. The right lung is divided into three lobes, and the left lung is divided into two lobes. Each lobe receives a separate branch of the bronchus. Can you think of why the right lung has one more lobe, and is therefore larger, than the left lung?

The bronchi subdivide into smaller *bronchial* (BRAHNG kee ul) *tubes*, which in turn branch to form *bronchioles*. The bronchioles do not have cartilaginous tissue in their walls or cilia in their lining. There are more than 250 000 bronchioles in the lungs. All bronchioles end in **alveoli** (al VEE uh LYE) (sing., alveolus), which are microscopic bubblelike sacs.

Most of the lung tissue, then, is composed of the alveoli—approximately 300 million of them. If it were possible to open and flatten all alveoli, bronchioles, and other tubular passageways of the respiratory system, they would cover about 70 m² (84 yd²). That's about half the size of a tennis court!

An extensive network of small blood vessels called capillaries surrounds each alveolus. The walls of the alveoli and the walls of the capillaries are each made of a single layer of epithelial cells. This close physical relationship of alveoli and blood capillaries provides for rapid exchange of carbon dioxide and oxygen.

The **pleura** (PLOOR uh) is a delicate membrane that lines the thoracic cavity and covers the lungs. The *pleural space* is between these two layers and contains *pleural fluid*. The pleural fluid provides lubrication to reduce the friction between the lungs and walls of the thorax. Imagine how painful breathing would be if the delicate tissues of the lungs rubbed against the chest cavity with every breath! If either air or fluid, such as blood, enters the pleural space, it prevents the lung from inflating. The air or fluid must be removed as soon as possible.

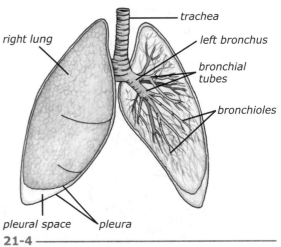

21-4
Structures of the lungs

alveolus: (L. *alveolus*—small cavity)
pleura: (Gk. *pleura*—rib)

Review Questions 21.1

1. Name, in order, the structures by or through which a molecule of air passes as it enters the nose and passes to the capillaries in the lungs.
2. What are the functions of the mucous membranes lining the respiratory system?
3. List the openings of the pharynx and tell where they lead.
4. Describe the pleura, the pleural space, and the pleural fluid.

21.2 Respiration

Why do you need oxygen? An easy answer is that you need oxygen to survive. But why is oxygen necessary for your survival? The answer to this question is actually found in the final phase of respiration.

Respiration is usually divided into three phases—external, internal, and cellular.

- *External respiration*: the exchange of gases in the air between the alveoli and the blood
- *Internal respiration*: the exchange of gases between the blood and body cells
- *Cellular respiration*: the use of oxygen by body cells in converting glucose or other energy-storing substances into water, carbon dioxide, and energy

21.2

Objectives

- Contrast the three phases of respiration
- Explain the mechanics of breathing
- Summarize the gas exchange process

Key Terms

inspiration
expiration
hemoglobin

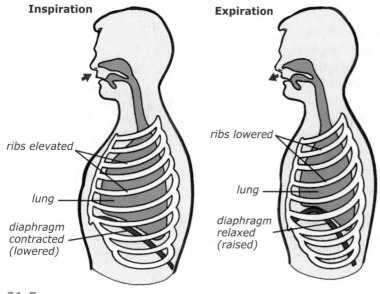

Inspiration

ribs elevated

lung

diaphragm
contracted
(lowered)

Expiration

ribs lowered

lung

diaphragm
relaxed
(raised)

21-5
The mechanics of breathing

inspiration: in- (in) + -spira- (to
breathe)
expiration: ex- (out) + -(s)pira- (to
breathe)

It is cellular respiration that requires the oxygen from the air we breathe in. Similarly, the carbon dioxide we breathe out comes from cellular respiration. Why is oxygen necessary for cellular respiration (see Subsection 4.3)?

Mechanics of Breathing

Breathing involves the movement of outside air into the alveoli and then out again. Filling the lungs is **inspiration** (inhalation), and forcing air out of the lungs is **expiration** (exhalation). During inspiration the dome-shaped diaphragm muscle contracts, moves down, and becomes flatter. At the same time, the ribs move upward and outward. These movements increase the chest cavity volume, reducing the pressure in the lungs so that air rushes into them.

When the diaphragm relaxes and the ribs are depressed following inspiration, the tissues of the lungs rebound to force air out. The abdominal organs that were slightly compressed during inspiration spring back and push the diaphragm upward. These body changes constitute expiration. When you need to control expiration, as in singing, the abdominal muscles squeeze the abdominal organs upward against the diaphragm to force out the air.

a Closer Look

Air Volumes Exchanged in Breathing

The amount of air that enters and leaves your lungs during a normal inspiration and expiration during sleep is about 0.5 L. This volume of air is known as the *tidal volume*.

If you breathe in as much air as you possibly can, you have accomplished forced inspiration. The air that enters your lungs in addition to the tidal volume is called the *inspiratory reserve volume*, or complemental air. Normal adults have about 3 L of inspiratory reserve volume.

Forced expiration is forcing out as much breath as you can, about 1 L of air in addition to the tidal volume. This quantity of air is known as the *expiratory reserve volume*, or supplemental air. The air that remains in your lungs after forced expiration is the *residual volume* (about 1 L).

The combined total of tidal volume, inspiratory reserve volume, and expiratory reserve volume is the *vital capacity*. This is the total amount of air you can exchange between your lungs and the environment. The sum of all lung volumes (vital

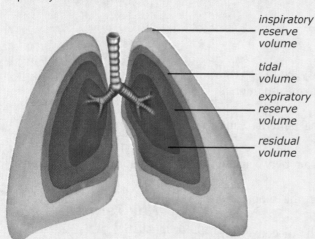

inspiratory
reserve
volume

tidal
volume

expiratory
reserve
volume

residual
volume

Lung volumes

A spirometer is used to accurately measure lung capacity.

capacity plus residual volume) is called the *total lung capacity* and is about 5.5 L.

Lung volumes and capacities are primarily dependent on the size and physique of the individual and vary considerably. Body position and posture also cause variation in air volumes. For example, the air volumes are greater when a person is standing than when he is sitting or lying down. This explains why it is easier to sing when standing. Diseases that restrict the breathing mechanism or occupy space inside the lungs will also affect lung volumes.

Lung Volumes

Name			Description	Approximate volume
total lung capacity (TLC) 5500 mL	vital capacity (VC) 4500 mL	tidal volume (TV)	amount of air that normally enters and leaves lungs at rest	500 mL
		inspiratory reserve volume (IRV)	air volume forcefully inspired after tidal volume	3000 mL
		expiratory reserve volume (ERV)	air volume forcefully expired after tidal volume	1000 mL
	residual volume (RV)		air remaining in lungs after forced expiration	1000 mL

Transport and Exchange of Oxygen and Carbon Dioxide

How are oxygen and carbon dioxide transported in the body? Do they travel as tiny bubbles alongside blood cells? No. The gases in the blood are either dissolved in or chemically combined with blood substances. Almost all the oxygen (98%) in the blood is combined with **hemoglobin**, the red oxygen-carrying pigment of the red blood cells.

Hemoglobin is a complex two-part molecule made of *heme* and *globin*. The heme portion of the molecule contains four atoms of iron, each of which can combine with one oxygen molecule. As oxygen moves from the alveoli into the blood, it combines rapidly with hemoglobin to form *oxyhemoglobin*. Oxyhemoglobin is a brilliant red, whereas hemoglobin without oxygen is a dull purplish red. The chemical bonds that form between oxygen and hemoglobin are rather unstable; therefore, the oxygen can be released quickly to the body cells for use in cellular respiration.

Oxygen diffuses through the alveolar walls into the blood because there is a higher concentration of oxygen in the air inside the alveoli than there is in deoxygenated blood. In the capillaries of the body, oxygen leaves the oxyhemoglobin to go to the body cells because there is a greater concentration of oxygen in the blood cells than in the cells of the body. The blood absorbs the carbon dioxide produced by cellular respiration because, when flowing through the capillaries of the body tissues, it has less carbon dioxide in it than do the tissues.

Blue Blood?

Is your blood ever blue? No! Why, then, might your veins appear blue when you look down at your wrist? It is because of the way light reflects off the tissues surrounding the veins. Blood is either bright red (oxygenated) or dull red (deoxygenated), never blue.

In illustrations, deoxygenated blood is usually diagrammed blue (see Fig. 21-6, for example). This is simply to distinguish the oxygenated from the deoxygenated blood; it is not because deoxygenated blood is actually blue.

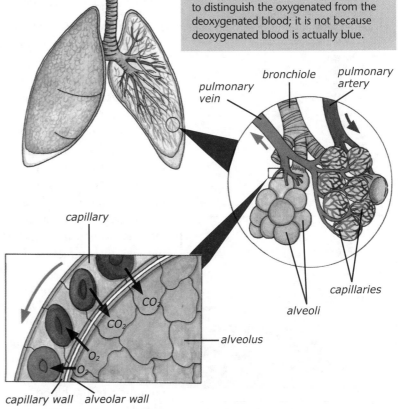

21-6
Oxygen exchange at the alveoli in the lungs

Carbon dioxide is transported to the lungs in three different ways. Most of the carbon dioxide (about 70%) is transported in the form of bicarbonate ions (HCO_3^-) in the fluid of the blood. About 7% of the carbon dioxide is dissolved in the fluid portion (plasma) of the blood. Only about 15%–25% of the carbon dioxide of the blood is combined with the hemoglobin. Hemoglobin can carry both oxygen and carbon dioxide at the same time since they do not compete for the same bonding sites.

As the blood flows through the capillaries that cover the alveoli, most of the carbon dioxide, which is dissolved or attached to the hemoglobin, diffuses into the lungs.

Review Questions 21.2

1. What are the two main phases of breathing? How do the diaphragm and ribs function during each of these phases?
2. Define and give the amounts of the respiratory system's various air volumes.
3. How are (a) oxygen and (b) carbon dioxide transported in the blood?

⊙4. Individuals that have had a tracheostomy breathe through an opening at the base of their neck, bypassing the upper airway. How might the air delivered to the lungs through this opening be different from air that passes through the upper airway?

21.3 Control of Breathing

The muscles involved in breathing are controlled voluntarily when a person sings, speaks, or prepares to hold his breath. Normal breathing, however, is a rhythmic, involuntary activity that continues even when a person is asleep or unconscious. The nervous system, in response to several chemical and environmental factors, regulates the rate and depth of respiration.

Respiratory Control Centers

The amount of oxygen required by the body depends on the level of activity. Playing racquetball or soccer increases the oxygen requirement, causing the player to breathe faster and more deeply. Conversely, his body requires much less oxygen while sitting at a desk doing homework; the breathing is much less intense.

The *respiratory center*, which controls involuntary respiration, is a group of nerve cells in the brain. The respiratory center monitors the level of carbon dioxide in the blood. When the level of carbon dioxide is elevated, the brain stimulates the respiratory muscles to increase the rate and depth of breathing in order to decrease the level of carbon dioxide and increase the oxygen level.

The respiratory center contains both an inspiratory and an expiratory center. The inspiratory center stimulates the diaphragm and external intercostal muscles to contract, which causes the lungs to draw in air. The expiratory center interrupts inspiration and thus permits air to be forced out of the lungs. The lung tissue contains specialized nerve cells called *stretch receptors*, which are stimulated as the alveoli expand during inspiration. These stretch receptors tell the brain when to stop inspiration and begin expiration.

During normal breathing, some of the alveoli are not ventilated, and some may be temporarily closed. The blood that passes over these alveoli, therefore, does not become very well oxygenated. This lowers the overall oxygen saturation of the blood. It is believed that a lower oxygen level in the blood stimulates long, deep inspirations that supply fresh air to the collapsed alveoli, as in a sigh or a yawn.

Objectives

- Explain the role of carbon dioxide in the regulation of breathing
- Describe the relationship between altitude and breathing

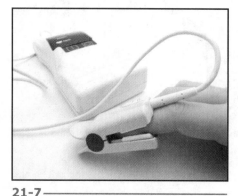

21-7

Pulse oximetry is used to monitor the blood's level of oxygen saturation.

Factors Affecting Breathing: Carbon Dioxide Level in Blood

Muscle cells produce carbon dioxide rapidly during active periods, raising the level of carbon dioxide in the blood above normal. The excess carbon dioxide forms quantities of carbonic acid that cause the blood to become too acidic. Special brain cells called *chemoreceptors* are very sensitive to carbon dioxide. When stimulated by the excess carbonic acid, they send messages to the brain's respiratory centers. The respiratory centers in turn send impulses to increase both the depth and rate of inspiration, permitting the blood to lose carbon dioxide more rapidly, and seeking to restore the blood's carbon dioxide level to normal.

Factors Affecting Breathing: Atmospheric Pressure

The "flatlander" who visits the mountains soon finds that he is out of breath during relatively mild exercise. The reason for this is the decrease in atmospheric (barometric) pressure at higher altitudes. Although approximately the same amount of oxygen is in the air at high altitudes as at sea level, the decrease in atmospheric pressure at high altitudes reduces the amount of oxygen that diffuses from the alveoli into the bloodstream.

This difficulty in breathing, however, can be overcome by *acclimatization* (uh KLYE muh tih ZAY shun). Acclimatization refers to the changes that occur in the body as a person becomes accustomed to a different atmospheric pressure. At higher altitudes the primary changes include an increase in the depth of respiration and an increase in the heart rate and the circulation of blood. These changes cause more red blood cells, with an increased amount of oxygen, to pass more rapidly through the blood vessels.

21-8

Above certain altitudes the body cannot acclimatize, and pressurized oxygen must be used.

acclimatization: ac- (to) + -clima- (Gk. *klima*—region of the earth)

Carbon Monoxide Poisoning

You have probably heard that you should never be near a car while the engine is running if the area is not properly ventilated. But why is this so dangerous? Whenever there is burning (such as with gasoline, chimneys or furnaces, gas stoves, generators, kerosene space heaters, and other sources that involve open flames), carbon monoxide (CO) is produced. It is a colorless, odorless gas that combines with hemoglobin (Hb) at the same site as oxygen. However, carbon monoxide forms a bond with hemoglobin that is two hundred times stronger than the bond with oxygen. Carbon monoxide thus prevents hemoglobin from carrying a normal amount of oxygen, and the person is stimulated to breathe more deeply. If he is still inhaling car exhaust, his heavier breathing will result in binding more carbon monoxide onto hemoglobin. The continuing cycle may result in death from carbon monoxide poisoning unless the person gets fresh air quickly. Breathing fresh air does not correct the damage done by carbon monoxide, but it prevents more damage from occurring. The oxygen-carrying capacity of the blood gradually improves as the poisoned red blood cells are replaced, a process which may take several weeks.

Home carbon monoxide monitors are readily available in stores. Since CO has about the same density as air, the sensors can be placed at any height; however, it is always best to follow the installation instructions included with the monitor.

Carbon Monoxide Poisoning	
Percentage of CO-bound hemoglobin (% COHb)	**Symptoms and consequences**
10%	no symptoms (Heavy smokers may have levels as high as 9% COHb.)
15%	mild headache
25%	nausea and serious headache; usually rapid recovery after exposed to oxygen or fresh air
30%	symptoms increase; potential for irreversible long-term effects
45%	unconsciousness
50% and above	death

Tobacco Smoking

Tobacco smoking, which contributes to more than 400 000 deaths every single year, is the major health problem affecting the respiratory system. The effects of smoking, however, are not limited to the respiratory system. Nicotine tars and other harmful substances in the smoke affect the entire body.

Inhaling tobacco smoke slows down the action of the cilia along the air passageways of the respiratory system. Continued smoking may eventually destroy these cilia. The mucus, normally moved along by the cilia, then collects and acts as an irritant, causing "smoker's cough." This nonmoving mucus allows smoke particles, bacteria, dust, and other substances to move into the lungs. As smoke particles collect in the alveoli, a person must breathe more deeply to obtain enough oxygen.

Though human lungs are much larger than is necessary for normal, quiet activity, an athlete who smokes experiences shortness of breath because his alveoli are filled with foreign matter. A long-time smoker has often filled so many

of his alveoli that he finds it difficult to breathe, especially when suffering from a respiratory disease such as pneumonia, flu, or even a cold. Emphysema, a crippling disease often caused by smoking, results in degenerated lungs that can no longer transfer oxygen and carbon dioxide between the air and the blood.

The nicotine in tobacco smoke causes the walls of the blood vessels to thicken, restricting the flow of the blood; therefore, nicotine increases the heart rate and blood pressure. Smokers have an increased risk of heart attack and stroke. In addition, tars and nicotine sometimes stimulate abnormal growth of cell layers in the air passageways. This growth, a type of lung cancer, restricts the air flow. Nicotine is an addictive drug (see p. 606). Once a person's body becomes used to nicotine, it craves more of the drug.

Smokers may also suffer a variety of symptoms such as shortness of breath; the tendency to tire easily; loss of appetite; inability to sleep; eye and skin irritations; discoloration of teeth, fingers, and lips; difficulty in performing precise movements with the hands; and impairment of vision and hearing. Smokers constantly have a low level of carbon monoxide poisoning. Some people are willing to live with these problems in order to get the "pleasure" of a cigarette, pipe, or cigar.

Ninety percent of smokers began smoking before age twenty-one. When asked why, people give responses such as the following: "My friends did, and it made me feel like a part of the group."

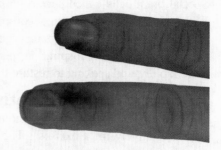

Nicotine-stained fingers and nails

"It made me feel more mature." "It was exciting to do something I knew I shouldn't."

When asked why a person who recognizes the hazards of smoking continues to smoke, common answers include the following: "I get nervous if I don't smoke." "I like the feeling of being accepted by my friends."

Nicotine produces feelings of relaxation in many people. Often, those who have the most difficulty stopping smoking are those who are most affected by the nicotine, feeling tense and irritable when deprived of the drug.

Tobacco itself is not bad. Several beneficial uses for tobacco are currently being used; these include treatments for certain diseases and as insecticides. However, a Christian cannot accept any of these as justification for smoking tobacco. The apostle Paul tells Christians to "not be brought under the power of any" (1 Cor. 6:12). A child of God must not knowingly subject his body to misuse.

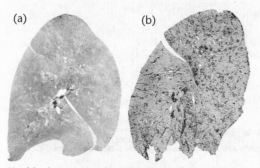

(a) (b)

Healthy lung (a) and smoker's lung (b)

Breathing Disorders

✦ *Asthma*: difficulty in breathing as a result of the bronchioles' becoming constricted. Asthma is sometimes caused by mold spores and other irritants.

✦ *Bronchitis*: an inflammation of the bronchi and bronchial tubes.

✦ *Collapsed lung*: failure of the alveoli of the lungs to inflate properly. Common causes are incomplete expansion of lungs in premature babies, accidental puncture of the chest wall, and fluid buildup in the chest cavity (such as sometimes occurs in elderly people).

✦ *Emphysema*: a degenerative condition in which the lungs overexpand and the walls of alveoli lose their elasticity and often rupture. Fluids fill the lungs.

This condition is often, but not always, a result of smoking.

✦ *Hiccups*: irregular contractions of the diaphragm while the glottis is closed. No useful function for hiccupping is known.

✦ *Lung cancer*: many symptoms. Breathing is difficult because abnormal growths of tissues block air passages in the lungs. Inhaling tobacco smoke is the most common cause.

✦ *Pleurisy*: an inflammation of the pleural membranes that causes painful breathing. Pleurisy usually occurs as a complication of pneumonia and tuberculosis.

✦ *Pneumonia*: an inflammation of the lungs caused by a fungal, bacterial, or viral infection. Usual symptoms are fever, chest pain, and a severe cough.

Other Factors That Affect Breathing

Different sensory stimuli can also affect the respiratory system. For example, severe pain usually increases respiration, and sudden exposure to cold can temporarily halt breathing. Physical condition, body size, and posture also affect the breathing rate.

Age also influences the respiration rate. The more rapid metabolic rate of children demands a higher respiratory rate. This rate gradually decreases until a person reaches old age unless health disorders and inactivity cause the breathing rate to increase slightly.

Respiration Rates

(breaths per minute)

At birth	40–70
1 year	35–40
10 years	19–24
25 years	16–18
Old age	20 or more

Review Questions 21.3

1. List the respiratory control centers. Tell what they measure and how they affect the breathing rate.
2. Describe how levels of (a) carbon dioxide, (b) carbon monoxide, and (c) atmospheric pressure affect the breathing rate.
⊙ 3. List several harmful effects of tobacco smoking.
⊙ 4. For what reasons do most people smoke, and why are these reasons unacceptable for Christians?

21B—The Digestive System

Beaumont and St. Martin by Dean Cornwell; Reproduced with permission of Wyeth Pharmaceuticals

21-9
U.S. Army surgeon William Beaumont (1785–1853) was one of the first researchers to study the physiology of digestion.

In order for your body cells to produce usable energy in the form of ATP, they need more than just oxygen. They must also have an energy source to use in cellular respiration. This energy source is usually glucose. The *digestive system* is responsible for obtaining the energy sources and other substances needed for the cells of the body.

Knowledge of human digestive processes was very limited until 1833. In that year William Beaumont published the results of the experiments that he had conducted on Alexis St. Martin, whose stomach and body wall had been accidentally perforated by a musket shot. Even after surgical repair and healing, the opening through the body wall and stomach persisted. Beaumont attempted to close it but failed.

Beaumont was able to conduct numerous experiments by inserting food (tied to a string) into the valvelike opening of St. Martin's stomach. Beaumont would then remove the food to determine the extent and speed of digestion. The permanent opening also allowed observation of the churning action of the stomach. Since 1833, scientists have learned much more about the functions of the digestive organs.

21.4 Organs of the Digestive System

The organs of the digestive system provide a means for **ingestion** (food intake) and provide structures and enzymes for **digestion** (breakdown of food to soluble substances). They also allow **absorption** of soluble food molecules into the bloodstream. Once the food is utilized for growth and repair, it is removed through **egestion** (elimination of undigested, unabsorbed material). Muscle tissue in the walls of the digestive system moves the food material along while these functions are being performed.

21.4

Objectives

- Describe the functions of the digestive system
- Describe and summarize the functions of each digestive system organ

Key Terms

ingestion	cementum
digestion	pulp cavity
absorption	esophagus
egestion	stomach
alimentary canal	chyme
peristalsis	small intestine
salivary gland	villus
crown	liver
root	bile
neck	gallbladder
dentin	pancreas
enamel	large intestine

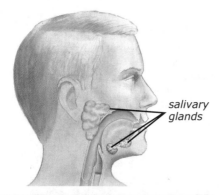

salivary glands

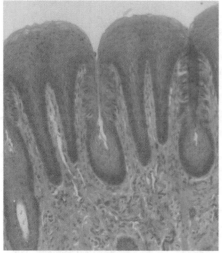

21-10

Location of the salivary glands (top) and a TEM of the papillae of the tongue (bottom)

alimentary: aliment- (L. *alere*—to nourish)

gastrointestinal: gastro- (belly) + -intestinal (L. *intestinus*—internal)

amylase: amyl- (Gk. *amulon*—starch) + -ase (E. *-ase*—enzyme)

mastication: (Gk. *mastikhan*—to grind the teeth)

The organs for digestion are usually divided into the following two groups:

- The **alimentary** (AL uh MEN tuh ree) **canal** includes the mouth, pharynx, esophagus, stomach, small intestine, and large intestine.
- Accessory organs include the teeth, tongue, salivary glands, liver, gallbladder, and pancreas.

The alimentary canal, or *gastrointestinal* (GAS troh in TES tuh nul) tract, is a continuous tube extending from mouth to anus and measuring about 9 m (30 ft) long. Food is moved along the canal by wavelike muscular contractions called **peristalsis** (PEHR ih STAHL sis).

The Mouth

Food taken into the mouth is mixed with saliva from the **salivary glands**. The saliva moistens and lubricates the food for easy swallowing. Three pairs of salivary glands, as well as numerous small glands distributed in the lining of the mouth, secrete saliva. All the salivary glands secrete *salivary amylase* (AM uh LASE), an enzyme that begins the chemical breakdown of starch to sugar. Taking food into the mouth or even the sight, smell, or thought of food triggers the production of saliva.

During *mastication* (MAS tih KAY shun), the chewing of food, the lips keep food in the mouth cavity, and the tongue pushes food toward the teeth. The taste sensations—sweet, salty, sour, and bitter—that are experienced as food is chewed originate with the *taste buds*, located in small bumplike structures called *papillae* on the surface of the tongue. Many scientists now recognize a fifth taste, umami (oo MAH mee), or savory, a meaty taste associated with proteins. However, because the sense of taste is closely associated with the sense of smell, much of what is defined as taste is actually the detection of odors from the food. That is why food does not taste the same to a person with a cold or some kinds of allergies.

The Teeth

Four different types of teeth are involved in mastication. The eight chisel-shaped *incisors* bite, the four cone-shaped *cuspids* (canines) tear, and the *premolars* (bicuspids) and *molars* have surfaces for crushing and grinding food.

A child's first set of teeth, called *deciduous teeth*, lacks premolars and has only eight molars. The set of *permanent teeth* in adults has the same number of incisors and cuspids as the deciduous set but has eight premolars and twelve molars. The last molars, or wisdom teeth, frequently fail to emerge into the mouth or, if they do, are often crooked. Sometimes the wisdom teeth must be removed.

Layers of the Alimentary Canal Wall

The walls that form the organs of the alimentary canal are typically composed of four layers. Starting from the inside of the canal (from deep to superficial), the layers are as follows:

✦ *Mucosa, or mucous membrane.* This layer is formed by epithelium and some connective tissue and comes into direct contact with food as it passes through the canal. This membrane secretes mucus that protects and lubricates the lining.

✦ *Submucosa.* The submucosa is composed of loose connective tissue, blood vessels, and many nerve endings. The blood

vessels carry away the nutrients that are absorbed, and the nerve endings stimulate the muscle fibers so that the food is continually moving by peristalsis.

✦ *Muscular layer.* This layer, consisting of a circular band and a longitudinal band of visceral muscle, is the thickest of the four layers. The main function of the muscular layer is peristalsis.

✦ *Serous layer.* The outermost portion of the canal wall is the serous layer. It is continuous with the *mesentery*, the connective tissues that attach to the posterior body wall and hold the digestive organs in their proper position.

Teeth are set in sockets in the bony ridges of the jaws. The *gingivae*, or gums, cover these ridges and merge with the connective tissue that attaches each tooth in its socket. The part of the tooth above the gingivae is called the **crown**, and the part that is anchored in the socket is called the **root**. Between the crown and the root is the **neck**, the part of the tooth surrounded by the gingiva.

Each tooth is composed of two layers of hard tissues. **Dentin**, which is similar to bone, forms most of the tooth structure. In the crown portion, the dentin is covered by the **enamel**, the hardest material in the body. The enamel protects the tooth from physical damage and chemical corrosion. A thin layer of bonelike **cementum** covers the dentin in the root region and is attached to the connective tissue that anchors the tooth. Inside each tooth is a **pulp cavity** filled with blood vessels, lymph vessels, nerves, and connective tissues collectively called the *pulp*.

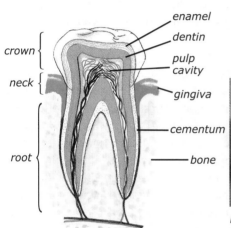

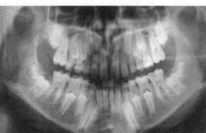

Radiograph showing permanent teeth forming in the bone

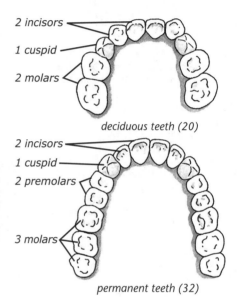

deciduous teeth (20)

permanent teeth (32)

21-11

Human teeth

Dental Caries

Dental caries—sometimes called dental decay or cavities—is the single most common childhood disease. Over 40% of children aged two to eleven have cavities in their "baby" teeth, and almost 60% of adolescents have at least one filling in their permanent teeth. This disease is caused by organic acids produced by bacteria that ferment the sugary film (plaque) on the surface of the tooth.

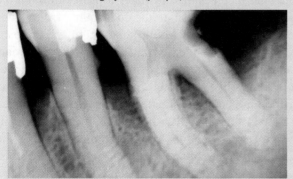

Radiograph showing caries (pink highlighted area)

After these acids break down the enamel, the bacteria and acids may enter the dentin. Spreading rapidly through the dentin, dental caries may invade the pulp and cause pus formation in the pulp and in the socket between the root of the tooth and the gum. Such a tooth is abscessed. A tooth with caries or an abscess should receive professional dental care to prevent the loss of the tooth and the spread of infection to other organs of the body.

People can prevent, or at least slow down, caries by using fluoride-containing toothpaste and by brushing and flossing properly after every meal. Eating less candy and drinking fewer sugary beverages can also decrease dental caries by reducing plaque formation.

Regular dental examinations can identify dental caries before the tooth becomes abscessed. Repairing dental caries often requires an injection of anesthetic, which allows the dentist to remove the caries painlessly using a dental handpiece, or "drill." Newer techniques for removing dental caries utilize special types of lasers or micro-air abrasion instruments, although they are not applicable in all situations. Sometimes these new techniques do not require an injection to anesthetize the teeth.

21-12 —————————

As the muscles contract and relax, the bolus is moved to the stomach.

The Pharynx, Esophagus, and Stomach

As the food is chewed, softened, and lubricated, it is formed into a ball-like mass called a *bolus* and is swallowed. During swallowing, the tongue pushes upward and back to direct the bolus into the *pharynx*. At the same time the *uvula* moves superiorly, closing the back of the nose to prevent the food from moving into the nasal cavities. The pharynx constricts after receiving the food in order to direct it into the esophagus, which is stretched open to receive the food. The *epiglottis* covers the *glottis* of the *larynx* as the food moves into the esophagus, thereby preventing the food from entering the trachea.

The **esophagus** is a muscular tube that is lubricated by many mucous glands so that the food can be easily moved downward by peristalsis. Swallowed food takes about three seconds to get to the stomach from the mouth.

The esophagus joins the stomach at the *cardiac sphincter*, a muscular valve that remains closed except when food passes into the stomach. If the cardiac sphincter does not remain properly closed, acidic liquid can move from the stomach back into the esophagus, causing a burning sensation called heartburn. What type of muscle (cardiac, visceral, or skeletal) do you think makes up the cardiac sphincter?

The **stomach** is a J-shaped muscular pouch that in most healthy adults can hold up to 2 L (2.1 qt) of food. The thick, muscular walls of the stomach churn and mix the food with the acidic gastric juices that are secreted by the epithelial lining of the stomach. The gastric juices contain enzymes for protein digestion and would digest the walls of the stomach if they were not protected by the slimy mucus also secreted by the mucous-membrane lining of the stomach. The gastric juices also contain hydrochloric acid (HCl), which aids in digestion and kills bacteria that are swallowed into the stomach. The duration of churning may be four hours for coarse foods like celery and spinach, while foods like oatmeal or pudding move rapidly through the stomach. Although the stomach lining does not absorb many food molecules, it does absorb alcohol, water, and certain drugs.

After the food is thoroughly broken down into a semiliquid state called **chyme** (KYME), the peristaltic contractions propel the food through the *pyloric sphincter* into the small intestine in small spurts. This pyloric valve usually remains closed until the food is converted to chyme.

The Small Intestine

Most of the digestion and absorption of food occurs in the **small intestine**. The entire small intestine is about 7 m (23 ft) long and 2.5 cm (1 in.) in diameter. The *duodenum* (doo AH di nuhm), the first section of the small intestine, is about 25 cm (10 in.) long and receives the chyme as it passes through the pyloric sphincter. As the acidic chyme passes into the duodenum, it stimulates the intestinal lining to secrete hormones. These hormones stimulate the pancreas, gallbladder, and intestinal lining to release enzymes and other materials into the duodenum to break down food into small molecules. **Villi** (VIL eye) (sing., villus) are microscopic fingerlike structures that line the small intestine and function to absorb the food molecules. The villi increase the surface area of the small intestine and contain small capillaries and lymph vessels; the increased surface area allows greater absorption. The food molecules cross the epithelial layer of the villi by diffusion and active transport and are absorbed through the vessels within the villi. These vessels then distribute the food molecules throughout the body.

21-13 —————————

SEM of small intestine villi

villus: (L. *villus*—shaggy hair)

The Liver, Gallbladder, and Pancreas

The **liver**, the largest internal organ, is located in the upper right part of the abdomen immediately inferior to the diaphragm. It is composed of soft tissue that contains many microscopic spaces called *sinusoids* (SYE nuh SOYDZ). The sinusoids can serve as blood reservoirs. The blood from all digestive organs (stomach, small and large intestines) flows through the liver sinusoids before going elsewhere in the body. As the blood passes through the sinusoids, it comes in contact with the liver cells. Each liver cell may perform more than five hundred separate functions. Several functions of the liver cells include the following:

- Engulfing bacteria and worn-out red blood cells
- Removing many drugs and poisons from the blood and detoxifying them
- Converting excess glucose into glycogen and storing it
- Changing excess glucose into fats
- Absorbing amino acids from the blood and using them to manufacture most of the proteins found in the blood
- Storing the fat-soluble vitamins, iron, and copper

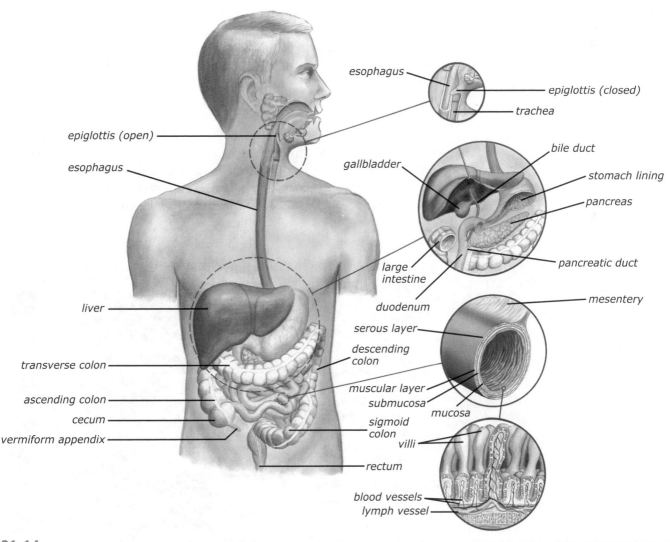

21-14
Structures of the human digestive system

Perhaps the most familiar function of the liver is the formation of **bile**, a greenish fluid that is necessary for the processing of fatty substances in the small intestine. The liver secretes about 0.5 L (1 pt) of bile a day.

The **gallbladder** is a 7.5–10 cm (3–4 in.) green, pear-shaped sac attached to the liver by the *bile duct*. The gallbladder serves as a reservoir for bile. During a meal, when fatty materials enter the duodenum, the gallbladder contracts and sends bile into the small intestine.

The **pancreas** is a soft, pinkish white gland 15–25 cm (6–10 in.) long and 2.5 cm (1 in.) wide. About 97% of the pancreas cells produce digestive juices that contain enzymes for digesting carbohydrates, fats, and proteins. These digestive juices flow into the small intestine through the *pancreatic duct*. Most of the remaining pancreas cells produce hormones involved in regulating the amount of sugar in the blood.

The Large Intestine

The **large intestine**, or colon, is 1.5–2 m (5–6 ft) long and about 6.5 cm (2.5 in.) in diameter. It has four major sections—ascending, transverse, descending, and sigmoid—named according to their shape and the direction in which they move waste.

Material from the small intestine moves into the *cecum* (SEE kum), which is a pouch about 6 cm (2 in.) long at the start of the ascending colon. The narrow end of the cecum is the *vermiform appendix* (VUR muh FORM • uh PEN diks), which secretes mucus for lubrication and harbors certain microorganisms that digest cellulose. It is also believed to function in immunity.

A major function of the large intestine is the removal of *feces* (waste materials) through the *anus*. The *rectum* is the straight, muscular portion that expels the feces.

A second function of the large intestine is the reabsorption of water. In other words, the water in food, saliva, bile, pancreatic juices, and secretions from the stomach and small intestine is recycled and used again. On a daily basis this amounts to about 7.5 L (2 gal). If this water were egested with the feces, a person would soon become severely dehydrated.

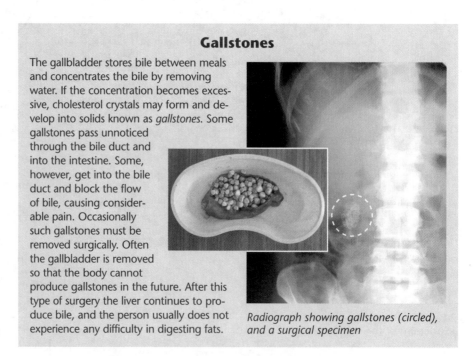

Gallstones

The gallbladder stores bile between meals and concentrates the bile by removing water. If the concentration becomes excessive, cholesterol crystals may form and develop into solids known as *gallstones*. Some gallstones pass unnoticed through the bile duct and into the intestine. Some, however, get into the bile duct and block the flow of bile, causing considerable pain. Occasionally such gallstones must be removed surgically. Often the gallbladder is removed so that the body cannot produce gallstones in the future. After this type of surgery the liver continues to produce bile, and the person usually does not experience any difficulty in digesting fats.

Radiograph showing gallstones (circled), and a surgical specimen

vermiform: vermi- (L. *vermis*—worm) + -form (*forma*—form or shape)

appendix: ap- (to, upon) + -pendix (to hang)

Intestinal Bacteria

The human intestines provide a home for millions of microbes called the "intestinal flora." Infants acquire these microbes from their mother and the environment during birth and for about a month after birth. Although people can survive without these microbes, they perform many beneficial functions. The large intestine serves as an ideal site for growth of bacteria and yeast that produce vitamins B_1, B_2, B_{12}, and K. Other bacteria in the intestines perform functions such as building the immune system and preventing the growth of harmful bacteria.

A loss of balance in the intestinal flora has been linked with various illnesses, including inflammatory bowel diseases, such as Crohn's disease. Such conditions are sometimes treated by having the patient take dietary supplements of live microbes to restore beneficial bacteria to the intestines. Sometimes these supplements can be obtained in the regular diet (e.g., in yogurt); other times they must be obtained by prescription.

Many yogurts contain live bacterial cultures that help restore beneficial bacteria to the intestines.

Review Questions 21.4

1. Name and describe the four main functions of the digestive system.
2. What are the four types of teeth in a human's mouth?
3. What is the difference in numbers and kinds of deciduous teeth and permanent teeth?
4. What are (a) the three regions of a tooth and (b) the layers of a tooth?
5. What structures keep the chyme within the stomach while it is being churned? Where is each structure located?
6. What prevents the stomach from digesting itself?
7. Which organ of the digestive system is the major site of digestion and absorption? What structures aid in absorption?
8. List several functions of the liver.
9. List and tell the functions of (a) the organs in the alimentary canal and (b) the accessory organs of the digestive system.
10. Describe the formation of gallstones.
11. Describe the location and functions of the vermiform appendix.
⊙12. Why is the colon called the "large" intestine, when it is significantly shorter than the small intestine?

21.5 Foods and Digestion

Food is any material that contributes to growth and repair of the body or supplies energy. According to this definition, minerals, vitamins, carbohydrates, fats, and proteins are all foods. Abundant water is also needed for the body to properly utilize these foods.

Most meals contain something from these groups. The kinds and amount of food a person needs, however, depend in part on his physical activities. For example, some people require more of certain vitamins than other people; athletes and lumberjacks require food with more calories than do office workers and laboratory technicians.

The mechanical breakdown of large food particles into smaller ones is known as *physical digestion*. Examples include the teeth breaking up food by mastication and the stomach churning food into chyme. Table 21-2 summarizes the major components of physical digestion.

Because the molecules of many of the carbohydrates, fats, and proteins that are eaten are too large for absorption, they must be broken down chemically by the action of *enzymes* (see pp. 39–40). This *chemical digestion* occurs on the molecular level, breaking food down into simpler chemicals. Most enzymes are proteins, and many of them require the presence of *coenzymes* in order to function. Many vitamins serve as coenzymes.

21.5

Objectives

- Distinguish between physical and chemical digestion
- Identify the primary digestive enzymes and which foods they digest
- Recognize the sources and functions of major vitamins and minerals
- Describe guidelines for a healthy diet

Key Terms

emulsification
fiber
mineral
vitamin
calorie

21-1 Summary of Chemical Digestion

Location	Mouth		Stomach		Small intestine					
Digestive fluid	Saliva from salivary gland		Gastric juices from stomach lining		Bile from liver		Pancreatic juices from pancreas		Intestinal juices from intestinal lining	
	Enzyme	Breakdown	Enzyme	Breakdown	Enzyme	Breakdown	Enzyme	Breakdown	Enzyme	Breakdown
Starch	AMYLASE	disaccharides (maltose) / starch					AMYLASE	disaccharides (maltose) (sucrose) (lactose)	MALTASE / SUCRASE / LACTASE	monosaccharides (glucose) (fructose; glucose) (galactose; glucose)
Fat						emulsified fats	LIPASE	fatty acids and glycerol	LIPASE	fatty acids and glycerol
Protein			PEPSIN	protein fragments			TRYPSIN and CHYMO-TRYPSIN	polypeptides	PEPTIDASE	amino acids

21-2 Summary of Physical Digestion

Structure	Activity
Mouth	
lips and cheeks	keep food between teeth during chewing
tongue	moves food during chewing and swallowing
salivary glands	produce saliva, which softens and moistens food, covering it with mucus for easier swallowing
teeth	masticate food for more rapid digestion and easier swallowing
Pharynx	moves food and liquids to esophagus
Esophagus	moves food and liquids to stomach
Stomach	churns food until semiliquid (chyme); moves chyme into small intestine
Small intestine	mixes chyme from stomach with juices from intestine and pancreas and with bile from liver; absorbs most of the food
Large intestine	prepares waste material for egestion; absorbs water, minerals, and vitamins

Water

Besides oxygen, water is the most important molecule for human survival. Most chemical reactions that occur in the body do so in an aqueous environment. Water is vital to the body because it dissolves certain foods for easier digestion. It also aids in regulating fluid balance, pH of the blood, and body temperature (through perspiration).

A person can live without food for five or more weeks, but without water he can survive only a few days. Most of the normal water loss from the body passes through the kidneys during urine formation. The body also loses water during expiration, through perspiration, and in feces. The amount of water loss varies depending on the level of activity, air temperature, and other environmental conditions. Lost water must be replaced as soon as possible.

Carbohydrates

The most economical and abundant foods are rich in carbohydrates. Carbohydrates include the sugars, which are a source of quick energy, and more complex molecules such as starch and glycogen.

Carbohydrate digestion begins in the mouth as the salivary enzyme *amylase* begins to break down starches into maltose, a disaccharide. Since the bolus of food is usually swallowed before starch digestion is complete, not all the starch is converted to maltose. The highly acidic stomach fluids soon stop the amylase digestion of starches.

When the food material moves into the small intestine, the pancreas secretes amylase to continue starch digestion. The intestinal lining secretes other enzymes that convert disaccharides (such as maltose, sucrose, and lactose) into monosaccharides (such as glucose, fructose, and galactose). These simple sugars can then be absorbed into the bloodstream and used for energy production, or they can be stored in the liver for future use.

Fats

Fat digestion in humans occurs in the small intestine. As the fats move from the stomach into the small intestine, a hormone stimulates the release of bile from the gallbladder. Flowing into the duodenum, the bile causes the fats to form droplets. The separation of fat into small droplets is called **emulsification** (ih MUL sih fih KAY shun). Bile is the *emulsifier*. Since enzymes can work only on the surface of a substance, emulsification greatly speeds fat digestion by increasing the surface area of the fats.

The enzyme *lipase*, secreted by the pancreas, digests fats into glycerol and fatty acids. The villi of the small intestine, aided by bile, then absorb these smaller molecules. After absorption into the bloodstream, glycerol and fatty acids recombine to form certain types of fat molecules. Two to four hours after a meal rich in fats, the blood reaches its highest concentration of fats, which can be 1%–2% of the total blood volume.

Fats are used to form cell membranes, form protective membranes around certain nerve cells, form certain hormones, cushion delicate structures, and provide energy. Many fats in the body do not come directly from the diet but are formed in liver cells from excess sugars or amino acids (that come from proteins).

Fat not used immediately by the body is stored in *adipose* (AD uh POHS) *cells*. The conversion of other foods to fats explains why eating any type of

lipase: lip- (fat) + -ase (enzyme)
adipose: (L. *adeps*—fat)

Food Fiber

Food **fiber**, often called roughage or bulk, is primarily the indigestible cellulose (a complex carbohydrate) of fruits and vegetables. It is the "stringy stuff" in roots, leaves, stems, and fruits that provides physical support for the plant. Some good sources of natural food fiber are whole-grain breads and cereals, fruits with the skin, vegetables, nuts, and seeds. The U.S. Food and Drug Administration (FDA) recommends 20–38 g/day; however, the average U.S. diet contains only 14–15 g/day.

Even though natural food fiber cannot be digested, it is important for moving material through the intestine. It also seems to shield the intestinal lining from irritation by certain harmful substances that may be in food. Some research studies link increased dietary fiber to a decreased incidence of colon cancer. Fiber also decreases blood cholesterol by binding to the cholesterol molecule and preventing it from being absorbed. Beta glucan, a specific component of oat bran, has been demonstrated to reduce blood cholesterol by decreasing the liver's cholesterol synthesis. Since elevated blood levels of cholesterol have been linked to an increased risk of cardiac disease, eating foods high in fiber, especially oat bran, can help reduce the risk of developing heart disease.

Minerals and Vitamins

The human body needs many different kinds of minerals. **Minerals** are inorganic substances found naturally in soil. They are absorbed by plants and then passed on to animals as they feed on the plants. Humans obtain minerals by eating foods that come from plants and animals.

Although the body needs only small amounts of minerals, they are necessary for it to produce certain compounds. Minerals are important for nerve and muscle function, blood coagulation and pH maintenance, enzyme function, and many other uses. The body does recycle some minerals but not enough to take care of its needs; therefore, minerals must be consumed daily in food. The minerals consumed, however, must be in usable forms. A person cannot get the iron he needs by eating rusty nails. Iron in the ferrous (Fe^{+2}) form is absorbed more readily than the ferric (Fe^{+3}) form. Calcium must be consumed with other materials to be absorbed and used.

The human body needs more than just the proper amounts of carbohydrates, fats, proteins, and minerals. Other things are needed for proper growth and development. **Vitamins** are organic substances necessary for normal metabolism.

There are two classes of vitamins.

✦ *Fat-soluble vitamins.* Vitamins A, D, E, and K are fat-soluble vitamins and can be stored in the body for future use. The accumulated effects of too much vitamin A or D, however, can be toxic. Many fat-soluble vitamins work with certain molecules in the cell membrane to regulate the movement of substances into and out of cells.

✦ *Water-soluble vitamins.* All other vitamins are water-soluble and must be taken in daily to ensure good health. If more water-soluble vitamins are consumed than the body can use, the excess vitamins will be excreted in the urine. Most of the water-soluble vitamins function as coenzymes, working together with enzymes to regulate certain metabolic reactions.

Under normal conditions, only small amounts of vitamins are needed. Even though all food groups contain certain types and amounts of vitamins, animal foods (such as milk, meat, fish, and eggs) are rich in fat-soluble vitamins, while plant foods contain more water-soluble vitamins.

Selected Minerals and Their Functions

Mineral	Sources	Functions	Comments
calcium (Ca)	dairy products, egg yolk, and leafy green vegetables	major part of bones and teeth; necessary for blood-clot formation, nerve cell functions, and muscle contractions	blood calcium level controlled by hormones; deficiency results in muscular twitching, spasms, poor blood clotting, fragile bones, and stunted growth (in children)
phosphorus (P)	dairy products, meat, fish, poultry, and nuts	important part of bones and teeth; part of ATP, DNA, RNA, and certain proteins	deficiency results in poor appetite, retarded growth, and general body weakness
iron (Fe)	meat, liver, egg yolk, beans, peas, dried fruits, nuts, and cereals	part of hemoglobin and many enzymes	deficiency results in anemia; mineral most likely to be deficient in diet
iodine (I)	iodized salt, seafood, cod-liver oil, and vegetables grown in iodine-rich soils	part of hormone from thyroid gland	deficiency results in goiter (abnormal growth of thyroid gland)
copper (Cu)	eggs, whole-wheat breads, beans, beef, liver, fish, spinach, and asparagus	involved in formation of melanin and hemoglobin	some stored in liver and spleen
sodium (Na)	table salt, meat, seafood, poultry, milk, and cheese	important in fluid balance, nerve cell function, and muscle contraction	excessive amounts may lead to high blood pressure; too much included in many American diets
potassium (K)	avocados, bananas, dried apricots, meats, nuts, and potatoes	important regulator of muscle contraction and nerve cell function	excessive or deficient amounts affect heart function and cause muscle weakness
chlorine (Cl)	same as sodium	important in fluid balance, acid-base balance, and formation of acids in stomach	closely associated with sodium; most abundant in fluids in brain and stomach
magnesium (Mg)	nuts, legumes, dairy products, and leafy green vegetables	needed for muscle and nerve cell functions; involved in bone formation; part of many coenzymes	most abundant in bones

Vitamin Myths

- *Vitamins from natural foods are better than vitamin tablets.* Synthetic vitamins manufactured in laboratories usually have the same effect on the body as natural vitamins obtained directly from plant and animal sources. However, if the molecular structure of a synthetic vitamin differs from its natural counterpart, the body may not be able to use the synthetic vitamin as effectively.
- *The more vitamins you consume, the healthier you will be.* Too much of certain vitamins can cause health problems.

For example, extreme excess of vitamin A can produce peeling skin or blurred vision, and too much vitamin D can cause vomiting and high blood pressure. However, it is much more common to have a deficiency of a vitamin than an excess.
- *Vitamins provide energy.* Vitamins are not a direct supply of energy. They do, however, aid in metabolism of carbohydrates, fats, and proteins. Therefore, they do affect the release of energy in the body.

Vitamins: Their Functions and Deficiency Disorders

Vitamin	Sources	Functions	Deficiency disorders
A	fish-liver oil, milk, butter, yellow and green vegetables	essential for normal vision, healthy epithelial tissue, and proper growth of teeth and bones	night blindness; scaly skin; poor growth
D	fish-liver oil, egg yolk, fortified milk, butter, and margarine	essential for absorbing calcium from intestines	bone disorders such as rickets in children
E	nuts and wheat germ, oils from seeds, fish, eggs, and liver	inhibits oxidation; involved in formation of DNA, RNA, and red blood cells	increased breakdown of fatty acids; may lead to anemia
K	spinach, cauliflower, cabbage, and liver	involved in forming prothrombin for blood clotting	delayed clotting time, resulting in excessive bleeding
B_1 (thiamine)	eggs, pork, nuts, whole-grain cereals, liver, yeast, and leafy green vegetables	coenzyme for many enzymes involved in carbohydrate and amino acid metabolism	beriberi—general body weakness, stunted growth, poor appetite
B_2 (riboflavin)	yeast, liver, cheese, milk, eggs, asparagus, peas, beets, and peanuts	used in metabolism of carbohydrates and proteins, especially in cells of eyes, skin, blood, and intestinal lining	eye and skin disorders; anemia; sores in intestines
niacin	yeast, liver, meat, fish, cereals, whole-grain breads, peas, beans, and nuts	essential part of a coenzyme involved in cellular respiration	pellagra—skin problems, diarrhea, mental problems
B_6 (pyridoxine)	salmon, liver, meat, cereals, legumes, yeast, tomatoes, spinach, and yogurt	serves as coenzyme for fat metabolism; involved in production of antibodies	sores of eyes, nose, and mouth
B_{12} (cyanocobalamin)	liver, meat, fish, milk, eggs, and cheese	serves as coenzyme for red blood cell formation; involved in nerve function	type of anemia; nerve disorders
pantothenic acid	kidney, liver, yeast, green vegetables, and cereals	part of coenzyme involved in cellular respiration and formation of certain hormones	general tiredness; nerve and muscle disorders
folic acid	leafy green vegetables, liver, and cereals; also produced by bacteria in large intestine	part of enzymes that form DNA, RNA, and red blood cells	type of anemia
biotin	yeast, liver, and egg yolk; produced by bacteria in large intestine	coenzyme in cellular metabolism	mental depression; muscular pains; tiredness; skin problems
C (ascorbic acid)	citrus fruits, tomatoes, green vegetables, and strawberries	needed for connective tissue formation; detoxifies body	scurvy—slow healing, weak blood vessels and bones

✦ *Cirrhosis of the liver*: a disease characterized by the formation of dense connective tissue and deposits of fatty tissue that destroy liver cells. It may be caused by drinking alcoholic beverages, infections by bacteria and viruses, or breathing certain gases.

✦ *Diverticulosis*: the condition in which sacs or pouches have formed in the wall of the large intestine. Usually no inflammation is experienced. It commonly occurs in persons who suffer from constipation for long periods. When inflammation occurs, the condition is called diverticulitis.

✦ *Flatus*: a condition of gas in the digestive tract that results from chemical digestion of foods. It may also result from

drinking too many carbonated beverages or from swallowing air while eating too fast.

✦ *Jaundice*: a condition in which the skin and white of the eye appear yellow because of excessive bile in the blood. It may be caused by diseases of the liver or blockage of bile ducts by gallstones.

✦ *Peptic ulcer*: an ulcer (an area that has lost the epithelial covering) occurring in the lower esophagus, stomach, or duodenum. A peptic ulcer is usually caused by too much acid in the stomach and certain bacteria.

food can increase the amount of adipose (fat) tissue in the body. Adipose tissue is found under the skin, between muscles, behind eyeballs, on the heart, and between most internal organs. In these locations fat tissue has three general functions: it is a future supply of energy, an insulation from cold, and a layer that protects from physical injury.

Proteins

Proteins are not only a part of all living cells but also a part of nonliving structures such as hair, nails, and the matrix of connective tissues. Most enzymes and some hormones are proteins. About 50% of the body's dry weight is protein, and most of that is in the muscle tissue. Proteins are most important for growth and repair, but they may also be used for energy production.

Protein digestion begins in the stomach. The enzyme *pepsin*, which is secreted by the cells lining the stomach, begins by breaking proteins into smaller fragments. Some proteins, however, are not digested by pepsin. When the food material moves from the stomach into the small intestine, these remaining proteins are exposed to *trypsin* and *chymotrypsin*, enzymes secreted by the pancreas. Other enzymes from the intestinal lining complete the digestion of the protein fragments into *amino acids*, which are absorbed and carried to various parts of the body by the blood.

pepsin: (to digest)

Amino acids are used within the body cells to form various proteins. Many amino acids can be produced by the body from other substances. However, if the body is not able to produce enough amino acids, the body may break down some skeletal muscle proteins into amino acids. If there is an excess of amino acids, they are broken down in the liver cells to release energy.

A Healthy Diet

There is often much confusion about the various diets advocated today. The first step in eliminating confusion is to define the term *diet*. In this context, *diet* means the usual food and drink a human consumes. It does not mean guidelines for weight loss or gain. In fact, most teenagers should not "diet" for weight loss since their bodies are still growing and maturing. If a person needs to gain or lose weight, professional treatment is recommended so that proper nutrition is maintained. That being said, there is disagreement among authorities regarding what the ideal human diet should be. The Bible

21-15

The Food Pyramid from the U.S. Department of Agriculture (USDA) provides guidelines for a healthy diet. Larger wedges indicate more servings per day from that category.

does give a general principle against gluttony, which is the sin of overeating (Prov. 23:20–21). Moderation should be at the center of the Christian's diet. There are, however, several guidelines nutritionists have devised that clarify what an appropriate diet includes.

✦ *Daily meals should include a variety of foods from all of the five basic food groups:* (1) milk and dairy products; (2) meat (including fish), legumes, nuts, and seeds; (3) fruits; (4) vegetables; and (5) whole grains (breads, pasta, rice, and cereals). Such meals will supply the body with most of the needed nutrients.

✦ *Eat a sufficient amount of raw fruits, vegetables, and seeds.* These foods, often lacking in a teenager's diet, are important because they are the best sources of vitamins and minerals. Many nuts and seeds (sunflower, sesame, flax, and poppy) are also good sources of protein and monounsaturated fats.

✦ *Drink more water, fruit juices, and milk than coffee, tea, soda, and other beverages.* Some fruit juices are good sources of vitamins and minerals, but you should always read the label to be certain. Milk is a good source of calcium, phosphorus, protein, and certain vitamins.

✦ *Restrict the intake of sugars in foods such as candy and soda.* Foods that contain large quantities of sugar and little nutritional value should be avoided.

✦ *No one food or group of foods should be overemphasized.* Many fad diets promote the elimination of entire food groups or focus on only one food group. This can be hazardous to your health. A variety from all the food groups is needed to supply all the nutrients, vitamins, and minerals necessary for a healthy diet.

✦ *Supplement the diet with a multivitamin and mineral tablet.* The supplement should include the water-soluble vitamins (especially B-complex and C) because they are not stored in body tissues.

The Calorie and Weight Gain or Loss

The term **calorie** refers to a measurement of the amount of energy in a substance. One calorie is the amount of heat required to raise the temperature of 1 g of water 1 degree Celsius. Since this is such a small value, the energy content of food is measured in kilocalories (Calorie, kcal). One Calorie is equal to 1000 calories. The caloric value of any specific amount of food is determined by burning the food in a special container. The increase in water temperature caused by the burning food is measured.

A healthy human body uses carbohydrates as the first source of energy, but fats and proteins are also used as energy sources. Fats, however, release more than twice as many calories as the same weight of carbohydrates or proteins. One gram of carbohydrate or protein releases 4.1 Cal, while 1 g of fat produces 9.3 Cal.

The calories needed to maintain normal body functions while at rest (but not asleep) is called the basal metabolic rate (BMR). A person with an abnormally high BMR uses many calories and has a faster metabolism. An abnormally low BMR indicates that a person is using few calories and has a slower metabolism.

The BMR is usually lower in females than in males since men usually have more muscle tissue per pound than women and therefore use more oxygen. Other factors that increase the BMR are height, exercise, and lean body mass. As a person ages, the BMR will probably decrease. Emotions such as anger and anxiety as well as physical factors such as a chill, a fever, or an infection increase the BMR.

The total number of daily calories a person needs depends on his BMR and the amount and type of activities he engages in. Scientists have determined that a pound of body fat is equal to 3500 Cal. When the diet supplies 3500 Calories more than is used for energy purposes, the excess may be converted to fat, resulting in an increase of one pound in body weight. When more calories are used than are taken in, the body normally breaks down stored fats to supply the energy. If this occurs for an extended period of time, a decrease in weight usually results. Dieting to lose weight without an increase in activity may actually lower your BMR, resulting in little if any weight loss.

calorie: (L. *calor*—heat)

Reading Food Labels

Have you ever wondered what all the information in the "Nutrition Facts" section of a packaged food label actually means? The Food and Drug Administration (FDA) requires all packaged foods to be labeled with certain nutritional information. This information will be very helpful to you as you try to maintain a healthy diet.

The first part of the food label tells how much food equals one serving and how many servings are in the package. Be sure to pay attention to how much of the package you are eating and how many servings you are actually consuming.

The second section tells how many Calories are in each serving and how many of these Calories are from fat. Generally, less than 30% of a person's Calorie intake should come from fat. Eating too many Calories a day will lead to unhealthy weight gain. The number of Calories a person needs each day varies from person to person, depending on size, age, activity level, genetics, and other factors. The information on nutrition labels is usually based on a 2000 Calorie per day diet, which is appropriate for many people.

The third section of the food label tells the amounts of certain key nutrients that are present in each serving of the food. These amounts are usually measured in grams (g) or milligrams (mg). The first three nutrients listed are fat, cholesterol, and sodium; these should be limited in a healthy diet. Excessive sodium intake has been linked to heart problems; an individual's sodium intake should not exceed 2400 mg per day, and less is better. This section of the food label also tells the amount of carbohydrate, protein, and sometimes other nutrients present in one serving.

After listing the amounts of key nutrients, the food label gives the percentage daily value (%DV) of various vitamins and minerals present in the food. For example, if the label reads, "Vitamin A 20%," that means that one serving contains 20% of the vitamin A you need for the day. Experts say that most Americans do not get enough fiber, vitamin A, vitamin C, calcium, and iron in their diets.

The rest of the nutrition label gives general recommendations for daily nutrient intake. The ingredients in the food are listed below the nutrition facts.

Nutrition Facts

Serving Size 1 cup (228g)
Servings Per Container 2

Amount Per Serving

Calories 250	Calories from Fat 110

	% Daily Value*
Total Fat 12g	18%
Saturated Fat 3g	15%
Trans Fat 3g	
Cholesterol 30mg	10%
Sodium 470mg	20%
Total Carbohydrate 31g	10%
Dietary Fiber 0g	0%
Sugars 5g	
Protein 5g	

Vitamin A	4%
Vitamin C	2%
Calcium	20%
Iron	4%

*Percent of daily values are based on a 2,000 calorie diet. Your daily values may be higher or lower depending on your calorie needs.

	Calories	2,000	2,500
Total Fat	Less than	65g	80g
Sat Fat	Less than	20g	25g
Cholesterol	Less than	300mg	300mg
Sodium	Less than	2,400mg	2,400mg
Total Carbohydrate		300g	375g
Dietary Fiber		25g	30g

Nutrition label on a package of macaroni and cheese

Review Questions 21.5

1. List the six basic nutritional substances normally found in food.

2. Describe the digestion processes for (a) a carbohydrate, (b) a fat, and (c) a protein. Tell where each of the steps takes place.

3. Refer to the nutrition label below, and answer the following questions.

   a. What percentage of the Calories in this food comes from fat? Is this an acceptable amount?

   b. This food is a good source of which vitamins and minerals?

   c. How many servings of this food would you have to eat in order to obtain 100% of your daily value of vitamin A? Would this be a good way to obtain all of your vitamin A for the day? Explain your answer.

Nutrition Facts

Serving Size 250g
Servings Per Container 2

Amount Per Serving

Calories 230   Calories from Fat 70

	% Daily Value*
Total Fat 8g	13%
Saturated Fat 3g	17%
Trans Fat 1g	
Cholesterol 8mg	3%
Sodium 960mg	40%
Total Carbohydrate 31g	10%
Dietary Fiber 3g	10%
Sugars 5g	
Protein 8g	

Vitamin A	10%	•	Vitamin C	0%
Calcium	3%	•	Iron	15%

*Percent of daily values are based on a 2,000 calorie diet. Your daily values may be higher or lower depending on your calorie needs.

Nutrition label for canned ravioli (Question 3)

d. Could this food item be a regular part of a healthy diet? Explain your reasoning.

⊙ 4. What is the difference between a vitamin and a mineral?

⊙ 5. What are the two main classes of vitamins?

⊙ 6. Why are fats a better unit for storing excess energy than carbohydrates?

⊙ 7. "Calories really do not count in losing or gaining weight." Present evidence either to defend or to contradict this statement.

22

Internal Transport

22A—The Circulatory System

The cardiovascular system consists of numerous blood vessels through which blood is pumped by the muscular heart. From the time of the Greek Galen (ca. AD 130–210), most scientists believed that the heart produced blood and then pumped it through the vessels to all the parts of the body.

Then William Harvey, an English physician, observed that the heart could not possibly produce as much blood as it pumped and that the structures of the heart and the blood vessels were not designed for blood production. Through careful observation and experimentation, Harvey clearly demonstrated the circulation of blood in 1628.

Since that time scientists have learned much more about the *circulatory system*, especially the heart. Many diseases of the circulatory system can now be prevented or controlled by proper nutrition, exercise, medication, and if needed, corrective surgery.

Galen   *William Harvey*

22-1
Galen's theory of circulation predominated for over 1400 years until disproved by William Harvey.

22.1 The Blood

The circulatory system continually transports blood throughout the body. **Blood** is a marvelous red "river" that transports oxygen, nutrients, and hormones to all body cells; carbon dioxide to the lungs for expiration; and waste

Objectives

- Describe the components of blood and summarize their functions
- Explain the clinical significance of blood typing

Key Terms

blood
blood plasma
erythrocyte
leukocyte
platelet

coagulation
atherosclerosis
antigen
agglutinate

Centrifuging

Centrifuging allows the components of a substance (such as blood) to be separated by density. The substance is placed in a tube, which is then placed in a centrifuge machine and spun rapidly around a central axis. Denser particles are concentrated at the bottom of the tube, whereas less dense substances remain at the top. Look at Figure 22-2. Which part of the blood is densest? Which is least dense?

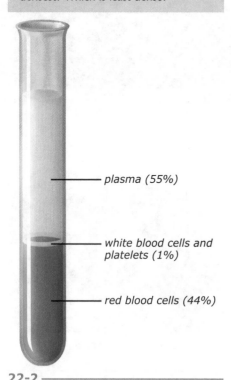

plasma (55%)

white blood cells and platelets (1%)

red blood cells (44%)

22-2

Blood separated by centrifuging has three layers.

hemocytoblast: hemo- (blood) + -cyto- (cell) + -blast (bud)

erythrocyte: erythro- (Gk. *eruthros*—red) + -cyte (cell)

molecules to the kidneys for excretion. A healthy adult has almost 5 L of blood and can lose 20%–30% of his blood volume without serious consequences. Nearly every cell in the body is located next to a tiny blood vessel so that it can exchange these materials with the blood.

Blood seems to be uniform in color, but examination under a microscope reveals that some cells are clear while others are red. When blood is centrifuged, it separates into two distinct layers that are separated by a thin third layer. The upper layer is the plasma, or liquid part of the blood; the red blood cells compose the bottom layer. The very thin layer that separates the plasma from the red blood cells is made of white blood cells and platelets.

Blood Plasma

Approximately 55% of the blood consists of **blood plasma**, a straw-colored liquid. Plasma is about 90% water. The remaining 10% contains mostly proteins, but it also has a variety of dissolved gases, minerals, vitamins, nutrients, hormones, and waste substances. Blood proteins perform many functions, including maintaining blood volume, clotting, transporting substances, and immunity. The water, minerals, vitamins, and nutrients in the plasma are absorbed from the digestive organs. Hormones are mostly proteins and lipids secreted into the blood by certain organs. The gases present in the plasma include small amounts of dissolved oxygen, carbon dioxide, and nitrogen. Waste substances, such as urea and uric acid that form during protein metabolism, are also present in the plasma. Wastes are continuously being filtered from the blood so that they do not reach toxic levels.

The composition of plasma varies somewhat, even in a healthy person. For example, after a meal the plasma may contain many tiny droplets of emulsified fat that were absorbed through the intestinal lining. The plasma composition may also vary if a person is suffering from an illness or a severe injury.

Formed Blood Components

About 45% of the total blood volume is made of formed components. This includes the red and white blood cells as well as the platelets. Although each of these components has a very different shape and function, they all originate from a single type of stem cell, the *hemocytoblast*. The hemocytoblasts are located in the bone marrow and respond to certain growth factors that affect their genes. Depending on which genes are turned on or off, the hemocytoblast forms different types of blood cells.

Erythrocytes

The term *cell* does not properly describe **erythrocytes** (ih RITH ruh SITES), even though they are often called *red blood cells* (*RBCs*). Early in their life cycle, erythrocytes contain a nucleus but very little hemoglobin. As they mature and shortly before they move into the bloodstream, the amount of hemoglobin increases. The nucleus decreases in size and is squeezed out. At this point the structure is no longer a true cell.

Erythrocytes contain few cellular structures and do not undergo mitosis. They may be more properly described as membrane-bound biconcave structures filled with hemoglobin. This shape provides the greatest possible surface area for the volume of the "cell." Greater surface area allows maximum diffusion of oxygen into and out of the erythrocyte. Erythrocytes are unable to move by themselves but are carried by the current that moves the plasma through the blood vessels.

When oxygen is bound loosely to the iron-containing heme group in hemoglobin, the oxyhemoglobin is a brilliant red color. After oxygen is released into other body cells, the blood becomes a dull purplish red. It changes back to a brilliant red after it absorbs oxygen while passing through the lungs.

Although the erythrocytes are the most numerous of all the formed blood components (99%), they can vary in number according to the condition of the body. Someone who exercises regularly has a greater need for oxygen. Chemical factors cause the hemocytoblasts to respond by forming more erythrocytes to meet this need. When a person changes his place of residence from a low elevation to a mountaintop, the erythrocytes and hemoglobin in his body increase to ensure that he absorbs enough oxygen into his blood.

If dietary iron is lacking, the body cannot form enough hemoglobin. The result is a general decrease in erythrocytes and hemoglobin, a disorder known as *anemia*, which is characterized by a lack of energy. Because certain B vitamins also aid in proper formation of hemoglobin, B vitamin deficiency also causes anemia. Proper diet, rest, and exercise usually correct such anemia. Some anemias, however, are caused by exposure to radiation or toxic substances such as benzene or arsenic. These types of anemias are difficult to correct and are sometimes fatal. Excessive bleeding, certain diseases, and drugs used in chemotherapy for cancer or other diseases can also cause certain types of anemia.

In a developing fetus, erythrocytes are formed primarily by the spleen, liver, and red bone marrow. At birth only the red bone marrow continues to actively produce erythrocytes. In adults erythrocyte production is restricted to the red bone marrow in the sternum, skull bones, vertebrae, ribs, pelvis, and ends of long bones. Red blood cells are continually being manufactured throughout your lifetime. Someone weighing about 59 kg (130 lb) produces more than one billion new erythrocytes each day.

The average life span of an erythrocyte is 90–120 days. Specialized cells in the liver and spleen break down old erythrocytes. Most of the remains of destroyed erythrocytes are recycled to form new erythrocytes.

Leukocytes

The leukocytes (LOO kuh SITES), or *white blood cells* (*WBCs*), do not contain hemoglobin. Leukocytes are about twice the size of erythrocytes; however, they possess no definite shape. Leukocytes have nuclei during their entire life span, and although carried along in the plasma, they are also able to move themselves. Leukocytes function in immunity.

There are various types of leukocytes. Some leukocytes manufacture substances for killing microorganisms or digesting toxins in the blood. Another type of leukocyte can phagocytize large amounts of bacteria and foreign material in the bloodstream and break them down in vacuoles. Many of the white blood cells leave the blood vessels through the capillary walls to enter an area of the body to "wage war" against foreign organisms or substances.

In a healthy person the ratio of leukocytes to erythrocytes is about 1:600. This difference in number makes sense when the functions of these blood cells are considered. When the body is in good health, it requires fewer leukocytes. However, when harmful microorganisms enter the body faster than they can be controlled, the number of leukocytes increases drastically. Physicians often seek to know a patient's blood count in order to determine whether they have an infection.

22-3

SEM of erythrocytes

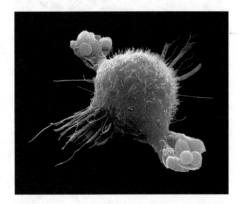

22-4

Leukocytes (blue) play an important role in fighting foreign invaders such as bacteria.

anemia: an- (without) + -hemia or -emia (blood)

leukocyte: leuko- (clear or white) + -cyte (cell)

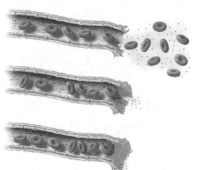

Vessel is ruptured and special chemicals are released.

Platelets begin to stick to each other.

Platelet plug seals ruptured vessel.

22-5

Platelet plug formation controls bleeding from small breaks.

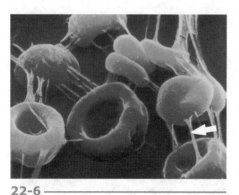

22-6

An electron micrograph of blood clotting. The arrow indicates fibrin threads.

An infection is the invasion of harmful organisms into the body. When the first leukocytes are not successful in stopping the infection, the invading organisms are free to multiply and injure body cells. Chemicals that are released from the injured cells initiate an inflammatory response. Additional leukocytes then move out of blood vessels to engulf and digest or to kill the harmful invaders. Leukocytes also digest and remove injured and dead body cells.

The accumulation of dead leukocytes, dead organisms, and broken cells forms a thick fluid called *pus*, which is characteristic of some infections. Infections may be rather minor, such as the common cold, or, as with pneumonia, may result in prolonged suffering and possibly death. In all infections, however, the leukocytes are necessary for ultimate victory over the invaders.

Platelets and Blood Clotting

Platelets, or *thrombocytes*, are cell fragments that lack a nucleus and are less than half the size of erythrocytes. When someone is cut or bruised, the broken blood vessels are plugged by platelets. The injured tissue releases a chemical that causes the platelets to stick to the broken edges of the vessels. The platelets release *serotonin* (SEHR uh TOH nin), which causes the smooth muscles of the vessel walls to contract. This reduces the blood loss from the damaged vessels. A platelet plug may control bleeding from small vessels or from small holes in large vessels; however, bleeding from larger vessels or large tears usually requires the formation of a clot.

Platelets also play an important role in **coagulation** (koh AG yuh LAY shun), the formation of a blood clot. Coagulation occurs as a result of complex biochemical pathways that involve multiple interdependent reactions, enzymes, and coenzymes. Through a cascading series of events, insoluble threads of the protein *fibrin* are produced.

These fibrin threads form a microscopic meshwork that entangles the blood cells to form a *blood clot*. The time required for coagulation varies from five to fifteen minutes. Chemicals in the blood gradually dissolve clots as the vessel wall heals.

thrombocyte: thrombo- (Gk. *thrombos*—clot) + -cyte (cell)

coagulation: co- (together) + -agulation (L. *agere*—to drive)

"The life of the flesh is in the blood" (Lev. 17:11). Biologically speaking, this is true—without an adequate supply of the right kind of blood to the flesh, life will be short.

Blood Types

Blood types (blood groups) are determined by the presence or absence of protein or carbohydrate molecules on the membranes of the erythrocytes. These molecules, or **antigens**, stimulate the formation of *antibodies* that cause blood with a different antigen to clump together—or **agglutinate**. If blood of the wrong type is transfused into someone, the recipient's antibodies will react with the donor's antigens, causing a transfusion reaction that can be fatal.

Injury, illness, surgery, and other problems can cause a loss of blood. Without transfusions, many people would die. Although it may seem odd, the idea of transfusing blood from one individual to another has been around for several centuries. In 1492, Pope Innocent VIII had a stroke and was deemed to be in need of blood. His transfusion was unsuccessful. In 1667, a successful blood transfusion was done between a sheep and a human. For many years, physicians did not know what caused some transfusions to fail and others to succeed.

In the early 1900s, Karl Landsteiner, an Austrian-born pathologist, began to study what caused transfusions to succeed or fail. He obtained blood samples from workers in his laboratory and mixed the blood in all possible combinations. From this he determined the number of blood groups and which types could be safely transfused. Landsteiner speculated that blood should be classified into groups, and in 1901 he proposed the ABO blood grouping system.

Determining Blood Types

The presence or absence of two antigens—A and B—on the membranes of erythrocytes determines the ABO blood type. If a person's erythrocytes have antigen A, then his blood is type A. A person with only antigen B has type B blood. If a person has both antigens A and B, his blood type is AB. A person who does not have either antigen A or B has type O blood.

In addition to antigens, three of these four blood types also possess antibodies. Unlike the antigens, which are located on red blood cell membranes, the antibodies are found only in the blood plasma. These particular antibodies, however, are not present at birth but are produced between the second and eighth months after birth as the body responds to antigens A and B in the person's food. Therefore, if a person has type A blood, his body produces anti-B antibodies. Anti-A antibodies do not form because they would react with his own antigens to destroy his erythrocytes. A person with type B blood produces anti-A antibodies. A person with type O blood has both anti-A and anti-B antibodies. A person with type AB blood does not produce any blood-type antibodies.

When blood is given to a patient, care must be taken to ensure that his

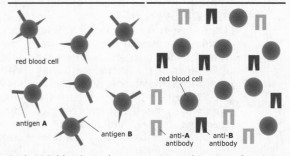

Each ABO blood type has a certain combination of antigens and antibodies. A person produces antibodies against antigens that are not his own.

antibodies will not attack the donor's erythrocytes and cause agglutination. Ideally, donor blood should be the same type as the recipient's blood. If the same blood type is not available, careful matching can determine which other blood type may be transfused. A person with type A (anti-B antibody) blood should never be given type B or AB blood because his anti-B antibodies would agglutinate both types of the donor erythrocytes. Similarly, a

agglutinate: ag- (to) + -glutinate (L. *glutinare*—to glue)

Karl Landsteiner developed the ABO blood grouping system.

Compatibility of Blood Types for Transfusion				
Blood type	**Antigen**	**Antibody**	**Can donate to**	**Can receive from**
A	A	anti-B	A and AB	A† and O
B	B	anti-A	B and AB	B† and O
AB	A and B	none	AB	AB†, A, B, and O
O	none	anti-A and anti-B	A, B, AB, and O	O

†This is the preferred blood type for transfusion, but the patient can receive other types listed if the preferred type is not available.

person with type O blood (with anti-A and anti-B antibodies) should never be given type A, B, or AB blood.

Someone with type AB blood can receive blood of any other type of the ABO grouping. For this reason, people with blood type AB are sometimes called *universal recipients*. Even though types A, B, and O contain antibodies that could cause agglutination of the recipient's type AB erythrocytes, there is usually no particular problem if the donor's blood (about 1 pt) is transfused slowly and therefore diluted by the large volume (about 5 qt) of the recipient's type AB blood.

Persons with type O blood are often referred to as *universal donors* because they can give to all types in the ABO grouping. Can you explain how this is possible?

The Rh System

The Rh system (named after the rhesus monkey from which the antigen was first isolated) involves the presence or absence of the Rh antigen on the erythrocyte membrane. Most people are Rh positive (Rh$^+$); they have the Rh antigen on the membranes of their erythrocytes. About 15% of all Americans lack the Rh antigen and are Rh negative (Rh$^-$).

Normally, human blood plasma does not contain anti-Rh antibodies, but these antibodies can be stimulated into production in an Rh$^-$ person. For example, when an Rh$^-$ person receives Rh$^+$ blood, his body begins to form anti-Rh antibodies. These antibodies remain in his blood plasma. Later, if a second transfusion of Rh$^+$ blood is administered, the anti-Rh antibodies react with the Rh antigens of the donor's blood.

Dr. Charles Drew, the first African American to graduate from Columbia Medical School, was also the first director of the first American Red Cross blood bank.

Problems with the Rh antigen may occur during pregnancy when blood leaks through the membranes between the unborn child and the mother. If the fetus is Rh$^+$ and the mother is Rh$^-$, the mother forms anti-Rh antibodies. The antibodies pose no danger until she becomes pregnant with a second Rh$^+$ baby. Then, if some of the mother's anti-Rh antibodies enter the unborn child's bloodstream, they will react with the child's Rh antigens, destroying the baby's erythrocytes.

Blood Transfusion Technology

In the 1930s, most of the blood transfusions were performed directly from the donor to the recipient. However, when World War II began, vast amounts of blood were needed to treat casualties, but storage was a major obstacle—the blood spoiled in just five to seven days. If blood were collected in the United States for use in either the European or Pacific theaters, it would spoil before it could be delivered.

Physician Charles Drew helped solve the problem. He determined that blood plasma (without the cells) was an effective substitute in emergency treatment. Also, the plasma could be stored easily and could be transfused without type testing. The use of plasma saved thousands of lives during World War II.

Although many improvements have been made, blood transfusions still carry some risks. Despite the precau-

Transfusions of the Future

For many years researchers have tried to develop artificial blood that can match the important properties of blood without the potential for complications. However, blood chemistry is very complex and the process has been slow. Some companies have developed an "artificial blood" that is derived from animal hemoglobin. One company has produced such a product that has been used in South Africa for several years. It can be stored at room temperature, and no testing is required for the donor; however, it is extremely expensive and quickly breaks down once transfused. Nonetheless, it is an important step in the search for risk-free blood transfusions.

tions, about 2% of all transfusions have complications. To be sure that a recipient's blood is completely compatible with the blood he is about to be given, a small amount of his blood is usually mixed with a sample of the donated blood. If there are any problems, other blood is sought. Today the naturally occurring substances in donated blood are not as big a problem as the various foreign substances it might contain.

Agglutinated blood (top) caused by incompatible donor. No agglutination occurs with compatible blood (bottom).

In the 1980s, some transfused blood was found to contain the virus that causes AIDS. Other diseases, such as hepatitis C, can also be passed to the recipient by contaminated blood. Researchers have developed tests to screen blood to detect contaminated blood, and all potential donors are screened with a series of questions before they are allowed to donate. Despite all of the precautions, blood supplies can still be contaminated.

Auto Transfusions

Many who face surgery in which they may need additional blood have decided to give their own blood before the surgery to be used if needed. But some people require blood in an emergency or are too ill to give blood before surgery. In these situations auto transfusion is sometimes used. The blood a person loses during surgery is collected, filtered, cleaned, and then replaced into his own bloodstream.

22-1 Summary of Formed Blood Components

Blood cell type	Quantity	Where produced	Description	Function
erythrocyte (red blood cell)	4.5–6 million/mm^3	red bone marrow	non-nucleated, biconcave disc; ≈8 µm in diameter; contains hemoglobin; red in color	transports gases for cellular metabolism
leukocyte (white blood cell)	6000–9000/mm^3	most in red bone marrow; some in lymphatic tissue	nucleated; 9–25 µm in diameter; varied shapes; cytoplasm appears granular in some; clear (white)	fights infection by phagocytosis and production of antibodies (depending on type of leukocyte)
thrombocyte (platelet)	150 000–450 000/mm^3	red bone marrow	non-nucleated, platelike disc; 2–4 µm in diameter	forms blood clots (along with other plasma proteins)

Review Questions 22.1

1. What are the components of blood?
2. List the components of blood plasma and tell their significance.
3. Describe the shape, size, structure, and function of (a) an erythrocyte, (b) a leukocyte, and (c) a platelet.
4. Describe the formation of a blood clot.
5. The presence or absence of what substances determines a person's blood type? Where are these substances located?
6. What is a blood transfusion, and why might a blood transfusion be needed?
7. Why should blood type compatibility be checked for the donor and the recipient?
8. What are a universal blood donor and a universal blood recipient?
9. What major problems of blood transfusions did Dr. Charles Drew overcome?
10. What steps are being taken to ensure a safe blood supply?
11. A person with type B blood needs a blood transfusion. What type(s) of blood could safely be given to this person? Why would it be unsafe for other blood types to be donated to this person?

22.2 The Heart

The *heart* is a hollow, muscular organ that has been described as a four-chambered, double-barreled blood pump. Approximately the size of a clenched fist and weighing about 340 g (12 oz), the heart lies in the thoracic cavity between the lungs, in a slanted position, with the *apex* (point) directed toward the left.

The heart muscle has greater endurance than any other muscle in the body; it contracts about 70–80 times per minute. At this rate, the heart beats over 100 000 times per day, about 37 million times per year. Every heartbeat pushes about 80 mL (2.7 fl oz) of blood from the heart, or 8000 L (2100 gal) per day.

Structure of the Heart

A fibrous sac, the **pericardium** (PEHR ih KAR dee um), loosely covers the heart and prevents it from rubbing against the lungs and inner chest wall. A slight space between the pericardium and the surface of the heart contains a slippery liquid, the *pericardial fluid*, which is secreted by the pericardium. This fluid reduces the friction between the heart and the surrounding structures.

The wall of the heart is made up of three layers.

✦ The **epicardium** is the outermost layer of the heart and is composed of connective tissue that is tightly attached to the muscular tissue of the heart. The epicardium keeps the heart muscle from becoming saturated with pericardial fluid.

✦ The **myocardium**, the thickest portion of the heart wall, is the cardiac muscle tissue that contracts and pumps blood.

22.2

Objectives

- Describe the structures of the heart and their functions
- Trace the blood flow through the heart
- Describe the cardiac cycle

Key Terms

pericardium	semilunar
epicardium	valve
myocardium	aorta
endocardium	systole
atrium	diastole
ventricle	sinoatrial node
atrioventricular	atrioventricular
valve	node

pericardium: peri- (around) + -cardium (heart)

epicardium: epi- (upon or over) + -cardium (heart)

myocardium: myo- (muscle) + -cardium (heart)

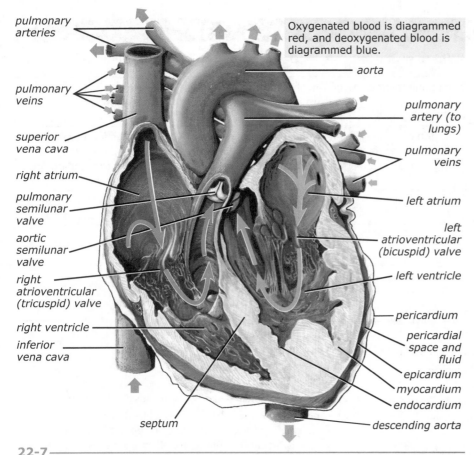

pulmonary arteries

Oxygenated blood is diagrammed red, and deoxygenated blood is diagrammed blue.

aorta

pulmonary veins

pulmonary artery (to lungs)

superior vena cava

pulmonary veins

right atrium

left atrium

pulmonary semilunar valve

left atrioventricular (bicuspid) valve

aortic semilunar valve

left ventricle

right atrioventricular (tricuspid) valve

pericardium

pericardial space and fluid

right ventricle

epicardium

inferior vena cava

myocardium

endocardium

septum

descending aorta

22-7

Anatomy of the human heart

Sounds of the Heart

The characteristic *lubb-dubb*, *lubb-dubb* sounds heard through a stethoscope are produced when the atrioventricular and semilunar valves close. As the ventricles contract, the atrioventricular valves snapping closed produce the *lubb* (first heart sound, or S_1). When the atria contract, the semilunar valves snap closed, producing *dubb* (the second heart sound, or S_2).

What happens if the heart valves do not close properly? This sometimes happens when the valves do not develop properly or become scarred by infection. Blood is then allowed to leak in the reverse direction. The backward flow of blood through the damaged heart valves creates turbulence in the blood flow, producing the abnormal hissing sounds known as *heart murmurs*. If the heart murmurs are heard only after strenuous exercise, they are usually not a threat to a person's health; however, if the murmurs are still heard after a restful night, there may be cause for concern. Sometimes it is necessary to replace a defective valve with an artificial valve. Many heart murmurs in children correct themselves as the children mature.

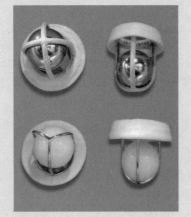

Types of artificial heart valves

endocardium: endo- (within) + -cardium (heart)

tricuspid: tri- (Gk. *tri*—three) + -cuspid (L. *cuspis*—point)

semilunar: semi- (half) + -lunar (L. *luna*—moon)

✦ The inner surface of the myocardium is lined with a thin layer of epithelial tissue, the endocardium. This layer prevents blood from saturating the myocardium.

A muscular wall called the *septum* separates the right and left sides of the heart. Each half is then divided into an upper chamber, the **atrium**, and a lower chamber, the **ventricle**. The atrial myocardium is thin because these chambers primarily receive blood. The ventricles, in contrast, have a thick myocardial layer because they are responsible for pushing the blood into the blood vessels of the body.

Between the atrium and ventricle in each half of the heart are the **atrioventricular** (AY tree oh ven TRIK yuh lur) **valves** (AV valves). The right AV valve is the *tricuspid valve*, and the left AV valve is the *bicuspid valve*. The tricuspid valve is composed of three flaps (or cusps, hence the name) of tissue, and the bicuspid is composed of two. These are one-way valves—they permit the flow of blood from the atria to the ventricles but prevent a reverse flow because the flaps close and are anchored to the ventricle walls by fibrous strands.

The semilunar **valves**, located at the exits of the ventricles, have three cup-shaped membranes. The *pulmonary semilunar valve* is located at the exit of the right ventricle and controls the flow of deoxygenated blood to the lungs. The *aortic semilunar valve* is located at the exit of the left ventricle and controls the flow of oxygenated blood to the rest of the body. Semilunar valves permit the blood to flow into the blood vessels but keep it from returning to the ventricles. All heart valves are passive—they do not move by themselves but are moved by the force of the blood during each heartbeat.

Blood Flow Through the Heart

The *superior vena cava* drains deoxygenated blood from body parts above the heart, including the arms and hands, and the *inferior vena cava* returns deoxygenated blood from body regions below the level of the heart. As the right atrium fills with blood, it contracts, squeezing the blood through the tricuspid valve and into the right ventricle.

When the right ventricle contracts, the tricuspid valve is forced shut, and the pulmonary semilunar valve opens to allow the blood to flow into the *pulmonary artery*. Thus, the right side of the heart pumps only deoxygenated blood.

Each of the two main branches of the pulmonary artery leads to a lung. As the blood flows through the capillaries surrounding the alveoli, oxygen is added to the hemoglobin of the blood.

The richly oxygenated blood returns to the left atrium through the *pulmonary veins*. The left atrium then contracts, squeezing the blood through the bicuspid valve and filling the left ventricle.

As the left ventricle contracts, the bicuspid valve shuts with considerable force, and blood rushes through the aortic semilunar valve into the **aorta**. All parts of the body receive blood from branches of the aorta. After the blood flows through the body organs, it eventually returns again to the right atrium by way of the superior and inferior venae cavae.

The Cardiac Cycle

The *cardiac cycle*, or heartbeat, is one complete contraction and relaxation of the heart muscle. A person can feel his heartbeat by placing his hand over the region of the heart's apex, which for most people is the level of the fifth and sixth ribs, about 7.6 cm (3 in.) left of the midline.

The contraction of the heart muscle is known as **systole** (SIS tuh lee). The heart's relaxing and filling with blood is called **diastole** (dye AS tuh lee).

The regular rhythm of systole and diastole is controlled by the *cardiac conduction system*, specialized tissue embedded in the myocardium. One part of this system, the **sinoatrial** (SYE noh AY tree ul) **node** (SA node), starts each systole and thus sets the pace. The SA node has its own rhythm of about eighty electrical impulses per minute. For this reason it has been called the *cardiac pacemaker*. The SA node rate, however, can be increased or decreased by input from the nervous system.

The electrical impulse from the SA node is transmitted through muscle tissue to both atria, causing them to contract together. About 0.1 s later, the impulse reaches the **atrioventricular node** (AV node), where there is a brief pause to allow for proper emptying of blood from the atria. When the AV node "fires," it sends an electrical impulse down the *bundle of His* (special cardiac muscle fibers) to the wall of each ventricle. The bundle of His divides into many small branches called *Purkinje fibers*, which spread throughout the myocardium. The fibers of the ventricle walls

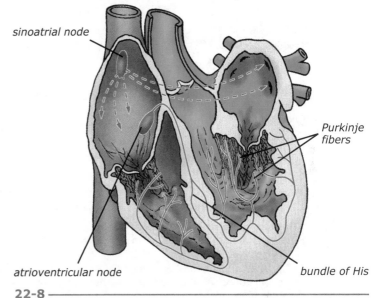

sinoatrial node

Purkinje fibers

atrioventricular node

bundle of His

22-8

The cardiac conduction system. The SA node is located in the right atrium near the entrance of the superior vena cava. The AV node is located at the base of the right atrium near the septum. The bundle of His originates from the AV node and enters the septum, where it divides into two branches. These two branches supply the ventricles. The terminal branches of the bundle of His are called Purkinje fibers.

systole: (Gk. *sustole*—contraction)
diastole: (Gk. *diastole*—expansion)

The Electrocardiogram (ECG)

A special apparatus can record on paper the electrical activity of the heart's conduction system. This graphic representation of heart impulses is known as an electrocardiogram (ih LEK tro KAR dee uh GRAM)—ECG or EKG. The various peaks or waves of an ECG, designated by the letters *P*, *Q*, *R*, *S*, and *T*, represent particular activities of the heart. Any deviation from a normal wave pattern indicates some defect in the conduction system or in the structure of the heart.

If a portion of the conduction system is injured or diseased, or if the nervous system does not properly regulate the SA node, the impulses do not pass from the atria to the ventricles. This condition is known as a *heart block*. There are different degrees and types of blocks depending on the particular physical defect. For example, in a partial heart block some impulses are missed between the SA node and AV node. This blockage causes the atria to beat somewhat faster than the ventricles.

In a complete heart block there is no communication between the SA and AV nodes, and the atria may beat twice as fast as the ventricles. In heart blocks, the *P* wave of an ECG will be present, but the *QRS* wave may be absent for several heartbeats.

Such abnormal rhythms (*arrhythmias*) may require surgical implantation of an artificial pacemaker. Such a device may be set electronically, or it may stimulate the heart only when the SA node slows in its function. Powered by batteries or nuclear energy, it consists of either a long wire inserted through a vein into the right ventricle or a device implanted in the ventricle wall. Modern artificial pacemakers may be as small as one inch in diameter.

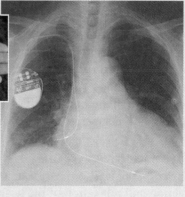

Pacemakers are implanted to control the heart rate. The white lines on the x-ray are the wires from the pacemaker.

A normal electrocardiogram. P wave: atria contracting; QRS wave: ventricles contracting; T wave: rest and preparation for next contraction.

electrocardiogram: electro- (electric) + -cardio- (heart) + -gram (letter)

Factors That Affect the Heart Rate

The following may increase the heart rate:

- Increase in body temperature (may increase the heart rate 3–4 bpm)
- Excessive amount of calcium (may cause irregular contractions)
- Certain hormones (may also increase the strength of contraction)
- Caffeine (as in coffee, tea, and many carbonated beverages)
- Exertion (such as during exercise; increased need for more blood to transfer oxygen to body cells in a given amount of time)
- Increased amount of blood returning to the heart (increased muscle contractions during exercise that force more blood through the veins into the heart)
- Increased acidity of the blood

The following may decrease the heart rate:

- Excessive amount of potassium
- Extreme deficiency of oxygen (may cause weaker heart contractions)
- Decreased amount of blood returning to heart

contract together and efficiently push the blood into the pulmonary artery and the aorta.

The Heart Rate

The typical resting heart rate of an adult is about 70 beats per minute (bpm); however, during moderate exercise it is commonly about 120 bpm. If the heartbeat is more than 140 bpm, ventricular diastole may be too short for the ventricles to fill with blood. Therefore, less blood is pumped at each heartbeat, and the person begins to tire.

If the heart rate is 72 bpm, about 0.8 s is required for one heartbeat. During this brief period of time, the atria are in systole for only 0.1 s, and then they relax and fill with blood for 0.7 s. The ventricles are in systole for nearly 0.3 s, and they relax and receive blood from the atria for about 0.5 s.

22-2 The Cardiac Cycle

Time (seconds)	Atria	Ventricles	AV valves	Semilunar valves	Sounds
0.1	systole (blood pumped into ventricles)	diastole (blood enters from atria)	open ↓	closed ↓	
0.2	diastole (blood enters from body)	systole (blood pumped into arteries)	closed	open	lubb
0.3					
0.4		↓			
0.5		diastole (blood enters from atria)	open	closed	dubb
0.6					
0.7					
0.8	↓	↓	↓	↓	

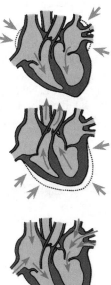

Heart Attack

Each year over 1.2 million people in the United States suffer a heart attack, and over 450 000 die from the attack. A heart attack, or *myocardial infarction*, is damage to or destruction of some of the myocardium because blood supply to that area is blocked. Heart disease was once thought to be a "man's disease." However, today more women die from heart disease than men, and heart disease is the second leading cause of death in middle-aged women. Since the causes and results of heart muscle damage can be vastly different, the term *heart attack* is somewhat vague. Treatments that help one person recover from a heart attack may cause one in another person.

Most men who experience a heart attack describe it as a crushing or squeezing pain in the chest. This painful sensation may spread to the shoulders, neck, and arms. It may come and go and may last for two minutes or longer. The person may not appear ill, but he may feel weak, nauseated, short of breath, and sweaty. No matter how ill he feels, a heart attack victim often denies that he is having a heart attack. In fact, denial is so common that it is often listed as a heart attack symptom.

Researchers have found that the symptoms of heart attack in women are different from those in men. Women often feel breathless, sometimes without any chest pain. Rather, the pain may be in the upper back, shoulders, or neck. Flu-like symptoms such as nausea, cold sweats, fatigue, and dizziness are often accompanied by feelings of anxiety. Heart attacks in women are sometimes misdiagnosed as colds or anxiety attacks. It is interesting to note that a woman is less likely to survive her initial heart attack than a man is.

When someone is suspected of having a heart attack, early treatment is the key to a positive outcome. If the person is conscious, he should lie down and try to relax. Someone should immediately seek medical attention. If the victim is unresponsive, someone trained in cardiopulmonary resuscitation (CPR) should begin CPR immediately.

Ventricular fibrillation is the most common arrhythmia (irregular heartbeat) associated with sudden death due to cardiac failure; in this condition, the ventricles begin quivering rapidly but fail to effectively pump blood. Thankfully, this condition responds well to electrical shock treatment (defibrillation). Automatic external defibrillators (AEDs) can automatically determine whether the victim has ventricular fibrillation and will deliver an electrical shock if needed.

The leading cause of heart attacks is *coronary thrombosis*, the blockage of one or more of the coronary arteries by a blood clot. The seriousness of a coronary thrombosis depends on how much of the heart's blood supply is cut off and which area of the heart is being affected. Sometimes medications can be used to dissolve the clot. Other times surgery is necessary to remove the blood clot. Other factors that may contribute to heart attacks include coronary atherosclerosis and heart block.

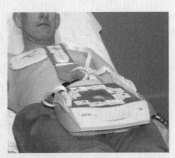

The availability and use of AEDs in public places such as shopping malls, airports, sports arenas, and churches has the potential to save hundreds of lives each year.

1. List and describe the three layers of the heart's wall, from superficial to deep.
2. List the heart's four valves. Describe the action of each.
3. Sketch the heart and trace the path of a drop of blood through the heart. Label all parts of the heart through which the blood passes. Label the arteries and veins passing into and out of the heart.

4. What causes the sounds of the heart?
5. Describe the cardiac conduction system, including all relevant structures in your answer.
6. What can be determined by an ECG? Sketch a normal ECG, and describe what happens during each wave of the cycle.

22.3

⮕ **Objectives**

- Describe the structural differences between arteries, veins, and capillaries
- Summarize the different circulations in the human body

⮕ **Key Terms**

artery
vein
capillary
pulse
pulmonary
  circulation
systemic
  circulation
blood pressure
hypertension

22.3 Blood Vessels and Circulations

Blood Vessels

The heart and blood vessels form a completely closed system. There are three types of blood vessels: **arteries**, which carry blood away from the heart; **veins**, which carry blood toward the heart; and **capillaries**, the tiny vessels which connect arteries to veins.

Usually the arteries of the body are in the deeper muscles and between muscles and bones. Some veins are located near the body surface, and many can be seen under the skin. There are almost 80 500 km (50 000 mi) of capillaries in an adult body. Every cell in the body is supplied with oxygen and nutrients by capillaries.

Blood leaves the heart through the strongest arteries in the body: the *pulmonary artery*, which leads to the lungs, and the *aorta*, which arches posterior to the heart and continues down to the rest of the body. The *aortic arch* has branches that take blood to the upper body. The *descending aorta* has many branches that carry blood to all organs of the body except the lungs. The walls of arteries have three layers: an outer elastic layer, a middle muscular layer, and an inner one-cell-thick layer of epithelial cells.

Arteries form smaller vessels known as *arterioles* (ar TIR ee OHLZ), which in turn branch into microscopic capillaries whose walls are only one cell thick. The *capillaries* are the functional units of the circulatory system—through them the diffusion of oxygen, glucose, hormones, and other essential substances occurs. As the blood flows through the capillaries, it also absorbs

The Pulse

As the heart beats and the left ventricle forces blood into the aorta, the elastic walls of the arteries expand. This wave of expansion moves down the aorta and along smaller arteries. As the wave passes, the elastic walls of the arteries resume their normal size. This dilation and rebounding of an artery is the **pulse**.

A person can measure his pulse rate wherever he can push an artery against a bone. Commonly, two fingers are placed at the wrist, pressing the radial artery against the radius. Fingers rather than a thumb should be used because the pulse can also be detected in arteries of the thumb.

There is no pulse in a vein because the force of the heart's contraction is absorbed as the blood flows through the numerous capillaries.

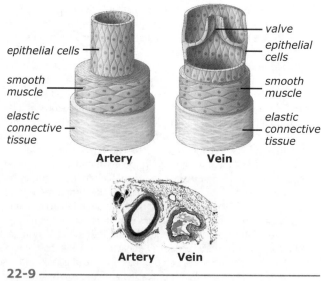

22-9
Comparison of artery and vein anatomy

carbon dioxide and other waste substances and carries them away to the lungs or kidneys for removal or to the liver, where many toxic substances are broken down.

The capillaries then merge to form *venules* (VEN yoolz), which join with other venules to form *veins*. Like arteries, all veins have walls with three layers, but the walls are thinner, less elastic, and less muscular. Veins possess semilunar valves that prevent reverse blood flow. Skeletal muscle contractions (which squeeze the vessels) and the semilunar valves help force the blood toward the heart. (See Figure 22-11.)

Paths of Circulation

For the benefit of study, it is convenient to group the entire circulatory system into two major divisions. Each division involves the circulation of blood from the heart through one or more vital organs and back to the heart. The two main divisions are the pulmonary circulation and the systemic circulation.

Pulmonary Circulation

Pulmonary circulation carries oxygen-poor blood from the right ventricle to the lungs for the purpose of absorbing oxygen and releasing carbon dioxide. After passing through the lungs, the oxygen-rich blood returns to the left side of the heart. At any moment about 0.5 L (0.5 qt) of blood is in pulmonary circulation. Circulation of this amount requires only a few seconds.

Systemic Circulation

Systemic circulation consists of the flow of blood from the left ventricle to all parts of the body (except the lungs) and then back to the right atrium. This circulation carries oxygen and nutrients to the body tissues and removes carbon dioxide and wastes from the tissues.

There are several smaller circulatory pathways within the systemic circulation; some of the familiar ones include the following:

✦ *Coronary circulation. Coronary circulation* carries blood into and out of the heart muscle. Two coronary arteries that branch off the aorta just past the aortic semilunar valve deliver nutrients and oxygen to the myocardium.

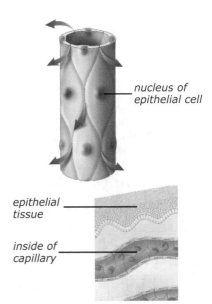

nucleus of epithelial cell

epithelial tissue

inside of capillary

22-10 ——————
Capillaries control the exchange of waste and nutrients between the blood and body cells.

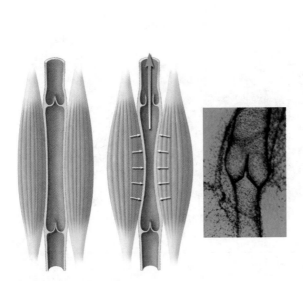

22-11 ——————
Venous circulation is aided by valves that prevent blood from flowing in reverse and by adjacent muscle groups that squeeze the veins during contraction. Photomicrograph of venous valve (inset).

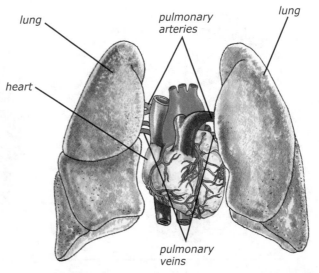

lung

pulmonary arteries

lung

heart

pulmonary veins

22-12 ——————
In pulmonary circulation, arteries carry oxygen-poor blood (blue), and veins carry oxygen-rich blood (red). Can you explain why the pulmonary arteries and veins are different from those of the rest of the body?

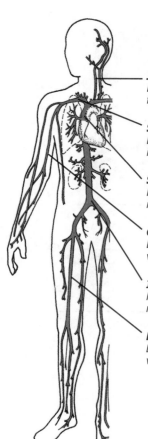

Veins

Jugular
Drains brain, skull, scalp, and face; located near surface of neck

Subclavian
Drains arms, shoulders, chest, and in females, mammary glands

Superior vena cava
Receives blood from all body regions above level of the heart

Cephalic
Drains blood from arms into subclavian vein; located in upper arm

Inferior vena cava
Receives blood from all body regions below the heart

Femoral
Drains blood from legs into inferior vena cava; located in thigh region

Arteries

Carotid
Supplies head, neck, and brain

Subclavian
Supplies neck, shoulders, and arms

Aorta
Largest artery in body; receives blood from left ventricle; extends from left ventricle to hip region

Brachial
Located on medial surface of arm; where blood pressure is usually measured

Radial
Located in lower arm region to the wrist; often used to measure pulse

Iliac
Supplies urinary bladder, reproductive organs, and lower limbs

Femoral
Located in medial thigh region; important "pressure point" to stop blood loss if leg is severely injured

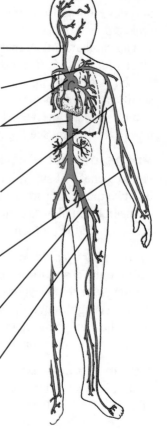

22-13
Major veins and arteries of the human body

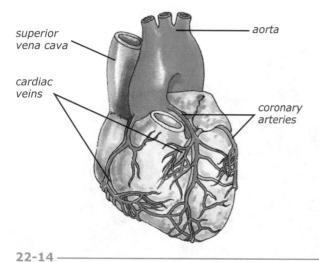

superior vena cava

cardiac veins

aorta

coronary arteries

22-14
Coronary circulation

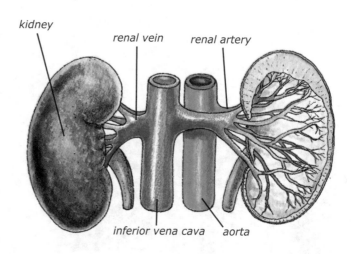

kidney

renal vein

renal artery

inferior vena cava

aorta

22-15
Renal circulation

Facets of Biology

Blood Pressure and Hypertension

High blood pressure is a common medical concern. In fact, it is known as the "silent killer" because it often has no symptoms. Although high blood pressure is usually thought of as a problem of older people (over 70 million adults in the United States), elevated blood pressure is being diagnosed in elementary children and high schoolers. Those who are at greatest risk are usually overweight and inactive, or they have a diet filled with processed food high in salt. Although many people have heard the term *blood pressure*, they have little understanding of how high blood pressure actually affects their health.

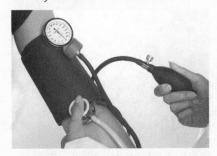

Blood pressure (BP) is defined as the pressure that the blood exerts against the walls of arteries. Healthy, elastic artery walls usually expand to absorb this pressure and then rebound to push the blood along. When the heart beats (systole of the ventricles), BP in the arteries is high; when at rest (diastole of the ventricles), the arterial blood pressure is lower. Because the standard measurement of blood pres-

sure initially involved pushing mercury up a thin tube, the units used in blood pressure are millimeters of mercury (mm Hg).

The normal pressure the blood of an adult puts on his blood vessels at rest is about 120 mm Hg during ventricular systole and about 80 mm Hg during ventricular diastole. For convenience, such measurements are expressed as 120/80.

Normally BP rises when more oxygen is needed, such as during strenuous activity. To raise the BP, the body increases the amount of blood pumped by the heart, decreases the size of the arteries, and, under extreme stress, thickens the blood by the addition of stored blood cells. After the need for high BP is over, a healthy person's BP quickly returns to the resting value.

Systolic pressures of 120–139 mm Hg and diastolic pressures of 80–89 mm Hg are termed *prehypertensive*. Blood pressures in this range are a warning that the person may develop high blood pressure in the future and should be monitored closely for any increase. A person whose BP measurements exceed one or both of these limits suffers from high blood pressure, or **hypertension**. If the systolic pressure exceeds 200 mm Hg, there is real danger that an artery may rupture. Such a rupture in the brain region is often fatal.

Hypertension is a condition caused by a complex of diseases, disorders, and condi-

tions. Some factors that contribute to hypertension include being overweight, eating too much salt (sodium), lack of exercise, anxiety, inability to relax, advanced age, certain drugs (usually taken to treat some other condition), and buildup of fatty substances inside the arteries (atherosclerosis).

In many cases the BP increases as the walls of the arteries become hardened and thicker. In such cases the arterioles and small arteries are decreased in diameter. Thus, the heart has to work harder as it attempts to push adequate amounts of blood through the partially closed, inelastic blood vessels.

Untreated hypertension can result in death by heart failure, stroke, kidney failure, or closure of the coronary arteries. Treatment or control of prehypertension includes proper diet, rest, and exercise; stages I and II hypertension may require one or more medications in addition to lifestyle changes.

Although the ability of the body to increase blood pressure is essential to maintaining homeostasis and responding to physical stress, continued high blood pressure can often result in severe health problems.

Blood Pressure Values†	
Prehypertensive	120–139/80–89
Stage I hypertension	140–159/90–99
Stage II hypertension	160/100 and above

†from the American Heart Association, 2009

The flow of blood in the capillaries of the myocardium nearly stops each time the heart contracts. Why? The contracted cardiac muscle fibers compress the adjacent coronary vessels as they contract. When the heart is in diastole, the cardiac muscle is relaxed—the capillaries open, and blood circulates. The blood that flows into the myocardium drains out of the heart through the cardiac veins. The cardiac veins join to form a large vein called the coronary sinus that returns the blood to the right atrium.

✦ *Renal circulation.* The circulation of blood into and out of the kidneys is known as *renal* (REE nul) *circulation.* This route begins with the renal

renal: (L. *renes*—kidneys)

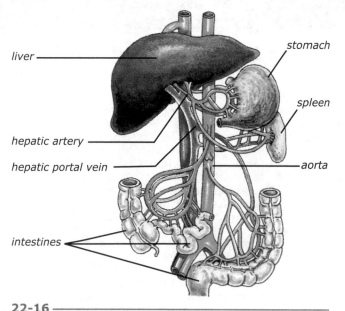

liver

stomach

spleen

hepatic artery

hepatic portal vein

aorta

intestines

22-16

Portal circulation

hepatic: (Gk. *hepatikos*—liver)

arteries that branch from the aorta. As the blood flows through the kidneys, the waste materials are removed for excretion as urine. The blood that leaves the kidneys in the *renal veins* is the cleanest blood in the body.

✦ *Portal circulation.* Blood flow through the liver and then to the inferior vena cava is called *portal circulation.* The blood from the digestive organs—rich in food molecules—is carried to the liver by the *hepatic portal vein.* The liver also receives oxygenated blood from the *hepatic artery,* a branch of the aorta. The blood from these two sources mixes as it enters small chambers of the liver, called sinusoids. The liver cells adjust the blood chemistry, taking out and storing materials when they are too abundant and then putting them in when they are needed. Liver cells also deal with many toxins and waste products.

Diseases and Disorders of the Circulatory System

✦ *Anemia*: a condition in which there is a decrease in hemoglobin or in the number of erythrocytes

  ✧ *Iron deficiency anemia*: results from a lack of usable iron or other nutrients necessary for the use of iron to make hemoglobin in the diet

  ✧ *Pernicious anemia*: the erythrocytes are fragile and abnormally large but usually contain less hemoglobin than normal; caused by a lack of certain proteins that are needed for vitamin B_{12} absorption in the stomach

  ✧ *Sickle cell anemia*: abnormal sickle- or crescent-shaped erythrocytes found in the bloodstream

✦ *Aneurysm*: a permanent stretching or bulging of an artery wall or heart chamber; caused by the pressure of blood on muscular walls weakened by disease or injury

✦ *Arrhythmia*: an irregularity of the heartbeat

✦ *Arteriosclerosis*: commonly called hardening of the arteries; a general term for a variety of conditions that cause the artery walls to become thick and hard and to lose their elasticity

✦ *Atherosclerosis*: a form of arteriosclerosis characterized by the accumulation of fatty material in artery walls

✦ *Endocarditis*: inflammation of the endocardium of the heart; usually caused by bacterial infection

✦ *Fibrillation*: the quivering, uncoordinated contraction of individual muscle fibers in atria, ventricles, or throughout the entire myocardium; may be caused by physical injury, a blood clot in coronary arteries, or an overdose of certain drugs

✦ *Hemorrhage*: a break in a blood vessel resulting in loss of blood; may be due to physical injury or a disease

✦ *Leukemia*: a type of cancer resulting in an increase in the number of leukocytes to about 1 million per cubic millimeter (mm^3); many immature leukocytes are present in blood; usually is fatal but in some cases has been controlled by drugs, proper diet, and marrow stem cell replacement

✦ *Leukocytosis*: an increase in leukocytes (more than 10 000 per mm^3) in response to an infection in the body

✦ *Pericarditis*: an inflammation of the pericardium; may be caused by bacterial infection, physical injury to the heart, or tumors

✦ *Phlebitis*: an inflammation of a vein; may be caused by varicose veins or irritations from intravenously administered medications or the IV cannula itself

✦ *Stroke*: loss of brain function caused by a hemorrhage or blood clot in arteries to the brain; often characterized by partial paralysis, slurred speech, loss of sensation to certain parts of the body, dizziness, or loss of consciousness

➥ Review Questions 22.3

1. Describe and compare arteries and veins.
2. How do pulmonary and systemic circulations differ?
3. List three divisions of systemic circulation and tell what happens to the blood in each.

⊙ 4. Compare pulse and blood pressure.

⊙ 5. What do the two numbers given in a blood pressure reading mean?

⊙ 6. Describe hypertension and list several factors that are believed to contribute to the condition.

22B—The Lymphatic System and Immunity

The human body exists in a hostile environment. Vast numbers of bacteria, viruses, and other microbes find the body an ideal place to grow and reproduce. Many of these invaders are pathogenic. That most humans are healthy is a tribute to God's design of the body to be able to resist infection. Some mechanisms that resist infection are *nonspecific* in that they do not distinguish one infectious agent from another. Many of these are discussed in Chapter 10, but some of the nonspecific responses involve the lymphatic system and are discussed here too.

Some body responses to foreign materials are highly *specific*. Each of these mechanisms responds only to a specific foreign substance, such as a particular protein in a certain virus's coat, a chemical produced by a specific bacterium, or cells that have a certain substance on their cell membranes. The specific defense mechanisms a person's body employs against invading pathogens is called his *immune system*.

The structures of the human immune system are not as easy to identify as are parts of the skeletal, circulatory, or other systems. Many of the structures that help the body fight off invading pathogens are tiny, and many are parts of other systems and carry on only certain immune system functions. Since much of the actual battle waged against pathogens must be carried on at sites where they invade the body, much of the immune system involves "mobile units" like chemicals and blood cells.

22.4 The Lymphatic System

As blood flows through the capillaries, water and some dissolved substances diffuse out of them into the spaces between cells. Some of this *interstitial fluid* between the cells enters the cells and is used in metabolic reactions, and some of the fluid is reabsorbed by the capillaries. However, most interstitial fluid passes into the *lymphatic* (lim FAT ik) *system* and is called lymph.

Lymph Vessels

The lymphatic system eventually returns the lymph to the bloodstream. Its tiny "dead-end" *lymph capillaries* are located in almost every region of the body. They absorb the excess fluids and proteins from most body tissues, keeping these tissues from becoming swollen.

The pressure of the interstitial fluid is usually greater than the pressure in the lymph capillaries. This forces the fluid into the lymph capillaries. The one-cell-thick walls of the lymph capillaries are constructed with overlapping cells so that the fluid can enter easily. This mechanism works somewhat like a door that opens inward only. As the volume of fluid increases within the lymph capillary, the cells are pushed together, preventing the outward flow of lymph.

The lymph capillaries combine and form *lymph vessels*, which are structurally similar to veins. The pressure of skeletal muscle contractions propels the lymph through the vessels as the body moves. Tiny one-way valves, similar to those in the veins, prevent the lymph from flowing backward. Lymph is also carried along the larger lymph vessels by a pumping action of smooth muscle fibers in their walls. After the lymph flows through the body, it finally drains into two main ducts that join the bloodstream where the jugular and subclavian veins meet in the shoulder region.

22.4

▶ Objectives

- List and describe the lymphatic system organs
- Describe the flow of lymph through the lymphatic system

▶ Key Terms

lymph
lymph node
lymphocyte
macrophage

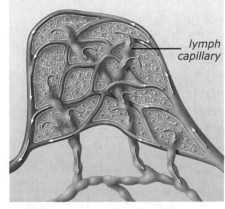

lymph capillary

22-17

Terminal end of lymphatic system. The arrows show the movement of blood and lymph fluid. How is the lymphatic system different from the circulatory system?

lymph: (L. *lympha*—water)

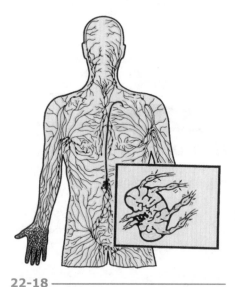

22-18

A portion of the lymphatic system. Note the distribution of lymph nodes. A lymph node (inset).

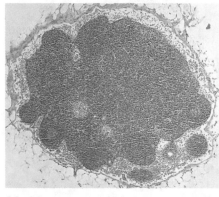

22-19

PMG of lymph node

lymphocyte: lympho- (water) + -cyte (cell)

macrophage: macro- (large) + -phage (to eat)

Lymph Nodes

The lymphatic system helps in combating infections in the body. Along the tiny lymph vessels are numerous small masses of tissue called **lymph nodes**. As the watery lymph flows into and through the nodes, bacteria, viruses, and other foreign matter are filtered out and destroyed.

Lymph nodes are composed of a layer of connective tissue surrounding small groups of cells. The size of the nodes may vary from that of a pinhead to that of a large lima bean. Lymph nodes are located in all body regions but are most numerous in the floor of the mouth, the neck region, the armpit, and the groin region.

There are two main types of cells found in lymph nodes: **lymphocytes** (which are primarily involved in specific immune responses and are discussed in the next section) and **macrophages**. Macrophages (sometimes called monocytes) are large phagocytic cells. They ingest virtually any substance that is foreign to the body. There are two basic types:

- *Wandering macrophages*, which travel through the blood and lymph
- *Fixed macrophages*, which are found embedded in various body tissues, including lymph nodes

As the lymph slowly flows through the lymph node, fixed macrophages engulf certain bacteria, viruses, and other foreign substances and digest them in vacuoles. For some minor infections, this is all that is necessary to rid the body of the pathogen. If, however, the foreign substance is in large quantity, it may slip by the first lymph node. Rarely, however, will a recognized foreign substance make it past the second or third lymph node along a lymph vessel.

The combat between bacteria and lymph node macrophages may result in swollen lymph nodes near the region of the body where the infection occurs. For example, if a person suffers from a sore throat, the lymph nodes in the neck region may become swollen, causing pain, especially when he swallows or moves his head.

In severe bacterial infections the bacteria that are carried into the node may become so abundant that they overwhelm the node. The node may swell and become a place where bacteria are growing rather than where they are being destroyed. This infection of the lymphatic system may result in bacteria entering the blood and being carried to other areas of the body.

The enzymes produced by the macrophages cannot digest some foreign substances. For example, macrophages are unable to break down many substances in tobacco smoke. In heavy smokers the lymph nodes fill with smoke particles and become dark gray or black. In time these foreign materials may interfere with the function of the nodes. Asbestos is another substance that macrophages cannot digest.

▶Review Questions 22.4

1. Where does lymph come from? Where does lymph go?
2. Name and describe two types of macrophages.
3. What happens to lymph as it passes through lymph nodes?
4. Describe three lymphatic organs.

22.5 Immunity

A person is **immune** to a disease when his body reacts to the presence of a pathogen by destroying it or rendering it harmless. Today, scientists recognize that the mechanisms involved in immunity are often complex. Frequently the body's defense mechanisms deal with the pathogen's products or with some aspect of the human body and not directly with the pathogen itself.

The human immune system's primary function is to search for, recognize, and eliminate antigens. An *antigen* is any foreign substance (usually a protein) to which the body responds by making chemicals or specialized cells that directly or indirectly eliminate that antigen. An antigen may be one of many different substances, including a portion of a virus's protein coat, a chemical released by a bacterium, an enzyme produced by a fungus, or even a toxin released by an infected body cell.

Where Immunity Begins

The cells responsible for the body's immune responses are collectively called *lymphocytes*, a kind of white blood cell. Lymphocytes, like all blood cells, originate in the red bone marrow from the hemocytoblasts. At first all lymphocytes are alike, but depending on where they mature, they develop into one of several different types.

During fetal development, some immature lymphocytes migrate to the thymus gland and become **T cells**. In time the T cells either will be directly involved in a type of immunity called cell-mediated immunity or will stimulate other cells to produce antibodies.

Other immature lymphocytes are believed to remain in the bone marrow and mature into B cells. **B cells** are involved with manufacturing antibodies and thus deal with another type of immunity, humoral immunity.

22.5

▶ Objectives

- Describe the difference between humoral and cell-mediated immunity
- Explain the function of B cells in the immune response
- List and explain the functions of the various types of T cells
- Compare and contrast passive and active immunity
- Describe the theories of auto-immune diseases

▶ Key Terms

immune	memory T cell
T cell	AIDS
B cell	HIV
humoral immunity	active immunity
antibody	autoimmune disease
cell-mediated immunity	vaccine
helper T cell	passive immunity
cytotoxic T cell	serum
suppressor T cell	allergy
	histamine

As the T and B cells mature, each develops the ability to recognize one of several specific antigens. Once mature, many of these cells migrate to the lymph nodes and spleen, where they await the arrival of their specific antigen. Others circulate in the bloodstream. T cells make up approximately 70%–80% of the circulating lymphocytes, while B cells make up the remaining 20%–30%. These cells may quietly wait for decades, but when their specific antigen comes, they rapidly respond to produce an army of immune system cells that wages war on the antigen.

Humoral Immunity

Humoral immunity involves activated B cells and the production of antibodies. **Antibodies** are proteins that are involved in the destruction of antigens. They circulate through the body by dissolving into the blood or lymph.

Before B cells can produce antibodies, they must become activated. Certain polypeptides released by other immune system cells activate B cells to form plasma cells and memory B cells. The presence of an antigen may also stimulate B cells to produce plasma cells and memory B cells. *Plasma cells* are responsible for producing antibodies. Some plasma cells produce 2000 antibody molecules per second for the four to five days they live. Even at that rate, the first buildup of plasma cells and antibodies needed to fight off a particular infection may take several days to several weeks.

Antibodies are capable of fighting antigens in several ways:

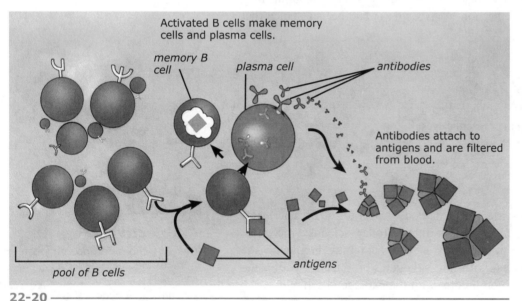

Activated B cells make memory cells and plasma cells.

memory B cell

plasma cell

antibodies

Antibodies attach to antigens and are filtered from blood.

pool of B cells

antigens

22-20

Humoral immunity involves B cells and the production of antibodies.

✦ *Neutralization*. Some antigens are toxins. Their toxic effect may be neutralized when the proper antibody combines with them. Some antibodies neutralize viruses by combining with a viral coat protein, thereby preventing the virus from entering its host cell.

✦ *Precipitation*. Some antibodies cause precipitation, the formation of clumps bonded together, of the antigen.

✦ *Agglutination*. In agglutination the antigen is part of the cell, and the antibody causes the cells to form clumps. The precipitated antigens and the agglutinated cells can be removed from the body by phagocytosis or other means.

✦ *Immobilization*. Some antibodies cause immobilization of cells by reacting with their flagella. Stopping cellular movement permits the body to deal with the cells more easily.

The *memory B cells* produced by a stimulated B cell may live for several dozen years. Subsequent exposure to the same antigen causes a rapid response by the memory B cells. They quickly divide, making plasma cells that produce antibodies and additional memory cells. This more rapid response usually

humoral: (L. *humor*—fluid)

permits the elimination of the antigen before the person suffers symptoms of the disease.

Cell-Mediated Immunity

T cells play the central role in **cell-mediated immunity**. In this kind of immunity, several types of cells are developed to target and destroy the antigen. In some cases the target may be body cells that have been invaded by a virus or have become cancerous. Other times, cell-mediated immunity acts against foreign cells or substances.

Cell-mediated immunity is activated when a macrophage ingests an antigen, forming an *antigen-macrophage complex*. When the receptors on a helper T cell match the antigen-macrophage complex, they bind together. Once they bind, the macrophage secretes polypeptides that cause helper T cells to rapidly divide and activate the other types of T cells to destroy the pathogen.

✦ *Helper T cells.* An activated **helper T cell** will make more of the same kind of helper T cells. A rapid buildup of a specific kind of helper T cell permits the body to fight the infection quickly. Stimulated helper T cells may also cause plasma cells to produce antibodies against the antigen and may produce other chemicals to help the body fight the infection. Some of these chemicals attract macrophages to the site, where they phagocytize the invaders.

✦ *Cytotoxic T cells* (also called *killer T cells*). **Cytotoxic T cells** attack other body cells that have been affected by the antigen. The cytotoxic T cell attaches itself to the affected cell and then releases a protein that forms a hole in the affected cell's membrane. This causes the cell's contents to spill (*lysis*), killing the affected cell.

✦ *Suppressor T cells.* Toward the end of the fight, **suppressor T cells** are produced. These cells stop the activities of the killer T cells, the helper T cells, and the plasma cells, returning the body to normal operations.

✦ *Memory T cells.* For many diseases, **memory T cells** made during an infection will circulate in the blood for dozens of years. When the antigen-macrophage complex of that disease again appears, these cells are ready to divide, rapidly forming cytotoxic and helper T cells. These can rapidly attack the antigen before the person experiences symptoms; thus, the person is immune to the disease.

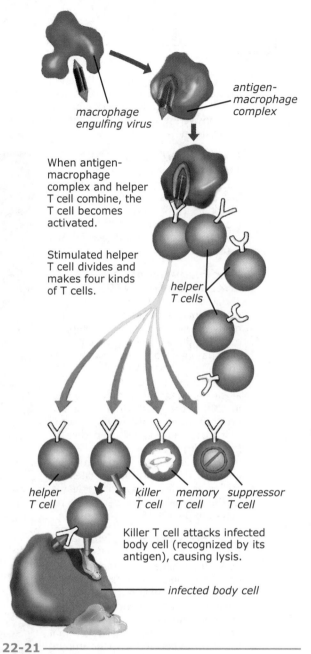

macrophage engulfing virus

antigen-macrophage complex

When antigen-macrophage complex and helper T cell combine, the T cell becomes activated.

Stimulated helper T cell divides and makes four kinds of T cells.

helper T cells

helper T cell

killer T cell

memory T cell

suppressor T cell

Killer T cell attacks infected body cell (recognized by its antigen), causing lysis.

infected body cell

22-21
Cell-mediated immunity involves T cells.

22-3 Comparison of T Cells and B Cells		
	T cells	**B cells**
Place of origin	bone marrow	bone marrow
Place of maturation	thymus gland	bone marrow
Type of immunity	cell-mediated	humoral
Activated cell type	helper, cytotoxic, suppressor, and memory T cells	plasma cells; memory B cells

cytotoxic: cyto- (cell) + -toxic (poison)

HIV and AIDS are terms heard often in the news. They also are terms that people often do not fully understand. This facet answers seven common questions about this devastating disease.

What are HIV and AIDS?

AIDS stands for *acquired immune deficiency syndrome*. This disease was first officially described in the United States in 1982, with 266 cases being diagnosed. **HIV** stands for *human immunodeficiency virus*, the virus that causes AIDS.

How many people are affected by HIV/AIDS?

Today HIV and AIDS are *pandemic* (spread over a large region, affecting many in the population), resulting in global health and social problems. The virus has infected people in almost every country, and it does not discriminate—it affects men and women of all races, ages, and economic levels. In 2007, the Joint United Nations Programme on HIV/AIDS reported that there were 33 million people infected with HIV worldwide; an almost equal number of men and women are infected. There were an estimated 2.7 million new cases and 2 million deaths. Of the 2.7 million new cases, 45% occurred in people 15–24 years of age.

How does the HIV virus work?

Researchers have learned a great deal about HIV since it was first described in 1982. It is now known that HIV is a retrovirus that attacks the T cells of the immune system. Initially the immune system may be able to adequately fight the HIV infection and keep the amount of virus in the system low. In cases such as this, the individual may show no symptoms of disease for several years. Once enough of the virus is produced, however, the immune system becomes overwhelmed and cannot effectively fight the HIV or any other infections. When a person with HIV has at least one of these "opportunistic infections" and has a lowered T cell count, he is considered to have AIDS.

What are the symptoms of HIV/AIDS?

Most people do not know that they have become infected with HIV because most do not show symptoms until several years after they become infected. The length of time between infection with HIV and development of AIDS varies but is usually 5–10 years.

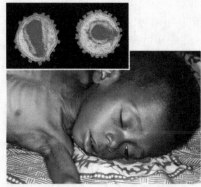

Over 2.1 million children worldwide have AIDS. HIV (inset).

As the disease progresses over the years and more T cells are destroyed, the person begins to develop symptoms, such as swollen lymph nodes, weight loss, diarrhea, and fever. Infections by other pathogens also become more frequent; these can include respiratory infections, tuberculosis, bacterial infections, and cancers.

It has been said that no one dies of AIDS. Rather, an AIDS victim usually dies from opportunistic infections caused by other organisms or viruses that his body cannot fight off due to the destruction of T cells by HIV. Medical science is able to help the

person fight many of these diseases. In time, however, the number of infections and the physical strain of the diseases and the medications slowly wear the person down. The immune systems of most healthy people easily fight off the pathogens that lead to death for AIDS victims.

Are there treatments or cures for HIV/AIDS?

There is no cure for HIV/AIDS, although there are several drugs that are used to fight HIV infection. Using their knowledge of the HIV life cycle, researchers have designed medications to disrupt the life cycle at various stages. These drugs can slow the progression of the infection but cannot stop it completely.

Unfortunately, many of these drugs have severe side effects and all of them are very expensive. Some of the drug companies have lowered the cost of their drugs to make them more affordable in Third-World countries. Nonetheless, these drugs are still unavailable to many people.

Vaccines against HIV are currently being developed and are being tested in several countries. However, when the HIV RNA is converted into DNA, there are often very subtle mutations that can produce a different strain of HIV. A vaccine may be effective against one strain of HIV but completely ineffective against other strains. Researchers believe it will be several more years before an effective, usable vaccine is available.

How is HIV transmitted?

HIV is normally found in the bodily fluids of an infected person: blood, semen (sperm-containing fluid), secretions of the female reproductive organs, and (to a much lesser extent) saliva and tears. For HIV to be transferred from one human to another, there must be a transfer of bodily fluids.

Anyone who receives bodily fluids— for example, through sexual contact or by sharing IV drug needles—from a person that has the virus can get the disease. Sadly, some people innocently contracted HIV as they were given transfusions of blood that contained

HIV/AIDS Statistics for the United States†	
estimated number living with HIV or AIDS	>700 000
annual new cases	>38 000
prevalence by sex	male 75%, female 25%
prevalence by race	black 49%, white 29%, Hispanic 20%, other 2%

†statistics from Centers for Disease Control and Prevention, 2007

pandemic: pan- (Gk. *pan*—all) + -demic (*demos*—people)

the virus. Today all donated blood is tested for HIV. The test is reliable, but like all such tests, it is not infallible. There is a slim chance a person can get HIV from a blood transfusion. The disease can also be transferred from an infected mother to her unborn child.

A person cannot become infected with HIV through casual physical contact (such as shaking hands, kissing the cheek, or hugging); by eating with an infected person or using the same cooking utensils; by using water fountains, bathroom facilities, or swimming pools; or by *donating* blood. The virus is actually somewhat fragile and can live outside the body only briefly.

What should be a Christian's response to HIV/AIDS?

Our response as Christians to a person who has HIV or AIDS should be no different from our response to any other victim of disease—compassion and love. First, we should be careful not to assume that an infected person has contracted HIV through immoral or illegal activity. Many people become infected through no fault of their own. Second, in dealing with those who have contracted the disease through sinful behavior, we should remember that the love and grace of our Lord Jesus Christ knows no boundaries. It is true that sexually transmitted diseases (STDs) serve in part as God's way of judging the sexually impure (Rom. 1:27). But it is also true that "Christ Jesus came into the world to save sinners" (1 Tim. 1:15). We should be ready and willing to share the gospel of Jesus' death and resurrection, which paid the price for man's sin, with all who will listen.

Acquired Immunity

When someone suffers from certain diseases, the body manufactures memory B cells and memory T cells. When the proper antigen stimulates them, these cells quickly divide to produce antibodies and reactivate the processes of cell-mediated immunity. Thus, in secondary infections (infections from a pathogen the body has previously identified), the body's defenses can attack more quickly than when the body first had the disease. In secondary infections the disease is usually conquered before a person experiences the symptoms. During secondary infection, more memory cells are made. This boosts immunity to later infections of that pathogen.

Kinds of Immunity

Immunity can be either active or passive. A person has an **active immunity** when he manufactures the antibodies himself or has activated T cells for a particular antigen. One way a person can acquire an active immunity is, of course, to

The History of Immunity

The Greek historian Thucydides (thoo SID i DEEZ) made one of the earliest known observations of acquired immunity. About 450 BC he noted that those who were sick and dying of the plague were cared for by those who had recovered from it, "for no one was ever attacked a second time." That people could develop immunity to certain diseases interested the medical profession for many years. Could a person become immune to a disease without having the disease? Could people be made immune to diseases that were normally contracted repeatedly, like the common cold?

Edward Jenner induced his patients to develop immunity against smallpox without having the disease (see p. 237). His success was based on the fact that the cowpox virus and the smallpox virus are quite similar. He had little idea about what the body was doing to become immune.

Nor did Louis Pasteur understand how immunity develops when he accidentally left a culture of chicken cholera bacteria on a shelf. After a two-week vacation, he remembered the bacteria and decided to see whether it was still pathogenic. He injected it into chickens, and they did not develop the disease. Later, when he exposed these chickens to the cholera pathogen, they did not develop the disease. Pasteur speculated that he had weakened the bacteria so that they did not cause the disease but did cause immunity. He experimented with several ways to weaken other pathogens and developed a vaccine for rabies.

The Common Cold—a Disease Without Immunity

If the human immune system is so effective that when exposed to a virus similar to the smallpox virus it can develop a permanent immunity, why is there no immunity for the common cold?

If there were only one common cold virus, it would probably be possible. But any one of over two hundred similar viruses can cause the common cold. Many of these are believed to be the result of minor mutations, changing only the protein coat of the virus. In the case of the common cold virus, the protein coat appears to be what the immune system recognizes as the antigen.

The many common cold viruses are just different enough to require a whole new reaction of the immune system. Therefore, though a person may be immune to some cold viruses, there are probably still many more for which he does not have immunity.

Autoimmune Diseases

Autoimmune diseases are those in which an individual's immune system cannot distinguish the body from foreign invaders. The immune system then produces autoantibodies and cytotoxic T cells that attack and damage cells of the body.

Researchers are unsure why this occurs, but they do have some theories. Part of the maturing process of the T cells in the thymus is to learn to distinguish host from foreign substances. If the T cells fail to mature properly, they never learn to make that distinction. Therefore, the T cells attack body cells. Another theory is that a virus might incorporate body membrane proteins into its protein coat. This "camouflage" fools the immune system, and it attacks both the virus and any other cells that have that antigen. A third theory is that the antigens on the invader are so close to those of host cells that the immune system cannot make the distinction. Once a foreign invader sensitizes it, the immune system attacks all cells that have similar membrane markers—foreign and host.

Autoimmune diseases can affect many different types of cells, causing a wide range of symptoms. The table gives some examples of autoimmune diseases, their symptoms, and the cells that are attacked by the host immune system.

Autoimmune Diseases

Disease	Symptoms	Antibodies against
systemic lupus erythematosus	fatigue, characteristic facial rash, prolonged fever, kidney damage, joint pain	blood cells, DNA, nerve cells
rheumatoid arthritis	joint deformity and pain	cells lining joints
ulcerative colitis	lower abdominal pain, diarrhea	colon epithelial cells
juvenile diabetes	weakness, hunger, thirst	pancreatic beta cells
myasthenia gravis	progressive muscle weakness	nerve receptors in skeletal muscle cells

Kinds of Immunity

Active immunity: The antibodies or the T cells needed for the immunity are produced by the person.

- Natural: The person has experienced the disease.
- Artificial: The person has been given a vaccine or toxoid.

Passive immunity: The antibodies needed for the immunity are supplied to the person.

- Natural: Antibodies are given by the mother to her child through the placenta or through her milk.
- Artificial: The antibodies are obtained from the blood serum of other humans or animals.

experience the disease. This is a *natural active immunity*. However, by being inoculated with the antigen that stimulates the production of antibodies or activates T cells, a person can gain immunity without having the disease. In the case of a serious disease such as polio, this method of acquiring an *artificial active immunity* is preferred.

A **vaccine** is a weakened form of a pathogen sometimes produced by exposing the pathogen to chemicals, acids, heat, x-rays, or other conditions. Some vaccines contain a *toxoid*—a weakened form of the toxin produced by a pathogen—rather than the pathogen itself. Vaccines are used to stimulate the body to produce antibodies or to activate T cells to provide just as much protection as those formed as a response to the disease. Some diseases require booster shots (which have the effect of secondary infections), causing the body to produce more memory cells and thereby increasing the level of immunity.

A person can develop a **passive immunity** by being given antibodies that have been formed by another person or an animal. Since the person is not producing the antibodies himself, these immunities are short-lived. A mother who has immunity to a specific disease can supply antibodies to an unborn infant through the placenta or to a newborn infant by her milk. In this way antibodies formed in the mother's body can protect the infant for the first six to twelve months of his life. This is called a *natural passive immunity*.

A passive immunity can also be acquired from an immune animal or human. Most often the blood of the organism that has the immunity is used to make a **serum**. A serum is made by removing the blood cells and most dissolved substances from the fluid portion of the blood, leaving specific antibodies. A serum injection results in an *artificial passive immunity* that generally lasts only a few weeks or months. These preformed antibodies can effect full protection from a disease, or in reduced amounts they can cause partial protection. In the latter case, if a person contracts the disease, he will have only a light case and will then develop his own active immunity to it.

Allergies

Sometimes good things can be carried too far, which is the case with an **allergy**. Normally, pollen, perfume, dog dander, feathers, strawberries, milk, wheat, dust mites, and other normal substances in the environment do not stimulate the body to produce antibodies. The immune systems of some people, however, after being exposed to these substances, begin to produce antibodies or helper T cells against them. Then, whenever that person is exposed to that substance, his body reacts against it. This is what is called an allergy.

The foreign substance that causes the reaction is called an *allergen*. When the antibodies that cause an allergy come in contact with the allergen, the two combine, causing certain cells to release **histamine** and other chemicals. These chemicals cause the symptoms of an allergic reaction. Many allergy sufferers find relief by taking *antihistamines* that prevent the release of histamine, thereby relieving allergies. There are many different antihistamines on the market. Some are potent; others, mild. Some work well for certain allergies but do very little for others.

Inhaled allergens (such as pollen, mold spores, animal dander, and house dust) usually cause a runny nose, sneezing, and itchy eyes. Skin contact with an allergen (such as poison ivy, cosmetics, perfumes, chemicals, and metals) usually causes redness, itching, swelling, and blisters. Reactions to ingested or injected allergens (such as foods, drugs, and venoms) can cause symptoms in almost any area of the body. Symptoms range from itching and sneezing to diarrhea and vomiting. In exceptionally severe cases the person's blood pressure drops and the air passages become constricted. This is called *anaphylactic shock*, and it may be fatal if not treated immediately.

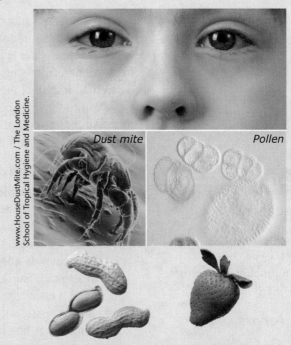

www.HouseDustMite.com / The London School of Tropical Hygiene and Medicine.

Dust mite *Pollen*

Red, watery eyes and a runny nose are common responses to pollen and dust mites. Peanuts and strawberries can cause life-threatening allergic reactions in some people.

histamine: hist- (Gk. *histos*—web) + -amine (E. *-amine*—a group of organic compounds of nitrogen)

antihistamine: anti- (opposite) + -hist- (web) + -amine (organic compounds of nitrogen)

⮕ Review Questions 22.5

1. List two responses of the body to an antigen.
2. Compare and contrast T cells and B cells.
3. How do antibodies react with antigens?
4. List several differences between humoral and cell-mediated immunity.
5. What do memory B cells and memory T cells have to do with acquired immunity?
6. What is the difference between active and passive immunity?
7. List and describe two ways of acquiring an active immunity and two ways of acquiring a passive immunity.
8. What is an autoimmune disease?
9. Describe the theories of how autoimmune diseases occur.
10. How are allergies and the body's immune system related?
⊙11. Why is HIV considered pandemic?
⊙12. Describe how HIV attacks the body.
⊙13. What treatments are available for HIV? Can these treatments cure an HIV infection?
⊙14. How is HIV spread? How can it be prevented?
⊙15. What is the proper Christian response to those who have contracted HIV/AIDS?
⊙16. Why would an antigen that attacks a person's red blood cells cause fatigue?
⊙17. Which kind of acquired immunity lasts longer? Explain. What are the advantages of other kinds of immunity that may not last as long?

22C—The Excretory System

Various organs remove body wastes. The skin's sweat glands (primarily responsible for regulating body temperature) excrete salts, urea, and other wastes. The lungs excrete water and carbon dioxide during breathing. The liver produces bile that contains pigments from broken-down erythrocytes. Because bile flows into the small intestine, these pigments are removed from the body with the feces.

Despite these excretory functions of other organs, the kidneys and associated structures compose the primary *excretory system* of the human body and are vitally important in regulating the blood's chemical composition.

22.6 The Kidneys

The **kidneys** resemble two large, purplish brown beans. Their function is to filter metabolic wastes from the blood and excrete these wastes in a liquid called *urine*. As the kidneys form urine, they regulate not only the composition of the blood but also the fluid balance throughout the body tissues.

Each kidney is about 11 cm (4.3 in.) long, 6 cm (2.4 in.) wide, and 3 cm (1.2 in.) thick. The kidneys are located posteriorly against the lower back body wall a little above the iliac bones. Protective layers of connective tissue and adipose tissue surround each kidney, holding it in place. An examination of a kidney that has been sectioned sagittally through the midline will show the outer *renal cortex* region distinguished from the inner *renal medulla*. The medulla consists of several cone-shaped masses of tissue called *pyramids*. The hollow center of the kidney is called the *renal pelvis*.

The cortex of each kidney is composed of approximately one million microscopic tubular units called **nephrons**, which are held together by thin layers of connective tissues that contain blood vessels, nerves, and lymph vessels. If all the nephrons in a single kidney could be untangled and placed end to end, they would be over 112 km (70 mi) long!

nephron: (kidney)

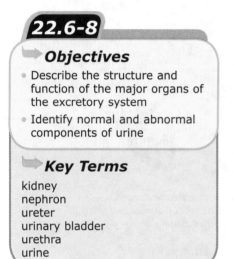

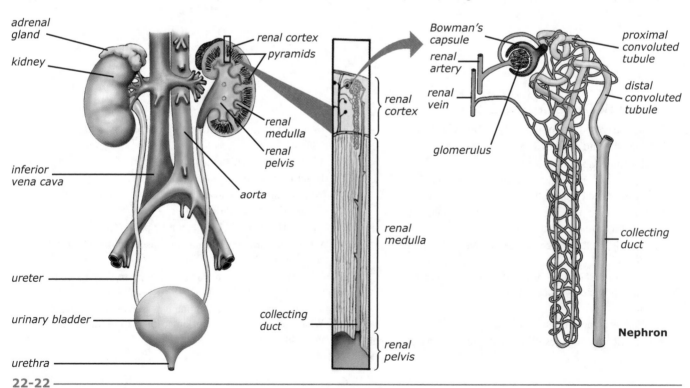

22-22
The excretory system

Kidney Function

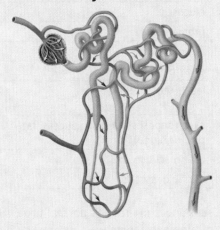

Yellow arrows indicate filtration, blue arrows show reabsorption, and red arrows show secretion.

To rid the body of wastes, the kidney performs three major functions: filtration, reabsorption, and secretion. The blood that flows into the kidney through the renal artery not only carries oxygen to the kidney tissues but also brings the waste substances from the rest of the body.

The blood flowing into the glomerulus is under higher pressure than the fluid in the adjacent Bowman's capsule. This pressure difference causes some of the fluids and other substances, such as urea, vitamins, salt, and glucose, to move from the glomerulus into the Bowman's capsule and then into the proximal convoluted tubule. This is the process of *filtration*. The fluid and substances that enter the Bowman's capsule are called the filtrate.

As the filtrate flows from the Bowman's capsule into the proximal convoluted tubule, many of the filtrate substances are reabsorbed into the capillaries that surround the proximal convoluted tubule. This is called *re-absorption*. During reabsorption most water diffuses back into the capillaries by osmosis. Glucose, sodium, potassium, and calcium ions are actively transported back into the capillaries. Some reabsorption also occurs in the distal convoluted tubule.

In the distal convoluted tubule, additional substances are moved from the blood to the tubule—this is *secretion*. The substances that are secreted include urea, various ions, drugs, and other waste products. Hydrogen ions may also be secreted into the tubule to adjust the pH of the blood.

As the fluid flows from the distal convoluted tubule into the collecting duct, water may be reabsorbed into the bloodstream, especially if the body is becoming dehydrated. The remaining fluid and substances constitute urine. The collecting ducts empty the urine into the central renal pelvis of the kidney.

As the blood flows into the nephron of a kidney, it passes into a collection of capillaries called the *glomerulus* (gloh MEHR yuh lus). Some of the plasma leaves the capillaries and passes into a *Bowman's capsule*, a cup-shaped end to the tubule of a nephron that surrounds the glomerulus. This plasma passes along the tubules of the nephron. Even beneficial molecules such as water, glucose, and amino acids are forced out of the blood into the Bowman's capsules. But most of these substances are reabsorbed into the blood as they travel along the nephron's tubules. The tubule is designed to contain the wastes and make sure that the beneficial molecules and minerals are returned to the bloodstream. After these processes are completed, the fluid that flows out the *collecting ducts* (which compose the medulla of the kidney) into the renal pelvis is urine.

22.7 Accessory Excretory Organs

From the kidneys, the urine moves down muscular tubes called the **ureters** by peristaltic contractions. About 60 mL of urine per hour is carried down the ureters and is emptied into the **urinary bladder**, the reservoir for urine. The elastic wall of the bladder is composed of muscle layers and an epithelial lining.

The **urethra** (yoo REE thruh) is the tube that leads from the bladder to the outside of the body. There are two sphincter muscles around the urethral opening of the bladder that remain closed when the bladder is relaxed. As the bladder prepares to empty, the muscles of the bladder wall contract. The increased pressure inside the bladder forces one sphincter open. However, before urine enters the urethra, the second sphincter must be opened voluntarily.

22.8 Urine

Normal **urine** is about 95% water, with the rest being solid solutes. *Urea* (yoo REE uh), formed from protein catabolism, is usually the most common organic substance in urine. *Uric acid*, which forms during nucleic acid breakdown, and *creatinine*, which is produced during muscular contraction, are also common

Diseases and Disorders of the Excretory System

+ *Floating kidney*: a kidney that has been jarred from its usual location and is left movable. It may be caused by an injury or blow to the lower back; it can also occur after an obese person reduces body fat.
+ *Kidney failure*: an abnormal condition in which the kidneys fail to form urine. Possible causes include physical injury, bacterial infections, or exposure to toxic chemicals.
+ *Kidney stones*: accumulations of various mineral crystals in the pelvis of the kidney. They may produce severe pain if they get wedged in the ureter. Many people with kidney stones pass them on their own, but others require surgery. They are believed to be prevented by drinking sufficient water.

glomerulus: (L. *glomus*—ball)

22-4 Normal Components of Urine	
Component	**Amount per day**
water	1200–2000 mL (g)
urea	25–30 g
sodium chloride	10–15 g
potassium chloride	2.5 g
sulfate	2.0 g
creatinine	1–2 g
phosphate	1.7 g
uric acid	0.7 g
ammonia	0.6 g
calcium	0.3 g
magnesium	0.2 g
minute amounts: fatty acids, amino acids, pigments, mucin, enzymes, hormones, and vitamins	

22-5 Abnormal Components of Urine	
Component	**Possible cause**
glucose	diabetes mellitus
proteins	kidney or heart disease
acetone	diabetes mellitus, starvation
erythrocytes	infection in urinary system
leukocytes	large numbers indicate an infection in the urinary system
casts (deposits of epithelial cells, fat, pus, or blood)	lesions (sores) in kidney
amino acid crystals	severe liver disease
uric acid crystals	gout

substances in urine. A normal urine sample also contains very small quantities of pigments, hormones, enzymes, vitamins, and various inorganic substances.

A *urinalysis* is a test that detects the amounts of normal substances and the presence of any abnormal substances in urine. For example, urine that contains glucose indicates that the pancreas is probably not producing enough insulin. Normal urine may contain small amounts of proteins, but if there is more than a trace, then there must be some tissue breakdown in the nephron that allows these large molecules to be filtered out of the blood. If blood or a large amount of bacteria are present in a urine sample, the person is probably suffering an infection somewhere in the urinary system.

Drinking enough fluids (6–8 glasses per day) is helpful in maintaining healthy kidneys. This water is absorbed into the bloodstream and later is used to dilute any toxins in the blood. An abundance of water makes it easy for the kidney to filter out toxins and other unwanted substances. This is one reason for the "drink plenty of liquids" prescription so often recommended when a person is sick.

Artificial Kidneys and Dialysis

If a person has to take certain drugs over long periods of time, one or both kidneys may cease to function. Alcoholic beverages and nicotine (from tobacco) as well as trauma and other causes may also contribute to kidney failure. Although a person can survive with fewer than 25% of his nephrons working, if his kidneys fail completely, an artificial kidney will be necessary for his survival. The body cannot tolerate a buildup of wastes or other substances in the blood. An artificial kidney, or *dialysis* (dye AL ih sis) machine, cannot fully duplicate the functions of a normal kidney, but it certainly can prolong life.

In a dialysis machine, a cellophane membrane substitutes for the glomerulus, and the substitute for the tubule is a large volume of fluid called *dialysate*. When the machine is in use, the patient's blood flows on one side of the membrane and the dialysate on the other. Thus, as both fluids circulate through the machine, wastes move from the blood into the dialysate, and some substances that are lacking in the blood are absorbed from the dialysate. In this manner the wastes are carried away by the circulating, fresh dialysate.

Although kidney transplantation is an option for those who have kidney failure, the number of potential recipients greatly outnumbers the available organs for transplantation.

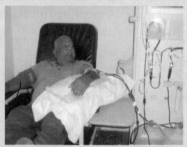
Patient undergoing dialysis

➡ *Review Questions 22.6–22.8*

1. List the various organs in the excretory system and tell their functions.
2. What are the three primary regions of the kidney?
3. Describe a nephron. Tell the function of each part.
4. What are the normal constituents of urine?
5. List several abnormal constituents of urine and tell what a physician might suspect if he found these substances in a patient's urine.

The Nervous System

Control Part I

Facets

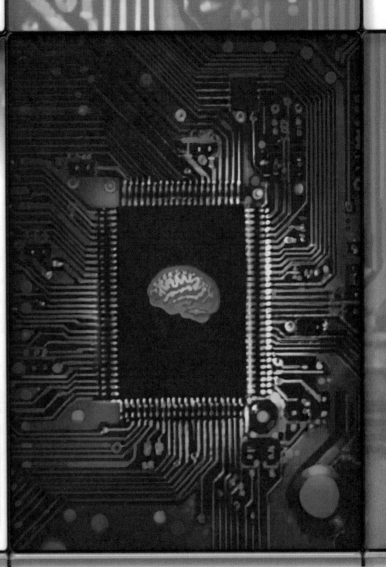

23A—The Structure of the Nervous System

One of the great marvels of the human body is that it all works together. When something hot is touched, the response of withdrawing the hand is immediate. If a person consciously had to feel, see, recognize, and think before drawing away his hand, he could be seriously burned. The body's internal workings, so often taken for granted, offer interesting examples of carefully designed control. As bodily activity increases, the heart automatically increases its pace; the respiratory system responds with deeper, more rapid breathing; but the digestive system slows down, while hundreds of other adjustments are made so that homeostasis can be maintained. When the activity stops, the systems automatically readjust.

Even a simple movement like taking a step involves thousands of cells, all of which must work with great precision. The intricate control mechanisms that are so vital to your well-being were programmed by the Creator. How else could they be adequately explained?

Two main systems control the body: the *nervous system* and the *endocrine system*. The brain, spinal cord, sensory organs (eyes, ears, taste buds, touch

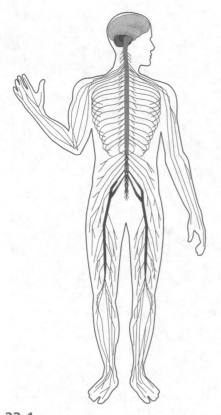

The brain and spinal cord compose the central nervous system. The nerves extending throughout the body are the peripheral nervous system.

23.1

Objectives

- Describe a typical neuron's structures and discuss their functions
- Distinguish between the three functional types of neurons
- Summarize the initiation and transmission of a nerve impulse

Key Terms

neuron	interneuron
cell body	motor neuron
dendrite	neurotransmitter
axon	reflex arc
myelin sheath	resting
synapse	potential
synaptic cleft	action potential
sensory neuron	

neuron: (nerve)
dendrite: (Gk. *dendron*—tree)

receptors, and others), and the nerves that supply them compose the nervous system. The endocrine system is discussed in the next chapter.

All voluntary activities, such as speaking and running, and many involuntary activities, such as heartbeat and digestion, are controlled by the nervous system. The nervous system receives stimulation (possibly from the environment or an internal source such as the level of a blood chemical), interprets it, and responds to it. Scientists understand, at least in part, how the nerve cells are able to perform some of these simple functions.

But the human nervous system also performs activities that scientists have not been able to explain. How does a cell or group of cells in the brain remember a beautiful mountain landscape seen two years ago? Or, for that matter, how does it "store" the concept and the word *mountain*? How does a person think? What is a dream? What is an emotion? The brain controls all of these. But what are these things, and how does a mass of cells inside the skull cause them? These questions continue to be the subjects of scientific studies, and many are not yet fully answered. However, many things about the nervous system are known. This study begins with the neuron—the basic cell of the nervous system—and how nerve impulses are transmitted.

23.1 Neurons

Neurons (NOOR AHNZ) are cells that serve as the functional units of the nervous system. Once mature, neurons are capable of living as long as the body does, although many do not. A typical neuron is designed to receive and distribute a nerve impulse. The number of impulses and whether or not a particular neuron is stimulated is the essence of the control of the body by the nervous system.

Neuron Anatomy

Neurons are composed of three basic parts.

✦ The **cell body** is the part of the neuron with the greatest diameter, and it contains the nucleus and cytoplasmic organelles. The cell body may also receive impulses from other neurons.

✦ The **dendrites** are multibranched fibers that project from the cell body. They receive and relay nerve impulses *toward* the cell body of the neuron.

✦ The **axons** are fibers that carry impulses *away* from the cell body. They are usually the longest portion of a neuron. Some axons in the legs can reach several feet in length; others are less than a millimeter. The axons of most neurons are covered by an additional lipid layer called the **myelin** (MYE uh lin) **sheath**. The myelin sheath functions to insulate the axon as well as increase the rate of transmission of nerve impulses.

A special structure called the *axon terminal* is located at the end of the axon. Axons terminate at another neuron, a muscle fiber, or a gland cell. It is important to note that neurons do not touch each other. The junction between two neurons is called a **synapse**, and the space between the axon terminal and another neuron is called the **synaptic cleft**.

Although neurons have many different functions, they may be grouped into three functional types.

✦ **Sensory neurons** carry impulses toward the central nervous system (the brain and spinal cord), and each responds only to particular stimuli. For example, the sensations of seeing, hearing, tasting, temperature changes, and so on result from stimulated sensory neurons that transmit impulses to other

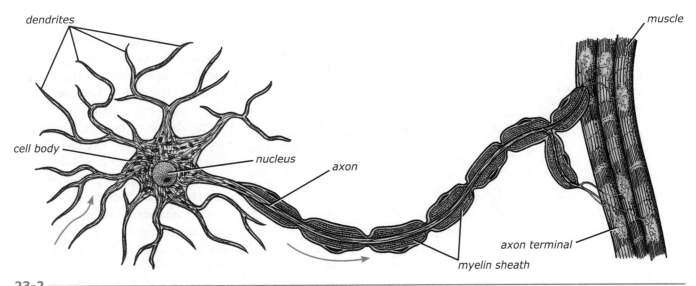

dendrites

muscle

cell body

nucleus

axon

axon terminal

myelin sheath

23-2 ─────────────────────────────────

A typical neuron. Arrows show the direction of nerve impulses.

neurons in the central nervous system. Usually, the cell body of a sensory neuron is part of the central nervous system, and the dendrites are part of the peripheral nervous system (the remainder of the nervous system outside the central nervous system).

✦ **Interneurons** are located within the central nervous system and receive impulses from sensory neurons. Interneurons distribute the impulse to other neurons.

✦ **Motor neurons** have dendrites and a cell body located in the central nervous system but axons in the peripheral nervous system. The ends of the motor neuron axon usually stimulate muscles or glands to cause a response.

Neuron Function

The primary function of neurons is to transmit nerve impulses from one area to another. A *nerve impulse* is an electrochemical pulse that moves along the membrane of a neuron. Neurons are *at rest* when they are neither receiving nor transmitting impulses. How is a nerve impulse transmitted?

When a neuron is stimulated, it becomes more permeable to positively charged ions, and they rush into the cytoplasm. That area of the cytoplasm now has a net positive charge. The increased positive charge causes adjacent areas of the neuron to become positively charged, and a nerve impulse is created. This is a chain reaction, very similar to the falling of a row of dominoes. However, as the nerve impulse travels along the membrane, positively charged ions stop moving into the cytoplasm, returning it once again to a net negative charge. It is almost as if someone is repositioning the dominoes as quickly as they fall down. The nerve impulse continues along the membrane until it reaches the axon terminal.

Once the nerve impulse arrives at the axon terminal, it must be transmitted across the synaptic cleft to the next neuron. The impulse cannot "jump" this space. The

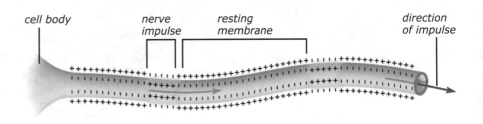

cell body

nerve impulse

resting membrane

direction of impulse

23-3 ─────────────────────────────────

Propagation of a nerve impulse. Does this illustration represent an axon or a dendrite? How can you tell?

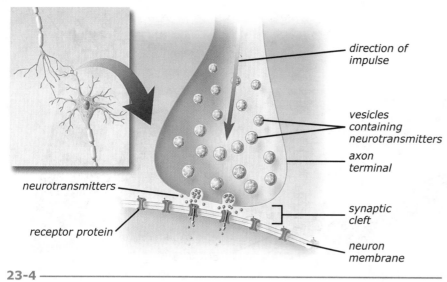

direction of impulse

vesicles containing neurotransmitters

axon terminal

synaptic cleft

neuron membrane

neurotransmitters

receptor protein

23-4

A synapse showing the transfer of a nerve impulse across a synaptic cleft

change in membrane permeability of the nerve impulse is believed to cause the end of the axon terminal to release chemicals called **neurotransmitters** into the synaptic cleft. The neurotransmitters stimulate receptor proteins in the membrane of the next neuron. A nerve impulse is created, and the impulse is carried along the neuron.

Once the impulse has crossed the synaptic cleft, special enzymes in the cleft rapidly inactivate the neurotransmitters. Since neurotransmitters are produced only at the axon terminals, nerve impulses can be passed only from the axon terminal of one neuron to the dendrite or cell body of the next neuron.

The Reflex Arc

The **reflex arc** is a good example of various types of neurons working together. A *reflex* is an involuntary response to a stimulus. The automatic, immediate jerk of a person's body in response to a stimulus such as a wasp sting or a cut is a reflex. The nerve impulses travel so rapidly that the body moves before the person has time to think.

Skin contains an abundance of receptors, which are actually the dendrites of sensory neurons. Once the sensory neuron is stimulated, the impulse is carried to the central nervous system. There the sensory neuron may form a synapse with only one interneuron, or it may branch and pass the impulse to many interneurons. What happens depends on the particular stimulus and the part of the body involved.

The interneuron passes the impulse to a motor neuron (or motor neurons), which carries the impulse to a muscle; body movements result. At the same time, the interneurons may send impulses to the brain, telling of the stimulus it received. But the

muscle action prompted by the motor neuron often happens before a person is conscious of the stimulus. Thus reflex arcs are involuntary.

The body part that responds to the stimulus is the *effector*. Effectors may be muscles or glands, depending on the reflex arc. For example, the eyelid muscles are effectors when something gets in the eye. Sweat glands may be the effectors in a reflex arc when a person is in a warm room. Many bodily movements and most internal responses are controlled by reflex arcs.

There may be only one sensory and one motor neuron or dozens of each in a reflex arc. If only two or three neurons are involved, a person is usually unaware of the stimuli, as is the case with those that control internal organs. No matter how many neurons are involved, the central nervous system is the center of each reflex arc. Parts of the brain and spinal cord are reflex centers for such actions as eye movement, sneezing, coughing, and breathing.

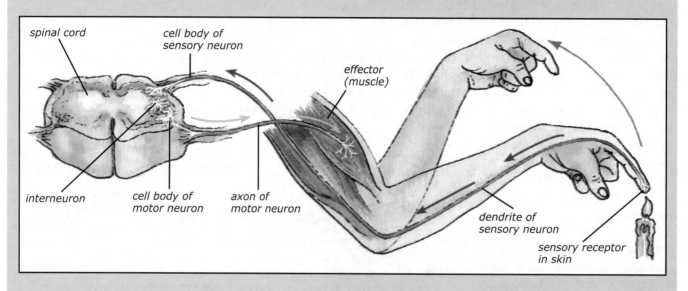

spinal cord

cell body of sensory neuron

effector (muscle)

interneuron

cell body of motor neuron

axon of motor neuron

dendrite of sensory neuron

sensory receptor in skin

effector: (L. *effectus*—to bring about)

The Nerve Impulse

To fully understand how a nerve impulse occurs, you must examine the chemical and electrical changes that occur on both sides of a neuron's membrane.

A resting neuron is *polarized*. This means that positive and negative charges are concentrated in different areas in relation to the neuron membrane. The difference in electrical charge between two areas is called a *potential*. Since the neuron is at rest, this relative charge difference between the inside and the outside of the neuron is called the **resting potential**.

Inside the neuron, there is a high concentration of positively charged potassium ions and negatively charged proteins. Outside the membrane there is a higher concentration of positive sodium ions. The negatively charged proteins cannot cross the membrane; however, the potassium ions can freely cross the membrane. As potassium ions leave the neuron, there is a net negative charge inside the neuron and a positive charge

outside. Cellular processes maintain this imbalance of various ions. The membrane is not very permeable to sodium ions; therefore, they remain concentrated outside the membrane.

When a neuron is stimulated, its membrane becomes more permeable to sodium ions, and they rush into the cytoplasm. That area of the cytoplasm now has a net positive charge—it is *depolarized*. The increased positive charge causes adjacent areas of the membrane to become depolarized, and an **action potential** (nerve impulse) is created.

As the action potential travels along the membrane, the sodium ions stop moving into the cytoplasm, and potassium ions rush out. This *repolarizes* the membrane, making the outside more positive than the cytoplasm. Once the membrane is repolarized, the action potential is over. The membrane then uses active transport to pump potassium back into the cell and sodium out to reestablish the resting ion concentrations.

 ### Review Questions 23.1

1. What are the two primary systems of body control?
2. List the three main parts of a neuron and describe how they differ.
3. Describe the three basic types of neurons.
4. Describe a nerve impulse.

5. Trace a nerve impulse through a simple reflex arc. Name each neuron involved and give its location in the body.
⊙ 6. Compare the electricity in a wire to a nerve impulse traveling along a neuron.

23.2 Divisions of the Nervous System

In general, the nervous system has two major divisions. The **central nervous system (CNS)** consists of the brain and spinal cord. The **peripheral nervous system (PNS)** is the remainder of the nervous system that carries information between the CNS and the rest of the body.

The Central Nervous System

The brain is one of the first organs to form in the human embryo. After about eighteen days of development, a group of cells forms the *neural plate*. This grows and eventually becomes a hollow, tubular mass of tissue. The anterior region enlarges and forms the major parts of the brain; the remainder becomes the spinal cord. The inside space increases in size and forms the *ventricles* (spaces inside the brain) and the central canal of the spinal cord.

Coverings of the Central Nervous System

The brain and spinal cord are covered by three protective coverings called **meninges** (muh NIN jeez). Starting with the external covering and moving inward, they are as follows:

- *Dura mater*—a thick, tough membrane that may contain blood vessels. In the skull it is tightly bound to the inner surface of the cranium.

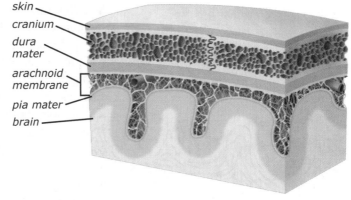

skin
cranium
dura mater
arachnoid membrane
pia mater
brain

23-5
The meninges are protective coverings of the brain and spinal cord.

meninges: (Gk. *menix*—membrane)

dura mater: dura (L. *dura*—hard) + mater (*mater*—mother)

Objectives

- Identify the two major divisions of the nervous system
- List and summarize the functions of the seven regions of the brain
- List and describe the functions of the peripheral nervous system's divisions
- Differentiate between spinal nerves and cranial nerves

Key Terms

central nervous
 system (CNS)
peripheral
 nervous
 system (PNS)
meninges
cerebrospinal
 fluid
brain
cerebrum
thalamus
hypothalamus
midbrain
pons

medulla
 oblongata
cerebral cortex
cerebellum
spinal cord
autonomic
 nervous
 system (ANS)
sympathetic
 nervous
 system
parasympa-
 thetic nervous
 system

arachnoid: (Gk. *arakhnoeides*—cobweblike)

pia mater: pia (L. *pia*—tender) + mater (mother)

cerebrospinal: cerebro- (brain) + -spinal (L. *spina*—the human spine)

thalamus: (Gk. *thalamus*—inner chamber)

hypothalamus: hypo- (beneath) + -thalamus (inner chamber)

- *Arachnoid* (uh RAK noyd) *membrane*—a thin, delicate, cobweblike membrane that forms many small spaces by its attachment to the pia mater.
- *Pia mater*—a thin membrane on the surface of the spinal cord and brain. This membrane contains many small blood vessels.

The **cerebrospinal** (seh REE bro SPY nul) **fluid** flows through the spaces between the pia mater and the arachnoid membrane. It also fills the ventricles of the brain and central canal of the spinal cord. The cerebrospinal fluid filters from the blood in the ventricles, passes over the central nervous system, and then returns to the blood. It nourishes the cells of the brain and spinal cord and protects these delicate organs by suspending them in a watery cushion.

The meninges surround the spinal cord down to the level of the coccyx. Examination of the cerebrospinal fluid can help identify infections in the central nervous system. Physicians obtain a sample by performing a spinal tap—an insertion of a hypodermic syringe between the lower lumbar vertebrae and withdrawal of 1–2 mL of cerebrospinal fluid. Since the spinal cord usually ends above this level, there is little chance of injury to the spinal cord by the syringe. The fluid is examined under a microscope; the presence of erythrocytes, bacteria, or viruses in the fluid clearly indicates infection or tissue damage.

The Brain

The adult human **brain** accounts for only about 2% of body weight (about 1.36 kg, or 3 lb), but it receives about 20% of the blood volume pumped by the heart. The natural color of the brain surface is pinkish gray and red. Blood gives the red color. The gray is characteristic of tissue composed primarily of cell bodies and dendrites and gives the *gray matter* its name. The deeper tissue of the brain is mostly *white matter* composed of myelinated axons.

The brain is subdivided into seven regions: the cerebrum, thalamus, hypothalamus, midbrain, pons, medulla oblongata, and cerebellum.

The **cerebrum** is divided sagittally into two large, irregularly folded lumps. Each half of the cerebrum is termed a *hemisphere*. A thick layer of white matter called the *corpus callosum* connects the hemispheres, allowing communication between them. The ridged or raised areas of the brain are *gyri* (JYE RYE) (sing., gyrus). The deeper, depressed areas are *fissures*.

The **thalamus** is composed of two oval gray masses near the center of the brain, connected at the middle by a bundle of white matter. The thalamus receives general sensations and quickly decides which impulses are important enough to be relayed to the parietal lobe. The thalamus is the first part of the brain that is aware of changes in the environment. The thalamus also helps keep a person awake and alert.

The **hypothalamus** weighs only about 4 g, yet it controls many involuntary activities, such as regulation of body temperature, blood volume, and fluid balance. It also has some control over appetite and emotional expression. Additionally, the hypothalamus controls hormone release by the *pituitary gland* and forms two hormones itself. Therefore, the hypothalamus functions in both the nervous system and the endocrine system.

The **midbrain** is located between the thalamus and the pons. Many motor impulses transmitted from the pons and the cerebellum to the cerebrum pass through the midbrain. In it are centers for controlling body movements and posture, especially controlling the head in relation to the rest of the body and space. Other midbrain centers are involved with vision and hearing reflexes.

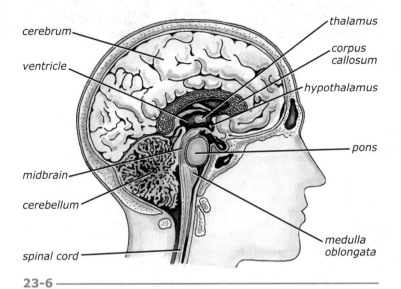

cerebrum
ventricle
midbrain
cerebellum
spinal cord

thalamus
corpus callosum
hypothalamus
pons
medulla oblongata

23-6
Cross section of the human brain

The **pons** is a bulging structure located between the medulla oblongata and the midbrain. It carries information from one side of the brain to the other. With the medulla oblongata, the pons controls involuntary respiration. It also contains reflex centers that control chewing, facial expression, response to sounds, and eyeball movement.

The **medulla oblongata** is continuous with the spinal cord and functions primarily as a relay center between the spinal cord and the brain. Centers for the control of respiration, blood vessel diameter (blood pressure), and heart rate are in the medulla. Therefore, injuries to the medulla may result in a *coma* (unconscious state) or death. Its reflex centers control sneezing,

pons: (L. *pons*—bridge)
coma: (Gk. *koma*—deep sleep)

Diseases and Disorders of the Nervous System

◆ *Alzheimer disease*: a progressive disease affecting the brain that causes problems with memory, thinking, and behavior
◆ *Cerebral palsy*: paralysis or lack of muscle coordination from improper development of the brain or from damage to the brain at birth
◆ *Encephalitis*: inflammation of the brain; often a symptom of other diseases
◆ *Epilepsy*: disturbed brain function often causing convulsions and loss of consciousness
◆ *Hydrocephalus*: abnormal increase in the amount of cerebrospinal fluid; may result in brain damage and the enlargement of the skull in children
◆ *Multiple sclerosis*: patches of hardened tissue in the brain and spinal cord with deterioration of myelin sheaths; may result in paralysis and jerking muscle contractions
◆ *Parkinson disease*: chronic nervous disorder that usually affects the elderly; characterized by rigidity of muscles and tremors of arms and legs
◆ *Shock*: sudden reduction in vital processes that control blood volume and blood pressure; results in critically low blood pressure; often caused by severe injuries or emotional trauma

The Cerebrum

The gray matter surfaces of the cerebral hemispheres appear "wrinkled," increasing the surface area and making it possible to have more neurons in the gray matter. The **cerebral cortex** (outer gray matter) is only 2–5 mm thick, but it contains 12–15 billion neurons (approximately 75% of all neuron cell bodies in the nervous system). Generally speaking, the cerebral cortex is responsible for conscious activities. Researchers have mapped the various lobes according to their specific functions. Neurosurgeons use such information to determine which regions are diseased or damaged. The lobes are named for the skull bones that cover those areas.

◆ The *frontal lobe* is responsible for mental functions, such as reasoning, planning, and memorizing. It also controls the ability to communicate verbally and starts the commands for voluntary body movements.

◆ The *parietal lobe* is responsible for sensations, such as pain, pressure,

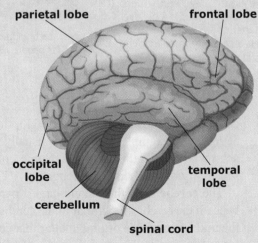

parietal lobe
frontal lobe
occipital lobe
cerebellum
temporal lobe
spinal cord

The lobes of the cerebrum

touch, and temperature. This information is directed to the frontal lobe, which determines what to do about it. The parietal lobe also responds to muscle tension, sensing the position of the body. For example, an expert skydiver knows his body position while he moves through the air because of impulses sent to his parietal lobe.

◆ The *occipital lobe* is involved primarily in vision and memory of objects and symbols. A severe blow to the head may cause the sensation of seeing "stars" because it stimulates the neurons leading from the eyeballs to the occipital lobe. Nerve impulses from these fibers are interpreted as visual impulses by the occipital lobe.

◆ The *temporal lobe* perceives the sensations of hearing and smell. It also provides the ability to remember the pronunciation of words and the melody of songs and stores memories of both sight and sound.

Each year approximately 11 000 people in the United States sustain injuries to their spinal cords. Of those injured, over 80% are males and over 55% are between the ages of sixteen and thirty. Sometimes the results are only temporary; other times the victim sustains permanent injuries that may drastically change his life. In the United States over 250 000 people are living with permanent spinal cord injuries (SCI).

Injury can result when the spinal cord is pinched, torn, or severed. The body responds to injury by inflammation. Since the spinal cord is enclosed within the vertebral column, inflammation increases the pressure placed against the spinal cord and actually reduces the blood flow. The neurons begin to die and there is a loss of function. The amount of damage and loss of function depends on the location and severity of the injury. Forty-five percent of SCI result in loss of function that is complete—there is no function below the level of the injury on both sides. A person with incomplete loss of function maintains some sensory and motor ability below the injury.

People who sustain spinal cord injuries must receive special care. At the scene of the accident, they must be carefully strapped to the stretcher to prevent further injury. If possible, they should be taken to hospitals that specialize in treating this type of injury. Medications are given to reduce the amount of swelling, and sometimes surgery is needed. Sometimes full function returns after the swelling has resolved. Researchers are investigating several potential treatments, such as the use of stem cells to regenerate damaged spinal cord cells, nerve cell transplants, and medications to enhance nerve growth.

As with many types of injuries, recovery depends on the extent of the injury. Today, 85% of those who survive the first twenty-four hours are still alive ten years later. Despite their physical limitations, many SCI survivors lead active lives. Technological advances in wheelchair materials, construction, and mobility have allowed even greater freedom of activity.

An injury such as this changes a person's life forever. How would you respond if you, someone in your family, or one of your friends sustained a spinal cord injury that resulted in permanent loss of function? It would probably test your faith in God and you would have many questions, especially, "Why did this happen?" While it is not necessarily a sin to ask God why, Christians must realize that God often calls His children to face difficulties without revealing His specific purposes. This truth is well illustrated in the life of Job. This man suffered a great deal in the will of God, yet God never told him why. Nevertheless, the suffering that Job experienced was very valuable. He did not come to know why he suffered. He came to know something far better: "I have heard of thee by the hearing of the ear: but now mine eye seeth thee" (Job 42:5). A believer who goes through this type of trial can take comfort in knowing that God is never taken by surprise. He is all-knowing and completely sovereign (Job 42:2). Knowing this, one should ask, "How can God be glorified by this injury and my testimony?" Even in the face of a life-changing injury, a believer can know that he can do all things through Christ Who strengthens him (Phil. 4:13) and that even this injury will in time prove to be an evidence of God's goodness and love (Rom. 8:28).

Causes of Spinal Cord Injuries†	
Vehicular	42%
Falls	27%
Violence	15%
Sports	8%
Other	8%

†from the National Spinal Cord Injury Statistical Center, 2009

New technology has improved the mobility and quality of life for people with SCI.

coughing, swallowing, and vomiting. Other factors, however, stimulate these actions. For example, vomiting is usually triggered by an irritant to the stomach or by some other stimulus. Some cough medicines work by dulling the sensitivity of the medulla oblongata.

The **cerebellum**, the second largest part of the brain, is located behind and beneath the cerebrum. Similar to the cerebrum, it has two hemispheres joined in the middle and folds of gray matter that cover a mass of white matter. The primary function of the cerebellum is coordination; it monitors and adjusts body activities that are stimulated by other brain regions. The cerebellum functions totally on an involuntary level; therefore, it stimulates no voluntary movements. Other brain regions initiate the movements, but the cerebellum regulates their quickness and force. It ensures that a movement goes where it should at the proper time and with the proper strength.

The Spinal Cord

The **spinal cord**, the second major division of the central nervous system, is continuous with the medulla and usually extends down to the first or second lumbar vertebra. It relays messages between the peripheral parts of the body and the brain. It is a cylindrical mass of nervous tissue composed of thirty-one segments with one pair of spinal nerves originating from each segment.

The spinal cord is made of white matter in the outer regions and gray matter in the center. The H- or butterfly-shaped central gray matter is mostly interneurons and the cell body portions of sensory and motor neurons. The white matter is myelinated axons arranged vertically. It is similar to a bundle of communication cables, all insulated by the myelin sheaths. Some of the axon bundles carry impulses to the brain; others carry impulses only away from the brain. For example, pain impulses pass along only certain axons, while impulses to muscles pass along other axons.

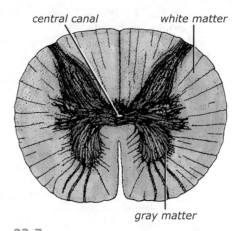

23-7
Cross section of the spinal cord

The Peripheral Nervous System

The peripheral nervous system is divided into two parts: the sensory and motor divisions. The *sensory division* contains all of the receptors and nerve fibers that carry information from the body to the central nervous system (brain and spinal cord).

The *motor division* carries information from the CNS to the rest of the body, allowing the body to react to the information received through the sensory system. It is composed of two independent divisions—the *somatic system* and the *autonomic system*. The somatic system is connected to the skeletal muscles and can be controlled. It is sometimes called the voluntary nervous system. This system also includes reflex arcs that can function automatically as protective reactions or to maintain balance. The autonomic system acts independently and without conscious effort. For example, no one needs to remind himself to breathe or digest his lunch—it happens automatically.

Twelve pairs of peripheral nerves, called *cranial nerves*, originate directly from the brain. Most of these nerves control the sensations and movements of the head and neck. Scientists classify cranial nerves by their general function— *sensory*, *motor*, or *mixed* (containing both sensory and motor fibers)—and name them according to their function or the general distribution of their fibers.

Thirty-one pairs of *spinal nerves* originate in the spinal cord and branch out to both sides of the body. These are also mixed nerves. The spinal nerves consist of a dorsal root that contains neurons that carry signals to the CNS, and a ventral root that carries information from the CNS to the effector organs. As they are relayed to the brain, all impulses enter the spinal cord through the dorsal roots of the spinal nerves. Impulses from the brain exit through the spinal cord's ventral roots as they deliver the stimuli to the rest of the body.

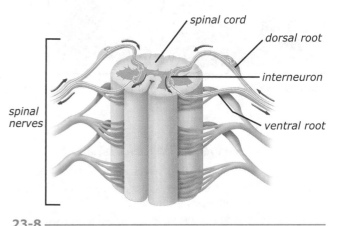

23-8
Basic spinal nerve anatomy

autonomic: auto- (self) + -nomic or -nomous (Gk. *nomos*—law)

cranial: (Gk. *kranion*—skull)

All internal organs (glands, blood vessels, heart, the smooth muscle tissue in the walls of internal organs, etc.) are under the control of the **autonomic nervous system (ANS)**, the involuntary part of the peripheral nervous system. These parts function automatically without conscious control.

The ANS helps maintain a steady internal condition in the body. For example, the ANS carefully regulates the internal body temperature (keeping it an almost constant 98.6 °F) and the pH of the blood (keeping it within the narrow range of 7.35–7.45). It also regulates the heartbeat, perspiration, breathing rate, and other body activities.

There are two divisions of the autonomic nervous system: the *sympathetic* and the *parasympathetic*. These divisions have generally opposing influences on body organs. While both divisions function at all times, usually one dominates at a time, depending on the environmental stimuli.

The **sympathetic nervous system** is composed of neurons whose

Sympathetic—"prepare for emergency"

fibers originate in the thoracic and lumbar portions of the spinal cord. It helps the body adjust to stressful or frightening experiences. In other words, it is "sympathetic" to problems during times of emergency—it helps a person "gear up" in order to survive and be successful.

For example, if a person comes across a grizzly bear while he is hiking, his sympathetic division instantly activates his body. Initially, the adrenal glands release a rapid surge of the hormone epinephrine (adrenaline). The epinephrine stimulates a quick increase in heart and breathing rates and causes a rise in the blood sugar needed for muscle action. The pupils of the eyes dilate quickly so that he can see where he is running. His digestive system is "shut off." ("I can finish digesting food later; now I must save the body.") His entire nervous system is put on emergency standby, making him jumpy.

The **parasympathetic nervous system** is composed of neurons from the brain stem and the sacral portion of the spinal cord. Its overall function opposes that of the sympathetic nervous system. It serves the "business as usual" functions of the body. It stimulates proper digestion, absorption of food, and elimination of wastes. It also maintains slower, normal heart and breathing rates as well as lower blood pressure. These functions are performed best when a person relaxes in a pleasant, peaceful environment.

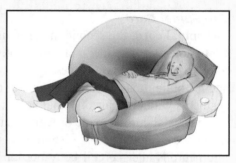

Parasympathetic—"business as usual"

The best example of the control of the parasympathetic system is its control over a person who sleeps after eating a delicious meal.

The sympathetic and parasympathetic nervous systems are both necessary, but both can be misused. A person who is almost constantly in stressful situations, stimulating the sympathetic nervous system and having the body on alert most of the time, may damage his body. If a person is constantly "tense," he may not be able to meet the demands of an emergency because he has exhausted his supply of reserves. Peace and rest are necessary for a properly functioning human body.

When many people are told that they must relax, they may respond that they cannot because their lives are too stressful. How can a person have peace in the high-pressure world in which he lives? The Creator knows the needs of the body and has supplied a "divine escape." The Bible says to "[cast] all your care upon him; for he careth for you" (1 Pet. 5:7).

➡ Review Questions 23.2

1. What are the main divisions of the central nervous system?
2. List and describe each of the meninges of the CNS and tell their functions.
3. What is the structural difference between gray matter and white matter in the CNS?
4. List the major divisions of the brain and tell what each controls.
5. List the lobes of the cerebrum and tell which bodily functions each controls.
6. List and describe the two main divisions of the peripheral nervous system.
7. List several factors that contribute to the amount of function loss after a spinal cord injury.
8. What are the two main divisions of the autonomic nervous system? What is the general function of each?
9. In our society many people suffer from stress-related illnesses. For example, stress is a contributing factor in many cases of high blood pressure (hypertension). Some, however, claim that no Christian should have these disorders. Is this a legitimate position? Why or why not?

23B—The Sensory Organs

It is commonly thought that humans have five senses: sight, smell, taste, touch, and hearing. If those were all, a person would have great difficulty. For example, a person not only sees his hand as he moves it in front of his face, but he also "senses" its movement. By extremely complicated mechanisms, one sees thousands of shades of color clearly. Skin not only relays the information to the brain that something has been touched, but it also senses the texture and temperature of the object.

The "sense of balance" seems to be two separate senses functioning together. A person is able to sense not only when *he* moves his body but also when some external force moves it. Further, he can sense the relative positions of parts of his body even when they are not moving. Scientists surmise that humans are somehow able to sense humidity, atmospheric pressure, and even the presence of static electricity.

All known sensory organs contain **sensory receptors**, specialized dendrites of sensory neurons. Receptors are stimulated by various external and internal conditions and inform the body of changes. Usually a receptor is sensitive to only one type of stimulus. The neurons of the eye, for example, are sensitive to light but not to sounds or odors.

The sensory organs also contain structures and tissues that support and assist the receptors. For example, the back of the eyeball contains light receptors. Other structures of the eyeball direct and focus the light entering the eye while they support and protect the receptors.

23.3 Minor Senses

Those senses that most people do not greatly depend on are the *minor senses*. They include the senses of the skin (such as temperature, touch, and pain) and the senses of smell and taste.

Often people who do not have use of one of their major senses (sight or hearing) are said to develop special abilities in their other senses. A person deprived of one sense may become more aware of the changes noted by his other senses, but there is little evidence to suggest that a person who is blind, for example, can smell or hear significantly better than other people.

Senses of the Skin

Cutaneous receptors sense cold, heat, pain, pressure, and touch in the skin. (Other regions of the body also have pain receptors.) Each sensation has different receptors, and certain receptors are more numerous in certain areas. For example, touch receptors are densely arranged in the tip of the tongue but are relatively few in number on the back of the neck. Other neurons transport impulses from the receptors to the brain. A stimulated heat receptor in a person's

23.3

Objectives
- Identify and describe the minor senses
- List the cutaneous receptors
- Explain how taste and smell are detected

Key Terms

sensory receptor
accommodation

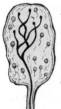
Touch
near surface of skin (most numerous in fingertips, palms, soles)

Heat
deeply embedded in dermis

Cold
dermis and subcutaneous areas of skin, cornea of eye, tip of tongue

Pressure
below the skin, membranes of abdominal cavity, around joints and tendons

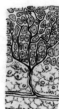
Pain
nearly every tissue of the body

23-9 —————
The five types of cutaneous receptors and their locations

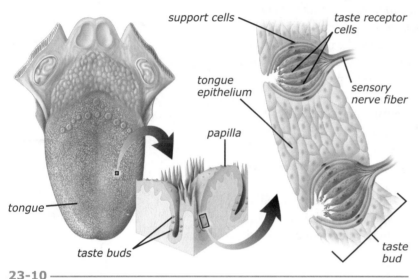

23-10
Location of taste buds

finger sends impulses up a set of neurons and stimulates a particular area of the brain. The brain then determines that what is felt in the finger is heat.

Taste and Smell

There are about 10 000 *taste buds*—the receptors for taste—on the average adult's tongue. Children, however, have considerably more. Most taste buds are on the *papillae* of the tongue, but some may be found on the inner surface of the cheek, roof of the mouth, tonsils, and epiglottis. All portions of the tongue respond to the five basic taste sensations—sweet, sour, bitter, salty, and umami. However, different areas of the tongue respond more to certain tastes.

The cells that form the taste buds are not neurons, but they are attached to nerve endings. The taste cells have chemoreceptors that stimulate a nerve impulse when they bind to a particular molecule. These cells live only about five to ten days and are replaced continually. The rate of replacement slows with age; therefore, many older people cannot taste substances as well as children. This fact may explain a child's objection to certain foods that he may enjoy as an adult. Exceptionally hot or spicy foods can damage taste buds. Consequently, some older people require more seasonings in order to enjoy foods.

The *olfactory sense* (sense of smell) is one of the least understood senses. The *olfactory receptors* consist of numerous cells in the mucous membranes that line the upper region of the nasal cavities. Olfactory receptors are neurons; therefore, if they are injured (as during infection or trauma), they do not regenerate. Even under normal conditions, these neurons decrease in number with age. As with taste, the older a person becomes, the less sensitive his sense of smell becomes.

Humans apparently can distinguish several thousand different odors. However, these odors can be detected only if the molecules stimulate the olfactory receptors. Since the olfactory receptor areas are poorly ventilated, the ability to detect odors is greatly increased by sniffing. The sense of smell is less acute when someone has a cold because the olfactory receptors are blocked by mucus, preventing odors from reaching them. A person's sense of taste is also affected because much of taste is actually smell.

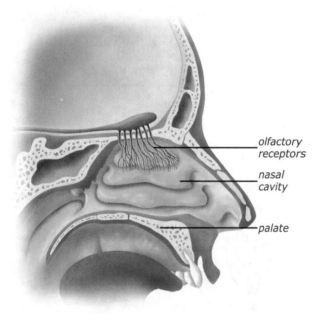

23-11
Location of olfactory receptors

Receptors for smell reach a point of **accommodation** quite rapidly. Accommodation means that after a short period of stimulation, the receptors become insensitive to that specific stimulus, and a person does not "smell" it any more. Accommodation also occurs with most other nerve sensations (except pain), although usually not as quickly as with smell. For example, it is common to become accustomed to the feel of what you are wearing or the background noise you are hearing (such as a fan or air conditioner) so that you no longer notice them.

Review Questions 23.3

1. List the five cutaneous receptors.
2. Why is the sense of taste more acute in children than in adults?
3. What is accommodation to a sensation?

⊙ 4. Occasionally when a body part is amputated, the person can still feel sensations that seem to come from the missing limb. This is called phantom limb pain. What could explain these sensations?

23.4 The Ear: Hearing and Balance

When discussing hearing, many people think first about the skin-covered flap of cartilage on either side of the head. However, this plays a rather small role in the sense of hearing. Embedded within the temporal bone is a small chamber, and beyond that is a tiny bony structure shaped like a snail's shell. These areas inside the bones of the skull are a small fraction of the size of the outer flaps, but they are actually responsible for the sense of hearing and most of the sense of balance.

Within the inner structures of a person's ear are thousands of nerve endings that are indirectly stimulated by sound waves (vibrations). These nerve endings transform an almost limitless variety of sound waves into nerve impulses, which travel along neurons to the brain where they are "heard."

The Ear

The *outer ear* consists of the *auricle*, an outer flap of tissue designed to collect sound waves, and the **external auditory canal**, a tube which goes into the head and ends at the **tympanic membrane** (eardrum). The skin lining the canal contains *ceruminous glands* (wax glands). The skin continually sheds dead surface cells, which combine with earwax and gradually move out of the ear. It is unnecessary and even dangerous to the delicate tissues of the ear to use any object other than a washcloth to clean the outer ear. (A physician should be consulted if the ear becomes clogged with wax.)

23.4

Objectives

- Identify the structures of the ear and describe their functions
- Trace the path of a sound wave from the external ear to the formation of a nerve impulse
- Identify the structures involved in the sense of balance and explain their functions

Key Terms

external auditory canal	static equilibrium
tympanic membrane	dynamic equilibrium
ossicle	semicircular canal
oval window	
cochlea	

Popping Your Ears

Have you ever felt your ears "popping" when driving up a mountain? What causes this sensation? The middle ear connects with the pharynx by the *eustachian tube*. This tube makes it possible for the air pressure within the middle-ear chamber to remain the same as the atmospheric pressure.

When driving high in the mountains, a person experiences a decrease in atmospheric pressure, but the pressure in the middle-ear chamber remains the same. The higher pressure in the ear causes the eardrum to bulge outward. When a person blows, yawns, or swallows, the eustachian tube opens and allows the air pressure in the middle-ear chamber to equalize with the atmospheric pressure. As the eardrum snaps back to a flat position, a "popping" sensation may be felt and/or heard in the ears.

auricle: (L. *auricula*—little ear)
auditory: (L. *audire*—to hear)

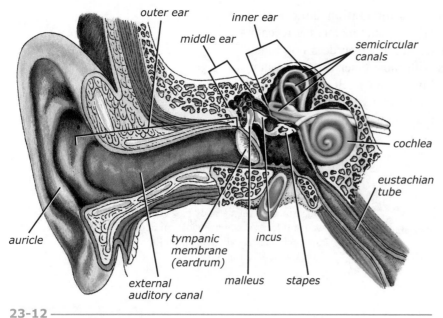

23-12

Human ear anatomy

ossicle: (L. *ossiculum*—small bone)
cochlea: (L. *cochlea*—snail shell)
static: (standing)
dynamic: (power)

① Stapes responds to sound waves.

② Movement of the oval window creates ripples in the cochlear fluid.

③ Ripples cause hair cells to bend.

④ Hair cells initiate a nerve impulse that is transmitted to the cerebrum.

nerve fibers

hair cells

oval window

23-13

The cochlea and the organ of Corti (inset)

The *middle ear* is a moist, air-filled chamber containing three tiny bones, or **ossicles** (AHS ih kulz): the *malleus* (MAL ee us, hammer), *incus* (ING kus, anvil), and *stapes* (STAY peez, stirrup). The joints between the bones are movable and form a lever system that picks up vibrations of sound waves that strike the tympanic membrane. The vibrations are transferred to the **oval window**, a membrane-covered opening of the inner ear.

Two small muscles attach to the ossicles and modify their movements. When exposed to sudden loud noises, they respond by limiting the vibrations, protecting the delicate membranes of the inner ear. In contrast, they also respond to soft sounds by amplifying the movement of the ossicles, enabling better hearing.

The *inner ear* consists of a *bony labyrinth* enclosing a *membranous labyrinth*. The bony labyrinth consists of channels and cavities within the temporal bone. The inner membranous labyrinth closely duplicates the shape of the bony channels. It is a tube-within-a-tube arrangement. The spaces between the bony and membranous labyrinths are filled with two slightly different fluids necessary for the sensations of hearing and equilibrium.

Hearing

The **cochlea** (KOH klee uh), a snail-shaped division of the inner-ear labyrinths, receives sound waves. Along the full length of the cochlea's inner surface runs the actual sound receptor, the *organ of Corti* (KOHR tee), which contains thousands of receptor cells called hair cells.

The sound waves in the atmosphere hit the eardrum, moving the ossicles in the middle ear. The vibrations of the stapes move the oval window in and out. This motion sets up "ripples" in the cochlear fluids, which cause some of the cilia-like projections of the hair cells to bend. The bending of the hair cells produces a nerve impulse that is transferred to the nerve cells. These impulses are carried to the temporal lobe of the cerebrum, where they are perceived as different sounds.

The Sense of Balance

The body maintains balance by two different senses: the sense of static equilibrium and the sense of dynamic equilibrium. **Static equilibrium** refers to the sense of body position when a person is not moving. **Dynamic equilibrium** is the ability of the body to respond automatically to positional changes when it is moving.

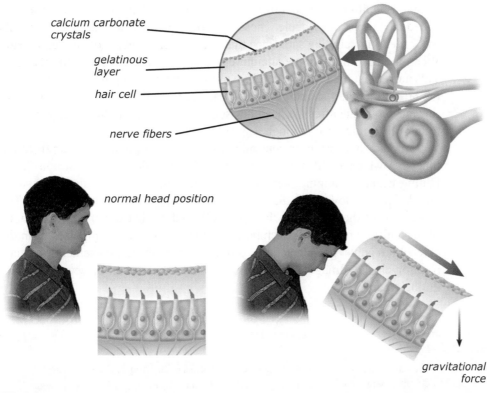

calcium carbonate
crystals

gelatinous
layer

hair cell

nerve fibers

normal head position

gravitational
force

23-14

When the head is depressed, gravity causes the calcium carbonate crystals to shift, stimulating the hair cells. Nerve impulses are sent to the brain, which interprets the new head position.

Two Kinds of Deafness

Most people define deafness as the inability to hear or the loss of the sense of hearing. Deafness can be a partial or complete hearing loss. Other symptoms include the inability to understand speech when there is background noise or difficulty hearing soft speech such as whispers. Loss of balance, dizziness, and ringing in the ears are other symptoms of hearing loss. There are two major types of deafness, distinguished by the anatomical structures that are affected.

✦ *Conductive deafness.* Anything that blocks the transmission of sound vibrations to the inner ear is called *conductive deafness*. This is the most common type of deafness (95%). Common causes include blockage of the auditory canal, stiffness or fusion of joints between ossicles, stiffness or tearing of the tympanic membrane, and middle-ear infections. Otosclerosis is the deposition of excess bone around the base of the stapes that decreases or totally prevents the transmission of sound vibrations. Some types of conductive deafness can be treated by surgery or by a hearing aid. A hearing aid is a small microphone that amplifies the sound. Some hearing aids have filters that reduce the amount of background noise.

✦ *Nerve deafness.* The other type of hearing loss is *nerve deafness*, or *sensorineural* hearing loss. This type of loss occurs when the cochlea, the auditory nerve, or the brain does not function properly. Although one in one thousand children is born with nerve deafness, this kind of deafness can also be caused by prolonged exposure to loud noise, such as

construction noise, loud music, and gunfire at close range. Some drugs, such as the antibiotic vancomycin, can cause nerve deafness.

Nerve deafness cannot be treated with the same type of surgery done for conductive loss or by the use of traditional hearing aids. However, in some people a special type of hearing aid called a cochlear implant can help. A cochlear implant is a surgically implanted device that converts sound waves into electrical signals that stimulate the neurons. It will not restore hearing but will enable the person to detect medium to loud sounds and speech rhythms. Therefore, cochlear implants are most successful in people who had normal hearing at one time. Nevertheless, some children who were born deaf have been able to learn to detect and to recognize certain sounds and have progressed quite well using these devices.

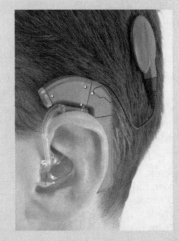

Cochlear implant

Static equilibrium is controlled by two small chambers in the inner ear called the utricle and saccule. These chambers are lined with sensory hair cells embedded in a jellylike substance that contains crystals of calcium carbonate. When the head moves, the calcium carbonate crystals slide, pulling the jelly. This movement bends the hair cells, stimulating nerve impulses that are sent to the brain. Violently shaking the head can set the jelly and crystals in motion; this may continue after the head has stopped moving, and the head may seem to be still "going" even though it is not moving.

Dynamic equilibrium is controlled by the three **semicircular canals** in the inner ear, which are also lined with hair cells embedded in a gelatinous layer. During walking, running, and other dynamic activities, the fluid in the semi-circular canals flows over the gelatinous material, bending the hair cells. The movement of the hairs stimulates nerve impulses that travel along a branch of the auditory nerve to the temporal lobe and cerebellum. Impulses from the cerebellum adjust muscle actions, producing coordinated movements for each position change. Imagine all the impulses that a basketball player experiences during a game!

⇒ Review Questions 23.4

1. What are the three main divisions of the ear? What structures are in each division? What structures separate each division from the next?

2. What is the primary function of the eustachian tube?

3. Describe the process of hearing in the inner ear.

4. List and describe the two main types of equilibrium. What are the primary organs for sensing each type?

23.5 The Eye and Sight

23.5

⇒ Objectives

- Identify the anatomical structures of the eye
- Summarize the functions of the structures involved in vision
- Explain the difference between rods and cones

⇒ Key Terms

sclera	retina
cornea	lens
choroid	rod
iris	cone
pupil	

sclera: (hard)
iris: (Gk. *iris*—rainbow)

The eyes are a pair of spheres a bit smaller than table tennis balls. They supply a continuous series of nerve impulses for about sixteen hours per day and then operate repair and maintenance systems while the body is sleeping. Normal eyes can focus on a hair as near as a few inches and on large objects as distant as several miles.

The brain controls muscles that move and focus the eyes so that they work harmoniously to provide a stereoscopic image that permits depth perception. While only silhouettes can be seen in near darkness, a person can distinguish minute variations in color in bright light. Scientists have not been able to explain completely how the eyes operate. The beauty and wonder of the eye is that it is perfectly designed by God to permit sight.

The Eye

The eyeball has three tissue layers. The outer layer, known as the sclera, is the "white of the eye." This white fibrous tissue maintains the shape of the eyeball. The transparent, anterior portion of the sclera, the **cornea** (KOHR nee uh), allows light to enter the eyeball. The cornea lacks blood vessels but receives nourishment from the fluid underneath it in the eyeball.

The middle layer, the **choroid** (KOHR OYD), is fragile and thin with many blood vessels for nourishing the innermost part of the eye. The anterior portion of the choroid, the iris (EYE ris), contains muscles and is the colored part of the eye. The circular opening in the iris is the **pupil**, which lets light into the eyeball. Muscles in the iris change the diameter of the pupil, regulating the amount of light entering the eye. In bright light the pupil is almost closed, protecting the eye from too much light. In dim light, however, it dilates (opens), permitting all the available light to enter.

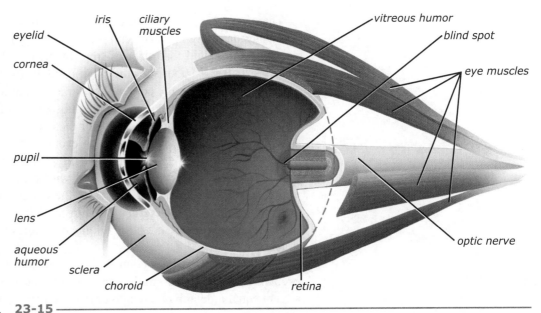

iris
ciliary muscles
eyelid
cornea
vitreous humor
blind spot
eye muscles
pupil
lens
aqueous humor
sclera
choroid
retina
optic nerve

23-15
Human eye anatomy

The third and innermost layer of the eyeball is the **retina**, composed of thousands of specialized neurons and their fibers. The neurons of the innermost layer are *photoreceptors* that can be stimulated by light. The impulses from the photoreceptors are transmitted to the occipital lobe of the brain by way of the *optic nerve*.

There are no photoreceptors where the optic nerve fibers leave the eye to form the optic nerve. This area is the *blind spot*. In people with normal eyes, the blind spot of each eye affects a different area of vision; therefore, the total field of vision is unbroken.

The **lens** of the eye is a biconvex, semisolid substance supported by the *ciliary* (SIL ee EHR ee) *muscles* and *suspensory ligaments*. The ciliary muscles and suspensory ligaments can change the shape of the lens. When looking at a close object, the muscles contract, making the lens thicker (more convex) and focusing the image on the retina. When looking at a distant object, the muscles relax and the lens flattens to focus the image.

The ability to focus on objects at different distances from the eye is *visual accommodation*. The lens is elastic in children but becomes more rigid with age. Therefore, about age forty some people begin having difficulty focusing on things closer to them; they hold reading material farther from their eyes. Lenses in glasses can compensate for the hardened natural lenses.

photoreceptor: photo- (light) + -receptor (L. *receptare*—to receive)
ciliary: cili- (eyelid) + -ary (related to)

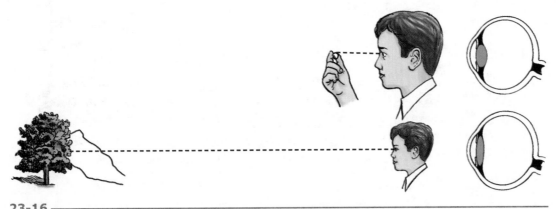

23-16
The thickness of the lens changes, permitting the eye to focus on objects at different distances.

There are two cavities in the eyeball: one in front of the lens and one behind. The anterior cavity is filled with the *aqueous* (AY kwee us) *humor*, a transparent, watery fluid that diffuses from blood vessels located near the ciliary muscles. This fluid nourishes the cornea and diffuses into the blood by way of canals on the edge of the cornea. The larger, posterior chamber of the eyeball contains a clear, permanent, jellylike substance called the *vitreous* (VIT ree us) *humor*, which provides support for the eye.

In the upper lateral region of each eyelid is the *lacrimal* (LAK ruh mul) *gland* (tear gland). It secretes about 1 mL of fluid each day, which is spread evenly over the surface of the eyeball with each blink. The fluid moistens and cleanses the cornea and lubricates the eyelid. It also contains *lysozyme*, an enzyme that kills bacteria. If the eyeball is irritated or if the person is under emotional stress, the lacrimal gland secretes more fluid, often resulting in tears.

Several sets of muscles control the movement of the eyeballs so that both eyeballs are directed toward the same object. In some individuals, muscles that are not equal in length or strength or that are paralyzed cause the eyes to cross.

Vision

There are more than 130 million photoreceptors in each eye, most of which are shaped like rods. **Rods**—responsible for night vision—are scattered over the retina. They are sensitive to low intensity light and produce a shadowed or silhouette image. Rods cannot determine color but can rapidly discern movements.

Cone-shaped photoreceptors detect colors. The **cones** are especially concentrated at the *fovea* (FOH vee uh), a small depression in the central region of the retina. Therefore, a person sees the sharpest color image of an object when he is looking directly at it in a well-lit environment; in dim light the image would not be clear because there are no rods at the fovea.

Both rods and cones contain light-sensitive pigments, and when the rods and cones are stimulated by light, the pigments decompose. These substances initiate a complex biochemical pathway that changes light energy into a nerve impulse that is carried to the brain by the optic nerve. The brain then interprets the sensory input as images and colors.

23-1 Photoreceptors	
Rods	**Cones**
night vision	color vision
produce general outlines of objects	produce sharp images
scattered over retina	concentrated at fovea
about 125 million per eye	about 7 million per eye

aqueous humor: aqueous (L. *aqua*—water) + humor (fluid)

vitreous: (L. *vitrum*—glass)

lacrimal: (L. *lacrima*—tear)

fovea: (L. *fovea*—pit)

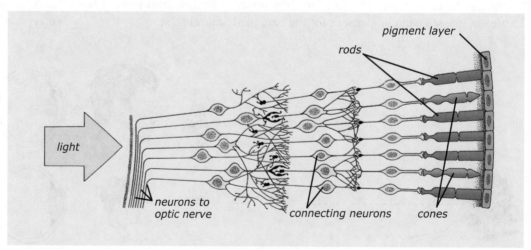

23-17
Cross section of the retina

Eye Disorders

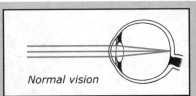

Normal vision

◆ *Myopia* (mye OH pee uh), or *nearsightedness*: a condition in which light rays from close objects can be focused on the retina, but those from distant objects are focused in front of the retina and therefore are not seen clearly. The problem is usually caused by an abnormally long eyeball, which is an inherited trait. Eyeglasses with biconcave lenses may compensate for myopia.

◆ *Hyperopia* (HYE puh ROH pee uh), or *farsightedness*: a condition in which light rays from far objects can be focused on the retina but those from near objects focus behind the retina. Hyperopia may be caused by an inherited short eyeball or by hardening lenses. Eyeglasses with

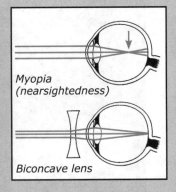

Myopia (nearsightedness)

Biconcave lens

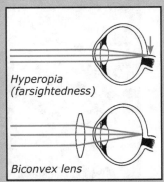

Hyperopia (farsightedness)

Biconvex lens

biconvex lenses may help a farsighted person to see close objects. Occasionally bifocal glasses, which have lenses for farsightedness in the lower portion and other lenses in the upper portion, are necessary.

◆ *Astigmatism* (uh STIG muh TIZ um): condition in which either the cornea or lens or both are uneven or unequally curved and the light rays from an object are not focused properly on the retina. This results in an area of the person's vision being out of focus. Sometimes the other eye will compensate for mild astigmatisms.

◆ *Cataracts*: clouded lenses that may be caused by old age, overexposure to bright sunlight, or diseases such as diabetes. Treatment may involve surgically removing the lens and implanting an artificial one. Sometimes special eyeglasses are used to compensate for the loss of the lens. Clouded corneas can also be removed and replaced with transplants.

◆ *Glaucoma*: the buildup of aqueous humor resulting in abnormal pressure within the eye. This causes a decrease of circulation to the retina and may damage the retina and cause blindness.

◆ *Night blindness*: lack of a certain pigment in the rods of the eye. Using cones, the person can see clearly in bright light, but in dim light he cannot see well. Vitamin A is needed to form this pigment; therefore, a diet with insufficient vitamin A can cause this condition.

astigmatism: a- (without) + -stigmat- (Gk. *stigma*—spot) + -ism (E. *-ism*—indicating a condition)

➥ Review Questions 23.5

1. List and describe the functions of the three main layers of the eye.
2. How is the amount of light that enters the eye regulated?
3. How does the eye focus?
4. Where are the lacrimal glands? What do they produce, and what are the functions of this substance?

5. Name and describe the two types of photoreceptors in the eye.
6. Describe myopia and hyperopia.
7. Differentiate between cataracts and astigmatisms.

Hormones and the Human Mind

Control Part II

24A—The Endocrine System

The *endocrine* (EN duh krin) *system* consists of the ductless glands of the body. These glands lack ducts (tubes) to transport the substances they secrete. Instead, the endocrine glands release these substances into the bloodstream, which transports them to the body's tissues and cells.

Endocrine glands help control the body, but they control it differently than the nervous system does. The nervous system connects directly with individual muscles and internal organs throughout the body to regulate movements. The endocrine system is composed of glands located throughout the body. The secretions of these glands affect cells, tissues, or organs in other areas of the body through the blood. The effects of the endocrine system are slower and longer lasting than the effects of the nervous system.

24.1 Hormones

Hormones are chemical messengers produced in one area of the body and carried by the blood to affect cells in other areas of the body. Hormones affect many different metabolic processes throughout the body: they regulate water balance in the blood, reproduction, growth, and blood pressure, to name a few.

24.1

Objectives

- Define and describe hormones
- Explain how the structure of hormones affects their mode of action on target cells
- Describe how a negative feedback system works

Key Terms

hormone
target cell
negative feedback system

endocrine: endo- (within) + -crine (separate)

hormone: (Gk. *hormon*—to stir up)

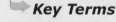

It is necessary to understand hormones in order to understand how the endocrine system works. Hormones, which are organic chemicals, can be described as follows:

- They are produced by the endocrine glands. There are two major categories of hormones based on their chemical structures. The steroid hormones are made from lipids. Nonsteroid hormones are made from amino acids and may be proteins, small peptides, or glycoproteins.
- They are carried and distributed by the blood. Small amounts of over thirty different hormones can be found in the blood at any time.
- They are specific chemical messengers. Most hormones stimulate or speed up processes in body cells, but some slow them down. As hormones pass through the blood, they are exposed to all body cells; however, they affect only specific **target cells**.
- They are secreted in small quantities. Usually only small amounts of hormones are produced, and only a tiny amount is needed to affect the target cell.
- They are constantly filtered out of the blood by the kidneys or deactivated by the liver or other glands. This forces the endocrine glands to constantly produce hormones.
- They are closely regulated. The quantity of the secretion of hormones is closely monitored to maintain homeostasis and regulate bodily functions.

Modes of Action

The target cells have specific receptors that recognize and bind specific hormones. The receptors are proteins located either in the cytoplasm or on the cell membranes of their target cells. The hormone structure determines to which type of receptor that hormone binds.

Since both steroid hormones and cell membranes have a high lipid content, steroid hormones diffuse through the membrane into the cytoplasm. There they bind to receptors, forming a *hormone-receptor complex*. The hormone-receptor

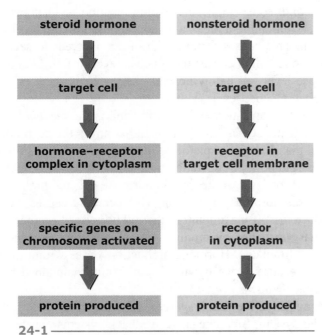

24-1
Steroid and nonsteroid hormone modes of action

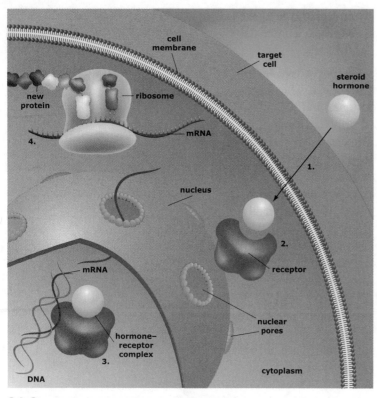

24-2

Steroid hormone action
1. Steroid hormone passes through cell membrane.
2. Steroid hormone binds with receptor in cytoplasm, forming a hormone-receptor complex.
3. Hormone-receptor complex enters nucleus and binds to DNA, stimulating formation of mRNA.
4. Ribosomes translate mRNA to produce proteins that will cause the desired effect in the cell.

24-3

In a negative feedback system, the product inhibits the system.

complex causes the target cell to alter its cellular processes. Many times the hormone-receptor complex binds to the DNA to activate specific genes on the chromosome. The activated genes form mRNA that binds to ribosomes. The ribosome translates the mRNA into proteins that cause the desired effect in the target cell.

Since the nonsteroid hormones do not have a large lipid content, they cannot diffuse across the target cell membrane. They require a two-receptor system—one receptor is located in the cell membrane and the other in the cytoplasm. When the hormone combines with the membrane-bound receptor, specific chemicals are released into the cytoplasm. These chemicals bind to specific receptors in the cytoplasm. The resulting cytoplasmic receptor complex releases substances that catalyze cellular processes to produce the desired changes in the target cell.

Control of Hormone Secretions

Since the endocrine system regulates many functions in the body, the secretion of hormones must be precisely controlled. God has designed intricate systems that respond to both external and internal stimuli, which control secretion to keep the body working properly.

First, the nervous system controls some glands directly. For example, the sympathetic nervous system has nerves that terminate in the adrenal gland. When someone is startled, the sympathetic system stimulates the adrenal glands to produce epinephrine (adrenaline).

Second, some endocrine glands respond to changes within the body. After eating a meal, a person's blood glucose level rises above normal. This causes certain cells in the pancreas to secrete insulin, a hormone which causes the glucose level to fall. Should the blood glucose level fall below normal, the pancreas secretes glucagon to raise the glucose level.

The third and most common method is the *feedback mechanism,* in which the secretion of one gland controls the secretion of a different gland. An example of a feedback mechanism is the control of the thyroid hormone secretion by a hormone secreted by the pituitary gland. If the level of the thyroid hormone gets too high, the pituitary decreases hormone production. This decrease causes the thyroid to produce less. If the amount of thyroid hormone becomes too low, the pituitary responds by producing more hormone, and the thyroid in turn is stimulated. Hormone production varies within rather narrow limits because of the dynamic equilibrium maintained by the feedback mechanisms.

A **negative feedback system** is one in which the product inhibits the system. Since negative feedback systems are the most common endocrine control systems, it is important to understand how they work.

An example of a negative feedback system is the heating system in a house. Suppose that the thermostat is set to maintain a temperature of 72 °F. Since the thermostat is sensitive to temperature changes, when the temperature drops below 72 °F, the thermostat will activate the heating system. As the warm air fills the house, the temperature begins to rise. When 72 °F is reached, the thermostat causes the heating system to shut down—the product of the system (the warm air) has controlled the system (the heating system).

The previous example of the pituitary and thyroid glands can also be used to illustrate a negative feedback mechanism. Special cells in the pituitary gland monitor the blood level of thyroid hormone. If the level increases, the pituitary gland produces a hormone that causes the thyroid gland to stop producing thyroid hormone.

Review Questions 24.1

1. List six characteristics of hormones.
2. Explain the significance of the nervous and circulatory systems in the function of the endocrine glands.
3. Explain how the structure of hormones affects the mode of action on their target cells.
4. Describe three methods of control of hormone secretions, and give an example of each.

24.2 The Endocrine Glands

Most of the endocrine glands produce hormones that are essential throughout a person's life. Some hormones are released only at specific times, such as puberty, pregnancy, or periods of stress. In a system as complex and sensitive as the endocrine system, it is not uncommon for hormone levels to have temporary fluctuations within the body or for levels to differ between individuals.

Occasionally there may be major problems. If a disease or injury affects an endocrine gland, the production of its hormone may decrease, resulting in **hyposecretion** (HYE poh sih KREE shun), or it may increase, resulting in **hypersecretion** (HYE pur sih KREE shun). These abnormal amounts of hormones can eventually produce physical and occasionally mental and psychological abnormalities, some of which can cause death.

Pituitary Gland

The **pituitary** (pih TOO ih TEHR ee) **gland** is about the size of a small marble. It is connected to the hypothalamus of the brain by the pituitary stalk. The pituitary gland has two lobes. The anterior lobe is epithelial tissue that is connected to the hypothalamus by special blood vessels. The posterior lobe is nerve tissue that communicates directly with the hypothalamus. Therefore, the hypothalamus controls both lobes. How? When the hypothalamus receives information from the body, special *neurosecretory cells* produce hormones that cause the pituitary to produce its hormones. These hormones then affect other glands and organs throughout the body.

Anterior Pituitary

Chemical secretions from the hypothalamus are released into the blood vessels that communicate directly with the anterior pituitary. These secretions are of two basic classes: releasing hormones and release-inhibiting hormones. The *releasing hormones* direct the anterior pituitary to release a particular hormone. Each hormone has a specific releasing hormone from the hypothalamus. The *release-inhibiting hormones* direct the anterior pituitary to stop producing a specific hormone.

24.2

Objectives

- List the major endocrine glands, the hormones they produce, and their effects on the body
- Explain why the pituitary gland is considered two glands in one
- Discuss the relationship between the hypothalamus and the pituitary gland

Key Terms

hyposecretion
hypersecretion
pituitary gland
growth hormone
thyroid gland
thyroxine
insulin
glucagon
diabetes mellitus
adrenal gland
epinephrine
norepinephrine
puberty
gonad
androgen
testosterone
estrogen

hyposecretion: hypo- (beneath) + -secretion (L. *secretio*—separation)

hypersecretion: hyper- (over) + -secretion (separation)

pituitary: (L. *pituita*—phlegm; from the belief that the gland secreted phlegm)

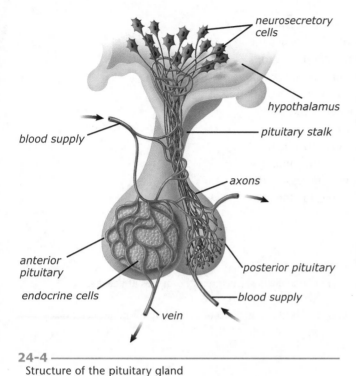

neurosecretory cells

hypothalamus

pituitary stalk

blood supply

axons

anterior pituitary

posterior pituitary

endocrine cells

blood supply

vein

24-4
Structure of the pituitary gland

The anterior pituitary secretes three groups of *tropic hormones* that stimulate growth and function of other endocrine glands.

✦ *Thyroid-stimulating hormone* (TSH) regulates the thyroid gland.

✦ *Adrenocorticotropic* (uh DREE noh KOHR ti koh TROH pik) *hormone* (ACTH) acts on the adrenal cortex.

✦ *Gonadotropins* (go NAD uh TROH penz) control the hormones secreted by the reproductive organs.

Body height and size are determined largely by the action of the **growth hormone** produced by the anterior pituitary. This hormone increases the uptake of amino acids for protein synthesis, stimulates the breakdown of fats, increases the level of sugar in the blood, and stimulates the formation of the periosteum and the epiphyseal plates of bones.

If the pituitary gland does not produce enough growth hormone during childhood, the person will be a dwarf. Usually pituitary dwarfs, unlike dwarfs by some other causes, possess normal intelligence and body proportions. Abnormally tall persons ("giants") develop if a hypersecretion of the growth hormone occurs during childhood. If hypersecretion of the growth hormone occurs after the epiphyseal plates have calcified, the person will not grow taller. However, the bones of the hands, feet, and face may become thicker, and the soft tissue may enlarge. This condition is acromegaly (ACK roh MEG uh lee), which means "large extremities."

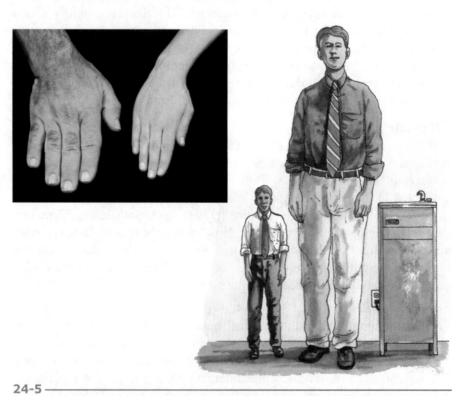

tropic: (adjective form of *tropism*)

adrenocorticotropic: ad- (L. *ad-*—near) + -reno- (kidneys) + -cortico- (bark) + -tropic (turn or change)

gonadotropin: gonado- (L. *gonad*—primary sex gland) + -tropin (change)

acromegaly: (Fr. *acromégalie*—enlargement of extremities)

24-5
Comparison of hands—one normal and one with acromegaly. Pituitary dwarf and giant with normal body proportions.

Posterior Pituitary

The posterior lobe is composed of nervous tissue and is actually an extension of the brain. The hypothalamus communicates directly with the posterior pituitary via the axons of the neurosecretory cells, whose nuclei are in the hypothalamus. The posterior pituitary stores and releases two hormones that are actually made by the neurosecretory cells in the hypothalamus. One of these hormones, *oxytocin* (AHK sih TOH sin), stimulates the smooth muscles of the body to contract. Its greatest effect is on the uterus, causing the birth of a baby.

The second hormone released by the posterior pituitary is *antidiuretic* (AN tee DYE uh RET ik) *hormone* (ADH), which regulates the volume and water content of the blood. If the amount of solutes in the blood increases, the hypothalamus stimulates the posterior lobe to release ADH. The target cells of ADH are tubules in the nephrons of the kidney. It causes them to reabsorb water from the tubules into the blood and decrease the volume of the urine. This helps the body conserve water. A hyposecretion of ADH causes an increase of water in the urine; this condition, called diabetes insipidis, increases urine volume and thirst.

Thyroid Gland

The **thyroid** (THY ROYD) **gland** is anterior to the trachea and just inferior to the larynx. It has two large lateral lobes connected by a wide strip of tissue. The thyroid produces the hormones **thyroxine** (thye RAHK sin) and *triiodothyronine* (TRY eye OH doh THY roh NEEN), both of which are derived from the amino acid tyrosine and iodine. Several weeks' supply of these hormones can be stored in the thyroid gland until needed.

These hormones increase oxygen use, stimulate heat production in most body tissues, and indirectly affect growth. Their secretion is regulated by a negative feedback mechanism between the anterior pituitary and the thyroid. The amount of thyroid-stimulating hormone (TSH) produced by the pituitary depends on the amount of thyroid hormones in the bloodstream, and vice versa. If the hypothalamus determines that the bloodstream contains too little

oxytocin: oxy- (Gk. *oxus*—sharp, quick) + -tocin (*tokos*—childbirth)

antidiuretic: anti- (against) + -diuretic (Gk. *diourein*—to pass urine)

thyroid: thyr- (Gk. *thureos*—oblong shield) + -oid (shape or form)

Too Much or Too Little Thyroid Hormone

A person who has hyposecretion of thyroid hormones during the early years of his life may be physically and mentally retarded. This type of person is called a *cretin* (KREET in). When there is extreme hyposecretion in an adult, the individual develops the condition called *myxedema* (MIK sih DEE muh). In this condition, a general swelling of body tissues (edema) develops, cholesterol levels increase (possibly leading to hardening of the arteries), and weight gain, lethargy, and sensitivity to cold can result. Cretinism and myxedema are both treated by taking thyroxine; early diagnosis of cretinism is critical for successful results.

Hypersecretion of thyroid hormones may increase the body's metabolism by 60% or more. The metabolic rate often exceeds the food intake, causing weight loss. Other symptoms include restlessness, protruding eyes, sensitivity to heat, and hyperactivity. The thyroid gland usually becomes enlarged, causing a swelling in the neck called a *goiter* (GOY tur).

Iodine deficiency also causes goiters. Without enough iodine, the thyroid gland cannot produce enough hormones, and the blood concentration decreases. The anterior pituitary secretes TSH in an attempt to correct the deficiency. However, since the thyroid cannot produce the hormones, the blood level never increases—the negative feedback mechanism fails. The anterior pituitary continues to produce TSH, causing the thyroid gland to grow, sometimes to a large size. The enlarged thyroid gland still cannot produce thyroxine until there is iodine in the diet. The use of iodized salt helps prevent a goiter, especially where the iodine content in the food supply is low. Often a goiter shrinks when iodine is added to a person's diet. Goiters may also result from other illnesses, most of which can be corrected.

cretin: (Fr. *cretin*—mentally retarded person)
goiter: (L. *guttur*—throat)

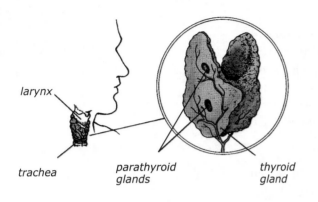

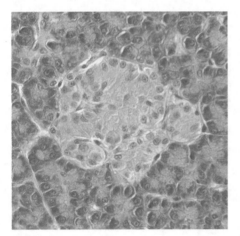

larynx

trachea   parathyroid   thyroid
           glands   gland

24-6
Location of thyroid and parathyroid glands (posterior surface of thyroid gland)

24-7
Light micrograph of pancreas tissue. Islets of Langerhans (lighter colored structure in center) produce insulin.

concentration of the thyroid hormones, it produces a releasing hormone that causes the anterior pituitary to secrete additional TSH. Exposure to cold weather increases metabolism by increasing production of TSH, causing an increase in the amount of thyroid hormones produced.

Calcitonin (KAL sih TOH nihn), another hormone secreted by the thyroid gland, is released when the calcium level of the blood gets too high. It causes calcium to move from the blood to the bone tissue, stabilizing the blood calcium level.

Parathyroid Glands

The *parathyroid* (PAR uh THY royd) *glands* are brownish yellow structures embedded in the connective tissue on the posterior surface of the thyroid gland. Parathyroid hormone regulates the calcium and phosphorus levels of the blood and bodily fluids. The function of the parathyroid hormone is opposite that of calcitonin: it moves calcium from the bone to the blood. It ensures there is sufficient calcium in the blood for normal nerve and muscle function and for blood clot formation. The blood calcium level may decrease as calcium is used by the bone tissue or when the diet is calcium deficient. In response, the parathyroid secretes more hormone to raise the calcium level of the blood.

Pancreas: Islets of Langerhans

There are more than a million small groups of cells known as *islets of Langerhans* (LAHNG ur HAHNZ) scattered throughout the pancreas. These cells secrete insulin and glucagon, which work together to control the level of sugar in the blood.

Insulin is a hormone that stimulates body cells to absorb glucose from the blood, thereby lowering the blood glucose level. Insulin also stimulates the conversion of glucose to glycogen, which is stored in the liver.

Glucagon performs functions that are basically opposite those of insulin. Primarily, it raises the blood sugar level by stimulating the liver to convert glycogen to glucose. It also stimulates the breakdown of fats to form glucose.

The secretion of insulin and glucagon is controlled by and helps to control the blood sugar level. Soon after a meal, the increased sugar content of the blood stimulates the secretion of insulin. Then, as the body cells take up the glucose, the blood sugar level decreases. This results in a decrease in insulin secretion. Several hours after a meal, the blood sugar level begins to drop. The lower blood sugar

Other Endocrine Glands

Some other organs produce hormones as well and therefore function in the endocrine system.

The *pineal gland*, a small oval structure located deep in the brain between the cerebral hemispheres near the thalamus, secretes the hormone melatonin (MEL uh TOH nin), but its function in humans is still debated. Experimental evidence from animal studies indicates that the pineal gland affects the regular rhythms of the body (e.g., sleep) and possibly pigmentation. In humans the pineal gland begins to decrease in size before puberty.

The *thymus* (THYE mus) *gland*, located between the lungs, is large in young children but shrinks after puberty. It secretes thymosin (THYE muh sin), a hormone that activates lymphocytes formed in the red bone marrow to differentiate into T cells.

The *placenta* (pluh SEN tuh), the tissue that joins the mother and fetus during pregnancy, produces estrogens and progesterone. The function of the placenta is covered more fully in Chapter 25.

parathyroid: para- (beside) + -thyr- (oblong shield) + -oid (shape or form)

insulin: (L. *insula*—island)

Diabetes Mellitus and Hypoglycemia

The most common disorder involving the hormones from the pancreas is **diabetes mellitus**. It is the seventh most common cause of death and is the leading cause of blindness, kidney disease, and gangrene. There are 1.6 million new cases diagnosed each year in adults twenty years old or older. Its usual cause is hyposecretion of insulin. There are two types—type 1 or insulin-dependent diabetes mellitus (IDDM) and type 2 or non-insulin-dependent diabetes mellitus (NIDDM).

Type 1 diabetes is primarily diagnosed in persons under forty years old. It is sometimes called juvenile-onset diabetes. In this type of diabetes, the pancreas eventually stops producing insulin. Although no one is sure of the etiology of type 1 diabetes, theories include a type of autoimmune disease in which certain pancreas cells are attacked by the immune system or a viral infection that gradually destroys these cells. People who have type 1 diabetes must monitor their blood sugar and inject insulin to control their carbohydrate metabolism.

Type 2 diabetes is the more common form and is usually found in persons over forty years old. Contributing factors in the development of type 2 diabetes are a diet continually high in carbohydrates, obesity, lack of exercise, and heredity. Treatment includes weight loss, dietary control, and medications that increase the production of insulin.

In both diseases, the decrease in insulin causes three major metabolic disturbances:

✦ Most of the glucose remains in the blood, largely unavailable to body cells. The liver cells are unable to store glucose, resulting in highly elevated blood glucose levels, or *hyperglycemia*.

✦ The body cells must rely on fats for energy. The release of fats into the blood may multiply the fat level five times. As fats are broken down, certain substances (ketone bodies) are produced that decrease the pH of the blood. Without insulin treatment, this condition may result in coma and death.

✦ The body breaks down proteins to amino acids. Since the glucose in the blood is not entering the body cells, they are "sugar starved." This signals the liver cells to then convert amino acids into glucose, compounding the blood sugar level problems. Normally, there is a balance of protein synthesis and protein breakdown, but an untreated diabetic may become thin and emaciated as his body uses its proteins for energy. Without the energy from glucose metabolism by cellular respiration, protein synthesis is greatly impaired.

The diabetic's blood sugar level often becomes so high that sugar is present in the urine. As the sugar in the urine increases, more water is also lost, and the diabetic has a constant thirst.

A person with diabetes mellitus suffers from too much blood sugar. A person with too little blood sugar has *hypoglycemia*. One cause of hypoglycemia is an overproduction of insulin in response to a meal. If the blood sugar level drops too low, the person will feel extremely drowsy and weak and may become uncoordinated or unconscious.

Hypoglycemia may also result if a diabetic injects too much insulin or skips a meal after injecting insulin. This severe reaction, known as insulin shock, requires a quick supply of sugar because the overdose of insulin has drastically reduced the glucose supply to the brain. If untreated, the person may lose consciousness, suffer brain damage, or even die.

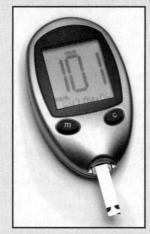

Diabetics must monitor their blood sugar level daily.

hyperglycemia: hyper- (over) + -glyc- (sweet) + -emia (blood)

hypoglycemia: hypo- (beneath) + -glyc- (sweet) + -emia (blood)

---

level stimulates the secretion of glucagon. This stimulates the liver to produce more glucose and return the blood sugar concentration to its normal level.

Adrenal Glands

The adrenal (uh DREE nul) **glands** are small yellow masses of tissue enclosed in fat tissue located on top of each kidney. Each gland consists of an inner portion, the *adrenal medulla*, and a larger outer region, the *adrenal cortex*. The two portions function as two different endocrine glands.

The cells of the adrenal medulla secrete **epinephrine** (EP uh NEF rin) and **norepinephrine** (NOHR ep uh NEF rin). The secretions of the adrenal medulla are controlled by the sympathetic nervous system and help prepare a person for emergencies or threatening situations.

When these hormones are released, the liver responds by breaking down glycogen into glucose that can be used for energy, the bronchial tubes dilate to increase oxygen in the blood, the rate of breathing increases, the heart beats faster, and the blood pressure increases. All these reactions provide the brain, heart, and muscles with a maximum supply of sugar, oxygen, and hormones so that the person is ready to respond immediately to any emergency.

The adrenal cortex secretes two groups of hormones. The first group helps regulate glucose metabolism during emergencies. The second group regulates the concentration of minerals, ions, and water in the blood and urine.

adrenal: ad- (near) + -renal (kidneys)

The time during which the secondary sex characteristics develop is called **puberty** (PYOO bur tee). Girls may show signs of puberty anytime between ten and sixteen years of age, but between twelve and thirteen years is average. In females the beginning of *menstruation* is considered the start of puberty. Menstruation, a monthly discharge of the uterine lining, is a normal part of a woman's life until she is about forty-five (see Ch. 25). Actually, the hormonal and bodily changes of a developing woman begin months or sometimes years before the first menstruation.

In males puberty usually begins sometime between eleven and seventeen years of age, but fourteen is average. The exact start of puberty in young men is not as obvious as it is in young women.

Puberty: Physical Growth

For many teenagers, growth during puberty is slow and regular. About 50% of all teenagers, however, have a growth spurt in which they grow about 15 cm (6 in.) in a year, and a few may even grow 20–25 cm (8–10 in.) in a year. Girls normally experience a prepubertal growth spurt and grow rapidly when they are about eleven years old. Boys usually have a pubertal growth spurt about age fourteen.

The brain does not always immediately adjust to the body's new height and strength during a growth spurt. The person is often clumsy when he wants most to be coordinated. Spilling things and running into doors can be embarrassing, but it is a normal part of growing up.

Puberty: Hormonal and Emotional Imbalance

Puberty is also a time of hormonal imbalance. The body develops in response to the new hormone levels, but the hormone release is not steady. This unsteadiness may cause temporary emotional imbalance.

An exciting activity (such as a big game, a big date, a performance, or a trip to an amusement park) stimulates the body to release various hormones. Levels of epinephrine and other hormones are increased, and the teenager may experience a "hormonal high": he feels good, as if he could conquer the world. This hormonal high, how-

ever, is short-lived. Negative feedback mechanisms begin to operate, and soon hormone levels drop; the person may have a "hormonal low." He is discouraged, feels everybody hates him, and is ready to either pout or lose his temper at the slightest provocation.

Puberty is not only a time of physical and emotional change but also a time of self-identification. A young child is concerned about himself and cares little about other children. The teen, however, begins to be concerned about what others think of him, as well as what he thinks of himself.

The desire for acceptance by friends of the same age reflects itself in many ways. For example, teens become concerned about their clothes. They want to dress to reflect themselves and may balk at school, church, or home dress codes. Actually, most reflect the styles of those whom they want to be accepted by. Dressing to imitate rock stars, athletes, or movie stars is not "being yourself" but is being like someone else to gain acceptance, something a teen may later regret.

Peer pressure (the compulsion of wanting to be accepted by the group) is strong during the teen years. Some teens will do things that they know are wrong just because their friends do them. For example, some teenagers steal, not because they need the items or money, but because their friends are stealing and they find it exciting. The excitement is in part a hormonal high. Later, when the depression comes, they feel it is time to steal again because it was "fun" (they were accepted by the group) and "exciting." The principle holds true in other areas, too. This is one reason that Christian teens need fellowship with Christian friends.

There are more subtle ways whereby the body's changes can lead a person to temptation. The body, especially during a time of change, greatly affects mental attitudes. Hormonal highs and lows, however, are not an excuse to sin. For example, suppose Bob's mother asks him to take out the trash, but Bob has had a rotten day (partly caused by a hormonal low) and he does not "feel like it." He tells her that if she wants the trash taken out, she can take it herself. She then scolds him for disrespect and disobedience. Bob feels like screaming, sock-

ing someone in the jaw, or running out of the house and never coming back. But he takes out the trash, goes to his room, and plops down, all the while thinking how bad his life is and what it is going to be like when he can "get out of here."

Some people would say that Bob's thoughts and actions were justified because he was having a hormonal low, a natural stage during puberty. However, Paul says, "I keep under my body, and bring it into subjection" (1 Cor. 9:27). Bob is tempted to the sin of rebellion (1 Sam. 15:23) and is using his body to get him to sin. Sin is sin, no matter how a Christian feels. Part of a Christian's maturing process is learning to submit himself to Christ's will, no matter how he may feel (1 Cor. 10:13; 2 Cor. 10:4–5).

Maturity

In some cultures, a teenage boy performs a task (such as hunting and killing a certain animal or traveling a distance alone), after which he is considered a man. Literature often portrays a teen going through a difficult experience; at the end he is "grown up."

Maturity, however, is not something that happens when a person reaches a certain height or age or accomplishes some feat. Circumstances work together to cause a person to mature, but maturity is not achieved all at once. In fact, some people never achieve it.

Maturity is a difficult concept to define. Some of the major characteristics are as follows:

✦ *The mature person can control himself.* He thinks things through, makes wise decisions, and then disciplines himself to do what he knows to be best, no matter what his whim or another person's wishes might be. A mature person has control of his anger, love, hate, and other emotions. He does not "fly off the handle" but tempers his actions with good judgment.

✦ *The mature person recognizes true value.* Mature people make good value judgments. For example, they waste little time or money on unprofitable activities. They do not buy what they

puberty: (L. *puber*—adult)

do not need or cannot afford. They give their time to valuable activities that are not necessarily "fun," such as doing homework, practicing the piano, or visiting the sick.

Mature Christians do not sacrifice prayer and Bible reading even for a good television program, nor do they skip church so that they can have a full day at the beach. Mature, spiritual Christians realize that thrills like those obtained by hormonal highs or purely physical excitement (such as watching a game or enjoying a ride at an amusement park) have their place. However, they are second-rate compared to the thrill of leading a person to Christ or spending time alone with the Lord.

✦ *The mature person works for deferred goals.* A mature person realizes that there is value in preparation for the future. Mature students study not just for grades but to prepare for their future goals. Mature people practice (sports, piano, or sewing, for instance) not only to have fun but also to acquire a valuable skill to use in the Lord's service.

Mature Christians realize that giving money to missionaries is not just performing a duty but is helping to advance their Lord's work. They are laying up treasures in heaven (Matt. 6:20). Witnessing is a normal part of the mature Christian's life, for he recognizes his long-range goal.

Acting Mature Versus Being Mature

Although it is easy to find people who act in mature ways, it is not always easy to find a person who is mature. All too often incentives cause him to do the right thing. One who responds only to a "big prize" (a pizza party for the class that brings the most food to the school's annual food drive, a gift certificate to the student who says the most verses in Bible club, and so forth) but does not seek "the prize of the high calling of God" by giving the same effort when there is no contest is not really mature. Usually, when a mature person is convinced of the value of an action, he is self-motivated in ways that are in keeping with Scripture.

How does a person become mature? The physical body grows and matures as a person exercises and eats proper food. A baby's share of food and exercise would be of little profit to an adult. A person matures emotionally as he stops depending on his parents to do everything for him and begins to assume responsibility wisely. If his parents still have to tell him to do the things he knows he should do, he lacks maturity.

Likewise, if the only spiritual food a Christian receives is supplied in Sunday school and church and the only spiritual exercise he engages in is an occasional prayer, he remains a babe in Christ. Hebrews 5:12–14 discusses Christian maturity. There are Christians who, because of the length of time they have known Christ, ought to be living examples and teachers of the Christian faith. But instead, they need to be taught the principles of Christianity again. They keep needing the "milk" of the Word rather than "strong meat," which "belongeth to them that are of full age, even those who by reason of use have their senses exercised to discern both good and evil" (Heb. 5:14).

Here is the key to Christian growth: "by reason of use." Only when a Christian begins to assimilate the Word of God and exercise his faith in prayer, witnessing, and a daily walk in the path of righteousness can he become a mature Christian.

The Gonads and Sex Hormones

From birth the **gonads** (the reproductive organs) produce small amounts of the sex hormones. In males the testes produce hormones called **androgens** (AN druh junz), the most familiar of which is **testosterone** (tes TAHS tuh RONE). In females the ovaries produce hormones called **estrogens** and another hormone called *progesterone* (proh JES tuh RONE).

Gonadotropic hormones (gonadotropins) secreted by the anterior pituitary act as a feedback mechanism, controlling the hormonal secretion of the gonads. During childhood, little or no gonadotropic hormones are produced, and the small amounts of sex hormones secreted by the gonads barely affect the body.

In the early teens, however, there are great increases in the secretion of gonadotropic hormones. Additional gonadotropins stimulate the testes to begin to produce sperm and the ovaries to continue the maturation of ova. The increase also causes comparatively large quantities of sex hormones to be produced and released.

Sex hormones cause the body to mature and activate genes that cause the development of *secondary sex characteristics*. Most secondary sex characteristics are sex-limited characteristics; that is, the genes for these characteristics are present in both sexes but are not expressed until the gonads produce adequate amounts of the necessary sex hormones. Tumors or hormone treatments for other disorders (such as cancer) can cause a hormonal imbalance, causing the opposite secondary sex characteristic to develop.

	Men	**Women**
Body hair	Hair on the scalp becomes coarser. The beard begins to grow. Hair grows on chest, arms, legs, under the arms, and in the pubic region.	Hair on the scalp becomes coarser. Hair on the legs, under the arms, and in the pubic area begins to grow.
Body fat	The child's subcutaneous layer of fat diminishes, resulting in a more angular, trim appearance.	The child's layer of subcutaneous fat is retained, resulting in a smooth appearance.
Muscular growth	Stimulated muscular growth causes broader shoulders, thicker arms and legs, and a deeper chest.	The muscles are stimulated to grow but to a lesser extent than in the male.
Bone growth	The person grows to his adult height.	The person grows to her adult height. The pelvis spreads and rotates to supply a wider base for supporting a child during pregnancy.
Voice	The vocal folds and other speech mechanisms become larger, causing the voice to become deeper. As this happens, difficulty in controlling the "new voice" often results in vocal "cracking."	The voice deepens slightly, but this change usually occurs over a long period. Young ladies, therefore, usually do not have the problem of their voices "cracking."
Skin	The skin is stimulated to become thicker and tougher. During the time of adjustment, the oil glands often overproduce, causing acne or other skin problems.	Changes are the same as in the male, except the retained layer of subcutaneous body fat gives the skin a smoother, somewhat translucent appearance.
Reproductive glands	The reproductive glands enlarge as they prepare for reproduction and producing functional gametes.	As they prepare for reproduction, the gonads and accessory sex organs enlarge. Menstruation begins and the breasts (mammary glands) develop.

➡ *Review Questions 24.2*

1. Describe the pituitary gland. What hormones does it produce, and what do they control?

2. Explain how the hypothalamus controls the pituitary gland.

3. Give examples of hyposecretion and hypersecretion of the pituitary growth hormone in (a) children and (b) adults.

4. Describe the location of the thyroid gland. What do its hormones control? What symptoms characterize hyposecretion and hypersecretion of thyroxine in (a) young children and (b) adults?

5. What do the parathyroid glands control?

6. Why would hyposecretion of insulin cause an increase in the amount of sugar in urine? What is the name of this disorder? What is the usual treatment of severe cases?

7. Explain the relationship between insulin and glucagon.

8. Describe the adrenal glands. What hormones do they produce, and what do they control?

9. List the secondary sex characteristics for (a) males and (b) females.

⊙10. List several characteristics of (a) physical maturity, (b) emotional maturity, and (c) spiritual maturity.

⊙⊙11. What are typical temptations during puberty? Does the fact that these are normal temptations justify yielding to them? Give scriptural support for your answer.

⊙⊙12. How does a person mature (a) emotionally and (b) spiritually?

24B—Drugs

The expression "give your heart to the Lord" is a familiar one. This phrase does not refer to someone's giving God the muscular pump in his chest; it refers to giving Him the very essence of his being, his will. The person who gives his will to the Lord has decided to do what God wants, not what he wants.

The Bible says that when a person does his own will rather than the Lord's will, he allows sin to control him (Rom. 6:16). When a Christian daily denies himself, refusing to allow sin to control him, and chooses instead to do the will of God, he becomes stronger and grows in the Lord. This daily dying to self is impossible without the strength God gives to do His will (Ps. 73:26;

Phil. 2:13). Salvation is the miracle of a moment, but living the Christian life obedient to the will of God is the work of a lifetime.

Although victory is assured for the Christian, there is a powerful adversary who "as a roaring lion, walketh about, seeking whom he may devour" (1 Pet. 5:8). One of Satan's most powerful weapons is tempting the Christian to take the control of his life back from God (James 1:14–15). Satan has won a victory if he can make a Christian believe that God's will is not good and can persuade him to violate it. Though God permits His children to be tempted, He does not permit them to be tempted beyond their ability to resist by His grace (1 Cor. 10:13). Those who claim God's power in times of temptation will know His victory.

24.3 Drugs and Addiction

The Greek word *pharmakon*, which can mean either a healing drink or a deadly drink, has supplied the English language with many words. *Pharmacy* (a place to purchase drugs) and *pharmacology* (the study of drugs and their effects) are both derived from *pharmakon*. So are the words *poison* and *potion*.

It is true that "all drugs are poisons, and all poisons are drugs." A **drug** can be defined as any chemical that causes an alteration in the function or structure of a living tissue. Any poison will also fit this definition. Even medicinal drugs, such as penicillin and aspirin, can be considered poisons in the sense that they can kill if taken in sufficient doses. This does not mean that medicine is bad; by this definition even water is a drug. Pure water, consumed in large amounts, can cause the body to reduce its salt content. Children, especially infants, who have rapidly consumed large quantities of water have died of "water poisoning."

Medicinal Drugs

A *medicinal drug* is one taken to prevent, treat, or cure a disease or disorder. Today, thousands of medicinal drugs are available. Some are legally obtained only by a prescription written by a physician, but some may be purchased "over the counter" without a prescription.

Miraculous Healing

The Bible teaches that God controls sickness and health and works His will through both. Sometimes He accomplishes His will through healing and other times through death. He sometimes accomplishes His will through healings that take a long time and other times through miraculous healings.

Scripture says that Christ healed people instantly and that apostles and prophets were instruments of miraculous healings. Several observations from Scripture about miraculous healing are possible:

- The faith of the person being healed is important, but it is not always necessary (Acts 3:1–11).
- The ability to heal is not a sign of salvation or of spiritual power (Matt. 7:21–23).
- Christians are told to pray for the sick (James 5:15), asking for God's will to be done.
- God sometimes chooses not to heal, but He always promises grace (2 Cor. 12:7–10).

Scripture does not condemn the use of medicines. If anything, it supports their use (James 5:14, olive oil was a principal medicine in New Testament times; see also Luke 10:34). Drugs have their place and purpose. If God wants to heal miraculously, He will. The will of God, however, often includes man's using his intelligence. Most medicinal drugs are tools that God uses to effect His will.

Christ at the Pool of Bethesda, Jean Restout, From the Bob Jones University Collection

Alcohol affects many body tissues in many undesirable ways. Although the body appears to be able to tolerate small amounts of alcohol, those who repeatedly consume even small amounts of alcohol may experience major physical problems. Alcohol's most immediate and pronounced effect is on the brain. Those who consume alcohol want their brains to be affected; they are willing to tolerate the physical effects of alcohol on other body tissues to gain the mental effects.

Alcohol acts as a *depressant*—it slows down brain activity. The degree of this effect depends on the dosage. In small quantities, alcohol produces a feeling of euphoria. It decreases tension by causing a disorganization of mental processes. A person who has had a few drinks cannot concentrate on his prob-

Alcohol is the leading cause of automobile-related accidents and deaths. Motorists suspected of drunk driving may be given breath tests to determine whether their alcohol level is above the legal limit.

lems. One reason that people drink at social gatherings is to become "loosened up" by decreasing their tension and becoming euphoric.

Increasing the dosage causes a loss of muscular control. The slowed mental activity results in a brain that is no longer able to control the body. The person staggers, cannot focus his eyes, and becomes sleepy. It is easy to see why alcohol is responsible for more traffic fatalities than any other cause. After only one or two drinks, the person feels he is capable of driving well and can usually pass police sobriety tests. Even so, his reaction time is slower and his judgment is poorer than when he is sober.

Larger doses of alcohol cause more disorganization: the person loses his ability to reason, and his mood changes from euphoria to irritation and anger. Even larger doses cause complete loss of bodily control, and sleep comes. If the dosage is large enough, the drinker will enter a coma. Alcohol consumption can also contribute to liver and brain damage, as well as heart disease and certain types of cancer.

If alcohol is consumed during pregnancy, it may affect the unborn child's development. *Fetal alcohol syndrome* (FAS) is a group of birth defects that are a result of a pregnant woman's drinking alcohol. Since there is no established safe level of alcohol during pregnancy, the damage may be caused before the mother ever realizes she is pregnant. Symptoms of FAS include low birth weight, malformed head,

Some of the characteristics of fetal alcohol syndrome may include low birth weight, small head circumference, flattened mid-face, wide-set eyes, and a thin upper lip. These characteristics may not be apparent at birth.

mental retardation, slow growth, and decreased motor coordination.

Alcoholism is the dependence upon the drug alcohol. A person who is dependent on alcohol is called an alcoholic. An alcoholic could be someone who must have just one drink to calm down after school or work or someone who starts drinking and does not stop until his body forces him to. Some researchers claim that alcoholism is a disease and, as such, alcoholics cannot help themselves—they have to drink. There is no conclusive scientific data to support this position. A person must make a conscious decision to take the first drink and willfully decide to keep drinking to be an alcoholic.

It is easy to understand why an unsaved person in this high-pressure

Medicinal drugs can be grouped according to their sources, the compounds they contain, and the body parts they affect. The following classifications are designed to help explain what medicinal drugs are. This simple grouping, however, does not include all the drugs in medical use today.

✦ *Dietary supplements.* If the diet is insufficient in a nutrient (such as a vitamin or mineral), that nutrient must be added to the diet as a supplement, most often in the form of a drug. Some people, because of injury, disease, or their metabolism, may require more of certain nutrients than their diet supplies and must take them as dietary supplements.

✦ *Supplements of body chemicals.* Some people's bodies cannot make necessary chemicals. Hemophiliacs, for example, cannot make certain blood chemicals that are needed for normal clotting. Therefore, the missing

Physical Effects of Alcohol

- Dehydrates and numbs membranes of the mouth and throat, causing thirst.
- May inflame and irritate the esophagus, stomach, and duodenum. This inflammation may extend to rawness, ulceration, hemorrhage (bleeding), and perforation (holes in the walls) of these organs.
- Inactivates the centers of judgment in the brain. Progressive anesthesia of the whole nervous system ensues, causing sleep and, depending on the amount consumed, coma. Large doses lead to depression of the respiratory and circulatory centers of the brain, causing death. Long-term alcohol abuse leads to brain atrophy.

- Inactivates the antidiuretic hormone (ADH) and causes excretion of excessive amounts of water.
- Lowers blood pressure by dilating (enlarging) blood vessels near the skin. This causes body heat to be brought to the surface. The immediate effect is a feeling of warmth. This effect can be harmful in cold climates because a person may feel warm even though he really isn't, and he may not recognize signs of hypothermia.
- Increases the workload of the heart.
- Decreases surface tension in the alveoli of the lungs, interfering with lung function.
- Poisons the sensitive cells of the liver that metabolize alcohol in order to eliminate it from the body. This elimination process is only temporary. Large or continual amounts of alcohol lead to fibrosis, fatty degeneration, and distortion of the liver—a condition known as alcoholic cirrhosis. Eventually, cirrhosis of the liver causes death. Food and vitamins help protect the liver (and other body parts as well) against some of alcohol's toxic effects.
- Causes physical and psychological addiction.

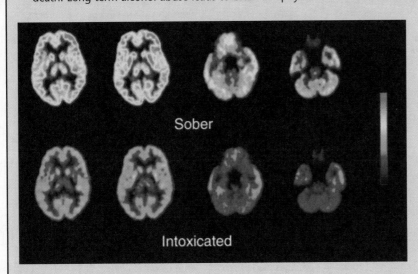

Sober

Intoxicated

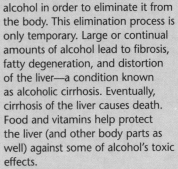

As these brain scans show, drinking alcohol significantly reduces brain metabolism.

---

society would turn to something that would permit him to forget, relax, and sleep. Friends and alcohol advertisements tell him that "this is the way to go." Many people know of no better way to escape their problems than to drown them in alcohol. Unfortunately, drowning problems does not remove them; it just preserves them until later.

At the same time, the alcohol creates many additional problems not only for the person but also for his family, friends, and co-workers—and sometimes even strangers.

Knowing these things, Solomon wrote in Proverbs that strong drink is deceitful (20:1). Scripture repeatedly condemns drunkenness (Luke 21:34;

Rom. 13:13; Gal. 5:19–21). Besides the fact that a Christian should not abuse his body, he is told to bring his burdens to the Lord, for He will either take them away or give the grace and strength to bear them. A Christian disobeys God if he depends on strong drink to relieve or solve his troubles.

---

substances must be injected into their blood for them to live a nearly normal life. Additionally, many diabetics are unable to make sufficient insulin and must receive it as a drug.

✦ *Psychoactive drugs.* A **psychoactive** (SY koh AK tiv) **drug** is one that alters the emotional state, the sense of reality, or both. Many of the more commonly used psychoactive drugs are listed in Table 24-2 (pp. 600–603). The psychoactive drugs include the following main groups.

✧ **Narcotics** act on the central nervous system to produce a feeling of *euphoria* (yoo FOHR ee uh), an emotional state of well-being and happiness. They are *physically addictive*.

✧ **Stimulants** increase the action of the central nervous system and often step up body metabolism. The person feels alert and has increased

psychoactive: psycho- (Gk. *psukho*—life) + -active (L. *agere*—to drive)

narcotic: (Gk. *narkotikos*—numbing)

euphoria: (Gk. *euphoros*—healthy)

stimulant: (L. *stimulare*—to excite)

24-2 Some Psychoactive Drugs and Their Effects

Class	Name	Common street names	Effective duration	Source	Short-term effects
Narcotics	morphine	M, morph, Emma, Number 13	6 hr	derived from opium (the dried sap of the opium poppy)	causes lack of feeling (painkiller) and euphoria followed by sleeplessness and anxiety
	heroin	H, brown sugar, mud, smack, junk	4 hr	synthetically derived from morphine	causes euphoria followed by drowsiness and anxiety
	cocaine†	C, coke, lady, rock, candy, crack, snow	varies	derived from leaves of the coca plant (not the cola nut used in beverages)	external application—deadens skin sensations internal application—stimulates nervous system, causing euphoria, excitement, talkativeness, and shakes
	methadone		24 hr	synthetically produced	no mental or emotional effects; same physical effects as opiates (therefore prevents withdrawal symptoms)
	codeine		4 hr	derived from the opium poppy	same as morphine but milder
Stimulants	amphetamines	meth, speed, ice, crystal, chalk, crank, uppers, stove top	4 hr	synthetically produced	increases heart rate and blood pressure; decreases appetite
	caffeine		4 hr	derived from many plants (and found in beverages made from them: cocoa, coffee, cola, tea) and synthetically produced	causes slight increases in heart rate; sense of wakefulness
Depressants	barbiturates	barbs, blues, downers, bluebirds	varies	synthetically produced	causes quiet stages, followed by sleep; impairs memory, physical coordination, and reflexes; curbs anxiety
	tranquilizers		varies	synthetically produced	dampens feelings a person has about life's situations; relieves anxiety; affects ability to concentrate on a situation
	alcohol		depends on dose	derived from fermentation	dampens feelings; increases blood pressure; produces mild euphoria, followed by sleep (Large doses can cause coma and death.)
	GHB (gamma hydroxybutyric acid)	G, Georgia home boy, grievous bodily harm, liquid ecstasy, goop	up to 4 hr	synthesized from other chemicals	has euphoric and sedative effects; can cause coma and seizures; when mixed with other drugs, such as alcohol, can produce breathing difficulties
	rohypnol	rofies, roach, rope, R2	8–12 hr	synthetically produced	causes drowsiness, lowered blood pressure, dizziness, confusion, and gastrointestinal problems; can incapacitate victims and produce amnesia; has been used as a date rape drug; when mixed with alcohol or other depressants, can be lethal

†Concentrated forms, such as crack, produce intensified effects more rapidly.

Long-term effects	Medicinal uses	Addiction characteristics
same as short-term	used as a powerful painkiller	As short-term effects wear off, physical withdrawal begins (cramps, internal pain, chills, shakes, vomiting, diarrhea). Severity depends on dosage and length of time the drug has been taken. Rarely fatal. Body does build up tolerance; larger doses must be taken to prevent withdrawal and maintain the euphoric state. Also psychologically addictive.
loss of appetite, constipation	none	
external—none known internal—after early symptoms, may cause depression and body convulsions (Large doses may cause other effects.)	sometimes used as a local anesthetic on mucous membranes (nose and throat)	
same as short-term	taken to prevent withdrawal symptoms from narcotics	
same as short-term	used as an easily administered mild painkiller; often found in cough syrups	Physical addiction with withdrawal symptoms is questionable. Body does build up tolerance.
False sense of energy causes people not to take care of themselves. Large doses over periods of time cause headaches, insomnia, confusion, panic, circulatory collapse, and nausea. Long-term use also leads to damage to the teeth ("meth mouth").	often prescribed as "pep pills" for those who lack energy or to counteract depression caused by other drugs; used to treat some mental disorders; for appetite control	Both physical and psychological addiction. Body does build up tolerance; increasingly larger doses must be taken to produce effects.
Doses of the equivalent of ten cups of coffee result in insomnia and the jitters in some people; larger doses may cause various other effects.	used as a stimulant in conjunction with other drugs; used in treatment of migraines	
Relaxing of the heart and breathing muscles can cause death; some people experience wakefulness after using.	sedatives; inducing sleep or hypnotic trance	Physical withdrawal begins 24 hr after last dose and lasts up to 7 days; weakness, visual distortion, sleeplessness, tremors, low blood pressure, delirium, and convulsions are common.
varies, often loss of interest in life and responsibilities	used to prevent some symptoms of mental illness; used to help people "cope with problems"	Usually no physical withdrawal; however, a person may develop a psychological addiction.
loss of motor and mental control; deterioration of brain, liver, and other organs	solvent for other drugs	Physical withdrawal from prolonged larger doses involves cramps, shakes, hallucinations, insomnia, and indigestion—a condition known as DTs (delirium tremens).
difficulty thinking, slurred speech, headaches, amnesia, coma, and death	none today (Until 1992, it was available over the counter to aid in fat reduction and muscle building.)	Withdrawal symptoms include insomnia, anxiety, tremors, and sweating. Has been used as a date rape drug.
memory loss, impaired motor movements, difficulty speaking	none in U.S. (in some countries, prescribed for insomnia)	Physical and psychological dependence is common.

Class	Name	Common street names	Effective duration	Source	Short-term effects
Hallucinogens	LSD (lysergic acid diethylamide)	acid, hits, microdot, trips, sugar cubes, window pane	10 hr	synthetically produced	senses amplified; judgment affected; hallucinations; person feels "out of himself" (experiencing himself and other phenomena in a distorted way); extreme emotions; may be a pleasing or terrifying experience, depending on mood and environment when taking the drug
	DMT (N,N-dimethyltryptamine)	businessman's trip	45 min	synthetically produced	
	DOM (2,5-dimethoxy-4-methylamphetamine)	STP	72 hr	synthetically produced	
	mescaline	magic mushrooms, shrooms	12 hr	derived from a cactus	senses amplified; hallucinations; pupils dilated; extreme emotions
	psilocybin	magic mushrooms	6–8 hr	derived from a mushroom	
	ecstasy (MDMA)‡	XTC, E, X	depends on dose	synthetically produced; related structurally to amphetamines	lowers inhibitions; affects judgment; bizarre behavior; hallucinations; may increase temperature and blood pressure, leading to death
	marijuana (cannabis)	grass, weed, pot, Mary Jane, MJ, dope, kush, hash	4 hr	derived from species of *Cannabis* leaves, shoots, etc.	senses amplified; slows speech; affects judgment; may make anxious or relaxed (Hallucinations are rare.)
	ketamine	special K, vitamin K, jet, super acid, cat, green	depends on dose; usually 4–6 hr	synthetically produced	in some cases, produces a dreamlike state and euphoria; in other cases, produces a terrifying feeling of sensory detachment ("K-hole"); can cause amnesia, delirium, impaired motor function, depression, and respiratory problems in high doses
Inhalants	volatile solvents, aerosols, gases, and nitrites	poppers, snappers, air blast, huff	rapid onset	natural and synthetic substances	causes euphoria, muscle incoordination, and dizziness that may last several minutes

‡3,4-methylenedioxymethamphetamine

physical energy. Usually these drugs mask the body's sense of fatigue, rather than removing it. The stimulants are not usually physically addictive, but some people rely on them for the "boost" that they think they need to face life.

✧ **Depressants** slow down the central nervous system, usually removing the ability to focus attention on problems. Often memory is impaired, and the person may go to sleep. Some depressants are physically addictive.

✧ **Hallucinogens** (huh LOO suh nuh junz) usually amplify a person's senses and distort his judgment. Some of the stronger drugs make him feel things and experience smells, tastes, and visions that are not really there (hallucinations). He is not in control of himself and cannot exercise his will. These drugs are not usually physically addictive.

✧ **Inhalants** are substances that form chemical vapors that can be inhaled. The inhaled vapors move rapidly from the lung alveoli to the bloodstream, producing psychoactive results such as euphoria and hallucinations.

hallucinogen: hallucino- (L. *alucinari*— to dream or be deceived) + -gen (born)

Long-term effects	Medicinal uses	Addiction characteristics
Can cause psychosis, panic; chromosomal damage (unconfirmed); time distortion. Flashbacks to experiences while taking the drug may happen weeks to years later.	not used today (historically was used to treat mentally ill)	No physical withdrawal; however, a person may develop a psychological addiction, feeling the dream world is better and more "real" than the real world. Body does build up a tolerance and may require increasingly larger doses to produce similar effects.
cramps, vomiting, increased heart rate, anxiety or euphoria; time distortion	none	
depression, sleep disorders, liver damage	none	Psychological addiction is common. Can cause paranoia; adversely affects thought and memory.
Can cause abnormal emotional behavior; various physical problems have been linked to sustained use.	Studies have been done on its use as a glaucoma treatment, nausea preventive, painkiller, and other uses. However, the wisdom of the medicinal use of marijuana is debated.	Psychological addiction is common. Body builds up tolerance and may require larger doses for similar effects.
Long-term cognitive and memory problems can result.	anesthetic for human and animal use	
same as short-term; additionally may result in antisocial behavior and impaired judgment	Some gases are used as general anesthetics; nitrites are used in many heart medicines.	Prolonged use can cause unconsciousness or death. Permanent brain damage can occur. Some studies show decreased immune response with nitrite use.

✦ *Agents to treat infections.* Some drugs either kill pathogens directly or help the body in its war against them. Antibiotics are drugs that help destroy bacteria and some other pathogens. Other chemotherapy treatments deal with pathogenic bacteria, fungi, protozoans, and worms. Some topical (surface) medications contain drugs to prevent infections.

✦ *Agents to induce immunity.* Physicians give vaccines or similar drugs to stimulate T cells or to cause the body to produce antibodies and thus develop immunity.

✦ *Substances that decrease a body function or relax muscles.* During an asthmatic attack, people are often given drugs that relax the muscles of the trachea, bronchial tubes, and bronchioles to permit breathing. Physicians use drugs to relax the muscles of the blood vessels to lower blood pressure. Some drugs slow the heart rate to restore a regular heartbeat. Others control the amounts of chemicals the body produces. Some young people take drugs to control the quantity of oil their skin produces to alleviate acne.

Good Drugs/Bad Drugs

No drug can be called truly good because all drugs can be misused and, thus, could be called bad. Some drugs can be considered good because they are useful and their wise use is acceptable. In the same sense, some drugs can be called bad because their use is neither wise nor acceptable. Here is a comparison between aspirin and alcohol.

- Aspirin is taken to relieve physical pain. Alcohol is a poor pain reliever except when taken in large doses.
- Aspirin does not significantly impair the mental or emotional processes. Alcohol produces multiple mental and emotional side effects.
- A dose of aspirin in sufficient quantity to relieve most minor pains produces only minor physical side effects in most people. Even a small dose of alcohol produces significant physical effects in most people.
- Aspirin has significant, potentially harmful side effects in some people (those who are allergic to it or consume too much of it over a long period of time, for instance). Alcohol has significant, harmful side effects in all people.
- Few circumstances besides great physical pain would cause a person to take large quantities of aspirin. Alcohol's depressant qualities and temporary euphoria-producing effect cause many to consume it in large quantities.

It cannot be said absolutely that aspirin is a good drug. Although it has good uses when taken in proper doses for the proper conditions, has few side effects, and does not affect the will, aspirin does have its drawbacks. Alcohol, on the other hand, is easily classified as a bad drug. Its harmful physical side effects, altering of emotions, and impairment of self-control are dangerous and undesirable.

Aspirin

In the early nineteenth century, a bitter fluid said to be effective in reducing fevers was produced from the bark of the willow tree. By 1852, salicylic (SAL ih SIL ik) acid was isolated, and methods of preparing it in large quantities became available. Doctors used it to relieve the pain of rheumatism and found it effective on types of gout (caused by excess uric acid in the body) because the drug increased the amount of uric acid in the urine.

Salicylic acid was given with great care, however, because it irritated membranes in the mouth, throat, and stomach. Pharmacists found that adding sodium to the acid made sodium salicylate (suh LIS uh LATE). The new drug was effective and not so irritating, but it tasted so bad that some patients could not take it.

In 1897, Felix Hoffman formulated acetylsalicylic (uh SEET ul SAL ih SIL ik) acid, which was much more effective and could be produced inexpensively. It reduced fever, pain, and swelling of the joints and increased the excretion of uric acid. In small doses it is relatively safe and can be purchased over the counter. Acetylsalicylic acid is the chemical name for aspirin.

Most people consider aspirin a relatively safe pain reliever. There are, however, more than two dozen possible dangerous side effects if aspirin is taken in large doses or for prolonged periods.

✦ *Substances that increase a body function or stimulate muscles.* Sometimes a heart attack patient needs drugs to stimulate his heart. Some people are able to make hormones but need drugs to stimulate their bodies to do so. Some drugs are used to stimulate the kidneys to remove extra quantities of a substance from the blood. Drugs are used to constrict the muscles of the blood vessels, increasing blood pressure.

✦ *Pain relievers.* The most common over-the-counter drug purchase is pain relievers. Several effective pain relievers (such as acetaminophen and ibuprofen) can be used to relieve minor pains. Several stronger pain relievers are available as prescription drugs when the pain is extreme. Minor pain relievers have few harmful side effects on most people. Some pain relievers, however, are psychoactive drugs, and some are addictive.

24.4 Addition

Psychoactive Drugs and Psychological Addiction

Although some psychoactive drugs cause physical withdrawal and addiction, most of them are psychologically addictive. In **psychological addiction** the person does not experience actual physical symptoms of withdrawal but may go through a period of emotional withdrawal. After a time, a person who has taken stimulants to be alert, tranquilizers to remain calm, and sedatives to go to sleep usually becomes dependent on the drugs to help him function "normally." If he is deprived of the drug, he thinks he cannot function normally and therefore cannot function normally without the drug. This thinking that he needs the drug is psychological addiction.

Thousands of people in the United States are psychologically addicted to tranquilizers, sleeping pills, or stimulants. These people are usually not criminals obtaining their drugs on the black market, although there is an active black market for these drugs. Many psychoactive drugs are available in over-the-counter forms. Thousands are prescribed by physicians to help their patients sleep, calm down, wake up, or "cope" with problems. Some of these prescriptions are made merely to satisfy the patient rather than for legitimate medical reasons.

Psychoactive Drugs and the Christian

Some psychoactive drugs have legitimate medical uses. Some of them are strong painkillers; others increase or lower blood pressure; some are the only known treatment for various physical conditions. The vast majority of the psychoactive drugs, however, are taken to alter the person's emotional state and sense of reality, to give him the euphoria, sleepiness, alertness, escape, or other "feeling" he thinks he needs.

Are drugs that are used to alter a person's emotional state acceptable for the Christian? There are some problems caused by injury or chemical imbalances that may be treated with psychoactive drugs. The Christian can accept these uses

24.4

Objectives
- Describe physical and psychological symptoms of drug addiction
- Discuss a Christian response to psychoactive drugs

Key Terms
psychological addiction
tolerance
physical withdrawal
addiction

24-3 Possible Symptoms of Drug Addiction

Psychological symptoms	Physical symptoms
seeing drugs as the solution to problems	changes in sleeping habits
spending a lot of time figuring out how to get drugs	changes in physical health caused by drugs
stealing money or selling belongings to be able to afford drugs	feeling shaky or ill when drugs are stopped
withdrawing from relationships with friends or family	a need to take more of the drug to get the same effect
losing interest in school, sports, or hobbies that used to be important; letting grades slip	changes in eating habits
experiencing anxiety or depression	weight loss or gain for no apparent reason
keeping secrets from friends or family	
hanging out with people who use drugs	
inability to stop using drugs	
inability to control moods	
having friends or family members express concern about mood swings	

of psychoactive drugs; but the Christian cannot support the recreational use of psychoactive drugs. God has placed the Christian in life situations for a reason, and if the Christian takes a drug for a non-medical reason to escape from the harsh reality of life, it is a sin (Eph. 5:18; Titus 2:12). Misusing drugs is also a way of saying that Christ cannot provide the strength to deal with the difficulties of life (Matt. 6:25–34). Recreational use of psychoactive drugs not only is a sinful escape from reality, but it can also lead to harmful addiction and even death.

Facets of Biology

Narcotics

A narcotic is a drug that dulls the senses, induces sleep, and becomes addictive with prolonged use. Most narcotics also soothe the mind, producing a chemically induced feeling of euphoria. Some narcotics are used as medicinal drugs to relieve pain. People who use narcotics for nonmedicinal reasons do so because the euphoria obscures their problems.

The body builds up a **tolerance** to narcotics; that is, the next time the drug is consumed, the size of the dose must be increased to obtain the same euphoria. No matter what the dosage, the euphoria lasts only a few hours. When the person no longer takes the drug, he enters **physical withdrawal**, experiencing actual physical symptoms due to lack of the drug. If he has taken a single small dose, the period of withdrawal may take place while he is asleep and may not even wake him. Estimated thousands of "weekenders" take narcotics occasionally on weekends and "sleep it off" before going to work on Monday.

If a person has taken narcotics over a longer period of time and has built up a tolerance to large doses, the physical withdrawal may be severe. Cramps, leg jerking (from which comes the expression "kicking the habit"), shakes, chills ("cold turkey"), vomiting, diarrhea, irregular sleep, insomnia, and other painful side effects are common symptoms of narcotics withdrawal. These symptoms stop once the body has become detoxified (no longer has the poison in it). The time needed for detoxification varies according to the type and amount of narcotics taken and the duration of use.

Although everyone agrees that narcotics are addictive drugs, few authorities agree on what **addiction** actually is. In addiction, the drug user becomes dependent on the drug and cannot comfortably function without it. Part of the addiction problem is that tolerance requires consuming ever-increasing amounts to maintain the desired effects and to prevent withdrawal.

There is more to addiction than just maintaining euphoria and preventing withdrawal. Even after detoxification, an unexplained craving for the drug remains. Many studies have shown that even after detoxification and years of "rehabilitation," over 95% of the people who have taken narcotics in large quantities for a period of time will return to using them. This return to drug abuse is scientifically unexplainable yet a very real part of addiction.

Some people feel that the high rate of return to the abuse of narcotics (as well as alcohol, barbiturates, nicotine, and other addictive drugs) is caused by some type of "biological need" created in the body by the drug itself. Other than the fact that most people return to taking drugs, there is little scientific evidence to support this view. Others think that the reason for returning to the drug is merely a desire for previous euphoric feelings (psychological reasons). Some feel that social reasons are the cause (when the person returns to his life among drug users, he returns to using the drug himself). No matter what the reason may be, the person consciously decides to take the drug.

The temptation to take a drug again is a powerful weapon in Satan's arsenal. Satan has convinced many people that these addictive drugs control them, while he is really the primary influence. Temptation comes from outside the will (although it may be inside the body), but it is the will that yields to temptation. Christians are promised strength from God so that they do not have to yield to temptation. Non-Christians do not have this strength and must rely on their own willpower to resist taking the drug again. Human willpower cannot match the strength of Satan's will, and it is easy to see why most unsaved people return to drugs.

detoxified: de- (reversal) + -toxi- (poison) + -fied (making)

addiction: (L. *addictus*—one awarded to another as a slave)

rehabilitation: re- (again) + -habilita- (L. *habilitas*—ability)

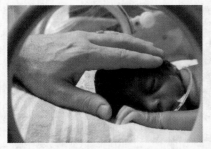 

Drugs can affect developing fetuses—when the baby is born, he may be addicted to the same drugs the mother took during pregnancy.

Narcotics and the Law

Before narcotics became widespread, alcohol was the more common drug. In 1920, the United States passed the Eighteenth Amendment to the Constitution and established Prohibition. During Prohibition the manufacture, transportation, and sale of alcoholic beverages was prohibited. A *black market* (illegal manufacturing, buying, and selling) for alcohol grew. Major criminal groups made fortunes in the black market for alcohol. The laws were difficult to enforce, and the criminal activities associated with the black market were rampant. In 1933, the Twenty-first Amendment to the Constitution repealed Prohibition.

In 1914, the legal supply of narcotics was stopped. A minor black market for narcotics quickly developed. Following Prohibition the narcotics black market started to grow. In the 1960s, the use of narcotics became rampant. Once a person becomes addicted, he is willing to pay high prices to support his habit. To obtain money, many become criminals because they are incapable of keeping most jobs while taking large doses of narcotics. By the 1980s, the narcotics black market had made fortunes for drug suppliers and had filled many streets with crime.

Today the narcotics black market affects the user in three major ways:

- The high price of narcotics leads many people to commit crimes to support their addiction.
- The purity of the narcotics is unknown. A person who purchases the drug on the black market has no idea what its strength is. Accidental lethal overdoses are not uncommon.
- Drug use is often unsanitary. Many narcotics users "mainline," injecting the drug into their veins. Using makeshift equipment and improper techniques, many drug users have contracted diseases such as hepatitis and AIDS.

Legislation against sin or man's weakness does not prevent him from sinning, but having such laws does set a certain moral standard and discourages some people from taking drugs who might otherwise do so.

Crack Cocaine

A concentrated form of cocaine called "crack" or "rock" cocaine is popular in the drug culture. Cocaine, which is usually sniffed, is a narcotic stimulant that slowly dissolves into the bloodstream and causes euphoria.

Crack, however, is smoked. This strong stimulant enters the bloodstream and reaches the brain in about ten seconds. Thus, the user gets an immediate reaction. The euphoria from crack lasts only five to twenty minutes.

Crack causes a person to be active, excited, and even aggressive. Since all of the drug taken enters the body at once, almost all of it passes from the body at one time. Withdrawal comes immediately after the euphoria, and because of the large amount of the drug involved, symptoms are severe. A person addicted to crack needs more of the drug within minutes. A powerful addiction can be built up within two weeks.

Because crack enters the blood quickly in a concentrated form, a lethal overdose is easy. Some people have died from a single smoke of crack, with over five times the lethal dose of cocaine in their bodies.

➡ Review Questions 24.3–24.4

1. Compare the terms *drug* and *medicinal drug*. List and give examples of the kinds of medicinal drugs.
2. List several points that should be considered about miraculous healings.
3. List the five groups of psychoactive drugs. List three drugs in each group.
4. Why can aspirin be considered a good drug? Why can it be considered a bad drug?
5. List several differences between psychological and physical addiction. List several similarities.
6. Describe some psychoactive drugs that have legitimate medical uses. Can these medical uses become unacceptable for a Christian? Explain your answer.
⊙ 7. What are some of the physical effects of alcohol?
⊙ 8. How does alcohol produce euphoria? Why does this euphoria fade as large quantities of alcohol are consumed?
⊙ 9. Why do many Christians consider it unacceptable to view alcoholism as a disease?
⊙ 10. What characteristics of narcotics make them so desirable that many people seek to obtain them on the black market?
⊙ 11. What are the three major results of narcotics being sold on the black market?
⊙ 12. What two main characteristics of narcotics contribute to their addictive qualities?
⊙ 13. List the characteristics of good drugs and bad drugs. What Christian principles can you cite to substantiate your position? How would penicillin and LSD fit into your classification of drugs? Why?
⊙ 14. "If a physician prescribes it, it's all right to take it." Why is this an unacceptable position for a Christian?

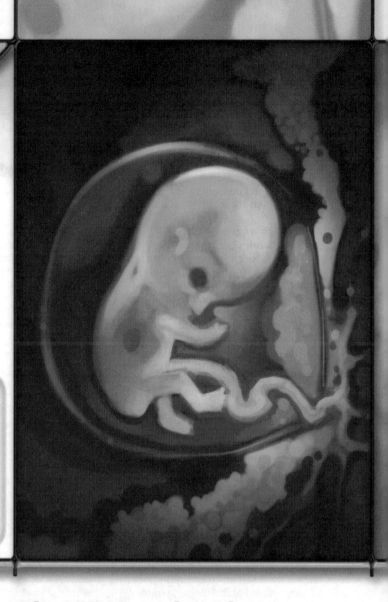

Human Reproduction

25A—The Reproductive System

It is natural for young people to be interested in the subject of human reproduction. As a child becomes a teenager, his interest changes from the curious "Where did I come from?" to an awareness that his body and mind are being prepared for possible marriage and parenthood. Such changes are part of God's plan. To ensure that mankind would continue, God created humans with strong feelings of love and desire and with a need for security and companionship (Gen. 2:18, 23–24).

These feelings motivate a man and a woman to unite in the fulfillment and joy—and even heartache—that are part of rearing a child. In part, God established marriage to fulfill these needs and accomplish this purpose. He ordained marriage and the intimate relationship between man and wife (Heb. 13:4) and commanded Adam and Eve to reproduce (Gen. 1:28). God created a beautiful example of the relationship between Christ and the church when He created the relationship between a man and his bride (Eph. 5:25).

As is often the case, humans took something good—sex—and perverted it. Humans make use of the normal drives and processes of the body for the evil purpose of rebelling against God. These sins have become increasingly accepted and widespread.

Christians must be careful as they stand against evil not to also condemn the good. Knowledge of human reproduction should not be suppressed simply because some people sinfully misuse reproductive processes. It is good to understand God's design and the basics of the human reproductive processes. An understanding of God's purposes for human reproduction enables a Christian to discern temptations to misuse his body. Being aware of the temptations set for God's children, the Christian can avoid the traps and be a good steward of the body God gave him.

Unlike any other body system, the functions of the *reproductive system* are not necessary for the survival of an individual. But its functions are vital for the continuation of the human race. Although many of the hormones and substances produced by the reproductive system make people "normal," life does not depend on them.

This chapter deals with the physical structures of the human reproductive system and their functions just as the other structures of the body were presented. Other related topics are examined using the Word of God as a guide.

25.1 Male Reproductive System

The organs of the male reproductive system produce sperm and transfer them into the wife's body. The primary organs of the male reproductive system are the **testes** (sing., testis). Normally a man has two testes that produce sperm and the male sex hormones continuously, though not always at a constant rate, from puberty to death. The other structures of the male reproductive system are accessory organs.

The Testes and Sperm

The testes begin their development in the abdominal cavity of the male embryo; however, about two months before birth they descend into the **scrotum** (SKROH tum). The scrotum is a skin pouch that develops from the abdominal wall. In the scrotum the testes are kept at the ideal temperature for sperm production—about 2 °C (3.6 °F) less than in the abdominal cavity.

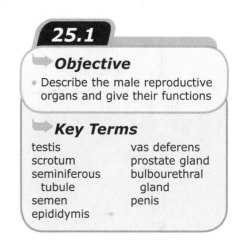

25.1

Objective
- Describe the male reproductive organs and give their functions

Key Terms

testis	vas deferens
scrotum	prostate gland
seminiferous	bulbourethral
tubule	gland
semen	penis
epididymis	

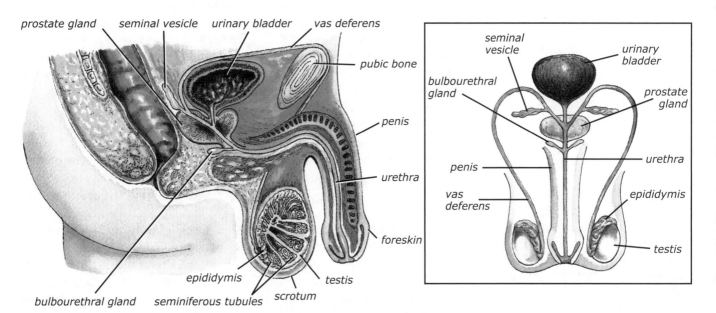

25-1

The male reproductive system, lateral view (left) and frontal view (right)

Each testis is shaped like a slightly flattened, 5 cm (2 in.) long egg and is covered by connective tissue. Most of the testis consists of tiny **seminiferous** (SEM uh NIF ur us) **tubules**, where the sperm are produced. Between the seminiferous tubules are clusters of cells that secrete the male sex hormones. Each day in the adult male, cells lining the seminiferous tubules carry on meiosis to produce millions of immature, nonmotile sperm. Other cells in the seminiferous tubules produce jellylike secretions that nourish the sperm.

Although sperm are able to live for over six weeks in the ducts of the male reproductive tract, they normally survive less than two days after they are released from the body. A sperm has three basic parts:

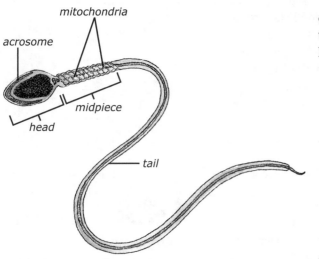

- The *head* contains the haploid nucleus. At its front tip is an *acrosome* (ACK ruh SOHM), which contains enzymes that help the sperm enter the ovum.
- The *midpiece* of the sperm contains many mitochondria that are wrapped into a tight coil. They provide the energy needed for the swimming action of the sperm.
- The *tail* is a modified flagellum capable of whiplike movements that propel the sperm. The flagellum is not activated until just prior to the sperm leaving the body.

Semen is the fluid containing the sperm. Semen is alkaline and rich in substances that nourish the sperm. The alkalinity activates the sperm and protects them from the acidic environment inside the female's body.

Accessory Reproductive Organs

The accessory organs of the male reproductive system carry the sperm from the testes to the outside of the body and produce about 95% of the semen. The sperm-containing jelly is carried through ducts from the testes into the sperm-storing **epididymis** (EP ih DID uh mis). Each epididymis is a thin, tightly coiled, 6 m (20 ft) long tube that is located on the outer surface of the testis. After at least eighteen hours in the epididymis, the sperm become mature. Mature sperm are motile and can fertilize an ovum.

Sperm exit the epididymis through a straight tube called the **vas deferens** (VAS • DEF ur enz), or *seminal duct*. The vas deferens travels from the scrotum upward into the abdominal cavity and around the urinary bladder before finally connecting to the urethra. Contractions of the smooth muscle in the walls of the vas deferens transport the sperm through the vas deferens.

The *seminal vesicle*, located near the urinary bladder, produces a fluid and adds it to the fluid of the semen. The vas deferens then passes through the doughnut-shaped **prostate gland**, which produces the largest part of the seminal fluid. The sperm can then be deposited into the urethra. The urethra serves as a common passageway for both sperm and urine. However, the action of certain muscles makes it impossible for the urine and sperm to mix together.

The **bulbourethral** (BUL boh yoo REE thrul) **glands** are two pea-sized structures located below the prostate gland. Their ducts also open into the urethra. These glands secrete a clear, sticky, alkaline fluid that neutralizes the normal acids along the urethra. The urethra passes through the penis and opens outside the body. The **penis** is designed to transfer the sperm into the wife's body during sexual relations.

seminiferous: semini- (seed) + -ferous (to carry, bear)

semen: (seed)

epididymis: epi- (near) + -didymis (Gk. *didumos*—testicle)

vas deferens: vas (L. *vas*—vessel) + deferens (*deferens*—carrying off)

25-2
Human sperm

Circumcision

When a boy is born, a loose-fitting fold of skin called the foreskin covers the tip of the penis. This is usually surgically removed soon after birth in a process known as circumcision. It is done today for medical reasons.

However, circumcision was practiced in biblical times for a different reason. For the Jews, circumcision is part of the Law (Gen. 17:10–14) and is to be performed on all male children eight days after they are born. Early

Jewish Christians felt that the Gentiles who converted to Christianity should be circumcised. In 1 Corinthians 7:19 Paul points out that circumcision is part of the Law and is not necessary for salvation. Circumcision is still practiced in most civilized areas for sanitary reasons.

circumcision: circum- (L. *circum*—around) + -cision (to cut)

Review Questions 25.1

1. List (a) the primary organs and (b) the accessory organs of the male reproductive system and give their functions.
2. What are the three main parts of a human sperm? What is the function of each part?
3. What is circumcision? What was the significance of circumcision to an Old Testament Jew? What was Paul's objection to circumcision for New Testament Gentiles?

25.2 Female Reproductive System

The female reproductive system is designed to produce ova (eggs), receive the sperm, and provide protection and nourishment for the developing embryo. The reproductive organs are also important in the birth of the baby and in nourishing the baby after birth. The principal organs of the female reproductive system are the ovaries. The accessory organs include the oviducts, uterus, vagina, and mammary glands (breasts).

The Ovary and Ovum

Human **ovaries** (sing., ovary) are solid organs approximately 2 cm (0.8 in.) long and 1 cm (0.4 in.) thick—about the size and shape of large almonds. They are attached to the lower lateral walls of the pelvic cavity by ligaments. Unlike sperm production, which begins at puberty and continues throughout life, each ovary already contains approximately 200 000 immature ova at birth. A few ova will mature each month for about thirty-five years of a woman's adult life. Usually the two ovaries alternate, one producing an ovum one month and the other ovary the next.

Each immature ovum with the surrounding cells constitutes a **follicle**. When puberty begins, the anterior pituitary gland begins secreting a gonadotropin called *follicle-stimulating hormone* (FSH), which stimulates several follicles to develop each month.

Accessory Reproductive Organs

The **oviduct**, or *fallopian tube*, is a tube with one end attached to the uterus. The other end is funnel shaped with fingerlike projections that partially encircle the ovary. The inner surface of the oviduct is lined with a ciliated mucous membrane. The cilia beat, creating a current that draws the released ovum, which cannot move by itself, into the oviduct. The action of the cilia and peristaltic contractions of the oviduct walls carry the ovum toward the uterus. Fertilization normally occurs if the ovum and sperm meet and fuse in the oviduct. The resulting zygote then begins its early development as it moves along toward the uterus. If fertilization does not occur, the ovum degenerates.

25.2

Objectives
- Describe the female reproductive organs and give their functions
- Summarize the female reproductive cycles

Key Terms

ovary	ovarian cycle
follicle	ovulation
oviduct	corpus luteum
uterus	menstrual cycle
cervix	menstruation
endometrium	menopause
vagina	

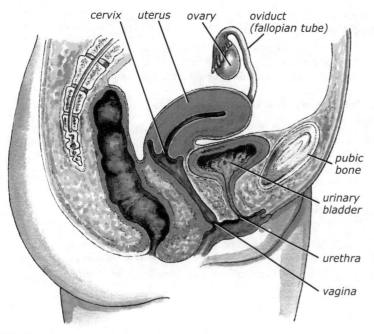

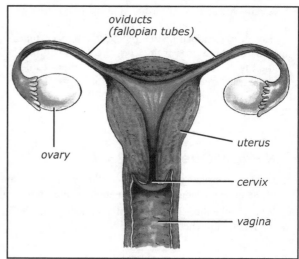

cervix   uterus   ovary   oviduct (fallopian tube)

pubic bone

urinary bladder

urethra

vagina

oviducts (fallopian tubes)

ovary

uterus

cervix

vagina

25-3

The female reproductive system, lateral view (left) and frontal view (right)

Cervical Cancer

A common disease of women is cancer of the cervix. It often spreads to the uterus and causes thousands of deaths each year. This type of cancer, however, can easily be detected at early stages by a simple test (Pap smear) in which a few cells are removed from the cervix and microscopically examined. Most cervical cancers are caused by the human papillomavirus (HPV).

Cancer of the uterus was once a major cause of death in females. If diagnosed early, the cure rate is almost 100%. Today, if cases are found early enough, only the cancerous portion of the cervix and uterus needs to be removed. Often the woman can still become pregnant. Sometimes the removal of the uterus, a hysterectomy (HISS tuh REHK tuh mee), is needed if the disease has spread.

The **uterus** is a 7 cm × 5 cm (3 in. × 2 in.) hollow, pear-shaped organ held in place by ligaments. The neck of the uterus, where it narrows and joins the vagina, is called the cervix (SUR viks). The wall of the uterus consists of muscle tissue, but the inner surface is lined with ciliated epithelial cells called the **endometrium**. The endometrium becomes thicker in preparation to receive the tiny embryo. If an embryo becomes implanted in the uterine lining, the membranes remain enlarged, and the uterus serves as the womb, a protective, nourishing place for the developing baby.

The vagina (vuh JYE nuh) is the short, elastic canal that leads from the cervix of the uterus to the outside of the body. It receives the semen and is the birth canal (the way the baby leaves the womb). Folds of tissue called the labia provide protection for the delicate tissues around the opening of the vagina and urethra.

Female Reproductive Cycles

After puberty, the female reproductive organs undergo periodic changes to prepare for a possible pregnancy. Both the ovary and the uterus must be prepared in the proper sequence if reproduction is to be successful. The changes occur through complex interactions between the endocrine and reproductive systems. Each organ is prepared by a separate yet dependent cycle. The ovarian cycle prepares and releases an ovum. The menstrual cycle prepares the lining of the uterus to receive an embryo. Although these two cycles are discussed separately in the text, they occur simultaneously in the body.

Hormones released by the anterior lobe of the pituitary gland control the ovarian cycle. These hormones stimulate the ovary to produce other hormones that control the menstrual cycle. Thus, a negative feedback mechanism regulates these two cycles. If an ovum is prepared but the uterus is not ready to receive it or vice versa, a pregnancy will not develop.

The Ovarian Cycle

The **ovarian cycle** refers to the changes happening in the ovaries. It can be divided into three phases—the follicular phase, ovulation, and the luteal phase. The *follicular phase* begins when the anterior pituitary releases a relatively

cervix: (L. *cervix*—neck)

vagina: (L. *vagina*—sheath)

large amount of follicle-stimulating hormone (FSH). A follicle is a small sac in the ovary that contains immature ova. The release of FSH causes the cells of the follicle to undergo mitosis. As the follicles grow, the process of meiosis (oogenesis) continues. At a certain point in their development, one or two developing ova are selected by unknown mechanisms to continue development. The others usually degenerate. The developing follicle enlarges and forms a cavity that fills with a clear fluid that protects and nourishes the ovum. Eventually the mature follicle appears like a blister on the surface of the ovary. At this stage the immature ovum has not yet completed the second division of meiosis.

As the follicle develops, it begins to secrete increasing amounts of estrogen. After about two weeks of development, the anterior pituitary responds by increasing the amount of luteinizing hormone (LH). This response causes the wall of the follicle to weaken and rupture, releasing the ovum. This process is called **ovulation** (OH vyuh LAY shun). Some cells from the follicle still surround the released ovum and secrete a gelatinous coating to protect it. When the ovum is released, it is barely visible to the unaided eye.

The *luteal phase* begins after ovulation. The luteinizing hormone causes the remaining cells of the ruptured follicle to develop into a temporary glandular structure called the **corpus luteum** (KOHR pus • LOO tee um). The corpus luteum produces progesterone, which affects the uterine wall and initiates the second female reproductive cycle.

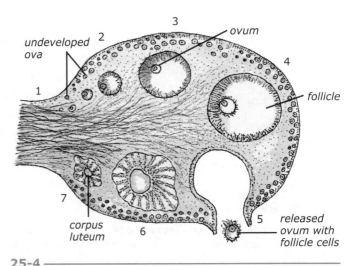

25-4

Cross section of an ovary showing the ovarian cycle: follicular phase (1–4), ovulation (5), luteal phase (6–7)

corpus: (L. *corpus*—body)
luteum: (L. *luteum*—yellow)
menstrual: (L. *mensis*—month)

The Menstrual Cycle

The **menstrual cycle** refers to the changes happening in the uterus. At the beginning of the menstrual cycle, the progesterone produced by the corpus luteum causes the endometrium to thicken, stimulates the growth of the blood vessels in the uterine wall, and causes glands of the membrane to secrete substances that prepare the uterus to receive a fertilized ovum. About seven days after ovulation, the amount of progesterone is at its peak, and about ten days after ovulation, the uterine lining is at its thickest.

At this point, one of two things happens. If an ovum has been fertilized in the oviduct and the tiny embryo has been carried to the uterus and implanted in its wall, the nourishing fluids produced by the uterine glands sustain the embryo as the *placenta* develops. The placenta produces a hormone that stops the degeneration of the corpus luteum, and pregnancy begins. If, however, no ovum has been fertilized, no placenta is formed. As the corpus luteum degenerates, it produces less and less progesterone. Decreasing progesterone causes the uterine lining to shrink, and about four weeks

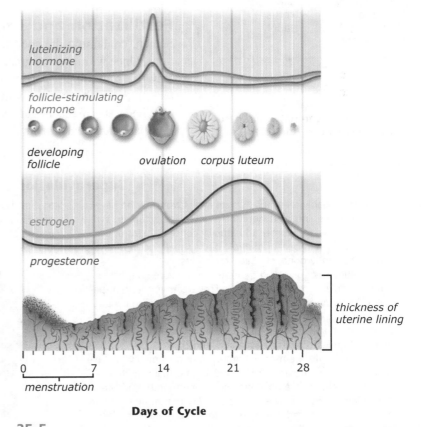

25-5

The menstrual cycle

after it started to grow (about fourteen days after ovulation), the uterine lining is shed from the body.

The several days that the uterine lining is being shed is known as **menstruation**. Hormone levels cause the uterine muscle to contract, often causing it to cramp, resulting in varying degrees of discomfort. The dramatic change in hormone levels just prior to and during menstruation, coupled with the discomfort, often temporarily affects the moods of the woman. The ovarian and menstrual cycles continue until a woman stops ovulating. This usually occurs around age 50 and is called menopause.

menopause: meno- (Gk. *men*—month) + -pause (*pausis*—pause or stop)

Review Questions 25.2

1. List the (a) primary organs and (b) accessory organs of the female reproductive system and give their functions.

2. What are the two primary female reproductive cycles? What happens during each cycle?

25.3

Objectives

- Summarize the processes of fertilization, cleavage, and implantation
- Describe the basic steps in embryonic and fetal development
- Describe the birth process

Key Terms

embryology
cleavage
blastocyst
implantation
placenta
fetus
labor

embryology: em- (in) + -bryo- (to grow) + -logy (the study of)

25.3 Human Embryology and Birth

The implantation of a human embryo and his development into a baby who is ready to be born is an orderly process of changes controlled by genes. Some genes are expressed only at certain stages of development and may be repressed during the remainder of the person's life. For example, certain genes stimulate the formation of the eyes, but after this process is completed, these genes may not be expressed again, even if an eye is removed or lost.

Although scientists are just beginning to understand the "hows" of human embryology—the study of human development from fertilization to birth—they already know a great deal about the physical development of an embryo, some of which is presented in the following sections.

Fertilization and Implantation

Semen, containing the sperm, is deposited in the vagina during sexual relations. Once within the vagina, the sperm travel through the uterus rapidly and may reach the upper portion of the oviducts in a few minutes. Fertilization takes place in the oviduct within 24 hours after ovulation.

When a sperm reaches an ovum, it moves through the follicular cells and the transparent gel that surrounds the ovum. An enzyme secreted by the acrosome

Superovulation

You have probably seen news stories about women who have given birth to quintuplets, septuplets, and even octuplets. While it is not unusual for women to naturally conceive twins or even triplets, larger numbers of multiple births usually occur after the mother has undergone infertility treatments.

Sometimes couples who desire to have children but have not been able to conceive naturally seek treatment from a specialist. Many procedures have been developed to help couples have children. Sometimes the doctor will recommend fertility drugs, which are designed to induce superovulation, causing the woman to release multiple ova each month.

Multiple births are more common among women who have undergone superovulation because under normal conditions, women usually release only one ovum each month. With superovulation, there is always the possibility that more than one ovum will be fertilized by a sperm. However, multiple births of large numbers—although widely publicized—are still very rare.

Statistics show that approximately 20% of women who conceive after undergoing superovulation treatment have twins (compared to about 3% in the general population), 2%–3% have triplets, and less than 1% have quadruplets or more.

in the sperm head helps the sperm penetrate these barriers. Many sperm may reach one ovum. But after one sperm penetrates the ovum, there is a change in the ovum membrane that prevents the entrance of other sperm.

During fertilization, the sperm loses its tail and its head swells. At this time the immature ovum nucleus divides, completing meiosis. The nuclei of the ovum and the sperm move toward the center of the cell, their nuclear membranes disappear, and their chromosomes combine. A new life has just begun.

Shortly after formation, the zygote undergoes mitosis, giving rise to two daughter cells. These cells in turn divide into four cells, which divide into eight cells, and so forth. This phase in development is termed **cleavage**.

It takes about three days for the young embryo to reach the uterus, and by then the embryo consists of a solid mass of about sixteen cells, called the *morula*. Once inside the uterus, the embryo remains free within the uterine cavity for about three additional days. Before the tiny embryo attaches to the mother's uterus, the embryo's own cytoplasm and nutrients secreted by the uterus supply the energy needed for growth.

morula: (L. *morum*—mulberry [which it greatly resembles])

blastocyst: blasto- (bud) + -cyst (bladder, pouch)

Cell divisions continue, resulting in a mass of cells contained within a fluid-filled sphere of cells. This stage is known as the blastocyst (BLAS tuh SIST). The blastocyst attaches, or implants, into the inner uterine wall. **Implantation** takes place as the outer cells of the blastocyst, together with the cells of the uterine wall, form the **placenta**. This organ is the link between mother and child where nutrients and wastes are transferred.

The baby makes his own blood, and under normal circumstances none of the mother's blood cells enter the child. Substances dissolved in the blood plasma, however, easily diffuse between the blood vessels of the mother and child in the placenta.

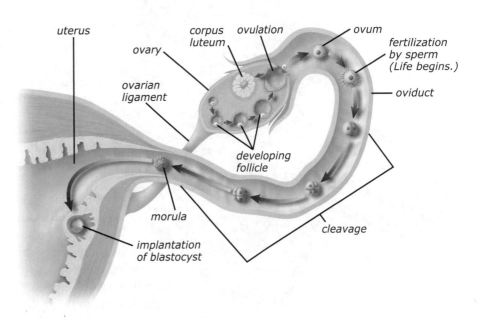

25-6

Ovulation, fertilization, and implantation of the embryo

4 weeks: 5 mm (0.2 in.)

6 weeks: 20 mm (0.8 in.)

8 weeks: 30 mm (1.2 in.)

17 weeks: 70 mm (2.8 in.)

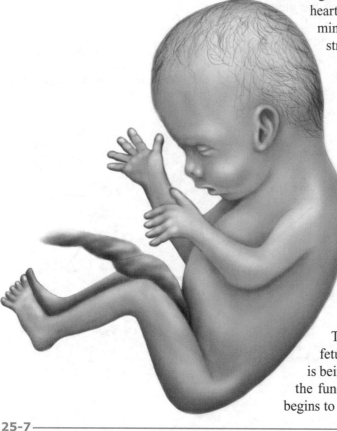

6 months: 400 mm (15.8 in.)

Gastrulation

By the third week of development, the cells of the embryo are arranged into three distinct layers during a process called *gastrulation*. These layers develop into the various adult tissues.

- The *ectoderm* (outer layer) forms the epidermis and associated structures such as nails and hair and also much of the nervous system.
- The *mesoderm* (middle layer) forms the muscles, blood, and bones.
- The *endoderm* (inner layer) gives rise to the epithelial lining of the digestive tract, respiratory tract, bladder, and urethra.

Thus, nutrients, oxygen, and antibodies diffuse into the baby's blood. Other substances dissolved in the mother's blood, such as drugs, alcohol, and toxins (for example, from diseases), can also diffuse into the child's blood. These substances affect the mother's body in relatively minor ways when compared to the effects they have on the rapidly growing and dividing cells of the unborn child. Babies born addicted to drugs or deformed by substances to which they have been unnecessarily exposed are the innocent victims of a mother's neglect.

The Embryo

At the end of the first month, the embryo is cylindrical in shape. He is about 5 mm (0.2 in.) long and weighs about 20 mg (0.0007 oz), approximately 10 000 times as heavy as the zygote. The embryo has a large head in proportion to the rest of his body because the brain develops rapidly. The head and jaws have begun to develop, and the arm and leg buds have appeared. The heart looks like an S-shaped tube and beats about sixty times per minute. The nervous system and the heart are the first major structures to develop.

At the eighth week, all major internal organs are present. The first eight weeks are the most critical stage of development. While he is less than an inch long and the major organs are forming, the embryo is susceptible to injury. Levels of drugs and disease-caused toxins in the mother's blood that would have little or no effect on the mother can greatly hamper the development of the child. At two months, the embryo is 30 mm (1.2 in.) long and weighs about 1 g (0.04 oz)—the weight of a paper clip.

The Fetus

By the end of the second month, the embryo is clearly recognizable as a human. From this point until birth, the developing child is called the **fetus**, which means "offspring." The facial features become more recognizable with the development of the nose, lips, and eyelids. The muscle tissues are beginning to differentiate, and the fetus is capable of some movement. The cartilaginous skeleton is being transformed into bone. The liver enlarges and takes over the function of blood cell formation from the spleen. The brain begins to send impulses to regulate the functions of other organs.

By the end of the third month, the fetus is about 56 mm (2.2 in.) long and weighs about 14 g (0.5 oz).

25-7
Stages in the development of a human fetus

Membranes

Special membranes develop around the embryo that protect, help obtain nourishment for, and eliminate wastes of the developing child. Although these membranes are vital to the survival of the baby inside the womb, they are not part of the embryo and are discarded at birth.

- The *amnion* begins to develop even before the first structures of the embryo take form and eventually expands to surround the entire embryo. The amniotic cavity is filled with a clear *amniotic fluid* that bathes the embryo. This sterile fluid serves to keep the embryo moist, functions as a shock absorber, maintains a uniform temperature, and provides room for the embryo's movements.
- The *yolk sac* forms during the second through the sixth week of development. There is no yolk in the human ovum, but this sac (along with the spleen and liver) functions in blood cell formation.
- The *allantois* consists of numerous blood vessels that converge to form the *umbilical cord*. This cord contains the umbilical arteries and veins that carry the blood of the fetus to the placenta, where food, wastes, oxygen, and carbon dioxide are exchanged with the mother's blood.
- The *chorion* grows rapidly along a section of the uterine wall after the embryo is implanted, forming the child's part of the placenta.

During the third month, reproductive organs are distinguishable as male or female. The ears and eyes have moved to their final positions, but the eyelids are fused. The fetus "practices" sucking and begins breathing movements by inhaling and exhaling amniotic fluid.

During the fourth month, the fetus grows rapidly, but the rate of growth decreases somewhat during the fifth month. During the fifth month, the baby becomes active and strong enough that the mother may feel the movements. At this stage the fetus has grown to about 250 mm (9.8 in.), about half the length the baby will be at birth, and weighs slightly more than 454 g (1 lb). The heart beats about 150 times per minute and can be heard with a stethoscope.

During the fifth and sixth months, hair begins to appear on the head, and eyelashes and eyebrows begin to grow. The skin becomes covered with a mixture of oil from the sebaceous (oil) glands and dead skin cells. This "cheesy" layer protects the baby's skin from the amniotic fluid. In the sixth month the arms and legs reach the relative proportions they will have at birth.

During the last three months of development, the fetus grows rapidly and gains a layer of fat tissue under the skin. If the baby is born at the beginning of the seventh month, he is able to move and cry, but unless given special help, the baby often dies because the brain is not developed enough to control breathing and maintain body temperature. The eyelids open during the seventh month, and the baby begins to appear chubby. When ready to be born, the fetus is usually positioned upside down with the head toward the cervix.

Birth

Toward the end of the pregnancy, the uterus is stretched to the maximum and the fetus has outgrown the womb. The stretching stimulates reflexes that cause the uterine muscles to contract occasionally. In the final weeks the contractions become more frequent but are still weak. A few hours before the birth of the baby, the contractions become quite strong and occur about every thirty minutes. This is the beginning of **labor**.

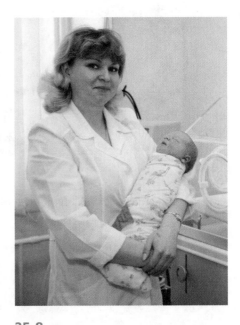

25-8

Today, babies born early can be nurtured in special units until they are strong enough to survive in the natural world.

During labor the stretching of the cervix causes the pituitary gland to release more *oxytocin*, which stimulates stronger uterine contractions. This cycle of positive feedback is repeated until strong contractions cause the baby to begin moving through the *birth canal*. The birth canal (uterine cervix and vagina) is elastic and stretches to accommodate the baby. Even the joints of the bones of the mother's pelvis at the pubic symphysis can be stretched to aid passage of the baby. The flexible bones of the baby and the fact that the skull bones of the child are not yet fused (permitting the head to be compressed slightly) also help the child in passing through the birth canal.

As the pressure increases in the uterus, the amnion bursts, releasing the amniotic fluid through the vagina. Often referred to as "breaking of the water," this sometimes occurs before labor. The strong uterine contractions, aided by the mother's pushing, move the baby through the birth canal, and he is born.

In about 96% of births, the head of the fetus is positioned downward. This position is ideal and allows the head to act as a wedge to open the birth canal as the baby emerges. This position also allows the baby to breathe safely even before he is completely born.

In about 4% of births, the baby is positioned with his posterior end toward the birth canal. This type of delivery is called a *breech birth*. Breech births are difficult because the baby's neck is sometimes trapped in the birth canal or the umbilical cord may become pinched or wrapped around the child's neck. Breech births require special procedures by the physician.

When the baby is born, the umbilical cord still connects the baby with the placenta. Most physicians clamp and cut the cord within the first ten minutes after the baby is born. The baby, deprived of oxygen, builds up carbon dioxide in his blood. This is a powerful stimulant to the respiratory center of the brain, and the baby begins to breathe on his own.

After delivery of the baby, the uterus begins to contract again, and the placenta and remains of the umbilical cord are expelled. Generally this "afterbirth" occurs within twenty minutes after the birth of the baby. About five to six weeks after the birth of a baby, the uterine wall has healed, the uterus has returned to its normal size, and the normal reproductive cycles resume.

If the space between the mother's pelvic bones (which can be measured before the time of labor) is too small, if the baby's head is unusually large, or if complications threaten the baby's life, the baby may need to be delivered by *cesarean* (sih ZEHR ee un) *section*. This is a surgical procedure, delivering the baby by surgically opening the mother's abdominal and uterine walls.

Beyond maintaining general good health and a proper diet, the mother has no control over the unborn child's development, much of which takes place before she is even aware of the child. At birth the baby leaves a protected place

25-9

Movement of the fetus through the birth canal

cesarean: (after Julius Caesar, who was reputedly born this way)

where all needs are supplied and enters a more demanding but more interesting world, where he can develop into an adult.

Now the parents must support the new life physically, mentally, emotionally, and spiritually. In God's sight the child is the "heritage of the Lord" (Ps. 127:3) and not the property of the parents. The child will grow, but it is the parents' responsibility to rear the child to love and fear the Lord (Deut. 6:2–7; Eph. 6:4).

Review Questions 25.3

1. In what organs do (a) fertilization and (b) implantation take place?
2. List the three layers formed during gastrulation. Into what adult structures do each of these embryonic tissues develop?
3. List the four embryonic membranes that develop around the embryo but are not considered part of the embryo. Give their functions.
4. What is the function of the umbilical cord?
5. How long is the embryonic stage of human development? What is the child like at the end of the embryonic stage?
6. List a few things that characterize the unborn child (a) during months 3–4, (b) during months 5–6, and (c) during months 7–9 of pregnancy.
7. List three conditions that help the baby pass through the birth canal.
8. What is a breech birth?
9. What is a cesarean section?
⊙10. Some have said that parental care begins long before pregnancy, is crucial during pregnancy, but is most important from birth on. Support each of these concepts from the physical, emotional, and spiritual standpoints.

25B—Human Relationships: A Christian Perspective

Sadly, some Christians have a degraded view of sexuality. However, the Bible indicates that God created Adam and Eve with sexual natures to be enjoyed as a married couple. The book of Song of Solomon delights in the wonderful relationship between man and wife.

Just as Christians should avoid the view that sex is evil, so Christians should avoid the world's view of sex. The world sees sex as a good thing, but they go too far in their pursuit of it. In the world's eyes (and unfortunately in some Christian's eyes), sex is a god to be worshiped at the expense of other responsibilities and relationships. Sex between a husband and wife in a committed marriage relationship is God's ideal, but humans have departed from God's ideal in many ways.

25.4 Fallen Human Sexuality

In Proverbs 5, Solomon, the wisest man who ever lived, advises his son regarding sexual relationships. In definite terms he tells him to stay away from the "strange woman." In this passage the strange woman is any woman who would entice him into sinful sexual relationships. Verses 4–13 tell of the consequences to the people who do not heed this advice, and verses 20–23 tell of God's condemnation of those who ignore this warning.

Adultery and Fornication

A sexual relationship between a married person and anyone other than that person's spouse is *adultery*. Scripture says repeatedly that adultery is sin (Matt. 5:27–28). Even one of the Ten Commandments is "Thou shalt not commit adultery" (Exod. 20:14; Rom. 13:9).

25.4

Objectives
- Define *sexual purity*
- Give Scripture references commanding Christians to maintain sexual purity
- Define *homosexuality* and give Scripture references condemning it

Key Term
sexually transmitted disease

adultery: (L. *adulterare*—to pollute)

25-10

The family is not only the basic unit of society but also the model God uses to illustrate His close personal relationship with believers (Rom. 8:14–17; Eph. 5:23–32). The bonds within the Christian family are ordained of God and should be strengthened.

Most people agree that unfaithfulness in marriage causes severe problems between the marriage partners, and therefore many agree that adultery is wrong. Although adultery may cause marital difficulties, this in itself does not make it sin. Adultery is not a sin merely because it harms a person or harms his relationship with others; it is sin because it is disobedience to God's Word.

Fornication as used in the Bible is a general term referring to any form of sex outside marriage. Fornication may be adultery, premarital sexual relationships, or any sexual relationships other than between a man and his wife. Fornication is condemned in Scripture (Gal. 5:19–21). Today, worldly people consider many forms of fornication permissible. However, if God says something is sin, no amount of human rationalizing or action can make it right. Man cannot remove God's condemnation.

Sexual purity, however, means more than just refraining from physical sexual relations with someone other than a spouse; it means also that one's thought life is pure. Proverbs 12:20 and Psalm 41:6 declare that sin is committed in the heart (mind, will). Therefore, even looking at others and wishing to have sexual relations with them is to have sinned already in the mind (Matt. 5:28). An evil thought life, in God's sight, is just as sinful as evil physical acts.

Some Christian young people, after having had impure thoughts about another person, feel that since they have sinned already, they may as well go ahead with the physical act. This is illogical; the person is sinning again. The Bible says to turn from sin, not to continue sinning (Rom. 6:1–2; 1 John 3:6–9).

Homosexuality

Homosexuality is sexual relations among members of the same sex: men with men, women with women. Romans 1:25–28 says,

> Who changed the truth of God into a lie, and worshipped and served the creature more than the Creator. . . . For this cause God gave them up unto vile affections: for even their women did change the natural use into that which is against nature: And likewise also the men, leaving the natural use of the woman, burned in their lust one toward another; men with men working that which is unseemly, and receiving in themselves that recompence of their error which was meet. And even as they did not like to retain God in their knowledge, God gave them over to a reprobate mind, to do those things which are not convenient.

God calls homosexuality a sin and condemns those who engage in it (Rom. 1:32). First Corinthians 6:9–10 places homosexuality ("abusers of themselves with mankind") with adultery, fornication, idolatry, and other sins.

Despite these and other passages condemning the sin of homosexuality, today many people claim that homosexuality is not a sin but an "alternative lifestyle" that must be tolerated by all people. Some lawmakers have attempted to give homosexuals special privileges and legal rights, and some states now recognize homosexual marriages. Extensive tests have not revealed a biological cause of homosexuality, nor are there any biological differences between homosexuals and heterosexuals (those who engage in normal sexual relations with members of the opposite sex).

homosexuality: homo- (same) + -sexuality (L. *sexus*—stem) [also called sodomy after Sodom—see Gen. 19:5]

heterosexual: hetero- (other) + -sexual (stem)

Even so, many people consider homosexuality to be a "natural" alternative lifestyle; such feelings permit homosexuals to justify their sin in their minds. Even though man is born with a sinful nature, that is no excuse for continuing in a sinful lifestyle. Christians are to follow God's precepts, not those of the world, and are to "walk in the Spirit, and . . . not fulfil the lust of the flesh" (Gal. 5:16).

Dealing with Sexual Temptations

Many Christian teenagers feel helpless about sexual sins. The combined forces of peer pressure and bodily desire can seem like undefeatable enemies. The strange woman of Proverbs 5 no longer just lurks in the street. She now prowls the television and the Internet looking to lead believers astray. Some Christian teens have already fallen into sexual sin and think that since they have failed already, there is no point in fighting against more sexual sin. But this is not the picture the Bible presents about the believer and sexual sin. Jesus has defeated sin, and it no longer has mastery over us (Rom. 6). Sexual sin, though it is a great enemy, does not rule the believer's life. There is forgiveness for past sexual sin and strength to fight against future temptation.

Christians use many methods to avoid sexual temptation. There are filters that can block out sexual temptation on the Internet. Many Christians choose

Facets of Biology — *Sexually Transmitted Diseases*

For many years people condemned sexual immorality because it spread certain diseases. These **sexually transmitted diseases** (STDs), or *venereal diseases* (VDs), are the most widespread contagious diseases in America. Some STDs are minor infections, but some can cripple a person for life, and, if not treated, can kill. Several STDs, such as herpes, human papillomavirus, and AIDS, have no known cure.

All STDs can be transferred by various means, but the vast majority of cases are the result of sexual relationships among infected people. Many STDs can be passed from an infected mother to an unborn child, some can be transferred during blood transfusions, and some can be transmitted by contact between a cut on an uninfected person's skin and the blood of an infected person. These other methods of transfer, however, account for only a tiny percentage of the cases of STDs.

Physicians can treat most STDs. The bacterial STDs, such as syphilis and gonorrhea, can be successfully treated with antibiotics. The damage they cause the body, however, often cannot be corrected even though the infection is cured. A few STDs cause only minor discomfort for the person with the disease, but when it is contracted by a pregnant woman, some of these cause problems for the child in the womb. Many babies are born suffering from the STDs their mothers have; others are born deformed because of their mothers' diseases.

Sexual purity before and during marriage virtually prevents the spread of all STDs. In times past, a person with an STD was considered to have been "caught in his sin." This changed as medical science found cures for many STDs. Prior to the AIDS epidemic, modern attitudes toward sexual relationships accepted various forms of fornication, in part because the STDs were not considered a ma-

jor problem. To people who did not know God, it appeared as though He no longer considered fornication a sin since there was no longer a punishment for it.

When the AIDS epidemic began, some people said that the disease was God's judgment on the sins of homosexuals and fornicators since they were the primary ones affected by the disease. Many were offended by such an analysis, claiming that it is unreasonably cruel to tell people in pain that they have caused their own disease. Nevertheless, the Bible does teach that diseases that result from sexual impurity are part of God's punishment of sin (Rom. 1:27). Such punishment is in fact evidence of God's grace. It allows the sinner to experience the offensiveness of his sin and points him to the need for a Savior—"the Lamb of God, which taketh away the sin of the world" (John 1:29).

Common STDs

Human Papillomavirus (HPV)

Human papillomavirus is the most common sexually transmitted disease. It is estimated that there are currently more than 20 million Americans infected with this virus, and over 6 million new cases are reported each year. There are over 40 kinds of HPV that can cause STDs. Some kinds cause genital warts, and others can cause cervical cancer in women. In fact, 80% of cervical cancers have been linked to the type of tissue changes caused by HPV. Unfortunately, the infection can be completely asymptomatic. Annual Pap smears are recommended to test for abnormal cervical cells. If warts are found, they can be treated with either topical medication or surgery. In 2006, the U.S. Food and Drug Administration approved a vaccine that is effective in preventing infection by the human papillomavirus. However, someone who becomes infected with HPV cannot be cured; once he has the virus, he has it for life.

Genital Herpes

Genital herpes is caused by the herpes simplex virus. It affects over 45 million people in the United States. Symptoms include genital sores (lesions, similar to cold sores, in the area of the reproductive organs) and fever. People infected with the virus typically go through asymptomatic periods interspersed with bouts of symptoms. The disease can be treated with certain antiviral drugs, which lessen the frequency and severity of attacks. However, there is no cure; the virus remains in the infected person's body.

Chlamydia

Chlamydia (kluh MID ee uh) is caused by the bacterium *Chlamydia trachomatis* and is the most common bacterial STD. There are an estimated 3 million new cases per year. Like an HPV infection, a chlamydia infection can be asymptomatic—75% of infected women and 50% of infected men never know they have it. Symptoms include painful urination, itching, and persistent abdominal pain. Chlamydia can be treated with antibiotics.

If untreated in women, chlamydia infection can lead to pelvic inflammatory disease (PID), infertility, ectopic pregnancy (the leading cause of death during pregnancy),

miscarriage, and chronic pelvic pain. Babies born to an infected mother can have conjunctivitis (inflammation of the conjunctiva of the eye) and pneumonia.

Syphilis

Syphilis (SIF uh lis), a common STD, is caused by the bacterial spirochete *Treponema pallidum*. From twenty-one to ninety days after a person has been exposed to the syphilis spirochete, he may develop a canker (an open sore or lesion producing a fluid discharge). The cankers usually form on the soft skin of the reproductive organs near the place where the spirochete entered the body; however, they may also form on the mouth or near the eyes. The fluid from these cankers contains the spirochetes. When the fluid comes in contact with other skin, the syphilis organism may enter the body, and the disease spreads.

In about three weeks the canker disappears, but unless medically treated, the person is not cured of the disease. A few days to a few weeks later, secondary symptoms, such as a general body rash, fever, headaches, or sore throat, develop. In time, these symptoms disappear, but the spirochetes still live in the bloodstream.

The third stage of syphilis may happen immediately after the second stage, or it may not be evident for ten to twenty-five years. During the third stage, the spirochete affects various internal organs. In untreated syphilitic patients, one in two hundred becomes blind, one in twenty-five becomes permanently crippled as the organism affects the brain, and one in thirteen develops heart or blood disorders. Syphilis may also affect the liver, bones, or almost any other organ, causing various symptoms. Infected women can pass the syphilis organism to their unborn children, causing the death of the child or causing him to be born blind, deaf, or with other disorders.

Gonorrhea

Gonorrhea (GAHN uh REE uh), another common STD, is caused by the bacterium *Neisseria gonorrhoeae*. Although these bacteria can spread through the body and cause serious complications, gonorrhea often infects only the reproductive organs, causing pain and sterility. An infected mother may give birth to deformed children.

(a)　　(b)　　(c)　　(d)　　(e)

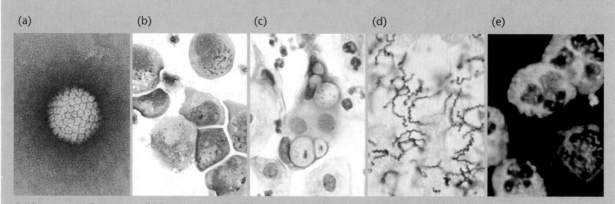

(a) human papillomavirus, (b) herpes virus, (c) Chlamydia, (d) syphilis spirochete, and (e) gonorrhea bacteria

not to watch television shows that have sexual themes. Some Christians join together in accountability groups to encourage each other to remain faithful. But these are just human methods. Internet filters can be avoided, television is ubiquitous, and you can lie to an accountability partner. Ultimately the Christian must rely on the transforming work of Jesus Christ, the grace of God, and the power of the Holy Spirit to keep from sinning. It is comforting to know that God promises every believer that there is no temptation in life that He has not promised an escape from (1 Cor. 10:13).

⮞ **Review Questions 25.4**

1. Distinguish between the terms *adultery* and *fornication*.
2. List two Scripture passages that condemn homosexuality.
⊙ 3. What separates sexually transmitted diseases from other forms of contagious diseases? What is another name for STDs?
⊙ 4. Give five examples of sexually transmitted diseases.

⊙ 5. In what way can unregenerate people consider homosexuality "natural"? Does this justify a person's being a homosexual?
⊙ 6. In a paragraph, summarize what you believe to be the biblical position on sexual relationships. Use Scripture references to support your position.

25.5 Sanctity of Human Life

The Bible clearly states that man is a special creation because he is made in God's image. The principle of the *sanctity of human life* (that human life is of high value) appears repeatedly in the Bible. After Abel was killed, God cursed Cain, his murderer. To make sure that there was no blood feud, God said He would inflict Cain's punishment sevenfold on whoever killed Cain (Gen. 4:15). "Thou shalt not kill" is one of the Ten Commandments written directly by the hand of God (Exod. 20:13; Rom. 13:9). God's condemnation falls on anyone who murders another human, and the Bible teaches that murderers are to be put to death (Gen. 9:6).

Christians agree that a man's spiritual life is important. After all, God's Son was sent to die to redeem man's soul. A Christian's physical life is special, too, because it belongs to Christ. A Christian who does not yield all of his physical self to Christ is sinning (Rom. 12:1–2). It is in his physical body that a Christian accomplishes his earthly ministry. A Christian's physical life belongs to God and is His to direct.

God holds human life as special, something of high value. This concept establishes the principle of the sanctity of human life. Deformity or illness does not negate the fact that man is created in God's image. Two aspects of medical practice—abortion and euthanasia—violate the scriptural principle of the sanctity of human life.

Abortion

Many people believe a pregnant woman has the right to decide whether she wants the child inside her to live. If she does not want the child, it is legal in many countries for her to have the child killed. The killing of an unborn child is called **abortion**.

Most abortions in the United States are performed before the baby is sixteen weeks old. The most common method involves a physician's placing sharp instruments into the womb through the vagina. As the walls of the womb are then scraped, the baby's body is cut into pieces. As the material is removed from the uterus, tiny arms, legs, or other parts of the unborn child are often recognizable.

25.5

⮞ **Objectives**
- Discuss what is meant by *sanctity of human life*
- Define *abortion* and give scriptural reasons why it is wrong

⮞ **Key Terms**
abortion
euthanasia

abortion: (L. *abortare*—to die, miscarry)

The Bible and Abortion

The Bible is quite clear that the life within a pregnant woman's womb is fully human. Psalm 139:13–16 uses personal pronouns for the unborn, indicating that unborn babies are considered legitimate people to God. Exodus 21:22–25 describes the punishment for harming a baby in his mother's womb, which shows that the unborn are to be valued just as any other person in Israel was. Luke 1:41–44 describes John the Baptist in Elisabeth's womb as having all the qualities of a human being: cognition, emotion, will, perception, and so on. These and many other verses teach that God considers a developing baby to be a human being. Taking any human life is wrong (Exod. 20:13; Rom. 13:9). No ruling, whether by a man or government agency, that says abortion is legal makes it lawful to God. In this and every case, the Word of God stands above the laws of sinful men and women.

In obeying the biblical command to defend the defenseless (Prov. 24:11–12; 31:8–9), Christians should seek a position that safeguards human life. Many Christians believe that life begins at conception. It is at that time that the gametes of the father and mother come together and form a new and distinct genetic code. To place the beginning of life at a later point is to risk calling a human being just a bundle of tissue—a risk no Christian should be willing to take.

Some people argue that God would not mind if a deformed baby were aborted. A growing number of diagnostic and imaging techniques can now show whether an unborn child is normal. Some people feel that God does not want "abnormal" babies. It is true that some children with deformities are naturally aborted. These natural abortions are often called miscarriages or stillbirths. But God is sovereign; He is in control of everything. He decides whether what would be called an abnormal baby will naturally abort or will be born.

Despite what humans may think is "best," God makes what people call "perfect" and also what people call "imperfect." Both are within the realm of His intention and, for Christians, will "work together for good" (Rom. 8:28). Although man may consider something "deformed" or "wrong," God's ways are beyond man's ability to understand (Isa. 55:8–9). For example, when God spoke to Moses in the burning bush, Moses claimed he was not fit to lead the children of Israel out of Egypt, for he could not speak well. The extent of Moses's handicap is unknown, but God's response is significant: "Who hath made man's mouth? or who maketh the dumb, or deaf, or the seeing, or the blind? have not I the Lord?" (Exod. 4:11).

Deliberately killing a human that someone feels is not physically or mentally "normal" is ignoring the Bible's teaching about the sovereignty of God and the sanctity of human life. Some parents who suspect a possible deformity (perhaps inherited) can use prenatal testing to determine whether the child is afflicted. This information can often be used to help the parents (and physicians) prepare wisely for the child, but never is it right to kill the child before or after birth.

Madonna and Child with St. Elizabeth and St. John the Baptist, Lubin Baugin (attr. to), From the Bob Jones University Collection

The unborn John the Baptist leaped for joy when his mother was greeted by Mary while she was carrying Jesus in her womb (Luke 1:39–44). Here the French painter Lubin Baugin (ca. 1610–1663) depicts a later meeting of these mothers and their sons.

In later stages of pregnancy, the physician will inject a strong salt solution into the fluid-filled amniotic sac that protects the baby. As the baby breathes and swallows the salt, his outer layer of skin is burned off by the salt, and the baby dies in about an hour. About a day later the woman enters labor and delivers a dead baby.

If the pregnancy is past the time when the child can be poisoned by salt, the baby can be removed in much the same way a baby is taken by cesarean section. The baby is delivered alive but is usually left unattended. He dies a few moments after he has entered the world. Often, when aborted in the final stages of pregnancy, the child could have lived if proper care had been given.

After the twentieth week, a baby can be aborted by a method called intact dilation and extraction, or partial-birth abortion. In this procedure, the child is partially delivered feet first. A cannula (hollow tube) is then inserted into the skull and the contents are suctioned out. This collapses the skull and the dead child is delivered.

Certain medications can also be used to cause abortions. Mifepristone, or RU 486, is administered to terminate a pregnancy. It counteracts the effect of progesterone (needed to maintain pregnancy) and causes a "spontaneous" abortion. RU 486 has been associated with severe bleeding in the mother as well as other medical complications. It is legal to use this drug up to the first nine weeks of pregnancy. RU 486 is also used to prevent the embryo from implanting; some advocates of this medication argue that no abortion occurred if the embryo was not implanted. However, life begins at fertilization (conception), and if an embryo is purposely prevented from implanting, a baby has been killed and an abortion has occurred.

Except under certain laboratory conditions, human sperm and ova have no chance of reproduction or even a lengthy survival outside the body. However, when they do unite, whether naturally or in a laboratory, the zygote they produce is fully human. A human's physical life is a continuing existence that begins when the sperm fertilizes the ovum and progresses through development, birth, growth, maturity, and death. Few people would argue that a newborn baby is not a complete human or that killing an infant is not murder. Tragically, because evolutionary philosophy holds that man is just a highly evolved animal, many believe that an unborn child is not human and that killing children before birth is not murder. These people may try to redefine terms and pass laws to the contrary, but abortion is killing a human—and that is murder.

End-of-Life Issues

Euthanasia (YOO thuh NAY zhuh) is the act of ending the life of a person for the person's "good." It can be done voluntarily as an act on the part of the patient, or it can be done involuntarily by a doctor or legal representative, who decides whether the person should live or not. It is often portrayed as an act of mercy toward someone suffering continual pain or a debilitating, terminal illness. It is a noble motive to want to end suffering, but it is always an evil act to end life, which is sacred in God's eyes. The reality in the end is that euthanasia is simply murder.

Legally in most of the United States, practicing euthanasia when a person is not clinically dead is murder. A physician who practices euthanasia may face a malpractice suit or criminal charges. A person is *clinically dead* if he has stopped breathing, has no heartbeat, and has no (or exceptionally subnormal) brain waves.

The issue of euthanasia is a clear one for the Christian, but other issues regarding the end of life are much more difficult to deal with. With modern life-sustaining apparatuses, drugs, and medical techniques, people who are about to die can often be kept alive for additional hours, days, months, or even years. Examples of life support are respirators, feeding tubes, and kidney dialysis machines. Some Christians accept "pulling the plug" on the life support of a person who is unconscious and terminally ill or is of considerable age and whose death is merely being postponed. These Christians feel that it is not always in God's will to use all the apparatuses, drugs, and techniques at the physician's disposal to keep alive a person in such advanced stages of age and illness.

Making decisions about ending life support asks people to make decisions regarding the quality of the patient's life. Factors such as the patient's age,

euthanasia: eu- (good) + -thanasia (Gk. *thanatos*—death)

possibility of regaining consciousness, and even spiritual condition must be taken into account. Some Christians feel that removing life support for any reason is morally wrong, so even the conscience of the person making the decision should be taken into account.

The brief comments made here oversimplify this involved topic, but an in-depth discussion is beyond the scope of this book. Various cases pose so many different situations that even Christian authorities disagree on appropriate guidelines.

Review Questions 25.5

1. Give at least four biblical supports for the sanctity of human life principle.
2. List three abortion methods.
3. List three Scripture references supporting the concept that an unborn child is a human.
4. What is clinical death?
5. List some types of life support.

Appendix A
Major Biomes of the World

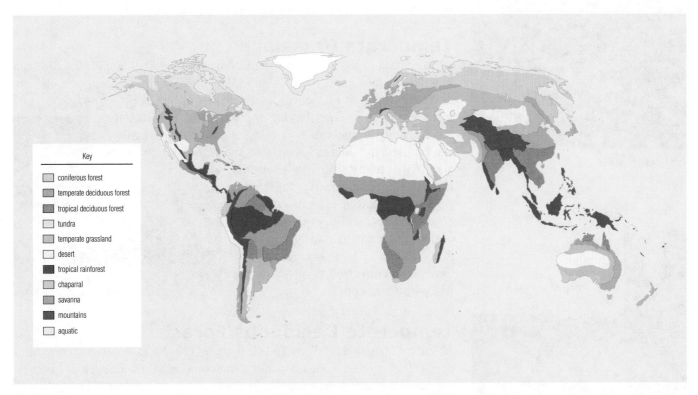

Key
- coniferous forest
- temperate deciduous forest
- tropical deciduous forest
- tundra
- temperate grassland
- desert
- tropical rainforest
- chaparral
- savanna
- mountains
- aquatic

Ecologists often divide the biosphere into large areas that have relatively uniform physical environments and therefore support similar biotic communities. These areas, which may cover major sections of continents, are called *biomes* (BYE ohmz). Usually the type of biome that exists in an area depends on the amount of water available and the seasonal range of temperatures. While the concept of different biomes is accepted by all, the division and naming of the regions varies. What follows is a list of the most commonly accepted biomes.

Tundra

In the tundra, the ground remains frozen except for two months during which the top few inches thaw, permitting a brief growing season. Beneath the thawed surface is the *permafrost*. The predominant vegetation includes lichens, grasses, and dwarf woody plants. The ground holds abundant dead organic material (which decomposes slowly because of the temperatures) and is often waterlogged during the growing season because the permafrost keeps the moisture on top. During the growing season, insects thrive, attracting various migratory birds to the area. Caribou, reindeer, musk oxen, lemmings, ptarmigans, and arctic foxes are tundra animals. Some tall mountains away from the North Pole also have regions known as alpine tundra that are quite similar but are usually better drained. Tundra covers one-fifth of the earth's land surface.

Reindeer

Coniferous Forest

Pines, spruces, and firs are predominant in coniferous forests; the prevailing temperatures and amount of rainfall determine the species. This biome usually has two major seasons, each lasting about half the year. Most of the precipitation comes in the summer. Coniferous forests support deer, elk, moose, snowshoe hares, squirrels, wolves, and many birds.

White-tailed deer

Red-breasted sapsucker

Temperate Grassland

In temperate regions, where the rainfall is too light to support a forest, a temperate grassland grows. Also called prairies, these areas are characterized by grasses that cover vast areas of the interior of many continents. Grasslands supply natural pastures for large grazing animals such as bison and pronghorn, or they can be used for domestic cattle or sheep. Most grasslands also have many rodents such as ground squirrels, prairie dogs, and gophers. Predators such as badgers, coyotes, foxes, and tigers also thrive in the grasslands. Because they have deep, fertile soils, many grasslands have been converted to grow crops like wheat and corn.

Prairie dog

Bison

Temperate Deciduous Forest

Abundant annual rainfall of 100–130 cm/yr (39–51 in.) and moderate temperatures characterize the temperate deciduous forest. Plants such as beeches, maples, oaks, hickories, chestnuts, mosses, and ferns dominate. The falling of leaves, a dormant period, the regrowth of leaves, and a growing period are the four different seasons. The decaying plant material creates a rich layer called humus. Animals such as deer, foxes, squirrels, bobcats, wild turkeys, woodpeckers, and wood thrushes are common.

Chaparral

A chaparral (SHAP uh RAL) is a region with mild temperatures, abundant winter rainfall but dry summers, and predominant vegetation including trees or shrubs with thick evergreen leaves. Often the plants are spiny. Chaparrals can be found extensively around the Mediterranean Sea and in California and Mexico. Coyotes, mule deer, rabbits, wood rats, chipmunks, and lizards are common animals. The abundant leafy plants also support many insects and spiders. Because the low-growing shrubs often contain flammable oils, fire is a frequent and important environmental limiting factor in the chaparrals.

Black-tailed jackrabbit

Tropical Deciduous Forest

In a tropical area with a lush rainy season and a severe, lengthy dry season, tropical deciduous forests thrive. The "bush" areas in Africa and Australia are tropical deciduous forests. The trees and bushes in these areas lose their leaves during the dry season, often revealing grotesquely shaped branches. These *thornwood forests* support large animals like tigers and bush elephants as well as diverse reptile and insect life.

Bush elephant

Savanna (Tropical Grassland)

A savanna (suh VAN uh) is a grassland with a few scattered trees. Savannas are found in warm regions with average rainfalls of 130 cm (51 in.) a year,

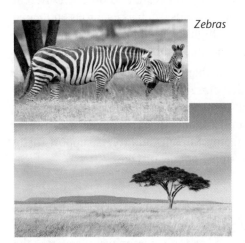

Zebras

but there is a major dry season when the animals are less active. Fire plays an important part in this biome. Fire-resistant trees and grasses are the predominant vegetation types. Antelope, wildebeest, zebras, giraffes, and their predators are common animals here. In some areas the savannas have been turned into farms and cattle ranches. Savannas are similar to the temperate grasslands but maintain a more consistent year-round temperature because they lie closer to the equator.

Macaw

Tropical Rainforest

Areas with rainfall over 250 cm (98 in.) a year and with high temperatures are termed tropical rainforests. These forests usually do not enter a dormant state, although some of the trees may lose their leaves in the dry seasons. Plants in these areas are greatly diversified; vines and epiphytes are common. Much of the animal life dwells in the trees. Monkeys, apes, chameleons, geckos, tree-climbing snakes, frogs, and such birds as toucans, hornbills, birds of paradise, and parrots are common. With their ideal, almost constant growing conditions, rainforests support more biodiversity than all other biomes combined. Living things are stratified into various levels from the forest floor to the emergent layer above the canopy.

Horned lizard

Desert

Regions where the rainfall is less than 25 cm (10 in.) a year are deserts. Most deserts do have sparse vegetation consisting mainly of succulents (such as the cactus), which store water; shrubs that have many short branches

from a thick trunk and shed their leaves during longer dry periods; and annuals that grow, bloom, form seeds, and die within the short growing periods. There are both hot deserts and cold deserts. Various insects, rodents, reptiles, and a few birds usually live in the desert, but in absolute deserts (such as the central parts of the Sahara), rainfall is so little and infrequent that almost no life exists there.

Lionfish (marine)

Rainbow trout (freshwater)

Aquatic Ecosystems

Many aquatic ecosystems, such as streams and ponds, are quite small. Others are temporary (such as rivers that change channels, and ponds and lakes that fill with sediment), but their significance cannot be ignored. Plankton, invertebrates, and fish are the predominant populations. Aquatic ecosystems are divided into two groups:

✦ *Marine ecosystems.* The ocean contains ecosystems such as shores, reefs, estuaries (where fresh and salt water mix), open ocean, and others. Different organisms grow in each. Some marine ecosystems, such as estuaries and reefs, are among the most productive and most highly populated ecosystems, while others are "watery deserts." Oceans cover three-fourths of the earth's surface.

✦ *Freshwater ecosystems.* Making up less than 2% of the earth's surface, the lakes, ponds, streams, and rivers are tied closely to the surrounding terrestrial habitats. The depth, temperature, bottom materials, and dissolved materials determine the type of biotic community that will exist in any freshwater ecosystem. Current is also a limiting factor. Streams and rivers support vastly different populations from those in swamps, ponds, or lakes.

Mountains

Those traveling from the equator to the North Pole will notice marked changes in the biomes. They will travel first through tropical forests and several other biomes, until they reach tundra and then ice cap. If a person were to start at the base of a tall mountain in the tropics and climb upward, he would pass through the same types of areas, reaching tundra and ice cap at the top. In some mountains, this zonation occurs in almost distinct lines, whereas between the equator and poles, changes are often gradual with large transitional areas between biomes. Zonation in mountains is caused by the predominant temperature, available water, and winds—the same conditions that cause other biomes.

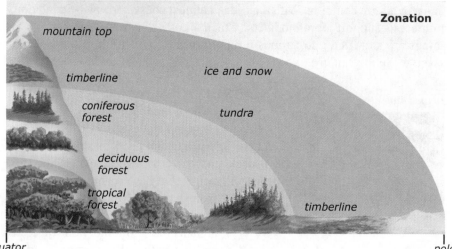

Zonation

mountain top

timberline

ice and snow

coniferous forest

tundra

deciduous forest

tropical forest

timberline

equator

pole

Appendix B
A Modern Classification System

The major divisions of the classification system, along with their major characteristics, examples, and their location in this book, are listed below. Not all of the orders, classes, or even phyla have been listed here. (At least twenty other phyla are recognized under the kingdom Animalia, but they are not listed here due to their obscurity and space constraints.) This listing is to be used for summary and review of the text material. Several additional divisions and examples are listed for reference purposes. The classifications are presented in the order they are covered in the text. The term *phylum* is used for uniformity throughout this classification system. The term *division* is technically correct in the kingdom Plantae and is sometimes also used in the kingdoms Archaebacteria, Eubacteria, Protista, and Fungi.

Kingdom Archaebacteria (10A)[1]
Prokaryotic (lack organized nucleus; lack membrane-bound organelles); most have cell walls lacking peptidoglycan; have capsules or sheaths and are unicellular or simple colonial organisms; exist in some of the most extreme conditions.

Examples: thermoacidophiles, halophiles, methanogens

Kingdom Eubacteria (10A)
Prokaryotic; have peptidoglycan (a special polymer) in their cell walls; unicellular.

Examples: "true" bacteria, like *Bacillus*, *Staphylococcus*, *Streptococcus*, and *Escherichia*, as well as cyanobacteria (the blue-green algae), mycoplasmas, spirochetes, and rickettsiae

Kingdom Protista (11)
Eukaryotic (have an organized nucleus and membrane-bound organelles); most are unicellular but some are colonial or form large multicellular structures; none have tissues; some are heterotrophic (either ingesting material or absorbing substances); some are photosynthetic (autotrophic); some can be either heterotrophic or photosynthetic.

Animal-like Protists (11A)[2]
Motile in at least one phase of the life cycle; heterotrophic (some are heterotrophic and photosynthetic); unicellular.

Example: protozoans

Phylum Sarcodina Sarcodines
Move by pseudopodia; heterotrophic (phagocytic); some parasitic, some pathogenic; some form tests or cysts; reproduction usually by fission; salt- and freshwater forms.

Examples: amoebas, foraminifers, radiolarians

Phylum Ciliophora Ciliates
Unicellular; have pellicle; many have macronucleus and micronucleus; move by cilia; heterotrophic; some parasitic, some pathogenic; some form cysts; salt- and freshwater forms.

Examples: *Paramecium*, *Stentor*, *Vorticella*

Phylum Sporozoa Sporozoans
Unicellular; rarely colonial; no locomotive structures; all parasitic; many pathogenic; form spores.

Example: *Plasmodium*

Phylum Zoomastigina Zooflagellates
Unicellular or colonial; have pellicle; move by flagella; heterotrophic, but some can be either photosynthetic or heterotrophic; some parasitic, some pathogenic; salt- and freshwater forms.

Example: *Trypanosoma*

Plantlike Protists (11B)[3]
Photosynthetic with chlorophyll and other pigments in plastids.

Example: algae

Phylum Euglenophyta Euglenophytes
Lack cell walls but are supported by pellicle; unicellular; two unequal flagella used for movement; mainly photosynthetic with chloroplasts.

Example: *Euglena*

Phylum Chlorophyta Green algae
Unicellular or colonial forms; motile, free-floating, and sessile forms; food stored as starch; cell walls of cellulose; salt- and freshwater forms.

Examples: *Spirogyra*, *Oedogonium*, *Chlamydomonas*

Phylum Chrysophyta Golden algae

Unicellular and colonial forms; cell walls often contain silicon; motile and free-floating forms; mainly photosynthetic with food stored as oils; salt- and freshwater forms.

Examples: *Vaucheria, Dinobryon*

Phylum Bacillariophyta Diatoms

Unicellular; secrete a silica shell designed like a box with a lid; photosynthetic; salt- and freshwater forms; major component of plankton.

Examples: *Navicula, Pinnularia*

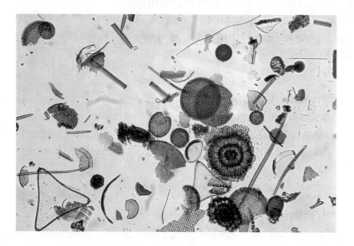

Phylum Phaeophyta Brown algae

Multicellular (usually large thallus); sessile; food stored as oil and complex carbohydrates; cell walls of cellulose; mostly shallow saltwater forms.

Examples: kelp, *Fucus, Sargassum, Laminaria*

Phylum Rhodophyta Red algae

Colonial; cell walls of cellulose; nonmotile, usually sessile; food stored as carbohydrates; mostly deep saltwater forms.

Examples: *Chondrus, Lemanea, Polysiphonia*

Phylum Dinoflagellata Dinoflagellates

Unicellular with two flagella; rigid body of two cellulose plates; heterotrophic and photosynthetic forms with food stored as starch or oil; mostly saltwater forms.

Example: *Karenia*

Funguslike Protists (11C)[4]

Phylum Oomycota Water molds

Unicellular or multicellular; cell walls contain cellulose, not chitin; heterotrophic parasites or decomposers; mostly aquatic.

Examples: downy mildews, *Phytophthora*

Phylum Acrasiomycota Cellular slime molds

Amoeboid cells that join (maintaining their own membranes) to form multicellular structure; heterotrophic; mass called a slug produces spores that form new organisms.

Example: *Dictyostelium*

Phylum Myxomycota Plasmodial slime molds

A plasmodium (multinucleate cellular structure) that often moves with pseudopodia and produces sporangia.

Example: *Physarum*

Kingdom Fungi (12)

Eukaryotic; some unicellular, some colonial; some form large structures of intertwined hyphae; none have tissues; cell walls usually contain chitin; all heterotrophic with external digestion; some carry on simple fission; many form various kinds of spores; most have sexual reproduction by conjugation or gametes; some produce motile gametes.

Phylum Zygomycota Common molds

Hyphae are nonseptate (lack cross walls); sexual reproduction by conjugation; terrestrial or parasitic.

Examples: *Rhizopus, Pilobolus*

Phylum Ascomycota Sac fungi[5, 6]

Hyphae are septate with perforated cross walls; form sexual spores in an ascus; many produce asexual spores on conidia.

Examples: yeasts, morels, *Penicillium*, athlete's foot fungus, ringworm, thrush

Phylum Basidiomycota Club fungi

Hyphae usually septate; mycelia often produce large fruiting bodies; produce spores on a club-shaped basidium.

Examples: mushrooms, puffballs, shelf fungi, rusts, smuts

Kingdom Plantae (13A)

Eukaryotic; multicellular; have tissues; most have organs; most have chlorophyll and other pigments in plastids; cell walls made of cellulose; most sessile, some float; most autotrophic (photosynthetic) with food stored as starch, some parasitic; reproduce sexually, many also with asexual forms of reproduction; sexual reproduction involves alternation between haploid and diploid generations.

Nonvascular Plants[7]

Phylum Bryophyta Mosses

Lack vascular tissues; plants are small; no true leaves, roots, or stems; gametophyte plant is dominant; sporophyte is a parasite on the gametophyte; moist habitats.

Example: *Sphagnum*

Phylum Hepatophyta Liverworts

Lack vascular tissues; lobe-shaped plants; gametophyte is the dominant generation; water required for reproduction.

Example: *Marchantia*

Phylum Anthocerophyta Hornworts

Characteristics similar to liverworts; named for horn-shaped dominant sporophyte generation.

Example: *Anthoceros*

Vascular Plants without Seeds

Phylum Pteridophyta[8] Ferns

Have vascular tissue (xylem and phloem), true roots, stems, and leaves; leaves have sori that produce spores; sexual reproduction by sperm swimming to ovum on small gametophyte plant; sporophyte plant is dominant; damp environments.

Examples: *Osmunda, Pteris, Polypodium*

Phylum Sphenophyta Horsetails

Thick perennial underground stem with erect annual stems containing silica; leaves in whorls from nodes on the stem; conelike spore-producing structures on top of stems.

Example: *Equisetum*

Phylum Lycophyta Club mosses

Have vascular tissue; usually have creeping and erect stems; reproduce by spores; water required for reproduction in gametophyte stage.

Example: *Lycopodium*

Phylum Psilotophyta Whisk ferns

Have vascular tissue; lack roots; have scales instead of leaves; reproduce by spores.

Example: *Psilotum*

Vascular Plants with Seeds[9]

Phylum Coniferophyta Conifers

Woody plants (mostly evergreen) with needles or scale-like leaves; produce cones containing seeds.

Examples: *Pinus, Taxus*

Phylum Cycadophyta Cycads

Evergreen shrubs resembling a palm tree; some have fiddleheads; produce seeds in a single large, conelike structure; sexes on separate plants.

Example: *Cycas*

Phylum Ginkgophyta Ginkgoes

Deciduous trees with fan-shaped, leathery leaves; sexes on separate trees, with females producing cherry-sized fruit.

Example: *Ginkgo*

Phylum Gnetophyta Gnetophytes

Sporophyte is dominant; seeds produced in conelike structures that can resemble flowers; often grow in dry habitats.

Examples: *Ephedra, Welwitschia*

Phylum Anthophyta Angiosperms

Have vascular tissue; produce flowers; seeds develop inside an ovary wall; ovary and seeds mature into fruit.

Class Monocotyledoneae Monocots

Seeds have one cotyledon and one seed leaf; parallel leaf venation; roots usually fibrous; flower parts usually in threes or sixes; young stems have scattered vascular bundles and usually become hollow as they mature.

Examples: grasses, *Zea, Lilium*

Class Dicotyledoneae Dicots

Seeds have two cotyledons and two seed leaves; netted leaf venation; usually have a taproot; flower parts usually in fours, fives, or multiples of these numbers; young stems have vascular bundles in a ring and are usually solid.

Examples: *Prunus, Quercus*

Kingdom Animalia (15–18)

Multicellular and eukaryotic; no cell walls; all heterotrophic with some pathogenic and some parasitic; most have specialized tissues; all reproduce sexually, but many reproduce both sexually and asexually; all have embryonic stages, but some have both embryonic and larval stages; most motile with muscles.

Phylum Porifera Sponges (15B)

Either asymmetrical or radially symmetrical; supported by a system of interlacing spicules and/or spongin fibers; body bears many pores; sessile adults; filter feeding accomplished by collar cells with intracellular digestion; no respiratory or excretory structures; response to stimuli mainly on the cellular level; asexual reproduction by budding, gemmules, or regeneration; sexual reproduction by eggs and sperm; aquatic.

Example: *Grantia*

Phylum Cnidaria Cnidarians (15B)[10]

Usually radially symmetrical; body consists of two cell layers forming a gastrovascular cavity with a single opening; movement using a system of musclelike fibers; jellylike tentacles for obtaining food; digestion is first extracellular within the gastrovascular cavity and then intracellular; no respiratory or excretory structures; nervous system consists of a nerve net with limited sensory capabilities; asexual reproduction by budding and regeneration; sexual reproduction by gametes; life cycle includes either a sessile polyp or a free-swimming stage or both.

Class Hydrozoa
Examples: hydras, obelias

Class Scyphozoa
Example: jellyfish

Class Anthozoa
Examples: corals, sea anemones

Phylum Platyhelminthes Flatworms (15C)

Bilateral symmetry; body consists of three cell layers; ribbonlike body; body covering of ciliated epidermis or tegument; movement using layers of muscles; digestion is extracellular and intracellular within an intestine that has a single opening; some adults lack digestive structures; slight cephalization; some have sensory organs such as eyespots; asexual reproduction by division with regeneration; sexual reproduction usually by cross-fertilization.

Examples: tapeworms, *Planaria*

Phylum Nematoda Roundworms (15C)

Body covering has a protective cuticle; tubular digestive system with mouth and anus; anterior nerve ring with dorsal and ventral nerve cords; longitudinal muscles only, permitting thrashing movements; sexes are separate; some parasitic on humans.

Examples: *Ascaris*, *Trichinella*

Phylum Annelida Segmented worms (15C)

Distinct external and internal segmentation; digestive system is complete (mouth and anus); respiration by skin or gills; closed circulatory system; nervous system has anterior ganglia and a ventral nerve cord with a ganglion in each segment; movement by two layers of muscles; asexual reproduction by regeneration in some; sexual reproduction involving hermaphroditic adults or separate sexes.

Examples: earthworms, leeches

Phylum Mollusca Mollusks (15D)

Bilaterally symmetrical or asymmetrical; soft bodied, many with a hard shell secreted by the mantle; movement by muscular foot; digestive system is complete (mouth and anus); many have sensory organs for vision, taste, touch, smell, and balance; no asexual reproduction; sexual reproduction involving a trochophore larval stage.

Class Bivalvia Bivalves
Examples: clams, oysters, scallops

Class Gastropoda Gastropods
Examples: snails, slugs, nudibranchs

Class Cephalopoda Cephalopods
Examples: squids, octopuses, nautiluses

Phylum Echinodermata Echinoderms (15D)

Radial symmetry for adults, bilateral for larvae; supported by a system of hardened plates beneath the epidermis; spiny skinned; locomotion by a water-vascular system and tube feet; asexual reproduction by regeneration; sexual reproduction involving separate sexes and external fertilization; marine.

Class Asteroidea
Example: starfish

Class Echinoidea
Examples: sea urchins, sand dollars

Class Holothuroidea
Example: sea cucumbers

Class Ophiuroidea
Examples: brittle stars, basket stars

Class Crinoidea
Examples: sea lilies, feather stars

Phylum Arthropoda Arthropods (16)

Exoskeleton usually of chitin; body segmented, usually with head, thorax, and abdomen; movement by jointed appendages moved by muscles; open circulatory system, with a dorsal heart; respiration by gills, tracheae, or book lungs; sensory organs include antennae, sensory hairs, compound and/or simple eyes; sexes usually separate; parthenogenesis in some.

Subphylum Trilobita *Trilobites (16A)*
All extinct.

Subphylum Crustacea *Crustaceans (16A)*
Cephalothorax and abdomen; two pairs of antennae; one pair of mandibles, two pairs of maxillae, and one to three pairs of maxillipeds (mouthparts); respiration by gills; usually have larval stages; usually aquatic, few terrestrial.

Examples: crayfish, crabs, pill bugs

Subphylum Chelicerata (16A)

Class Arachnida
Cephalothorax and abdomen; no antennae; chelicerae and pedipalps (mouthparts); four pairs of walking legs attached to thorax; respiration by book lungs or tracheae; no larval stages (except for ticks); usually terrestrial.

Examples: spiders, ticks, mites, scorpions

Subphylum Uniramia (16B)

Class Insecta
Head, thorax, and abdomen; one pair of antennae; three pairs of legs on thorax; may have one or two pairs of wings on thorax; respiration by tracheae; usually have larval stages with metamorphosis; mainly terrestrial.

Examples: ants, flies, wasps

Class Chilopoda
Head and many body segments; one pair of antennae; have poison glands; carnivorous; one pair of walking legs per body segment; no larval stages; terrestrial.

Example: centipedes

Class Diplopoda
Long body with head, short thorax, and many body segments; one pair of antennae; two pairs of walking legs per body segment; one pair of mandibles and one pair of maxillae (mouthparts); herbivorous, eating dead plant material; no larval stages; terrestrial.

Example: millipedes

Phylum Chordata Chordates (17–18)
Dorsal notochord, which is replaced by vertebrae in many; dorsal tubular nerve cord; have pharyngeal pouches during embryonic development.

Subphylum Cephalochordata (17A)
Fishlike; retain notochord throughout life; no skeleton; marine filter feeders.

Example: lancelets

Subphylum Urochordata (17A)
Notochord as larvae only; saclike covering called the tunic; many adults sessile; marine.

Example: tunicates (sea squirts)

Subphylum Vertebrata Vertebrates (17B–18B)
Usually a vertebral column (backbone) protecting a dorsal nerve cord; distinct head with brain.

Class Agnatha Jawless fish (17B)
Eel-like fish with no true jaws and no scales; circular sucking mouth; some scavengers and some parasites; no stomach; skeleton of cartilage; two-chambered heart; ectothermic; external fertilization; oviparous development.

Examples: lampreys, hagfish

Class Chondrichthyes Cartilaginous fish (17B)
Skin with mucus-producing glands and placoid scales; mouth usually on ventral surface; jaws with teeth; skeleton of cartilage; fins both paired and unpaired; respiration by gills; two-chambered heart; ectothermic; fertilization generally internal; development usually ovoviviparous or viviparous; mostly marine.

Examples: sharks, rays, skates

Class Osteichthyes Bony fish (17B)
Skin with many mucus-producing glands; usually covered by scales; mouth with jaw located in the front of the head; skeleton chiefly of bone with some cartilaginous parts; fins both paired and unpaired; respiration by gills covered with opercula; two-chambered heart; often a swim bladder for controlling depth in water; ectothermic; fertilization generally external; development usually oviparous, but some ovoviviparous and viviparous; eggs usually minute and numerous; salt- and freshwater forms.

Examples: perch, eels, salmon

Class Amphibia Amphibians (17C)
Skin smooth with many glands and chromatophores (pigment cells); no scales; large mouth; nostrils open into mouth cavity; skeleton chiefly of bone; two pairs of limbs usually present (some lack limbs); no claws; feet often webbed; respiration by gills, lungs, mouth lining, and skin—either individually or in combination; three-chambered heart as adults, two-chambered heart as larvae; ectothermic; fertilization generally external (except salamanders); development usually oviparous; eggs surrounded by jellylike coat.

Order Apoda
Example: caecilians

Order Caudata
Examples: salamanders, newts

Order Anura
Examples: frogs, toads

Class Reptilia Reptiles (17D)
Body covering of tough, dry scales; two pairs of limbs with clawed toes (some lack limbs); respiration by lungs; most have three-chambered heart with partial division of ventricle, but some have four-chambered heart; predominantly ectothermic; internal fertilization; oviparous, producing amniotic eggs laid on land; some ovoviviparous or viviparous; terrestrial or aquatic.

Order Rhynchocephalia
Example: tuataras (one living species: *Sphenodon*)

Order Squamata
Examples: snakes, lizards

Order Testudinata
Example: turtles

Order Crocodilia
Examples: alligators, crocodiles, caimans

Class Aves Birds (18A)
Body covering of feathers; skeleton composed of porous, lightweight bones; forelimbs (wings) used for flight in most species; body supported on hind limbs only; toothless bill; four-chambered heart; respiratory system with air sacs; endothermic; oviparous reproduction with egg enclosed in lime-containing shell; female with only one developed ovary and oviduct.

Class Mammalia Mammals (18B)[11]
Body covering of hair; usually have teeth in both jaws; two pairs of limbs (one pair in some); respiration by lungs; diaphragm used to assist in breathing; four-chambered heart; endothermic; reproduction viviparous with young sustained within mother by a placenta (except Monotremata and marsupials); some ovoviviparous (Monotremata); some lack placenta (marsupials); all have young nourished by milk from mammary glands.

Egg-Laying Mammals

Order Monotremata Monotremes
Lay eggs in leathery shell; lack nipples.
Examples: duck-billed platypuses, spiny anteaters

Pouched Mammals Marsupials[12]
Terrestrial, arboreal, and aquatic species; no placenta; young born immature and develop in an abdominal pouch where nipples are located; as many as seven different orders.
Examples: kangaroos, koalas, opossums

Placental Mammals
As many as twenty different orders. See Subsection 18.5 for coverage of eleven of the most studied.

Footnotes

1. The five-kingdom classification, which still persists in some books, combines the two prokaryotic kingdoms, Archaebacteria and Eubacteria, into the single kingdom Monera.
2. The animal-like protists are often classified as subkingdom Protozoa.
3. The plantlike protists are often classified as subkingdom Protophyta.
4. The funguslike protists were at one time in kingdom Fungi but are now part of kingdom Protista.
5. The imperfect fungi, those that lack a known form of sexual reproduction, are placed in phylum Ascomycota, awaiting observation of the sexual reproduction many scientists believe they have.
6. Many species of fungi, usually ascomycetes, form symbiotic relationships with green algae and/or cyanobacteria to form lichens. Because more than one species of alga or bacterium can exist in a lichen, these partnerships are usually classified on the basis of their fungal component.
7. At one time, the three phyla of nonvascular plants were combined under phylum Bryophyta, but each now has its own phylum.
8. The phylum Pteridophyta sometimes appears as Pterophyta.
9. The conifers, cycads, ginkgoes, and gnetophytes are sometimes collectively referred to as gymnosperms.
10. Phylum Cnidaria was formerly Coelenterata.
11. Many taxonomies place monotremes in the subclass Prototheria and marsupials and placentals into subclass Theria.
12. Marsupials were formerly organized into families under the order Marsupialia but now have been reclassified into as many as seven different orders.

Appendix C
Metric System Conversions and Unit Abbreviations

Metric System Conversion Table

	Metric unit	Appropriate English system equivalent	Metric conversion	Handy comparisons
Length	kilometer (km)	1090 yards	1000 m	11 football fields
	meter (m)	1 yard 3 inches; 39 inches	0.001 km	a yardstick and 3 inches
	centimeter (cm)	0.4 inch	0.01 m; 10 mm	a nickel is about 2 cm in diameter
	millimeter (mm)	0.04 inch	0.1 cm	the thickness of a penny
	micrometer (µm)	0.000 04 inch	0.001 mm	a red blood cell is 7.5 µm
	nanometer (nm)	0.000 000 04 inch	0.001 µm	a polio virus is 25 nm
	angstrom (Å)	0.000 000 004 inch	0.1 nm	½ size of a hydrogen atom
Weight	kilogram (kg)	2.2 pounds; 35 ounces	1000 g	half of a bag of sugar
	gram (g)	0.035 ounce	0.001 kg	the weight of 24 drops of water; a dime weighs 2.3 g
	milligram (mg)	0.000 035 ounce	0.001 g	the weight of 0.024 of a drop of water
Volume	liter (L)	1 quart and ¼ cup	1000 mL	¼ gallon
	milliliter (mL)	0.004 cup; 1000 µL	0.001 L	the volume of 24 drops of water, about ⅓ teaspoon
	microliter (µL)	0.000 004 cup	0.001 mL	the volume of 0.024 of a drop of water
	liter, dry (L)	0.03 bushel		the volume of about one quart

Temperature

On the Celsius (centigrade) scale, 0 °C is the freezing point of water and 100 °C is the boiling point of water at sea level. On the Fahrenheit scale, 32 °F is the freezing point of water and 212 °F is the boiling point of water at sea level. Normal body temperature is 37 °C, or 98.6 °F. The following formulas can be used to convert one scale to the other:

°F to °C: subtract 32, multiply by 5, divide by 9.
°C to °F: multiply by 9, divide by 5, add 32.

Unit Abbreviations

Å	angstrom	**kcal**	kilocalorie	**nm**	nanometer
AD	anno Domini (in the year of our Lord)	**kg**	kilogram	**oz**	ounce
		km	kilometer	**pH**	measure of acidity
BC	before the birth of Christ	**km/h**	kilometers per hour	**pt**	pint
bpa	bushels per acre	**L**	liter	**qt**	quart
bpm	beats per minute	**lb**	pound	**sec**	second
bu	bushel	**m**	meter	**m²**	square meter
°C	degree Celsius	**mg**	milligram	**sq mi**	square mile
Cal	kilocalorie	**mi**	mile	**sq yd**	square yard
cm	centimeter	**mL**	milliliter	**tn**	ton
°F	degree Fahrenheit	**mm**	millimeter	**w**	week
ft	foot	**mm Hg**	millimeters of mercury	**×**	times larger, power (magnification)
g	gram	**mph**	miles per hour		
gal	gallon	**µL**	microliter	**yd**	yard
in.	inch	**µm**	micrometer	**y**	year

Appendix D
Index to Combining Forms

aer-, aero- (air) 4A
albus (white) 6A
amphi- (on both sides) 17C
an-, a- (without) 4A
ana- (up) 4B
ante- (before) 8B
anti-, ant-, anta- (opposite, against) 10C
ap-, ad-, ac-, ag- (to, near) 15C
arthro-, arthr- (joint) 16A
-ase (enzyme) 21B
aster-, astero-, astr- (star) 15D
auto- (self) 4A

bacter-, bacterio-, bactero-, -bacteria (small rod) 9A
bi- (two) 13A
bio-, bi-, -bic, -be, -biosis (life) 2B
-bolism (to throw) 4B

cardi-, cardia-, cor-, cord- (heart) 16A
carn- (flesh) 17A
cauda- (tail) 15A
cell- (cell, chamber) 3A
centi-, cent-, -cent (hundred) 1C
centri-, centr-, centro- (center) 3A
cephal- (head, brain) 15A
cereb- (brain) 17A
chem-, chemic-, chemo- (chemical) 4A
chloro- (greenish yellow) 3A
chondr-, chondro- (cartilage) 17B
-chord (cord) 17A
chrom-, chromo-, -chrome (color) 3A
cili- (eyelash, eyelid, cilia) 3A
cis-, -cide, -cision (to kill or cut) 16B
con-, com-, co- (with, together) 10A
crani- (skull) 23A
-cretion (to separate) 3A
cutan-, cuti- (skin) 13B
cyano-, -cyanin (dark blue) 10A
-cycle, cyclo-, cycl- (circle) 13B
cyst (bladder, pouch) 11A
-cyte, cyto-, -cytic (cell) 3A

-dactyla (finger) 18B
derm-, derma-, dermi-, -derm, -dermis (skin) 13B

di-, diplo- (two, double) 2B
dia- (through) 20C
dif-, dis-, di-, de- (apart) 2A
dors-, dorso-, dorsi- (back) 15A

echin- (spiny) 15D
eco- (house) 19A
ecto- (outside, outer, external) 11A
-emia (blood) 17A
endo-, en-, em- (within) 2A
epi- (upon, over) 13A
equi- (equal) 2A
erythro- (red) 22A
eu- (good, true) 3A
ex-, exo-, ef-, e- (out) 2A
extra- (outside) 4B

-fer, -fera (to bear) 11A (see also -phora)
fibro-, fibr-, -fibril (fiber) 13B
flagell- (little whip) 3A
foli- (leaf) 12C

-gamete, gameto- (marriage, spouse) 5A
gastro-, gastri-, gastr- (belly) 15B
-geneous (stock, race, or kind) 2A
-genesis, gene-, geno- (birth, origin, family, beginning) 5A
germ (seed) 6A
-gestion (to carry) 3A
-glia, -glutinate (glue) 15B
gluco-, glyc- (sweet) 4A
gonado- (primary sex gland) 24A
-gonium (seed) 11B
-graph, -graphy (to write) 1C
gymno- (naked) 13A

halo- (salt, sea) 10A
hemo-, hema-, hem- (blood) 17A (see also -emia)
hetero-, heter- (other, different) 4A
homo-, homeo- (same) 2A
-hybrid (offspring of different parents) 5B
hydro-, hydra- (water) 2B
hyper- (over, beyond) 3B
hypo- (under, beneath) 3B

-ichthyes (fish) 17B
in-, il-, im-, ir- (into, within) 3A
in- (not) 15A
intra- (within) 4B
iso- (equal) 2A

-jug- (see zygo-)

-karyotic (with a central part) 3A
-kinesis (to move) 2A

later- (side) 15A
leuco-, leuko- (clear, white) 3A
lip-, lipo- (fat) 2B
-logy (word, study of) 3A
-lunar (moon) 22A
lyso-, -lysis, -lyst, -lytic (loosening, break apart) 2A

macro- (large) 11A
mater (mother) 23A
media-, medi- (middle) 15A
melan-, melano- (black) 20B
-mensal- (table) 19A
-mer (part) 2B
meso-, mes- (middle) 13B
meta- (change) 4B
-meter, metri- (measure) 1C
micro- (small) 1C
milli- (thousand) 1C
mono- (one) 2B
-morph, morpho-, -morphosis (form, shape) 6A
myco-, myci- (fungus) 12C
myo-, my- (muscle) 20D

nano- (dwarf) 1C
nastic (pressed closed) 14A
nemato-, -nema (thread) 15B
nephro-, nephr- (kidney) 15C
neur-, neuro- (nerve) 17D
-nomy (system of rules) 9A
nuc-, nucle-, nucleo-, -nucleus (central part) 2B

ocular (eye) 1C
-oid (shape, form) 5A
omni- (all) 17A

oo-, ovo-, ovi-, ova- (egg) 5A
optic-, -optic (visible) 17A
or-, os- (mouth) 15B
ortho-, orth- (straight, correct, right) 16B
ossi-, osteo-, oste- (bone) 17B

para- (beside) 10A
-parous (to give birth) 17A
per- (through) 2A
peri- (around) 13B
phago-, -phage, -phagy (to eat) 3B
phil-, -phile, -philic, philo- (beloved, loving) 2B
-phobia, -phobe, -phobic (fear) 2B
-phora, -phore, -phera, -fera (to bear) 11A
photo- (light) 4A
-phyll (leaf) 4A
-physis (growth) 20C
-phyta, -phytic, phyto-, -phyte (plant) 10A
plasm-, -plasm (material forming cells, to mold) 3A
platy-, plan-, plano- (flat) 15C
-pod, -ped, -pus, -p (foot) 15B
poly- (many) 2B

pro- (before, in front) 3A
prot-, proto- (first) 6A
-ptera (wing, fan) 16B
pulmo-, pneumo- (lung, wind, breath) 17C

radio- (to emit) 8B
re- (back) 2B
-renal (kidney) 22A
rhizo-, rhiza-, rad- (root) 12B

-saccharide (sugar) 2B
-scope, -scopy (to view) 1C
-script, -scribe (to write) 2B
-secretion (separation) 24A
semi- (half) 2A
semin-, semen- (seed) 15C
-sexual (stem) 25B
somato-, somatic, -some (body) 3A
sperm-, sperma-, spermato-, -sperm (seed) 5A
spir- (coil or twist) 10A
-spire, -spiration (to breathe) 14A
sporo-, spore, -spore (seed) 5A
stato-, -stasis, -stat, -status (standstill) 3B
stomat-, -stomium (mouth) 13B

syn-, sym-, syl-, sys-, sy- (same, with, together) 2B

thermo-, -therm (heat) 2A
thigmo- (to touch) 14A
thora-, -thorax (breastplate) 16A
-tomous, -tomy (to cut) 9A
-tonic (tone or tension) 3B
-tope (place) 2A
toxi-, toxo-, toxic-, -toxin (poison) 10C
trans- (across) 2B
tri- (three) 22A
trop-, -tropism (turn) 14A
-troph, -trophic, tropho- (nourishment, feeder) 4A
-type (impression) 5B

uro-, -ura (tail) 17A

vas-, vaso- (vessel, duct) 13A
ventr- (belly) 15A
-vert (to turn) 15A
-vorous (to devour, eat) 17A

zoo-, zo- (animal) 11B
zygo-, -zygous, -jug- (yoke, join) 5A

Glossary

A

abdomen (16.1) A body region posterior to the thorax.

abiogenesis (8.5) The concept that life can arise from nonliving substances.

abiotic factor (19.2) A nonliving aspect of an ecosystem.

abortion (25.4) The killing of an unborn child.

abscisic acid (14.3) A plant hormone that affects plant growth by inhibiting the action of other hormones. It also promotes dormancy.

abscission layer (13.5) A layer of cells, located at the base of leaf petioles and fruits, that die, causing the separation of the leaf or fruit from the stem.

absorption (21.4) The movement of food molecules from the alimentary canal into the bloodstream.

accommodation (23.3) The desensitization of sensory receptors after a period of stimulation.

acetyl CoA (acetyl coenzyme A) (4.3) A two-carbon substance found in many cellular metabolisms.

acid (2.3) Any substance that yields hydrogen ions when dissolved in water; a substance that neutralizes a base.

acquired immune deficiency syndrome (AIDS) (22.5) A disease caused by the human immunodeficiency virus that affects the human immune system.

actin (20.8) One of the two types of protein found in muscle fibers.

activation energy (2.2) The initial energy necessary to start a chemical reaction.

active immunity (22.5) An immunity in which a person manufactures the antibodies himself or has activated T cells for a particular antigen.

active potential (23.1) The nerve impulse that results when sodium ions rush into the cytoplasm after stimulation.

active site (2.2) The portion of an enzyme's surface that is believed to bond to a particular substrate during the action of an enzyme.

active transport (3.5, 14.2) The movement of molecules across cellular membranes against the concentration gradient; requires cellular energy expenditure.

adaptation (9.3) Any inheritable characteristic that gives a survival advantage to the organism; according to evolutionists, the change of an organism that enables it to survive in a new environment.

addiction (24.3) The continued use of habit-forming drugs; the inability to stop using a drug for physical or psychological reasons or both.

addition See *nucleotide addition.*

adenine (4.4) A base in a nucleic acid molecule.

adenosine diphosphate (ADP) (4.1) The molecule that is produced when ATP is split to yield energy.

adenosine triphosphate (ATP) (4.1) A compound that serves as a temporary energy storage molecule in all cells.

adhesion (2.3) The force that holds molecules of different substances together.

ADP See *adenosine diphosphate.*

adrenal cortex (24.2) The larger outer region of the adrenal glands; secretes the sex hormones, mineralocorticoids, and glucocorticoids.

adrenal gland (24.2) Endocrine gland located on each kidney; composed of cortex and medulla regions.

adrenal medulla (24.2) Inner portion of adrenal gland; produces epinephrine and norepinephrine.

adultery (25.5) The act of sexual relations between a married person and anyone other than that person's marriage partner.

aerial hypha (12.4) One of the hyphae of a fungus that grows above the substrate.

aerobe, obligate (10.2) An organism that can live only in the presence of free oxygen.

aerobic (4.3) Requiring oxygen.

aerobic cellular respiration (4.3) The oxygen-requiring process of breaking down a food substance to obtain cellular energy.

agar (11.4) A gelatinous substance obtained from red algae used as a culture medium in microbiology.

agglutinate (22.1) To cause the clumping together of blood cells when blood types are not matched properly in a transfusion.

AIDS See *acquired immune deficiency syndrome.*

air sac (18.2) One of a series of hollow chambers connected to the respiratory system of birds.

albinism (6.3) A genetic abnormality resulting in a lack of pigmentation.

albumen (18.3) The white of an egg.

alcoholic fermentation (4.3) The formation of alcohol and carbon dioxide from glucose; performed by yeast cells.

alcoholism (24.3) A dependence on the substance alcohol.

alga (pl., **algae**) (11.3) An organism in one of the seven primarily photosynthetic phyla of kingdom Protista.

alimentary canal (21.4) A group of digestive organs arranged in a continuous tube extending from mouth to anus.

allantois (17.8) An embryonic membrane in an amniotic egg that serves for respiration and excretion for the embryo; in humans, becomes part of the umbilical cord.

allele (5.4) One of a pair of genes that have the same position on homologous chromosomes.

allergen (22.5) Any foreign substance that triggers an allergic reaction.

allergy (22.5) A disorder caused by the body's producing antibodies when stimulated by natural, nonpathogenic substances.

alternation of generations (13.1) The reproductive cycle in which the asexual reproductive stages give rise to sexual reproductive stages that, in turn, give rise to asexual reproductive stages.

altricial chick (18.3) A young bird that is naked and helpless when hatched.

alveolus (pl., **alveoli**) (21.1) One of the small, bubblelike structures of the lungs where gases are exchanged between the atmospheric air and the blood.

amensalism (19.2) The situation in which one population in an environment is inhibited by another, while the other is not affected by the first.

amino acid (2.6) The basic "building block" of a protein molecule.

amniocentesis (6.5) A medical test using amniotic fluid to determine the sex and health of an unborn child, including possible genetic defects.

amnion (17.8) A thin, membranous sac enclosing a developing reptile, bird, or mammal.

amniotic egg (17.8) An egg that has a leathery or hard shell in which the embryo is enclosed by an amnion.

amoebocyte (15.3) An amoeba-like cell in a sponge's mesenchyme that produces spicules, transports food, and eliminates waste.

amoeboid movement (11.2) A constant change in shape by an amoeba or similar cell by the formation of pseudopodia.

amphibian (17.6) A member of the class Amphibia.

amplexus (17.7) The physical contact of a male and a female amphibian that stimulates the female to release eggs into the water.

amylase (21.4, 21.5) An enzyme secreted by the salivary glands and pancreas to digest starches into sugars.

anabolism (4.5) The phase of metabolism that builds molecules and stores energy; the constructive part of metabolism.

anaerobe, obligate (10.2) An organism that cannot live in the presence of free oxygen.

anaerobic (4.3) Not requiring oxygen.

anaphase (5.2) The third phase of mitosis; paired chromatids separate and begin to migrate toward opposite poles of the cell.

anatomical position (20.2) A standing position of the human body with the arms at the sides and the palms turned forward.

anatomy (20.2) The science that deals with the structure of organisms.

androgen (24.2) A male sex hormone produced by the testes.

aneuploid (6.1) An organism in which the chromosome number is not an exact multiple of the haploid number.

angiosperm (13.3) A flowering, seed-producing plant; seeds enclosed in an ovary when mature.

annelid (15.7) A member of the phylum Annelida; a segmented worm.

annual (13.1) A plant that grows from a seed, produces more seeds, and dies during one growing season or within one year.

annual ring (13.7) In woody stems, one layer of xylem that forms during one year.

ANS See *autonomic nervous system.*

antagonist (20.10) The muscle that performs an action opposite to that of the muscle acting as the prime mover.

antediluvian (8.7) Before the Flood.

antenna (pl., **antennae**) (16.1) An elongated, movable sensory append-age on the head of various invertebrates.

antennule (16.2) The sensory appendage responsible for the sense of balance in some arthropods.

anther (14.6) The structure on a flower's stamen in which pollen is produced.

antheridium (pl., **antheridia**) (11.3) The reproductive structure that produces sperm in certain algae and (13.1) nonvascular plants.

antibiotic (10.6) A chemical produced by living organisms that naturally kills or inhibits the growth of other organisms.

antibody (10.6, 22.5) A protein substance produced to eliminate antigens that have entered the body.

anticodon (4.4) The triplet of nucleotides on transfer RNA that will pair with the codon of the messenger RNA to line up amino acids during protein synthesis.

antigen (22.1, 22.5) Foreign material in the body that stimulates anti-body production or begins cell-mediated immunity.

anus (15.6) The posterior opening of the alimentary canal for egestion of feces.

aorta (22.2) The large vessel that carries blood from the left ventricle out to the body.

apical dominance (13.7) A condition in plants in which the terminal bud suppresses the growth of lateral buds.

apical meristem (13.7) A region of active cell division in the tip of a woody stem; cells may differentiate into tissues for leaves, stems, or flowers.

apparent age (8.3) A feature of God's creation; man, plants, and animals created in their mature forms; earth created with the appearance of age.

appendage (16.1) An extension of an animal's body; a limb.

appendicular skeleton (17.2, 20.6) The bones of the pelvic and pectoral girdles and their appendages.

applied science (1.5) The use of pure science to solve practical problems.

aquaculture (11.3) The farming of aquatic environments.

aqueous humor (23.5) A transparent, watery fluid in the eyeball that nourishes the cornea.

aqueous solution (2.3) A mixture in which the solvent is water.

arachnid (16.3) A member of the class Arachnida.

arachnoid membrane (23.2) A thin membrane surrounding the brain and spinal cord; between the dura mater and pia mater.

archegonium (pl., **archegonia**) (13.1) Female reproductive structure in algae, fungi, and some plants.

artery (17.2, 22.3) Any blood vessel that carries blood away from the heart.

arthropod (16.1) A member of the phylum Arthropoda.

articular cartilage (20.6) A cartilage layer covering the epiphyses of bones; provides cushioning and smooth movement at the joints.

artificial classification system (9.1) A classification system based on observable characteristics.

artificial insemination (6.5) The mechanical placement of a male's sperm into a female's reproductive organs.

artificial selection (8.4) Man's controlled breeding of organisms in an attempt to influence characteristics.

ascus (12.7) The structure in which the haploid ascospores are formed in the molds of the phylum Ascomycota.

asexual reproduction (5.2) The production of a new organism without the fusion of a sperm and an ovum; involves only mitotic cell divisions.

assimilation (1.6) The conversion of nutrients into living cells; a process of growth.

astigmatism (23.5) A condition in which the cornea or lens or both are unequally curved and the light rays from an object are not equally focused on the retina.

atherosclerosis (22.1) A disease causing the lining of blood vessels to be narrowed by the deposition of lipids and rough from extra connective tissue.

atom (2.1) The smallest unit of an element that can exist either alone or in combination.

ATP See *adenosine triphosphate.*

ATP synthase (4.2) An enzyme necessary for the conversion of phos-phate and adenosine diphosphate (ADP) into adenosine triphosphate (ATP).

atrioventricular node (AV node) (22.2) A mass of specialized cardiac tissue located in the right atrium; responsible for the contraction of the ventricles.

atrioventricular valve (22.2) One of the membranous structures be-tween the atria and ventricles in the heart that prevents backflow of blood into the atria.

atrium (pl., **atria**) (17.4, 22.2) One of the heart chambers that receives blood from different parts of the body.

attenuated vaccine (10.4) A vaccine produced from live viruses.

auricle (23.4) An outer flap of ear tissue leading to the auditory canal.

autoimmune disease (22.5) A disease in which the immune system pro-duces antibodies and cytotoxic T cells that attack the normal host cells.

autonomic nervous system (ANS) (23.2) The involuntary portion of the peripheral nervous system.

autophagy (4.5) A process whereby a cell forms a membrane around some of its own cellular parts and digests them.

autosome (5.6) Any chromosome other than a sex (*X* or *Y*) chromosome.

autotroph (4.1) An organism that is able to make its own food.

auxin (14.3) A growth-regulating hormone in plants.

avascular (20.2) Describing human and animal tissues in which there are no blood vessels.

AV node See *atrioventricular node.*

axial skeleton (17.2, 20.6) The portion of the skeleton that supports and protects the organs of the head, neck, and trunk.

axon (23.1) The portion of a neuron that carries impulses away from the cell body.

bacillus (pl., **bacilli**) (10.1) A rod-shaped bacterium.

bactericidal (10.6) Capable of killing bacteria.

bacteriophage (10.2) A virus that parasitizes a bacterial cell.

bacteriostatic (10.6) Preventing the multiplication of bacteria.

bacterium (10.1) An extremely small, unicellular, prokaryotic organism found in the kingdoms Archaebacteria and Eubacteria.

baraminology (9.4) A movement among Creationists to create a taxo-nomic system based on the biblical kinds of the original Creation.

bark (13.7) The outer protective covering of a mature woody stem; com-posed of all the tissues outside the vascular cambium.

basal disc (15.4) The flattened structure at the lower end of certain cni-darians; used for attaching to objects and for locomotion.

basal metabolic rate (BMR) (21.5) The amount of calories needed to maintain normal body functions while at rest.

base (2.3) A substance that releases hydroxyl ions when dissolved in water; neutralizes an acid.

basidium (pl., **basidia**) (12.8) A microscopic structure in basidiomycetes that produces the asexual basidiospores.

B cell (22.5) Cell involved in humoral immunity.

behavior (17.3) The way an animal responds to its environment.

behavioral isolation (9.3) The inability of two subgroups of a popula-tion to interbreed due to differences in their behaviors, especially their courtship and breeding rituals.

benign (6.3) Describing abnormal cellular growth most often character-ized by localized, nonaggressive growth.

bias (7.4) A preference or inclination in thinking, especially a belief that causes one to be partial.

biblical kind (9.4) Any of the natural groupings of organisms established by God.

biennial (13.1) A plant that sprouts and grows in one season but does not flower and produce seeds until the following growing season.

bilateral symmetry (15.2) A body pattern that can be divided into equal halves only by a cut that passes longitudinally through the body, resulting in mirror images on each side.

bile (21.4) A greenish fluid produced by the liver; necessary for the breakdown and absorption of fatty substances.

binary fission (10.2) A method of asexual reproduction in which the nuclear material is copied and the parent cell divides into two equal cells.

binomial nomenclature (9.2) A system of naming organisms in which each organism is given a genus and species name.

biodegradable (19.6) Capable of being broken down by the environment and returned to the normal cycling of substances.

biodiversity (19.2) A measure of the scope or range of living organisms in an environment.

biogenesis (1.3) The concept that life comes only from pre-existing life.

biogeochemical cycle (19.3) The movement of a particular chemical substance through the earth's system.

biological father (6.5) The man who supplies the sperm for the zygote of a child.

biological key (9.2) An arrangement of descriptions, questions, and illustrations used to identify an organism.

biological magnification (19.6) The process that concentrates small quantities of a substance into larger quantities as it is passed in a food chain.

biological mother (6.5) A woman who supplies the ovum for the zygote of a child.

biological species (9.3) A population of organisms that are similar; those organisms that interbreed and produce fertile offspring.

biology (1.6) The science that deals with living organisms and vital life processes.

biomass (19.2) The total mass of tissue of a population or species (usually measured in a dried form).

biome (App. A) A major biotic community with populations of climax species.

biopsy (10.7) The removal of a sample of tissue to be observed for abnormalities.

biosphere (19.3) The part of the earth in which life can exist.

biosynthesis (2.4) The formation of a chemical compound by a living organism.

biotechnology (7.1) The use of technology to enhance living organisms and processes.

biotic factor (19.2) A living thing (population) in an ecosystem.

bipedalism (18.5) The characteristic of walking upright on two legs.

birth canal (25.3) The cervix and vagina; very elastic at time of delivery to allow passage of the baby.

bivalve (15.8) A mollusk from the class Bivalvia.

blade (13.5) The large, flattened area on most leaves.

blastocyst (25.3) A fluid-filled sphere of embryonic cells.

blind spot (optic disc) (23.5) The area where the nerve fibers leave the eye to form the optic nerve; contains no photoreceptors.

blood (22.1) The liquid carrier of nutrients, hormones, and gases that moves through the circulatory system.

blood plasma (22.1) The liquid portion of the blood.

blood pressure (BP) (22.3) The pressure of the blood against the walls of arteries caused by contraction of the heart ventricles.

bloom condition (11.4) A rapid, seasonal reproduction of a particular organism in an optimal environment.

blowhole (18.5) The single breathing opening on the dorsal surface of the head of a cetacean.

BMR See *basal metabolic rate*.

body tube (1.7) The cylindrical part of the microscope between the eyepiece and the objectives.

bone marrow, red (20.6) A tissue that makes red blood cells and that is located in the marrow cavities of some bones.

bone marrow, yellow (20.6) Fatty tissue that gradually replaces red bone marrow as humans become older.

book lung (16.3) The platelike respiratory structure found in arachnids.

botany (13.1) The study of plants.

Bowman's capsule (22.6) The end of the nephrons of the kidney where blood plasma is absorbed from the blood vessels.

bract (13.7) Brightly colored leaf on plants that appears to be a petal; leaflike structure.

brain (23.2) The portion of the central nervous system of vertebrates that is encased in the skull; composed mainly of neurons.

breech birth (25.3) A delivery in which the baby is positioned with his posterior end toward the birth canal.

bronchus (pl., **bronchi**; adj., **bronchial**) (21.1) One of the two branches of the trachea that carries air to the lungs.

budding (5.2, 12.7, 15.3) A type of asexual reproduction in which portions from the parent organism or unicellular organism (fungi) form a new individual. (14.5) In plants, a method of grafting in which a bud is placed under the bark of another plant.

buffer (2.3) A dissolved substance that makes a solution resistant to a change in its pH (the concentration of hydrogen ions).

bulbourethral gland (25.1) One of the paired structures beneath the prostate gland that contributes a neutralizing alkaline substance to the semen as it flows through the urethra.

bundle of His (22.2) The specialized heart muscle tissue that transports impulses from the AV node in the heart to each ventricle.

bursa (20.7) A saclike structure between tendons, ligaments, and bones that cushions and reduces friction.

C

calcitonin (24.2) A thyroid hormone that lowers the blood calcium level.

calorie (21.5) A measurement of heat produced during the oxidation of food (Calorie); the amount of heat required to raise 1 g of water 1 Celsius degree (calorie).

Calvin cycle (4.2) The most common pathway of photosynthesis; also called the *carbon fixation cycle* or *light-independent phase*.

cancer (6.3, 10.7) A disorder in which the cell is unable to control cell division.

canine (18.4) A pointed tooth generally used for tearing; name comes from general term for members of the dog family Canidae.

cap (12.8) The top portion of a mushroom, the final fruiting structure of many basidiomycetes.

capillarity (14.1) The property of water that causes it to cling to surfaces.

capillary (17.2, 22.3) A blood vessel that has walls one cell thick where diffusion of nutrients and exchange of gases occurs.

capsid (10.3) The outer covering of a virus; made from protein.

capsule (3.3, 10.1) A cellular secretion surrounding certain algae and bacteria.

carapace (16.2) The portion of the exoskeleton that covers the cephalothorax in some arthropods. (17.9) The dorsal part of a tortoise's body shell.

carbohydrate (2.5) An organic compound that contains only carbon, hydrogen, and oxygen.

carbon backbone (2.4) The basic carbon chain or ring upon which the remainder of the organic molecule is built.

carbon fixation (4.2) The process in photosynthesis whereby a carbon molecule from carbon dioxide is added (fixed) to another organic molecule.

carcinogen (6.3) A cancer-causing substance.

cardiac cycle (heartbeat) (22.2) The series of physical events that transports blood through all four heart chambers during one heartbeat.

cardiac muscle (20.8) Muscle tissue found only in the heart; striated and involuntary.

carnivore (adj., **carnivorous**) (17.2) An animal that eats other animals.

carotenoids (13.5) The yellow, orange, and red fat-soluble pigments found in the plastids of some plant cells.

carpel (14.6) The female reproductive structure that produces seeds in a flower.

carrier (4.2) A molecule that carries one or more electrons and passes them on during photosynthesis. (5.6) A heterozygous organism that is normal for a trait but has a recessive gene for an undesirable trait and can transmit that gene to offspring. (10.5) An individual who spreads disease without showing signs of the disease himself.

carrier protein (3.5) A transport protein embedded in the cellular membrane that functions to transport specific molecules across the cellular membrane.

carrying capacity (19.3) The maximum population size that a given area can sustain.

cartilage (20.6) The soft, fibrous matrix often associated with the skeletal system of vertebrates.

caste (16.5) Among social insects, one of the groups in a colony that performs a particular function.

catabolism (4.5) The phase of metabolism that breaks down a molecule or releases energy; the destructive phase of metabolism.

catalyst (2.2) A substance that affects the rate of a reaction but is not changed in the reaction.

cecum (18.4) A pouch of the digestive system that helps some herbivores digest plant material. (21.4) A blind pouch that forms the first portion of the large intestine; its lower end forms the appendix.

cell (3.1) The functional and structural unit of life.

cell body (23.1) The part of the neuron with the greatest diameter; contains the nucleus and cytoplasmic organelles.

cell cycle (5.2) The repeating cycle of events in the life of a cell; composed of interphase, mitosis, and cytokinesis.

cell mediated immunity (22.5) An immunity to disease involving activated cells.

cell membrane (3.3) The cellular membrane that forms the outermost boundary of a cell's cytoplasm and also encloses the membrane-bound organelles within the cell.

cell plate (5.2) The precursor to the cell wall during cytokinesis in plant cells.

cell theory (3.1) The theory that all living organisms are made up of microscopic units called cells and that these cells perform all the functions of living things.

cellular differentiation (6.3) The specialization of cells that occurs during embryological development under the control of proteins.

cellular fermentation (4.3) See *fermentation*.

cellular respiration (4.3) The breakdown of foods (glucose) to release energy, including both aerobic and anaerobic cellular respiration.

cellular slime mold (11.6) A slime mold of the phylum Acrasiomycota in which cells maintain their individual membranes but appear to be one single organism.

cellulose (2.5) Chains of glucose molecules; found in plant cell walls.

cell wall (3.3) A rigid structure manufactured by the cell; located outside the cell membrane; often made of cellulose, silica, or other substances.

cementum (21.4) An external bony layer on the roots and the neck of a tooth; it anchors the tooth in the socket.

central nervous system (23.2) The part of the nervous system consisting of the brain and spinal cord.

central vacuole (3.3) A large vacuole found in some cells that regulates the water content of the cell.

centriole (3.3) An organelle composed of microtubules and located near the nucleus; doubles before cell division to establish the poles.

centromere (5.1) The attachment point of two sister chromatids; also serves as point of attachment of spindle fibers during mitosis.

centrosome (3.3) An area near the nucleus of a cell that functions in the production of microtubules.

cephalization (15.5) The presence of a "head" region, usually containing nerve tissue and supplied with sensory organs.

cephalopod (15.8) A mollusk from the class Cephalopoda.

cephalothorax (16.1) A body region in some arthropods consisting of a fused head and thorax.

cerebellum (17.2, 23.2) A part of the brain; monitors and adjusts body activities involving muscle tone, body posture, and equilibrium.

cerebral cortex (23.2) The gray matter of the cerebrum.

cerebrospinal fluid (CSF) (23.2) The fluid that nourishes and protects the brain and spinal cord and that flows between the two inner meninges.

cerebrum (17.2, 23.2) The part of the brain containing major motor and sensory centers; controls voluntary muscle activity; the area of conscious activity.

cervix (25.2) The neck of the uterus where it narrows and joins the vagina.

chelicerae (16.3) The first pair of appendages in arachnids; used for feeding; poisonous fangs in certain spiders.

chemical change (2.1) A change in which a substance loses its characteristics and changes into one or more new substances.

chemistry (2.1) The science of matter, its properties, and interactions.

chemosynthesis (4.2, 10.2) A process whereby certain organisms obtain cellular energy from the breakdown of inorganic chemicals.

chemotherapy (10.6) The use of chemical agents to treat a disease.

chemotropism (14.3) Growth movement of a plant toward or away from certain chemicals.

chitin (12.3, 16.1) A complex polysaccharide that is a component in the exoskeletons of arthropods and in the cell walls of fungi.

chlorophyll (4.2) The green pigment of plant cells that is necessary for photosynthesis.

chloroplast (3.3) An organelle that contains chlorophyll for photosynthesis.

chordate (17.1) A member of the phylum Chordata.

chorion (17.8, 25.3) In an amniotic egg, an embryonic membrane that becomes closely joined to the inner surface of the egg membrane; in humans, becomes part of the placenta.

choroid (23.5) The thin middle layer of the eyeball; contains blood vessels for nourishing the retina.

Christian worldview (8.2) A way of looking at life based on three key principles of the Bible: God made the world and placed humans at the center, the world is fallen because of man's sin, and God is working to redeem the world to Himself.

chromatid (5.1) One of the two DNA duplicates that compose one chromosome (when the chromosome is not separated).

chromatin material (3.3) A complex of DNA and surrounding proteins in the nucleus of a cell.

chromatophore (17.4) A skin cell that contains pigments.

chromosomal change (6.1) A genetic change involving either the number of chromosomes or the arrangement of genes on a chromosome.

chromosome (5.1) A strand of DNA entwined with proteins; usually found within the cell's nucleus.

chrysalis (16.5) A protective case found in the pupal stage of metamorphosis of some insects, especially lepidopterans.

chyme (21.4) The semiliquid mixture of partly digested food and digestive juices in the stomach and small intestine.

ciliary muscle (23.5) A muscle that supports the lens of the eye and can change the shape of the lens.

ciliate (11.2) A protozoan that possesses cilia.

cilium (pl., **cilia**) (3.3) One of numerous short extensions of a cell membrane; aids in movement.

circumcision (25.1) The surgical removal of the foreskin.

clade (9.4) A group of related organisms, usually depicted as a branch on a phylogenetic tree.

class (9.1) A taxonomic division of a phylum; composed of one or more orders.

classify (9.2) To assign an organism to a particular classification group.

cleavage (25.3) A series of cell divisions in the development of an embryo during which the zygote forms a multicellular mass.

climax vegetation (19.4) The predictable plant community that would normally be found in an area if it were not disturbed.

clinically dead (10.7) The state of an individual who shows no brain waves for 24–48 hours.

clitellum (15.7) The swollen region in the anterior of an earthworm; secretes the cocoon that contains the eggs.

cloaca (18.2) The terminal portion of the digestive tract in certain vertebrates that serves as a common passageway for the elimination of urine and feces.

clone (7.1) Any one of a group of organisms produced asexually from an individual organism; to reproduce organisms asexually.

closed circulatory system (15.7) A system of fluid circulation in an animal where the fluid (typically blood) remains within vessels.

cnidarian (15.4) A member of the phylum Cnidaria.

cnidocyte (15.4) A stinging cell containing a nematocyst; used by cnidarians for defense.

coagulation (22.1) The biochemical process that forms a blood clot.

coccus (pl., **cocci**) (10.1) A spherical bacterium.

cochlea (23.4) A snail-shaped division of the inner ear that functions in sound perception.

codominance (5.5) The expression (but not blending) of both alleles in a heterozygous offspring.

codon (4.4) A triplet of bases (on nucleotides) that forms the code for a particular amino acid on messenger RNA.

coelom (15.9) The body cavity of many animals.

coenzyme (2.2) A nonprotein substance that helps form the active portion of an enzyme.

cohesion (2.3, 14.1) The force that holds molecules of the same substance together; characteristic that causes water to move up plant stems.

colchicine (6.1) A poison that disrupts the spindle fibers during cell division, resulting in polyploid cells.

collar cell (15.3) One of the flagellated cells that lines the inner cavity of a sponge.

collenchyma (13.4) A ground tissue that provides strengthening in plants, especially in growing regions.

colloid (2.3) A mixture of fine particles, often including protein molecules; these particles do not settle out.

colonial organism (3.2) An organism that consists of a group of similar cells living together. Each cell functions like a unicellular organism.

columnar (13.7) A growth pattern in plants typified by a crown of leaves atop a nonbranching stem, as in palm trees.

commensalism (19.2) A relationship in which one population benefits from a second population, but the second population is not harmed or helped by the first.

common ancestor (8.4) A hypothetical organism that supposedly gave rise to two or more types of organisms.

communicable disease (10.5) A disease that can be spread from one organism to another.

companion cell (13.4) A cell paired with a sieve tube cell to provide metabolic functions for the sieve tube cell.

competition (19.2) A relationship in which two populations inhibit each other because they both depend on the same resource.

complete metamorphosis (16.4) A common type of insect development; four stages are egg, larva, pupa, and adult.

compound (2.1) A substance composed of two or more elements chemically combined in definite proportions.

compound eye (16.1) An eye composed of many individual lenses.

compound leaf (13.5) A leaf having several blades on a single petiole.

compound light microscope (1.7) A magnification device with two sets of lenses, one set for magnification and a second set to allow the observer to view the image from a convenient distance.

concentration (2.3) An expression of the proportions of solute to solvent in a solution.

concentration gradient (2.3) The difference between the number of molecules in one area and the number of the same molecules in an area nearby.

conditioned behavior (17.3) A behavioral response learned by experience.

cone (23.5) One of the cone-shaped receptors in the eye; sensitive to colors.

conidiophore (12.5) A fungus sporophore that forms asexual conidia by repeated divisions at its tip and not in an enclosure.

conjugation (10.2, 11.2) A temporary union of two organisms or cells for the one-way transfer of genetic material; type of sexual reproduction.

conjugation tube (10.2) A tube used for the transfer of genetic material between bacteria; may also function in attachment to surfaces.

connective tissue (20.2) Any of the tissues of the body that connect, support, cushion, and fill spaces around organs.

conservation, law of (2.2) The concept that the quantities of matter and energy in the universe are constant, not being created or destroyed.

constriction (17.9) A method of suffocating prey by squeezing it.

consumer (19.2) An organism that takes materials from the ecosystem.

contour feather (18.1) Any of the outer feathers on a bird; includes the flight feathers.

contractile vacuole (11.2) A vacuole, found in some cells, that collects water and expels it from the cell.

control group (1.3) The group in an experiment that is not exposed to the experimental variable.

controlled experiment (1.3) An experiment testing two identical groups for a single variable.

convergent evolution (8.5) The development of similarities in groups of organisms assumed to be unrelated, attributed to adaptation to similar environmental conditions and pressures.

core (10.3) The inner structure of a virus, composed of either DNA or RNA.

cork (13.4) The waterproof, protective layer on the outside of a stem or root that forms from dead epidermal cells.

cork cambium (13.7) A layer of cells under the epidermis that produces cork cells for protecting the stem of woody plants.

corm (13.7) A short, upright, underground stem that produces aerial leaves at its top.

cornea (23.5) The transparent anterior portion of the sclera of the eye.

coronary circulation (22.3) The circulatory pathway that supplies the heart with oxygenated blood.

corpus luteum (25.2) The structure formed in the follicle of the ovary after ovulation; forms several hormones.

cortex (13.6) The region of thin-walled parenchyma cells that stores food just inside the epidermis of a root or stem. (22.6) The outer region of the kidneys containing the nephrons.

cosmological beginnings, theory of (8.2) The attempts to explain how the universe, the earth, and even matter and energy came into being.

cotyledon (13.3) An area of stored food in a seed.

countershading (17.4) A form of color camouflage in certain animals; one color on the top side of the animal and another color on the bottom side of the animal.

courtship (18.3) Animal behavior that promotes mate selection and breeding.

covalent bond (2.1) A chemical bond formed between atoms as a result of sharing a pair of electrons.

cranial nerve (23.2) One of the nerves originating from the brain.

Creationist (8.2) One who believes the Bible's account of Creation.

Creation Mandate (1.1) God's command in Genesis 1:28 requiring that man "subdue" or have stewardship over the earth.

cristae (3.3) The folds of the inner membrane of the mitochondria; contain enzymes necessary for cellular respiration.

critical dark period (14.4) The period of uninterrupted darkness that is required for a plant to flower.

crop (15.7, 18.2) A portion of the digestive tract that temporarily stores food.

cross-pollinate (5.4) To fertilize a flower with the pollen from another flower.

crown (21.4) The visible portion of a tooth, above the gingiva.

crustacean (16.2) A member of the subphylum Crustacea.

cud (18.4) In ruminants, partially digested food regurgitated from the rumen for additional chewing to aid in digestion.

cuspid (21.4) A cone-shaped tooth used for tearing food; sometimes called a canine tooth.

cutaneous receptor (23.3) One of the receptors of cold, heat, pain, pressure, and touch located primarily in the skin.

cuticle (13.4) The protective, waxy covering found on the outer surface of the epidermis of plants; a dead layer of skin. (15.6) The noncellular covering of certain parasitic invertebrates.

cyclic (19.2) Recurring in a series; that which can be used and reused in a recurring series of events.

cyst (11.1) A structure similar to a spore that is formed as part of the life cycle of some organisms or when conditions become unfavorable. (15.5) A protective shell formed by parasite larvae.

cytokinesis (5.2) The division of the cytoplasm in a dividing cell.

cytokinin (14.3) A substance that affects plant cells in many ways, including stimulation of the division of cells.

cytology (3.1) The study of cells.

cytolysis (3.4) The bursting or disintegration of a cell.

cytoplasm (3.3) All the material inside the cell membrane of a cell, excluding the nucleus.

cytoplasmic streaming (3.3) A flowing of the cytoplasm inside the boundaries of the cell for moving the cell's contents.

cytosine (4.4) A base in a nucleic acid molecule.

cytoskeleton (3.3) The internal structure of cytoplasm; made of microfilaments and microtubules.

cytotoxic (killer) T cell (22.5) A cell that functions in immunity by attacking and destroying cells that have been affected by a particular antigen.

D

data (1.3) The recorded information from an experiment or survey.

daughter cells (5.2) The two cells that result from a mitotic division.

day-age theory (8.6) An attempt to interpret the days of Genesis 1 as ages rather than as literal 24-hour days.

day-neutral plant (14.4) A plant that flowers independently of the photoperiod.

deciduous (13.5) Describing plants that shed their leaves before a period of dormancy.

deciduous teeth (21.4) A child's first set of teeth.

decomposer organism (10.1, 19.2) An organism that breaks down dead organic matter into forms that can be used by other organisms.

deductive reasoning (1.1) The process of beginning with known facts and predicting a new fact.

degeneration, law of (2.2) The statement that in all natural processes there is a net increase in disorder and a net loss of usable energy.

dehydration synthesis (2.5) The process whereby two molecules combine and a water molecule is released.

deletion (6.2) A mutation involving the loss of a segment of chromosome during replication. See also *nucleotide deletion.*

deliquescent (13.7) A growth pattern of shrubs and trees in which the main stem branches repeatedly to form many small branches.

Deluge (8.7) The Genesis Flood (Gen. 6–8).

deluge fossil formation theory (8.7) The belief that most fossils were formed by the Genesis Flood.

dendrite (23.1) Part of the neuron that receives nerve impulses and transmits them toward the cell body.

density (19.2) A measure of the number of individuals from a population in a defined area or space.

density-dependent factor (19.3) A particular measure of the environment that becomes more limiting as a population increases.

density-independent factor (19.3) A particular measure of the environment that limits population growth regardless of the size of the population.

dentin (21.4) The primary component of teeth.

deoxyribonucleic acid (DNA) (2.6) The nucleic acid that is located primarily in the nucleus; carrier of genetic information.

dependent variable (1.3) The variable in a controlled experiment that shows the effect of the treatment.

depressant (24.3) A drug that slows down the central nervous system; may cause drowsiness and sleep.

derived characters (9.4) The shared unique features by which organisms are assigned to a particular clade.

dermal tissue (13.4) Any of the tissues that form the outer covering of a plant or other organism.

dermis (20.3) The thick inner layer of the skin.

detritivore (19.2) An animal that feeds on dead organic matter.

detritus (19.2) Dead organic matter.

diaphragm (1.7) The device under the stage of a microscope that regulates the amount of light on the specimen. (18.4) The muscle that separates the thoracic and abdominal cavities in mammals and man.

diaphysis (20.6) The shaft of a long bone.

diastole (22.2) The phase of the cardiac cycle during which the myocardium is relaxed and the heart chamber fills with blood.

diatom (11.4) One of the unicellular algae of phylum Bacillariophyta; has silicon in its cell walls.

dichotomous key (9.2) A series of paired statements used to identify a specimen.

dicot (13.3) A plant in the class Dicotyledoneae.

diffusion (2.3) The random movement of atoms, ions, or molecules from an area of higher concentration to an area of lower concentration.

diffusion pressure (2.3) The pressure for diffusion that is produced by the concentration gradient.

digestion (21.4) The process of breaking a large molecule down into its component parts.

dihybrid cross (5.5) A genetic cross dealing with two characteristics at the same time.

dimorphism (12.3) The ability of some fungi to change form in order to survive in different environments.

dinoflagellate (11.4) A member of the phylum Dinoflagellata; a unicellular alga usually possessing two flagella of unequal length and having cell walls made of cellulose plates.

diploid (5.1) Having homologous pairs of chromosomes.

disaccharide (2.5) A sugar composed of two monosaccharides.

disease (10.1) Any change, except for those caused by injuries, that affects an organism's normal function.

disorder (10.7) An affliction that is not caused by a pathogen; includes inherited conditions, injuries, and organic disorders.

DNA See *deoxyribonucleic acid.*

DNA fingerprint (7.5) A distinctive pattern of bands composed of fragments of an individual's DNA that is sufficiently specific to that individual to be used as a means of identification.

Doctrine of Humors (1.1) The belief that living things are composed of four different humors, or fluids, and that the ratio between these humors affected the function of the organism.

Doctrine of Signatures (1.1) The belief that the Creator left signs (signatures) in plants and other organisms that showed which ailments or organs they were intended to treat.

domain (9.1) The taxonomic division above the kingdom level that groups organisms based on biochemical and RNA analyses.

dominant generation (13.1) The stage that is most often seen in the life cycle of a plant.

dominant trait (5.4) The characteristic that is expressed even in the presence of the recessive genes.

dormancy (14.4) A period of greatly reduced activity in organisms.

dorsal (15.2) On or near the upper surface (the backs of bilaterally symmetrical animals and humans).

double fertilization (14.6) The union of gametes in angiosperms in which two nuclei within the ovary must be fertilized by two separate sperm nuclei.

double helix (2.6) The shape of a DNA molecule, characterized by two parallel, spiral strands.

down feather (18.1) One of the initial feathers of young birds; one of the underlying, insulating feathers in adult birds.

Down syndrome (6.1) A genetic disorder caused by a trisomy of the twenty-first chromosome.

drug (24.3) A chemical that causes a change in the function or structure of a living tissue.

duodenum (21.4) The first section of the small intestine.

dura mater (23.2) A tough membrane covering the brain and spinal cord.

dynamic equilibrium (23.4) An organism's maintaining a steady, balanced living state by expending energy; the ability of the body to respond automatically to position changes while it is moving.

echinoderm (15.9) A member of the phylum Echinodermata.

ecological pyramid (19.2) A diagram showing the quantitative relationships between the biomass or the quantities of organisms in an ecosystem.

ecological succession (19.4) The predictable, gradual change of a biotic community over a period of time.

ecology (19.1) The science of the relationships between an organism and its environment.

ecosystem (19.2) The total system of interactions between living organisms and nonliving things and factors within a limited area.

ectoderm (25.3) The outermost of the three embryonic germ layers.

ectoplasm (11.2) The thin cytoplasm on the perimeter of a cell.

ectothermic (17.1) Not able to maintain a constant body temperature; body temperature varies with the temperature of the environment.

effector (23.1) A body part (such as a muscle or gland) that responds to a stimulus as a result of a nerve impulse transmitted along neurons.

egestion (21.4) The elimination of nonsoluble, undigested wastes.

electrocardiogram (ECG or EKG) (22.2) A tracing made by an apparatus that records the impulses produced by the conduction system of the heart.

electroencephalogram (EEG) (10.7) A tracing made by an instrument used to measure brain waves.

electron (2.1) The part of an atom that has a negative charge and that moves in a shell-like orbit around the nucleus.

electron microscope (1.7) A high-powered magnification device that uses streams of electrons rather than light rays to form an image.

electron transport chain (4.2) A series of aerobic reactions that release energy as they combine hydrogen and oxygen to form water.

element (2.1) A substance that cannot be broken down into simpler substances by chemical reactions.

elongation region (13.6) The area of a plant where cells extend in length.

embolus (22.1) A blood clot, debris, or foreign matter floating in the bloodstream.

embryo (14.6) In plants, the plant within the seed. (18.3, 18.4) In animals, the young of a multicellular organism in early stages of development; the young before hatching. (25.3) In humans, the developmental time frame between three weeks and eight weeks.

embryology (25.3) The study of the development of the embryo.

embryonic membranes (17.8) The amnion, yolk sac, chorion, and allantois.

embryonic stage (15.1) One of the stages that a fertilized embryo goes through during its development.

embryonic stem cell (7.3) An undifferentiated cell, taken from an embryo, that has the potential to give rise to various other cell or tissue types.

embryo sac (14.6) The sac in a plant ovule that contains the haploid cells resulting from the division of the megaspore mother cell.

emerging disease (10.4) A viral illness that originated in animals but has been transferred and spread by humans.

emulsification (21.5) The process by which liquid fats are made into small droplets within another liquid.

enamel (21.4) Hard, durable material that covers the exposed surface of the teeth.

endocardium (22.2) The inner lining of the heart chambers.

endocrine system (24.1) A system of ductless glands that secrete hormones.

endocytosis (3.5) The movement of materials across a cell's membrane and into the cell by phagocytosis or pinocytosis.

endoderm (25.3) The inner germ layer in an embryo; forms the lining of the digestive and respiratory tracts, bladder, and urethra.

endodermis (13.6) The single-cell layer inside the cortex of a young root or stem; regulates the passage of substances into the vascular tissues.

endometrium (25.2) The lining of the uterus; formed from ciliated epithelial cells.

endoplasm (11.2) The dense cytoplasm found in the interior of a cell.

endoplasmic reticulum (3.3) A cellular structure consisting of a complex network of fine, branching tubules and interconnected folded membranes.

endoskeleton (15.1, 17.2) An internal skeleton usually composed of bone and cartilage; characteristic of vertebrates.

endosperm (14.6) The stored food that is used by the embryo in a mature seed.

endospore (10.2) An asexual spore that forms within a bacterium.

endosteum (20.6) The lining of the marrow cavity in a long bone; involved in bone growth and repair.

endothermic (2.2) An energy-consuming chemical reaction. (17.1) Able to maintain a constant body temperature.

endotoxin (10.5) A toxin produced by a cell; released only after the death and disintegration of the cell.

energy (2.2) The ability to do work.

entropy (2.2) A measure of the unusable energy that escapes when energy is being converted from one form to another; an increase in disorder and degeneration.

envelope (10.3) A membrane-like structure that forms an outer covering on some kinds of viruses.

environmental determinism (8.4) The concept that the environment determines an individual's characteristics.

enzyme (2.2, 21.5) A protein molecule that is produced by living cells to catalyze specific reactions.

epicardium (22.2) The outer connective tissue covering the muscular tissue of the heart.

epicotyl (plumule) (14.6) The portion of the plant embryo above the point of attachment to the cotyledon(s) that becomes the stem and leaves.

epidermis (13.4) In plants, the outer layer of cells that usually lack chlorophyll and serve for protection. (15.3, 20.3) In animals and humans, a tissue that usually covers or lines a structure.

epididymis (pl., **epididymides**) (25.1) A coiled tube on the outer surface of each testis that stores sperm.

epiglottis (21.1) The flap that closes the glottis during swallowing.

epinephrine (24.2) Adrenaline; a hormone secreted by the adrenal medulla that stimulates reactions needed in an emergency.

epiphysis (20.6) The end of a long bone.

epiphyte (13.2) A plant that grows on another plant or nonliving structure but is not parasitic.

epithelial tissue (20.2) A tissue of the body that covers or lines a body part; functions in absorption, secretion, and protection.

equilibrium (2.3) A state of balance as is seen when diffusion has progressed to the point where there are no longer regions of higher and lower concentrations.

erythrocyte (22.1) A red blood cell.

esophagus (15.7, 21.4) The tube of the alimentary canal connecting the pharynx and stomach.

estivation (17.6) A period of inactivity and slowed metabolism whereby some animals escape unfavorably hot weather conditions.

estrogen (24.2) A class of female sex hormones that stimulate the development of secondary sex characteristics.

estrus (18.4) The period when a female mammal is ovulating and receptive to mating.

ethylene (14.3) A gaseous plant hormone that causes fruit to ripen.

etiolated (14.4) A plant's growing in the absence of light; having thin, elongated stems with small, pale leaves.

etiology (10.5) The study of the causes of disease.

eugenics (6.5) The science that deals with improvement of the human race by applying principles of genetics.

euglenoid movement (11.4) A modified type of amoeboid movement seen in euglena.

eukaryotic cell (3.3) A cell that possesses both organelles and a nucleus that is surrounded by a nuclear membrane.

euphoria (24.3) A feeling of well-being often induced by drugs.

euploidy (6.1) A chromosome number that is an exact multiple of the haploid number for that organism.

eustachian tube (17.7, 23.4) One of the tubes leading from the pharynx to the middle-ear space to equalize air pressure.

euthanasia (25.4) The act of ending a person's life to relieve him from suffering.

evolutionary classification (9.4) A taxonomy based on characteristics such as genetic similarities and reproductive capabilities rather than physical characteristics.

evolutionist (8.2) One who believes in the three components of evolutionary theory.

evolution, philosophy of (8.2) The belief that all things are progressing toward a future perfection.

evolution, theory of (8.2) A composite of ideas involving the philosophy of evolution, the theory of cosmological beginnings, and the theory of biological evolution.

evolution, theory of biological (8.2) The theory of the beginning of life and the slow process of organisms becoming more complex.

excretion (3.1, 20.3, 22C) The elimination of soluble wastes.

excurrent (13.7) A cone-shaped branching pattern in trees and shrubs.

excurrent siphon (15.8) The tube that expels water from the body of a mollusk.

ex nihilo (8.2) Latin for "out of nothing"; descriptive of God's creative acts.

exocrine gland (20.5) A gland that releases its secretions through a duct.

exocytosis (3.5) The process in which vacuoles or vesicles fuse with the cell membrane to release particles or substances from the cell.

exon (4.4) A section of RNA that is kept when forming messenger RNA.

exoskeleton (15.1, 16.1) A system of external plates that protect and support.

exotoxin (soluble toxin) (10.5) A toxin that diffuses from a microorganism into the surrounding tissue while the pathogen is still alive.

experiment (1.3) See *controlled experiment*.

experimental group (1.3) The group in an experiment that is exposed to the experimental, or independent, variable.

expiration (21.2) Breathing air out of the lungs; exhalation.

expiratory reserve volume (21.2) The amount of air that can be forced out of the lungs after a normal expiration.

exponential growth (19.3) The rate of population growth in which the population size multiplies at a constant rate at regular intervals.

external auditory canal (23.4) The canal from the outer ear to the eardrum.

external digestion (12.3) The process in which enzymes are secreted to digest food outside the organism.

external fertilization (17.2) The uniting of the sperm and egg outside the organism.

external respiration (21.2) The transfer of oxygen from the air in the alveoli into the blood.

extracellular digestion (4.5) The breakdown of substances that occurs in spaces outside the cells such as within the stomach or intestine.

eyespot (11.1) A light-sensitive area of some organisms.

facilitated diffusion (3.5) Passive transport that requires the presence of a protein factor in the cellular membrane.

facultative anaerobe (10.2) An organism that can grow with or without the presence of oxygen.

family (9.1) A taxonomic division of an order; composed of one or more genera.

fang (17.9) A hollow, needlelike tooth of a reptile used to inject venom into prey.

fascicle (20.8) A small bundle of muscle fibers.

fatty acid (2.5) A common lipid composed of a chain of 14–28 carbon atoms with a carboxyl group on the end.

feather (18.1) One of the proteinaceous structures that covers birds and enables flight.

feces (21.4) Solid waste material of an organism.

feedback mechanism (24.1) A process in which the secretions of one gland either stimulate or inhibit the secretions of a different gland.

fermentation (4.3) The anaerobic breakdown of carbohydrates to pyruvic acid, and then to alcohol and carbon dioxide or lactic acid.

fern (13.2) A vascular plant of the phylum Pteridophyta; sporophyte generation is dominant.

fertilization (5.3) The process of forming a zygote; the union of gametes.

fetoscope (6.5) A fiber-optic device which can be inserted into the womb to take pictures of the unborn child.

fetus (25.3) The term applied to an embryo; in humans, the unborn child from the second month of development until birth.

fever (10.6) Elevated body temperature.

fiber (21.5) The indigestible cellulose and other roughage from fruits and vegetables that is essential to good health.

fibrin (22.1) An insoluble protein involved in blood clotting; forms a fine, interlacing network of filaments that traps blood cells.

fibrous root system (13.6) The branching root system of vascular plants that lack a taproot.

fibrovascular bundle (13.7) In herbaceous plants, a bundle composed of vascular tissues surrounded by fibrous tissue.

filament (10.1) A chain of cells. (14.6) The stalk of the stamen.

first law of thermodynamics (2.2) The concept that the quantities of matter and energy in the universe are constant, not being created or destroyed.

flagellate (11.2) An organism with one or more flagella.

flagellum (pl., **flagella**) (3.3) A long, tubular extension of a cell's membrane that aids in movement.

flame cell (15.5) A cell that possesses tufts of cilia; part of the planarian's excretory system.

flatworm (15.5) A member of the phylum Platyhelminthes.

fluke (15.5) A parasitic flatworm from the class Trematoda.

follicle (20.4) The tube from which a hair grows. (25.2) A small saclike structure in the ovary that encloses an immature ovum.

follicular phase (25.2) The first phase of the ovarian cycle; characterized by the development of certain ovarian follicles into ova.

fontanel (20.7) One of the fibrous membranes that fills the spaces between the skull bones at birth, forming the "soft spot."

food (1.6) Organic, energy-containing substances required by all living things.

food chain (19.2) A representation of the nutritional relationships between organisms in an ecosystem.

food web (19.2) A method of illustrating multiple nutritional relationships and interactions between populations in an ecosystem.

foot (15.8) Fleshy, muscular organ of locomotion for most mollusks.

fornication (25.5) Any act of sexual relations between people not married to each other.

fossil (8.3) Any evidence or remains of an organism preserved in the earth's crust.

fovea (23.5) A small depression in the central region of the retina; contains a concentration of cones.

fragmentation (11.2) A form of reproduction caused by the breaking of a colonial organism by a physical disturbance.

frame shift (6.2) A DNA mutation in which a nucleotide or nucleotides (other than three in number) are added to or taken from the chain so that subsequent codons are read incorrectly.

frond (13.2) A leaf of a fern or a palm; a thallus that resembles a leaf.

frontal lobe (23.2) The region of the brain that controls reasoning, communication, and commands for voluntary body movements.

fruit (13.3, 14.6) A ripened plant ovary with or without seeds.

fruiting body (11.6) The spore-producing reproductive structure of a slime mold.

G

gallbladder (21.4) The pear-shaped sac on the underside of the liver; concentrates and stores bile.

gamete (5.3) A haploid cell which can unite with another gamete to form a zygote.

gametophyte (13.1) The stage that produces gametes in the life cycle of a plant.

ganglion (pl., **ganglia**) (15.7) A mass of nerve tissue.

gap theory (8.6) An interpretation of the Genesis Creation account that states that there was a long period of time between Genesis 1:1 and 1:2.

gastric cecum (16.4) One of the finger-shaped organs, attached to an insect's stomach, that secretes enzymes for digestion of foods.

gastrodermis (15.4) The inner cellular layer of the digestive tract of cnidarians.

gastropod (15.8) A mollusk from the class Gastropoda.

gastrovascular cavity (15.4) The internal cavity of cnidarians where digestion and food circulation occur.

gel (2.3) The semisolid state of a colloid.

gel electrophoresis (7.5) The stage of DNA fingerprinting in which DNA fragments are separated and sorted by passing an electric current through gel.

gemmule (15.3) In a sponge, an internal dormant cluster of cells encased in a tough spicule-reinforced covering.

gene (5.1) A segment of DNA capable of producing a specific amino acid chain (polypeptide) resulting in a particular characteristic.

gene expression (6.3) The activation or turning on of a gene that results in its transcription and the production of a specific protein.

gene mutation (6.2) A changing of the gene itself, which alters the sequence of nucleotide bases within a gene.

gene pool (6.4) The sum of all the alleles that every member of a species' population could possess at a given time.

gene therapy (7.4) A technique to correct a specific disease or disorder caused by a defective gene by replacing the defective gene with a properly functioning gene.

genetically modified (GM) plant (7.6) A plant that has undergone genetic engineering to improve its characteristics.

genetic disorder (inherited disorder) (6.3) Any undesirable phenotype caused by genetic defects in an individual.

genetic drift (6.4) The genetic change due to chance, especially seen in small populations.

genetic engineering (7.2) The manipulation of chromosomes or genes by methods other than normal reproduction.

genetic equilibrium (6.4) A point at which the allele frequencies in a given population do not change.

genetic load (8.5) The number of mutations within an organism or its gene pool.

genetics (5.1) The study of heredity.

genetic screening (6.5) Medical and nonmedical methods used to find information on an individual's genetic makeup.

genome (7.1) The hereditary information encoded in an organism's DNA; a complete haploid set of an organism's chromosomes.

genotype (5.4) The genetic makeup of an individual organism.

genus (9.1) A group of organisms that has one or more common characteristics; includes one or more species; taxonomic division of a family.

genus-species name (9.2) The scientific name for an organism.

geographic isolation (9.3) The inability of two subgroups of a population to interbreed due to physical separation.

geologic column (8.5) The sequence of fossil-containing rock layers that shows the supposed record of the history of life on the earth.

germination (14.6) The beginning of growth by a seed, spore, bud, or other structure following a state of dormancy.

germ mutation (6.2) A mutation that affects the gamete-producing cells.

gerontology (10.7) The science of aging.

gestation (18.4) The period of pregnancy.

gibberellin (14.3) A plant hormone that causes rapid elongation of stems.

gill (12.8) The thin, spore-producing membrane of certain fungi. (15.8, 17.4, 17.6) A respiratory structure in aquatic organisms through which oxygen and carbon dioxide are exchanged.

gill slit (17.5) One of the external openings into gills found in some jawless and cartilaginous fish.

girdling (13.7) Removing a ring of bark from a woody stem.

gizzard (15.7, 18.2) A thick-walled digestive organ that grinds food.

global warming (19.6) The rise in temperature that has been noted over the last century on the earth.

glomerulus (22.6) A collection of blood capillaries surrounded by the Bowman's capsule of the nephron.

glottis (17.7, 21.1) The space between the vocal folds.

glucagon (24.2) A hormone that raises the blood sugar level; also stimulates breakdown of fats to form glucose.

glucose (2.5) A common six-carbon simple sugar.

glycogen (2.5) A polysaccharide; animal starch; branching chains of glucose molecules.

glycolysis (4.3) The breakdown of glucose to pyruvic acid during cellular respiration.

goiter (24.2) A swelling of the thyroid gland.

Golgi apparatus (3.3) A membrane-bound organelle that deals with synthesis and packaging of materials.

gonad (24.2) A reproductive organ.

grafting (14.5) The joining of two plant parts, usually stems, so that their tissues grow together.

granum (pl., **grana**) (3.3) One of the structures within chloroplasts that contains the chlorophyll and other pigments involved in photosynthesis.

gravitropism (14.3) Growth movement of a plant in response to gravity.

gray matter (23.2) Nerve tissue composed mainly of neuron cell bodies and dendrites.

green gland (16.2) An organ that excretes wastes in some crustaceans.

greenhouse effect (19.6) The phenomenon in which gases in the atmosphere prevent some of the sun's radiation from returning to space, thus maintaining a warm temperature on the earth.

ground tissue (13.4) The tissue in leaves, young roots, and stems that is not epidermal or vascular in function.

growth hormone (24.2) A hormone produced by the anterior pituitary gland; primarily affects bone growth.

guanine (4.4) A base in a nucleic acid molecule.

guard cell (13.5) One of the cells surrounding a stoma that controls the opening and closing of the stoma.

gullet (11.2, 17.7) A food passageway into the digestive tract of an animal.

guttation (14.1) The process whereby drops of water are forced through pores at the tip and edges of a leaf.

gymnosperm (13.3) A nonflowering, seed-producing plant; seeds are not contained in an ovary when mature.

habitat (19.3) Where an organism lives; the "address" of an organism.

half-life (8.5) The length of time necessary for a radioactive substance to decay to half its original amount.

hallucinogen (24.3) A drug that amplifies the senses, affects a person's judgment, and can produce hallucinations (visions).

halophile (10.1) An organism that thrives in a salty environment.

halophyte (14.2) A plant that grows in soil that has high salt concentrations.

haploid (5.1) Having only one member of each homologous pair of chromosomes; characteristic of gametes.

Hardy-Weinberg principle (6.4) The concept that the frequency of alleles and the ratio of heterozygous to homozygous individuals in a given population will remain constant unless the population is affected by outside factors.

haustoria (12.4) Hyphae of parasitic fungi which enter the host's cells to obtain nourishment.

Haversian system (20.6) A unit of bone in compact bone tissue.

helper T cell (22.5) A cell that functions in immunity by rapidly reproducing itself and causing the production of particular antibodies.

hemoglobin (17.2, 21.2) The red pigment of erythrocytes that transports oxygen and carbon dioxide.

hemophilia (5.6) Bleeder's disease; a genetic disorder in which a blood chemical for blood clotting is not produced.

hemotoxin (17.9) Venom that affects blood cells.

hepatic portal vein (22.3) The vein that carries food-rich blood from the digestive organs to the liver.

herbivore (adj., **herbivorous**) (17.2) An animal that eats plants.

hermaphrodite (adj., **hermaphroditic**) (15.5) An organism that has both male and female reproductive organs.

heterocyst (10.1) A large, colorless cell in the filaments of certain cyanobacteria.

heterogamete (5.3, 11.3) A gamete that differs in size and shape.

heterosexuality (25.5) Sexual desire for members of the opposite sex.

heterosis (**hybrid vigor**) (6.4) An increased capacity for growth or strength in a hybrid.

heterotroph (4.1) An organism that depends on other organisms for food.

heterozygous (5.4) Having two different alleles at the same position (locus) on homologous chromosomes.

hibernation (17.6, 18.4) A state of extremely slow metabolism by which certain animals survive unfavorable conditions.

hilum (14.6) The point at which a seed (ovule) is attached to the ovary wall.

histamine (10.6, 22.5) A compound produced by the body and released as part of an allergic reaction, resulting in inflammation.

histology (20.2) The study of tissues.

histone (5.1) A protein that supports, protects, and helps maintain the tightly coiled structure of the DNA in a chromosome.

HIV See *human immunodeficiency virus*.

holdfast (11.3) The special structure that anchors an organism.

homeostasis (3.4) The equilibrium or internal "steady state" that every living organism must maintain.

homeotic gene (6.3) A section of the DNA molecule that regulates morphogenesis by controlling gene expression.

homologous pair of chromosomes (5.1) Two chromosomes that have the same kinds of genes (alleles) in the same order.

homologous structures (8.5) Organs that are similar in structure between two organisms; once thought to show evolutionary relationships.

homologue (5.1) One member of a homologous pair of chromosomes.

homosexuality (25.5) Sexual desire for members of the same sex.

homozygous (5.4) Having the same two alleles at the same position (locus) on homologous chromosomes.

hormone (14.3) In plants, a chemical regulator produced in meristematic tissues, affecting cell maturation. (24.1) In animals, a chemical regulator produced in ductless glands and carried in blood, affecting metabolism.

host (19.2) An organism in or on which a parasite lives.

Human Genome Project (7.3) The attempt to decode the human genome, begun in 1990 and completed in 2003.

human immunodeficiency virus (HIV) (22.5) The virus that causes AIDS.

humoral immunity (22.5) An immunity to disease involving antibodies.

hybridization (6.4) The crossbreeding of two genetically unrelated individuals.

hydrogen bond (2.1) Weak intermolecular attraction between a hydrogen atom of one molecule and a nitrogen, oxygen, or fluorine atom of another molecule; especially significant in affecting the characteristics of water.

hydrolysis (2.5) A reaction whereby a compound is split apart by the addition of a molecule of water.

hydrophilic (2.5) Attracted to or having an affinity for water.

hydrophobic (2.5) Not having an affinity for water.

hydrostatic skeleton (15.4) A support system in some soft-bodied animals in which water pressure keeps the body firm and enables movement through the flexing of muscles.

hyperopia (23.5) An eye condition in which it is easy to focus on distant objects but not on near objects.

hypersecretion (24.2) Excessive production of a substance.

hypertension (22.3) High blood pressure.

hypertonic solution (3.4) A solution in which the concentration of solutes is greater than in the cytoplasm of living cells.

hypha (pl., **hyphae**) (12.4) One of the slender filaments that composes the mycelium of a fungus.

hypocotyl (14.6) The stem portion of an embryonic plant in a seed.

hypoglycemia (24.2) Low blood sugar; often caused by an overproduction of insulin.

hyposecretion (24.2) Insufficient production of a substance.

hypothalamus (23.2) The region of the brain that controls involuntary activities, emotional expressions, appetite for food, and release of certain hormones.

hypothesis (1.3) An educated guess about the solution to a problem; helps define and direct an experiment.

hypotonic solution (3.4) A solution in which the concentration of solutes is less than in the cytoplasm of living cells.

identify (9.2) To determine the group in which an organism belongs.

immunity (adj., **immune**) (10.3, 22.5) The ability to resist infection or to overcome the effects of infection.

imperfect fungi (12.8) Any fungi in which sexual reproduction has not been observed.

implantation (25.3) The process whereby the tiny embryo attaches to the uterine wall and forms the placenta.

inborn behavior (17.3) A pattern of reaction and response that the organism has inherited and does not need to learn.

inbreeding (6.4) The mating of closely related organisms.

incisor (18.4, 21.4) A flat, thin tooth used in gnawing, biting, and cutting food.

incomplete dominance (5.5) The type of inheritance in which the alleles for expressing characteristics are neither dominant nor recessive.

incomplete flower (14.6) A flower that lacks petals or sepals or either the stamen or carpel.

incomplete metamorphosis (16.4) A type of insect development in which eggs hatch into nymphs that eventually grow into adults.

incubation period (10.5) The time between infection by a pathogen and the appearance of the first symptoms.

incurrent siphon (15.8) The tube that draws water into a mollusk.

independent assortment (5.5) The Mendelian idea that the separation of one set of alleles during gamete formation is not affected by the separation of another set of alleles.

independent variable (1.3) The variable in a controlled experiment that is manipulated by the researcher; also called the experimental variable.

individual characteristic (5.1) A trait that differs among members of a species; variation.

inductive reasoning (1.1) The process of beginning with many facts or assumptions in order to reach a general conclusion.

infection (10.5) The condition of the body after it has been invaded by harmful organisms.

infectious disease (10.5) A disease caused by a pathogen.

inflammation (10.6) The reaction of tissues to injury or infection; characterized by increased blood flow, redness, pain, and swelling.

infusion (1.1) A nutrient-rich solution in which microorganisms can live.

ingestion (3.3, 21.4) The intake of food.

inhalant (24.3) A substance that releases vapors that are then inhaled and taken into the bloodstream.

inorganic (1.6, 2.4) Describing objects that are not alive and that have never been alive; also, substances that lack carbon (with few exceptions); usually derived from nonliving material.

insect (16.4) A member of the class Insecta.

insecticide (16.5) A chemical or agent that destroys insects.

insectivorous plant (14.2) A plant that captures and digests insects.

insertion (20.10) The point of attachment of a muscle's tendon to a more movable bone.

inspiration (21.2) Filling the lungs with air; inhalation.

inspiratory reserve volume (21.2) The amount of air that can be forced into the lungs beyond a normal inspiration.

instinct (17.3) Elaborate, often highly complex inborn behavior.

insulin (24.2) A hormone produced by the pancreas; helps control glucose level in blood.

intelligence (17.3, 20.1) Behavior marked by the ability to use knowledge to manipulate the environment or to communicate.

intelligent design (ID) (8.6) A movement that asserts that living things show evidence of design that cannot be totally explained by the random processes of Darwinian evolution.

interferon (10.4) A protein substance or substances produced by cells exposed to viruses; acts to slow the spread of a virus.

intermediate host (15.5) An animal that temporarily harbors an immature form of a parasite.

internal fertilization (17.2) The fertilization of the ovum inside the female's body.

internal movement (1.6) An attribute of life that includes movement of fluids, organs, and even structures within cells.

internal respiration (21.2) The exchange of gases between the blood and the body cells.

interneuron (23.1) A neuron located in the central nervous system; transmits an impulse from a sensory neuron to another neuron.

interphase (5.2) The period of time between cellular divisions.

intestine (15.5) A section of the alimentary canal where most of the digestion and absorption of foods usually occurs.

intracellular digestion (4.5) The breakdown of substances within cells.

intron (4.4) A section of RNA that is cut out when forming messenger RNA.

invasive species (19.5) An organism that, when introduced to an area outside its original range, becomes a nuisance due to excessive growth or reproduction.

inversion (6.2) A chromosomal mutation in which genes break off a section of DNA and reattach but in the opposite order.

invertebrate (15.2) An animal that lacks a backbone or vertebral column.

in vitro fertilization (6.5) The union of sperm and egg in an artificial setting, usually in a laboratory.

ion (2.1) An atom or group of atoms that has a positive or negative charge as a result of losing or gaining electrons.

ionic bonding (2.1) The formation of a chemical bond between ions of opposite charge.

iris (17.7, 23.5) The colored portion of the eye.

irreducible complexity (8.6) A tenet of the intelligent design movement claiming that many structures and processes in living things have so many interdependent parts and steps in their actions that they could not operate unless all the components were present simultaneously in their finished state.

irritability (1.6) The ability to respond to changes in the environment.

islets of Langerhans (24.2) Endocrine glands that consist of small groups of cells in the pancreas that secrete the hormones glucagon and insulin.

isogamete (5.3) A gamete that is similar in shape and size to another gamete.

isotonic solution (3.4) A solution that has the same concentration of solutes as the cytoplasm of living cells.

isotope (2.1) One of the forms of an atom produced by having different numbers of neutrons but the same number of protons in the nucleus.

Jacobson's organ (17.8) Sensory pit used by a reptile to detect chemicals in the air.

joint (20.7) The point where two bones come together.

karyotype (5.1) An illustration in which the chromosomes of a cell are arranged according to their size.

keel (18.1) The ridge on a bird's sternum.

keratin (17.8, 20.3) A tough, fibrous protein found in reptile scales, bird feathers, and mammal and human hair.

kidney (17.2, 22.6) The organ in most vertebrates that filters waste from the blood and excretes it in a liquid called urine.

kinetic energy (2.2) The energy of motion; may take the form of heat, light, electricity, etc.

kingdom (9.1) One of the six broad groups of organisms to which all species are assigned.

Koch's Postulates (10.5) A set of laws developed by Robert Koch to conclusively determine the cause of a disease.

Krebs cycle (4.3) The stage of aerobic cellular respiration in which pyruvic acid reacts with an enzyme that removes a carbon from the pyruvic acid to produce acetyl CoA, CO_2, hydrogen ions, ATP, and electrons; also called the *citric acid cycle*.

labor (25.3) The period immediately preceding the birth of a baby during which contractions expel the child from the uterus through the birth canal.

lacrimal gland (23.5) The tear gland in the eye.

lactic acid fermentation (4.3) A series of reactions resulting in the formation of lactic acid from glucose.

large intestine (colon) (21.4) The part of the alimentary canal that extends from the small intestine to the anus.

larva (pl., **larvae**) (15.1, 16.4) An immature stage in the life cycle of many animals, usually different from the adult.

larynx (18.4, 21.1) The short passageway that leads from the pharynx to the trachea; the sound-producing organ; the voice box.

lateral bud (13.7) A bud at the base of the petiole; a bud on the side of a branch.

lateral line (17.4) A canal running the length of a fish's body that detects vibrations in the water.

layering (14.5) A method of vegetative reproduction in which a branch is exposed to the soil, allowed to form roots, and then separated from the parent plant.

leaflet (13.5) One of the blades of a compound leaf.

leaf scar (13.5) A layer of cork cells left on the stem after a leaf falls.

leaf venation (13.5) The pattern of the veins of a leaf.

leafy shoot (13.1) A stem-and-leaf-like arrangement of mosses that lacks water-conducting tissues.

learned behavior See *conditioned behavior*.

lens (23.5) The transparent structure within the eye that focuses light rays on the retina.

lenticel (13.7) A small opening in the cork layer of older woody roots and stems through which air is admitted into the plant.

lethal mutation (6.2) A mutation that causes the death of the organism.

leucoplast (3.3) A colorless plastid used as a storehouse in a cell.

leukocyte (22.1) A white blood cell.

lichen (12.8) A fungus and an alga or cyanobacterium living together in a symbiotic relationship.

life (living condition) (1.6) A highly organized cellular condition that is derived from preexisting life and that faces death; requires energy to carry on processes such as growth, movement, reproduction, and response.

ligament (20.7) A band of connective tissues that holds a joint together.

light-dependent phase (4.2) The first phase of photosynthesis; requires light energy to energize electrons in pigments.

light-independent phase (4.2) See *Calvin cycle*.

limiting factor (19.3) Something that in some way restricts the growth or existence of an organism.

lipid (2.5) An organic compound that is insoluble in water but soluble in certain organic solvents.

lipid bilayer (3.3) Description of the two layers of phospholipids that make up the cell membrane.

literal interpretation (8.6) The most obvious meaning of a word or phrase.

liver (21.4) The largest organ in the body; secretes bile, purifies blood, metabolizes food molecules, and stores minerals and vitamins.

living condition (1.6) See *life*.

locus (pl., **loci**) (5.5) The specific location of a gene on a chromosome.

logical reasoning (1.1) The process of arriving at a conclusion through a series of ordered steps.

long-day plant (14.4) A plant that flowers when the period of light is more than twelve hours.

long day theory See *day-age theory*.

lung (17.6, 21.1) A structure for the exchange of gases between the atmosphere and the blood of an organism.

luteal phase (25.2) The phase of the ovarian cycle from ovulation through the development of the corpus luteum.

lymph (22.4) The fluid found between body cells; absorbed in the lymphatic system and returned to the bloodstream.

lymphatic system (22.4) A network of vessels and nodes carrying lymph in a vertebrate's body.

lymph node (22.4) A small mass of tissue through which lymph passes and in which lymphocytes are found.

lymphocyte (22.4) A type of white blood cell that functions in immunity.

lysis (10.3) The rupturing of a cell.

lysogenic cycle (10.3) The process in which a virus remains latent in cells but spreads by becoming part of the host cell genome. Factors may then trigger these viruses to become lytic.

lysosome (3.3) A membrane-bound organelle that contains various hydrolytic enzymes.

lytic cycle (10.3) The sequence of events whereby a virus replicates within a cell and eventually destroys the cell.

M

macronucleus (11.2) An organelle found in certain protozoans; contains multiple copies of the cell's genetic material.

macrophage (22.4) A large, phagocytic cell found in the lymphatic system and surrounding tissues.

madreporite (15.9) A sievelike structure of the starfish's water-vascular system that helps vary the pressure.

malaria (11.2) A disease affecting the human's circulatory system that is caused by the blood parasite *Plasmodium* and spread by *Anopheles* mosquitoes.

malignant (6.3) Characterized by rapid and chaotic growth; often spreads and may be fatal.

Malpighian tubule (16.4) One of numerous threadlike tubules in insects that extracts wastes from the blood and empties them into the intestine.

mammal (18.4) A member of the class Mammalia.

mammary gland (18.4) Organ of female mammals and humans that produces milk to nourish the young.

manager (19.5) Man's role in having dominion in which he changes the environment to better meet his needs.

mandible (16.4) A paired chewing mouthpart found in insects and some other arthropods.

mantle (15.8) The sheath of tissue that covers the body of a mollusk; also secretes the shell.

mantle cavity (15.8) The space between the mantle and the body of a mollusk.

marsupial (18.4) A mammal that possesses a marsupium; a pouched mammal.

mass selection (6.4) The method for selecting breeding stock in which only the desirable organisms are selected.

mastication (21.4) The chewing of food.

matrix (20.2) Nonliving material in a tissue; secreted by the tissue's cells.

matter (2.1) Anything that occupies space and has mass.

maturation region (13.6) The area of a young root or stem in which the primary tissues are developed; area of cell differentiation.

medulla (22.6) The inner region of an organ.

medulla oblongata (17.2, 23.2) A part of the brain; the relay center between spinal cord and brain; contains several reflex centers.

medusa (15.4) The free-swimming, umbrella-shaped stage in the life cycle of cnidarians; reproduces sexually.

meiosis (5.3) Cell division in which the chromosome number is reduced from the diploid to the haploid state.

melanin (20.5) A dark brown or black pigment.

melanocyte (20.5) A cell in the human epidermis that produces melanin.

memory T cell (22.5) A cell that functions in immunity by remaining in the circulatory system and dividing rapidly to form cytotoxic and helper T cells when an antigen reappears.

meninges (23.2) The protective coverings of the brain and spinal cord.

menopause (25.2) The point in a woman's life when the ovarian and menstrual cycles cease.

menstrual cycle (25.2) The process by which the uterine lining is prepared to receive an embryo.

menstruation (25.2) The time during which the uterine lining is shed.

meristematic region (13.6) An area in a plant containing young, rapidly dividing cells.

meristematic tissue (13.4) A plant tissue that is able to reproduce and become other plant tissues.

mesenchyme (15.3) A noncellular, jellylike matrix between cell layers of a sponge; contains the amoebocytes.

mesentery (17.7, 21.4) Any of the transparent membranes that surround body organs and attach them to the body wall.

mesoderm (15.5, 25.3) The middle tissue layer in some animals; the middle germ layer in an embryo.

mesoglea (15.4) The jellylike layer found between the epidermis and the gastrodermis of the cnidarians.

messenger RNA (mRNA) (4.4) The RNA molecule that carries the code for a polypeptide chain from the DNA.

metabolism (4.4) The sum of all reactions that occur in a living organism.

metamorphosis (16.4, 17.6) A change in shape or form that an animal undergoes in its development from egg to adult.

metaphase (5.2) The second phase of mitosis; chromosomes congregate along the equatorial plane of the cell.

metastasize (10.7) The ability of cancer cells to separate from the tumor and travel by the circulatory or lymphatic system to different parts of the body.

microbe (1.1) A microscopic organism.

microfilament (3.3) A flexible, rodlike assembly of protein molecules found in cells.

micronucleus (11.2) The small reproductive nucleus in some protozoans.

micropyle (14.6) A small opening between the integuments where the pollen tube may enter the ovule.

microscope (1.7) A scientific instrument that magnifies objects for more detailed study; usually contains a series of lenses.

microtubule (3.3) A hollow, spiral assembly of protein molecules that composes flagella, cilia, mitotic spindles, and other cellular structures.

midbrain (23.2) The structure between the thalamus and pons that controls body movements and posture as well as vision and hearing reflexes.

midline (median) (15.2) Dividing into right and left halves (only for animals with bilateral symmetry).

migration (9.3) The movement of an organism from one location to another.

milt (17.4) The sperm and seminal fluids of certain aquatic animals; released into the water.

mimicry (19.2) An organism's appearing like another organism; form of protective coloration.

mineral (21.5) An inorganic substance found in soil; many obtained from food and essential for good health.

missing link See *common ancestor*.

mitochondrion (pl., **mitochondria**) (3.3) A membrane-bound cellular organelle responsible for the respiration of foods to release usable energy.

mitosis (5.2) The duplicating and separating of a cell's chromosomes.

mixture (2.1) A material that contains two or more substances.

model (1.2) An explanation or representation of how something works.

molar (18.4, 21.4) A cheek tooth in humans and some mammals used for grinding.

molecular formula (2.1) A description of a molecule that indicates the number and kinds of atoms in the molecule.

molecule (2.1) The smallest possible unit of a substance that consists of two or more atoms.

mollusk (15.8) A member of the phylum Mollusca.

molt (16.1, 17.8, 18.1) To shed an exoskeleton, scales, feathers, or fur.

monocot (13.3) A member of the class Monocotyledoneae; any flowering plant whose embryo has only one cotyledon.

monohybrid cross (5.4) A genetic cross that deals with only one set of characteristics.

monomer (2.5) One of the repeating units within a polymer or macro-molecule.

monosaccharide (2.5) A simple sugar.

monosomy (6.1) A condition in which there is only one of a homologous chromosome pair.

monotreme (18.4) A mammal that reproduces by laying eggs.

morphogenesis (6.3) The change in form that occurs in an organism.

mother cell (5.2) Any cell that is ready to begin cell division.

motor neuron (23.1) A neuron that receives impulses from the central nervous system and stimulates muscles or glands.

mRNA See *messenger RNA*.

mucus (17.4) A slimy substance on the surface of mucous membranes and on the exterior of many fish and aquatic animals.

mulch (14.2) Decomposing organic matter; often added to the soil to enrich its mineral content or texture and to preserve soil moisture.

multicellular organism (3.2) An organism that consists of many cells.

multiple alleles (5.5) The possible arrangement of three or more genes (alleles) for a trait at a single locus.

muscle fatigue (20.9) The inability of muscle fibers to respond (contract) after prolonged use.

muscle fiber (20.8) A muscle cell.

muscle tissue (20.2) Tissue made of cells that can contract to cause movement.

mutagen (6.2) A substance that induces mutation.

mutation (6.2) A random change in a DNA molecule.

mutation-selection theory See *Neo-Darwinism*.

mutualism (19.2) A form of symbiosis in which the organisms depend on each other for protection and nourishment.

mycelium (pl., **mycelia**) (12.4) One of the hyphae in a fungal organism.

mycology (12.1) The study of fungi.

mycoplasma (10.2) A type of bacterium that lacks a cell wall.

mycorrhiza (12.2) Symbiotic relationship between fungal hyphae and plant roots.

myelin sheath (23.1) The white, fatty membrane that protects neurons.

myocardium (22.2) The muscular tissue of the heart.

myofibril (20.8) One of the functional fibers within a muscle that causes contraction by the movement of actin and myosin filaments.

myopia (23.5) An eye condition in which only light rays from close objects can be focused accurately on the retina.

myosin (20.8) One of the two types of protein found in muscle fibers.

naiad (16.5) An aquatic insect nymph that possesses gills.

narcotic (24.3) An addictive drug that induces a sense of euphoria and sleepiness, followed by anxiety; derived from opium or manufactured synthetically.

nastic movement (14.1) Any of the movements of some plants due to the loss of turgor pressure in cells, such as the opening and closing of petals.

natural system of classification (9.4) See *evolutionary classification*.

neck (21.4) The region of a tooth that is surrounded by the gingiva.

negative feedback system (24.1) A biochemical system in which the products inhibit the operation of the process.

nematocyst (15.4) A stinging cell, characteristic of cnidarians, that contains poisonous barbs, coiled threads, or a sticky substance.

Neo-Darwinism (mutation-selection theory) (8.5) An evolutionary theory proposing that mutations produce variations and that natural selection determines which variations will survive in order to produce biological evolution.

nephridium (pl., **nephridia**) (15.7) A tubelike structure that filters wastes from blood.

nephron (22.6) A microscopic tubular unit of a kidney.

nerve cord (17.1) Length of nerve tissue that connects the brain to the rest of a chordate's body.

nerve impulse (23.1) An electrochemical pulse that moves along the membrane of a neuron.

nerve net (15.4) A nervous system that lacks a brain and major ganglia.

nervous tissue (20.2) Body tissue capable of responding to changes and conducting electrical impulses.

netted venation (13.5) Vein arrangement in a leaf where the veins form a branching network throughout the leaf.

neuron (23.1) The functional unit of the nervous system; the cell that receives and distributes nerve impulses.

neurotoxin (17.9) Venom that affects the nervous system.

neurotransmitter (23.1) A chemical that a neuron releases into the synaptic cleft for the purpose of stimulating receptor proteins in the membrane of the next neuron.

neutralism (19.2) A situation in which there is no direct relationship between populations in an environment.

neutron (2.1) The non-charged particle in an atom's nucleus.

niche (19.3) What an organism does, including its relationship to and effect on its habitat.

nictitating membrane (17.7) A thin, transparent membrane that protects the eye and keeps it moist.

nitrogen fixation (19.3) A process in which certain bacteria capture atmospheric nitrogen and convert it to stored nitrogen compounds.

node (13.7) The place where a leaf, root, or flower attaches to the stem.

nonbiodegradable (19.6) Not capable of being broken down naturally in the environment.

noncyclic (19.3) Not recurring in a series; that which is not used again.

nondisjunction (6.1) The failure of a pair of homologous chromosomes to separate during meiosis.

norepinephrine (noradrenaline) (24.2) A hormone secreted by the adrenal medulla; functions with adrenaline during stressful situations.

notochord (17.1) A tough, flexible rod of cartilage, usually located along the dorsal side of an animal; supports the animal's body.

nuclear envelope (membrane) (3.3) The double membrane forming the surface of the nucleus in eukaryotic cells.

nucleic acid (2.6) An organic compound in living cells that is responsible for passing on hereditary information; DNA and RNA.

nucleoid region (10.1) A non-membrane-bound mass of DNA and proteins in a prokaryotic cell.

nucleolus (3.3) A spherical body in the nucleus that has a high concentration of RNA and proteins.

nucleotide (2.6) The basic component of a DNA or RNA molecule; each is made up of a sugar, a phosphate, and a base.

nucleotide addition (6.2) A genetic mutation in which an additional nucleotide is inserted into the sequence.

nucleotide deletion (6.2) A genetic mutation in which a nucleotide is lost from a DNA chain.

nucleus (2.1) The positively charged central portion of an atom. (3.3) The region of a eukaryotic cell that contains the chromosomes.

nyctinastic movement (14.1) A movement in plants caused by a gradual change in turgor pressure; sometimes called "sleep movements."

nymph (16.4) One of the stages of incomplete metamorphosis in an insect.

objective (1.7) The part of a light microscope that is near the specimen and that contains lenses; that which forms an image of an object.

obligate aerobe See *aerobe, obligate.*

obligate anaerobe See *anaerobe, obligate.*

occipital lobe (23.2) The brain region responsible for vision and memory.

ocular (1.7) The eyepiece of a microscope; contains lenses.

olfactory lobe (17.2) A part of the brain that receives impulses from smell receptors in the nostrils.

omnivore (adj., **omnivorous**) (17.2) An animal that eats both plants and animals.

oncogene (6.3) A mutated gene that causes cells to divide uncontrollably.

oogenesis (5.3) The meiotic process that forms ova.

oogonium (pl., **oogonia**) (11.3) The structure that produces the ovum.

operculum (pl., **opercula**) (17.4) A plate that covers the gills of a fish.

opposable thumb (18.5) A digit able to be placed opposite the other digits (fingers or toes).

optic lobe (17.2) A division of the brain that receives impulses from the eyes.

oral groove (11.2) The funnel-shaped indentation in the body of the paramecium; lined with cilia to sweep food into the mouth pore.

order (9.1) A taxonomic division of a class; composed of one or more families.

organ (3.2) A group of tissues that perform a specific function.

organelle (3.3) A specialized structure within a cell that performs a specific function.

organic (1.6) Naturally derived from living organisms; organic compounds are those that contain carbon.

organ system (3.2) A group of organs that work together to accomplish a life function.

origin (20.10) The point of attachment of a muscle's tendon to a more stationary bone.

osculum (pl., **oscula**) (15.3) The opening of the sponge's body that expels water.

osmosis (2.3) Diffusion of water molecules through a semipermeable membrane.

ossicles (23.4) The bones in the middle ear: malleus, incus, stapes.

ossification (20.6) The process of converting cartilage tissue into bone.

osteocyte (20.6) A living bone cell.

ostium (pl., **ostia**) (15.3) A tiny opening in the sponge's body for the intake of water.

ova See *ovum.*

oval window (23.4) A membrane-covered opening of the inner ear which receives vibrations transmitted from the ossicles.

ovarian cycle (25.2) The process by which the ovary prepares and releases an ovum.

ovary (14.6) In plants, the part of the carpel containing the ovules that mature into the fruit containing the seeds. (25.2) In animals and humans, the primary sexual reproductive organ in females; produces ova.

oviduct (25.2) Fallopian tube; tube transporting the ovum from the ovary to the uterus.

oviparous (17.2) Describing a method of reproduction in which young develop within eggs that are laid and hatched outside the body of the parent.

ovipositor (16.4) An insect organ used to deposit eggs.

ovoviviparous (17.2) Describing a method of reproduction in which young develop within the egg that hatches in the body of the parent.

ovulation (25.2) The release of ova from the ovary.

ovule (14.6) A structure in a plant ovary that contains the egg cell and will mature into a seed.

ovum (pl., **ova**) (5.3) A gamete formed by a female; usually nonmotile and larger than a sperm.

oxygen debt (20.9) The amount of oxygen that must be supplied to change lactic acid to glucose during physical exercise.

oxyhemoglobin (21.2, 22.1) A molecule that forms when oxygen combines with hemoglobin in the blood.

oxytocin (24.2, 25.3) A hormone released by the posterior pituitary; stimulates the smooth muscle of the uterus to contract during birth.

ozone (19.6) The triatomic form of oxygen that is found in the atmosphere.

P

palate (21.1) The structure serving as the floor of the nose and the roof of the mouth.

palisade mesophyll (13.5) The primary photosynthetic tissue in plant leaves that has the cells lined up side by side.

palmate (13.5) Having veins, leaflets, or lobes that originate from a common point.

pancreas (21.4, 24.2) An organ that secretes enzymes into the duodenum to perform digestion (also secretes hormones).

pandemic (22.5) An epidemic over a widespread geographic area.

pantheism (19.5) The worship of the universe and its phenomena as god.

papilla (pl., **papillae**) (18.1) A tiny bump on the skin of a bird from which a feather protrudes. (21.4) A small bump on the tongue surface in which a taste bud is located.

parallel venation (13.5) Vein arrangement in a leaf where the veins originate from the stem and remain nearly parallel to the tip.

parasite (10.2, 15.5, 19.2) An organism that obtains its nourishment by living in or on another organism.

parasympathetic nervous system (23.2) A system of neurons that helps the body return to normal processes after a stressful situation.

parathyroid gland (24.2) A gland of the endocrine system that regulates calcium and phosphorus levels in the body.

parenchyma (13.4) The tissue in plants that composes the pith, cortex, spongy tissue of leaves, and major parts of fruits.

parietal lobe (23.2) The brain region responsible for most sensations, such as pain, pressure, touch, and temperature.

parthenogenesis (6.1) Reproduction in which organisms develop from unfertilized ova.

passive immunity (22.5) An immunity in which an individual has been given antibodies that have been formed by another individual or an animal.

passive transport (3.5) The movement of substances through a cellular membrane without the expenditure of cellular energy.

pathogen (10.5) An agent that causes a disease.

pathologist (10.7) One who studies diseased tissues.

pectin (3.3) A jellylike substance that helps solidify a cell wall.

pectoral girdle (17.2, 20.6) The part of the appendicular skeleton designed to support and provide attachment for the upper limbs.

pedicel (14.6) The stalk that supports the flower.

pedigree (5.4) A diagram that shows the characteristics of several generations of organisms.

pedipalp (16.3) The second pair of arachnid appendages; used for sensory perception and sperm transfer.

pellicle (11.2) A firm yet flexible covering outside the cell membrane of certain protozoans.

pelvic girdle (17.2, 20.6) The hip bones; designed to support and provide attachment for the lower limbs.

penis (25.1) The organ that transfers sperm from male to female in humans and many vertebrates.

peptidoglycan (10.1) A compound found in the cell walls of eubacteria; important in bacterial classification due to Gram staining.

perennial (13.1) A plant that lives for many years.

pericardium (22.2) A fibrous sac covering the heart and protecting it from rubbing against the lungs and chest wall.

pericycle (13.6) A layer of meristematic tissue in a root.

periosteum (20.6) A layer of fibrous tissue covering the diaphysis of a long bone; serves for muscle attachment and bone growth and repair.

peripheral nervous system (23.2) The division of the nervous system containing the nerves that originate in the central nervous system and the sensory organs.

peristalsis (21.4) Muscular movements that move food in the alimentary canal.

persistent foliage (13.5) Leaves that do not fall until pushed off by the new growth in spring.

petal (14.6) One of the flower structures, just inside the sepals; often large and conspicuous.

petiole (13.5) The stalk connecting the blade of a leaf to the stem.

PGA See *phosphoglyceric acid.*

PGAL See *phosphoglyceraldehyde.*

pH (2.3) The measure of the concentration of hydrogen ions in a solution using values from 0 to 14.

phagocytosis (3.5) The process of a cell engulfing a substance.

pharyngeal pouches (17.1) Folds of skin along the neck region of vertebrate embryos that develop either into structures of the lower face, neck, and upper chest or into gill openings.

pharynx (15.5, 21.1) The portion of the digestive tract that connects the mouth cavity and the esophagus; also serves as the passageway for air from nose to larynx.

phenomenon (1.4) Something that can be observed or measured; may be an object, property, or process.

phenotype (5.4) The physical expression of an organism's gene.

pheromone (16.4) A chemical released by an animal which influences the behavior of another animal of the same species.

phloem (13.4) A vascular tissue that usually carries water and dissolved foods downward in plants.

phosphoglyceraldehyde (PGAL) (4.2) A three-carbon sugar produced during the Calvin cycle of photosynthesis; also found in glycolysis.

phosphoglyceric acid (PGA) (4.2) A three-carbon acid that forms from RuBP during photosynthesis; also forms during glycolysis.

phospholipid (2.5) A molecule consisting of two fatty acid molecules and a phosphate group attached to a glycerol molecule.

photolysis (4.2) The breaking apart of a water molecule by energized chlorophyll.

photoperiodism (14.4) The response of a plant to changes in light intensity and length of days.

photosynthesis (4.2) The process whereby simple sugars are formed from carbon dioxide and water in the presence of light and chlorophyll.

phototropism (14.3) Growth movement of a plant in response to light.

phylogenetic tree (9.4) A diagram that demonstrates the supposed stages of evolution.

phylum (9.1) A taxonomic division of a kingdom; composed of one or more classes.

physical change (2.1) Altering a substance in its state of matter and appearance without changing it into a new substance.

physical withdrawal (24.3) Physical symptoms that occur when a person stops taking a physically addictive drug.

physiology (20.2) The science that deals with the various processes and activities that occur within a living organism.

phytochrome (14.4) A protein plant pigment that regulates a plant's response to photoperiod changes.

phytoplankton (11.3) Plankton that are photosynthetic organisms.

pigment (4.2) A light-absorbing molecule that functions in photosynthesis.

pineal gland (24.2) A small structure in the brain that secretes melatonin.

pinnate (13.5) Having parts originating from a central axis.

pinocytosis (3.5) The process whereby a cell takes in fluid by forming vesicles.

pith (13.7) The central area of a woody stem, composed mainly of parenchyma cells.

pituitary gland (24.2) An endocrine gland attached to the lower part of the brain.

placenta (14.6) In plants, the structure that holds the ovule to the ovary wall. (18.4, 24.2, 25.3) In animals and humans, the structure that consists of a portion of the uterine wall and chorion of the embryo; allows nutrient and waste exchange between mother and embryo.

planarian (15.5) Common free-living flatworm of the class Turbellaria.

plankton (11.3) A tiny floating aquatic organism.

plasmid (7.2, 10.1) A small, circular piece of DNA; most often found in bacteria.

plasmodial slime mold (11.6) An acellular slime mold of the phylum Myxomycota.

plasmodium (11.6) A multinucleate mass of cellular material that forms the vegetative body of a slime mold.

plasmolysis (3.4) The shrinking of a cell's contents when the cell loses water.

plastid (3.3) A membrane-bound organelle found in plants, algae, and a few other organisms, but not in animals. (See also *leucoplast* and *chloroplast.*)

plastron (17.9) The ventral part of the turtle's bony shell.

platelet (22.1) A small, colorless body found in the blood; lacks hemoglobin and a nucleus; involved in blood clot formation; a thrombocyte.

pleura (21.1) A delicate membrane that lines the thoracic cavity and covers the lungs.

point mutation (6.2) A gene mutation involving only one or a few nucleotides.

polar molecule (2.3) A molecule with charged poles (not balanced by symmetry).

pollen cone (13.3) A structure on a conifer that produces pollen.

pollen grain (14.6) The structure of plant reproduction that contains the tube and sperm nuclei of plants; produced in the anther.

pollen tube (14.6) The structure that grows from a pollen grain down through the style into the ovule and through which the sperm travel.

pollination (14.6) The process in which pollen is transferred from the anther to the carpel or from the male cone to the female cone.

pollution (19.6) Contamination of the environment with substances or factors that change the environment significantly.

polygenic inheritance (5.5) The cumulative effect of two or more genes on the same trait.

polymer (2.5) A macromolecule made up of a chain of monomers, sometimes identical.

polymerase chain reaction (PCR) (7.5) A technique often used in DNA fingerprinting that takes a tiny amount of DNA and quickly produces millions of copies.

polyp (15.4) A sessile, tubular cnidarian with a mouth and tentacles at one end and a basal disc at the other; reproduces asexually.

polypeptide chain (2.6) A chain formed by many peptide bonds, as in the formation of a protein by many amino acids being bonded by peptide bonds.

polyploid (6.1) Having three or more complete sets of chromosomes.

polysaccharide (2.5) A large, complex carbohydrate composed of many monosaccharides.

pons (23.2) A rounded portion of the lower brain that relays information from one side of the brain to the other; contains reflex centers.

population (6.4, 19.2) All the members of the same type of living thing within an area.

population genetics (6.4) The study of the types and frequencies of genes in a given population.

poriferan (15.3) A member of the phylum Porifera; a sponge.

portal circulation (22.3) The flow of blood from the digestive organs to the liver.

postdiluvian (8.7) After the Flood.

potential energy (2.2) Energy that is stored until being released.

precocial chick (18.3) A bird that has a long incubation period and is usually able to care for itself when it hatches.

predation (19.2) Situation in which an organism (predator) kills and eats another organism (prey).

preformationist (5.1) One who believes that completely formed organisms exist within sperm.

prehensile tail (18.5) A tail that can be flexed and is used for gripping or holding things.

premolar (21.4) A tooth in front of the molars used for crushing and grinding food.

pressure-flow model (14.1) An explanation for how carbohydrates are translocated in plants.

primary growth (13.6, 13.7) The increase in length of a root or stem.

primary productivity (19.2) The rate of photosynthesis carried on by an ecosystem's producers.

primary succession (19.4) The type of ecological succession that starts from bare rock and must first form soil.

primary tissue (13.6) A plant tissue formed by the apical meristem; the plant tissue that results from primary growth.

prion (10.3) An abnormal form of protein found in some cells and linked to disease.

producer (19.2) An organism that produces its own food; a photosynthetic or chemosynthetic organism; an autotroph.

product (2.2) The physical result of a chemical reaction.

proglottid (15.5) A segment of a tapeworm's body.

progressive creationism (8.6) The belief that the physical universe is old and that at certain points in time God created certain organisms.

prokaryotic cell (3.3) A cell that lacks a nuclear membrane and has only non-membrane-bound organelles; found in kingdoms Archaebacteria and Eubacteria.

prophase (5.2) The first phase of mitosis; centromeres migrate to poles in the cytoplasm and chromosomes develop from chromatin material in the nucleus.

prostate gland (25.1) A structure that produces a portion of the semen.

protein (2.6) An organic compound that is composed of amino acids.

prothallus (13.2) The heart-shaped gametophyte generation in ferns.

protist (11.1) A member of the kingdom Protista; eukaryotic but not a true animal, plant, or fungus.

proto-oncogene (6.3) A gene that codes for the development of proteins affecting cell division.

protozoan (11.1) A member of the kingdom Protista that exhibits animal-like characteristics; most are motile and non-photosynthetic.

pseudopodium (pl., **pseudopodia**) (3.3, 11.2) A cytoplasmic extension of a cell; used for locomotion or engulfing substances.

psychoactive drug (24.3) A drug that alters the emotional state of the user and often alters his sense of reality.

psychological addiction (24.4) An emotional dependence on a substance after a period of use.

puberty (24.2) The period of hormonal-induced change during which the secondary sex characteristics develop.

pulmonary circulation (22.3) The flow of blood from the right ventricle to the lungs and back to the left atrium.

pulp cavity (21.4) The central region of a tooth; filled with blood vessels, lymph vessels, nerves, and connective tissues.

pulse (22.3) The dilation and rebounding of an artery's wall in response to the surge of blood from the heart's contraction.

punctuated equilibrium (8.5) The theory that evolution occurs rapidly for a period of time followed by a long period of nonevolving before another period of rapid evolution.

Punnett square (5.4) A diagram used to visualize genetic crosses.

pupa (pl., **pupae**) (16.4) One of the stages of complete metamorphosis of an insect; the resting or inactive stage.

pupil (17.7, 23.5) The circular opening in the iris of the eye.

pure science (1.5) Knowledge obtained through scientific activities.

pure strain (6.4) An organism that is homozygous for certain traits.

pus (10.6) A thick yellowish fluid composed of leukocytes, bacteria, and broken cells; characteristic of infections.

pyrenoid (11.4) A protein-containing structure present in the chloroplasts of algae; center for starch storage.

pyruvic acid (4.3) The organic acid formed during glycolysis.

Q

quadrate bone (17.9) A snake bone loosely attached to the skull and the jaw that enables the snake to open its mouth wide.

quarantine (16.5) Strict isolation to prevent the spread of disease or pests.

quill (18.1) The shaft portion of a feather that extends below the vane; the part of the feather in the skin.

R

rachis (18.1) The slender, central part of a bird's feather from which the barbs protrude.

radial symmetry (15.2) A body pattern that can be divided into equal halves through the center point of the body and along its length.

radicle (14.6) The portion of a plant embryo that will become the root.

radiocarbon dating method (carbon-14 dating method) (8.5) A method of determining the age of fossils using the half-life of carbon-14 as a basis.

radiometric dating method (8.5) A method of determining the age of an object by measuring the amount of a radioactive substance that is part of the object.

radula (15.8) A platelike structure in the pharynx of certain mollusks; composed of rows of tiny teeth.

ray (15.9) The arm of a starfish.

reactant (2.2) The starting substance of a chemical reaction.

receptacle (14.6) The enlarged end of the pedicel; bears the flower parts.

recessive trait (5.4) The characteristic that is expressed only in the homozygous recessive condition.

recombinant DNA (7.2) DNA which has had a new section of DNA that contains specific gene(s) spliced into it.

rectum (21.4) The muscular portion of the large intestine that contracts in order to rid the body of feces.

red bone marrow See *bone marrow, red.*

referred pain (23.3) Pain that seems to be in one area of the body but actually originates in a different area.

reflection (1.7) The image caused by light rays bouncing off an object.

reflex (17.3) An automatic, involuntary response to a stimulus.

reflex arc (23.1) A series of neurons that produces a single reaction in response to a stimulus.

refraction (1.7) The bending of a light ray when it passes from one medium to another at an oblique angle.

renal circulation (22.3) The flow of blood into and out of the kidneys.

renal medulla (22.6) The inner region of a kidney.

replication (2.6) The process whereby a DNA molecule duplicates itself and forms a new DNA molecule.

reproduction (1.6) The formation of another organism that has characteristics and limitations similar to the original.

reptile (17.8) A member of the class Reptilia.

research (1.3) An investigation into a topic often carried on by reading, inquiry, or scientific observation.

research method (1.5) The use of the scientific method to obtain knowledge.

residual volume (21.2) The amount of air in the lungs after all the vital capacity has been expired.

resolution (1.7) The characteristic that allows a microscope to form a clear image of detailed structures.

resting potential (23.1) The relative charge difference between the inside and the outside of a neuron.

restriction enzyme (7.2) An enzyme that is able to cut and separate DNA strands.

retina (23.5) The innermost layer of the eyeball; composed of specialized neurons and their fibers.

retrovirus (10.3) A special type of RNA virus that contains the enzyme reverse transcriptase.

rhizoid (12.4, 13.1) A rootlike structure that lacks water-conducting tissue.

rhizome (13.2, 13.7) A thick, fleshy, horizontal underground stem that produces leaves or leaf-bearing branches.

ribonucleic acid (RNA) (2.6) The type of nucleic acid that forms from DNA and functions with ribosomes to form protein molecules. (See also *messenger RNA, ribosomal RNA,* and *transfer RNA.*)

ribosomal RNA (rRNA) (4.4) The RNA molecule that combines with proteins to form a ribosome.

ribosome (3.3) A non-membrane-bound cellular organelle associated with protein formation.

ribulose biphosphate (RuBP) (4.2) A five-carbon sugar biphosphate that serves as a carbon dioxide acceptor in photosynthesis and then splits to form two molecules of PGA.

rickettsia (10.2) A group of obligate parasites in the kingdom Eubacteria.

RNA See *ribonucleic acid.*

RNA polymerase (2.6) An enzyme that separates the DNA double helix to initiate transcription.

rod (23.5) One of the sensory receptors in the eye that is not sensitive to color but can rapidly discern movements and is important in night vision.

root (13.6) The organ of a vascular plant that absorbs water and minerals necessary for growth; often serves to anchor the plant. (20.4) The part of the tooth that anchors it in the socket.

root cap (13.6) A layer of thick-walled cells that cover and protect the delicate root tip.

root hair (13.6) An outgrowth of epidermal cells of the root.

root pressure (14.1) The force exerted on the water in the vascular cylinder that results from the movement of water into a root.

roundworm (15.6) A member of the phylum Nematoda.

rRNA See *ribosomal RNA.*

RuBP See *ribulose biphosphate.*

rumen (18.4) The first of four chambers of the stomach of a ruminant.

rust (12.8) A fungus of the phylum Basidiomycota with a complex life cycle; most rusts are harmful plant diseases.

salivary gland (16.4, 21.4) A gland that secretes saliva to break down starches.

sanctity of human life (25.4) The biblical principle that human life has high value.

saprophytic (10.2, 11.4) Relationship in which an organism obtains its nourishment from dead organic matter.

sarcomere (20.8) One of the functional segments of a muscle.

saturated (2.5) Describing a fatty acid molecule with only single bonds between carbon atoms.

scale (13.3) One of the small overlapping leaves of some gymnosperms. (17.4, 17.8) A small body-covering plate on fish and reptiles.

science (1.2) A body of facts that man has repeatedly observed about the physical universe around him.

scientific method (1.3) A logical method of problem solving that involves observing and reaching a conclusion.

scientism (8.2) The belief that science is the only way to find truth.

scion (14.5) The unrooted portion of a plant that is grafted into the stock; may be a bud or twig.

sclera (23.5) The outer layer of the eye; the "white of the eye."

sclerenchyma (13.4) A rigid type of ground tissue that provides support for nongrowing areas of a plant.

scolex (15.5) The anterior end of a tapeworm.

scrotum (25.1) A skin pouch within which the testes are located.

scute (17.9) One of the broad scales on a snake's belly that aids in movement.

seaweed (11.3) A term often applied to the larger multicellular types of algae.

sebaceous gland (20.5) A gland of the skin that produces oil.

sebum (20.5) The material secreted by sebaceous glands.

secondary growth (13.6, 13.7) The increase in diameter of stems and roots.

secondary sex characteristic (24.2) A body characteristic caused by the sex hormones.

secondary succession (19.4) The type of ecological succession that starts at a point where soil and some plants are already present.

secondary tissue (13.6) Any tissue that is manufactured in a plant after primary growth.

secular (8.2) Free from any religious influences or beliefs.

sedimentary rock (8.7) Layers of rock formed by sedimentation.

sedimentation (8.7) The settling out of materials due to the action of water or wind.

seed (13.1, 14.6) A mature plant ovule that consists of an embryo and stored food enclosed by a coat.

seed cone (13.3) The structure on conifers that produces the seeds.

segregation, concept of (5.4) The Mendelian concept that only one gene for each characteristic may be carried in a particular gamete.

selective breeding (5.1) The process in which man crosses animals or plants to produce desirable traits.

selectively permeable membrane See *semipermeable membrane*.

self-pollination (5.4) The process by which a plant's structure allows its own pollen to provide the sperm for fertilization of ova.

semen (25.1) The fluid that contains sperm.

semicircular canal (23.4) A structure in the inner ear that maintains dynamic equilibrium.

semilunar valve (22.2) A membranous structure located at the exit of each ventricle; permits one-way flow of blood; pulmonary and aortic.

seminiferous tubule (25.1) One of the tiny tubes that produces sperm within the testes.

semipermeable membrane (2.3) A membrane that is permeable to certain molecules or ions but not to others.

sensory neuron (23.1) A neuron that carries impulses toward the spinal cord or the brain.

sensory receptor (23.3) A specialized dendrite of sensory neurons found in sensory organs.

sepal (14.6) The outermost flower structure; usually encloses the other floral parts in the bud.

septum (pl., **septa**) (12.4) A wall that separates individual cells in fungi. (15.7) One of the inner divisions of a segmented worm. (17.8) The muscular wall separating the chambers of either side of the heart.

sequencing (7.3) A technical process for determining the order of nucleotides in an organism's DNA.

serum (22.5) The clear fluid (obtained from blood) that contains antibodies; used to transfer immunity to another person or animal.

sessile (11.3) In organisms, growing while attached to something else; nonmotile. (13.5) In leaves, lacking a petiole.

seta (pl., **setae**) (15.7) One of the stiff bristles on a segmented worm; used for locomotion and sensation.

sex chromosome (5.6) A special chromosome (in humans, X or Y) that determines whether an organism will be male or female.

sex-linked trait (5.6) An inherited characteristic for which there is a gene on the X or Y chromosome but not on the other.

sexually transmitted disease (STD) (25.5) A contagious disease spread by sexual contact with an infected person.

sexual reproduction (5.3) The union of haploid gametes that results in a diploid zygote that develops into a new individual.

shaft (20.4) The visible portion of a hair.

shell (15.8) The mineralized covering secreted by the mantle of some mollusks; supports and protects the body. (17.8) The outer covering of an amniotic egg.

short-day plant (14.4) A plant that flowers when the period of light is less than twelve hours.

sieve tube cell (13.4) One of the hollow cells that forms the vessels of phloem for conduction through plants.

simple eye (16.1) An eye with only one lens.

sinoatrial node (SA node) (22.2) A small mass of specialized cardiac muscle located in the right atrium; performs the job of starting each systole; the cardiac pacemaker.

siphon (15.8) A tube in a mollusk used to draw in or expel water.

skeletal muscle (20.8) Muscle tissue that is attached to and moves the skeleton; striated and voluntary.

skin gill (15.9) A fingerlike projection on the surface of echinoderms; used for respiration and excretion.

slime mold (11.6) A funguslike protist that is classified as either a cellular or plasmodial species.

small intestine (21.4) The digestive organ where most of the digestion and absorption of food occurs.

smut (12.8) A harmful plant pathogen of phylum Basidiomycota, usually producing several different kinds of spores, but having only one host species.

solute (2.3) The dissolved substance in a solution.

solution (2.3) The uniform dissolving of one substance into another substance.

solvent (2.3) The substance (often a liquid) into which a solute is dissolved.

somatic (adult) stem cell (7.3) A stem cell obtained from the differentiated body tissue of adults or children.

somatic mutation (6.2) A mutation that affects only body cells (not gametes).

sonography See *ultrasonic scanning*.

soredium (pl., **soredia**) (12.8) An asexual reproductive structure in lichens; consists of a group of algal cells enclosed in fungal hyphae.

sorus (pl., **sori**) (13.2) A group of sporangia attached to the underside of fern fronds.

spawn (17.4) In aquatic animals, to lay eggs.

speciation (9.3) The process by which new species develop.

species (9.1, 9.3) A population of organisms that are structurally similar but do have a degree of variation; a group of organisms that interbreed and produce fertile offspring.

species characteristic (5.1) A trait that applies to all members of a species.

specified complexity (8.6) A tenet of the intelligent design movement attempting to show whether a particular structure or process is truly the result of design; it must be both specific and complex.

sperm (5.3) A gamete formed by a male; often motile.

spermatogenesis (5.3) The meiotic process of sperm formation.

spherical symmetry (15.2) A body pattern that can be divided into equal halves by a cut in any direction as long as it passes through the center point of the body.

spicule (15.3) A sharp, pointed, supporting structure in sponges; composed of silicon or calcium compounds.

spinal cord (23.2) The nervous tissue that conducts messages between the brain and the peripheral body parts.

spinal nerve (23.2) A mixed nerve attached to the spinal cord.

spine (13.7) In plants, a stiff, hard, pointed outgrowth on stems. (17.1) In animals, the backbone of a vertebrate. (17.4) In fish, a bony support in fins.

spinneret (16.3) An organ in spiders and some insects that spins web from the secretions of silk glands.

spiracle (16.4) One of the small pores in an insect's body that function in breathing.

spirillum (pl., **spirilla**) (10.1) A spiral-shaped bacterium.

spirochete (10.2) The common name for a group of spiral-shaped organisms in the kingdom Eubacteria.

spleen (22.1, 22.4) A lymphatic organ that filters blood, stores red blood cells, and destroys old red blood cells.

spongy bone (20.6) The type of bone that contains many small spaces; located in the ends of a long bone.

spongy mesophyll (13.5) The photosynthetic tissue in leaves that is formed of irregularly shaped cells and many air spaces.

spontaneous generation (1.1) The formation of living organisms from nonliving materials.

spontaneous mutation (6.2) A chromosomal or gene mutation that occurs naturally.

sporangiophore (12.5) A fungus sporophore that produces its asexual spores within an enclosure.

sporangium (pl., **sporangia**) (11.5, 12.5, 13.2) A structure in which spores are produced.

spore (5.2) A cell with a hard protective covering that is capable of producing a new organism.

sporophore (12.4) A spore-producing hypha of a fungus mycelium; spore-producing stage in the life cycle of a plant.

sporophyte (13.1) The stage that produces spores in the life cycle of a plant.

stage (1.7) The flat surface of a microscope on which a slide or specimen is placed.

stamen (14.6) The male reproductive structure of a flower.

starch (2.5) A polysaccharide; often used for energy storage by plants.

start codon (4.4) The triplet AUG on mRNA. It signifies the starting point for translation.

static equilibrium (23.4) The sense of body position a person has when not moving.

statocyst (16.2) An organ of equilibrium in crustaceans.

STD See *sexually transmitted disease*.

stem cell (7.3) A generalized cell that has the ability to differentiate into cells for specific functions.

steroid (2.5) A lipid composed of a carbon backbone of four carbon rings and a side chain of carbon atoms; many function as hormones.

stigma (14.6) The sticky tip of the carpel that receives the pollen.

stimulant (24.3) A substance that increases the action of the central nervous system, often masking fatigue.

stipe (12.8) The stalk of a mushroom.

stipule (13.5) A structure at the base of the petiole of a leaf; often leaflike.

stock (14.5) The rooted plant onto which a scion is grafted.

stolon (12.4) In fungi, an aerial, horizontal hypha that produces new fungi asexually. (13.7) In plants, a slender, branched, underground stem that produces new shoots.

stoma (pl., **stomata**) (13.5) An opening between the guard cells of a leaf that permits exchange of gases.

stomach (21.4) The muscular pouch of the digestive system that comes after the esophagus.

stop codon (4.4) Any of the codons UAA, UAG, and UGA. These codons do not code for amino acids but instead signal the end of the translation process.

stratum corneum (20.3) The dead outer cell layer of the epidermis, sloughed off by the body.

stratum germinativum (20.3) The deepest layer of the epidermis; carries on cell division.

stroma (4.2) Material within the chloroplast that surrounds the grana of thylakoids.

structural defense (10.6) The body's first line of defense in preventing pathogens from entering the body; includes the skin, mucous membranes, tears, and stomach acid.

structural formula (2.1) An expanded drawing that shows the arrangement of atoms and bonds within the molecule.

style (14.6) The elongated portion of the carpel that supports the stigma.

subcutaneous layer (20.3) The layer of fat and connective tissues below the dermis of the skin.

substitution (6.2) A gene mutation that occurs when a nucleotide in the DNA sequence is removed and replaced with a different nucleotide or when two nucleotides are inverted.

substrate (2.2) The chemical or chemicals an enzyme will affect.

succession (19.4) See *ecological succession*.

succulent leaf (13.7) A thick leaf with a tough cuticle; capable of storing water.

sucrose (2.5) A common disaccharide; table sugar; a substance made of one glucose molecule and one fructose molecule.

surrogate mother (6.5) A woman who nurtures and brings to birth from her womb a child for which she did not provide the ovum.

survey (1.3) A scientific study used to determine what exists or what is a common practice in an area.

survival of the fittest (8.4) Part of Darwin's evolutionary theory; only the organisms best suited to their environment will survive.

suspension (2.3) The state of a substance when its particles are mixed but are not dissolved.

suture (20.7) An interlocking margin of skull bones.

sweat gland (20.5) One of the glands in the skin that releases perspiration to cool the body and release wastes.

swim bladder (17.4) A structure found in many fish that enables them to float.

swimmeret (16.2) One of the small appendages on some crustaceans that is used for swimming and reproduction.

symbiosis (12.8, 19.2) Situation in which two organisms of different species live together in close association; lichens are an example.

symmetry (15.2) A likeness in size, shape, or structure of parts of an organism.

sympathetic nervous system (23.2) A system of neurons that helps the body adjust to stressful situations.

synapse (23.1) A space between an axon and a dendrite or between the end of an axon and the body structure it affects.

synaptic cleft (23.1) The space between the axon terminal and a neuron.

synovial fluid (20.7) A lubricating fluid in joints.

synovial membrane (20.7) The lining of a joint cavity that fills the cavity with a lubricating fluid.

syrinx (18.2) The song box of a bird.

system (3.2, 20.2) A group of body organs that works together to perform one or more vital functions.

systematics (9.1) The science of classifying organisms.

systemic circulation (22.3) The flow of blood from the left ventricle to all parts of the body, except the lungs, and back to the right atrium.

systole (22.2) The phase of the cardiac cycle when the myocardium contracts and the heart chamber pumps blood.

tadpole (17.6) The larval stage of most frogs and toads.

talon (18.1) One of the long curved claws found on birds of prey.

tapeworm (15.5) A flatworm from the class Cestoda.

taproot system (13.6) The plant root system in which the primary root continues to grow as the main root.

target cell (24.1) Cell that is activated by hormones.

taste bud (21.4, 23.3) A group of cells near the tongue surface that produces taste sensations.

taxis (pl., **taxes**) (11.2) An organism's response to a single stimulus.

taxonomy (9.1) The science of classifying organisms.

T cell (22.5) Any of several types of cells involved in cell-mediated immunity.

technical method (1.5) The use of prescribed techniques to gain knowledge about a specific case.

tegument (15.5) A protective body covering.

telophase (5.2) The final phase of mitosis; chromosomes have reached opposite poles of the cell and two distinct nuclei form.

temporal lobe (23.2) The brain region responsible for hearing and smelling.

tendon (20.7) The connective tissue that attaches muscles to bone.

test cross (5.4) The mating of an organism that possesses a dominant phenotype but unknown genotype with an organism that possesses a recessive phenotype to determine the genotype of the dominant individual.

testes (sing., **testis**) (25.1) The primary male reproductive organs, which produce sperm and male sex hormones.

testosterone (24.2) The male sex hormone that promotes the development of secondary sex characteristics; one of the androgens.

test-tube fertilization See *in vitro fertilization.*

tetrad (5.3) A group of four joined chromatids during meiosis.

tetraploid (6.1) Having four complete sets of chromosomes (genomes).

thalamus (23.2) The brain region that receives general sensations and relays impulses to the parietal lobe.

thallus (pl., **thalli**) (11.3) A leaflike or plantlike structure in algae, fungi, and plants; not differentiated into true leaves, roots, or stems.

theistic evolution (8.6) The interpretation of the Bible according to evolutionary theories, purporting that God used evolution as His means of creation.

theory of descent with modification (8.4) Part of Darwin's evolutionary theory that states that newer organisms are modified versions of older organisms.

theory of evolution See *evolution, theory of.*

theory of inheritance of acquired characteristics (8.3) An evolutionary theory that states that a characteristic acquired by an organism can be passed on to its offspring.

theory of natural selection (8.4) A process that supposedly results in the survival of the organisms that are best suited for their environment.

theory of need (8.3) An evolutionary theory that states that an organism must have a need for a structure in order to evolve it.

theory of use and disuse (8.3) An evolutionary theory that states that if an organ is used, it will keep evolving; if it is not used, it will degenerate.

thermoacidophile (10.1) Any of the members of the kingdom Archaebacteria that live in highly acidic soils and hot springs.

thigmonastic movement (14.1) A movement in plants caused by a rapid loss of turgor pressure in response to touch.

thigmotropism (14.3) Growth movement of a plant in response to contact.

thorax (16.1) The body region between the head and the abdomen.

thrombus (22.1) A blood clot lodged in an unbroken blood vessel or the heart.

thylakoid (3.3) A flattened sac in a chloroplast; arranged into stacks called grana.

thymine (4.4) A base in a nucleic acid molecule.

thymus gland (24.2) A mass of lymphatic tissue thought to produce hormones; degenerates prior to puberty.

thyroid gland (24.2) Endocrine gland that affects bone development and the body's overall rate of metabolism.

thyroxine (24.2) The thyroid hormone that regulates the metabolic rate.

tidal volume (21.2) The amount of air that enters the lungs during a normal inspiration or leaves the lungs during a normal expiration.

tissue (3.2) A group of many similar cells that perform a similar function.

tolerance (24.3) The body's response to prolonged use of narcotics in which increasingly greater amounts are required to obtain the same response.

tonsil (22.4) One of the masses of lymph nodes in the pharynx region.

topography (8.7, 19.2) Land features.

torsion (15.8) A twisted body arrangement unique to gastropods.

total lung capacity (21.2) The sum of the vital capacity and the residual volume.

toxin (10.5) A poisonous substance produced by a bacterium, another microorganism, or a plant or animal cell.

trachea (16.3, 16.4, 17.7, 21.1) A tube that extends from the larynx to the bronchi.

tracheid (13.4) A long, slender cell that forms one of the elements of xylem.

transcription (2.6) The process of forming messenger RNA from DNA.

transduction (10.2) The transfer of genetic material from one bacterium to another by a virus called a bacteriophage that attaches to bacteria.

transfer RNA (tRNA) (4.4) The RNA molecule that carries (transfers) a specific amino acid to the ribosome during protein synthesis.

transformation (10.2) The genetic change that occurs when DNA from one bacterium is taken up through the membrane of another bacterium.

transgenic (7.2) Containing genes from more than one species.

translation (4.4) The process of manufacturing polypeptides.

translocation (6.2) The transfer of a chromosome segment to a non-homologous chromosome. (14.1) The movement of carbohydrates throughout a plant.

transpiration (14.1) The release of water through the leaves of a plant.

transpiration-cohesion theory (14.1) A possible explanation for water movement in a plant; as water is released from the leaves, additional water molecules must enter the roots.

transport protein (3.5) A protein molecule embedded in the cell membrane that mediates the passage of certain molecules.

trichocyst (11.2) An organelle in a paramecium that discharges a thread-like filament in response to stimuli.

triplet See *codon.*

triploid (6.1) Having three complete sets of chromosomes (genomes).

trisomy (6.1) An abnormal condition in which there are three homologous chromosomes in a set instead of two.

tRNA See *transfer RNA.*

trochophore (15.8) The ciliated, free-swimming, larval stage of mollusks.

trophic level (19.2) A particular step in an ecological pyramid that shows the flow of energy through a food chain.

tropic hormone (24.2) A hormone that stimulates the growth and function of other endocrine glands.

tropism (14.3) A growth response of plants to external stimuli such as light, gravity, chemicals, and touch.

true fungus (12.8) A fungus that has a known form of sexual reproduction and can thus be classified into one of the three fungal phyla.

tube feet (15.9) Small, soft, tubular structures in echinoderms; used for locomotion and food capture; part of water-vascular system.

tuber (13.7) A storage stem which produces roots and often has "eyes" (buds) to produce aerial stems.

tumor (6.3, 10.7) An abnormal growth of cells.

tumor suppressor gene (6.3) A gene that produces proteins to prevent uncontrolled cell growth.

tunicate (17.1) A sessile urochordate, sometimes called a sea squirt, that pumps water through its body to obtain food.

turgor pressure (3.3) The added pressure within a plant cell that results from the movement of water into the central vacuole.

tympanic membrane (17.7, 23.4) A circular membranous structure that serves to transmit sound vibrations to an ear cavity; the eardrum.

tympanum (16.4) Tympanic membrane of some insects.

ultrasonic scanning (ultrasonography) (6.5) The use of sound waves to produce a computer-generated picture of internal structures.

umbilical cord (18.4, 25.3) The flexible structure that contains blood vessels and that conducts the blood of the fetus to the placenta for exchange of food, wastes, oxygen, and carbon dioxide.

understory (19.4) Plants that grow in the shade of trees.

ungulate (18.5) A hoofed mammal.

unicellular organism (3.2) An organism that consists of only one cell.

unit characteristics, concept of (5.4) One of the Mendelian theories that states that each characteristic of an organism is determined by a single gene.

universal flood (8.7) A flood that covers the entire earth; the Genesis Flood, Noah's Flood.

unsaturated fat (2.5) A fatty-acid molecule in which some of the carbon atoms are double-bonded to each other.

uracil (4.4) A base in a nucleic acid molecule.

ureter (22.7) One of the tubes that carries urine from the kidney into the bladder.

urethra (22.7) The passageway for urine from the bladder externally; also the passageway for sperm in the male.

urinary bladder (22.7) An organ designed as a reservoir for urine.

urine (22.8) A liquid excreted from the body; contains the metabolic wastes from the blood.

uterus (18.4, 25.2) A reproductive organ for storing ova until they are fertilized or laid; the womb; in most mammals and in humans, the organ in which the embryo develops.

vaccination (10.3) A method of exposing a person to a controlled amount of a disease-causing factor to develop an immunity.

vaccine (22.5) A weakened form of a pathogen used to build immunity by stimulating the body to produce antibodies or activate T cells.

vacuole (3.3) A membrane-bound organelle in a cell; stores materials.

vagina (25.2) The elastic canal that leads from the outside of the body to the cervix of the uterus.

valve (15.8) A mollusk shell.

vane (18.1) The flat part of a feather; composed of parallel rows of interlocking barbs.

variation (1.6, 6.4) The differences between individual organisms of the same kind; differences based on genotype; the expression of different individual characteristics in organisms of the same kind.

variety (9.2) Different form or type of organism within a species.

vascular cambium (13.6) A layer of meristematic tissue between the xylem and phloem; produces secondary xylem and phloem.

vascular cylinder (13.6) The central area of the young root or stem; contains xylem and phloem.

vascular ray (13.7) Tissue that extends from the central pith region to the outer areas of the stem and that permits the horizontal movement of water and dissolved substances.

vascular tissue (13.1, 13.4) The group of tissues (xylem and phloem) that conduct water and dissolved materials in a plant.

vas deferens (seminal duct) (25.1) The tube which carries the sperm from the testes to the urethra.

vector (7.2) An agent used to carry a gene from one organism to another. (10.5) An insect or other arthropod that carries pathogens to other host organisms.

vegetative reproduction (14.5) Asexual reproduction in plants.

vein (13.3, 13.5) A structure for strength and conduction in plants. (17.2, 22.3) Any blood vessel that carries blood toward the heart.

ventral (15.2, 20.2) On or near the lower surface (or at the front of some organisms).

ventricle (17.4, 22.2) The chamber of the heart that pushes blood into the arteries. (23.2) Space within the brain.

vertebra (pl., **vertebrae**) (17.1) One of the bones or cartilaginous segments of the vertebral column.

vertebral column (17.1) Bony or cartilaginous structures that are used in supporting an organism; the backbone.

vertebrate (15.2) An animal that possesses a backbone or vertebral column; a member of the subphylum Vertebrata.

vesicle (3.3) A small vacuole.

vessel cell (13.4) A cell type that forms the water-conducting tubes of the xylem.

vestigial structure (8.5) An organ that supposedly no longer has any function.

viable (14.6) Capable of growing and developing.

villus (pl., **villi**) (21.4) One of the microscopic fingerlike structures that lines the small intestine; aids in absorption of food molecules.

viroid (10.3) A short, single strand of circular RNA that has no capsid or envelope yet is still able to replicate once inside a host.

virology (10.3) The study of viruses.

virulence (10.3) The ability of a virus or other pathogen to cause a disease.

virus (10.3) A submicroscopic, noncellular particle, composed of a nucleic acid core and a protein coat called the capsid; an obligate parasite.

visceral mass (15.8) The portion of a mollusk's body that contains internal organs.

visceral muscle (20.8) Muscle tissue that forms the walls of internal organs; involuntary and smooth.

visual accommodation (23.5) The ability of the lens of the eye to focus on objects at different distances.

vital capacity (21.2) The sum of the tidal volume, inspiratory reserve volume, and expiratory reserve volume.

vitamin (21.5) An organic substance other than proteins, fats, and carbohydrates that is necessary for normal metabolism, growth, and development.

vitreous humor (23.5) A clear, jellylike substance in the eyeball.

viviparous (17.2) Describing a method of reproduction in which the young are born alive after being nourished in the uterus through a placenta.

vocal sac (17.7) One of a pair of sacs in the mouth region of male frogs; used in sound production.

water mold (11.5) A funguslike protist that reproduces both asexually and sexually through cells called sporangia; blights are common examples.

water-vascular system (15.9) A series of canals and tubules that are used for locomotion and food capture in echinoderms.

womb (25.2) A place where the unborn baby is protected and nourished; the uterus.

wood (13.7) The secondary xylem tissues in a stem or root.

workability (1.4) A characteristic of scientific knowledge that allows its practical application.

worldview (8.1) The belief system that a person uses to interpret and understand the world around him.

worship (1.5) Man's recognition of his insignificance and the resulting dependence on and praise to the almighty God.

xylem (13.4) A vascular tissue that carries water and dissolved minerals upward in a plant.

yeast (12.7) A unicellular fungus from the phylum Ascomycota that reproduces asexually by budding.

yellow bone marrow See *bone marrow, yellow*.

yolk (17.4, 17.8) Food material stored in an egg to nourish the embryo.

yolk sac (17.8) The membrane that contains the yolk in an amniotic egg.

Z line (20.8) The line where actin filaments are attached between sarcomeres.

zonation (App. A) Arrangement of organisms into their ideal biogeographic zones.

zooflagellate (11.2) A protozoan that possesses one or more flagella for propulsion; found in phylum Zoomastigina.

zoology (15.1) The study of animals.

zooplankton (11.3) Plankton that are tiny animals or protozoans.

zoospore (11.3) A motile, swimming spore; possesses cilia or flagella.

zygosporangium (pl., **zygosporangia**) (12.6) One of the thick-walled, spore-producing reproductive structures of molds in phylum Zygomycota.

zygospore (11.3) A zygote surrounded by a hard, protective covering to survive unfavorable conditions.

zygote (5.3) A diploid cell formed by the union of two haploid gametes.

Index

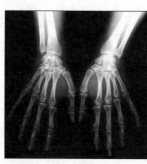
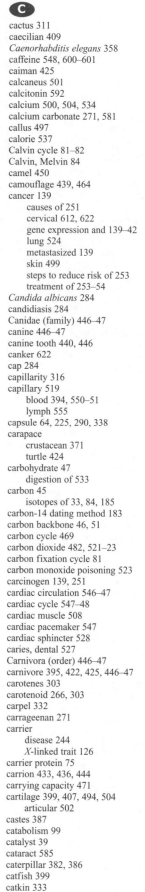

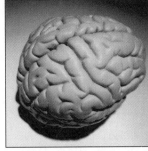

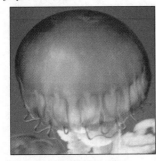

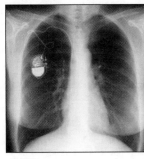

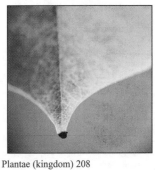

Photograph Credits

The following agencies and individuals have furnished materials to meet the photographic needs of this textbook. We wish to express our gratitude to them for their important contribution.

Cover
© iStockphoto.com/Sascha Burkard (front cover, spine); © iStockphoto.com/Tomasz Zachariasz (back cover)

Front Matter
Ryan McVay/Lifesize/Thinkstock iv

Chapter 1
CDC/Cheryl Tyron 3; Robert Thom, American (Grand Rapids, MI, 1915–1979, Michigan) *Hippocrates: Medicine Becomes a Science*, Collection of the University of Michigan Health System, Gift of Pfizer, Inc. UMHS.7 4 (top); National Library of Medicine, Washington, DC 4 (bottom); Suzanne Altizer 5 (far left); Stephen Sharnoff 5 (left), 17; Olympic National Park 5 (right); BJU Photo Services 5 (far right); Bob McCoy, Museum of Questionable Medical Devices 8; © Stanislav Freidin/Creative Commons ShareAlike 1.0 Generic 9; PhotoDisc/Getty Images 11, 16 (both top right), 29 (center left); Robert Thom, American (Grand Rapids, MI, 1915–1979, Michigan) *Pasteur: The Chemist Who Transformed Medicine*, Collection of the University of Michigan Health System, Gift of Pfizer, Inc. (detail of original) UMHS.32 13; © 2004 Hemera Technologies/Getty Images 16 (top left), 18 (left), 19 (top left), 24 (bottom both); Brian D. Johnson 16 (bottom left); Department of Energy 18 (right); U.S. Department of Agriculture 19 (top right, bottom); JupiterImages/Thinkstock 20; George R. Collins 21 (top); © Scientifica/Visuals Unlimited, Inc. 21 (bottom); © iStockphoto.com/Lee Pettet 22; © iStockphoto.com/Sam Gluzberg 23; Robert Thom, American (Grand Rapids, MI, 1915–1979, Michigan) *Leeuwenhoek and the "Little Animals"*, Collection of the University of Michigan Health System, Gift of Pfizer, Inc. UMHS.15 24 (center); © Inga Spence/Visuals Unlimited, Inc. 25 (top right); Joyce Landis 25 (bottom left); Mary Chapman 25 (bottom right); © iStockphoto.com/bojan fatur 28 (top right); Barge Memorial Hospital 28 (center left); © iStockphoto.com/Joel Johndro 28 (center right); Mediscan/Visuals Unlimited, Inc. 28 (bottom); Dr. Parsons, MD, Carolina ENT 29 (top right); © SIU/Visuals Unlimited, Inc. 29 (center bottom); © Carolina Biological/Visuals Unlimited, Inc. 30 (top); © Dr. Donald W. Fawcett/Visuals Unlimited, Inc. 30 (bottom 4)

Chapter 2
IBM Research Center, Almaden Research Center 32; BJU Photo Services 36 (top left), 49 (top); Susan Perry 36 (top right); © iStockphoto.com/Bonita Hein 36 (bottom right); © 2009 JupiterImages Corporation 37, 38, 42 (center left), 44 (top bar), 48; George R. Collins 41; PhotoDisc/Getty Images 42 (top bar), 49 (bottom), 53; © iStockphoto.com/Katrina Leigh 42 (bottom); © Dr. Ken MacDonald/Photo Researchers, Inc. 46; © Barrington Brown/Photo Researchers, Inc. 52

Chapter 3
Public Domain 56 (both left); © Biophoto Associates/Photo Researchers, Inc. 56 (bottom right); © Photo Researchers, Inc. 57; © Wim van Egmond/Visuals Unlimited, Inc. 59 (center right); George R. Collins 59 (bottom left); Richard A. Altizer 59 (bottom right); © Dr. Donald W. Fawcett/Visuals Unlimited, Inc. 65 (top), 69, 70, 76 (right); © R. Bolender & Dr. Donald W. Fawcett/Visuals Unlimited, Inc. 65 (center top); © Dr. Dennis Kunkel Microscopy, Inc./Visuals Unlimited, Inc. 65 (bottom center, bottom left), 66 (bottom), 76 (left); © Manfred Schliwa/Visuals Unlimited, Inc. 66 (top); © Dr. Richard Kessel & Dr. Gene Shih/Visuals Unlimited, Inc. 66 (center); © George Chapman/Visuals Unlimited, Inc. 67 (top left); © Henry Aldrich/Visuals Unlimited, Inc. 67 (bottom right); © RMF/Visuals Unlimited, Inc. 68 (top bar, center left); Larry Albee/Longwood Gardens 71 (both)

Chapter 4
© 2009 JupiterImages Corporation 78; Courtesy North Carolina State University Floriculture Program, Dept. of Horticultural Science/www.floricultureinfo.com 80; © micrographia.com 83; PhotoDisc/Getty Images 84 (top bar), 96 (both); Courtesy of the Chemical Heritage Foundation Collections 84 (top left); Edgar Fahs Smith Collection, University of Pennsylvania Library 84 (top right); Lawrence Berkley National Laboratory 84 (bottom); Susan Perry 89 (both); © 2004 Wisconsin Milk Marketing Board, Inc. 90; Stockbyte/Thinkstock 91; © Dr. Donald W. Fawcett/Visuals Unlimited, Inc. 95; © iStockphoto.com/rotofrank 98 (top right both); Vasco Gomes 98 (bottom left); uwe kils/Wikipedia/GNU Free Documentation License, Version 1.2 98 (bottom right)

Chapter 5
© Science VU/Visuals Unlimited, Inc. 104 (top); © Dr. Dennis Kunkel Microscopy, Inc./Phototake 104 (bottom); BJU Photo Services 106 (six on right), 115 (both); Ward's Natural Science Establishment, Inc. 107 (top); © Dr. David M. Phillips/Visuals Unlimited, Inc. 107 (bottom); Philip Gladstone 108 (top left); Sam Laterza 108 (top center); BJU Photo Services 109; Jiri Parizek/Abbey of St. Thomas 112; National Library of Medicine 112 (inset); John Innes Foundation Historical Collections, courtesy of the John Innes Trustees 116 (bottom); George R. Collins 116 (bottom); Joyce Landis 117; Ball Horticultural Company 118 (top); Max Planck Institute for Molecular Genetics 118 (bottom); Scott Bauer/USDA/ARS 119 (top); © My Pet Chicken 119 (bottom left); Courtesy Richland Ranch, photo by Don Trout 119 (bottom right); © 2009 JupiterImages Corporation 120, 122 (walnut comb); © Fotolia/Margo Harrison 122 (rose comb); © Fotolia/Craig Hanson 122 (pea comb); Muhammad Mahdi Karim/Wikipedia/GNU Free Documentation License, Version 1.2 122 (single comb); © 2004 Hemera Technologies/Getty Images 122 (bottom 4); Art Today/Clipart.com/© 2010 Getty Images 123 (top left, top right); © C.N.R.I./Phototake 123 (center left); © Phototake 123 (center right); HRR plates from www.RichmondProducts.com 123 (bottom); Public Domain 124 (bottom); Suzanne Altizer 126 (both)

Chapter 6
Scott Bauer/USDA/ARS 129 (left); Suzanne Altizer 129 (right); PhotoDisc/Getty Images 130 (center left), 131 (center left, center right), 140 (top right), 144 (top bar, center, bottom left); R&M Gaited Mules, Marie Lanier 130 (right); © 2004 Hemera Technologies/Getty Images 131 (top left), 144 (bottom right); Florida Department of Agriculture and Consumer Services 131 (top right); The Florida Citrus Growers 131 (top inset); Carlos Quiros 131 (bottom inset); Joyce Landis 132 (top); BJU Photo Services 132 (bottom left); © Dr. Dennis Kunkel Microscopy, Inc./Phototake 132 (bottom right); www.cridchat.u-net.com/Courtesy of Jeff Noneley 135; © K. H. Kjeldsen/Photo Researchers, Inc. 137 (center left); © Eye of Science/Photo Researchers, Inc. 137 (top right); © 2009 JupiterImages Corporation 138; Photo Courtesy House Rabbit Society 139 (top); © iStockphoto.com/Yuriy Sukhovenko 139 (center); © Science Source/Photo Researchers, Inc. 139 (bottom); © Fotolia/Oleg Golovnev 140 (top left); © Medical-on-Line / Alamy Images 140 (center left); © Dr. Ken Greer/Visuals Unlimited, Inc. 140 (center right); Brian D. Johnson 141 (left); © iStockphoto.com/Ann Marie Kurtz 141 (right); William M. Ciesla, Forest Health Management International, www.forestryimages.org 143 (center); University of Florida Department of Entomology and Nematology 143 (bottom); Stephen Christopher 146 (left); Cohocton Clydesdales 146 (right); GE Systems 148 (top); Anne Rayner, Vanderbilt University 148 (bottom); Brenda Hansen 149

Chapter 7
© Dr. Dennis Kunkel Microscopy, Inc./Visuals Unlimited, Inc. 153 (top right); Tom Adams/Visuals Unlimited, Inc. 153 (bottom right); Susan Perry 153 (bottom left); R. C. Ploetz, University of Florida 153 (bottom center); Advanced Cell Technology 154 (all); PhotoDisc/Getty Images 160 (top bar); Ernie Branson, National Institutes of Health 160 (top right); © 2004 Hemera Technologies/Getty Images 163 (top left); © SIU/Visuals Unlimited, Inc. 163 (top right); Agricultural Research Service/USDA 165; Paul Darrow/The New York Times/Redux 166

Chapter 8
BJU Photo Services 169, 192 (top bar); NASA and the Hubble Heritage Team 172 (top left); Wolfgang Wander/Wikimedia/GNU Free Documentation License, Version 1.2 172 (bottom left); Mdf/Wikimedia/GNU Free Documentation License, Version 1.2 172 (bottom right); © iStockphoto.com/Jello5700 173; © iStockphoto.com/konradlew 175; © 2009 JupiterImages Corporation 176; © iStockphoto.com/Eric Isselée 177 (bottom); © iStockphoto.com/Mark Kostich 178 (top); National Archives 178 (bottom); PhotoDisc/Getty Images 182 (top bar), 196 (top), 198 (top left); © iStockphoto.com/David Hills 182 (top right); © iStockphoto.com/Geoffrey Stradling 184; National Library of Medicine 185 (top); Ball Horticultural Company 185 (bottom); LadyofHats/Wikimedia/Public Domain 192 (bottom); Ward's Natural Science Establishment, Inc. 196 (center); © iStockphoto.com/Falk Kienas 196 (bottom); © 1982 Glen J. Kuban 197 (top left); Brian D. Johnson 197 (bottom left); AP Photo/Augustin Ochsenreiter 197 (bottom right); Breck Kent 198 (bottom left); Ian Juby, ianjuby.com 198 (bottom right)

Chapter 9
© 2004 Hemera Technologies/Getty Images 203 (all), 205 (spider, butterfly, moth, frog, bird), 210 (top center); PhotoDisc/Getty Images 204 (snake, bird, bat, butterfly), 205 (bony fish), 208 (Plantae, Animalia), 210 (top bar); USDA 204 (ant); © iStockphoto.com/Tammy Peluso 204 (fish); © iStockphoto.com/Torsten Wittmann 204 (lobster); JupiterImages/Photos.com/Thinkstock 205 (centipede, wolf); © iStockphoto.com/Lezh 205 (tick); © iStockphoto.com/Aleksander Trankov 205 (scorpion); © iStockphoto.com/Mike Sonnenberg 205 (cricket); © iStockphoto.com/Vladimir Krivsun 205 (grasshopper); © iStockphoto.com/Tomasz Zachariasz 205 (fly); JupiterImages/Stockxpert 205 (mosquito, chipmunk); © Fotolia/Fotografik/Foto©Daniel Leśniak 205 (salamander); © Fotolia/Ilia Shcherbakov 205 (rat); © iStockphoto.com/Mark Wilson 205 (cheetah); © iStockphoto.com/Dongfan Wang 207; © Ralph Robinson/Visuals Unlimited, Inc. 208 (Archaebacteria); © Dr. T. J. Beveridge/Visuals Unlimited, Inc. 208 (Eubacteria); © Carolina Biological/Visuals Unlimited, Inc. 208 (Protista); Suzanne Altizer 208 (Fungi); © Fotolia/Colin Hanrahan 209; Photo by Lawrence Steppanowicz / Alamy Images 210 (center right); © Ron Dengler/Visuals Unlimited, Inc. 210 (center right); Mdf/Wikimedia/GNU Free Documentation License, Version 1.2 211; © Image Source/Picture Quest 212 (top left); Carla Thomas 212 (top center); Comstock/Thinkstock 212 (top right); Smithsonian Tropical Research Institute 212 (bottom); Francesco Rovero, Trento Museum of Natural Sciences, Italy 213 (top); Breck Kent 213 (bottom); Francois Bourgeot 214; Kevin L. Cole/Wikipedia/Creative Commons Attribution 2.0 License 215 (bottom left); © iStockphoto.com/Robert Blanchard 215 (bottom right); © iStockphoto.com/Scott Orr 217 (left); © iStockphoto.com/ShaneKato 217 (bottom right); Rich Yasick/JupiterImages/Stockxpert 218 (top bar)

Chapter 10

© iStockphoto.com/Katie Fletcher 222; © iStockphoto.com/Rhienna Cutler 223 (top); © Ralph Robinson/Visuals Unlimited, Inc. 223 (center); © Dr. Dennis Kunkel Microscopy, Inc./Visuals Unlimited, Inc. 223 (bottom), 224 (bottom), 227, 233 (center left inset), 251; Ann Arbor Biological Center 224 (top left); Mike Clayton/University of Wisconsin, Madison 224 (top right); © 2009 JupiterImages Corporation 225, 240; Dr. George Chapman/Visuals Unlimited, Inc. 229; National Archives 230 (top left); CDC/World Health Organization 230 (bottom left); © 2009 Edward H. Gill/Custom Medical Stock Photo 230 (center); © Dr. David Phillips/Visuals Unlimited, Inc. 230 (bottom right); © iStockphoto .com/Alina555 231; Courtesy of The Bancroft Library, University of California,13:583 233 (top); R. J. Reynolds Tobacco Company slide set, R. J. Reynolds Tobacco Company 233 (center); © Scott Camazine/Photo Researchers, Inc. 233 (bottom left); © CAMR/A. B. Dowsett/Photo Researchers, Inc. 233 (bottom right); © Carolina Biological/Visuals Unlimited, Inc. 234; USDA/Barry Fitzgerald 235 (bottom left); USDA 235 (bottom right inset); USDA/APHIS photo by Dr. Al Jenny 236 (top left); AP Photo/Tito Alabiso 236 (top right); © Mediscan/Visuals Unlimited, Inc. 236 (bottom), 249 (bottom left); PhotoDisc/ Getty Images 237 (top bar), 239 (bottom), 252 (top bar), 254 (center right); Wellcome Library, London 237 (top right); Robert Thom, American (Grand Rapids, MI, 1915–1979, Michigan) *Jenner: Smallpox is Stemmed*, Collection of the University of Michigan Health System, Gift of Pfizer, Inc. (detail of original) UMHS.23 237 (center left); Centers for Disease Control 239 (top); Library of the College of Physicians of Philadelphia 241; Sue Renault/Christine Nichols, American Leprosy Missions 242; *Christ Healing the Blind Man*, (detail), Cornelis Cornelisz, van Haarlem, From the Bob Jones University Collection 243 (top); Susan Perry 243 (bottom); BJU Photo Services 244 (top); Clipart.com/JupiterImages 244 (bottom left); © Carolina Biological/Visuals Unlimited, Inc. 244 (bottom right); © 2009 JupiterImages Corporation 246; *The Era of Antibiotics* by Robert Thom. Printed with Permission of American Pharmacists Association Foundation. Copyright 2007 APhA Foundation. 249 (bottom right); Courtesy of Tom Porch 252 (bottom all); Comstock Images 254 (left)

Chapter 11

MedicalRF.com/Visuals Unlimited, Inc. 256; Suzanne Altizer 257, 273 (center right); Wim van Egmond 258 (top both); Courtesy Museum of Science Boston 258 (bottom); © Dr. David M. Phillips/Visuals Unlimited, Inc. 259, 267, 271 (bottom left); © Eric Grave/Photo Researchers, Inc. 260 (top left); © Dr. Dennis Kunkel Microsopy, Inc./Visuals Unlimited, Inc. 260 (top right); Dr. Margene Ranieri 260 (center left), 265; National Archives 261; Dr. Tom Coss 262 (top); © Michael Abbey/Visuals Unlimited, Inc. 262 (bottom); PhotoDisc/ Getty Images 264 (both); © Tom Adams/Visuals Unlimited, Inc. 268 (top left); George R. Collins 268 (center left); Joyce Landis 268 (center right); Comstock/Thinkstock 268 (bottom); M. D. Guiry/www.algaebase.org 269 (center left); John Fitzhugh/The Sun Herald 269 (center right); Amanda Cotton/JupiterImages/Stockxpert 269 (bottom); JupiterImages/ Thinkstock 270 (top bar); © Marufish/Wikimedia Commons/Creative Commons Attribution 2.0 Generic 270 (bottom); © Robert De Goursey/Visuals Unlimited, Inc. 271 (top); JupiterImages/Stockxpert 271 (center); NASA 271 (bottom right); © catalano82/Wikimedia Commons/Creative Commons 272 (top); © James Richardson/Visuals Unlimited, Inc. 272 (bottom left); USDA/ARS 272 (bottom right); Library of Congress 273 (top left); Scott Bauer/USDA 273 (top right); John H. Ghent, USDA Forest Service/www.forestryimages. org 273 (bottom right); Mike Clayton/University of Wisconsin, Madison 274 (left); © Carolina Biological/Visuals Unlimited, Inc. 274 (right)

Chapter 12

Ken Wagner/Visuals Unlimited, Inc. 276 (top); © Renee Morris / Alamy Images 276 (center); Dr. Ken Walker/www.padil.gov.au 276 (inset); Courtesy M. F. Heimann. Printed with Permission from Plant Disease 78(2). American Phytopathological Society, St. Paul, MN 276 (bottom); © Scott Camazine / Alamy Images 277 (top); JupiterImages/Stockxpert 277 (center), 282 (top), 284 (top left), 286 (center, letter a); © Gerald Van Dyke/Visuals Unlimited, Inc. 277 (bottom); PhotoDisc/Getty Images 278 (top bar); Pieria/Wikimedia Commons/Public Domain 278 (top left); Cato Edvardsen/Wikimedia Commons/GNU Free Documentation License 278 (top right); Grmica/Wikimedia Commons/GNU General Public License 278 (center); © Bon Appetit / Alamy Images 278 (bottom left); James E. Gordon 278 (bottom right); Centers for Disease Control 279 (top right), 284 (center right); Courtesy of www.doctorfungus.org © 2004 279 (center right, bottom right), 280 (bottom both); Strobilomyces/Wikimedia Commons/GNU Free Documentation License 279 (bottom left); Ejdzej/Wikimedia Commons/GNU Free Documentation License/Creative Commons Attribution ShareAlike 3.0 License 282 (bottom left); Velela/Wikimedia Commons/Public Domain 282 (bottom right); © Dr. Dennis Kunkel Microscopy, Inc./Visuals Unlimited, Inc. 283; Photograph by Ken Harris 284 (top right); Hemera Technologies/Photos.com/ Thinkstock 284 (center left); Centers for Disease Control/Dr. Lucille K. Georg 284 (bottom right); Clemson University, USDA Cooperative Extension Slide Series, www.forestryimages.org 285 (left); USDA/ARS 285 (right); © imagebroker / Alamy Images 286 (far left); © yogesh more / Alamy Images 286 (letter b); © Peter Arnold, Inc. / Alamy Images 286 (letter c)

Chapter 13

JupiterImages/Stockxpert 288 (left), 289 (center bottom), 292 (center top), 296 (letter c), 297 (bottom), 301 (right letter b), 311 (center right, bottom left, bottom center), 312 (bottom left); JupiterImages/Photos.com/Thinkstock 288 (center), 296 (letter a), 301 (left letters b and c), 311 (bottom right ivy); John H. Ghent, USDA Forest Service, www.forestryimages.com 288 (right); © 2004 Hemera Technologies/Getty Images 289 (top left); Ball Horticultural Company 289 (top center); PhotoDisc/Getty Images 289 (top right), 295 (right), 303 (right), 311 (top bar, top center); Vaelta/Wikipedia/GNU Free Documentation License/Creative Commons Attribution ShareAlike 3.0 License 290; J. F. Gaffard Jeffdelonge/fr.wikipedia/GNU Free Documentation License/Creative Commons Attribution ShareAlike 3.0 License 291 (left); © Henry Robison/Visuals Unlimited, Inc. 291 (right); Susan Perry 292 (top), 302 (left); © J. S. Peterson@Plants.usda.gov 293 (left); Forest & Kim Starr/Wikimedia Commons/ Creative Commons Attribution 3.0 Unported License 293 (right); Dr. Stewart Custer 294 (top); © Gerald and Buffy Corsi/Visuals Unlimited, Inc. 294 (bottom); © Jerome Wexler/ Visuals Unlimited, Inc. 295 (left); Erich G. Vallery, www.forestryimages.com 295 (center); © iStockphoto.com/Roger Whiteway 296 (letter b); © John Mead/SPL/Photo Researchers, Inc. 296 (letter d); © George Loun/Visuals Unlimited, Inc. 296 (cycad); © Ed Webber/ Visuals Unlimited, Inc. 296 (ginkgo); © Joe and Mary Ann McDonald/Visuals Unlimited, Inc. 296 (bottom); USDA 297 (top); tezzstock © Fotolia 297 (center top); © iStockphoto. com/bluebird13 297 (center bottom); © iStockphoto.com/asiseeit 301 (left letter a); © Elena Elisseeva. Image from BigStockPhoto.com 301 (right letter a); David W. Boyd Jr. 302 (right); © Carolina Biological/Visuals Unlimited, Inc. 303 (left), 306, 310; BJU Photo Services 305 (both); Dr. Margene Ranieri 308; Scott Bauer/USDA 311 (top right); © Margo Harrison. Image from BigStockPhoto.com 311 (center left); Raul654/Wikimedia Commons/ GNU Free Documentation License/Creative Commons Attribution ShareAlike 3.0 License 311 (center); © Bill Beatty/Visuals Unlimited, Inc. 311 (bottom far right); Dr. Mohammed Fayyaz/The University of Wisconsin 312 (top left); Sam Laterza 312 (top center); Emily Earp and Josh Hillman, FloridaNature.org 312 (center left); © Michael P. Gadomski/Photo Researchers, Inc. 312 (bottom left inset)

Chapter 14

George R. Collins 316; Mary Chapman 317 (top right both); Suzanne Altizer 317 (center right, all 4), 324 (bottom); © David Sieren/Visuals Unlimited, Inc. 317 (bottom left both); © iStockphoto.com/Kary Nieuwenhuis 318; USDA 319; The Scotts Company, Inc. 320 (left); BJU Photo Services 320 (inset), 321 (top right); Hemera Technologies/AbleStock.com/ Thinkstock 321 (top bar); © 2009 JupiterImages Corporation 321 (top and bottom left); Hemera Technologies/Photos.com/Thinkstock 321 (center left); Noah Elhardt/Wikimedia Commons/GNU Free Documentation License/Creative Commons Attribution ShareAlike 3.0 License 321 (center); Ward's Natural Science Establishment, Inc. 321 (center right); Goodshoot/Thinkstock 321 (bottom right); Courtesy of the Archives, California Institute of Technology 322; The International Rice Research Institute, Los Banos, Philippines 323; © Jack Bostrack/Visuals Unlimited, Inc. 324 (top both); © iStockphoto.com/heizfrosch 324 (center left); © iStockphoto.com/Matjaz Boncina 324 (center right); Mike Clayton/ University of Wisconsin, Madison 325 (top), 326 (bottom); © iStockphoto.com/Peter Garbet 325 (bottom); © iStockphoto.com/AtWaG 326 (top); PhotoDisc/Getty Images 328 (top bar), 330 (top bar), 337 (bottom left); © Mike Sieren/Visuals Unlimited, Inc. 329 (top); © iStockphoto.com/Pauline Mills 329 (bottom right); © Bill Beatty/Visuals Unlimited, Inc. 329 (inset); Stark Bro's Nurseries & Orchards Co. 330 (top right); © Wally Eberhart/ Visuals Unlimited, Inc. 331 (far right); Dr. Mohammed Fayyaz/The University of Wisconsin 331 (inset); © Rosenfeld Images, Ltd./Science Photo Library 331 (left); Susan Perry 332 (top right), 333 (center both); © iStockphoto.com/Scott Vickers 332 (bottom); Paul Wray/ www.forestryimages.org 333 (top), 337 (bottom right); © iStockphoto.com/Gary Ferguson 333 (bottom); Robert H. Mohlenbrock @ USDA-NRCS PLANTS Database/USDA SCS. 1989 337 (top)

Chapter 15

PhotoDisc/Getty Images 341 (top), 346 (bottom left), 350 (bottom right); JupiterImages/ Photos.com/Thinkstock 341 (bottom), 353; age fotostock/SuperStock 345; © Dr. John D. Cunningham/Visuals Unlimited, Inc. 346 (top); Susan Perry 346 (bottom right); NOAA 349 (top), 368 (top); YOMIURI SHIMBUN/AFP/Getty Images 349 (bottom); Wikimedia Commons/Public Domain 350 (top); COREL Corporation 350 (bottom left, center right), 365 (top), 366 (bottom left); © Tom Adams/Visuals Unlimited, Inc. 352; © Centers for Disease Control 356 (top bar, bottom left); © Cleve Hickman/Visuals Unlimited, Inc. 356 (bottom right); Bob Goldstein, UNC Chapel Hill/Wikimedia Commons/Creative Commons Attribution ShareAlike 3.0 License 358; Breck Kent 359, 362, 364; © iStockphoto.com/ Vinicius Ramalho Tupinamba 360; Paul Osmond/Deep Sea Images 365 (bottom); © Mark Norman/Visuals Unlimited, Inc. 366 (top left); © Reinhard Dirscherl/Visuals Unlimited, Inc. 366 (top right); © Alex Kirstitch/Visuals Unlimited, Inc. 368 (bottom)

Chapter 16

COREL Corporation 370, 384 (bottom left, bottom right), 386 (bottom left, center right); BJU Photo Services 371; PhotoDisc/Getty Images 372 (top), 381 (top bar), 386 (center left), 388 (top bar); NOAA 372 (bottom), 374; Justin Montemarano/Wikimedia Commons/ Creative Commons Attribution 3.0 Unported License 373; Centers for Disease Control 375 (top), 390 (all); © iStockphoto.com/William Howe 375 (center left); © Larry Jensen/ Visuals Unlimited, Inc. 375 (center right); Digital Vision 375 (bottom), 376 (top), 380; © iStockphoto.com/ChartChai MeeSangNin 376 (center left), 378 (left); © Bill Beatty/ Visuals Unlimited, Inc. 376 (center right); Phototake 376 (bottom); E. L. Manigault slide collection, Department of Entomology, Soils and Plant Sciences, Clemson University/www. forestryimages.org 377 (both); JupiterImages/liquidlibrary/Thinkstock 378 (right); David R. Lance, USDA APHIS PPQ Archives, www.forestryimages.org 381 (top center); Alfred Molon 381 (center left); © iStockphoto.com/zbindere 381 (bottom); The University of Georgia Archives, www.forestryimages.org 382; Clemson University, USDA Cooperative Extension Slide Series, www.forestryimages.org 384 (top); Dr. Picker 386 (top right); Hemera Technologies/PhotoObjects.net/Thinkstock 387 (top left); © iStockphoto.com/ Brian Evans 387 (top right); Susan Perry 387 (center); © Dr. Dennis Kunkel Microscopy, Inc./Visuals Unlimited, Inc. 387 (bottom); USDA 388 (bottom); Mark Robinson, USDA

Forest Service, www.forestryimages.org 389 (top left); USDA Forest Service Archives, www.forestryimages.org 389 (top left inset); John Ruberson, The University of Georgia, www.forestryimages.com 389 (center); USDA APHIS PPQ Archives, www.forestryimages.org 389 (top right)

Chapter 17

Hans Hillewaert/Wikimedia Commons/Creative Commons Attribution ShareAlike 2.5 License 392 (top); Nick Hobgood/Wikimedia Commons/Creative Commons Attribution ShareAlike 3.0 License, GNU Free Documentation License 392 (bottom); JupiterImages/Photos.com/Thinkstock 393, 402 (top); © iStockphoto.com/Maurice van der Velden 395; PhotoDisc/Getty Images 397 (top, center right, bottom), 408 (both), 418 (top); JupiterImages/Stockxpert 397 (center left); © iStockphoto.com/Jeremy Lang 399 (top); © Andrew J. Martinez / Photo Researchers, Inc. 399 (bottom); Dean A. Hendrickson, Texas Memorial Museum, University of Texas at Austin 400 (top bar, bottom); © David Fleetham/Visuals Unlimited, Inc. 400 (top right); © Luc Viatour/Wikimedia Commons/GFDL/CC 401 (top); U.S. Fish and Wildlife Service 402 (bottom left); University of Washington/Thomas Quinn 402 (bottom right); EyeKarma/Wikimedia Commons/Public Domain 405; © David Wrobel/Visuals Unlimited, Inc. 406 (top); Courtesy of EPA/U.S. Fish and Wildlife Service 406 (bottom); Courtesy of EPA 406 (bottom inset); © Dr. Dennis Kunkel, Inc./Visuals Unlimited, Inc. 407; © 2004 Henk Wallays 409, 410 (salamander eggs and adult); Brian Gratwicke/Wikimedia Commons/Creative Commons Attribution 2.0 License 410 (top left); © 2004 Peter Weish 410 (top right); Carey James Balboa/Wikimedia Commons/Public Domain 410 (tree frog); Patrick Gijsbers/Wikimedia Commons/GNU Free Documentation License/Creative Commons Attribution ShareAlike 3.0 License 410 (bottom); © Carolina Biological/Visuals Unlimited, Inc. 411; © iStockphoto.com/Rex Lisman 412 (top left); COREL Corporation 412 (center left); Clinton & Charles Robertson/Wikimedia Commons/Creative Commons Attribution 2.0 License 412 (bottom left); © iStockphoto.com/Kevin Snair 412 (bottom right); © iStockphoto.com/Alasdair Thomson 415; © Jim Merli/Visuals Unlimited, Inc. 417; © Nigel J. Dennis/ Photo Researchers, Inc. 418 (bottom); © 2004 Hemera Technologies/Getty Images 419 (top bar); Breck Kent 419 (bottom); 2010 © E-I-E-I-O Fotos. Image from BigStockPhoto.com 423 (top); © iStockphoto.com/Arnaud Weisser 423 (bottom); Brenda Hansen 424 (top); © iStockphoto.com/Daniel Budiman 424 (bottom); Digital Vision 425 (both top and center); Andy Heyward/JupiterImages/Stockxpert 426

Chapter 18

Brian D. Johnson 428, 435 (center), 441 (top right, bottom left); PhotoDisc/Getty Images 430, 434 (top bar), 437 (bottom right), 439 (center bottom), 442 (both), 444 (top, center, bottom inset), 446 (top, bottom left), 449 (top, center left); U.S. Fish and Wildlife Service 433, 437 (bottom left), 439 (top), 452; COREL Corporation 435 (top left); Guillaume Dargaud 435 (top right); U.S. Fish and Wildlife Service 435 (bottom both); © iStockphoto.com/Jill Lang 436 (top right); © iStockphoto.com/Julianna Tilton 436 (bottom); Kenneth M. Gale/www.forestryimages.org 436 (center left); Alan D. Wilson, www.naturespicsonline.com/Wikimedia Commons/CC 437 (top left); © 2010 JupiterImages/Gino Santa Maria 438 (top); © iStockphoto.com/Ewan Chesser 440; © iStockphoto.com/Len Tillim 441 (top right); Kenneth M. Gale/www.forestryimages.org 441 (bottom right); Cpl. Nicholas Tremblay/USMC 443; © iStockphoto.com/Mark Higgins 444 (bottom left); © iStockphoto.com/Donna Heatfield 444 (bottom right); © Fotolia/Susan Flashman 445 (top); © 2010 JupiterImages Corporation 445 (bottom), 449 (center right), 451 (top bar); Tom Brakefield/Stockbyte/Thinkstock 446 (bottom right); © iStockphoto.com/Ryan Saul 448; © Fotolia/Hii Boh Teck 449 (bottom); G. Keith Douce, The University of Georgia, www.forestryimages.org 450 (top right inset); William M. Ciesla, Forestry Health Management International, www.forestryimages.org 450 (top left); © iStockphoto.com/Pablo J Yoder 450 (center); Rexford Lord/Photo Researchers 450 (bottom); © Fotolia/ewanc 451 (top left); © Brian Jackson. Image from BigStockPhoto.com 451 (top right)

Chapter 19

NOAA 454 (top), 475; © 2009 JupiterImages Corporation 454 (bottom), 467 (bottom left), 470 (left); © iStockphoto.com/peterleabo 457 (top); David J. Moorhead, www.forestryimages.org 457 (bottom); NASA 458 (globe); © Oliver Trapp. Image from BigStockPhoto.com 461; © Fotolia/Clarence Alford 462; PhotoDisc/Getty Images 464 (top bar), 467 (all 3 on right), 469, 470 (right), 474, 476 (top), 478 (top bar, center), 481 (both), 482 (bottom), 484 (both); E. L. Manigault slide collection, Department of Entomology, Soils and Plant Sciences, Clemson University/www.forestryimages.org 464 (center left); William M. Ciesla, Forestry Health Management International, www.forestryimages.org 464 (bottom right); Breck Kent 465 (both), 472 (both); P. Frenzen, USDA Forest Service 473 (center left); © Melisa Taylor. Image from BigStockPhoto.com 473 (right); © iStockphoto.com/smartstock 476 (center); Friends of the Los Angeles River/Thea Wang 476 (bottom); Chuck Meyers, Office of Surface Mining 477 (both); Robert Thom, American (Grand Rapids, MI, 1915–1979, Michigan) *Banting, Best and Diabetes*, Collection of the University of Michigan Health System, Gift of Pfizer, Inc. UMHS.43 478 (bottom); Gail Pennewell 482 (top)

Chapter 20

© iStockphoto.com/Dean Mitchell 489 (top); BJU Photo Services 489 (bottom); *God the Father in a Glory of Angels*, Francesco Montemezzano, From the Bob Jones University Collection 490; © Christopher Meade/Shutterstock 493 (top); Image Source/Getty Images 493 (center top); © Dr. John D. Cunningham/Visuals Unlimited, Inc. 493 (center bottom); © Wim van Egmond/Visuals Unlimited, Inc. 493 (bottom); Brenda Hansen 497 (top); PhotoDisc/Getty Images 497 (bottom), 502; © Dr. John D. Cunningham/Visuals Unlimited, Inc. 498; COREL Corporation 499 (center left); Neil W. Scroggins 499 (right); © 2004

Hemera Technologies/Getty Images 503 (top bar); © iStockphoto.com/emin ozkan 503 (center); © iStockphoto.com/Ryan Lane 503 (bottom); Barge Memorial Hospital 504; © Ralph Hutchings/Visuals Unlimited, Inc. 508 (top); © Scientifica/Visuals Unlimited, Inc. 508 (center); © Nephron/Wikimedia Commons/Creative Commons Attribution ShareAlike 3.0 License 508 (bottom)

Chapter 21

© Dr. Dennis Kunkel Microscopy, Inc./Visuals Unlimited, Inc. 517; Dr. Parsons, MD, Carolina ENT 518; IAN HOOTON/Science Photo Library/Getty Images 520; © iStockphoto.com/bojan fatur 522; © iStockphoto.com/Christian Nasca 523 (top); © 2009 JupiterImages Corporation 523 (bottom), 527 (center); PhotoDisc/Getty Images 524 (top bar), 534 (top bar); Jmh649/Wikimedia Commons/Creative Commons Attribution ShareAlike 3.0 Unported 524 (top right); © James Steveson/Photo Researchers, Inc. 524 (left both); Reproduced with Permission of Wyeth Pharmaceuticals 525; Jpogi/Wikipedia/Public Domain 526; © Science VU/Visuals Unlimited, Inc. 527 (bottom); © Dr. David M. Phillips/Visuals Unlimited, Inc. 528; © Medscan/Visuals Unlimited, Inc. 530; Alex Khimich/Wikimedia Commons/Public Domain 530 (inset); © 2009 JupiterImages/Marc Dietrich 531; © Fotolia/monamakela.com 531 (inset); USDA 536

Chapter 22

© Sheila Terry/Photo Researchers, Inc. 539 (left); © Image Select/Art Resource 539 (right); Clemson University EM lab 541 (top); © Dr. David M. Phillips/Visuals Unlimited, Inc. 541 (bottom); Stem Labs 542; PhotoDisc/Getty Images 543 (top bar), 553 (both), 560 (top bar), 563 (bottom both); Biographical Memoirs of the National Academy of Sciences, vol. 40 543 (bottom); Apers0n/Wikipedia/Creative Commons Attribution 2.0 Generic 544 (top); © National Portrait Gallery, Smithsonian Institution/Art Resource 544 (bottom); Dr. Mirco Junge/Wikimedia Commons/Attribution 3.0 License 546; Steven Fruitsmaak/Wikimedia Commons/GNU Free Documentation License 548 (left); Lucien Monfils/Wikimedia Commons/GNU Free Documentation License 548 (right); © Scott Camazine/Photo Researchers, Inc. 549, 560 (top center); © Jubal Harshaw/Shutterstock 550; © Science VU/Visuals Unlimited, Inc. 551 (bottom center); Ed Uthman, MD/Wikimedia Commons/Creative Commons Attribution ShareAlike 2.0 License 556; © Dr. Hans Gelderblom/Visuals Unlimited, Inc. 560 (top center inset); © Steinhagen Artur/Shutterstock 563 (top); www.HouseDustMite.com/The London School of Tropical Hygiene and Medicine 563 (center left); BJU Photo Services 563 (center right); Shanel/Wikimedia Commons/GNU Free Documentation License 566

Chapter 23

PhotoDisc/Getty Images 574 (top bar), 576 (top bar); Karl Weatherly/PhotoDisc/Thinkstock 574 (bottom left); © iStockphoto.com/Kim Gunkel 574 (bottom right); Zipfer/Wikimedia Commons/Public Domain 581

Chapter 24

© Medscan/Visuals Unlimited, Inc. 590; © Jubal Harshaw/Shutterstock 592; © iStockphoto.com/Andrew Gentry 593; PhotoDisc/Getty Images 594 (top bar), 598 (top bar), 606 (top bar); *Christ at the Pool of Bethesda*, From the Bob Jones University Collection 597; © George Steinmetz - www.georgesteinmetz.com 598 (right); © iStockphoto.com/Marjan Laznik 598 (left); NIAAA/National Institutes of Health 599; © Andy Easthope. Image from BigStockPhoto.com 606 (bottom); Photo by Ted Thai/Time Life Pictures/Getty Images 607 (left); © iStockphoto.com/Ziga Lisjak 607 (right)

Chapter 25

Photo by Taro Yamasaki/Time Life Pictures/Getty Images 614; Hemera Technologies/Thinkstock 617; JupiterImages/Comstock/Thinkstock 620; Centers for Disease Control 621 (top bar); National Cancer Institute/Laboratory of Tumor Virus Biology 622 (letter a); National Cancer Institute/Zaki Salahuddin/Laboratory of Tumor Cell Biology 622 (letter b); National Cancer Institute/Dr. Lance Liotta Laboratory 622 (letter c); CDC/Dr. Edwin P. Ewing, Jr. 622 (letter d); CDC/Dr. M. S. Ferguson 622 (letter e); *Madonna and Child with St. Elizabeth and St. John the Baptist*, From the Bob Jones University Collection 624

Appendix A

PhotoDisc/Getty Images 627 (top inset), 628 (top left, top right, center left, bottom right), 629 (top right, top left inset); © iStockphoto.com/Lars Johansson 627 (bottom); U.S. Fish and Wildlife Service 628 (top right inset, center left inset, bottom right inset); © iStockphoto.com/Brian Raisbeck 629 (top left); Digital Stock/Corbis 629 (center right inset); © iStockphoto.com/Peter Short 629 (center right); National Park Service/Richard Frear 629 (bottom left inset); © 2004 Hemera Technologies/Getty Images 629 (bottom left); COREL Corporation 630 (top); © iStockphoto.com/eb33 630 (bottom)

Appendix B

PhotoDisc/Getty Images 632, 633, 634; PhotoObjects.net/Thinkstock 635; U.S. Fish and Wildlife Service 636

Index

PhotoDisc/Getty Images 659–671 (all)